Molecular Biology Intelligence Unit

Serpins: Structure, Function and Biology

Peter G.W. Gettins, M.A., D.Phil.

University of Illinois-Chicago
College of Medicine
Chicago, Illinois, USA

Philip A. Patston, B.Sc., D.Phil.

Steven T. Olson, B.S., M.S., Ph.D.

University of Illinois-Chicago
College of Dentistry
Chicago, Illinois, USA

CHAPMAN & HALL
I(T)P An International Thomson Publishing Company

New York • Albany • Bonn • Boston • Cincinnati • Detroit • London • Madrid • Melbourne •
Mexico City • Pacific Grove • Paris • San Francisco • Singapore • Tokyo • Toronto • Washington

R.G. Landes Company
Austin

Molecular Biology Intelligence Unit

SERPINS: STRUCTURE, FUNCTION AND BIOLOGY

R.G. LANDES COMPANY
Austin, Texas, U.S.A.

Printed in the U.S.A.

Please address all inquiries to the Publishers:
R.G. Landes Company, 909 Pine Street, Georgetown, Texas, U.S.A. 78626
Phone: 512/ 863 7762; FAX: 512/ 863 0081

North American distribution:
Chapman & Hall, 115 Fifth Avenue, New York, New York, U.S.A. 10003

International distribution (except North America):
Springer-Verlag GmbH & Co. KG, Tiergartenstrasse 17, D-69121 Heidelberg, Germany

U.S. and Canada ISBN: 0-412-10351-6
International ISBN: 3-540-60729-3

Library of Congress Cataloging-in-Publication Data

Gettins, Peter G.W., 1953-
Serpins:structure, function and biology/Peter G.W. Gettins, Phillip A. Patston, Steven T. Olson.
p.cm.—(Molecular biology intelligence unit)
Includes bibliographical references and index.
ISBN 0-412-10351-6 (alk. paper)
1. Serpins. I. Patston, Phillip A. II. Olson, Steven T. III. Title. IV. Series.
[DNLM: 1. Serpins—physiology. 2. Serine Proteinases—physiology.
QU 136 G394s 1996]
QP609.S47G48 1996
612'.01575—dc20
DNLM/DLC 95-50022
for Library of Congress CIP

Publisher's Note

R.G. Landes Company publishes six book series: *Medical Intelligence Unit, Molecular Biology Intelligence Unit, Neuroscience Intelligence Unit, Tissue Engineering Intelligence Unit, Environmental Intelligence Unit* and *Biotechnology Intelligence Unit.* The authors of our books are acknowledged leaders in their fields and the topics are unique. Almost without exception, no other similar books exist on these topics.

Our goal is to publish books in important and rapidly changing areas of bioscience for sophisticated researchers and clinicians. To achieve this goal, we have accelerated our publishing program to conform to the fast pace in which information grows in bioscience. Most of our books are published within 90 to 120 days of receipt of the manuscript. We would like to thank our readers for their continuing interest and welcome any comments or suggestions they may have for future books.

Deborah Muir Molsberry
Publications Director
R.G. Landes Company

CONTENTS

1. Introduction **1**
1. The Serpin Superfamily of Proteins 1
2. Characteristic Properties of Serpins 1
3. Occurrence of Serpins 2
4. Size and Composition of Serpins 2
5. Gene Structure, Chromosomal Localization and Evolution 4

2. Three Dimensional Structures of Serpins **15**
1. Structures Completed 15
2. The Core Serpin Domain 17
3. The Reactive Center Loop 19
4. Conformational Lability and Mobility of the Reactive Center Loop 23
5. Modeled Serpin Structures 23
6. Ligand Binding Regions 27

3. Mechanism of Action of Serpins as Suicide Substrate Inhibitors **33**
1. Comparison of Serpins with Other Classes of Serine Proteinase Inhibitors 33
2. Outcome of Serpin-Proteinase Interactions Substrate Versus Inhibition Reactions and the Branched Pathway Mechanism 40
3. Loop Insertion as a Requirement for Inhibition by Serpins 44
4. Influence of Reactive Center Loop Conformation on Reactivity 46
5. Nature of Complex Stabilization and Mechanism of Serpin Inhibition 47
6. Reaction of Serpins with Cysteine Proteinases 51
7. Stabilization of the Non-Loop Inserted “Active” Form of Serpins 51
8. Advantages and Disadvantages of the Serpin Mechanism 54

4. Serpins Activated by Heparin and Other Glycosaminoglycans **65**
1. Antithrombin 65
2. Heparin Cofactor II 79
3. Protein C Inhibitor 84
4. Protease Nexin 1 86
5. Plasminogen Activator Inhibitor-1 (PAI-1) 89
6. Comparisons Between Activation Mechanisms 92

5. Other Inhibitory Serpins 111
1. α_1-Proteinase Inhibitor 111
2. α_1-Antichymotrypsin 113
3. α_2-Antiplasmin 113
4. C1-Inhibitor 114
5. Plasminogen Activator Inhibitor-2 (PAI-2) 117
6. Kallistatin 117
7. Squamous Cell Carcinoma Antigen 117
8. Rodent Serpins 118
9. Intracellular Serpins 119
10. Insect Serpins 120
11. Parasite Serpins 121
12. Plant Serpins 121
13. Viral Serpins 121

6. Non-Inhibitory Serpins 135
1. Ovalbumin 135
2. Gene Y Protein 137
3. Hormone-Binding Serpins 137
4. Angiotensinogen 144
5. Pigment Epithelium-Derived Factor 147
6. Uterine Milk Proteins 149
7. *Xenopus* Serpin 150
8. Maspin 150
9. Collagen Binding Protein 151

7. Serpin Biosynthesis, Clearance and Receptors 161
1. Serpin Biosynthesis 161
2. Clearance of Serpin-Proteinase Complexes 162
3. The SEC Receptor 163
4. The α_2-Macroglobulin/LDL Related Protein (LRP) Receptor and the Urokinase Receptor 165
5. Other Serpin Receptors 165
6. Biological Activity of Cleaved and Complexed Serpins 167

8. Serpins and Disease 177
1. a_1-Proteinase Inhibitor Deficiency. 177
2. Antithrombin Deficiency 180
3. C1-Inhibitor Deficiency 180
4. α_2-Antiplasmin Deficiency 182
5. The Role of PAI-1 in Vascular Disease 182
6. Serpins and Alzheimer's Disease 182
7. Serpins and Cancer 182
8. Serpins as Targets for Therapeutic Intervention 183

9. Unanswered Questions and Directions for Future Research ... 193
1. Crystal Structure of Serpin-Proteinase Complexes 193
2. The Mechanism of Activation of Antithrombin by Heparin 194
3. Additional Structures of Native Serpins 194
4. Class Specificity 194
5. Protein Folding and Function 195
6. Evolutionary Relationships 195
7. Involvement of PEDF, Maspin and Other Serpins in Diverse Biological Processes 195
8. Human Disease and Biotechnology 195
9. The In Vivo Regulation of Proteolytic Pathways 196
10. Concluding Remarks 196

Index 199

PREFACE

From small beginnings a few years ago, the field of serpin biochemistry, biology, and genetics has been growing at an ever increasing rate. Serpins are now known from mammals, including man, from insects, from plants, and even from viruses. In man they are known to be involved in coagulation, in fibrinolysis, in the immune response and in inflammation and may be involved in many other processes such as cellular differentiation and cancer metastasis. In many of these situations they function as regulatable inhibitors of serine proteinases. Some serpins, such as antithrombin, ovalbumin and angiotensinogen, have been considered important objects of study for decades, but only with the identification of the serpin superfamily have they been examined in the light of them being structurally and mechanistically related to one another. As a result there are two distinct bodies of literature for some serpins, one related to them as serpins and one treating each as an isolated protein with a particular set of biochemical and biological properties.

The intent of this book is a selective one. It is to provide an overview of common mechanistic, structural and genetic features of serpins that make them such an unusual class of proteins, as well as to be a source of core biochemical and biological information related to individual serpins. Rather than attempt to include all literature on proteins that are serpins, we have concentrated on what is essential from a structural or mechanistic point of view and have cited reviews where the details are not critical to an understanding of the properties of a particular serpin. The book is organized into four sections.

The first section, Chapters 1-3, deals with common structural and mechanistic properties of serpins. In Chapter 1 we briefly review the history of serpins and provide an indication of their widespread distribution, some of their characteristic properties, and what is known about their gene organization and location and their evolution. In Chapter 2 we consider the common structural features of serpins that have been elucidated in the last few years through X-ray crystallography and examine the dramatic structural interconversions that serpins can undergo as a necessary part of their functioning. In Chapter 3 we contrast the simple "lock-and-key" mechanism of proteinase inhibition by non-serpin inhibitors with the much more sophisticated serpin inhibition mechanism. We examine what is needed for effective inhibition and point out the advantages and disadvantages of such a conformationally-dependent mechanism.

The second section, Chapters 4-6, deals with individual serpins grouped according to certain common features. Chapter 4 considers those serpins that bind to and are activated by glycosaminoglycans such as heparin and heparan sulfate. Such interactions are of great clinical importance since they form the basis for the anticoagulant nature of heparin and related molecules. A major focus of the chapter is therefore the different mechanisms by which such activation is achieved. Chapter 5 deals with other inhibitory serpins from a variety of sources and summarizes key properties of each. Chapter 6 covers those serpins that are known not to, or are suspected of being unable to, inhibit proteinases, but have other biological properties. We deal with both long known serpins, e.g. ovalbumin, and recently discovered serpins, e.g. maspin and pigment epithelium-derived factor.

The third section, Chapters 7 and 8, returns to common serpin properties, but from a more biological and physiological perspective. Chapter 7 deals with the biosynthesis and receptor-mediated clearance of serpins and their complexes with proteinase. Chapter 8 considers the role of serpins in disease, both deficiency states and involvement in cancer and Alzheimer's disease.

The final section, Chapter 9, is a brief summary of those remaining questions in all areas of serpin structure, function, and biology that are likely to be the most actively investigated in the coming years.

The authors of this book have been actively engaged in research on serpins for many years and have contributed particularly in the areas of structure and mechanism of serpins and of their activation by glycosaminoglycans. The book will be of use to a wide range of researchers; from those basic researchers interested generally in serpin mechanism and structure, through those with more clinical interests in the serpin-mediated anticoagulant properties of heparins, to those who have identified a new protein that turns out to be a serpin and who want to obtain an overview of the field.

though striking, has not yet been demonstrated to have any specific function.[19]

Secreted serpins typically are synthesized as a longer pre-protein containing a cleavable hydrophobic N-terminal signal sequence. Some serpins, such as plasminogen activator inhibitor-2 (PAI-2) exist in both secreted and non-secreted forms and may rely on an internal non-cleavable signal sequence, which is a characteristic of the ov-serpin subfamily, for export from the cell.

Within the core serpin domain, sequence identity between a pair of serpins can be as low as 30% (e.g., ovalbumin and α_1-proteinase inhibitor) or as high as 92% (e.g., two squamous cell carcinoma antigens occurring at the same gene locus[20]). Although some serpins contain disulfides, others do not. Those that do contain disulfides usually do not show conservation of their positions, unless the proteins are highly homologous. Huber and Carrell[21] have noted 51 residues that are either absolutely conserved in 20 serpins or involve at most change of phenylalanine to tyrosine. All of these conserved residues are either internal or in niches on the surface.[21] The reactive center loop region, which is the interaction site with proteinase, has a region with considerable sequence conservation and another in which sequences are highly dissimilar (Fig. 1.2). Using the nomenclature of Schechter and Berger[22] to designate residues N- and C-terminal to the scissile bond in proteinase substrates and inhibitors (the scissile bond is defined as between P1-P1'; N-terminal residues are designated successively P2, P3 etc. and C-terminal residues P2', P3' etc.), the highly conserved residues are in the "hinge region" from P17 to P9, which is considered very important in the functioning of the mechanistically crucial loop-insertion process (see Chapters 2 and 3), whereas the region of high sequence variation is centered on the scissile bond and presumably reflects the wide differences in target proteinase specificity.

Table 1.1. Human plasma serpins, their concentrations, target and other proteinases, and pathologies resulting from deficiency (d) or excess (e) states

Serpin	Concentration		Proteinases inhibited		Pathology	Ref.
	μg/ml	μM	Physiological	Other		
α_1-Proteinase inhibitor	1300	24.5	Neutrophil elastase	Trypsin, pancreatic elastase	Emphysema (d)	45,46
α_1-Antichymotrypsin	250	4.0	Cathepsin G	Chymotrypsin	Alzheimer's (?)	47
α_2-Antiplasmin	79	1.0	Plasmin	Trypsin, chymotrypsin	Hemorrhage (d)	48
Antithrombin	130	2.3	Thrombin, factor Xa	Trypsin, plasmin, other coagulation proteinases	Thrombosis (d)	49
C1-inhibitor	180	1.7	C1s, C1r, kallikrein h, factor XIIa	tPA, factor XIa, chymotrypsin	Angioedema (d)	50
Kallistatin	13.5	0.3	Tissue kallikrein	?	Hypo-, hypertension (?)	51
Heparin cofactor II	80	1.2	Thrombin	Chymotrypsin, cathepsin G	Thrombosis (?) (d)	52
PAI-1	0.02	0.0004	tPA, uPA	Trypsin	Hemorrhage(d), thrombosis (e)	53
PAI-2	<0.005	$<7 \times 10^{-5}$	tPA, uPA	?	Cancer ?	20
Protein C inhibitor	2-5	0.1	Activated protein C, uPA, tPA, acrosin	Thrombin, factor Xa, chymotrypsin	?	54

5. GENE STRUCTURE, CHROMOSOMAL LOCALIZATION, AND EVOLUTION

Based on number, size and phase of exons and introns, at least seven distinct serpin gene structures have been shown to exist. These are exemplified by the ov-serpins (genes characterized for ovalbumin,[23] SCCA,[20] PAI-2[24] and gene Y[25]), the α_1-PI family (genes characterized for α_1-PI,[26] α_1-antichymotrypsin,[27] corticosteroid binding globulin (CBG)[28] and angiotensinogen[29]), antithrombin,[30] plasminogen activator inhibitor 1,[31] PI-6,[32] C1-inhibitor[33] and α_2-antiplasmin[34] (Fig. 1.3). Among these groups of serpins, introns occur at both equivalent and unrelated positions. Whereas some of the exon/intron boundaries occur at positions corresponding

Table 1.2 Species distribution of serpins

Human	Fish
α_1-Antichymotrypsin	α_1-Proteinase inhibitor (carp)
α_2-Antiplasmin	
α_1-Proteinase inhibitor	Amphibians
Angiotensinogen	Ep-45 (*Xenopus laevis*)
Antithrombin	
C1-inhibitor	Birds
Corticosteroid binding globulin	Antithrombin (chicken)
Heparin cofactor II	Collagen binding protein (chicken)
Kallistatin	Ovalbumin (chicken, quail)
Leukocyte elastase inhibitor	
Maspin	Arthropods
Pigment epithelium-derived factor	α_1-Antichymotrypsin (*Bombyx mori*)
Plasminogen activator inhibitor-1	Ala-serpin (*Manduca sexta*)
Plasminogen activator inhibitor-2	Unidentified function (*Drosophila melanogaster*)
Protease nexin 1	Coagulation inhibitor (Horseshoe crab)
Protein C inhibitor	
Squamous cell carcinoma antigen	Nematodes
Thyroxine binding globulin	Unidentified function (*Schistosoma haematobium, S. mansoni, Brugia malayi*)
Other mammals	Plants
α_1-Proteinase inhibitor (baboon, chinchilla, chipmunk, dog, donkey, eastern barred bandicoot, fat-tailed insectivore, gerbil, goat, guinea pig, horse, mink, mongolian gerbil, monkey, mouse, opossum, pig, rabbit, rat, sheep, wallaby)	Barley Z protein Chymotrypsin inhibitor (wheat) Z_4 (barley) Z_7 (barley)
α_1-Antichymotrypsin (cow, goat, pig)	
Angiotensinogen (cow, mouse, rat, rabbit, sheep)	Fungi
Antithrombin (cow, mouse, pig, rabbit, ra, sheep)	Corticosteroid binding globulin (yeast)
C1 inhibitor (guinea pig, rat)	
Collagen binding protein (mouse, rat)	Viruses
Contrapsin (rat, mouse)	crmA
Corticosteroid binding globulin (rat, rabbit, squirrel monkey)	SPI-3 (vaccinia virus)
	SERP1 (myxoma virus)
Heparin cofactor II (rabbit)	
Kallistatin (rat)	
Leukocyte elastase inhibitor (horse, pig)	Bacteria
PEDF (cow)	None known
Uterine milk protein (cow, pig, sheep)	

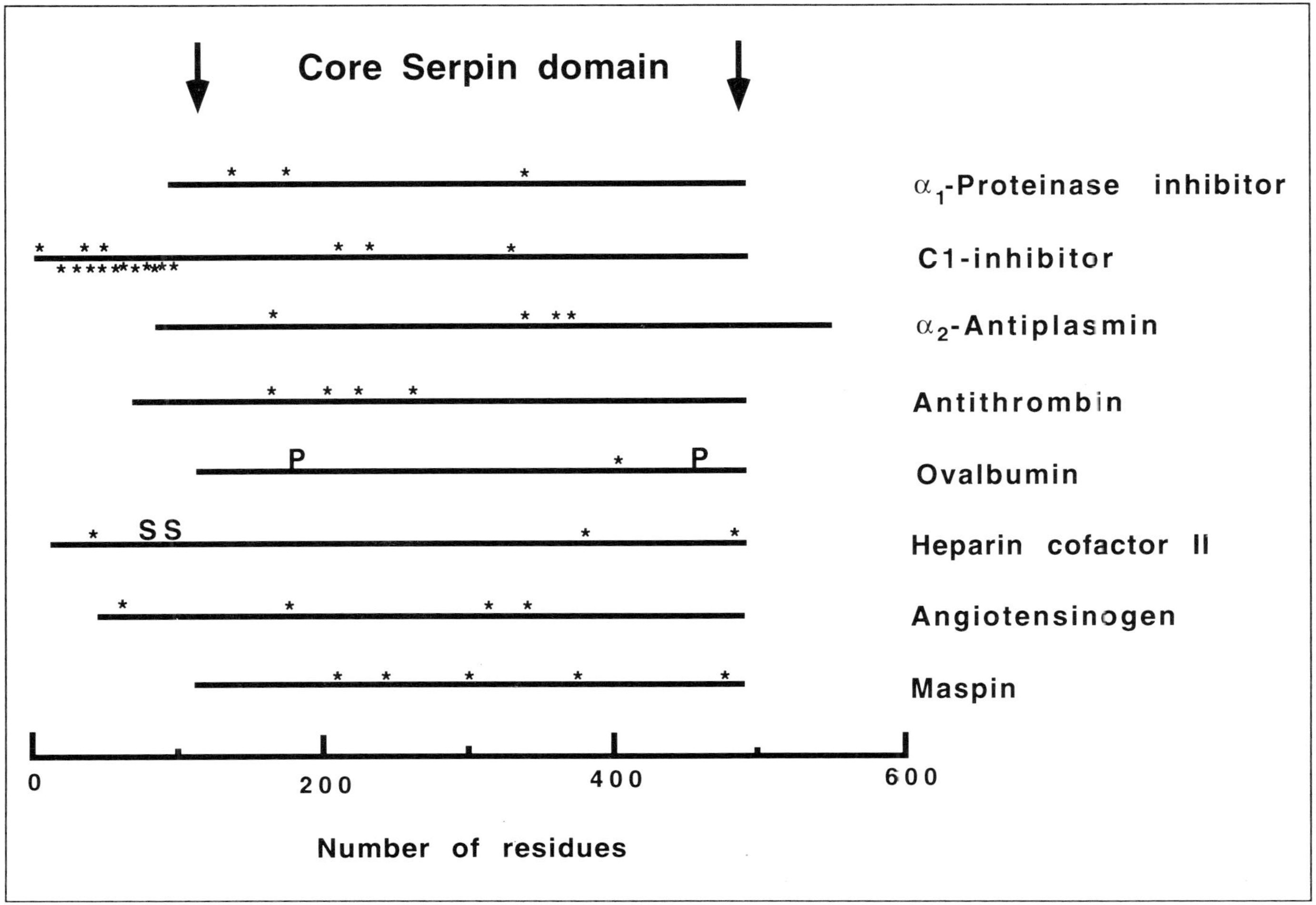

*Fig. 1.1. Schematic representation of the primary structures of selected serpins to illustrate the size and location of the core serpin domain, and the presence of non-conserved post-translational modifications. Glycosylation sites, either demonstrated to exist or putative, are indicated by *; above the line for N-linked and below the line for O-linked; P represents a phosphoserine and S represents a tyrosine sulfate. For C1-inhibitor, the high density of glycosylation sites makes it difficult to accurately represent all sites present in the N-terminal region.*

Fig. 1.2. Aligned primary structures of the reactive center loops (P17 to P3') of both inhibitory and non-inhibitory serpins. The P and P' residue designation is that of Schechter and Berger.[22] Note the frequent presence in inhibitory serpins of small side chains from P12 to P9, of uncharged side chains at P14, of glycine at P15, and glutamate at P17, P16 and P13. The P1 and P1' residues on either side of the scissile bond are underlined for clarity. The scissile bond is indicated with a down arrow. Abbreviations used are: ELEI, equine leukocyte elastase inhibitor; LICI, Limulus *intracellular coagulation inhibitor; PAI, plasminogen activator inhibitor; PI-6, proteinase inhibitor 6, also known as placental thrombin inhibitor; SCCA, squamous cell carcinoma antigen; CBG, corticosteroid binding globulin, TBG, thyroxine binding globulin; PEDF, pigment epithelium-derived factor.*

	P17	P16	P15	P14	P13	P12	P11	P10	P9	P8	P7	P6	P5	P4	P3	P2	P1 ↓	P1'	P2'	P3'	P4'	P5'	Ref
INHIBITORY SERPINS																							
α_1-Antichymotrypsin	Glu	Glu	Gly	Thr	Glu	Ala	Ser	Ala	Ala	Thr	Ala	Val	Lys	Ile	Thr	Leu	Leu	Ser	Ala	Leu	Val	Glu	62
α_1-Antitrypsin	Glu	Lys	Gly	Thr	Glu	Ala	Ala	Gly	Ala	Met	Phe	Leu	Glu	Ala	Ile	Pro	Met	Ser	Ile	Pro	Pro	Glu	63
Antithrombin	Glu	Glu	Gly	Ser	Glu	Ala	Ala	Ala	Ser	Thr	Ala	Val	Val	Ile	Ala	Gly	Arg	Ser	Leu	Asn	Pro	Asn	64
α_2-Antiplasmin	Glu	Val	Gly	Val	Glu	Ala	Ala	Ala	Ala	Thr	Ser	Ile	Ala	Met	Ser		Arg	Met	Ser	Ser	Ser	Phe	65
Barley Z protein	Glu	Glu	Gly	Thr	Glu	Ala	Gly	Ala	Ala	Thr	Val	Ala	Met	Gly	Val	Ala	Met	Ser	Met	Pro	Leu	Lys	66
C1-inhibitor	Glu	Thr	Gly	Val	Glu	Ala	Ala	Ala	Ala	Ser	Ala	Ile	Ser	Val	Ala		Arg	Thr	Leu	Leu	Val	Phe	67
crmA	Glu	Glu	Tyr	Thr	Glu	Ala	Ala	Ala	Ala	Thr	Cys	Ala	Leu	Val	Ala		Asp	Cys	Ala	Ser	Thr	Val	68
ELEI	Glu	Glu	Gly	Thr	Glu	Ala	Ala	Ala	Ala	Thr	Ala	Gly	Thr	Ile	Met	Leu	Ala	Met	Pro	Glu	Glu	Asn	69
Heparin cofactor II	Glu	Glu	Gly	Thr	Gln	Ala	Thr	Thr	Val	Thr	Thr	Ala	Gly	Phe	Met	Pro	Leu	Ser	Thr	Gln	Val	Arg	70
Kallistatin	Asp	Glu	Gly	Thr	Glu	Ala	Ala	Ala	Ala	Thr	Thr	Phe	Ala	Ile	Lys	Phe	Phe	Ser	Ala	Gln	Thr	Asn	71
LICI	Glu	Glu	Gly	Thr	Glu	Ala	Ser	Gly	Ile	Ser	Ser	Val	Val	Ala	Gly	Val	Arg	Ser	Gly	Trp	Lys	Arg	72
LICI2	Glu	Glu	Gly	Ser	Glu	Ala	Ala	Ala	Val	Ser	Gly	Val	Val	Val	Met	Leu	Lys	Ser	Ala	Ser	His	Asn	73
Manduca sexta Ala-serpin	Glu	Glu	Gly	Ala	Glu	Ala	Ala	Ala	Ala	Asn	Ala	Phe	Gly	Ile	Val	Pro	Ala	Ser	Leu	Ile	Leu	Thr	74
PAI-1	Glu	Ser	Gly	Thr	Val	Ala	Ser	Ser	Ser	Thr	Ala	Val	Ile	Val	Ser	Ala	Arg	Met	Ala	Pro	Glu	Glu	75
PAI2	Glu	Glu	Gly	Thr	Glu	Ala	Ala	Ala	Gly	Thr	Gly	Gly	Val	Met	Thr	Gly	Arg	Thr	Gly	His	Gly	Gly	76
Protein C inhibitor	Glu	Ser	Gly	Thr	Arg	Ala	Ala	Ala	Ala	Thr	Gly	Thr	Ile	Phe	Thr	Phe	Arg	Ser	Ala	Arg	Leu	Asn	77
Protease nexin-1	Glu	Asp	Gly	Thr	Lys	Ala	Ser	Ala	Ala	Thr	Thr	Ala	Ile	Leu	Ile	Ala	Arg	Ser	Ser	Pro	Pro	Trp	78
PI-6	Glu	Glu	Gly	Thr	Glu	Ala	Ala	Ala	Ala	Thr	Ala	Ala	Ile	Met	Met	Met	Arg	Cys	Ala	Arg	Phe	Val	79
SCCA-1	Glu	Glu	Gly	Ala	Glu	Ala	Ala	Ala	Ala	Thr	Ala	Val	Val	Gly	Phe	Gly	Ser	Ser	Pro	Ala	Ser	Thr	20
SCCA-2	Glu	Glu	Gly	Val	Glu	Ala	Ala	Ala	Ala	Thr	Ala	Val	Val	Val	Phe	Glu	Leu	Ser	Ser	Pro	Ser	Thr	20
Schistosome serpin	Glu	Ser	Gly	Ile	Glu	Ala	Thr	Thr	Val	Thr	Ser	Pro	Ile	Phe	Val	Pro	Phe	Ser	Ala	Ile	Ile	Pro	80
NON-INHIBITORY SERPINS																							
Ovalbumin	Glu	Ala	Gly	Arg	Glu	Val	Val	Gly	Ser	Ala	Glu	Ala	Gly	Val	Asp	Ala	Ala	Ser	Val	Ser	Glu	Glu	81
CBG	Glu	Glu	Gly	Val	Asp	Thr	Ala	Gly	Ser	Thr	Gly	Val	Thr	Leu	Asn	Leu	Thr	Ser	Lys	Pro	Ile	Ile	82
TBG	Glu	Lys	Gly	Thr	Glu	Ala	Ala	Ala	Val	Pro	Glu	Val	Glu	Leu	Ser	Asp	Gln	Pro	Glu	Asn	Thr	Phe	83
Angiotensinogen (h)	Ala	Asp	Glu	Arg	Glu	Pro	Thr	Glu	Ser	Thr	Gln	Gln	Leu	Asn	Lys	Pro	Glu	Val	Leu	Glu	Val	Thr	84
PEDF	Glu	Asp	Gly	Ala	Gly	Thr	Thr	Pro	Ser	Pro	Gly	Leu	Gln	Pro	Ala	His	Leu	Thr	Phe	Pro	Leu	Asp	6

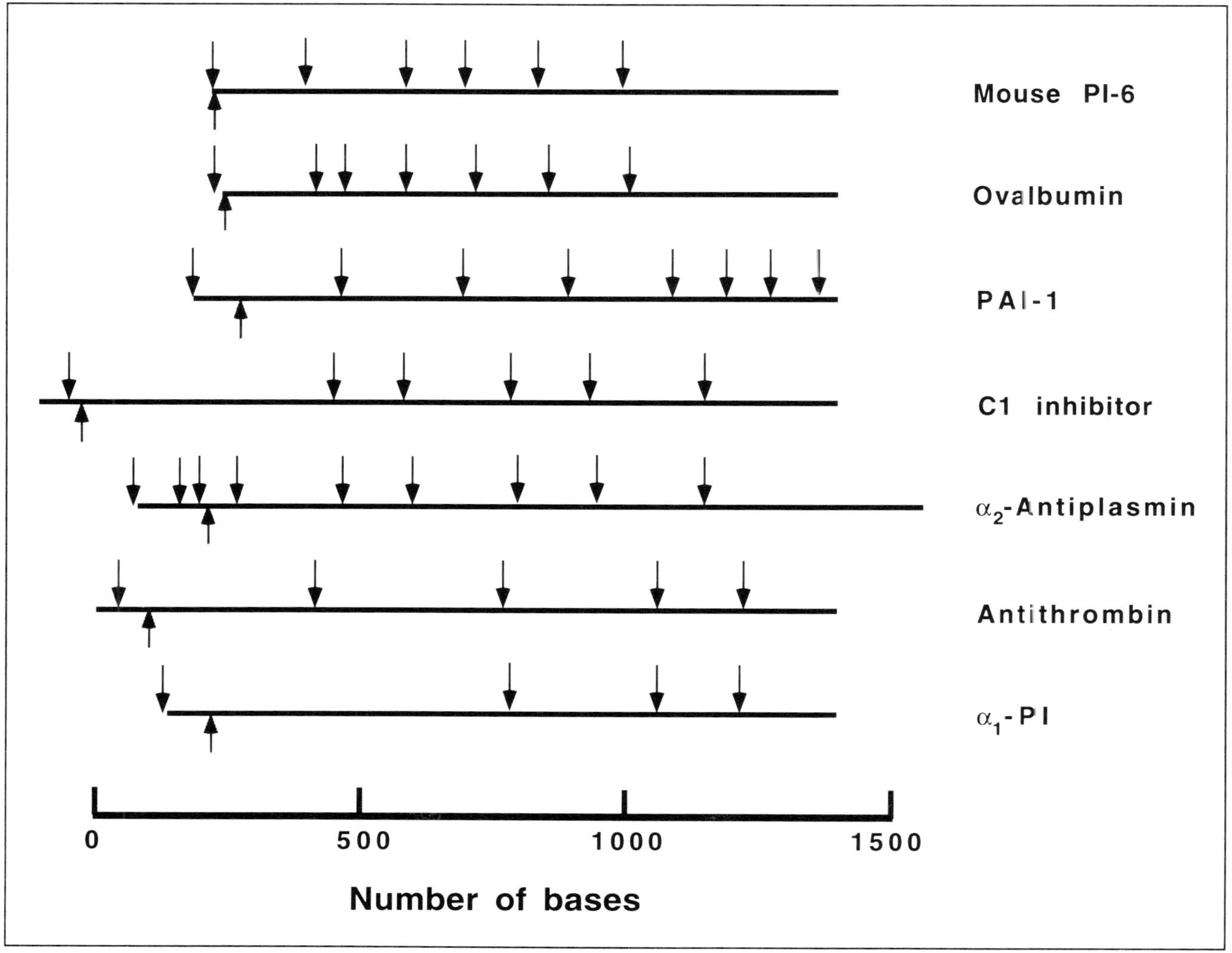

Fig. 1.3. Gene organization of selected serpins to illustrate the seven different patterns of exons and introns so far identified. The horizontal bars represent the exon coding regions, down arrows indicate the positions of introns, up arrows indicate the start position of the mature protein after processing of any signal peptide.

to non-structured parts of the serpin polypeptide, others occur within stretches of secondary structure, e.g., in the α_1-antichymotrypsin gene the second intron occurs in the middle of the portion of the gene coding for strand 3 of β-sheet A and the fourth intron in the middle of the portion of the gene coding for strand 5 of β-sheet A.

In proposing a model for the evolution of serpins from a common ancestral gene both the above differences in gene organization and divergence of amino acid sequence must be taken into account. Thus, whereas angiotensinogen and α_1-antichymotrypsin have the same gene organization, these serpins are amongst the most widely separated based on protein primary structure. Woo and colleagues[27] have proposed that a hybrid of the intron deletion[35] and intron insertion models[26] could account for the relationships between serpins. In this model the ancestral gene resembled the α_1-antichymotrypsin gene. Gene duplication over 500 million years ago (estimated from evolutionary distances based on protein primary structure data) led to the formation of angiotensinogen, whose protein sequence then diverged from that of the precursor of α_1-antichymotrypsin (Fig. 1.4). More recently (between 500 and 250 million years ago) a second gene duplication led to the precursors of (i) α_1-antichymotrypsin and α_1-proteinase inhibitor and (ii) ovalbumin and antithrombin. For ovalbumin and antithrombin, the precursor gene must have undergone several intron insertions and deletions prior to the gene duplication that led, about 250 million years ago, to anti-thrombin and the precursor of the ov-serpins. Even more recent gene duplication of the ovalbumin precursor (~100 million years ago) led to the distinct ovalbumin and gene Y genes. A recent analysis of the evolutionary relationships among serpins has been carried out, based on protein primary structures.[36] This found high percentages of conservative amino acid similarities, even between distantly-related serpins, reflecting the constraints placed on the serpin architecture by the need to retain the properties peculiar to serpins of folding into a metastable intermediate state that is capable of subsequent relaxation upon interaction with proteinase.

Some serpin gene duplication events have led to gene clusters (Table 1.3), either of variants of a specific inhibitor or of distinct, but perhaps coordinately expressed, serpin genes. An example of multiple variants occurs for the α_1-proteinase inhibitor gene in certain species of mouse and other rodents. Although humans, baboons, sheep, rats and the mouse strain *Mus caroli* contain only a single copy of the α_1-proteinase inhibitor gene, other *Mus* strains contain multiple copies.[37] The deduced protein sequences of four *M. saxicola* α_1-proteinase inhibitors have been compared and shown to exhibit high variability specifically localized to the reactive center region that interacts with proteinase,[38] suggesting an accelerated evolutionary rate stimulated by the need to resist novel proteinases from parasitic infections.[39] This hypervariability in the reactive center is disproportionate to the number of base changes within this region compared to other regions of the coding sequence[37] (Fig. 1.5). In humans the genes for α_1-antichymotrypsin, α_1-proteinase inhibitor, protein C inhibitor, and corticosteroid binding globulin, are despite different functions localized to the same region of chromosome 14 (14q32.1)[40] (Table 1.3) and may be expressed coordinately during inflammation,[41] since the distinct functions of proteinase inhibition and corticosteroid delivery may be required in concert. The gene for kallistatin is also very close by.[42] Another human serpin gene cluster has been identified on chromosome 18 (18q21.3).[20] This involves four ov-serpin genes; maspin, a non-inhibitory serpin identified from a mammary carcinoma cell line,[5] PAI-2, and squamous cell carcinoma antigen (2 distinct copies).[20] No normal coordinate functioning of this gene cluster has yet been proposed. However, it is possible that in pathological states involving chromosomal changes at 18q21, which have been shown to be associated with certain forms of

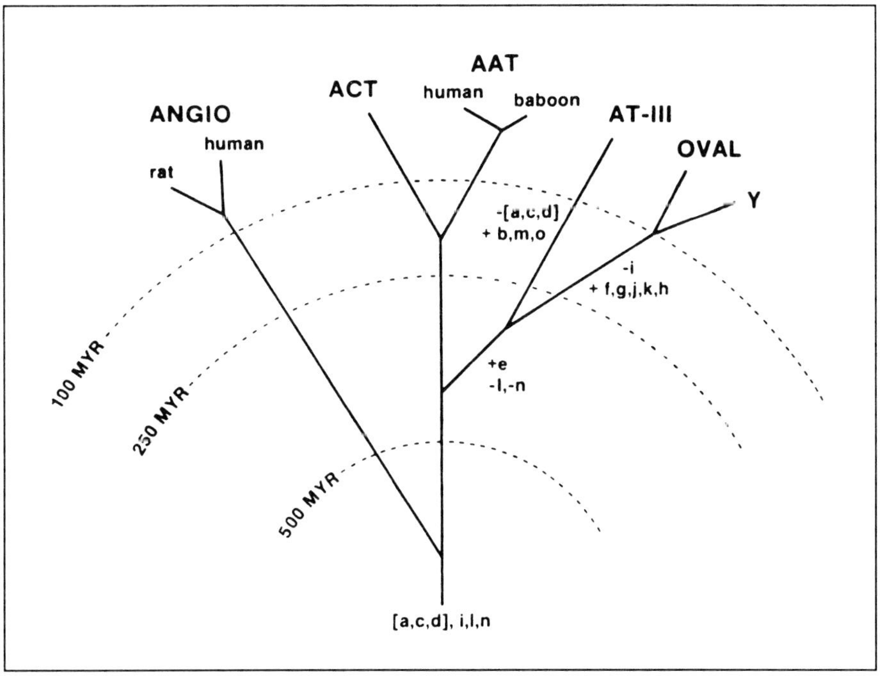

Fig. 1.4. Possible evolutionary tree for selected serpins based on gene organization and protein sequence divergence. The letters a through o represent a composite of intron positions such that [a, c, d] i, l, and n are assumed to be present in the primordial gene and the designation "+" or "–" indicates insertion or deletion respectively of the indicated intron. Reprinted with permission from Biochemistry.[27] *Copyright 1987 by the American Chemical Society.*

Table 1.3. Chromosomal location and gene type of human serpins

Serpin	Chromosomal location	Gene type*	Ref.
Angiotensinogen	1q4	α_1-PI	55
Antithrombin	1q23-25	Antithrombin	56
Protease nexin 1	2q33-35	?	57
Monocyte/neutrophil elastase inhibitor	6p24-pter	Ovalbumin	43
Placental thrombin inhibitor (PI-6)	6p25	? Mouse PI-6	44
Plasminogen activator inhibitor-1	7q	PAI-1	58
C1-inhibitor	11p11.2-q13	C1-inhibitor	33
Kallistatin	14q31-q32.1	α_1-PI	42
α_1-Antichymotrypsin, α_1-PI, protein C inhibitor, corticosteroid binding globulin	14q32.1	α_1-PI	40
α_2-Antiplasmin (α_2-AP)	17p13	α_2-AP	59
Maspin, plasminogen activator inhibitor-2, squamous cell carcinoma antigens 1 and 2	18q21.3	Ovalbumin	20
Heparin cofactor II	22q11	α_1-PI	60
Thyroxine binding globulin	Xq21-22	α_1-PI	61

* Serpin group to which the gene belongs based on exon/intron number and spacing rather than function or chromosomal organization

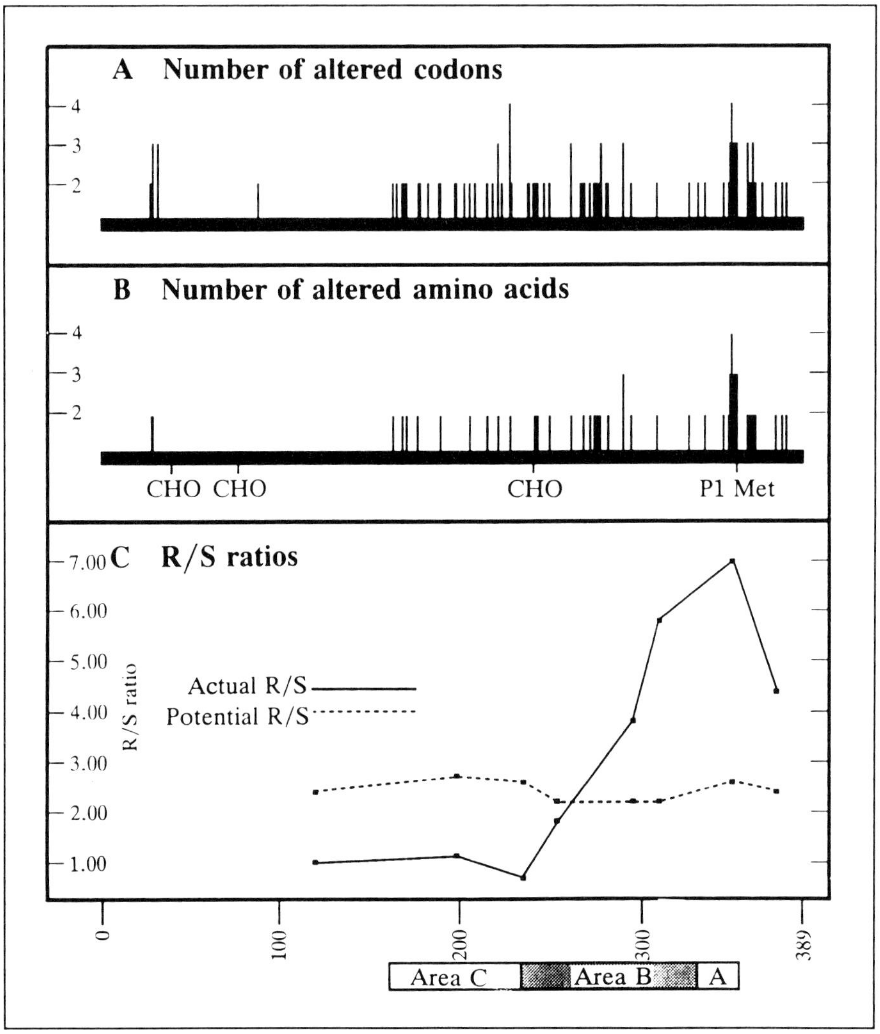

Fig. 1.5. Concentration of mutations in the reactive center region of mouse α_1-PI genes.[37] Panel A, the distribution of altered codons; panel B the distribution of altered amino acids; panel C comparison of observed and potential ratios of non-silent (R) to silent (S) changes, calculated for a window of 18 variant nucleotides. Each succeeding window begins 10 variant positions further downstream. Random nucleotide substitutions should lead to amino acid substitutions in about 75% of cases, giving a potential R/S value of about 2.2. the actual values calculated for α_1-PI (dotted line) are close to this and show little variation with sequence position. Values higher or lower than this indicate evolutionary pressure favoring or suppressing respectively such amino acid replacements. For α_1-PI the observed values (solid line) indicate suppression of amino acid changes (R/S < potential) outside of the reactive center region and a strong favoring of such changes (R/S > potential) in the reactive center region. Reprinted from reference 37 with kind permission of Dr. Krauter.

human cancer, mutations in one or more of these serpin genes may be involved.[20] A third possible close association exists for monocyte/neutrophil elastase inhibitor[43] and PI-6 (placental thrombin inhibitor),[44] both of which occur in the same region of chromosome 6 (Table 1.3).

REFERENCES

1. Hunt LT and Dayhoff MO. A surprising new protein superfamily containing ovalbumin, antithrombin III and α_1-proteinase inhibitor. Biochem Biophys Res Commun 1980; 95:864-871.
2. Doolittle RF. Angiotensinogen is related to the antitrypsin-antithrombin-ovalbumin family. Science 1983; 222:417-419.
3. Hill RE, Shaw PH, Boyd PA, et al. Plasma protease inhibitors in mouse and man: divergence within the reactive center regions. Nature 1984; 311:175-177.
4. Carrell RW and Travis J. α_1-Antitrypsin and the serpins. Variation and countervariation. Trends Biol Sci 1985; 10:20-24.
5. Zou Z, Anisowicz A, Hendrix MJC, et al. Maspin, a serpin with tumor-supressing activity in human mammary epithelial cells. Science 1994; 263:526-529.
6. Steele FR, Chader GJ, Johnson LV, et al. Pigment epithelium-derived factor: Neurotrophic activity and identification as a member of the serine protease inhibitor gene family. Proc Natl Acad Sci USA 1993; 90:1526-1530.
7. Ray CA, Black RA, Kronheim SR, et al. Viral inhibition of inflammation. Cowpox virus encodes an inhibitor of the interleukin-1β converting enzyme. Cell 1992; 69: 597-604.
8. Takeda A, Yamamoto T, Nakamura Y, et al. Squamous cell carcinoma antigen is a potent inhibitor of cysteine proteinase cathepsin L. FEBS Lett 1995; 359:78-80.
9. Remold-O'Donnell E. The ovalbumin family of serpin proteins. FEBS Lett 1993; 315:105-108.
10. Hubbard SJ, Campbell SF and Thornton JM. Conformational analysis of limited proteolytic sites and serine proteinase protein inhibitors. J Mol Biol 1991; 220:507-530.
11. Gettins P, Patston PA and Schapira M. The role of conformational change in serpin structure and function. BioEssays 1993; 15:461-467.
12. Creighton TE. Up the kinetic pathway. Nature 1992; 356:194-195.
13. Olson ST, Bock PE, Kvassman J, et al. Role of the catalytic serine in the interactions of serine proteinases with protein inhibitors of the serpin family. Contribution of a covalent interaction to the binding energy of the serpin-proteinase complexes. J Biol Chem 1995; 270:(In Press)
14. Abraham CR, Kanemaru K and Mucke L. Expression of cathepsin G-like and alpha(1)-antichymotrypsin-like proteins in reactive astrocytes. Brain Res 1993; 621:222-232.
15. Pasternack JM, Abraham CR, Van Dyke BJ, et al. Astrocytes in Alzheimer's disease gray matter express α_1-antichymotrypsin mRNA. Am J Pathol 1989; 135:827-834.
16. Jensen PJ, Wu Q, Janowitz P, et al. Plasminogen activator inhibitor type 2: an intracellular keratinocyte differentiation product that is incorporated into the cornified envelope. Exp Cell Res 1995; 217:65-71.
17. Nisbet AD, Saundry RH, Moir AJG, et al. The complete amino acid sequence of hen ovalbumin. Eur J Biochem 1981; 115: 335-345.
18. Hortin G, Tollefsen DM and Strauss AW. Identification of two sites of sulfation of human heparin cofactor II. J Biol Chem 1986; 261:15827-15830.
19. Coutinho M, Aulak KS and Davis AE III. Functional analysis of the serpin domain of C1 inhibitor. J Immunol 1994; 153: 3648-3654.
20. Schneider SS, Schick C, Fish KE, et al. A serine proteinase inhibitor locus at 18q21.3 contains a tandem duplication of the human squamous cell carcinoma antigen gene. Proc Natl Acad Sci USA 1995; 92: 3147-3151.
21. Huber R and Carrell RW. Implications of the three-dimensional structure of α_1-antitrypsin for structure and function of serpins. Biochemistry 1989; 28:8951-8966.
22. Schechter I and Berger A. On the size of the active site of proteases. Biochem Biophys Res Commun 1967; 27:157-162.
23. Woo SLC, Beattie WG, Catterall JF, et al.

Complete nucleotide sequence of the chicken chromosomal ovalbumin gene and its biological significance. Biochemistry 1981; 20:6437-6446.

24. Ye RD, Ahern SM, Le Beau MM, et al. Structure of the gene for human plasminogen activator inhibitor-2. The nearest mammalian homologue of chicken ovalbumin. J Biol Chem 1989; 264:5495-5502.
25. Heilig R, Muraskowsky R, Kloepfer C, et al. The ovalbumin gene family: complete sequence and structure of the Y gene. Nucl Acids Res 1982; 10:4363-4382.
26. Leicht LM, Long GL, Chandra T, et al. Sequence homology and structural comparison between the chromosomal human a_1-antitrypsin and chicken ovalbumin genes. Nature 1982; 297:655-659.
27. Bao J, Sifers RN, Kidd VJ, et al. Molecular evolution of serpins: homologous structure of the human α_1-antichymotrypsin and α_1-antitrypsin genes. Biochemistry 1987; 26:7755-7759.
28. Underhill DA and Hammond GL. Organization of the human corticosteroid binding globulin gene and analysis of its 5' flanking region. Mol Endocrinol 1989; 3:1448-1454.
29. Jeunemaitre X, Soubrier F, Kotelevtsev YV, et al. Molecular basis of human hypertension: role of angiotensinogen. Cell 1992; 71:169-180.
30. Prochownik EV, Bock SC and Orkin SH. Intron structure of the human antithrombin III gene differs from that of other members of the serine protease inhibitor superfamily. J Biol Chem 1985; 260:9608-9612.
31. Loskutoff DJ, Linders M, Keijer J, et al. Structure of the human plasminogen activator inhibitor 1 gene: nonrandom distribution of introns. Biochemistry 1987; 26:3763-3768.
32. Sun J, Rose JB and Bird P. Gene structure, chromosomal localization, and expression of the murine homologue of human proteinase inhibitor 6 (PI-6) suggests divergence of PI-6 from the ovalbumin serpins. J Biol Chem 1995; 270:16089-16096.
33. Bock SC, Skriver K, Nielsen E, et al. Human C1 inhibitor: primary structure, cDNA cloning, and chromosomal localization. Biochemistry 1986; 25:4292-4301.
34. Hirosawa S, Nakamura Y, Miura O, et al. Organization of the human α_2-plasmin inhibitor gene. Proc Natl Acad Sci USA 1988; 85:6836-6840.
35. Cornish-Bowden A. Related genes can have unrelated introns. Nature 1982; 297: 625-626.
36. Marshall CJ. Evolutionary relationships among the serpins. Phil Trans R Soc Lond B 1993; 101-119.
37. Borriello F and Krauter K. Multiple murine α_1-protease inhibitor genes show unusual evolutionary divergence. Proc Natl Acad Sci USA 1991; 88:9417-9421.
38. Rheaume C, Goodwin RL, Latimer JJ, et al. Evolution of murine α_1-proteinase inhibitors: gene amplification and reactive center divergence. J Mol Evolution 1994; 38 :121-131.
39. Hill RE and Hastie ND. Accelerated evolution in the reactive centre regions of serine protease inhibitors. Nature 1987; 326: 96-99.
40. Billingsley GD, Walter MA, Hammond GL, et al. Physical mapping of four serpin genes: α_1-Antitrypsin, α_1-antichymotrypsin, corticosteroid-binding globulin, and protein C inhibitor, within a 280-kb region on chromosome 14q32.1. Am J Hum Genet 1993; 52:334-353.
41. Hammond GL, Smith CL, Paterson NAM, et al. A role for corticosteroid-binding globulin in delivery of cortisol to activated neutrophils. J Clin Endocrinol Metab 1990; 71:34-39.
42. Chai KX, Ward DC, Chao J, et al. Molecular cloning, sequence analysis, and chromosomal localization of the human proteinase inhibitor 4 (kallistatin) gene. Genomics 1994; 23:370-378.
43. Evans E, Cooley J and Remold-O'Donnell E. Characterization and chromosomal localization of $ELANH_2$, the gene encoding human monocyte/neutrophil elastase inhibitor. Genomics 1995; 28:235-240.
44. Coughlin P, Nicholl J, Sun H, et al. Chromosomal mapping of the human proteinase inhibitor-6 (PI6) gene to 6p25 by fluorescence in situ hybridization. Genomics 1995; 26:431-433.
45. Pannell R, Johnson D and Travis J. Isola-

tion and properties of human plasma α-1-proteinase inhibitor. Biochemistry 1974; 13:5439-5445.

46. Courtney M, Jallat S, Tessier L-H, et al. Synthesis in *E. coli* of α_1-antitrypsin variants of therapeutic potential for emphysema and thrombosis. Nature 1985; 313:149-151.

47. Abraham CR, Selkoe DJ and Potter H. Immunochemical identification of the serine protease inhibitor α_1-antichymotrypsin in the brain amyloid deposits of Alzheimer's disease. Cell 1988; 52:487-501.

48. Griffin GC, Mammen EF, Sokol RJ, et al. α_2-Antiplasmin deficiency. An overlooked cause of hemorrhage. Am J Ped Heme/Onc 1993; 15:328-330.

49. Lane DA, Olds RJ, Boisclair M, et al. Antithrombin III mutation database: First update. Thromb Haemostas 1993; 70: 361-369.

50. Donaldson VH and Evans RR. A biochemical abnormality in hereditary angioneurotic edema. Absence of serum inhibitor C1-esterase. Am J Med 1963; 35:37-44.

51. Chen LM, Chao L, Mayfield RY, et al. Differential interactions of human kallikrein-binding protein and α_1-antitrypsin with human tissue kallikrein. Biochem J 1990; 267:79-84.

52. Tollefsen DM and Pestka CA. Modulation of heparin cofactor II activity by histidine-rich glycoprotein and platelet factor 4. J Clin Invest 1985; 75:496-501.

53. Erickson LA, Fici GJ, Lund JE, et al. Development of venous occlusions in mice transgenic for the plasminogen activator inhibitor 1 gene. Nature 1990; 346:74-76.

54. Suzuki K, Nishioka J and Hashimoto S. Protein C inhibitor purification from human plasma and characterization. J Biol Chem 1983; 258:163-168.

55. Gaillard-Sanchez I, Mattei MG, Clauser E, et al. Assignment by in situ hybridization of the angiotensinogen gene to chromosome band 1q4, the same region as the human renin gene. Hum Genet 1990; 84:341-343.

56. Bock SC, Harris JF, Balazs I, et al. Assignment of the human antithrombin III structural gene to chromosome 1q23-25. Cytogenet Cell Genet 1985; 39:67-69.

57. Carter RE, Cerosaletti KM, Burkin DJ, et al. The gene for the serpin thrombin inhibitor (P17), protease nexin I, is located on human chromosome 2q33-q35 and on syntenic regions in the mouse and sheep genomes. Genomics 1995; 27:196-199.

58. Ginsburg D, Zeheb R, Yang AY, et al. cDNA cloning of human plasminogen activator-inhibitor from endothelial cells. J Clin Invest 1986; 78:1673-1680.

59. Kato A, Hirosawa S, Toyota S, et al. Localization of the human α_2-plasmin inhibitor gene (PLI) to 17p13. Cytogenet Cell Genet 1993; 62:190-192.

60. Herzog R, Lutz S, Blin N, et al. Complete nucleotide sequence of the gene for human heparin cofactor II and mapping to chromosome band 22q11. Biochemistry 1991; 30:1350-1357.

61. Trent JM, Flink IL, Morkin E, et al. Localization of the human thyroxine-binding globulin gene to the long arm of the X chromosome (Xq21-22). Am J Hum Genet 1987; 41:428-435.

62. Chandra T, Stackhouse R, Kidd VJ, et al. Sequence homology between human α_1-antichymotrypsin, α_1-antitrypsin and antithrombin III. Biochemistry 1983; 266:5055-5061.

63. Long GL, Chandra T, Woo SLC, et al. Complete sequence of the cDNA for human α_1-antitrypsin and the S variant. Biochemistry 1984; 23:4824-4837.

64. Bock SC, Wion KL, Vehar GA, et al. Cloning and expression of the cDNA for human antithrombin III. Nucl Acids Res 1982; 10:8113-8125.

65. Holmes WE, Nelles L, Lijnen HR, et al. Primary structure of human α_2-antiplasmin, a serine protease inhibitor (serpin). J Biol Chem 1987; 262:1659-1664.

66. Hejgaard J, Rasmussen SK, Brandt A, et al. Sequence homology between barley endosperm protein Z and protease inhibitors of the α_1-antitrypsin family. FEBS Lett 1985; 180:89-94.

67. Carter PE, Dunbar BA and Fothergill JE. Genomic and cDNA cloning of the human C1 inhibitor. Intron-exon junctions and comparison with other serpins. Eur J Biochem 1988; 173:163-169.

68. Pickup DJ, Ink BS, Hu W, et al. Hemor-

rhage in lesions caused by cowpox virus is induced by a viral protein that is related to plasma protein inhibitors of serine proteases. Proc Natl Acad Sci USA 1986; 83: 7698-7702.
69. Dubin A, Travis J, Enghild JJ, et al. Equine leukocyte elastase inhibitor. Primary structure and identification as a thymosin-binding protein. J Biol Chem 1992; 267: 6576-6583.
70. Ragg H. A new member of the plasma protease inhibitor gene family. Nucl Acids Res 1986; 14:1073-1088.
71. Chai KX, Chen LM, Chao J, et al. Kallistatin—a novel human serine proteinase inhibitor—molecular cloning, tissue distribution, and expression in *Escherichia coli*. J Biol Chem 1993; 268:24498-24505.
72. Miura Y, Kawabata S and Iwanaga S. A *Limulus* intracellular coagulation inhibitor with characteristics of the serpin superfamily. Purification, characterization and cDNA cloning. J Biol Chem 1994; 269:542-547.
73. Miura Y, Kawabata S, Wakamiya Y, et al. A limulus intracellular coagulation inhibitor type 2. Purification, characterization, cDNA cloning and tissue localization. J Biol Chem 1995; 270:558-565.
74. Kanost MR, Prasad SV and Wells MA. Primary structure of a member of the serpin superfamily of proteinase inhibitors from an insect, *Manduca sexta*. J Biol Chem 1989; 264:965-972.
75. Pannekoek H, Veerman H, Lambers H, et al. Endothelial plasminogen activator inhibitor (PAI): a new member of the serpin gene family. EMBO J 1986; 5:2539-2544.
76. Ye RD, Wun T and Sadler JE. cDNA cloning and expression in *Escherichia coli* of a plasminogen activator inhibitor from human placenta. J Biol Chem 1987; 262: 3718-3725.
77. Suzuki K, Deyashiki Y, Nishioka J, et al. Characterization of a cDNA for human protein C inhibitor. A new member of the plasma serine protease inhibitor superfamily. J Biol Chem 1987; 262:611-616.
78. Gloor S, Odink K, Guenther J, et al. A glia-derived neurite promoting factor with protease inhibitory activity belongs to the protease nexins. Cell 1986; 47:687-693.
79. Coughlin PB, Tetaz T and Salem HH. Identification and purification of a novel serine proteinase inhibitor. J Biol Chem 1993; 268:9541-9547.
80. Blanton RE, Licate LS and Aman RA. Characterization of a native and recombinant *Schistosoma haematobium* serine protease inhibitor gene product. Mol Biochem Parasitol 1994; 63:1-11.
81. McReynolds L, O'Malley BW, Nisbet AD, et al. Sequence of chicken ovalbumin mRNA. Nature 1978; 273:723-728.
82. Hammond GL, Smith CL, Goping IS, et al. Primary structure of human corticosteroid binding globulin, deduced from hepatic and pulmonary cDNAs, exhibits homology with serine protease inhibitors. Proc Natl Acad Sci USA 1987; 84:5153-5157.
83. Flink IL, Bailey TJ, Gustafson TA, et al. Complete amino acid sequence of human thyroxine-binding globulin deduced from cloned DNA: close homology to the serine antiproteases. Proc Natl Acad Sci USA 1986; 83:7708-7712.
84. Kageyama R, Ohkubo H and Nakanishi S. Primary structure of human preangiotensinogen deduced from the cloned cDNA sequence. Biochemistry 1984; 23: 3603-3609.

CHAPTER 2

THREE DIMENSIONAL STRUCTURES OF SERPINS

The first X-ray structure of a serpin was determined as recently as 1984 and was that of reactive center-cleaved, and therefore inactive, α_1-proteinase inhibitor.[1] It was a further six years before the next serpin structure was published. That was also of a reactive center-cleaved serpin, but of ovalbumin, a serpin with no known inhibitory properties.[2] This structure was quickly followed by that of uncleaved ovalbumin.[3] After this, new structures appeared much more frequently and included reactive center-cleaved forms of other inhibitory serpins,[4-7] as well as of an uncleaved, but still non-functional, latent form of plasminogen activator inhibitor-1 (PAI-1).[8] It was only in 1994 that structures finally appeared of inhibitory serpins (antithrombin and α_1-antichymotrypsin) in uncleaved form.[9-11] Together these structures of cleaved and uncleaved inhibitory and non-inhibitory serpins have greatly advanced our understanding not only of serpin structure but also of the physical mechanism of their functioning. They have convincingly shown that the similarities in primary structure, which were originally used to identify serpins, indeed reflect very similar tertiary structures in the core serpin domain (Fig. 1.1) that represents the major part of the molecule. They have also revealed the dramatic structural interconversions that serpins can undergo,[12] which involve the reactive center region and two of the β-sheets, and which are almost certainly critical in the functioning of serpins as inhibitors and in determining whether a given serpin behaves as an inhibitor, as a substrate or as both upon interaction with proteinase.

1. STRUCTURES COMPLETED

Three-dimensional X-ray crystallographic structural information is now available for ovalbumin in native[3] and cleaved[2] forms, for α_1-antichymotrypsin in cleaved form for the plasma protein[6] and in the native state for a recombinant variant,[11] for antithrombin in native (two structures)[9,10] and cleaved[4,7] forms, for α_1-proteinase inhibitor in cleaved form,[1] for equine leukocyte elastase inhibitor in cleaved form[5] and for plasminogen activator inhibitor 1 in the latent conformation[8] (Table 2.1). As yet there is no structure of a serpin in complex with proteinase, which may reflect the instability of such complexes, on the time scale of days, with respect to completion of the proteolytic cleavage process and subsequent release of active proteinase and cleaved serpin (see Chapter 3), thereby making it difficult to maintain a homogeneous protein population during the long times needed for crystallization. Indeed, the crystals used in determination of the first serpin structure (cleaved α_1-proteinase inhibitor) resulted from an attempt to crystallize a complex between α_1-proteinase inhibitor and chymotrypsinogen. Microheterogeneity resulting from post-translational modifications might also be responsible for the relative paucity of structures obtained. Thus ovalbumin was one of the very first proteins to be crystallized,[13] yet satisfactory crystals for X-ray diffraction were only obtained when a single isoform, of the four that result from heterogeneity in phosphorylation of two serine residues, was used for crystallization trials.[3] Even here, the unit cell was found to contain four ovalbumin molecules, each with a slightly different orientation of the α-helical reactive center region relative to the body of the protein. Thus the conformational plasticity of the reactive center loop and the irreversibility of complex formation with proteinase may be intrinsically inimical to obtaining structures of the most interesting serpin species.

Reflecting the great similarities between serpin structures (see below), it has been possible to greatly simplify analysis of the diffraction data by using molecular replacement methods for almost all of the serpins examined to date, instead of preparation of heavy atom derivatives. Molecular replacement has most frequently used the structure of α_1-proteinase inhibitor. The exceptions are the original structure of α_1-proteinase inhibitor, for which no starting structure was available[1] and the structures

Table 2.1. Structures of serpins determined by X-ray crystallography

Serpin	Resolution	Method[1]	R-factor[2]	Coordinate ID. Brookhaven	Ref
α_1-Proteinase inhibitor - human, cleaved	3Å	MIR	0.251	7api	1
Ovalbumin - chicken, cleaved	2.8Å	SIR	0.25	ND[5]	2
Ovalbumin - chicken, native	1.95Å	MR	0.174	1ova	3
Antithrombin - bovine, cleaved	3.2Å	MR/SIR	0.212	1att	4,7
Antithrombin - human, native[3]	3.2Å	MR	0.179	ND	9
Antithrombin - human, native[4]	3Å	MR	0.322	ND	10,46
α_1-Antichymotrypsin - human, cleaved	2.7Å	MR	0.18	2ach	6
α_1-Antichymotrypsin - recombinant human variant, native	2.5Å	MR	0.207	ND	11,47
Plasminogen activator inhibitor-1 - human, latent	2.3Å	MR	0.198	ND	8,48
Leurocyte elastase inhibitor - equine, cleaved	1.95Å	MR	0.176	ND	5

1. Means of determining phase angles. MIR, multiple isomorphous replacement; SIR, single isomorphous replacement; MR, molecular replacement.
2. The R-factor is defined as $(\Sigma|F_0 - F_c|/\Sigma F_0)$.
3. The molecular unit is a dimer composed of one molecule of uncleaved and one molecule of cleaved antithrombin.
4. The molecular unit is a dimer composed of two uncleaved antithrombin molecules with different conformations for the reactive center loop and the A and C β-sheets.
5. ND, coordinates have not yet been deposited in the Brookhaven Database.

of cleaved ovalbumin[2] and cleaved bovine antithrombin.[4,7] For cleaved bovine antithrombin a combination of molecular replacement and single isomorphous replacement gave an improved R-factor over molecular replacement alone, whereas for cleaved ovalbumin, one of the principal reasons for determining this structure was for comparison with that of cleaved α_1-proteinase inhibitor. Using isomorphous replacement rather than molecular replacement avoided possible bias in the structure determined.

2. THE CORE SERPIN DOMAIN

Although there are striking differences between different serpin structures they are all related to the reactive center loop conformation and its interactions with β-sheets A and C (see section 3). Even more striking, however, is the almost invariant structure of the core serpin domain, which was already apparent with the determination of the second serpin structure, that of cleaved ovalbumin. The core domain covers approximately residues 21 to 389 (using the α_1-proteinase inhibitor numbering) and is composed of three β-sheets and nine α-helices (Figs. 2.1 and 2.2). The β-sheets are designated A, B, and C, and the α-helices A through I. β-Sheet A involves the most residues and is composed of 5 strands of polypeptide, followed by β-sheet B which has 6 strands and β-sheet C which has three strands. These β-sheets all involve antiparallel orientations of strands within the

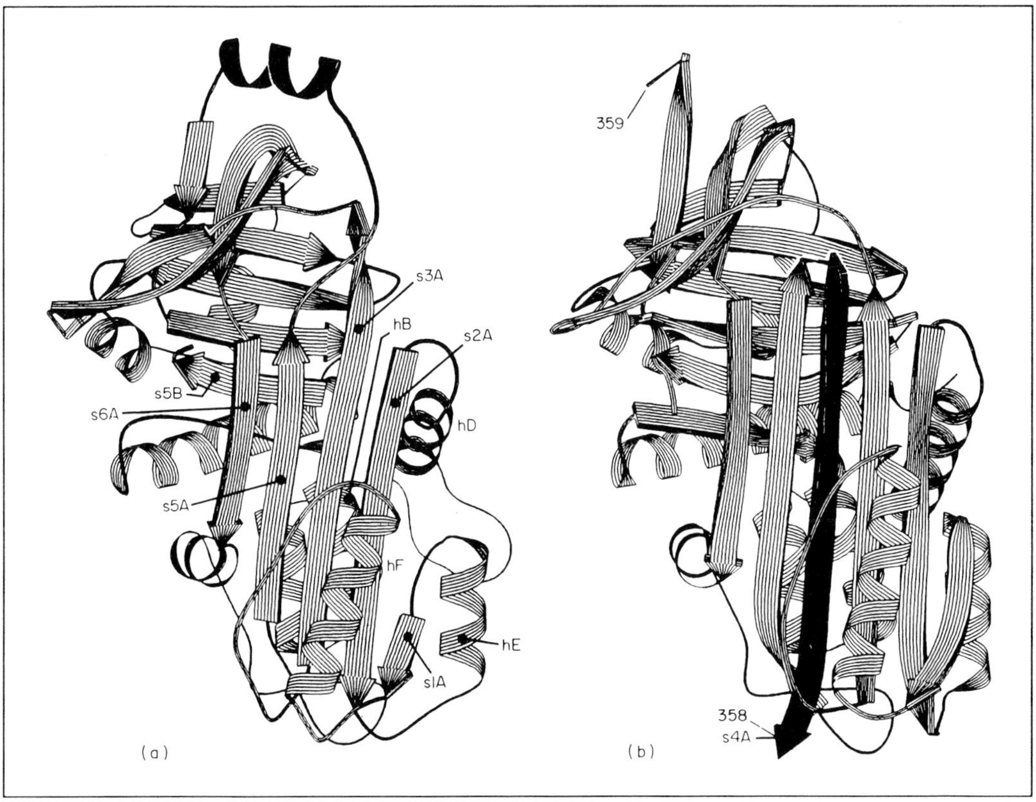

Fig. 2.1. Ribbon diagram of (a) native ovalbumin and (b) cleaved α_1-proteinase inhibitor showing the structural transformation that occurs upon cleavage of the reactive center loop of inhibitory serpins. The reactive center loop of native ovalbumin is the α-helix, shown in black, at the top of the molecule. The equivalent region of cleaved α_1-proteinase inhibitor, also shown in black, is integrated into β-sheet A as strand s4A. The cleaved ends of α_1-proteinase inhibitor, Met 358 (P1) and Ser 359 (P1') are indicated. Reprinted with permission from the Journal of Molecular Biology.[12] *Copyright 1991 by Academic Press Inc.*

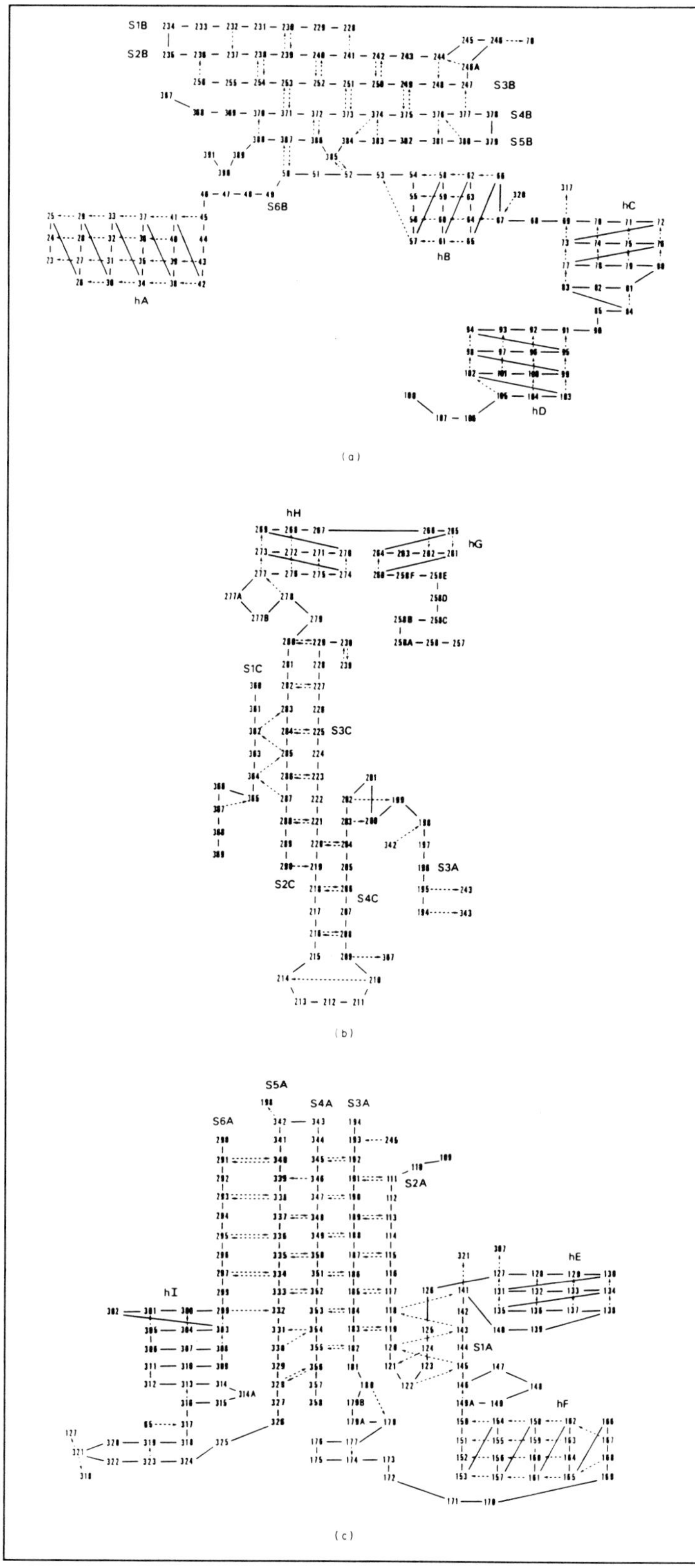

Fig. 2.2. Main chain hydrogen-bonding scheme of cleaved equine leukocyte elastase inhibitor, with arrows directed from amide NH towards O. (a) Helices A-D and β-sheet B; (b) helices G and H and β-sheet C; (c) helices E, F, and I and β-sheet A. Reprinted with permission from the Journal of Molecular Biology.[5] Copyright 1992 by Academic Press Inc.

of β-sheet A. It is possible that the dimers that formed in both these studies were initially identical and involved two uncleaved molecules, one in latent and one in partially loop-inserted conformation. Given the exposed and extended conformation of the P3'-P11' region in the latent subunit, this region may have been cleaved by contaminating proteinase, with loss of these residues.

4. CONFORMATIONAL LABILITY AND MOBILITY OF THE REACTIVE CENTER LOOP

It is clear from the above discussion of very different conformations of the reactive center loop region of serpins and the associated changes in structure of β-sheets A and C that the reactive center loop is capable of undergoing remarkable transformations in structure according to circumstances. Apart from the mechanistic questions that relate to these interconversions, which are addressed in Chapter 3, these properties also raise the thermodynamic questions of how serpins can fold in a metastable non-loop inserted form and whether the reactive center loop is mobile in addition to being conformationally labile.

Although the loop inserted forms of cleaved inhibitory serpins have been shown to be very much more stable than the uncleaved native forms, it could be argued that, since they have different chemical structures (2 chain versus one chain), the increased stability of the cleaved form does not necessarily imply that the uncleaved form is not in its most stable conformation. However, measurements on the stability of active and latent PAI-1, both wild-type and variant forms, indicate that the normal conformation of serpins represents a metastable state and that formation of the loop-inserted latent state is thermodynamically favorable.[23] This kinetic trapping of serpins in a metastable conformation appears to be an essential part of the inhibitory mechanism and is discussed in Chapter 3. The question of the mobility of the reactive center loop is most appropriately discussed here.

Only two X-ray structures of monomeric intact non-latent serpins exist; of ovalbumin and of variant α_1-anti chymotrypsin. Each of these shows the reactive center region to be in helical conformation. The temperature factors for the main chain of this region of α_1-antichymotrypsin are not significantly different from those of the remainder of the molecule[11] (Fig. 2.6), indicating no unusual mobility for this region of the protein. This is also true for ovalbumin,[3] though for ovalbumin there are four different protein molecules in the unit cell, each with a different orientation of the reactive center helix to the body of the protein. This lack of mobility is also what has been found from an NMR study of native ovalbumin and α_1-proteinase inhibitor.[24] The NMR approach is capable of distinguishing small domain motions on the time-scale of 1-50 ns. A distinction must therefore be made between relatively slow conformational lability of the reactive center region, i.e., its ready ability to change conformation, which it clearly possesses, and rapid domain mobility of this region within a given conformation, which it appears to lack.

5. MODELED SERPIN STRUCTURES

The conservation of the structure of the core serpin domain and an awareness of the conformational changes possible for the reactive center loop and β-sheets A and C have led a number of investigators to make up for the relative paucity of X-ray structures of serpins and their ligand complexes by modeling them.

The first serpin model was an attempt to predict the structure of the intact form of α_1-proteinase inhibitor.[25] This predicted structure was derived from the structure of cleaved α_1-proteinase inhibitor[1] by extraction of the reactive center strand from β-sheet A followed by energy minimization and reconnection of Met 358 (P1) to Ser 359 (P1'). The energy minimization procedure resulted in closure of the gap in β-sheet A caused by removal of the reactive center region and gave a core structure very similar to that of plakalbumin. The final structure of the intact reactive center loop was somewhat more problematic, since the conformations open to it are too many and the imposition of a "canonical" structure is not necessarily justified (see Chapter 3). More recently, the structure of latent PAI-1[8] has been used to predict the structure of active PAI-1[26] by removal of the sheet-inserted reactive center region and energy minimization of a structure in which strand s1C was restored to β-sheet C and the reactive center loop from P16 to P3' was given a starting structure based on a secondary structure prediction algorithm. In contrast to the molecular dynamics experiment with α_1-proteinase inhibitor described above, strands s3A and s5A did not come together, which, if a realistic phenomenon, might help explain the tendency of PAI-1 to convert so readily to the latent form.

Three other models were aimed at predicting the heparin binding regions of protein C inhibitor[27] and antithrombin.[28,29] Each prediction again used the cleaved α_1-proteinase inhibitor structure as a starting point, but required replacement of the α_1-proteinase inhibitor polypeptide with the primary structures of protein C inhibitor or antithrombin. A complication was that

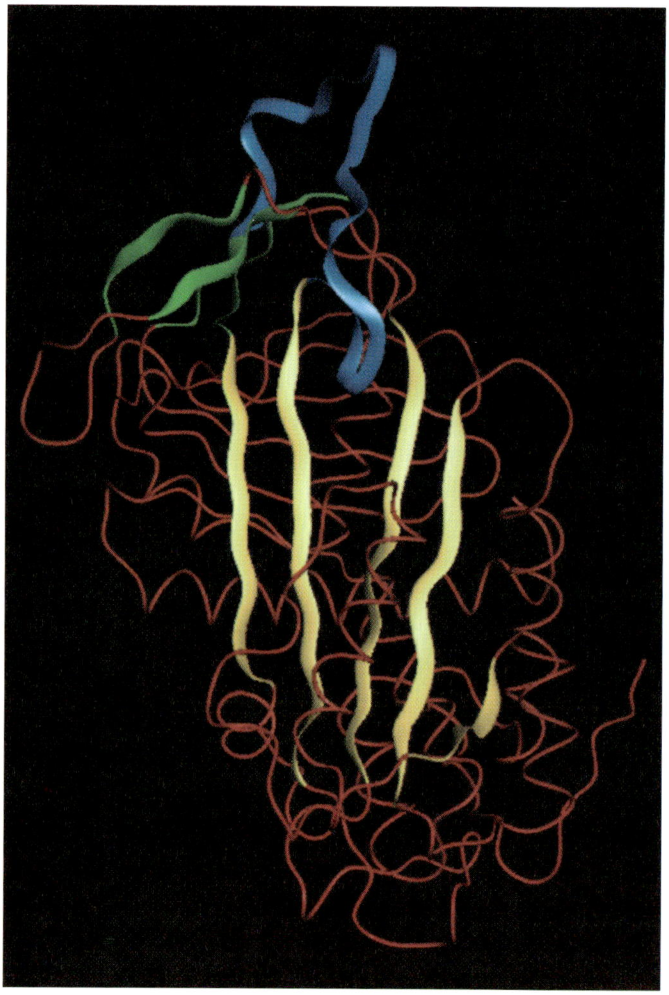

Fig. 2.5. Schematic structures of the two component monomers, (a) and (b), of the antithrombin heterodimer (c). Only one monomer, (a), retains inhibitory activity and has a reactive center region that is extended, whereas the second, (b) corresponds to an inactive, loop-inserted, latent conformation. The interaction between the two monomers (c) involves the association of the extended reactive center of the active molecule with β-sheet C of the second, latent molecule, replacing the missing strand 1. Reprinted with permission from Structure.[10] *Copyright 1994 Current Biology Ltd. Figure kindly provided by Dr. Mark Wardell.*

Fig. 2.5a

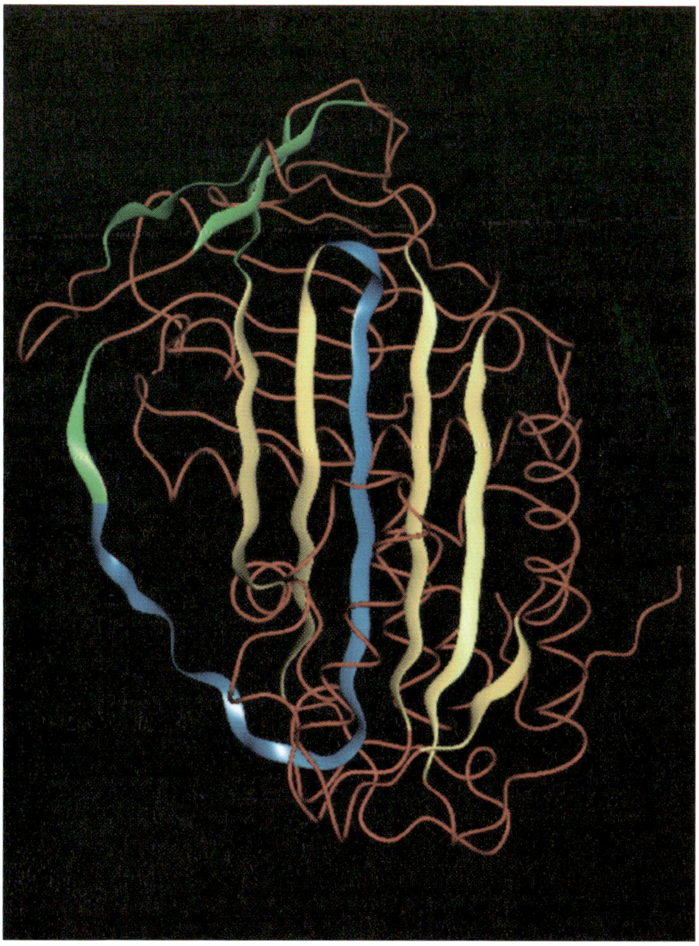

Fig. 2.5b

Fig. 2.5c

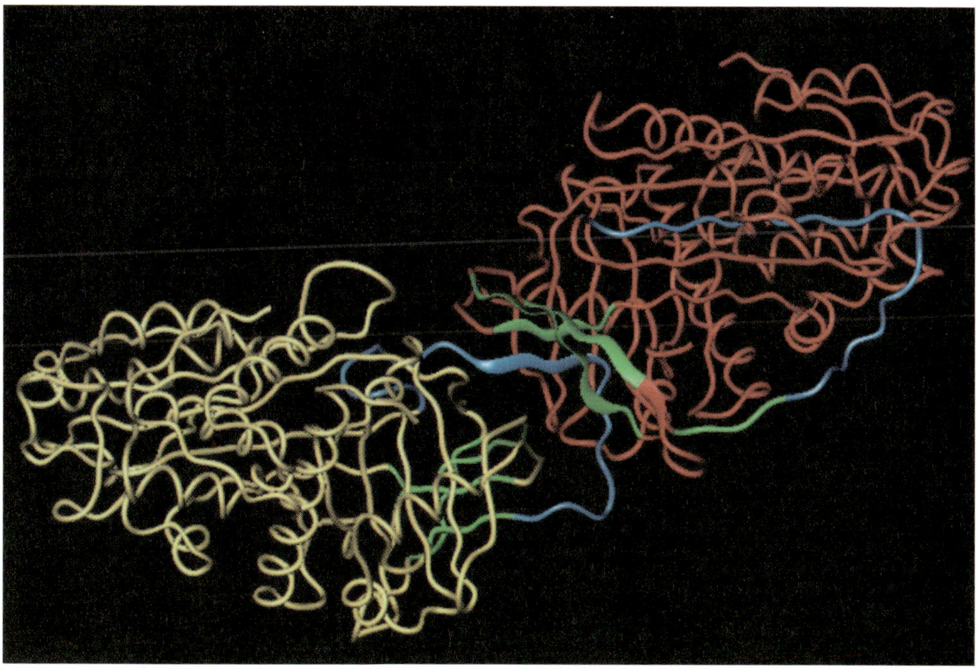

the N-terminal regions are not homologous to the corresponding region of α_1-proteinase inhibitor. In the case of protein C inhibitor this region was modeled separately and predicted to form an α-helix (designated helix A^+). In the case of one of the antithrombin predictions,[29] the modeled region omitted the N-terminal 44 residues. In neither case was an attempt made to reform the P1-P1' bond. Energy minimization of the protein C inhibitor model gave a structure that contained a highly positively charged region formed by the A^+ and H helices, suggesting that this region is the heparin binding site (see below). The earlier antithrombin-heparin model noted the presence of a band of positively charged residues on the D and A helices.[28] In the second antithrombin-heparin model[29] an attempt was made to dock a heparin structure corresponding to the high affinity pentasaccharide[30] with the predicted heparin binding site in the D and A helices. This prediction indicated a single complex in which the binding involved complementary charge-charge interactions. The protein side chains involved were Arg 46, Arg 47, Lys 125, Arg 129, Arg 132, and Lys 133 (see below).

The final two models have also been attempts to define ligand binding sites in these cases for the hormone binding sites of corticosteroid binding globulin (CBG)[31] and thyroxine binding globulin (TBG).[32] These predictions have used the parallel between the β-clam or β-barrel-like structures

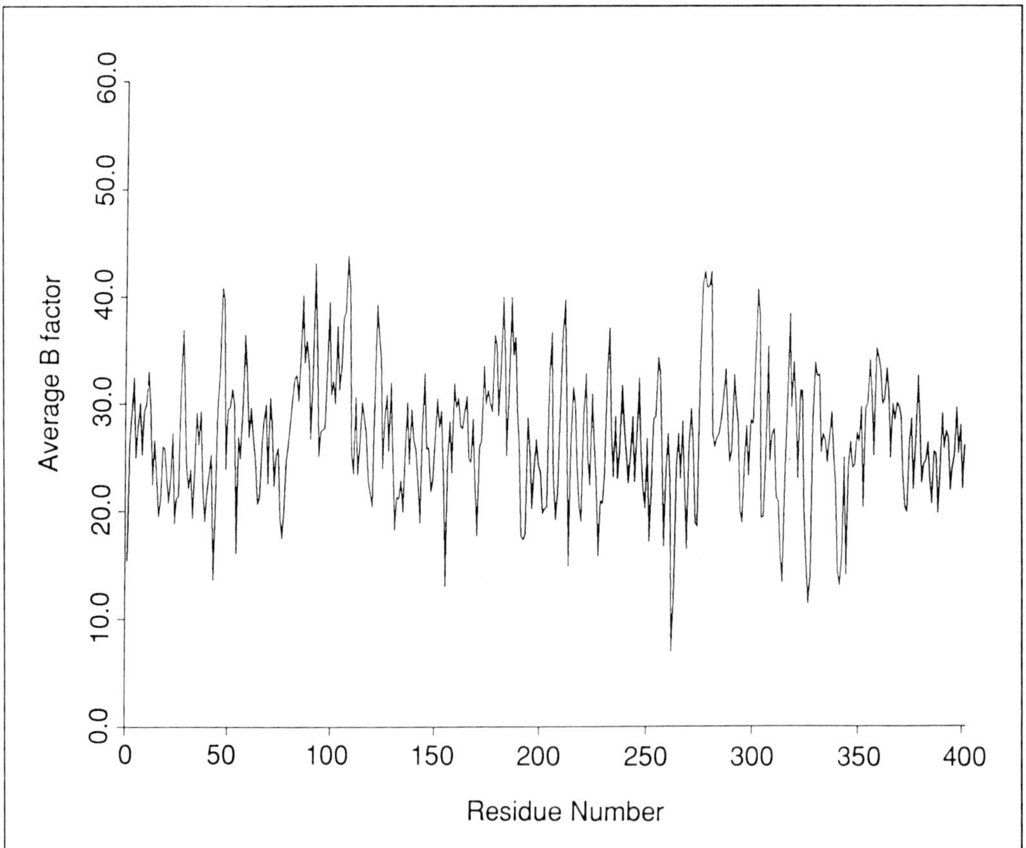

Fig. 2.6. Average main chain temperature factors (B factor) of variant recombinant α_1-antichymotrypsin showing that the reactive center region (residues 342-361, representing P17 to P3') does not have unusually high values that would indicate mobility of that region. Reprinted with permission from Nature Structural Biology.[11] *Copyright 1994, Macmillan Magazines Limited. Figure kindly provided by Dr. David Christianson.*

of a number of small hydrophobic ligand-binding proteins and the hydrophobic pocket formed by β-sheets B and C of serpins to obtain a model of the complexes by docking the hormone ligands with this region of predicted CBG and TBG structures. More details of the proposed binding sites are given in Chapter 6.

6. LIGAND BINDING REGIONS

If the binding of a regulatory protein is included as a ligand binding process, there are at present four types of serpin-ligand binding that are known. The four types of ligand are glycosaminoglycans including heparin, thyroid and corticosteroid hormones, DNA and the protein vitronectin. Where information is available (i.e., for glycosaminoglycan, hormone and vitronectin interactions) details of the role and mechanism of the interactions are given elsewhere. Here we consider only the location of the binding sites. Although there are no structures of complexes of any serpin with its ligand, the structures of the unliganded serpins, in conjunction with mutagenesis and other data, provide a good picture of the heparin binding region of antithrombin, the hormone binding regions of CBG and thyroxine binding globulin, the DNA binding region of α_1-antichymotrypsin, and the vitronectin binding site of PAI-1.

From many studies it is known that antithrombin has a specific high affinity heparin binding site that is involved in the

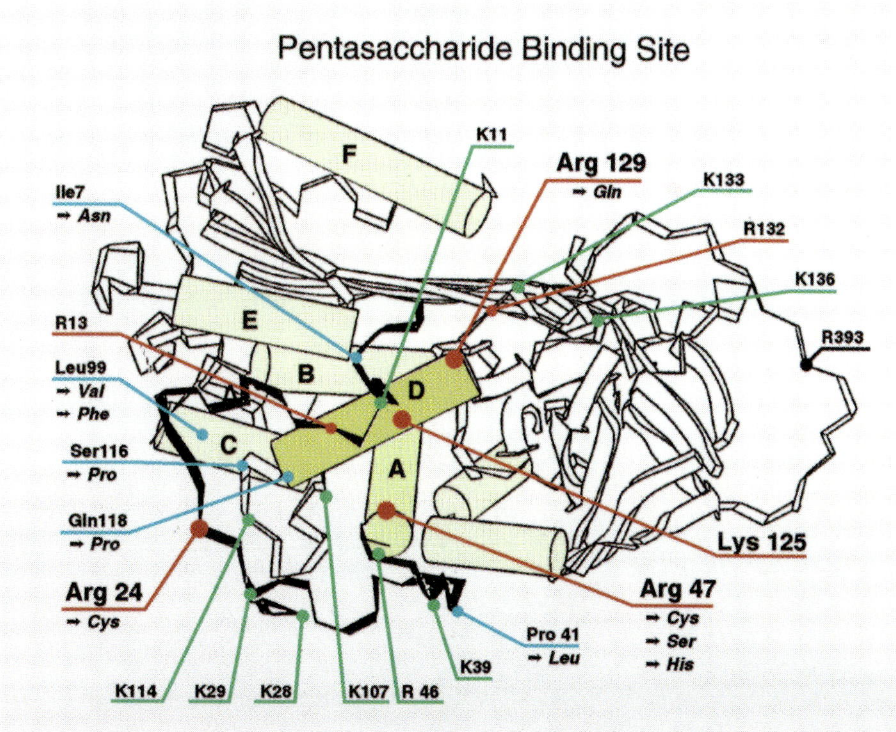

Fig. 2.7. Proposed localization of the heparin pentasaccharide binding site of inhibitory human antithrombin to a region with a high density of positively charged residues. Residues shown in red are basic residues implicated in heparin binding from the properties of natural variants. Residues shown in blue are non-basic residues implicated by natural variants, whereas residues in green are other basic residues in the vicinity. The reactive center conformation is that of the inhibitory monomer of the heterodimeric antithrombin structure. Modified with permission from Structure.[10] *Copyright 1994 by Current Biology Ltd. Figure kindly provided by Dr. Mark Wardell.*

inhibition of serine proteinases. Heparin binding is also a property of heparin cofactor II, protease nexin 1, PAI-1 and protein C inhibitor (see Chapter 4). The heparin binding site of antithrombin involves arginine and lysine residues in the two helices D and A, both of which are located in the N-terminal region of the molecule (Fig. 2.7). From chemical modification studies[33] and from the properties of natural[34] and recombinant variants[35] it has been shown that arginine 47, lysine 125 and arginine 129 all contribute to ionic interactions with the negatively-charged heparin. We have recently shown that arginine 132 and lysine 133 (both on helix D) are also involved in such interactions with heparin (J. Meagher and P.G.W. Gettins, unpublished results). These charged species appear to contribute about half of the ligand binding energy.[36] This cluster of positively charged residues on the outer face of the D helix is conserved in PAI-1, protease nexin 1 and heparin cofactor II and has been proposed to be a common heparin-binding motif.[37-39] In protein C inhibitor this cluster is missing. Instead, charged residues on the H and A⁺ helices (A⁺ is a postulated amphipathic helix that precedes the A helix of α_1-proteinase

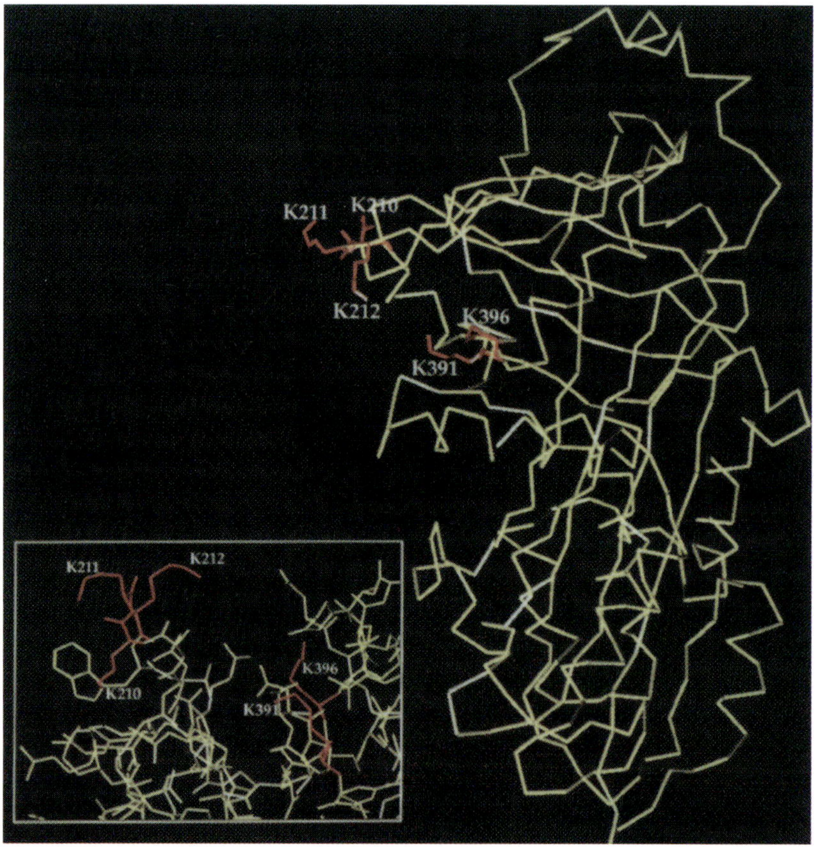

Fig. 2.8. Location of the lysines shown by mutagenesis studies to be involved in the DNA binding site of α_1-antichymotrypsin. The inset is an expansion of the DNA binding region. The backbone structure is that of the P3-P3' variant of α_1-antichymotrypsin. The reactive center region is the distorted helix at the top center and right of the picture. Reprinted with permission from the Journal of Biological Chemistry.[42] Copyright 1995, American Society for Biochemistry and Molecular Biology. Figure kindly provided by Dr. Harvey Rubin.

inhibitor) have been proposed to be involved in heparin binding.[27,40] The apex of the D helix in antithrombin leads into one of the strands of β-sheet A (s2A) and may provide the route for the transmission of heparin-induced conformational change from the heparin binding region to the reactive center region of antithrombin.[41] This mechanism is discussed in more detail in Chapter 4.

α_1-Antichymotrypsin has been reported to have a binding site for DNA,[6] but one without a demonstrated function. The location of the DNA binding region of α_1-antichymotrypsin is less well defined than the heparin binding region of antithrom-bin. Mutagenesis studies have implicated a triplet of lysines (212-214)[42] that occur in an exposed turn between strands 3 and 4 of β-sheet C and that are thus located relatively close to the reactive center region, as well as two other lysines that are more distant in primary structure (391 and 396), but still close in space to the other lysines (Fig. 2.8). This disposition of lysines represents a DNA-binding motif not previously seen in DNA-binding proteins.

As mentioned above (section 2.5), the hormone binding sites of CBG[31] and TBG[32] are thought to occur in the hydrophobic pocket formed by the interior surfaces of β-sheets B and C of the serpin (see Chapter 6).

The vitronectin binding site of PAI-1 has been localized by four different approaches that have used monoclonal antibodies,[43] random mutagenesis,[44] PAI-1/PAI-2 chimeras[45] and peptide mapping.[45] The consensus is that residues within the region 110 to 145 constitute an important determinant for vitronectin binding. Random mutagenesis specifically implicated the clustered group of residues 109, 110, 116 and 123 within this region of the primary structure as well as residue 55,[44] which is close in space to the cluster when modeled on the structure of cleaved PAI-1. These residues are located in helices C and D, which are in contact with one face of β-sheet A. Vitronectin binding at this site may stabilize the active conformation of PAI-1 by limiting the ability of β-sheet A to "breathe" and thus to undergo sheet expansion.

References

1. Löbermann H, Tokuoka R, Deisenhofer J, et al. Human α_1-proteinase inhibitor. Crystal structure analysis of two crystal modifications and preliminary analysis of the implications for function. J Mol Biol 1984; 177:731-757.
2. Wright HT, Qian HX and Huber R. Crystal structure of plakalbumin, a proteolytically nicked form of ovalbumin. Its relationship to the structure of cleaved α-1-proteinase inhibitor. J Mol Biol 1990; 213:513-528.
3. Stein PE, Leslie AGW, Finch JT, et al. Crystal structure of uncleaved ovalbumin at 1.95Å resolution. J Mol Biol 1991; 221:941-959.
4. Mourey L, Samama JP, Delarue M, et al. Antithrombin III: Structural and functional aspects. Biochimie 1990; 72:599-608.
5. Baumann U, Bode W, Huber R, et al. Crystal structure of cleaved equine leucocyte elastase inhibitor determined at 1.95Å resolution. J Mol Biol 1992; 226:1207-1218.
6. Baumann U, Huber R, Bode W, et al. Crystal structure of cleaved α_1-antichymotrypsin at 2.7Å resolution and its comparison with other serpins. J Mol Biol 1991; 218:595-606.
7. Mourey L, Samama JP, Delarue M, et al. Crystal structure of cleaved bovine antithrombin-III at 3.2-Ångstrom resolution. J Mol Biol 1993; 232:223-241.
8. Mottonen J, Strand A, Symersky J, et al. Structural basis of latency in plasminogen activator inhibitor 1. Nature 1992; 355: 270-273.
9. Schreuder HA, de Boer B, Dijkema R, et al. The intact and cleaved human antithrombin III complex as a model for serpin-proteinase interactions. Nature Struct Biol 1994; 1:48-54.
10. Carrell RW, Stein PE, Fermi G, et al. Biological implications of a 3 Å structure of dimeric antithrombin. Structure 1994; 2:257-270.

11. Wei A, Rubin H, Cooperman BS, et al. Crystal structure of an uncleaved serpin reveals the conformation of an inhibitory reactive loop. Nature Struct Biol 1994; 1:251-258.
12. Stein PE and Chothia C. Serpin tertiary structure transformation. J Mol Biol 1991; 221:615-621.
13. Hofmeister F. Über die Darstellung von krystallisirtem Eieralbumin und die Krystallisirbarkeit colloider Stoffe. Z Physiol Chem 1892; 14:165-172.
14. Franzén L-E, Svensson S and Larm O. Structural studies on the carbohydrate portion of human antithrombin III. J Biol Chem 1980; 255:5090-5093.
15. Mizuochi T, Fujii J, Kurachi K, et al. Structural studies of the carbohydrate moiety of human antithrombin III. Arch Biochem Biophys 1980; 203:458-465.
16. Bruch M, Weiss V and Engel J. Plasma serine proteinase inhibitors (serpins) exhibit major conformational changes and a large increase in conformational stability upon cleavage at their reactive sites. J Biol Chem 1988; 263:16626-16630.
17. Hood DB, Huntington JA and Gettins PGW. α_1-Proteinase inhibitor variant T345R. Influence of P14 residue on substrate and inhibitory pathways. Biochemistry 1994; 33:8538-8547.
18. Shitamori S and Nakamura R. Plakalbumin converts to heat-stable form under the same condition as an ovalbumin to S-ovalbumin transformation. J Agric Food Chemistry 1983; 31:513-516.
19. Gettins P and Harten B. Properties of thrombin- and elastase-modified human antithrombin III. Biochemistry 1988; 27:3634-3639.
20. Keijer J, Ehrlich HJ, Linders M, et al. Vitronectin governs the interaction between plasminogen activator inhibitor-1 and tissue-type plasminogen activator. J Biol Chem 1991; 266:10700-10707.
21. Lawrence D, Strandberg L, Grundström T, et al. Purification of active human plasminogen activator inhibitor 1 from *Escherichia coli*. Comparison with natural and recombinant forms purified from eucaryotic cells. Eur J Biochem 1989; 186:523-533.
22. Evans DL, Marshall CJ, Christey PB, et al. Heparin binding site, conformational change, and activation of antithrombin. Biochemistry 1992; 31:12629-12642.
23. Lawrence DA, Olson ST, Palaniappan S, et al. Engineering plasminogen activator inhibitor 1 mutants with increased functional stability. Biochemistry 1994; 33:3643-48.
24. Hood DB and Gettins P. A ^{1}H NMR probe for mobility in the reactive center loops of serpins: Spin echo studies of native and modified forms of ovalbumin and α_1-proteinase inhibitor. Biochemistry 1991; 30:9054-9060.
25. Engh RA, Wright HT and Huber R. Modelling the intact form of α_1-proteinase inhibitor. Prot Eng 1990; 3:469-477.
26. Aertgeerts K, De Bondt HL, De Ranter C, et al. A model of the reactive form of plasminogen activator inhibitor-1. J Struct Biol 1994; 113:239-245.
27. Kuhn LA, Griffin JH, Fisher CL, et al. Elucidating the structural chemistry of glycosaminoglycan recognition by protein C inhibitor. Proc Natl Acad Sci USA 1990; 87:8506-8510.
28. Huber R and Carrell RW. Implications of the three-dimensional structure of α_1-antitrypsin for structure and function of serpins. Biochemistry 1989; 28:8951-8966.
29. Grootenhuis PDJ and van Boeckel CAA. Constructing a molecular model of the interaction between antithrombin III and a potent heparin analogue. J Am Chem Soc 1991; 113:2743-2747.
30. Choay J, Petitou M, Lormeau JC, et al. Structure-activity relationship in heparin: A synthetic pentasaccharide with high-affinity for antithrombin III and eliciting high anti-factor Xa activity. Biochem Biophys Res Commun 1983; 116:492-499.
31. Edgar P and Stein PE. Hormone binding site of corticosteroid binding globulin. Nature Struct Biol 1995; 2:196-197.
32. Jarvis JA, Munro SLA and Craik DJ. Homology model of thyroxine binding globulin and elucidation of the thyroid binding site. Prot Eng 1992; 5:61-67.
33. Peterson CB, Noyes CM, Pecon JM, et al. Identification of a lysyl residue in antithrombin which is essential for heparin

binding. J Biol Chem 1987; 262: 8061-8065.

34. Lane DA, Olds RJ, Boisclair M, et al. Antithrombin III mutation database: First update. Thromb Haemostas 1993; 70: 361-369.
35. Fan B, Turko IV and Gettins PGW. Lysine-heparin interactions in antithrombin. Properties of K125M and K290M, K294M, K297M variants. Biochemistry 1994; 33:14156-14161.
36. Olson ST, Srinivasan KR, Björk I, et al. Binding of high affinity heparin to antithrombin III. Stopped flow kinetic studies of the binding interaction. J Biol Chem 1981; 256:11073-11079.
37. Cardin AD and Weintraub HJR. Molecular modeling of protein-glycosaminoglycan interactions. Arteriosclerosis 1989; 9:21-32.
38. Pratt CW and Church FC. General features of the heparin-binding serpins antithrombin, heparin cofactor-II and protein-C inhibitor. Blood Coag Fibrinolysis 1993; 4:479-490.
39. Ehrlich HJ, Keijer J, Preissner KT, et al. Functional interaction of plasminogen activator inhibitor type 1 (PAI-1) and heparin. Biochemistry 1991; 30:1021-1028.
40. Shirk RA, Elisen MGLM, Meijers JCM, et al. Role of the H helix in heparin binding to protein C inhibitor. J Biol Chem 1994; 269:28690-28695.
41. van Boeckel CAA, Grootenhuis PDJ and Visser A. A mechanism for heparin-induced potentiation of antithrombin III. Nature Struct Biol 1994; 1:423-425.
42. Naidoo N, Cooperman BS, Wang Z, et al. Identification of lysines within α1-antichymotrypsin important for DNA binding. An unusual combination of DNA-binding elements. J Biol Chem 1995; 270: 14548-555.
43. Van Meijer M, Gebbink RK, Preissner KT, et al. Determination of the vitronectin binding site on plasminogen activator inhibitor 1 (PAI-1). FEBS Lett 1994; 352:342-346.
44. Lawrence DA, Berkenpas MB, Palaniappan S, et al. Localization of vitronectin binding domain in plasminogen activator inhibitor-1. J Biol Chem 1994; 269:15223-15228.
45. Padmanabhan J and Sane DC. Localization of a vitronectin binding region of plasminogen activator inhibitor-1. Thromb Haemost 1995; 73:829-834.
46. Wardell MR, Abrahams J-P, Bruce D, et al. Crystallization and preliminary X-ray diffraction analysis of two conformations of intact human antithrombin. J Mol Biol 1993; 234:1253-1258.
47. Wei A, Rubin H, Cooperman BS, et al. Crystallization, activity assay and preliminary X-ray diffraction analysis of the uncleaved form of the serpin antichymotrypsin. J Mol Biol 1992; 226:273-276.
48. Goldsmith EJ, Sheng-Chang C, Danley DE, et al. Preliminary X-ray analysis of crystals of plasminogen activator inhibitor-1. Proteins 1991; 9:225-227.

CHAPTER 3

Mechanism of Action of Serpins as Suicide Substrate Inhibitors

1. COMPARISON OF SERPINS WITH OTHER CLASSES OF SERINE PROTEINASE INHIBITORS

Serpins represent only one of several classes of protein inhibitors of serine proteinases, but one that is distinct from the other classes in a number of important respects.[1] Before considering in detail what is known of the mechanism of serpins, it is instructive to compare serpin and non-serpin inhibitors to illustrate the fundamentally different mechanism of action of serpins.

(i) Gross Structural and Thermodynamic Differences

Some of the most studied classes of non-serpin serine proteinase inhibitors are listed in Table 3.1 together with other structural information. One striking difference from serpins, which have molecular weights of 45-100 kDa, is that these proteinase inhibitors are all comparatively small. Indeed, some of the squash trypsin inhibitors contain only about 30 residues. Even in the case of the much larger ovomucoid inhibitors, which contain about 190 residues, the inhibitory domain is relatively small and represents only one of three Kazal-type inhibitory units present in the whole protein. The possible significance of the size difference is discussed below. Since some of these low M.W. inhibitors share the same target enzymes as some of the serpins, a question that must be answered is why it is advantageous to the organism to use a high molecular weight serpin inhibitor of a given proteinase when a much lower molecular weight inhibitor is capable of tight complex formation and thereby of proteinase inhibition.

A second major difference between serpins and other classes of serine proteinase inhibitor is in the reversibility of complex formation with the target proteinase. Whereas all of the inhibitors in Table 3.1 form reversible complexes with proteinases, those with serpins are considered to be irreversible, with the possible exception of α_2-antiplasmin interacting with trypsin and chymotrypsin, for which there is evidence for recovery of active α_2-antiplasmin from complex with proteinase by incubation in the presence of α_2-macroglobulin (α_2M).[2] Other studies of serpin-proteinase encounters that have claimed to demonstrate reversibility by virtue of reappearance of proteolytic activity from complexes have failed to show that this coincides with reappearance of intact serpin, as would be required by true reversibility, rather than recovery of the cleaved serpin that is expected from slow continuation of the serpin-proteinase interaction as a substrate.[3,4] In keeping with the need to demonstrate concomitant regeneration of uncleaved serpin and free proteinase, a study that demonstrated the ability of α_2M to trap chymotrypsin released from chymotrypsin-α_1-antichymotrypsin complexes showed that this was unlikely to be by reversal of the pathway, since no significant increase in reformed serpin occurred at the same time.[5] Similarly, spontaneous dissociation of antithrombin-thrombin complexes released only cleaved serpin.[6]

Related to this apparent irreversibility of most serpin-proteinase complexes is the energy difference between the native and cleaved forms of serpins. For non-serpin inhibitors the energy difference between the cleaved and native inhibitor is close to zero,[7] and both cleaved and uncleaved forms of the inhibitor are effective in forming stable complexes with proteinase and thus acting as proteinase inhibitors. In contrast, not only is there a very large energy difference between cleaved and uncleaved serpins, but the cleaved forms are incapable of acting as inhibitors.[8] Quantitation of the energy differences between native and cleaved serpins is made difficult by the very much greater stability of the cleaved forms, which results in them being stable even at 100°C.[9]

This marked difference in thermodynamic stability between cleaved and uncleaved serpins is a manifestation of the more general difference between serpins and other protein proteinase inhibitors, namely that native serpins fold in a kinetically-trapped metastable state. Indeed serpins belong to a small group of proteins that fold in a native conformation that is a kinetically-determined local energy minimum rather than a thermodynamically-determined global energy minimum.[10] The latter had previously been thought to occur for all proteins.[11] It has already been pointed out (Chapter 2 section 4) that this metastability can lead to spontaneous change to a more stable conformation such

Table 3.1. Major families of non-serpin serine proteinase inhibitors

Family	Number of example	Number of residues	S-S bonds	Ref
BPTI (Kunitz)	Bovine pancreatic trypsin inhibitor	58	3	107
Pancreatic secretory trypsin inhibitor (Kazal)	Bovine pancreatic secretory trypsin inhibitor	56	3	108
Streptomyces subtilisin	*Streptomyces*-subtilisin inhibitor (S-SI)	113	2	109
Soybean trypsin (Kunitz)	Soybean trypsin inhibitor (Kunitz)	181	2	110
Soybean (Bowman-Birk)	Lima bean trypsin inhibitor	70	7	111
Potato inhibitor I	Eglin C	69	0[a]	112
Potato inhibitor II	Chymotrypsin inhibitor-1	51	4	113
Chelonianin	Mucus proteinase inhibitor	107	8	114
Squash seed	*Cucurbita maxima* trypsin inhibitor	29	3	115
Ascaris	*Ascaris* trypsin inhibitor	62	5	116

[a] Although eglin C and other inhibitors of this family do not have disulfide bonds, they are very stable as a result of strong electrostatic and hydrogen bonding interactions with the reactive site loop.[117]

as in the formation of the "latent" form of PAI-1. It is discussed below how this metastability plays an essential role in the inhibitory mechanism. Since native serpins are not in their most stable conformation, it is not surprising that they have relatively low unfolding temperatures, typically about 60°C. This contrasts with non-serpin inhibitors which have unusually high denaturation temperatures, which is due in part to the presence of one or more disulfide bonds in relatively small structures. The best documented example of a kinetically-trapped metastable protein structure is α-lytic proteinase, which is synthesized in a zymogen form.[12] The N-terminal pro portion of the zymogen serves to promote folding of the catalytic domain in the energetically lowest, catalytically active state. In the absence of the pro region, the unfolded form of the catalytic domain refolds as a metastable intermediate on the pathway to the lower energy active form. The failure of the catalytic domain to fold to its lowest energy active state in the absence of the pro-region is due to the size of the kinetic barrier between these conformations. This energy barrier has been estimated as greater than 27 kcal mol^{-1}, but is reduced by at least 23 kcal mol^{-1} by the pro-region[10] (Fig. 3.1). The parallel with serpin-proteinase interactions may be that the proteinase serves a role equivalent to the pro-region of α-lytic proteinase to promote conversion of the metastable serpin conformation to a more stable form by lowering the energy barrier for interconversion.

(II) Structural Differences in the Reactive Center Loop

The binding interaction between non-serpin serine proteinase inhibitors and proteinases is well understood from a combination of kinetic and X-ray structural studies. X-ray structures have been determined of complexes formed between

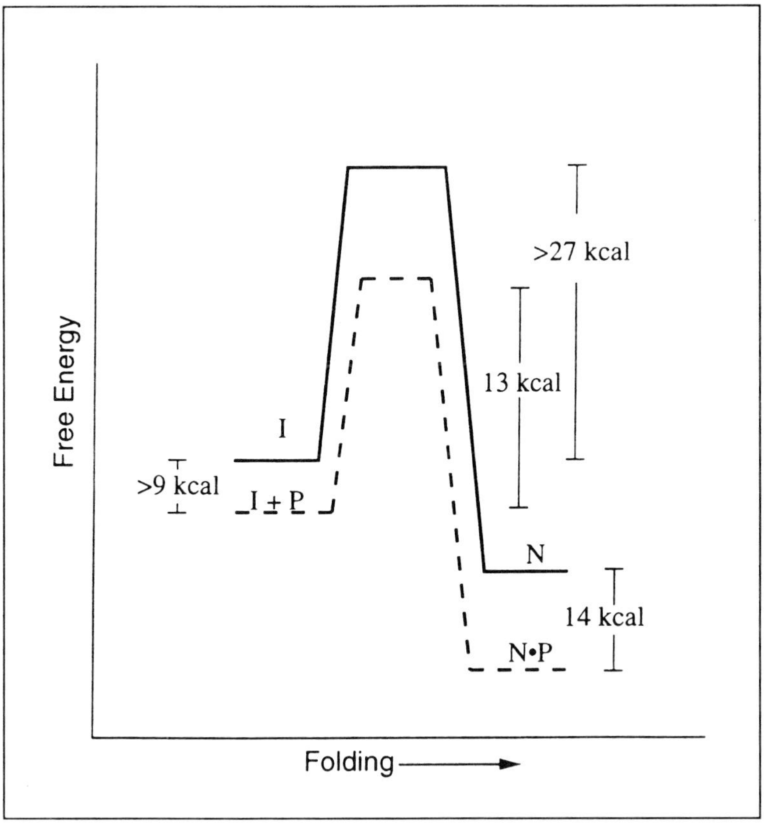

Fig. 3.1. Free energy diagram for the interconversion of the trapped metastable conformation of alpha-lytic proteinase (form I) into the more stable active conformation (N) in the absence (solid line) or presence (dashed line) of the pro-region (P). The effect of binding the pro-region is to lower the barrier for conversion of I into N by > 23 kcal mol^{-1}, resulting in an acceleration of the conversion rate of >10^7-fold. Reprinted with permission from Biochemistry.[10] Copyright 1994 by the American Chemical Society. Figure kindly provided by Dr. David Agard.

inhibitors from several of the classes given in Table 3.1 and proteinases from mammalian, amphibian, plant and bacterial sources (Table 3.2). Although there are differences in the precise position of the contact interactions between inhibitor and proteinase, the very high affinity of the complex results in all cases from multiple non-covalent interactions. Most commonly these involve complementarity between residues in the P3-P3' region interacting with the S3-S3' subsites on the proteinase. In these instances the portion of the inhibitor reactive center that interacts with the proteinase adopts an extended β conformation that has been termed the "canonical conformation".[13] This is true irrespective of the class of the inhibitor (Table 3.3) and of the large differences in structure of the protein "scaffolds" that hold the reactive center loops in place.[14] Since there are also X-ray structures of uncomplexed inhibitor for a number of these interactions, it is possible to determine the extent of the conformational change necessary for optimal interaction with the proteinase. In all cases little change is necessary within this region of

Table 3.2. Selected X-ray structures of serine proteinases with non-serpin inhibitors

Inhibitor Family	Proteinase Inhibitor:Proteinase Complex	Resolution (Å)	Ref.
BPTI	BPTI:bovine trypsin	1.9	119
	BPTI:bovine trypsinogen	1.9	120
	BPTI:porcine glandular kallikrein	2.3	121
Kazal	Porcine pancreatic secretory trypsin inhibitor:bovine trypsinogen	1.8	122
	Turkey ovomucoid third domain inhibitor:*Streptomyces griseus* proteinase B	1.8	123
	Turkey ovomucoid third domain inhibitor:bovine α-chymotrypsin	1.8	124
	Turkey ovomucoid third domain inhibitor:human leukocyte elastase	1.8	125
	Indian peafowl ovomucoid third domain inhibitor:human leukocyte elastase	1.8	118
STI-Kunitz	Soybean trypsin inhibitor:porcine trypsin	2.6	21
	Proteinase K:wheat α-amylase inhibitor	2.5	126
S-SI	*Streptomyces* subtilisin inhibitor:subtilisin BPN'	2.2	127
Potato I	Barley chymotrypsin inhibitor-2:subtilisin Novo	2.1	128
	Eglin C:subtilisin Carlsberg	1.2	13
	Eglin C:thermitase	2.2	129
Potato II	Chymotrypsin inhibitor-1:*Streptomyces griseus* proteinase B	2.1	130
Chelonianin	Human mucus inhibitor:bovine α-chymotrypsin	2.5	131
Bowman-Birk	Azuki bean proteinase inhibitor:bovine trypsin	3.0	132
	Mung bean trypsin inhibitor:porcine trypsin	2.5	133
Squash seed	*Cucurbita maxima* trypsin inhibitor:bovine trypsin	2.0	134
Ascaris	*Ascaris* chymotrypsin/elastase inhibitor:porcine elastase	2.4	135

Adapted from Bode and Huber[118] and updated

the protein, so that not much of the energy of interaction is wasted on promoting the optimal conformation for binding. For some inhibitors, such as chymotrypsin inhibitor-2[15] and *Streptomyces* subtilisin inhibitor,[16] there are high temperature factors associated with the reactive center loop regions, reflecting some mobility that might be beneficial for adapting to the active site structures of different enzymes, with consequently lowered affinity for proteinase. However, even in these cases, the change in structure of the reactive center loop itself upon complexation is minor.[17] These structurally different classes of serine proteinase inhibitor thus present a common conformation of reactive center loop residues to the proteinase that could be said to represent an optimized substrate-binding conformation.[17] Ironically analysis of actual substrate cleavage sites by trypsin in a number of proteins has shown that the conformation at, and surrounding, the cleavage site is far from this optimal conformation, thus requiring considerable change in local conformation and large changes in loop position relative to the remainder of the structure to interact with the proteinase active site.[17] On the basis of this comparison the (non-serpin) inhibitor-proteinase interaction has been described as the classic "lock and key" fit[17] of Fischer,[18] whereas true substrate-proteinase interactions are better described as "induced fits." Such flexibility in substrates but not in inhibitors is required to permit adoption of a conformation in which the transition state rather than the ground state is stabilized and thus to promote the proteolytic reaction rather than inhibition. For the rigid reactive site loop of the inhibitors only ground state stabilization is favored, with consequent inhibition.

Much more limited X-ray data are available on the structure of the reactive center loop of inhibitory serpins, and none are available for a serpin-proteinase complex. In the structure of an inhibitory α_1-antichymotrypsin variant,[19] the reactive center loop is a distorted helix. In ovalbumin, admittedly not an inhibitory serpin, it is α-helical. If such conformations are general to native serpins, major conformational changes involving the reactive center loop are likely to occur upon interaction with target proteinase to provide extended interactions with specificity sites on the proteinase.

(III) Involvement of the Proteinase Catalytic Triad in Inhibitor Binding

The view of the nature of the interaction between the catalytic site serine and the P1-P1' carbonyl in complexes between proteinase and non-serpin inhibitors has changed markedly over the years. Based on the original X-ray structures of trypsin complexes with Kunitz inhibitors (bovine pancreatic trypsin inhibitor and soybean trypsin inhibitor) it was thought that the active site serine formed a tetrahedral intermediate with the peptide carbonyl as a result of attack of the serine γ-O.[20,21] Several

Table 3.3. Backbone conformation of reactive site residues, P3-P2', of non-serpin inhibitors

	P3		P2		P1		P1'		P2'	
	Φ	Ψ	Φ	Ψ	Φ	Ψ	Φ	Ψ	Φ	Ψ
Bovine pancreatic trypsin inhibitor	-86	-28	-64	152	-116	87	-135	166	-116	87
Soybean trypsin inhibitor	-46	-21	-77	136	-89	85	-135	-115	-175	-179
Subtilisin inhibitor	-123	141	-52	134	-92	89	-117	169	-120	78
Hypothetical subtilisin substrate	-157	172	-85	162	-116	26	-62	136	-112	70

subsequent pieces of data made this interpretation untenable. Thus a refinement of the X-ray structure gave a separation of 2.6Å between the carbonyl carbon and serine oxygen, which is much too long for a normal covalent bond.[22] ^{13}C NMR measurements, using the sensitivity of the carbon chemical shift to changes in carbon hybridization, indicated minimal change in the nature of the carbonyl carbon.[23] Finally it has been shown that anhydrotrypsin, in which the active site serine has been changed by β-elimination to a dehydro-alanine residue, binds to trypsin inhibitor with similar affinity, thereby ruling out covalent interaction between the active site serine and the inhibitor peptide as a necessary part of the binding process.[24] The situation with serpins is the opposite of that with the non-serpin inhibitors. Thus anhydrotrypsin binds to PAI-1, to α_1-proteinase inhibitor and to antithrombin with apparent K_A at least four to six orders of magnitude smaller than does trypsin, showing that the active site serine hydroxyl is needed for tight binding.[25] The authors of a recent study that appears to contradict this need for a functional active site may have misinterpreted their data.[26] In this study human neutrophil elastase, trypsin and cathepsin G were reacted with the active site-directed inhibitor 3,4 dichloroisocoumarin (DCI) and subsequently shown to be able to form SDS-stable complexes. However, we have found that the amount and kinetics of complex formation paralleled the rate of decay of the DCI-modified proteinase to regenerate active enzyme (S.T. Olson, unpublished results). SDS-stable complex formation therefore almost certainly resulted from normal serpin reaction with regenerated functional proteinase.

(iv) Importance of Reactive Center Residues in Determining Specificity

Both serpin and non-serpin serine proteinase inhibitors usually show primary proteinase specificity determined by the P1 residue (Table 3.4). A consequence of this is that changes at this position, whether through species variation, natural mutation or in vitro site-directed mutagenesis, can result in loss of inhibitory properties or a change in inhibitory specificity (Table 3.4). An example of this is the methionine to arginine change in the serpin α_1-proteinase inhibitor (Pittsburgh variant), which converts the serpin from an inhibitor of elastase to an inhibitor of proteinases involved in blood coagulation.[27-30] In vivo this mutation led to hemorrhage and ultimately death in the boy found to have this mutation.[27] Similarly there are a number of antithrombin P1 variants in which replacement of the normal arginine with cysteine,[31,32] proline,[33] or histidine[33-37] resulted in reduction or loss of thrombin inhibitory activity. In the non-serpin inhibitors, change of P1 from lysine or arginine to phenylalanine or tryptophan resulted in change from a good inhibitor of trypsin to a good inhibitor of chymotrypsin.[38] However, P1 alone can not be the sole determinant of inhibitory specificity, since there are examples of inhibition of proteinase despite an apparently inappropriate residue at P1. Thus amongst the serpins heparin cofactor II is principally an inhibitor of the arginine-specific proteinase thrombin, despite having a leucine at P1[39,40] (hence an earlier name for heparin cofactor II of leuserpin[41]). However, mutation of the P1 leucine of heparin cofactor II to arginine did produce a faster inhibitor of thrombin.[42] A second example is α_1-proteinase inhibitor, which inhibits trypsin (hence the alternative, but somewhat inappropriate name of α_1-antitrypsin), despite having methionine at P1.[43]

In addition to the role of P1 in both types of inhibitor, interactions beyond this are also important for effective inhibition. In the case of non-serpin inhibitors there is an extended contact region that can involve residues on either side of the P1 residue, from as far away as P6 to P3'.[14] In serpins it has also been shown that changes beyond P1 can change the proteinase

Table 3.4. Influence of P1 and P1 mutations on proteinase inhibitory specificity

Inhibitor	P1 residue	Proteinase specificity	Ref.
NON-SERPIN INHIBITORS			
BPTI	Lys^{wt}	trypsin	136
	Val	elastase	136
Turkey ovomucoid	Leu	chymotrypsin	137
	Glu	*S. griseus* proteinase GluSGP	137
	Lys	trypsin	46
SSI	Met^{wt}	*Streptomyces* subtilisin	138
	Asp	10^3 reduction in subtilisin K_a	138
SERPINS			
Antithrombin	Arg^{wt}	thrombin, factor Xa	139
	Cys	not an inhibitor of thrombin	140
α_1-Proteinase inhibitor	Met^{wt}	elastase	141
	Arg	thrombin	27,142
	Leu	elastase	143
α_1-Antichymotrypsin	Leu^{wt}	cathepsin G	144
	Met	elastase	145
	Arg	trypsin	57
	Trp	chymase	146
C1-inhibitor	Arg^{wt}	C1s, kallikrein	147,148
	Glu	not an inhibitor of C1s	149
	His	not an inhibitor of C1s	150
Heparin cofactor II	Leu^{wt}	thrombin, chymotrypsin	151
	Arg	thrombin, factor Xa	42
PAI-1	Arg^{wt}	uPA, tPA	152
	Lys	uPA, tPA	153
	other	some inhibit tPA but not uPA	153,154
Protein C inhibitor	Arg^{wt}	protein C, thrombin	155
	Met	chymotrypsin	156

specificity.[44,45] However, whereas other mutations have additive effects in non-serpin inhibitors[46] (reviewed in Wells[47]), this appears not to be the case for serpins.[44] Even more distant interactions can also influence specificity between inhibitor and proteinase. Thus an insertion loop of seven residues in tissue-plasminogen activator (tPA) has been shown to affect the interaction with the serpin PAI-1, consistent with a favorable interaction between this loop and residues 350-355 of the serpin.[48] Deletion of the seven residue tPA loop made the proteinase resistant to inhibition by the serpin, though without affecting its ability to activate plasminogen.[49] Specific lysine residues in the carboxy-terminal end of α_2-antiplasmin have also been shown to interact with kringle domains of plasmin.[50,51]

(V) Mechanism of Inhibition by Non-Serpin Inhibitors

From the above discussion we now summarize the essential features of the mechanism of inhibition of non-serpin inhibitors, to provide a sharp contrast between their mode of action and that of serpins. Non-serpin inhibitors have rigid reactive center loops in an extended β-conformation that can interact with target proteinases by extended non-covalent interactions involving residues on both sides of the P1 primary specificity-determining residue. Complementarity involves both the side chains and the peptide backbone. These interactions result in extremely high affinities that stabilize the initial encounter complex, which accordingly resembles a Michaelis complex. This stabilization of the ground state and the inability of the reactive center to readily change conformation to provide optimum complementarity in the transition state, traps the inhibitor in the Michaelis complex and ensures that a substrate cleavage reaction is not favored. In contrast, good substrates of the same proteinases are more flexible and use their proteinase binding energy to stabilize the transition state. As a result of the non-covalent Michaelis-like nature of these non-serpin-proteinase complexes, they are readily reversible. In contrast to these properties, serpins have a conformationally-labile reactive center loop, which is therefore more substrate-like, require covalent interaction with the catalytic residues of the proteinase, and as a consequence almost always give irreversible complex formation.

2. OUTCOME OF SERPIN-PROTEINASE INTERACTIONS. SUBSTRATE VERSUS INHIBITION REACTIONS AND THE BRANCHED PATHWAY MECHANISM

Next to the irreversibility of complex formation, the clearest indication that serpins are mechanistically distinct from other protein inhibitors of serine proteinases is the range of outcomes that are possible for a serpin-proteinase encounter even though the encounter involves recognition of the P1-P1' bond of the serpin in each case. The possible outcomes are: (i) the serpin acts completely as an inhibitor, e.g., α_1-proteinase inhibitor with elastase; (ii) the serpin acts completely as a substrate, e.g., ovalbumin with elastase; (iii) for variant serpins, a complete switch from inhibitor to substrate, e.g., C1 inhibitor Ma (P12 variant ala→glu) with C1s;[52] and (iv) for variant serpins a partial change from inhibitor to substrate such that both outcomes occur side by side, e.g., T345R α_1-proteinase inhibitor (P14 threonine→arginine).[53]

The simplest mechanism consistent with the range of outcomes outlined above for serpin-proteinase encounters and with kinetic data for a number of variant serpins compared to wild type is a branched pathway mechanism (Fig. 3.2) in which the initial Michaelis complex and the subsequent, though as yet not well defined, intermediate EI' are common to both substrate and inhibitory outcomes. Steps leading up to formation of EI' are typically rate limiting, so that values of k_3 and k_4, which determine the *partitioning* of EI', do not affect the overall rate of reaction of serpin and proteinase. However, the fate of the EI' complex determines the actual

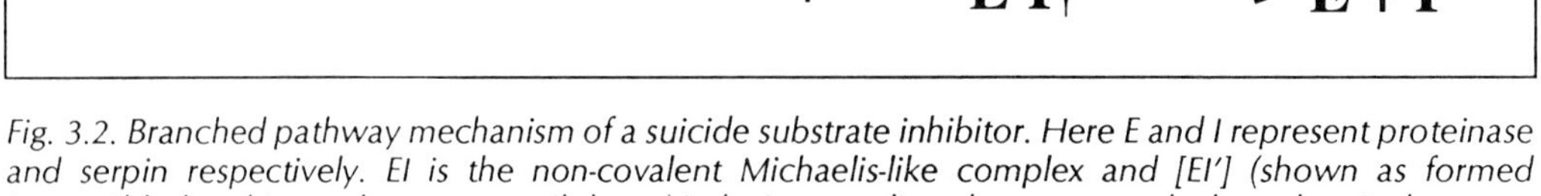

Fig. 3.2. Branched pathway mechanism of a suicide substrate inhibitor. Here E and I represent proteinase and serpin respectively. EI is the non-covalent Michaelis-like complex and [EI'] (shown as formed irreversibly, but this need not necessarily be so) is the intermediate that occurs at the branch point between substrate (upper) and inhibitory (lower) pathways. E-I† represents the long-term stable covalent intermediate between serpin and proteinase. However, this can decay at a very slow rate (k_5) to regenerate free proteinase, E, and release the same cleaved serpin (I) as is formed directly and much more quickly by the substrate pathway. Steps leading up to the branch point are considered rate-limiting here, so that alterations in k_3 and k_4, while altering the product outcome of the reaction, do not alter the overall rate of disappearance of I. No attempt has been made in this general scheme to assign particular structures to [EI'] and E-I†. However, see the text and Figure 3.4 for a discussion and representation of how the real reaction may proceed.*

outcome of the serpin-proteinase encounter. This in turn depends on the relative values of the rate constants k_3 and k_4. If $k_3 >> k_4$ the outcome is predominantly a substrate reaction. If $k_3 << k_4$ the reaction is predominantly inhibitory. If $k_3 \sim k_4$ both outcomes are observed, with product ratios dependent on the relative values of k_3 and k_4. Because the route to substrate reaction represents a normal encounter of a proteinase with its substrate, the alternative route to stable serpin-proteinase complex (EI†) can be considered to be a subversion of the normal substrate pathway. In this way the serpin is behaving as a "suicide substrate inhibitor" in which the ability to act as an inhibitor appears only after interaction with the proteinase.[54]

In considering product outcomes, it is useful to define a "stoichiometry of inhibition" or SI as the ratio of mols of serpin needed to inhibit one mol of proteinase. Thus an SI value of 1 represents complete inhibition reaction, whereas a value of infinity represents complete substrate reaction and a value in between represents mixed substrate and inhibition reactions. For the branched pathway given in Figure 3.2, the value of SI can also be represented as $(k_3 + k_4)/k_4$.

The evidence in support of a branched pathway mechanism rather than an alternative parallel pathway mechanism, in which independent substrate and inhibitory pathways co-exist, comes in part from measurements of the rates of reaction for variant serpin-proteinase encounters compared with the rates of inhibition for the wild-type serpin. According to the branched pathway mechanism, the apparent second order rate constant for inhibition is given by eq 1, under conditions where $k_5 << k_4$ so that k_5 can be ignored on the time scale of the experiment :

$$k_{app} = \frac{k_2}{K_S} \times \frac{k_4}{(k_3 + k_4)} \qquad \text{(eq. 1)}$$

Under conditions where the substrate pathway is insignificant relative to the inhibitory pathway (i.e., $k_3 << k_4$) this simplifies to:

$$k_{app} = \frac{k_2}{K_S} \qquad \text{(eq. 2)}$$

However, when k_3 is significant the value of k_{app} diminishes. Since this is not a result of alterations in the inhibitory branch of the pathway per se, it gives a misleading impression of the rate of the actual inhibitory reaction. As long as formation of EI' is rate limiting, an apparent rate constant that represents only the rate constant for the inhibitory branch of the pathway, k'_{app}, can be obtained by correcting k_{app} for the effect of partitioning of the EI' complex such that only a fraction of EI' leads to stable complex. This is given by:

$$k'_{app} = k_{app} \times \frac{(k_3 + k_4)}{k_4} = \frac{k_2}{K_S} \qquad \text{(eq. 3)}$$

where the factor $(k_3 + k_4)/k_4$ represents SI. This modified rate constant is very useful for evaluating the effects of mutations or other perturbations on different steps in the pathway. Several serpin species have been examined in this way. They include P14,[53] P12 and P10[55] variants of α_1-proteinase inhibitor, a P14 variant of anti-thrombin[56] and P1 variants of α_1-antichymotrypsin.[57] For each of these there was a change in SI; small in some cases but very large in others (Table 3.5). Nevertheless, the effect was largely restricted to the outcome of the reaction rather than the reaction rate at which serpin and proteinase interacted, as judged by the approximate constancy of the k_2/K_S values for a given proteinase-serpin pair (Table 3.5). Although k_{app} values showed large decreases as a result of the diminution in the inhibitory pathway, the k'_{app} values showed little change, indicating that the mutations did not significantly affect the rate of formation of the EI' complex that precedes the branch point. This is what is expected from the branched pathway mechanism since this mechanism predicts that if a mutation does not affect the rate-limiting steps leading to formation of the intermediate that precedes the branch point, even though it may greatly affect the product distribution (SI), there

Table 3.5. SI values and apparent second order rate constants for inhibition for selected serpin variants

Variant	Proteinase	SI	k_{app}[b]	k'_{app} (k_2/K_S)	$k_{app\ w.t.}$	Ref.
P14						
Antithrombin S→W	thrombin	10	3×10^2	3×10^3	8.5×10^3	56
	factor Xa	50				56
	trypsin	40	5×10^3	2×10^5	1.3×10^5	56
α_1-PI T→R	HNE[a]	45				53
	PPE	70	1.6×10^3	1.1×10^5	2.8×10^5	53
	trypsin	9.5				53
P12						
α_1-PI A→T	HNE	1.0	1.0×10^7	1.0×10^7	1.2×10^7	55
	trypsin	1.2	5.4×10^5	6.5×10^5	2.2×10^5	55
P10						
α_1-PI G→P	HNE	> 100	1×10^5	1.0×10^7	1.2×10^7	55
	trypsin	5.6	1.0×10^5	5.4×10^5	2.2×10^5	55
P1						
α_1-Antichymotrypsin L→R	chymotrypsin	4.4				5
α_1-Antichymotrypsin L→M	chymotrypsin	1.0	3×10^5	3×10^5	6×10^5	5, 57
	chymase	2.0	0.5×10^5	0.9×10^5	1.1×10^5 [c]	146
	HNE	5.0				145
α_1-Antichymotrypsin L→F	chymotrypsin	1.0				5
	chymase	7.0	0.2×10^5	1.7×10^5	1.1×10^5 [c]	146
α_1-Antichymotrypsin L→W	chymase	1.5	1.5×10^5	2.3×10^5	1.1×10^5 [c]	146

a, Abbreviations: HNE, human neutrophil elastase; PPE, porcine pancreatic elastase. b, Units of $M^{-1}s^{-1}$. c α_1-Antichymotrypsin gives an SI of 4.0 for reaction with chymase, therefore the values for $k_{app\ w.t.}$ have been corrected for SI by multiplying by 4.0.

should be no significant change in the overall flux through the reaction pathway (i.e., the rate of consumption of serpin will be unaffected). In contrast, the observed constancy of the rate of consumption of serpin is very hard to explain by a parallel pathway scheme and provides very strong evidence against it. Thus, if a mutation results in a reduction in the rate of an independent inhibitory pathway but does not affect the rate of the substrate pathway, blockage of the inhibitory path will reduce the overall consumption of serpin (the summed flux through both pathways) to that of the substrate branch of the pathway, which is known from the wild-type reaction to be very slow relative to the inhibitory rate (i.e., $k_4 >> k_3$ since SI~1). The effect will therefore be to greatly reduce the rate of consumption of serpin. As stated above, the opposite is observed, namely little or no change in serpin consumption.

A second kinetic demonstration of the operation of the branched pathway mechanism has been afforded by a characterization of the rate of substrate cleavage of an antithrombin-P14-P1 peptide binary complex.[58] Such complexes are thought to involve insertion of the tetradecapeptide into β-sheet A between strands s3A and s5A, thereby preventing insertion of the reactive center loop of the serpin and thus effectively blocking the inhibitory branch of the pathway. The expectation for parallel pathways would again be a large reduction in the rate of consumption of serpin, corresponding to loss of reaction through the inhibitory pathway, whereas the expectation for a branched pathway is that the flux would not be reduced. For the reaction of the binary antithrombin-peptide complex with thrombin or factor IXa, the substrate rate constant was indistinguishable from the second order rate constant for the inhibitory reaction of these proteinases with antithrombin, thus strongly supporting the branched pathway mechanism. The substrate reactions with plasmin and factor Xa were even higher with the binary complex than for antithrombin alone, which is again difficult to reconcile with parallel pathways.[58]

Other related evidence in favor of the branched pathway mechanism is provided by changes in SI for a given serpin-proteinase pair as a function of temperature or the binding of an effector molecule. Thus the reaction of C1-inhibitor with α-kallikrein, kallikrein light chain, plasmin and C1s showed changes in SI for a given reaction pair upon changing the temperature from 4°C to 38°C.[59] In each case the SI dropped upon increasing the temperature, as a result of an increasing fraction of inhibition at the higher temperature (Table 3.6). This is as expected for a higher activation energy for the inhibitory branch of the pathway than for the substrate branch, such that the rate of the former is increased more for a given temperature increase than the rate of the latter. Analogous changes have more recently been seen for reaction of the serine proteinase acrosin with the serpin protease nexin 1.[4] For the reaction of antithrombin with thrombin the effect of binding heparin is to change the SI from 1 to 1.5 and to cause even larger changes in SI, to about 10, when the ionic strength (I) is reduced to 0.01.[60]

It is possible that such a branched pathway operates generally for all serpin-proteinase encounters for which there is recognition of the serpin P1-P1' bond by the proteinase, even where SI appears to be indistinguishable from 1.0 (i.e., solely inhibitory) or to approach infinity (i.e., solely substrate). α_1-Proteinase inhibitor might thus be regarded as a substrate of elastase, but one for which $k_4 >>> k_3$. Equally ovalbumin might be considered as an inhibitor of elastase, but one for which $k_4 <<< k_3$. The experimental difficulty of determining the small percentage of substrate reaction in the former or of inhibition in the latter, however, makes it difficult to prove such a proposal. Many examples of reactive center loop variants exist for which the effect of the mutation has been qualitatively described as a conversion from "inhibitor" to "substrate" (see Table 3.7 for selected examples). It is possible that such statements are only approximately true and that careful examination

Table 3.6. Alteration in SI caused by changes in environment

Serpin-proteinase pair	Normal SI	Nature of environmental change	Effect on SI	Ref
C1-inhibitor-C1s	1.05 at 38°C	Temperature reduction to 4°C	Increase in SI to 1.2	59
C1-inhibitor-kallikrein	2.5 at 38°C	Temperature reduction to 4°C	Increase in SI to 6.3	59
Protease nexin I-acrosin	37 at 37°C	Temperature reduction to 12°C	Increase in SI to 3000	4
α_1-Antichymotrypsin-chymase	5 at pH 8	pH reduction to 6.0	Decrease in SI to 1.3	157
PAI-1 (P14 T→H)	~1 at pH 8	pH reduction to 5.0	Increase in SI to 40	158
α_1-Antichymotrypsin-chymase	5 at pH 8.0 and I 0.15	Increase in ionic stregth	Decrease in SI	157
Antithrombin-thrombin	1.0	Decrease in ionic strength in the presence of heparin	Increase in SI to 9.8 at I 0.01	60

Table 3.7. Selected serpin variants with mutations in the hinge region, ordered by mutation position

Serpin	Position of mutation	Mutation	Effect	Ref.
Antithrombin	P14	S→W	Still inhibitory, but with increased SI	56
C1-inhibitor	P14	V→E	Converted to substrate	159
α_1-Proteinase inhibitor	P14	T→R	Still inhibitory, but with increased SI	53
PAI-1	P14	T→H	pH-dependent SI (40→1)	158
Antithrombin	P12	A→T	Converted to substrate, but still loop-insertable	160, 161
α_1-Proteinase inhibitor	P12	A→T	Small increase in SI, but mostly inhibitory	55
C1-inhibitor	P12	A→E	Converted to substrate	52
Antithrombin	P10	A→P	Still inhibitory, but with increased SI	162
Antithrombin	P10	A→S	Still inhibitory, but with loss of heparin acceleration for thrombin	163
α_1-Proteinase inhibitor	P10	G→P	Still inhibitory, but with greatly increased SI	55
C1-inhibitor	P10	A→V	No inhibition of target proteinases	164
Antithrombin	P7	A→V	Thrombosis in vivo	165

of evidence for residual inhibitory capacity would show that SI, while large, is not infinite and that formation of some stable complex does occur.

3. LOOP INSERTION AS A REQUIREMENT FOR INHIBITION BY SERPINS

Given that the outcome of a serpin-proteinase reaction depends on the relative values of the substrate (k_3) and inhibitory branch (k_4) rate constants, the next question to be addressed is what determines the relative values of these rate constants. We consider here how k_4 may depend on the ease of insertion of the reactive center loop into β sheet A.

The clearest demonstration that insertion of some or all of the reactive center loop into β-sheet A is a prerequisite for operation of the inhibitory branch of the pathway is the conversion of an otherwise inhibitory serpin into an exclusive substrate by annealing a peptide corresponding to a portion of the reactive center loop with the serpin prior to reaction with the proteinase. This has been done for antithrombin[58,61] and α_1-proteinase inhibitor.[62] The peptide in the complex is thought, from spectroscopic and stability studies, to occupy the strand 4 position of the expanded β sheet A that is present in cleaved, loop-inserted inhibitory serpins.[61,63] For each serpin, formation of a binary complex with tetradecapeptide corresponding to the respective P14-P1 reactive center loop sequences converted the serpin into a substrate for proteinase. This was examined for thrombin, plasmin, factor IXa and factor Xa in the case of antithrombin[58] and for trypsin, thrombin and porcine pancreatic elastase in the case of α_1-proteinase inhibitor.[62,63] By using two series of nested peptides of different length, one of the studies with α_1-proteinase inhibitor also attempted to define the degree of reactive center loop insertion that is sufficient to permit operation of the inhibitory pathway.

The conclusion that insertion of only one residue (P14) is sufficient for inhibition to occur[62] is somewhat surprising (see below). Since the presence of the peptide in the putative ternary (serpin-peptide-proteinase) complex was not demonstrated, it may be that complexes with any peptide not containing P14 permitted the reactive center loop to start the insertion process, but with subsequent displacement of all or part of the peptide from β-sheet A during complex formation with the proteinase.

A recent study using a fluorescently-labeled PAI-1 variant also provides convincing evidence that loop insertion is an integral part of the inhibitory mechanism.[64] Mutation of the P9 residue in the reactive center loop of PAI-1 to cysteine permitted the introduction of a fluorescent reporter group (NBD) to allow evaluation of the environmental changes that occur under a number of conditions. Conversion of the labeled PAI-1 to either latent or cleaved forms, with concomitant loop insertion into β-sheet A resulted in the same large enhancement of NBD fluorescence as occurred when stable complex was formed with proteinase.[64] This suggests not only that loop insertion occurs upon complex formation, but that it involves residues at least up to and including P9. This labeled PAI-1 has also been used to examine the relationship between the covalent interaction involving the active site serine of the proteinase and the occurrence of loop insertion. By comparing the changes in NBD fluorescence upon complex formation with trypsin or with anhydrotrypsin, it was found that anhydrotrypsin, which is incapable of covalent interaction with the serpin reactive center loop, caused minimal changes in fluorescence of the NBD label at P9.[25]

Immunological evidence has also been provided for loop insertion in C1-inhibitor-proteinase complexes and in antithrombin-proteinase complexes. Using monoclonal antibodies in the case of C1-inhibitor[65] and polyclonal antibodies in the case of antithrombin[66] it was shown that an epitope that was absent from the native serpin was present on cleaved and complexed serpin. Antithrombin-reactive center loop peptide complexes also contained this neoepitope, suggesting that in each case expansion of β-sheet A had occurred with associated structural changes.

Finally, the partial or complete absence of inhibitory properties for certain reactive center loop variants can be most readily explained by the deleterious effect that particular side chains might have on the rate of loop insertion, and thus on the rate constant k_4. The more thermodynamically unfavorable that insertion of the first residue or residues is, the higher the kinetic barrier for the insertion process is likely to be. This will strongly depend on the side chain to be inserted, but may also be influenced by residues in the flanking strands of the β-sheet (s3A and s5A) and by the underlying residues against which the inserted side chains would pack. The extreme examples might be non-inhibitory serpins such as ovalbumin which contain unfavorable residues not only at P14 (arginine) but also at additional even-numbered P sites, all of which would be expected to either become buried residues upon insertion of the reactive center loop into β-sheet A or to cause major distortions in the inserted strand. Less extreme might be non-inhibitory serpins that nevertheless undergo sufficiently facile loop insertion that the cleaved serpin product of reaction with proteinase is loop inserted, but not fast enough to give proteinase trapping. Examples of these are corticosteroid binding globulin and thyroxine binding globulin.[67] Least extreme are variants of otherwise inhibitory serpins in which only a single reactive center loop residue has been changed, with a less than complete conversion from inhibitor to substrate. Examples of this are the T345R (P14) variant of α_1-proteinase inhibitor, which has SI values of 9.5, 45 and 70 for trypsin, human neutrophil elastase and porcine pancreatic elastase respectively,[53] or the A347T (P12) variant of α_1-proteinase inhibitor which shows only a small decrease in the relative flux along the inhibitory pathway[55] and thus only a small increase in SI.

4. INFLUENCE OF REACTIVE CENTER LOOP CONFORMATION ON REACTIVITY

The structure of the reactive center loop of a monomeric serpin is only known for the non-inhibitory serpin ovalbumin[68] and an inhibitory variant of the serpin α_1-antichymotrypsin.[19] In both cases the central part of the reactive center loop, from about P9 to P1 is involved in a stable helix; α-helix in the case of ovalbumin, distorted 3_{10} helix in the case of the proline-containing α_1-antichymotrypsin variant. If such helical structures are general for serpins it is unlikely that the P1-P1' bond and the immediately adjacent residues have the optimal Φ and Ψ angles for binding to the proteinase. Thus some of the binding energy would be wasted in changing the conformation of the reactive center loop. To determine whether this occurs generally in serpins it would be desirable to have data for the k_{cat}/K_M values for substrate cleavage of peptides containing the reactive center loop sequences of serpins, to be compared with data for the intact serpin-proteinase interaction. Such a study has been carried out for peptides derived from the reactive center of α_1-proteinase inhibitor (P4-P4').[69] Values of k_{cat}/K_M were orders of magnitude smaller for the peptide than were observed in the intact serpin. Although the reverse order of reactivity might have been anticipated, the study at least demonstrates the importance of conformational context for the recognition sequence. This is reinforced by the observation that, when the same peptides were constrained, much higher affinities to neutrophil elastase were obtained.[69]

The only structural data that bear on the question of conformational change in the reactive center loop are the effect of the heparin-induced conformational change in the reactive center loop of antithrombin on the rate of inhibition of factor Xa. From fluorescence studies on a P1 R393C variant of antithrombin labeled with the fluorophore NBD it has been shown that heparin binding caused a specific conformational change in the reactive center loop involving P1.[70] The magnitude of the change varied with the type of heparin used (i.e., low affinity or high affinity penta-saccharide-containing). The different degrees of conformational change for different heparin species correlated with different degrees of enhancement of the rate at which factor Xa was inhibited.[70] This is an indication that the normal conformation of the reactive center loop of antithrombin gives poor interaction with the factor Xa binding site, whereas heparin-induced structural changes apparently lead to improvements in complementarity and consequent enhancements in the rate of inhibition. The nature of the conformational changes is conjectural at this point, but may involve a partial unfolding of a helical structure. However, even these data on antithrombin may not be generally applicable to other serpins, since it can be argued that antithrombin is a special case of a serpin with a deliberately unfavorable conformation for the reactive center loop, designed to give a low rate of reaction with factor Xa in the absence of heparin, but, "normal" rate of reaction when bound to heparin. Equally, there may be differences in the structures of the reactive center loop of other serpins, such that some are α-helical and require large structural changes to give good interaction with proteinase, whereas others are already more open. Such differences might be reflected in the rate of proteinase inhibition, with the highest rates of reaction resulting where the least structural change needs to occur.

Also of questionable general applicability is the structure of the reactive center loop of the one active subunit of dimeric antithrombin present in the two recently determined X-ray structures.[71,72] Dimerization involves interaction between the reactive center loop of the active monomer with the remainder of β-sheet C of the latent conformation monomer (to permit conversion to the latent form, strand 1 of β-sheet C had been removed: see Chapter 2 section 3). Conversion to latency necessarily results in loss of activity, so that such dimers do not represent the situation in solution. Equally, the loop conformation of the active monomer depends for its stability on hydrogen-bonding interactions with β-sheet C of the second monomer. Whereas this conformation almost certainly does not represent the resting state structure of antithrombin in solution, it may represent the structure of the reactive center loop as it interacts with proteinase, with the second antithrombin molecule as stand-in for the proteinase.[71,72]

5. NATURE OF COMPLEX STABILIZATION AND MECHANISM OF SERPIN INHIBITION

There are two major questions with regard to the nature of the stabilization of the serpin-proteinase complex that results from loop insertion. (i) What is the nature of the interaction between the scissile bond and the proteinase active site in the complex and (ii) how does loop insertion bring about the stabilization of the reaction intermediate and thus of the serpin-proteinase complex? In considering both questions it is useful to recall that the branched pathway mechanism requires that the initial steps of an ultimately inhibitory interaction are the same as if the outcome were substrate cleavage. Thus, examination of the normal substrate reaction pathway of serine proteinases is helpful in trying to decide what the nature of the intermediate at the branch point is and what the nature of the trapped intermediate is. The pathway for substrate reaction (Fig. 3.3) shows that, after formation of the initial Michaelis complex, there is attack of the active site serine, assisted by the active site histidine, on the scissile peptide carbonyl to form a tetrahedral intermediate. The negatively charged tetrahedral intermediate is stabilized by hydrogen bonding to two backbone amides that form the "oxyanion hole." With the active site histidine acting as a general acid to donate a proton to the leaving group, the peptide bond is cleaved to produce the first product,

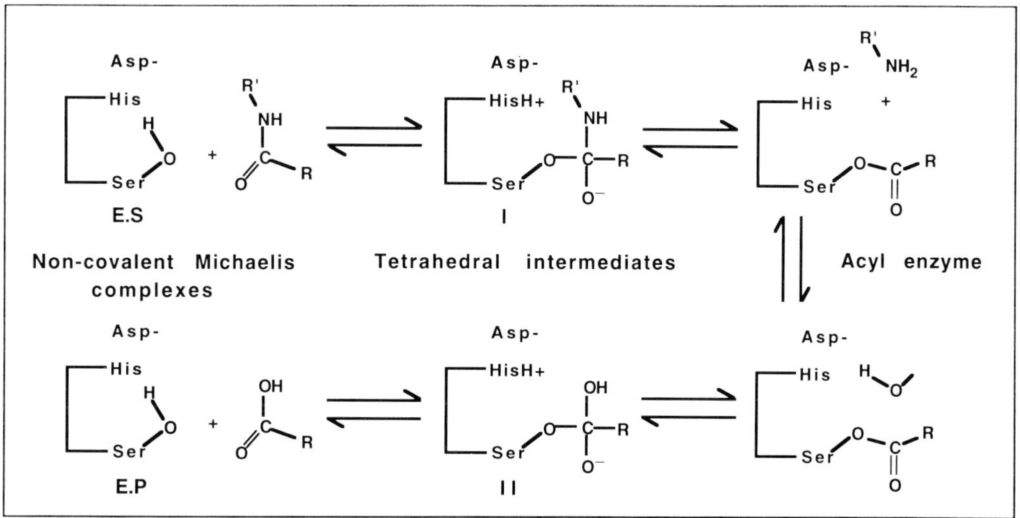

Fig. 3.3. Schematic representation of the mechanism of peptide cleavage by serine proteinases, i.e., the normal substrate pathway. The catalytic triad residues of aspartate 102, histidine 57 and serine 195 are shown, together with a general substrate R'NHCOR. The first-formed complex (the non-covalent Michaelis complex) is formed by association of proteinase and substrate. In helping to form the acyl intermediate, histidine 57 acts first as general base and then as general acid to promote attack of the serine on the peptide carbonyl and then protonate the leaving amine. In deacylating the acyl intermediate the histidine again acts in turn as general base and acid to promote attack by water and subsequently to reprotonate the regenerated serine. Note that two tetrahedral intermediates are expected to occur along this pathway, one preceding and one following the acyl intermediate.

RNH_2, and an acyl enzyme intermediate linking the N-terminal portion of the serpin to the active site serine. Here RNH_2 represents the carboxy-terminal portion of the serpin, containing P1'-P2'..., etc. Deacylation and release of the second product involves attack by water, facilitated by the active site histidine acting as a general base to abstract a proton from the water and give a second tetrahedral intermediate, stabilized in the same way as the first. The serine γO then accepts a proton from the histidine, acting as general acid, to regenerate the serine OH and release the second product.[73] Throughout this process, the reactivities of the serine γOH as a nucleophile, and of the histidine as a general base or acid, depend on the correct alignment of the "catalytic triad" of serine-histidine-aspartate residues, both among themselves, so that the aspartate can serve to form a possible very strong low-barrier hydrogen bond to the histidine imidazole,[74,75] and with respect to the substrate. Indeed, the importance of such coordinate arrangement of these three residues is illustrated by its existence in all serine proteinases, whether they belong to the chymotrypsin-like, subtilisin-type or wheat carboxypeptidase II-type families of proteinase.[76] Since these families differ markedly in their structures, the close homology of the catalytic triad represents an excellent example of convergent evolution. In addition the correct orientation of the oxyanion hole to the acyl or tetrahedral intermediates must be maintained.

(I) The Nature of the Catalytic Intermediate in the Serpin-Proteinase Complex

Candidates for the structure of the stable complex include (i) non-covalent Michaelis complex, (ii) tetrahedral intermediate I, (iii) acyl enzyme intermediate, (iv) tetrahedral intermediate II. Firm evidence for a particular structure is at present relatively limited. However, one of the main characteristics of the stable serpin-proteinase complex is that, after denaturation, the serpin appears as a two chain protein covalently bound to the proteinase by an acyl bond between the α-carboxyl of the scissile bond and the active site serine γO.[77-79] The complex thus appears on SDS-PAGE as a band of higher molecular weight than the serpin alone. Together with the recent demonstration that the apparent affinity of trypsin and anhydrotrypsin for PAI-1, α_1-proteinase inhibitor or antithrombin differed by over four orders of magnitude, indicating that the serine γO is required for strong interaction, this tends to rule out a non-covalent complex.[25] The observation that some serpin-proteinase complexes (notably those of α_2-antiplasmin) may be slowly reversible in the presence of a proteinase trap (i.e., α_2-macroglobulin)[2] is not necessarily inconsistent with this (see section (II) below).

Although there is general agreement that the serpin-proteinase complex involves a covalent intermediate, the nature of this intermediate is more controversial. Both the behavior of serpin-proteinase complexes on SDS-PAGE and the requirement for strong nucleophiles to bring about dissociation of such complexes are consistent with the presence of an acyl intermediate.[77-81] Further support, under non-denaturing conditions, comes from a recent study that demonstrated that several serpin-proteinase complexes contained a new amino terminus which was as accessible and reactive to N-hydroxysuccinimide as the α-amino group of the original N-terminus.[82] This strongly suggests that the peptide bond has been fully cleaved to form an acyl enzyme with concomitant generation of a new α-amino group. In contrast to these findings, a ^{13}C NMR study carried out with α_1-proteinase inhibitor labeled with ^{13}C at the Cα of methionine residues favors the stable complex being a tetrahedral intermediate.[83]

(II) The Mechanism of Complex Stabilization

Little is known about the mechanism by which the putative tetrahedral or acyl enzyme intermediate is stabilized to give a long-lived serpin-proteinase complex.

Since the complex ultimately, though slowly, decays to the normal substrate-like products with rate constant k_5, a process that is thermodynamically favorable but extremely slow,[5,6,45,80,84] the stabilization of the complex must be considered kinetic. For reasons not well understood the final steps of the proteolysis reaction are slowed by many orders of magnitude. The most straightforward explanation for this is that the position of the intermediate with respect to the active site histidine and/or the amides of the oxyanion hole may be unfavorably changed such that the kinetic barrier to continuation of the catalytic reaction is greatly increased. The stability of the complex can be further increased by limited proteolysis. With chymotrypsin-α_1-antichymotrypsin complexes, the presence of free chymotrypsin results in cleavage of the complex either within the chymotrypsin or the α_1-antichymotrypsin to give a modified complex from which no active chymotrypsin was seen to be released even after several days.[5] Similar stabilization may occur for other serpin-proteinase complexes, since excess proteinase typically results in additional cleavage within the complex.

We propose the following scheme (Fig. 3.4) as a possible mechanism that can accommodate much of the observed, and sometimes apparently contradictory, data on the pathway of serpin-proteinase encounters. This is based on the branched pathway mechanism of Figure 3.2 and the need to include loop insertion as the watershed for commitment to one or the other pathway. We have tried to accommodate the possibility of reversibility under certain circumstances, to incorporate the need for attack on the peptide carbonyl to precipitate loop insertion and also to provide a way in which the "final complex" EI† can be kinetically stabilized. In addition, we have used some of the limited structural information that shows that the ultimate product of either branch (cleaved

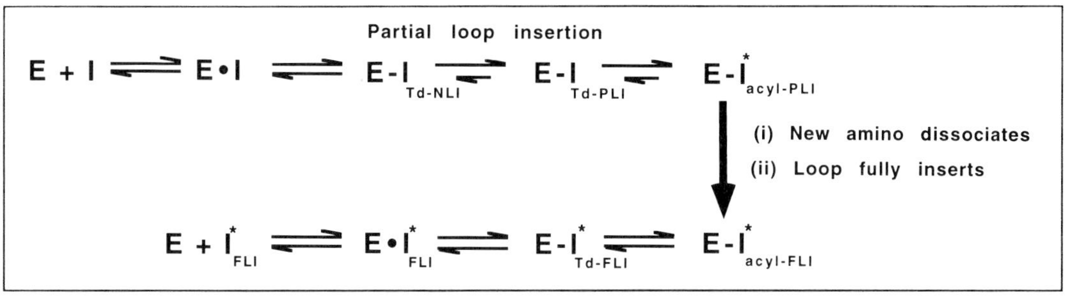

Fig. 3.4. Proposed mechanism of action of serpins as suicide substrate inhibitors, showing only the inhibitory pathway. The first two steps represent formation of the Michaelis complex (E • I) and the initial covalent attack on the peptide carbonyl to give a tetrahedral intermediate $E\text{-}I_{Td\text{-}NLI}$, in which loop insertion has not yet taken place but has been provoked by formation of the tetrahedral carbon. This represents the branch point of the reaction between inhibitory and substrate pathways. If loop insertion occurs too slowly the normal substrate reaction continues without interference. If loop insertion is rapid, most or all of the serpin-proteinase complexes become trapped in stable complexes; either as $EI_{Td\text{-}PLI}$ or more likely as one of the two forms of acyl intermediate ($E\text{-}I^*_{acyl\text{-}PLI}$ or $E\text{-}I^*_{acyl\text{-}FLI}$). Here the asterisk is used to indicate that the peptide bond has been cleaved and NLI, PLI and FLI signify not loop inserted, partially loop-inserted or fully loop-inserted respectively. The conversion of $E\text{-}I^*_{acyl\text{-}PLI}$ to $E\text{-}I^*_{acyl\text{-}FLI}$ requires the release of the first product RNH_2 from the active site and from the subsites of interaction (P1'-P2' etc.). This release permits loop insertion of the P8-P1 residues to occur, with a consequently very favorable free energy change signified by the bold down arrow. Reversal of this step is considered unlikely. Insertion of the remainder of the reactive center loop necessitates concomitant movement of the attached proteinase. Conformational changes involving P1-P2-P3 etc. subsite interactions, as well as possible conformational changes in the proteinase as a result of the movement, result in disruption of the appropriate alignment of the catalytic residues and the acyl intermediate and result in a large reduction in deacylation rate (to k_5), effectively trapping the complex. Slow hydrolysis can occur to regenerate proteinase and release cleaved, fully-loop inserted serpin(I^*_{FLI}).

serpin I*) is fully loop inserted, whereas the native serpin has an exposed reactive center loop. In this mechanism the serpin, with the reactive center in an exposed but not optimized conformation, forms an initial loose Michaelis complex (E•I) with the proteinase that requires recognition of the P1 residue. At this point the active site serine attacks the P1 carbonyl and forms a tetrahedral intermediate ($E\text{-}I_{Td\text{-}NLI}$; the subscript NLI, not loop inserted, signifies that the reactive center loop has not yet started inserting into β-sheet A). This change in geometry at the P1-P1' peptide provides the mechanical impetus to initiate loop insertion. This is the branch point for the substrate and inhibitory branches of the pathway. If loop insertion is slow, the normal substrate reaction out-competes the loop-insertion-dependent inhibitory pathway, and vice versa for fast insertion (no more needs to be said about the substrate pathway other than that it proceeds by the normal intermediates given in Fig. 3.3). For the inhibitory pathway the first product of loop insertion is still a tetrahedral complex, but with partial loop insertion from P14 to residue P10 or P9 ($E\text{-}I_{Td\text{-}PLI}$; PLI stands for partially loop inserted). The critical consequence of partial loop insertion is to alter the conformation of the reactive center from an incompletely extended to a fully extended conformation, such as has been suggested by Carrell.[85] A substrate-like reaction can continue, with conversion of the tetrahedral to an acyl enzyme intermediate ($E\text{-}I_{acyl\text{-}PLI}$) in which no additional loop insertion has taken place and in which the newly formed N-terminus is still held close to the acyl intermediate by subsite interactions involving the P' residues, with interactions improved over those that were present in the Michaelis complex by the extended conformation of the loop. All of the steps up to this point are reversible, since, if the newly-formed amino group is rigidly held, reformation of the peptide bond is not unfavorable.[86,87] At this point the irreversible step occurs. This involves the release of the amino product (containing the C-terminal P1'- - -Pn' residues) which thereby removes the constraint that had prevented further insertion of the reactive center loop. Full loop insertion can now occur, and involves large scale movement of both the reactive center loop as it inserts, and the covalently attached proteinase, with burial of reactive center residues up to P2 or even P1, depending on the constraints imposed by the attached proteinase. The proteinase will thus finish at the opposite end of the serpin from that at which it first attacked the P1-P1' bond. Of necessity, this burial of the P-side residues will involve disruption of the subsite interactions between these residues and the proteinase. This in turn will alter the previously fixed geometrical relationship between the acyl intermediate and the catalytic triad and result in a large reduction in catalytic efficiency, perhaps by removing the acyl intermediate from the vicinity of the oxyanion hole. This is the kinetic trap in all cases where irreversible complex formation is observed. The energetic impetus for formation of this structure is the greatly increased stability of the fully loop inserted serpin, which more than offsets the loss of energy from the disrupted P-S subsite interactions. Since reversal of the pathway would require removal of the reactive center loop from β-sheet A as well as a coming together of the $P1\text{-}COO^-$ and the $P1'\text{-}NH_3^+$, all in a thermodynamically uphill process, the reversal does not occur. Instead the acyl enzyme intermediate ($E\text{-}I_{acyl\text{-}FLI}$; FLI stands for fully loop inserted) can deacylate at a slow rate determined by the greatly reduced efficiency of the inappropriate relationship of the acyl group to the active site. The energy difference between this intermediate and the final products of cleaved, fully-loop inserted serpin (I*) and proteinase is dominated mostly by the energy gained from hydrolysis of the acyl intermediate and formation of two molecules from one complex, since little further energy change will occur for the serpin moiety. In keeping with the occurrence of conformational changes in the proteinase while in complex with serpin is a recent study on an Asp 189→Ser trypsin-

α_1-proteinase inhibitor complex, in which it was shown that Asp 189→Ser trypsin in complex was much more susceptible to cleavage by free bovine trypsin than was free Asp 189→Ser trypsin[88] (note that Asp 189 is not the aspartate of the catalytic triad).

This mechanism satisfies the major requirements for serpin inhibition, including the need for the serpin to be in a metastable state so that it can promote firstly partial and then full loop-insertion as well as the requirement for fast loop insertion to promote the inhibition pathway. At the same time it shows how some serpin-proteinase complexes may be reversible whereas others are not. This will depend on how well the $E\text{-}I_{acyl\text{-}PLI}$ intermediate is stabilized with respect to release of the amino-containing peptide. In turn this will depend on the strength of subsite interactions, which will be serpin- and proteinase-dependent. For those complexes stabilized at the $E\text{-}I_{acyl\text{-}PLI}$ intermediate stage, the equilibrium position with the preceding tetrahedral intermediate may favor either the acyl or tetrahedral species, depending on the stabilizing interactions of the subsites.

6. REACTION OF SERPINS WITH CYSTEINE PROTEINASES

The several reports of the inhibition of cysteine proteinases by serpins[89-91] pose no particular problems for the serpin inhibition mechanism discussed above. Both serine and cysteine proteinases share much in common in their mechanism of substrate cleavage, involving nucleophilic attack on the peptide carbonyl to form an acyl enzyme intermediate in both cases. The details of how this is accomplished are different for the two types of proteinase, but the nature of the pathway and the types of intermediate are the same (of course with the appropriate replacement of oxygen by sulfur in the cysteine proteinases). Since the most critical attributes for serpins to function as inhibitors rather than as substrates are (i) the ability to loop insert to give more stable structures and (ii) the ability to change conformation in the reactive center loop in response to covalent intermediate formation and as the trigger for loop insertion, serpins should, in principle, be capable of inhibiting both cysteine and serine proteinases. For the interaction of the viral serpin CrmA with interleukin-1β converting enzyme it has been shown that several of the characteristic serpin properties are present. Thus the active site cysteine (cf serine) is required for inhibition and the cleaved CrmA that dissociates from complexes or is produced by the substrate reaction is very stable towards denaturation, and the normal products of reaction are a mixture of substrate and inhibitory pathways with SI ~5.[92] A difference from most serpin-serine proteinase complexes, though not inconsistent with the proposed mechanism, is that the complexes are not stable to SDS and release cleaved CrmA and proteinase, which implies that k_5 is greater than for most serpin-proteinase complexes. However, the complex of α_1-antichymotrypsin with the prohormone thiol proteinase is SDS-stable,[90] showing that instability is not a general characteristic of cyteine proteinase-serpin complexes.

7. STABILIZATION OF THE NON-LOOP INSERTED "ACTIVE" FORM OF SERPINS

Implicit in the requirement for loop insertion to be a necessary part of the inhibitory mechanism is that native serpins fold with the reactive center loop exposed rather than inserted. Since the loop-inserted form is thermodynamically more stable than the native form, as shown by the greater stability of latent PAI-1 compared with native PAI-1 and the very much greater stability of cleaved, loop-inserted serpins compared with native serpins, this requires that serpins fold in a kinetically-trapped metastable state. This metastable state must be achieved in a way that does not place too high an energetic barrier between the loop-exposed and loop-inserted conformations, since loop insertion must only be disfavored in the absence of proteinase. How is this achieved?

Recent computer simulations of protein folding by Rose and colleagues have suggested that initial α-helix formation occurs readily in the newly synthesized polypeptide, whereas β sheet is much more dependent on interactions with other parts of the protein and thus may form later.[93] One could thus envision a situation in serpins where the reactive center loop is saved from incorporation into β-sheet A by a strong tendency to form α-helix. An examination of the helix- and sheet-forming propensities of the reactive center loop and flanking sequences in a number of serpins revealed that serpins that undergo a loop-to sheet insertion upon reaction with proteinase and are also inhibitory are strongly predicted to prefer α-helical conformation (Figs. 3.5 and 3.6). However, neither those serpins that loop-insert but are non-inhibitory (corticosteroid binding globulin and thyroxine binding globulin), nor those that do not spontaneously loop-insert upon reaction with proteinase showed the same prediction, despite all serpins apparently having an exposed conformation for the reactive center loop. Thus stabilization of an early α-helical conformation is not likely to be the mechanism for ensuring the metastable serpin conformation. Instead, it may be that formation of β-sheets A and C occurs early in the folding process and determines the final folding. The available X-ray structures indicate that, in a single chain form of a serpin, it is possible to have either strand s1C incorporated into β-sheet

	STRAND 5A									REACTIVE CENTER LOOP															
P	25	24	23	22	21	20	19	18	17	16	15	14	13	12	11	10	9	8	7	6	5	4	3	2	1
INHIBITORY SERPINS																									
Variant α_1-Antichymotrypsin			S	S	S	S	S									S	S	S	S	S	S	S	S		
α_1-Antichymotrypsin			S	S	S	S	S							H	H	H	H		S	S	S	S	S	S	S
α_1-Proteinase Inhibitor			S	S	S	S							H	H	H	H	H	H	H	H	H	H	S		
α_2-Antiplasmin				S		S	S					H	H	H	H	H	H	H	H	H	H	H	H	H	H
Antithrombin			S	S	S	S	S					H	H	H	H	H	H	H	H	H	S	S	S	S	S
Barley Z Protein			S	S	S	S	S				H	H	H	H	H	H	H	H	H	H	S	S	S	S	S
C1 Inhibitor				S	S	S	S	H	H	H	H	H	H	H	H	H	H	H	H	H	H	H	H	H	H
CrmA				S	S	S	S			H	H	H	H	H	H	H	H	H	H	H	H	S	S	S	
Equine leukocyte elastase inhibitor				S	S						H	H	H	H	H	H	H	H	H	H	H	H	H	H	H
Heparin Cofactor II			S	S	S	S	S	S					S	S	S	S	S	S	S	S	S	S	S	S	
Kallistatin			S	S	S	S						H	H	H	H	H	H	H	H	H	H	H	H		
Limulus coagulation inhibitor			S	S	S	S	S							S				S	S	S	S	S	S	S	S
Manduca sexta Ala-serpin			S	S	S	S	S					H	H	H	H	H	H	H	H	H	H	H	H	H	H
Placental thrombin inhibitor				S	S	S					H	H	H	H	H	H	H	H	H	H	H	H	S		
PAI-1			S	S	S	S	S						H	H	H	H	H	H	H	H	H	H	S	S	S
PAI-2			S	S	S	S	S					H	H	H	H	H	H	H	H	S	S	S	S	S	S
Proteinase Nexin I			S	S	S	S	S					H	H	H	H	H	H	H	H	H	S	S	S	S	
Protein C Inhibitor			S	S	S	S	S						H	H	H	H	H	H	H	S	S	S	S	S	S
Squamous carcinoma cell antigen			S	S	S	S	S				H	H	H	H	H	H	H	H	H	H	H	H	H	H	H
NON-INHIBITORY SERPINS																									
(A)																									
Corticosteroid binding globulin			S	S	S	S	S															S	S	S	
Thyroxine binding globulin			S	S	S	S	S													S					
(B)																									
Angiotensinogen				H	H	H	H	H									S	S							
Maspin			S	S	S	S	S	S	S	S	S	S						S	S	S	S			S	S
Ovalbumin			S	S	S	S	S					H	H	H	H	H	H	H	H	H	H				
PEDF																									

Fig. 3.5 Secondary structure predictions for the region P25 to P1 of selected serpins determined by the "winner takes all" algorithm of Rost and Sander.[166-168] "S" signifies a prediction of β-strand and "H" signifies a prediction of α-helix. The predictions were performed on 40 residue segments from P25 to P15', but only data for P25 to P1' are shown in this figure.[169] Serpins are grouped according to whether they are inhibitory or not, and if non-inhibitory by whether they spontaneously loop insert (A) or not (B).

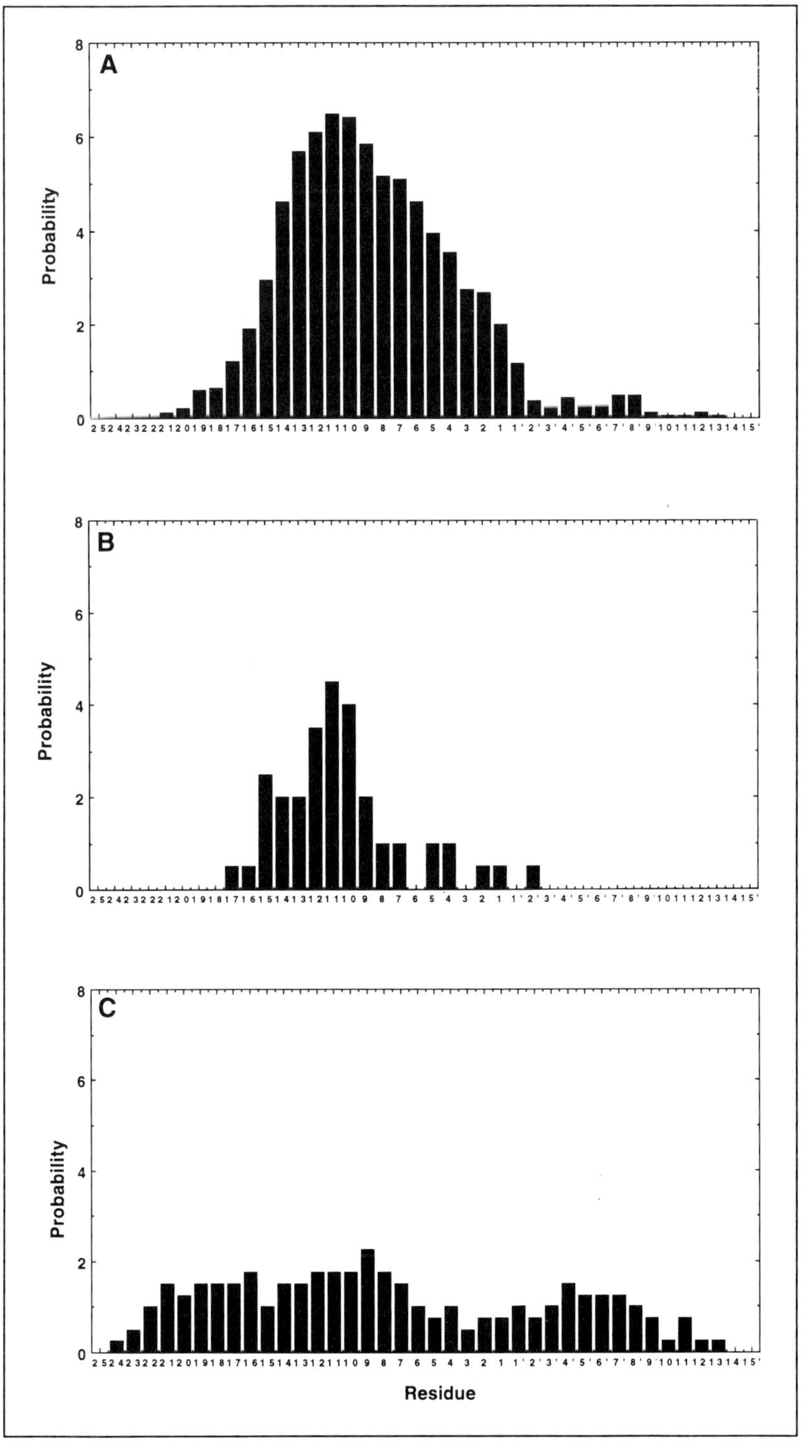

Fig. 3.6. Secondary structure predictions for P25 to P15' of the same serpins as treated in Figure 3.5 and obtained in the same way, but shown as the normalized sum of the numerical predictions of helix-forming propensity.[169] Panel A, inhibitory serpins; panel B, non-inhibitory, but spontaneously loop-insertable serpins (CBG and TBG); panel C non-inhibitory, non spontaneously loop-insertable serpins.

C or strand s4A (the reactive center loop) incorporated into β-sheet A, but not both simultaneously. If the isolated β-sheet C adopts a stable conformation incorporating strand s1C, concomitant incorporation of the reactive center loop into β-sheet A as s4A would be disfavored. The reactive center loop (putative s4A) would then become exposed by default. Since the ends of the reactive center loop are attached to strands of the A and C sheets, with a separation dictated by the positions of these two β-sheets (about 18Å), the conformation adopted by the reactive center loop in the folded serpin may then be determined primarily by the anchoring of the ends and the relatively large number of intervening residues. Under such circumstances a condensed (e.g., helical) conformation may represent the most stable conformation for the loop itself, even though in the context of the whole serpin its more stable conformation is as a strand of β-sheet A.

8. ADVANTAGES AND DISADVANTAGES OF THE SERPIN MECHANISM

From the above discussion it is clear that serpins not only function very differently as inhibitors than do other protein proteinase inhibitors, but that they must also possess the unusual ability to adopt multiple conformations, including the preference to fold in a metastable state. The fulfillment of these stringent demands is probably one reason that serpins are so much larger than other proteinase inhibitors, in that the mechanism of action demands the presence of extensive β-sheets, both to ensure adoption of the metastable state and to permit subsequent change in the presence of proteinase. These requirements may impose a minimum size on the protein. In compensation there must, however, be advantages to the organism that result from utilizing serpins rather than smaller proteins as inhibitors. One possible advantage is that, because of the competitive nature of the branched pathway mechanism of serpins, there are opportunities for regulation of serpin inhibitory activity for a given concentration of serpin. In contrast, a "lock-and-key" inhibitor has inhibitory potential determined only by its concentration and K_D for the target proteinase. By altering the ratio of k_4/k_3 it is possible, at a fixed serpin concentration, to effectively change the serpin from an inhibitor to a substrate or vice versa (Table 3.8).[94] In addition there are several examples where allosteric effector molecules, particularly heparin and heparan sulfate, can affect the rate of reaction with proteinase, without necessarily altering the SI by a large amount. Since such rate enhancements can be very large, e.g., heparin enhancements of factor Xa inhibition, they can effectively alter the milieu from

Table 3.8. Dependence of stoichiometry of inhibition (SI) on relative values of k_3 (substrate branch rate constant) and k_4 (inhibitory branch rate constant)

k_4	k_3						
	1	2	5	10	100	1000	10000
1	**2**	3	6	11	101	1001	10001
2	1.5	**2**	3.5	6	51	501	5001
5	1.2	1.4	**2**	3	21	201	2001
10	1.1	1.2	1.5	**2**	11	101	1001
100	1.01	1.02	1.05	1.1	**2**	11	101
1000	1.001	1.002	1.005	1.01	1.1	**2**	11
10000	1.0001	1.0002	1.0005	1.001	1.01	1.1	**2**

Large changes in one rate constant can have either a large or small effect on SI depending on the value of the other rate constant. Adapted from Patston and Gettins[94] and reprinted with permission. Copyright 1994, Thieme Medical Publishers Inc.

non-inhibitory, in the sense of the inhibition being too slow, to inhibitory.

Another means of regulating the activity of serpins that may have either positive or negative consequences is proteolytic inactivation by cleavage in the reactive center loop by a non-target proteinase. Thus human neutrophil elastase, in the presence of heparin, is an effective inactivator of antithrombin[95,96] through cleavage at the P4-P3 bond.[97] In pathological states where neutrophil activation leads to secretion of neutrophil elastase, such inactivation of antithrombin may lead to a large reduction in the nonthrombogenic properties of the vascular surface.[96] Of potential benefit in regulation of tissue remodeling and vascularization, α_1-proteinase inhibitor can be inactivated by matrix metalloproteinases such as matrilysin.[98] The susceptibility of the reactive center loop of serpins such as antithrombin, C1 inhibitor, α_2-antiplasmin and α_1-proteinase inhibitor to cleavage by snake proteinases, including metalloproteinases,[99-102] may also be utilized as a means of reducing the inhibitory potential in the plasma of attacked animals. In this regard it is interesting that the reactive center loop of α_1-proteinase inhibitor from opossum, an animal that tolerates rattlesnake envenomation, has apparently evolved to be resistant to inactivation by snake venom proteinases while still retaining the ability to inhibit neutrophil elastase.[103]

A major disadvantage of the serpin mechanism results from the very conformational lability that enables serpins to function as inhibitors. The need to fold as a metastable state and the subsequent ability to undergo loop-to-sheet transitions makes serpins sensitive to mutations that affect one or both of these processes. One of the most dramatic examples of such a variant is the Z variant of α_1-proteinase inhibitor which, as a result of problems during folding,[104] undergoes polymerization into sheets of linked α_1-proteinase inhibitor molecules, which are accordingly not secreted into the circulation and instead build up inside hepatocytes.[105] Although the mechanism of polymerization is still speculative, it is likely to involve insertion of the reactive center loop of one molecule into β-sheet A of the next, but also with possible involvement of strand 1 of β-sheet C. In the simple case of reactive center-cleaved forms of α_1-proteinase inhibitor, oligomeric forms were seen, including some closed circles, consistent with daisy-chain-like insertion of one cleaved reactive center loop into the strand 4 position of β-sheet A of the next molecule.[106]

References

1. Laskowski M Jr. and Kato I. Protein inhibitors of proteinases. Annu Rev Biochem 1980; 49:593-626.
2. Shieh B-H, Potempa J and Travis J. The use of α_2-antiplasmin as a model for the demonstration of complex reversibility in serpins. J Biol Chem 1989; 264: 13420-13423.
3. Longstaff C and Gaffney PJ. Serpin-serine protease binding kinetics: α_2-antiplasmin as a model inhibitor. Biochemistry 1991; 30:979-986.
4. Hermans JM, Monard D, Jones R, et al. Inhibition of acrosin by serpins. A suicide substrate mechanism. Biochemistry 1995; 34:3678-3685.
5. Cooperman BS, Stavridi E, Nickbarg E, et al. Antichymotrypsin interaction with chymotrypsin—partitioning of the complex. J Biol Chem 1993; 268:23616-23625.
6. Danielsson Å and Björk I. Properties of antithrombin-thrombin complexes formed in the presence and in the absence of heparin. Biochem J 1983; 213:345-353.
7. Estell DA, Wilson KA and Laskowski M. Thermodynamics and kinetics of the hydrolysis of the reactive-site peptide bond in pancreatic trypsin inhibitor (Kunitz) by *Dermasterias imbricata* trypsin 1. Biochemistry 1980; 19:131-137.
8. Fish WW, Orre K and Björk I. The production of an inactive form of antithrombin through limited proteolysis by thrombin. FEBS Lett 1979; 98:103-106.
9. Gettins P and Harten B. Properties of thrombin- and elastase-modified human antithrombin III. Biochemistry 1988; 27:3634-3639.

10. Baker D and Agard DA. Kinetics versus thermodynamics in protein folding. Biochemistry 1994; 33:7505-7509.
11. Anfinsen C. Principles that govern the folding of protein chains. Science 1973; 181:223-230.
12. Baker D, Sohl J and Agard DA. A protein-folding reaction under kinetic control. Nature 1992; 356:263-265.
13. Bode W, Papamokos E and Musil D. The high-resolution X-ray crystal structure of the complex formed between subtilisin Carlsberg and eglin c, an elastase inhibitor from the leech *Hirudo medicinalis*. Structural analysis, subtilisin structure and interface geometry. Eur J Biochem 1987; 166: 673-692.
14. Read RJ and James MNG. Introduction to the proteinase inhibitors: X-ray crystallography. In: Barrett AJ, Salvesen GS. Proteinase inhibitors. Amsterdam: Elsevier Science Publishers, 1986:301-336.
15. McPhalen CA and James MNG. Crystal and molecular structure of the serine proteinase inhibitor CI-2 from barley seeds. Biochemistry 1987; 26:261-269.
16. Mitsui Y, Satow Y, Watanabe Y, et al. Crystal structure of a bacterial protein proteinase inhibitor (*Streptomyces* subtilisin inhibitor) at 2.6Å resolution. J Mol Biol 1979; 131:697-724.
17. Hubbard SJ, Campbell SF and Thornton JM. Conformational analysis of limited proteolytic sites and serine proteinase protein inhibitors. J Mol Biol 1991; 220:507-530.
18. Fischer E. Einfluss der Configuration auf die Wirkung der Enzyme. Berichte der Deutschen Chemischen Gesellschaft 1894; 27:2985-2993.
19. Wei A, Rubin H, Cooperman BS, et al. Crystal structure of an uncleaved serpin reveals the conformation of an inhibitory reactive loop. Nature Struct Biol 1994; 1:251-258.
20. Rühlmann A, Kukla D, Schwager P, et al. Structure of the complex formed by bovine trypsin and bovine pancreatic trypsin inhibitor. Crystal structure determination and stereochemistry of the contact region. J Mol Biol 1973; 77:417.
21. Sweet RM, Wright HT, Janin J, et al. Crystal structure of the complex of porcine trypsin with soybean trypsin inhibitor (Kunitz) at 2.6Å resolution. Biochemistry 1974; 13:4212-4228.
22. Huber R and Bode W. Structural basis of the activation and action of trypsin. Adv Carb Chem 1978; 11:114-122.
23. Richarz R, Tschesche H and Wüthrich K. Carbon-13 nuclear magnetic resonance studies of the selectively isotope-labeled reactive site peptide bond of the basic pancreatic trypsin inhibitor in the complexes with trypsin, trypsinogen, and anhydrotrypsin. Biochemistry 1980; 19:5715-5720.
24. Ako H, Foster RJ and Ryan CA. Mechanism of action of naturally occurring proteinase inhibitors. Studies with anhydrotrypsin and anhydrochymotrypsin purified by affinity chromatography. Biochemistry 1974; 13:132-139.
25. Olson ST, Bock PE, Kvassman J, et al. Role of the catalytic serine in the interactions of serine proteinases with protein inhibitors of the serpin family. Contribution of a covalent interaction to the binding energy of serpin-proteinase complexes. J Biol Chem 1995; 270:(In Press).
26. Christensen S, Valnickova Z, Thogersen IB, et al. Sodium dodecyl sulfate-stable complexes between serpins and active or inactive proteinases contain the region COOH-terminal to the reactive site loop. J Biol Chem 1995; 270:14859-14862.
27. Owen MC, Brennan SO, Lewis JH, et al. Mutation of antitrypsin to antithrombin. α_1-Antitrypsin Pittsburgh (358 Met Arg), a fatal bleeding disorder. N Engl J Med 1983; 309:694-698.
28. Scott CF, Carrell RW, Glaser CB, et al. Alpha-1-antitrypsin-Pittsburgh. A potent inhibitor of human plasma factor XIa, kallikrein, and factor XII. J Clin Invest 1986; 77:631-634.
29. Schapira M, Ramus M-A, Jallat S, et al. Recombinant α-antitrypsin Pittsburgh ($Met^{358} \rightarrow Arg$) is a potent inhibitor of plasma kallikrein and activated factor XII fragment. J Clin Invest 1986; 77:635-637.
30. Wachtfogel YT, Bischoff R, Bauer R, et al. α_1-Antitrypsin Pittsburgh ($Met^{358} \rightarrow Arg$) inhibits the contact pathway of intrinsic

coagulation and alters the release of human neutrophil elastase during simulated extracorporeal circulation. Thromb Haemost 1994; 72:843-847.

31. Howarth DJ, Samson D, Stirling Y, et al. Antithrombin III Northwick Park: a variant antithrombin with normal affinity for heparin but reduced heparin cofactor activity. Thromb Haemostas 1985; 53:314-319.
32. Erdjument H, Lane DA, Ireland H, et al. Formation of a covalent disulfide-linked antithrombin-albumin complex by an antithrombin variant, antithrombin "Northwick Park." J Biol Chem 1987; 262: 13381-13384.
33. Owen MC, George PM, Lane DA, et al. P1 variant antithrombins Glasgow (393 Arg to His) and Pescara (393 Arg to Pro) have increased heparin affinity and are resistant to catalytic cleavage by elastase. FEBS Lett 1991; 280:216-220.
34. Owen MC, Beresford CH and Carrell RW. Antithrombin Glasgow, 393 Arg to His: a P1 reactive site variant with increased heparin affinity but no thrombin inhibitory activity. FEBS Lett 1988; 231:317-320.
35. Lane DA, Erdjument H, Di Marzo V, et al. Antithrombin Sheffield: Amino acid substitution at the reactive site (Arg 393 to His) causing thrombosis. Br J Haematol 1989; 71:91-96.
36. Erdjument H, Lane DA, Panico M, et al. Antithrombin Chicago, amino acid substitution of arginine 393 to histidine. Thromb Res 1989; 54:613-619.
37. Okajima K, Abe H, Wagatsuma M, et al. Antithrombin III Kumamoto II; A single mutation at Arg 393-His increased the affinity of antithrombin III for heparin. Am J Hematol 1995; 48:12-18.
38. Jering H and Tschesche H. Replacement of lysine by arginine, phenylalanine and tryptophan in the reactive site of the bovine trypsin-kallikrein inhibitor (Kunitz) and change of the inhibitor properties. Eur J Biochem 1976; 61:453-463.
39. Griffith MJ, Noyes CM, Tyndall JA, et al. Structural evidence for leucine at the reactive site of heparin cofactor II. Biochemistry 1985; 24:6777-6782.
40. Ragg H and Preibisch G. Structure and expression of the gene coding for the human serpin hLS2. J Biol Chem 1988; 263:12129-12134.
41. Ragg H. A new member of the plasma protease inhibitor gene family. Nucl Acids Res 1986; 14:1073-1088.
42. Derechin VM, Blinder MA and Tollefsen DM. Substitution of arginine for Leu^{444} in the reactive site of heparin cofactor II enhances the rate of thrombin inhibition. J Biol Chem 1990; 265:5623-5628.
43. Johnson D and Travis J. The oxidative inactivation of human α 1 proteinase inhibitor. Further evidence for methionine at the reactive center. J Biol Chem 1979; 254: 4022-4026.
44. Hopkins PCR, Crowther DC, Carrell RW, et al. Development of a novel recombinant serpin with potential antithrombotic properties. J Biol Chem 1995; 270: 11866-11871.
45. Olson ST, Stephens AW, Hirs CHW, et al. Kinetic characterization of the proteinase binding defect in a reactive site variant of the serpin, antithrombin. Role of the P1' residue in transition-state stabilization of antithrombin-proteinase complex formation. J Biol Chem 1995; 270:9717-9724.
46. Bigler TL, Lu WL, Park SJ, et al. Binding of amino acid side chains to preformed cavities: Interaction of serine proteinases with turkey ovomucoid third domains with coded and non-coded P1 residues. Prot Sci 1993; 2:786-799.
47. Wells JA. Additivity of mutational effects in proteins. Biochemistry 1990; 29: 8509-8517.
48. Madison EL, Goldsmith EJ, Gerard RD, et al. Amino acid residues that affect interaction of tissue-type plasminogen activator with plasminogen activator inhibitor 1. Proc Natl Acad Sci USA 1990; 87:3530-3533.
49. Madison EL, Goldsmith EJ, Gerard RD, et al. Serpin-resistant mutants of human tissue-type plasminogen activator. Nature 1989; 339:721-724.
50. Hortin GL, Trimpe BL and Fok KF. Plasmin's peptide-binding specificity: Characterization of ligand sites in α_2-antiplasmin. Thromb Res 1989; 54:621-632.
51. Wiman B, Almqvist A and Ranby M. The

non-covalent interaction between plasmin and α_2-antiplasmin. Fibrinolysis 1989; 3:231-235.
52. Skriver K, Wikoff WR, Patston PA, et al. Substrate properties of C1 inhibitor Ma (alanine→glutamic acid). Genetic and structural evidence suggesting that the P12 region contains critical determinants of serine protease inhibitor inhibitor/substrate status. J Biol Chem 1991; 266:9216-9221.
53. Hood DB, Huntington JA and Gettins PGW. α_1-Proteinase inhibitor variant T345R. Influence of P14 residue on substrate and inhibitory pathways. Biochemistry 1994; 33:8538-8547.
54. Waley SG. The kinetics of substrate-induced inactivation. Biochem J 1991; 279:87-94.
55. Hopkins PCR, Carrell RW and Stone SR. Effects of mutations in the hinge region of serpins. Biochemistry 1993; 32:7650-7657.
56. Huntington JA, Fan B and Gettins PGW. Inhibitory properties of S380W (P14) antithrombin. FASEB J 1995; 9:A1440. (Abstract).
57. Rubin H, Wang ZM, Nickbarg EB, et al. Cloning, expression, purification, and biological activity of recombinant native and variant human α_1-antichymotrypsin. J Biol Chem 1990; 265:1199-1207.
58. Björk I, Nordling K, Larsson I, et al. Kinetic characterization of the substrate reaction between a complex of antithrombin with a synthetic reactive-bond-loop tetradecapeptide and four target proteinases of the inhibitor. J Biol Chem 1992; 267: 19047-19050.
59. Patston PA, Gettins P, Beechem J, et al. Mechanism of serpin action: Evidence that C1 inhibitor functions as a suicide substrate. Biochemistry 1991; 30:8876-8882.
60. Olson ST. Heparin and ionic strength-dependent conversion of antithrombin III from an inhibitor to a substrate of α-thrombin. J Biol Chem 1985; 260:10153-10160.
61. Björk I, Ylinenjärvi K, Olson ST, et al. Conversion of antithrombin from an inhibitor of thrombin to a substrate with reduced heparin affinity and enhanced conformational stability by binding of a tetradecapeptide corresponding to the P1 to P14 region of the putative reactive bond loop of the inhibitor. J Biol Chem 1992; 267: 1976-1982.
62. Schulze AJ, Frohnert PW, Engh RA, et al. Evidence for the extent of insertion of the active site loop of intact α_1-proteinase inhibitor in β-sheet A. Biochemistry 1992; 31:7560-7565.
63. Schulze AJ, Baumann U, Knof S, et al. Structural transition of α_1-antitrypsin by a peptide sequentially similar to β-strand s4A. Eur J Biochem 1990; 194:51-56.
64. Shore JD, Day DE, Francis-Chmura AM, et al. A fluorescent probe study of plasminogen activator inhibitor-1. Evidence for reactive center loop insertion and its role in the inhibitory mechanism. J Biol Chem 1995; 270:5395-5398.
65. de Agostini A, Patston PA, Marottoli V, et al. A common epitope is created when the reactive center of C1 inhibitor is cleaved by plasma kallikrein, activated factor XII, C1 esterase, or neutrophil elastase. J Clin Invest 1988; 82:700-705.
66. Björk I, Nordling K and Olson ST. Immunologic evidence for insertion of the reactive-bond loop of antithrombin into the A β-sheet of the inhibitor during trapping of target proteinases. Biochemistry 1993; 32:6501-6505.
67. Pemberton PA, Stein PE, Pepys MB, et al. Hormone binding globulins undergo serpin conformational change in inflammation. Nature 1988; 336:257-258.
68. Stein PE, Leslie AGW, Finch JT, et al. Crystal structure of uncleaved ovalbumin at 1.95Å resolution. J Mol Biol 1991; 221:941-959.
69. McRae B, Nakajima K, Travis J, et al. Studies on reactivity of human leukocyte elastase, cathepsin G, and porcine pancreatic elastase toward peptides including sequences related to the reactive site of α_1-protease inhibitor (α_1-antitrypsin). Biochemistry 1980; 19:3973-3978.
70. Gettins PGW, Fan B, Crews BC, et al. Transmission of conformational change from the heparin binding site to the reactive center of antithrombin. Biochemistry 1993; 32:8385-8389.
71. Schreuder HA, de Boer B, Dijkema R, et al. The intact and cleaved human anti-

thrombin III complex as a model for serpin-proteinase interactions. Nature Struct Biol 1994; 1:48-54.

72. Carrell RW, Stein PE, Fermi G, et al. Biological implications of a 3 Å structure of dimeric antithrombin. Structure 1994; 2:257-270.
73. Polgár L. Serine proteases. In: Mechanisms of protease action, Boca Raton: CRC Press, 1989; 87-122.
74. Cleland WW and Kreevoy MM. Low-barrier hydrogen bonds and enzymic catalysis. Science 1994; 264:1887-1890.
75. Frey PA, Whitt SA and Tobin JB. A low-barrier hydrogen bond in the catalytic triad of serine proteases. Science 1994; 264: 1927-1930.
76. Perona JJ and Craik CS. Structural basis of substrate specificity in the serine proteases. Prot Sci 1995; 4:337-360.
77. Jörnvall H, Fish WW and Björk I. The thrombin cleavage site in bovine antithrombin. FEBS Lett 1979; 106:358-362.
78. Jesty J. The kinetics of formation and dissociation of the bovine thrombin-antithrombin III complex. J Biol Chem 1979; 254:10044-10050.
79. Longas MO and Finlay TH. The covalent nature of the human antithrombin III-thrombin bond. Biochem J 1980; 189: 481-489.
80. Fish WW and Björk I. Release of a two-chain form of antithrombin from the antithrombin-thrombin complex. Eur J Biochem 1979; 101:31-38.
81. Björk I, Jackson CM, Jörnvall H, et al. The active site of antithrombin. Release of the same proteolytically cleaved form of the inhibitor from complexes with factor IXa, factor Xa, and thrombin. J Biol Chem 1982; 257:2406-2411.
82. Lawrence DA, Ginsburg D, Day DE, et al. Serpin-protease complexes are trapped as stable acyl-enzyme intermediates. J Biol Chem 1995; 270:(In Press)
83. Matheson NR, van Halbeek H and Travis J. Evidence for a tetrahedral intermediate complex during serpin-proteinase interactions. J Biol Chem 1991; 266:13489-13491.
84. Vercaigne-Marko D, Carrere J, Ducourouble MP, et al. "In vivo" and "in vitro" inhibition of human pancreatic chymotrypsin A by serum inhibitors. Biol Chem Hoppe-Seyler 1987; 368:37-45.
85. Carrell RW and Evans DL. Serpins: mobile conformations in a family of proteinase inhibitors. Curr Op Struct Biol 1992; 2:438-446.
86. Fastrez J and Fersht AR. Demonstration of the acyl-enzyme mechanism for the hydrolysis of peptides and anilides by chymotrypsin. Biochemistry 1973; 12:2025-2034.
87. Longstaff C, Campbell AF and Fersht AR. Recombinant chymotrypsin inhibitor 2: Expression, kinetic analysis of inhibition with α-chymotrypsin and wild-type and mutant subtilisin BPN', and protein engineering to investigate inhibitory specificity and mechanism. Biochemistry 1990; 29:7339-7347.
88. Kaslik G, Patthy A, Bálint M, et al. Trypsin complexed with α_1-proteinase inhibitor has an increased structural flexibility. FEBS Lett 1995; 370:179-183.
89. Ray CA, Black RA, Kronheim SR, et al. Viral inhibition of inflammation. Cowpox virus encodes an inhibitor of the interleukin-1β converting enzyme. Cell 1992; 69: 597-604.
90. Hook VYH, Purviance RT, Azaryan AV, et al. Purification and characterization of α_1-antichymotrypsin-like protease inhibitor that regulates thiol protease involved in enkephalin precursor processing. J Biol Chem 1993; 268:20570-20577.
91. Takeda A, Yamamoto T, Nakamura Y, et al. Squamous cell carcinoma antigen is a potent inhibitor of cysteine proteinase cathepsin L. FEBS Lett 1995; 359:78-80.
92. Komiyama T, Ray CA, Pickup DJ, et al. Inhibition of interleukin-1β converting enzyme by the cowpox virus serpin CrmA. An example of cross-class inhibition. J Biol Chem 1994; 269:19331-19337.
93. Srinivasan R and Rose GD. LINUS: A hierarchic procedure to predict the fold of a protein. Proteins 1995; 22:81-99.
94. Patston PA, Gettins PGW and Schapira M. Serpins are suicide substrates: Implications for regulation of proteolytic pathways. Sem Thromb Hemostas 1994; 20:410-416.
95. Jordan RE, Kilpatrick J and Nelson RM.

Heparin promotes the inactivation of antithrombin by neutrophil elastase. Science 1987; 237:777-779.
96. Jordan RE, Nelson RM, Kilpatrick J, et al. Antithrombin inactivation by neutrophil elastase requires heparin. Am J Med 1989; 87 suppl 3B:3B-19S.
97. Carrell RW and Owen MC. Plakalbumin, α_1-antitrypsin, antithrombin and the mechanism of inflammatory thrombosis. Nature 1985; 317:730-732.
98. Sires UI, Murphy G, Baragi VM, et al. Matrilysin is much more efficient than other matrix metalloproteinases in the proteolytic inactivation of α_1-antitrypsin. Biochem Biophys Res Commun 1994; 204:613-620.
99. Kress LF, Kurecki T, Chan SK, et al. Characterization of the inactive fragment resulting from limited proteolysis of human α_1-proteinase inhibitor by *Crotalus adamanteus* proteinase II. J Biol Chem 1979; 254: 5317-5320.
100. Kress LF and Catanese JJ. Identification of the cleavage sites resulting from enzymatic inactivation of human antithrombin III by *Crotalus adamanteus* proteinase II in the presence and absence of heparin. Biochemistry 1981; 20:7432-7438.
101. Kress LF, Catanese J and Hirayama T. Analysis of the effects of snake venom proteinases on the activity of human plasma C1 esterase inhibitor, α_1-antichymotrypsin, and α_2-antiplasmin. Biochim Biophys Acta 1984; 745:113-120.
102. Janssen M, Meier J and Freyvogel TA. Purification and characterization of an antithrombin III inactivating enzyme from the venom of the African night adder (*Causus rhombeatus*). Toxicon 1992; 30:985-999.
103. Catanese JJ and Kress LF. Opossum serum α_1-proteinase inhibitor: Purification, linear sequence, and resistance to inactivation by rattlesnake venom metalloproteinases. Biochemistry 1993; 32:509-515.
104. Yu M-H, Lee KN and Kim J. The Z type variation of human α_1-antitrypsin causes a protein folding defect. Nature Struct Biol 1995; 2:363-367.
105. Lomas DA, Evans DL, Finch JT, et al. The mechanism of Z α_1-antitrypsin accumulation in the liver. Nature 1992; 357:605-607.
106. Mast AE, Enghild JJ and Salvesen GS. Conformation of the reactive site loop of α_1-proteinase inhibitor probed by limited proteolysis. Biochemistry 1992; 31: 2720-2728.
107. Kassell B, Radicevic M, Ansfield MJ, et al. The basic trypsin inhibitor of bovine pancreas. IV. The linear sequence of the 58 amino acids. Biochem Biophys Res Commun 1965; 18:255-258.
108. Greene LJ and Bartelt DC. The structure of the bovine pancreatic secretory trypsin inhibitor-Kazal's inhibitor. II. The order of the tryptic peptides. J Biol Chem 1969; 244:2646-2657.
109. Ikenaka T, Odani T, Sakai M, et al. Amino acid sequence of an alkaline proteinase inhibitor (*Streptomyces* Subtilisin inhibitor) from *Streptomyces albogriseolus* S-3253. J Biochem 1974; 76:1191-1209.
110. Koide T and Ikenaka T. Studies on soybean trypsin inhibitors. 3. Amino-acid sequence of carboxy-terminal region and complete amino acid sequence of soybean trypsin inhibitor (Kunitz). Eur J Biochem 1973; 32:417-431.
111. Tan CGL and Stevens FC. Amino acid sequence of lima bean protease inhibitor component IV. 2. Isolation and sequence determination of the chymotryptic peptides and the complete amino acid sequence. Eur J Biochem 1971; 18:515-523.
112. Seemüller U, Fritz H and Eulitz M. Eglin: elastase-cathepsin G inhibitor from leeches. Methods Enzymol 1981; 80:804-816.
113. Hass GM, Hermodson MA, Ryan CA, et al. Primary structure of two low molecular weight proteinase inhibitors from potatoes. Biochemistry 1982; 21:752-756.
114. Thompson RC and Ohlsson K. Isolation, properties, and complete amino acid sequence of human secretory leukocyte protease inhibitor, a protease inhibitor of leukocyte elastase. Proc Natl Acad Sci USA 1986; 83:6692-6696.
115. Hojima Y, Pierce JV and Pisano JJ. Pumpkin seed inhibitor of human factor XIIa (activated Hageman factor) and bovine trypsin. Biochemistry 1982; 21:3741-3746.
116. Bernard VD and Peanasky RJ. The serine proteinase inhibitor family from *Ascaris*

suum. Chemical determination of the 5 disulfides. Arch Biochem Biophys 1993; 303:367-376.

117. Jackson SE and Fersht AR. Contribution of residues in the reactive site loop of chymotrypsin inhibitor 2 to protein stability and activity. Biochemistry 1994; 33: 13880-13887.
118. Bode W and Huber R. Natural protein proteinase inhibitors and their interaction with proteinases. Eur J Biochem 1992; 204:433-451.
119. Huber R, Kukla D, Bode W, et al. Structure of the complex formed between bovine trypsin and bovine pancreatic trypsin inhibitor. II Crystallographic refinement at 1.9Å resolution. J Mol Biol 1974; 89: 73-101.
120. Bode W, Schwager P and Huber R. The transition of bovine trypsinogen to trypsin-like state upon strong ligand binding. The refined crystal structures of the trypsinogen-pancreatic trypsin complex and of its ternary complex with Ile-Val at 1.9Å resolution. J Mol Biol 1978; 118:99-112.
121. Chen Z and Bode W. Refined 2.5Å X-ray crystal structure of the complex formed by porcine kallikrein A and the bovine pancreatic trypsin inhibitor. Crystallization, Patterson search, structure determination, refinement, structure and comparison with its components and with bovine trypsin—pancreatic trypsin inhibitor complex. J Mol Biol 1983; 164:283-311.
122. Bolognesi M, Gatti G, Menegatti E, et al. Three-dimensional structure of the complex between pancreatic secretory trypsin inhibitor (Kazal type) and trypsinogen at 1.8Å resolution. Structure solution, crystallographic refinement and protein structure interpretation. J Mol Biol 1982; 162: 839-868.
123. Read RJ, Fujinaga M, Sielecki AR, et al. Structure of the complex of *Streptomyces griseus* proteinase B and the third domain of the turkey ovomucoid inhibitor at 1.8Å resolution. Biochemistry 1983; 22:4420-4433.
124. Fujinaga M, Sielecki AR, Read RJ, et al. Crystal and molecular structures of the complex of α-chymotrypsin with its inhibitor the ovomucoid third domain at 1.8Å resolution. J Mol Biol 1987; 195:397-418.
125. Bode W, Wei A-Z, Huber R, et al. X-ray crystal structure of the complex of human leukocyte elastase (PMN elastase) and the third domain of the turkey ovomucoid inhibitor. EMBO J 1986; 5:2453-2458.
126. Onesti S, Brick P and Blow DM. Crystal structure of a Kunitz-type trypsin inhibitor from *Erythrina caffra* seeds. J Mol Biol 1991; 217:153-176.
127. Hirono S, Agawa H, Iitaka Y, et al. Crystal structure at 2.6Å resolution of the complex of Subtilisin BPN' with *Streptomyces* subtilisin inhibitor. J Mol Biol 1984; 178: 389-413.
128. McPhalen CA, Svendsen I, Janassen I, et al. Crystal and molecular structures of chymotrypsin inhibitor 2 from barley seeds in complex with subtilisin Novo. Proc Natl Acad Sci USA 1985; 82:7242-7246.
129. Gros P, Betzel C, Dauter Z, et al. Molecular dynamics refinement of a thermitase-eglin C complex at 1.98Å resolution and comparison of two crystal forms that differ in calcium content. J Mol Biol 1989; 210:347-367.
130. Greenblatt HM, Ryan CA and James MNG. Structure of the complex of *Streptomyces griseus* proteinase B and polypeptide chymotrypsin inhibitor-1 from Russet Burbank Potato tubers at 2.1Å resolution. J Mol Biol 1989; 205:201-225.
131. Grütter MG, Fendrich G, Huber R, et al. The 2.5Å X-ray crystal structure of the acid-stable proteinase inhibitor from human mucous secretions analysed in its complex with bovine α-chymotrypsin. EMBO J 1988; 7:345-351.
132. Tsunogae Y, Tanaka I, Yamane T, et al. Structure of the trypsin-binding domain of Bowman-Birk type protease inhibitor and its interactions with trypsin. J Biochem 1986; 100:1637-1646.
133. Lin GD, Bode W, Huber R, et al. The 0.25 nm X-ray structure of the Bowman-Birk-type inhibitor from mung bean in ternary complex with porcine trypsin. Eur J Biochem 1993; 212:549-555.
134. Bode W, Greyling HJ, Huber R, et al. The refined 2.0Å X-ray crystal structure of the complex formed betweem bovine β-trypsin

and CMTI-I, a trypsin inhibitor from squash seeds (*Cucurbita maxima*). Topographical similarity of the squash seed inhibitors with the carboxypeptidase A inhibitors from potatoes. FEBS Lett 1989; 242:285-292.
135. Huang K, Strynanka NCJ, Bernard VD, et al. The molecular structure of the complex of *Ascaris* chymotrypsin elastase inhibitor with porcine elastase. Structure 1994; 2:679-689.
136. Beckmann J, Mehlich A, Schröder W, et al. Preparation of chemically 'mutated' aprotinin homologues by semisynthesis. Eur J Biochem 1988; 176:675-682.
137. Komiyama T, Bigler TL, Yoshida N, et al. Replacement of P1 Leu^{18} by Glu^{18} in the reactive site of turkey ovomucoid third domain converts it into a strong inhibitor of Glu-specific *Streptomyces griseus* proteinase (GluSGP). J Biol Chem 1991; 266: 10727-10730.
138. Kojima S, Nishiyama Y, Kumagai I, et al. Inhibition of subtilisin BPN' by reaction site P1 mutants of *Streptomyces* subtilisin inhibitor. J Biochem 1991; 109:377-382.
139. Rosenberg RD. Chemistry of the hemostatic mechanism and its relationship to the action of heparin. Fed Proc 1977; 36:10-18.
140. Erdjument H, Lane DA, Panico M, et al. Single amino acid substitutions in the reactive site of antithrombin leading to thrombosis. Congenital substitution of arginine 393 to cysteine in antithrombin Northwick Park and to histidine in antithrombin Glasgow. J Biol Chem 1988; 263:5589-5593.
141. Pannell R, Johnson D and Travis J. Isolation and properties of human plasma α-1-proteinase inhibitor. Biochemistry 1974; 13:5439-5445.
142. Travis J, Matheson NR, George PM, et al. Kinetic studies on the interaction of α_1-proteinase inhibitor (Pittsburgh) with trypsin-like serine proteinases. Biol Chem Hoppe-Seyler 1986; 367:853-859.
143. Jallat S, Carvallo D, Tessier LH, et al. Altered specificities of genetically engineered α_1-antitrypsin variants. Prot Eng 1986; 1:29-35.
144. Beatty K, Bieth J and Travis J. Kinetics of association of serine proteinases with native and oxidized α-1-proteinase inhibitor and α-1-antichymotrypsin. J Biol Chem 1980; 255:3931-3934.
145. Rubin H, Plotnick M, Wang Z, et al. Conversion of α_1-antichymotrypsin into a human neutrophil elastase inhibitor: Demonstration of variants with different association rate constants, stoichiometries of inhibition, and complex stabilities. Biochemistry 1994; 33:7627-7633.
146. Schechter NM, Jordan LM, James AM, et al. Reaction of human chymase with reactive site variants of alpha-1-antichymotrypsin—modulation of inhibitor versus substrate properties. J Biol Chem 1993; 268:23626-23633.
147. Sim RB, Arlaud G and Colomb M. Kinetics of reaction of human C1-inhibitor with the human complement system proteases C1r and C1s. Biochim Biophys Acta 1980; 612:433-449.
148. Schapira M, deAgostini A and Colman RW. Contribution of plasma protease inhibitors to the inactivation of kallikrein in plasma. J Clin Invest 1982; 69:462-468.
149. Eldering E, Huijbregts CCM, Lubbers YTP, et al. Characterization of recombinant C1 inhibitor P1 variants. J Biol Chem 1992; 267:7013-7020.
150. Aulak KS, Pemberton PA, Rosen FS, et al. Dysfunctional C1-inhibitor (At), isolated from a type II hereditary angio-oedema plasma, contains a P1' reactive centre ($Arg^{444} \rightarrow His$) mutation. Biochem J 1988; 253:615-618.
151. Church FC, Noyes CM and Griffith MJ. Inhibition of chymotrypsin by heparin cofactor II. Proc Natl Acad Sci USA 1985; 82:6431-6434.
152. Ehrlich HJ, Gebbink RK, Keijer J, et al. Alteration of serpin specificity by a protein cofactor. Vitronectin endows plasminogen activator inhibitor 1 with thrombin inhibitory properties. J Biol Chem 1990; 265: 13029-13035.
153. Sherman PM, Lawrence DA, Yang AY, et al. Saturation mutagenesis of the plasminogen activator inhibitor-1 reactive center. J Biol Chem 1992; 267:7588-7595.
154. Sherman PM, Lawrence DA, Verhamme IM, et al. Identification of tissue-type plasmi-

nogen activator-specific plasminogen activator inhibitor-1 mutants. Evidence that second sites of interaction contribute to target specificity. J Biol Chem 1995; 270:9301-9306.

155. Esmon CT. The regulation of natural anticoagulant pathways. Science 1987; 235: 1348-1352.

156. Phillips JE, Cooper ST, Potter EE, et al. Mutagenesis of recombinant protein C inhibitor reactive site residues alters target proteinase specificity. J Biol Chem 1994; 269:16696-16700.

157. Schechter NM, Sprows JL, Schoenberger OL, et al. Reaction of human skin chymotrypsin-like proteinase chymase with plasma proteinase inhibitors. J Biol Chem 1989; 264:21308-21315.

158. Lawrence DA, Olson ST, Palaniappan S, et al. Serpin reactive center loop mobility is required for inhibitor function but not for enzyme recognition. J Biol Chem 1994; 269:27657-27662.

159. Davis AE III, Aulak K, Parad RB, et al. C1 inhibitor hinge region mutations produce dysfunction by different mechanisms. Nature Genetics 1992; 1:354-358.

160. Devraj-Kizuk R, Chui DHK, Prochownik EV, et al. Antithrombin III Hamilton: a gene with a point mutation (guanine to adenine) in codon 382 causing impaired serine protease activity. Blood 1988; 72:1518-1523.

161. Wright HT and Blajchman MA. Proteolytically cleaved mutant antithrombin-Hamilton has high stability to denaturation characteristic of wild type inhibitor serpins. FEBS Lett 1994; 348:14-16.

162. Perry DJ, Harper PL, Fairham S, et al. Antithrombin Cambridge, 384 Ala to Pro: A new variant identified using the polymerase chain reaction. FEBS Lett 1989; 254:174-176.

163. Perry DJ, Daly M, Harper PL, et al. Antithrombin Cambridge II, 384 Ala to Ser. Further evidence of the role of the reactive center loop in the inhibitory function of the serpins. FEBS Lett 1991; 285:248-250.

164. Siddique Z, Mcphaden AR and Whaley K. C1-Inhibitor gene nucleotide insertion causes Type-II hereditary angioedema. Hum Genet 1993; 92:189-190.

165. White D, Abraham G, Carter C, et al. A novel missense mutation in the antithrombin III gene (Ala 387 to Val) causing recurrent venous thrombosis. Hum Genet 1992; 90:472-473.

166. Rost B and Sander C. Prediction of protein secondary structure at better than 70% accuracy. J Mol Biol 1993; 232:584-599.

167. Rost B and Sander C. Improved prediction of protein secondary structure by use of sequence profiles and neural networks. Proc Natl Acad Sci USA 1993; 90:7558-7562.

168. Rost B and Sander C. Combining evolutionary information and neural networks to predict protein secondary structure. Proteins 1994; 19:55-72.

169. Patston, PA, Gettins, PGW. Significance of secondary structure predictions on the reactive center loop region of serpins. A model for the folding of serpins into a metastable state, 1995. (submitted for publication)

CHAPTER 4

Serpins Activated by Heparin and Other Glycosaminoglycans

A number of serpins that are mostly found in the blood plasma of vertebrates exhibit the unique feature of being activated by heparin and certain other glycosaminoglycans. The activation results from the binding of the glycosaminoglycan to the inhibitor and is manifested by enhancements of up to several thousand-fold in the rate at which these inhibitors inactivate their target proteinases. This class of serpins act, largely to regulate proteinases that function in blood coagulation and fibrinolysis and, in particular, have the multifunctional proteinase thrombin as a common target (Fig. 4.1). This accounts for the presence of most of these inhibitors in blood plasma. The extravascular localization of other serpins in this group is thought to control extravascular thrombin-mediated wound healing and deposition of fibrin. Some of these serpins additionally regulate proteinases of seminal fluid, sperm and certain other tissue-specific sites. The abundance of the glycosaminoglycans that activate this group of serpins at the sites of serpin action has led to the suggestion that glycosaminoglycans are physiological regulators of the activity of these inhibitors. The known heparin-activatable serpins include antithrombin, heparin cofactor II, protein C inhibitor, protease nexin 1 and plasminogen activator inhibitor-1. The general properties of this class of serpins and their regulation by glycosaminoglycans are the subject of this chapter.

1. ANTITHROMBIN

Antithrombin is the primary inhibitor of the procoagulant proteinases of the intrinsic and extrinsic blood coagulation pathways (Fig. 4.1). The inhibitor is synthesized in the liver[1] and circulates in blood plasma at a concentration of 0.13 mg/ml or 2.3 μM[2] with a half-life of ~3 days[3] (Table 4.1). It is a single chain glycoprotein with a molecular weight of 58 kDa of which 16% is carbohydrate.[4] Protein and DNA sequencing

Serpins: Structure, Function and Biology, by Peter Gettins, Philip Patston and Steven Olson.

indicate that the human protein contains 432 amino acids and six cysteine residues all linked in disulfide bridges.[5-8] Bovine, sheep, rabbit and mouse antithrombins are all one residue longer and have 84-89% homology with the human inhibitor.[9-12] Four Asn-X-Ser/Thr consensus sequences are all glycosylated with identical biantennary complex carbohydrate chains terminating in sialic acid in the predominant α-form of antithrombin,[13,14] which constitutes 90% of the total circulating inhibitor.

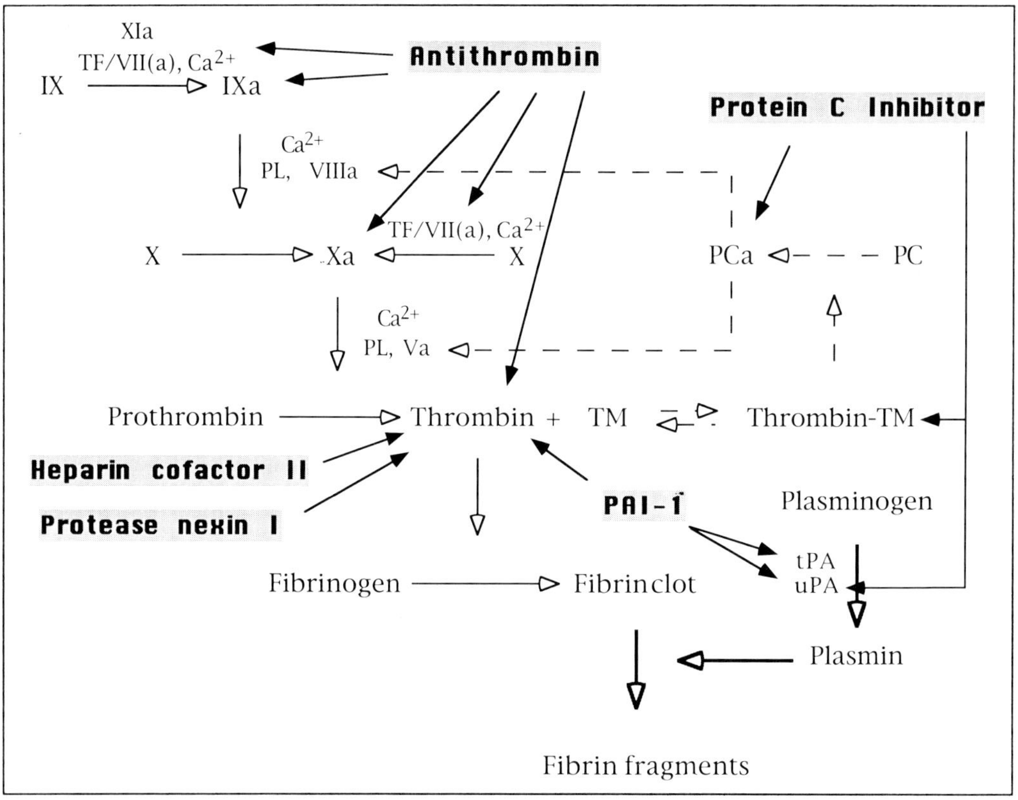

Fig. 4.1. Regulation of proteinases of blood coagulation and fibrinolysis pathways by heparin-activatable serpins. The pathways for the activation of proteinases of blood coagulation (solid and dashed arrows with open arrowheads) and fibrinolysis (bold solid arrows with open arrowheads) are shown along with the heparin-activatable serpins which regulate their activity (shaded, with solid arrows and closed arrowheads pointing to proteinase targets). Activation of clotting is initiated by the exposure of tissue factor (TF) receptors in the subendothelium to the blood. Binding of factor VII or traces of VIIa produced by autoactivation to tissue factor produces a calcium ion-dependent activator of factor X and factor IX. Factor Xa and IXa in turn activate more factor VII to factor VIIa. Factor IX can also be activated by factor XIa, which may become activated by thrombin through a feedback loop. Factor IXa amplifies the activation of factor X in a reaction which requires calcium ions, factor VIIIa and a phospholipid (PL) surface, most likely a platelet, endothelial cell or monocyte membrane. Factor Xa then activates prothrombin to thrombin in the presence of calcium ions, factor Va and a phospholipid membrane. Thrombin finally catalyzes the conversion of the soluble protein, fibrinogen, to insoluble fibrin which constitutes the framework of the blood clot. Dissolution of the fibrin clot is catalyzed by plasmin which is generated from plasminogen through the action of either tissue-type or urokinase-type plasminogen activators (tPA and uPA), reactions which constitute fibrinolysis. Binding of thrombin to its endothelial cell receptor, thrombomodulin (TM), blocks thrombin procoagulant functions and converts thrombin into an anticoagulant activator of protein C (PC). Protein Ca functions as an anticoagulant by catalyzing the proteolytic destruction of the cofactor proteins, factors VIIIa and Va, which are essential for activating factors IX and X. These reactions may occur both within the blood or in the extravascular space due to the expression of coagulant proteins by certain extravascular cell-types.

A minor glycoform, termed β-antithrombin, which makes up the remaining 10% of the inhibitor, lacks carbohydrate at the N135 glycosylation site.[15-17] Incomplete glycosylation results from a reduced efficiency of incorporating a carbohydrate chain at the Asn-X-Ser sequence of residues 135-137 as compared to the Asn-X-Thr sequences comprising the other three glycosylation sites.[18] This heterogeneity was eliminated in a recombinant S137T variant antithrombin. The pI of the inhibitor ranges from 4.9-5.3, reflecting charge heterogeneity in the terminal sialic acid residues.[19] The gene for antithrombin has been cloned and localized to chromosome 1q23-25.[20-24] The gene consists of 13,477 base pairs from the 5' transcription start point to the 3' polyA tail and contains 7 exons and 6 introns. Ten Alu repeats are present in the introns. The cDNA sequence indicates that the protein is expressed with a 32 residue signal peptide, consistent with it being a secreted protein. Recombinant antithrombin has been expressed in a cell-free system in a nonglycosylated form[25] as well as in yeast,[26] slime mold,[27] insect cells[28] and mammalian cells[29-31] in a glycosylated form. The glycosylation in these latter cases differs from that of the plasma protein, although the mammalian cell expressed protein shows a close similarity with the plasma protein in its glycosylation, including incomplete glycosylation at Asn 135.[32,33]

Antithrombin inhibits all the proteinases of the coagulation cascade as well as a number of nontarget proteinases.[34,35] However, its main targets appear to be thrombin, factor Xa and possibly factor IXa. Inactivation of these major target proteinases is slow in the absence of heparin with second order inhibition rate constants ranging from 4.8 x 10^2 $M^{-1}s^{-1}$ to 1.1 x 10^4 $M^{-1}s^{-1}$ (Table 4.2). Heparin accelerates these reactions several-thousand-fold, although the acceleration of factor IXa inhibition using direct assays of stable complex formation was found to be considerably less than that determined in an earlier report by indirect assay of enzyme inactivation.[36] The

Table 4.1. Properties of glycosaminoglycan-activated serpins

Serpin	M_r (kDa)	% CHO	# CHO	Plasma conc (µM)	pI	Number of cysteines	Number of disulfides	Number of amino acids	Signal peptide size	Reactive bond	GAG activators	[NaCl] from heparin agarose (M)
Antithrombin	58	16	4	2.3	4.9-5.3	6	3	432	32	R393-S394	Heparin, HS[a]	0.9
Heparin cofactor II	66	10	3	1.2	4.9-5.3	3	0	480	19	L444-S445	Heparin, HS, DS	0.4
Protein C inhibitor	57	23	5	0.1	4.5-6.0	1	0	387	19	R354-S355	Heparin, HS	0.5
Protease nexin 1	42-43	6	3	–	7.5-7.8	3	0	378 (α) 379 (β)	19	R345-S346	Heparin, HS, CS-A, CS-C	0.55-0.6
PAI-1	52	15	3	.0004	4.5-5.0	0	0	379	23	R346-M347	Heparin, HS, DS, C-4-S	0.4-0.5

a. HS, heparan sulfate; DS, dermatan sulfate; CS-A, chondroitin sulfate A; CS-C, chondroitin sulfate C; C-4-S, chondroitin-4-sulfate CHO, carbohydrate, GAG, glycosaminoglycan.

acceleration of other nontarget enzymes is also significantly less than that of target enzymes. The reactive bond of antithrombin occurs at Arg 393-Ser 394, consistent with the substrate specificity of the target enzymes of the inhibitor.[37] Natural and engineered variants with mutations in the P1 and P1' residues have indicated that both residues contribute to inhibitor specificity for these enzymes.[36,38-42]

The activating effect of heparin on antithrombin is responsible for the potent anticoagulant activity of the polysaccharide and its widespread clinical use for the prophylaxis of venous thrombosis. Heparin is a highly sulfated linear polysaccharide consisting of a repeating disaccharide unit composed of a hexuronic acid linked to glucosamine (Fig. 4.2). Heterogeneity in heparin structure results from the type of hexuronic acid, which may be glucuronic or iduronic acid, from variable sulfation of the disaccharide and from the molecular weight of the polysaccharide chain, which ranges from 5,000-30,000, or 15 to 100 saccharide units.[43] Antithrombin is unique among the heparin-activatable serpins in requiring a sequence-specific pentasaccharide in the polysaccharide for its activation[44-48] (Fig. 4.3). About one-third of heparin molecules in commercial preparations contain one such pentasaccharide sequence distributed randomly in the chains,[49] although multiple sequences are found in a small fraction of these chains.[50-52] Heparin molecules containing the pentasaccharide bind the inhibitor with high affinity and account for the bulk of the anticoagulant activity of the preparation. The remaining two-thirds of commercial heparin preparations lack the pentasaccharide and have an ~1,000-fold reduced affinity for antithrombin and < 5% of the anticoagulant activity.[53-55] The heparin fraction with high affinity for antithrombin can be separated from the low-affinity fraction by chromatography on matrix-bound antithrombin[56,57] or other similar methods.[58] Both heparin

Table 4.2. Second order rate constants for antithrombin-proteinase reactions measured in the absence and presence of glycosaminoglycans at pH 7.4 and the indicated temperatures

Target proteinase	Temp(°C)	Rate constant ($M^{-1}s^{-1}$) - Heparin	+ Heparin	Enhancement	Ref.
Acrosin	37	4.4×10^4	1.1×10^7	250	217
Activated protein C	37	0.15	1.9	13	228
β-Trypsin	25	1.5×10^5	1.1×10^6	7	35
Factor VIIa	37	-	1.7×10^2		116
Factor VIIa-TF[a]	37	4.5×10^2	5.6×10^3	12	116
Factor IXa	25	3.2×10^2	1.0×10^4	30	36
Factor IXa	37	4.9×10^2	3.3×10^5	680	89
Factor Xa	25	$2.1\text{-}2.5 \times 10^3$	1.3×10^6	500-600	95
Factor Xa	37	3.1×10^3	4.0×10^6	1300	89
Factor XIa	25	3.0×10^2	1.5×10^4	30	118
Factor XIa-HK[b]	25	3.0×10^2	5.0×10^5	1700	118
Factor XIIa	37	3.6×10^1	4.2×10^2	12	326
Plasma kallikrein	25	1.6×10^2	2.1×10^3	13	117
Kallikrein-HK[b]	25	1.6×10^2	1.9×10^5	1200	117
Plasmin[c]	25	4.5×10^2	1.0×10^4	22	36
Plasmin	37	6.7×10^2	6.7×10^4	100	89
Thrombin	25	8.9×10^3	3.7×10^7	~4000	94
Thrombin	37	1.1×10^4	4.0×10^7	3600	228

a, TF, tissue factor; b, HK, high molecular weight kininogen; c, + 20 mM ε-aminocaproic acid and the related glycosaminoglycan heparan sulfate, which is composed of the same disaccharide repeat but is less sulfated, can bind and activate antithrombin[59] whereas other glycosaminoglycans appear to have little or no activating ability.[60] Only ~1% of heparan sulfate chains appear active in this regard and appear to involve a similar or identical activating sequence, suggesting a common mechanism of activation.[61]

The dissociation constant for the binding of high-affinity heparin to antithrombin has been measured to be 10-20 nM at physiological ionic strength and pH.[54,55,62] Synthetic pentasaccharide binds with only a 3- to 4-fold reduced affinity, implying that this sequence accounts for the major part (~95%) of the binding energy of the interaction.[63] The affinity decreases with increasing ionic strength, indicating that ionic interactions are involved.[64] A quantitative analysis of this dependence revealed that 4-5 ionic interactions contribute to the binding,[63] consistent with the number of sulfates in the pentasaccharide implicated in the interaction.[43] A large fraction of the binding energy is also non-ionic in origin, suggesting that hydrophobic, hydrogen bond and van der Waals interactions make a significant contribution to the binding. An additional ionic interaction outside the pentasaccharide site has been suggested to account for the slightly higher affinity of full-length heparin chains.[63]

Binding of heparin to antithrombin induces a conformational change in the

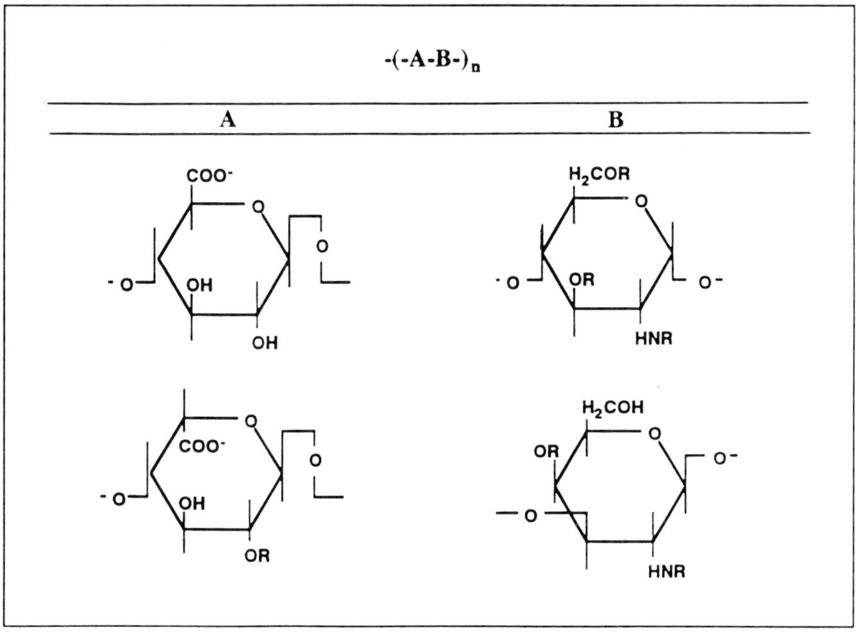

Fig. 4.2. General disaccharide repeat unit of glycosaminoglycans. Shown are the structures of the hexuronic acid (A) and glycosamine (B) residues, which comprise the repeating disaccharide of glycosaminoglycan polymers. The hexuronic acid residue can be either D-glucuronic acid (top) or L-iduronic acid (bottom) and the glycosamine residue can be either D-glucosamine (top) or D-galactosamine (bottom). R groups indicate positions of sulfation except for N-linkages which can be either sulfate or acetyl groups. In heparin the A unit can be either of the hexuronic acids, whereas the B unit is always glucosamine, usually N-sulfated. Heparan sulfate has the same disaccharide repeat as heparin, but is less sulfated, mostly N-acetylated and enriched in glucuronic acid. In dermatan sulfate, the A unit is mostly iduronic acid and the B unit is always N-acetyl-galactosamine. In chondroitin sulfate the A unit is invariably glucuronic acid and the B unit N-acetylgalactosamine. Reproduced with permission from Perspectives in Drug Discovery and Design.[327] *Copyright 1994 by ESCOM.*

inhibitor as evidenced by the significant changes in the spectroscopic properties of antithrombin which accompany heparin binding. These changes include a red-shift in the near UV spectrum,[53] a 40% enhancement in the protein fluorescence,[54,55] large increases in ellipticity in the near UV circular dichroism spectrum,[53] and substantial perturbations of the proton NMR spectrum.[65-67] Whereas such observations do not prove a conformational change, other evidence strongly supports this conclusion. Rapid kinetic studies indicate that heparin binding is a two-step process involving the formation of an initial complex followed by a subsequent conformational change in which the fluorescence of a buried tryptophan residue is enhanced.[62,68] A conformational change is also indicated by the observations that heparin binding to antithrombin is accompanied by an increased reactivity of Lys 236 to chemical modifying agents[69] and the appearance of a new N-terminal cleavage site for a non-target proteinase.[70] Further evidence for a conformational change and its transmission to the reactive center region has been provided by studies of an R393C variant of antithrombin specifically labeled at the P1 residue with a nitrobenzofuran (NBD) fluorescent reporter group.[71] High-affinity heparin or pentasaccharide binding to the labeled variant was normal and was accompanied by a saturable 100% enhancement in the fluorescence of the reactive center reporter group, whereas low-affinity heparin and dextran sulfate showed much smaller fluorescence changes requiring higher polysaccharide concentrations. A conformational change in the reactive center loop of antithrombin induced by heparin binding is also suggested from the observation that a monoclonal antibody which binds to the strand 1C/4B region immediately following the loop blocks thrombin inhibition in the absence of heparin but not in the presence of the polysaccharide.[72]

The heparin binding site on antithrombin has been mapped by studies of variant antithrombins with natural and site-directed mutations which have a diminished ability to bind heparin. The residues implicated in binding heparin map to the amino-terminal tail and to the A and D α-helices[73-75] (Fig. 4.4). An early proposal that Lys residues 290, 294 and 297 are part of a primary or secondary site

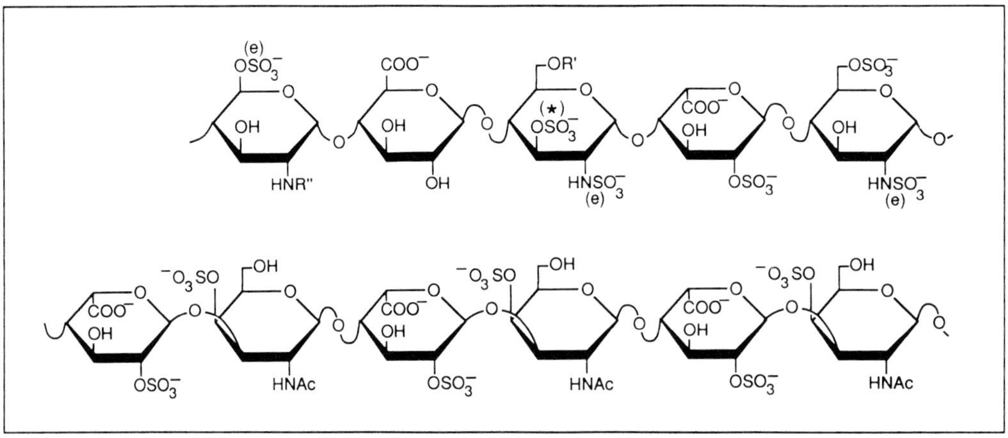

Fig. 4.3. Structures of the heparin pentasaccharide binding sequence for antithrombin (top) and the dermatan sulfate hexasaccharide binding sequence for heparin cofactor II (bottom). Structural variants in the heparin pentasaccharide are indicated by R′ (-H or $-SO_3^-$). The 3-O-sulfate group marked by an asterisk is a distinguishing structural feature of the antithrombin-binding region of heparin. This sulfate group and the three sulfate groups marked (e) are essential for high-affinity binding of the polysaccharide to antithrombin. Reproduced with permission from the Annual Review of Biochemistry volume 60,[328] Copyright 1991 by Annual Review Inc. Figure kindly provided by Dr. Ulf Lindahl.

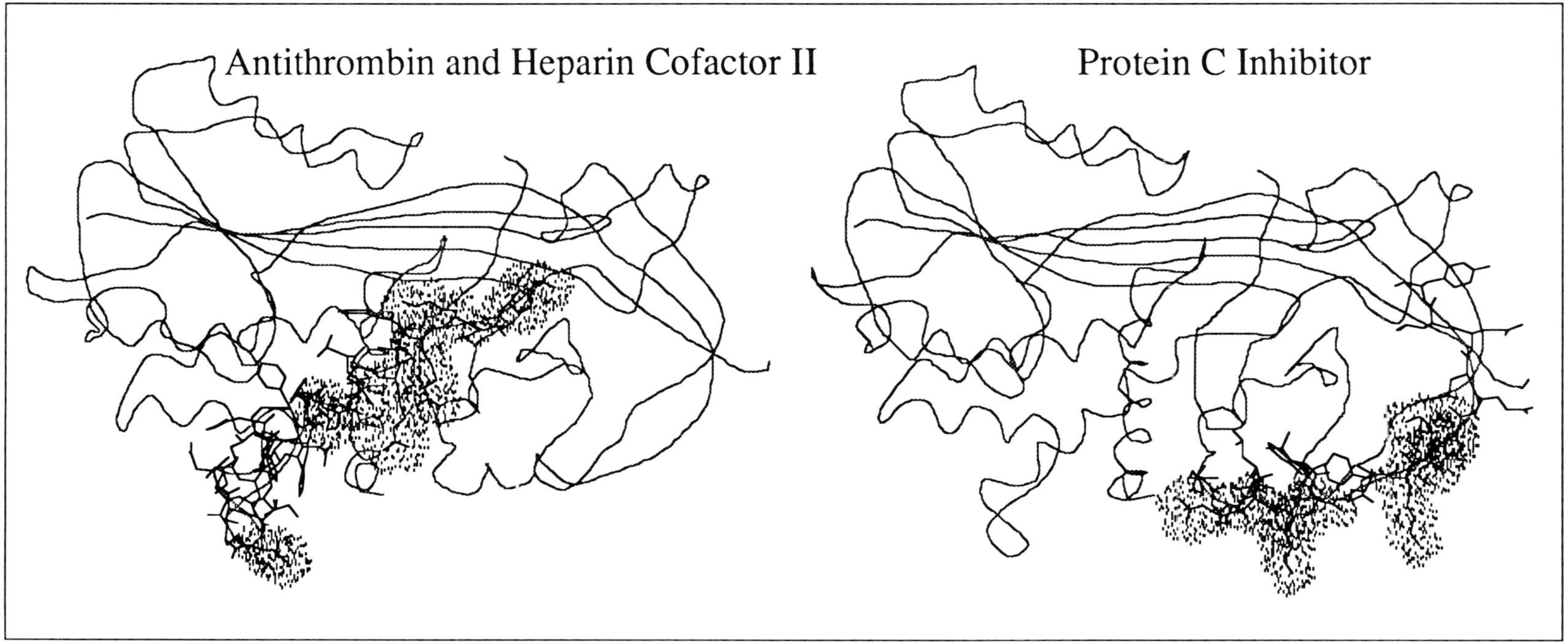

Fig. 4.4. Molecular models of the putative heparin binding domains of heparin-activatable serpins. Shown on the left is a three-dimensional model of antithrombin and heparin cofactor II constructed from the crystal structure of α_1-proteinase inhibitor. The model highlights the D helix region thought to be involved in binding heparin in these serpins as well as in protease nexin 1 and PAI-1 by depicting the van der Waals radii of basic residues in this region. The D helix region is contrasted on the right with the H helix region implicated in binding heparin to protein C inhibitor. The H helix is again highlighted in a model of protein C inhibitor similarly constructed from an α_1-proteinase inhibitor template by showing van der Waals radii of basic residues. Reprinted with permission from Blood Coagulation and Fibrinolysis.[329] *Copyright 1994 by Rapid Science Inc. Figure kindly provided by Dr. Frank Church.*

for heparin binding[76] has been conclusively shown by mutagenesis to be incorrect.[77] The D helix in particular contains a number of charged basic residues conserved among the heparin-activatable serpins, consistent with this being a common site for heparin binding and activation (Fig. 4.5). Molecular modeling of the antithrombin-heparin interaction suggests that the pentasaccharide makes a complementary interaction with R46 and R47 in helix A and with K125, R129, R132, K133 and K136 in helix D[78] (Fig. 4.6). A recent proposal for heparin activation of antithrombin based on this modeling suggests that heparin binding transmits conformational changes to β-sheet A and reactive center loop regions of the inhibitor by inducing an elongation of helix D[79] (Fig. 4.7). Since strand 2 of β-sheet A follows helix D in the primary structure, the elongation of helix D is thought to produce a contraction of the A sheet, which expels the reactive center loop from its partial loop inserted state observed in the X-ray structure of native antithrombin.[74,80] Such a model provides an explanation of how heparin binding affinity is modulated by loop insertion, since partial or full insertion of the reactive center loop into β-sheet A in antithrombin-proteinase complexes, cleaved antithrombin or latent antithrombin is associated with a large loss in heparin affinity.[55,81-83] However, an alternative model suggests that the loop is completely exposed in the native structure. In this model heparin binding activates the inhibitor by breaking salt bridges involving helix D residues, which induces A sheet expansion and partial insertion of the loop, leading to an activated loop conformation.[84]

The accelerating effect of heparin on antithrombin-proteinase reactions can result from two different mechanisms[63] (Fig. 4.8).

Antithrombin	107	K	F	D	T	I	S	E	K	-	-	-	-	T	S	D	Q	I
Heparin cofactor II	163	H	F	K	D	F	V	N	A	S	S	**K**	Y	E	I	T	T	I
Protease nexin 1	63	R	Y	-	-	-	-	-	-	-	-	-	-	G	V	N	G	V
PAI-1	63	G	F	**K**	-	-	-	-	-	-	-	-	I	D	D	**K**	G	M
Protein C inhibitor	77	G	L	N	-	L	Q	K	S	-	-	-	-	S	E	K	E	L
α1-Proteinase inhibitor	81	N	F	N	-	L	T	E	I	-	-	-	-	P	E	A	Q	I

Antithrombin	120	H	F	F	F	A	**K**	L	N	C	**R**	L	Y	**R**	**K**	A	N	**K**
Heparin cofactor II	180	H	N	L	F	**R**	**K**	L	T	H	**R**	L	F	**R**	**R**	N	F	G
Protease nexin 1	70	G	**K**	I	L	**K**	**K**	I	N	**K**	A	I	V	S	**K**	**K**	N	**K**
PAI-1	72	A	P	A	L	**R**	H	L	Y	**K**	E	L	M	G	P	W	N	**K**
Protein C inhibitor	89	H	R	G	F	Q	Q	L	L	Q	E	L	N	Q	P	R	D	G
α1-Proteinase inhibitor	93	H	E	G	F	Q	E	L	L	R	T	L	N	Q	P	D	S	Q

Fig. 4.5. Amino acid sequences of the helix D region of the heparin-activatable serpins. The sequences have been aligned with the prototypical serpin, α_1-proteinase inhibitor, according to Huber and Carrell.[73] The helix D region is underlined and residues implicated in heparin binding by mutagenesis or molecular dynamics studies are indicated in bold. Note that the conserved basic residues in antithrombin, heparin cofactor II, protease nexin 1 and PAI-1 are absent in protein C inhibitor, whose heparin binding site is thought to lie in the H helix.

In one of these mechanisms, heparin activates antithrombin by inducing a conformational change in the reactive center loop which increases the reactivity of the inhibitor toward proteinase. In the second mechanism, heparin acts as a surface or bridge to bring the two proteins in close proximity by their binding together on the same polysaccharide chain. The conformational activation mechanism predominates when factor Xa is the proteinase. Thus the pentasaccharide, which is incapable of binding both antithrombin and proteinase, is nevertheless sufficient to produce the bulk of the heparin rate-enhancing effect.[47,63,85,86] By contrast, the bridging mechanism mostly accounts for heparin's rate-enhancing effect on the antithrombin-thrombin reaction, since the pentasaccharide alone produces an insubstantial rate enhancement and heparin chains at least 18 saccharide residues in length are required to produce an enhancement characteristic of full-length heparins.[63,85,87] This minimal chain length requirement is in agreement with the smallest heparin chain able to bind antithrombin and thrombin simultaneously[86] and with recent molecular modeling studies of the ternary heparin-antithrombin-thrombin complex[88] (Fig. 4.9). The molecular modeling studies suggest that, in the ternary complex, the pentasaccharide sequence that binds specifically to antithrombin is bridged on its non-reducing end to a 4-6 saccharide thrombin-binding region by an 8-10 saccharide heparin spacer which makes few interactions with either protein (Fig. 4.9). This model is supported by the ability of synthetic oligosaccharides consisting of the antithrombin-binding pentasaccharide linked to a non-specific sulfated pentasaccharide by a non-sulfated spacer to accelerate antithrombin inhibition of thrombin to an extent exceeding that of heparin.[88]

Kinetic studies support two separate mechanisms for heparin activation of antithrombin. The dependence of heparin rate enhancement on heparin concentration thus parallels heparin binding to antithrombin for all proteinase reactions.[89,90] However, the rate enhancement is abolished at high heparin concentrations which favor the binding of antithrombin and proteinase to separate polysaccharide chains when thrombin is the target proteinase, but not when factor Xa is the target, in keeping with a bridging mechanism for the former but not the latter proteinase[55,89-92] (Fig. 4.10). Rapid kinetic studies have

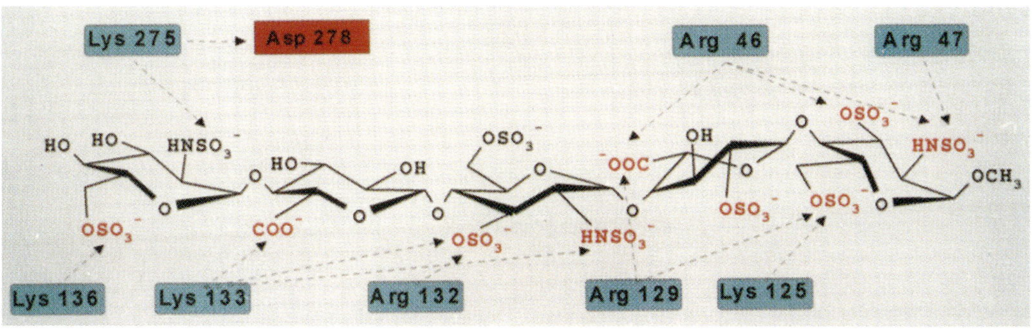

Fig. 4.6. Model of the antithrombin-heparin pentasaccharide interaction. Shown are the interactions resulting from the docking of pentasaccharide with the experimentally determined crystal structure of intact antithrombin by molecular dynamics simulations. Essential sulfates and carboxylate groups of the pentasaccharide are shown in red and interacting basic amino acid residues of antithrombin are in blue. Predominant in the latter group are residues in helix A (Arg 46 and Arg 47) and helix D (Lys 125, Arg 129, Arg 132, Lys 133 and Lys 136). Reprinted with permission from Nature Structural Biology.[79] *Copyright 1994 by Macmillan Magazines Limited. Figure kindly provided by Dr. C. A. A. van Boeckel.*

shown that a thrombin-heparin interaction in an intermediate heparin-antithrombin-thrombin ternary complex provides additional binding energy to promote the encounter of thrombin with antithrombin in this complex and thereby generate heparin rate enhancement.[93,94] In contrast, heparin interaction with just antithrombin and not factor Xa in a similar ternary complex produces the bulk of the rate enhancement presumably through an activated antithrombin-factor Xa interaction.[63,95] Finally, chemical modification or mutagenesis of basic residues in thrombin which diminish heparin binding coordinately diminish heparin rate enhancement of its reaction with antithrombin,[96-99] whereas a similar chemically-induced abrogation of factor Xa binding to heparin does not influence the rate-enhancing effect of heparin on the antithrombin-factor Xa reaction.[100] Together, such results support a bridging mechanism for heparin acceleration of thrombin inactivation but a conformational activation mechanism for the acceleration of factor Xa inactivation. Low-affinity heparin accelerates the reactions of antithrombin with thrombin and factor Xa also by these two distinct mechanisms, although the accelerations are much lower

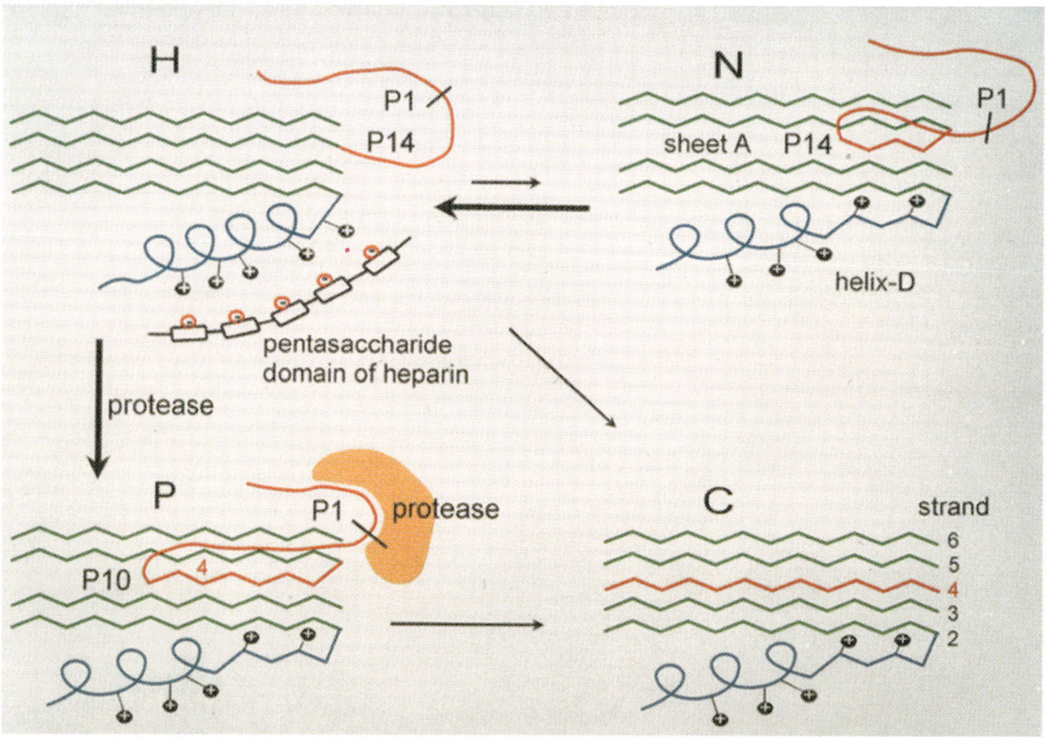

Fig. 4.7 Proposed model for heparin activation of antithrombin. The model postulates that in the native antithrombin structure (state N), the reactive center loop is partially inserted in β-sheet A resulting in a low heparin binding affinity. Pentasaccharide binding to helix D induces an elongation of the helix on the carboxyl-terminal end to produce additional interactions which result in high-affinity heparin binding (state H). This elongation induces a contraction of the A-sheet due to the transmission of the D-helix structural change to strand 2 of the sheet, which immediately follows the helix in the polypeptide chain sequence. The contraction of the A sheet expels the reactive center loop from the sheet, allowing the loop to adopt an activated inhibitory conformation (state H). Cleavage in the reactive center loop (state C) or formation of a stable complex with proteinase (state P) results in deep reinsertion of the loop into the A sheet which disrupts the extended helical interaction and induces the inhibitor back into a low heparin affinity state. Reproduced with permission from Nature Structural Biology.[79] *Copyright 1994 by Macmillan Magazines Limited. Figure kindly provided by Dr. C. A. A. van Boeckel.*

than those of high-affinity heparin[101] (Fig. 4.10). The lower accelerations result from the reduced affinity of antithrombin for the low affinity polysaccharide diminishing the ternary-complex bridging-effect, the formation of an antithrombin-heparin binary-complex and the conformational activation of the inhibitor in the binary complex. The specificity of the heparin-induced conformational change in the reactive center loop for factor Xa is consistent with the Ile-Ala-Gly-Arg sequence of the P4 to P1 residues of the loop representing a factor Xa and not a thrombin recognition sequence.[75] Supporting this assertion, replacing the reactive center loop sequence of α_1-proteinase inhibitor with that of antithrombin produces an inhibitor which inactivates thrombin at a slow rate similar to that of antithrombin, whereas optimal thrombin recognition sequences placed in the loop produce inhibitors with substantially higher rates of thrombin inactivation.[102]

Heparin acceleration of antithrombin-proteinase reactions is catalytic rather than stoichiomeric, implying that heparin is recycled during the reaction.[103] This catalytic action results from a ~1,000-fold decreased affinity of heparin for antithrombin following the reaction of inhibitor with proteinase to form a stable complex.[55,104] The reduced affinity induces heparin to dissociate once the stable complex has formed, allowing it to bind another antithrombin

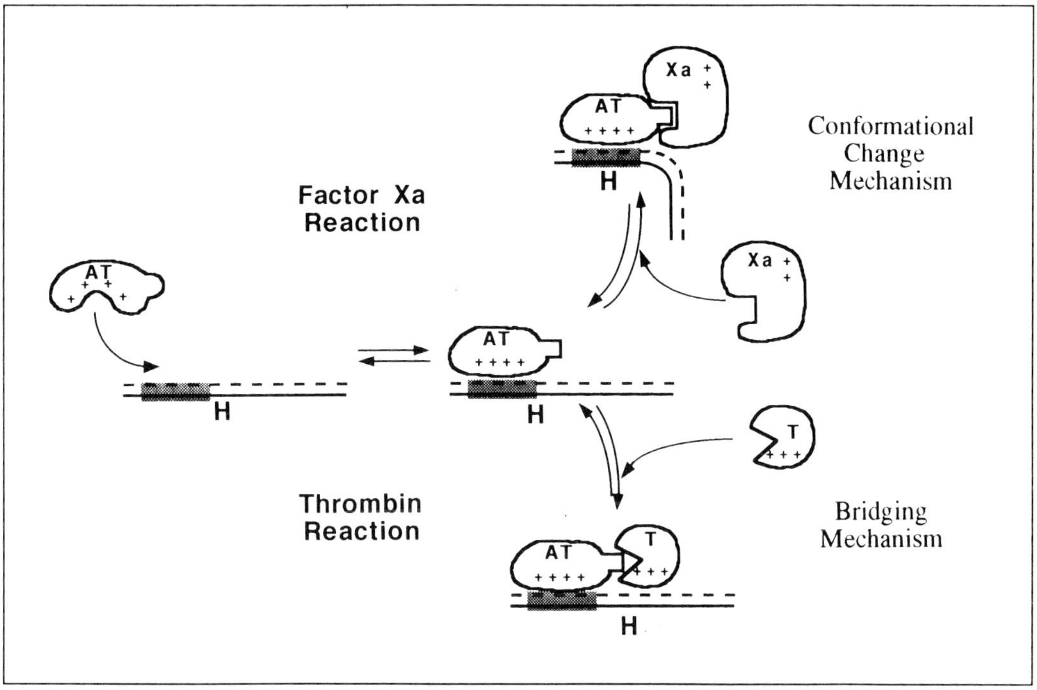

Fig. 4.8. Mechanism of heparin activation of antithrombin-proteinase reactions. The scheme depicts two component mechanisms which contribute to heparin activation of antithrombin reactions with proteinases. In both mechanisms heparin (H) first binds to antithrombin (AT) through a unique pentasaccharide binding region. This binding induces a conformational change in the inhibitor which greatly enhances its affinity for the polysaccharide and alters the structure of the reactive center region. The reactive center structural change enhances the recognition of antithrombin by factor Xa to produce rate enhancement but does not affect inhibitor recognition by thrombin (T). Thrombin recognition is instead enhanced by a proteinase binding site on heparin adjacent to bound antithrombin, which promotes the proteinase-inhibitor interaction through a ternary complex bridging effect. Reproduced with permission from Perspectives in Drug Discovery and Design.[327] *Copyright 1994 by ESCOM.*

molecule.[105] A similar decrease in heparin affinity occurs when antithrombin is cleaved in the reactive center loop,[81] is exposed to mild denaturants[82] or forms a complex with a reactive center loop peptide.[106] As discussed above, what is common to all these low heparin affinity forms of antithrombin is that they all have their reactive center loops inserted into β-sheet A, implying that loop insertion disrupts the high-affinity interaction between pentasaccharide and antithrombin.

Heparin also enhances a substrate reaction of antithrombin with proteinases in which an inactive, reactive-site cleaved form of the inhibitor is produced concomitant with the formation of a stable proteinase-inhibitor complex.[81,107,108] About 30% of antithrombin reacts with thrombin or factor Xa as a substrate and 70% as an inhibitor at physiological ionic strength and pH, in contrast to the ~5-10% of substrate reaction in the absence of polysaccharide. The substrate reaction is further enhanced at lower ionic strength in the presence but not in the absence of heparin, with up to 95% substrate reaction observable at I 0.02 when thrombin is the proteinase.[108] Such findings support the branched pathway for serpin-proteinase reactions in which partitioning from a common intermediate leads to either a stable inhibited complex or turnover of the inhibitor as a regular substrate (Chapter 3). Heparin binding to antithrombin is thought to alter the partitioning in favor of the substrate reaction by interfering with loop insertion into the A sheet which is

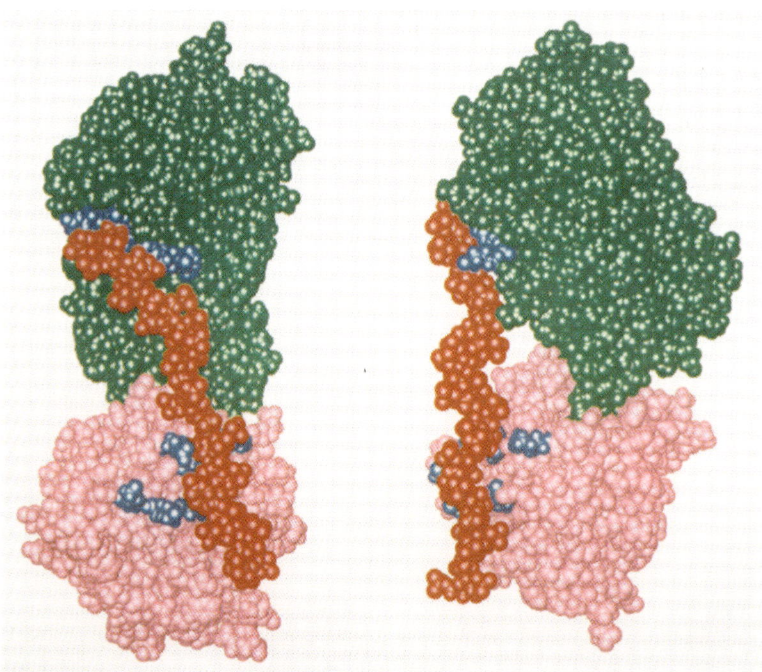

Fig. 4.9. Model of the ternary heparin-antithrombin-thrombin complex. Front (top) and side (bottom) views have antithrombin in green, thrombin in magenta and heparin in red. The heparin binding residues on antithrombin and thrombin are shown in blue. The model was constructed by docking X-ray structures of antithrombin[80] and thrombin[330] with an NMR structure of heparin.[331] Reproduced with permission from Nature Structural Biology.[88] *Copyright 1995 by Macmillan Magazines. Figure kindly provided by Dr. C.A.A. van Boeckel.*

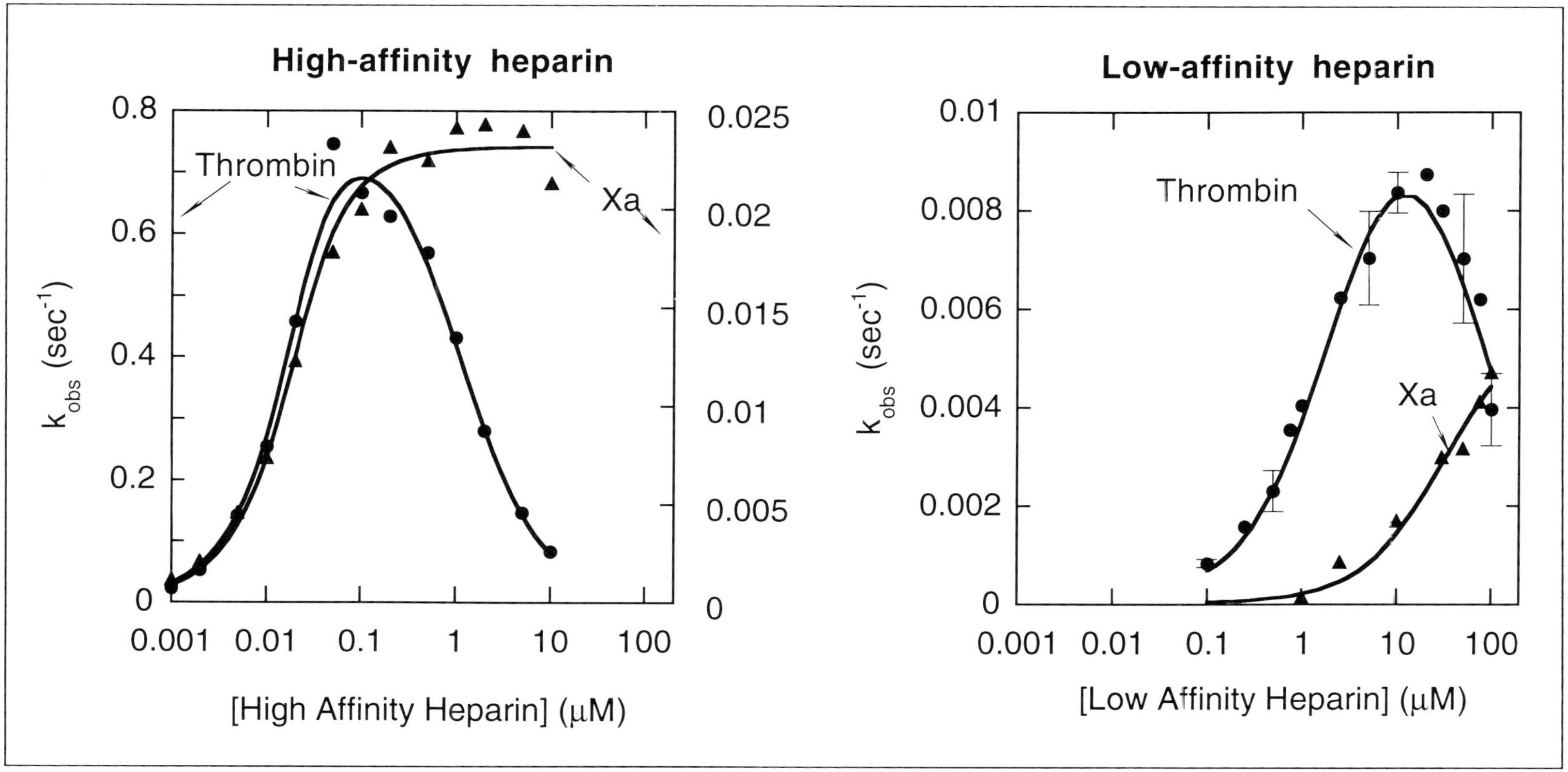

Fig. 4.10. Heparin concentration dependence of the accelerating effect of high- and low-affinity heparins on antithrombin-proteinase reactions. Pseudo first order rate constants (k_{obs}) for the inhibition of 1 nM thrombin or 2 nM factor Xa by 20 nM antithrombin are shown as a function of high-affinity heparin concentration (left panel) or low-affinity heparin concentration (right panel) at pH 7.4, I 0.15, 25°C. The bell-shaped curves are fits of thrombin reaction rate constants by a ternary complex bridging mechanism requiring the binding of both antithrombin and proteinase to the same heparin chain. The sigmoid curves are fits of factor Xa reaction rate constants by a conformational activation mechanism requiring the binding of just antithrombin to heparin. The higher heparin concentrations and lower maximal rate enhancements for low-affinity as compared to high-affinity heparin-catalyzed reactions result from the absence of the antithrombin binding pentasaccharide in low-affinity heparin reducing the antithrombin affinity and the ability of heparin to fully activate the inhibitor (see text). Adapted with permission from the Journal of Biological Chemistry.[101] Copyright 1995 by the American Society for Biochemistry and Molecular Biology.

required to trap proteinase in a stable complex.[106] However, proteinase binding and attack on the scissile bond may induce a decrease in heparin binding affinity, which could promote dissociation and allow insertion of the loop to proceed.[75]

A role for antithrombin in regulating blood clotting enzymes is indicated by the predisposition to thrombosis of individuals with inherited or acquired deficiencies of the inhibitor.[109,110] In agreement with such a role, antithrombin is the major inactivator of thrombin,[111,112] factor Xa[113,114] and factor IXa[115] when these enzymes are added to plasma. Factor VIIa, factor XIa and plasma kallikrein may be additional targets of antithrombin when these proteinases associate with certain cofactor proteins[116-118] (Fig. 4.1). Inherited deficiencies of antithrombin can result from mutations that yield low circulating levels of protein or normal levels of a defective protein. Mutations of the latter type may be due to defects in proteinase binding or heparin binding.[119,120] Certain pleiotropic-type mutations lead to an abnormal protein with defects in both proteinase and heparin binding functions and also to low circulating levels of the variant protein.[121] Heterozygotes with an abnormal antithrombin defective in proteinase binding are at risk for thrombosis whereas homozygotes with such defects are not known, suggesting that antithrombin is essential for survival. By contrast, both homozygotes and heterozygotes with defects in heparin binding are known, with only the former showing a predisposition to thrombotic disease.[119] Heparin activation of antithrombin thus appears to be important for regulating clotting enzyme activity in vivo.

Heparin does not circulate freely in the blood stream, but rather is localized extravascularly in mast cells closely associated with the vasculature, suggesting that heparin is not involved in maintaining normal hemostasis.[122] Anticoagulantly active heparan sulfate glycosaminoglycans, however, are present in the sub-endothelial basement membrane and possibly on the luminal endothelial surface, where they could play a role in activating circulating antithrombin by plasma protein exchange with the sub-endothelial compartment.[123,124] Perfusion of thrombin or factor Xa through isolated regions of the vasculature of experimental animals results in accelerated enzyme inhibition by co-perfused or endogenous antithrombin, consistent with such a role.[124,125] However, the acceleration of thrombin inhibition may partly result from thrombin binding to its endothelial cell receptor, thrombomodulin.[126] Antithrombin inactivation of receptor-bound thrombin is thus enhanced 3- to 9-fold by a covalently associated chondroitin sulfate glycosaminoglycan.[127-131] The chondroitin sulfate-dependent enhancement of receptor-bound thrombin inhibition also blocks heparin acceleration of proteinase inhibition,[132,133] implying different modes of regulation of the free procoagulant and bound anticoagulant forms of thrombin.

Extravascular heparin and heparan sulfate may both contribute to the regulation of clotting enzymes when hemostasis is perturbed by vascular injury. This regulation may primarily be a scavenging-type function, since clotting enzymes appear to be protected from antithrombin when assembled on activated platelets or endothelial cells where clotting takes place.[134,135] The rate of thrombin inactivation by antithrombin is decreased ~3-fold and heparin acceleration is blocked when thrombin is generated from its precursor, prothrombin, by the action of the prothrombinase complex.[136,137] This effect appears to be due to interaction of thrombin with the fragment 2 activation peptide[138] which is bound in the heparin binding site of the proteinase.[139] The rate of factor Xa inactivation by antithrombin is also greatly diminished when the enzyme is bound in the prothrombinase complex,[140-142] consistent with antithrombin acting as a scavenger of enzyme which escapes from its site of action.

The binding of thrombin and factor Xa to soluble fibrinogen and insoluble fibrin clots additionally affects the rate of antithrombin inactivation of these proteinases.

Plasma levels of fibrinogen decrease the rate of thrombin inhibition ~2-fold, i.e., to the extent expected for fibrinogen acting as a competing substrate of the enzyme.[112] Thrombin binding to fibrin I polymer, in which fibrinopeptide A but not fibrinopeptide B has been cleaved from fibrinogen, can occur in modes which block antithrombin inactivation or which promote this inactivation.[143] The former mode appears to involve binding through both the active site and anion binding exosite I, while the latter mode involves only exosite binding with the active site available for antithrombin binding. Thrombin binding to fibrin II appears to involve binding solely to thrombin exosite I, resulting in a slight reduction (1.6-fold) in inactivation rate. Heparin acceleration of the inactivation of the bound thrombin, however, is reduced 300- to 500-fold due to the formation of thrombin-fibrin-heparin ternary complexes that block heparin bridging of thrombin and antithrombin.[144,145] Factor Xa bound to fibrin II clots is similarly greatly protected from antithrombin-heparin or antithrombin-pentasaccharide inhibition, suggesting that bound factor Xa is inaccessible to the inhibitor or undergoes conformational changes that alter its recognition by free or heparin-complexed antithrombin.[146] Such protection of these procoagulant enzymes is thought to explain the inability of heparin to prevent vessel reocclusion in patients undergoing thrombolytic therapy.

Heparin activation of antithrombin is regulated in vivo by a number of other factors. Plasma contains several proteins which bind heparin electrostatically and which thereby can compete with antithrombin for binding the polysaccharide.[147] The presence of such proteins has suggested that low-affinity-type heparins lacking an antithrombin binding site may function to bind these proteins and through mass action make more high-affinity heparin available to activate antithrombin.[148] Elevation of heparin binding proteins in thromboembolic disease may contribute to the phenomenon of heparin resistance wherein substantial clinical doses of the polysaccharide are required to achieve an anticoagulant effect.[149] The binding of some of these proteins to heparin appears to be important for the regulation of antithrombin function. Binding of vitronectin to cell surface heparan sulfate glycosaminoglycans mediates the clearance of antithrombin-thrombin complexes by endocytosis.[150] Binding of high molecular weight kininogen to heparin serves to promote the binding of factor XIa and plasma kallikrein to the polysaccharide and thereby to greatly increase the rate of antithrombin inactivation of these proteinases by a bridging-type mechanism.[118,151]

Antithrombin activity may additionally be regulated by inactivation of the inhibitor by nontarget enzymes. An elastase released by neutrophils at inflammatory sites cleaves antithrombin in the reactive center loop and inactivates the inhibitor without getting trapped and is thereby protected from this and other serpins at its site of action.[152,153] The cleavage does not occur at the reactive center which may explain the inability to trap such enzymes in stable complexes. The cleavage is accelerated by full-length heparin but minimally by pentasaccharide, consistent with a bridging mechanism for the accelerating effect.[154] Matrix metalloproteinases have also been reported to cleave and inactivate antithrombin in a similar fashion, although the effect of heparin on the rates of these cleavages has not been examined.[155] Whether the inactivation of antithrombin by such nontarget proteinases is a general mechanism for regulation of the inhibitor remains to be determined.

2. HEPARIN COFACTOR II

Heparin cofactor II is a 66 kDa molecular weight serpin inhibitor of thrombin, contains 10% carbohydrate,[156] and circulates in plasma at a level of 80 μg/ml or 1.2 μM.[157] DNA sequencing indicates that the protein contains 480 amino acids and is synthesized with a 19 amino acid signal peptide.[158-160] There are three potential N-linked glycosylation sites, three free cysteine residues,[161] and two sulfated tyrosines.[162] The pI is 4.9-5.3, reflecting glycosylation heterogeneity.[156] The gene is localized on chromosome 22q11[163] and contains 5 exons and 4 introns with 10 Alu repeats.[163,164] The protein has been expressed in COS cells[165] and in a non-glycosylated form in *E. coli*.[166]

The reactive center P1-P1' residues are Leu 444-Ser 445,[158,160,167] identical to those of α_1-antichymotrypsin.[168] Whereas chymotrypsin and the chymotrypsin-like proteinase, cathepsin G, are inhibited by heparin cofactor II as would be expected from the P1 residue specificity of these proteinases,[169,170] the physiological target proteinase is paradoxically thrombin, which does not match this specificity[171] (Fig. 4.1). Thrombin is inhibited with a slow second order rate constant of 1-8 x 10^3 $M^{-1}s^{-1}$. However, heparin and dermatan sulfate accelerate the inhibition 10^3- to 10^4-fold; i.e., with a rate constant of ~1 x 10^7 $M^{-1}s^{-1}$ at optimal levels of glycosaminoglycan[60,99,156,172,173] (Table 4.3). Other trypsin-like proteinases including factors Xa, IXa, XIa, XIIa, VIIa, kallikrein, activated protein C, plasmin, urokinase and tissue plasminogen activator are not significantly inhibited.[171] Consistent with the P1 residue affecting the inhibitor specificity, a Leu 444→Arg variant inhibits thrombin with a 100-fold increased rate constant, is active in inhibiting other trypsin-like proteinases and loses its ability to inhibit chymotrypsin.[172] However, the glycosaminoglycan acceleration of the reaction of thrombin with this variant inhibitor is correspondingly reduced, so that native and variant inhibitors inactivate thrombin at similar maximal rates in the presence of glycosaminoglycans.

Heparin cofactor II is exceptional among the heparin-activatable serpins in that it can be activated equally well by either heparin or dermatan sulfate, although optimal activation requires ~100-fold higher levels of polysaccharide than those required to activate antithrombin.[60] The dermatan sulfate disaccharide repeat differs from that heparin in consisting predominantly of iduronic acid linked to N-acetylgalactosamine, with variable extents of O-sulfation (Fig. 4.2). The heparin-related glycosaminoglycan, heparan sulfate, is also an activator as it is with antithrombin. However, chondroitin 4-sulfate, chondroitin 6-sulfate, keratan sulfate and hyaluronic acid do not activate the inhibitor.[60] Unlike the case for antithrombin, a number of other natural and semisynthetic or synthetic polyanions are also effective stimulators of the serpin, including pentosan polysulfate,[174] lactobionic acid,[175] fucoidan,[176] dextran sulfate,[177] polyphosphates[178] and polycarboxylates.[179]

The accelerating activity of heparin does not appear to involve a sequence-specific site as it does with antithrombin, but rather correlates with the charge density of the heparin chain.[180,181] Heparin oligosaccharides containing 24-30 saccharides are

Table 4.3. Second order rate constants for heparin cofactor II-proteinase reactions measured in the absence and presence of heparin and dermatan sulfate at pH 7.4 and the indicated temperatures

		Second order rate constant ($M^{-1}s^{-1}$)			
Target proteinase	**Temp(°C)**	**– GAG[a]**	**+ Heparin**	**+ DS[b]**	**Ref.**
Cathepsin G	37	2.3 x 10^2	-	1.4 x 10^3	171
Chymotrypsin	23	3.7 x 10^3	4.0 x 10^3	6.2 x 10^3	172
Leukocyte elastase	37	8 x 10^1	-	-	171
Thrombin	23	0.7-8.3 x 10^3	0.8-1.1 x 10^7	0.4-1.1 x 10^7	60, 99 156,172 173

a, GAG, glycosaminoglycan; b, DS, Dermatan sulfate

required for a significant accelerating effect, suggesting a mechanism involving heparin bridging the inhibitor and proteinase in a ternary complex.[182] The bell-shaped dependence of the accelerating effect on heparin concentration is consistent with such a bridging mechanism.[60] However, other evidence argues against this being the predominant accelerating mechanism (see below). Activation by dermatan sulfate does show evidence of sequence-specificity in that the smallest oligosaccharide with high affinity for the inhibitor is a specific hexasaccharide containing six sulfates[183] (Fig. 4.3). Hexasaccharides with five or four sulfates also bind, although with weaker affinity, whereas those with fewer sulfates do not bind at all. The positions of these sulfates may be critical for activity.[184] The high-affinity hexasaccharide activates the inhibitor with a 50-fold acceleration of the inhibition rate constant.[185] However, longer oligosaccharides bind with greater affinity and are more active.

The binding sites for heparin and dermatan sulfate are distinct, although they do appear to overlap. Mutations of Arg 189, Arg 192 and Arg 193 thus decrease dermatan sulfate affinity,[165,166] mutation of Lys 173 decreases heparin affinity,[186] and mutations of Arg 184 and Lys 185 decrease the affinity for both glycosaminoglycans.[165,187] These residues are all localized in the D helix, with Lys 173, Lys 185, Arg 189, Arg 192 and Arg 193 being homologous to Lys 114, Lys 125, Arg 129, Arg 132 and Lys 133 residues in the heparin binding region of antithrombin (Figs. 4.4 and 4.5). Unlike the case for antithrombin, there is no evidence for involvement of the A helix in glycosaminoglycan binding to heparin cofactor II. Mutation of Arg 103, homologous to the A-helix residue Arg 47 which mediates heparin binding to antithrombin, thus does not affect the binding of heparin to the inhibitor or its activation by either heparin or dermatan sulfate.[187]

Mutagenesis studies of heparin cofactor II and thrombin have indicated that the amino-terminal region of heparin cofactor II and the anion binding exosite I of thrombin play important roles in determining the specificity of this serpin for thrombin and in the mechanism of glycosaminoglycan activation.[185,188] Deletion of two repeated acidic sequences in the amino terminal region, which are homologous to the C-terminal tail of the specific thrombin inhibitor, hirudin, and which contain the tyrosine sulfates (Fig. 4.11), abolishes most of the glycosaminoglycan acceleration of thrombin inhibition without affecting the inhibition in the absence of glycosaminoglycan or the inhibition of chymotrypsin. The deletion of these acidic sequences also greatly enhances the binding of the variant inhibitor to heparin, suggesting that the basic glycosaminoglycan binding region in helix D makes an intramolecular interaction with the acidic region in the native inhibitor which is disrupted by glycosaminoglycan binding.[185,188] This disruption is thought to free the hirudin-like sequence and make it available for interaction with anion binding exosite I of thrombin (Fig. 4.12). Such an allosteric mechanism for glycosaminoglycan activation of heparin cofactor II is supported by other observations. The hirudin C-terminal dodecapeptide which binds to thrombin exosite I thus blocks glycosaminoglycan activation of heparin cofactor II inhibition of thrombin without affecting the unactivated reaction.[185] Similarly, the inhibition of the proteolyzed derivative, γ-thrombin, in which exosite I is no longer intact, or of engineered variants of thrombin with mutations in exosite I is only weakly activated by glycosaminoglycans.[185,189,190] Together, these observations strongly suggest that an interaction between the amino-terminal hirudin-like sequences of heparin cofactor II and anion binding exosite I of thrombin makes an important contribution to glycosaminoglycan rate enhancements and to the specificity of the inhibitor for thrombin.

The role of heparin in bridging the inhibitor and proteinase in a ternary

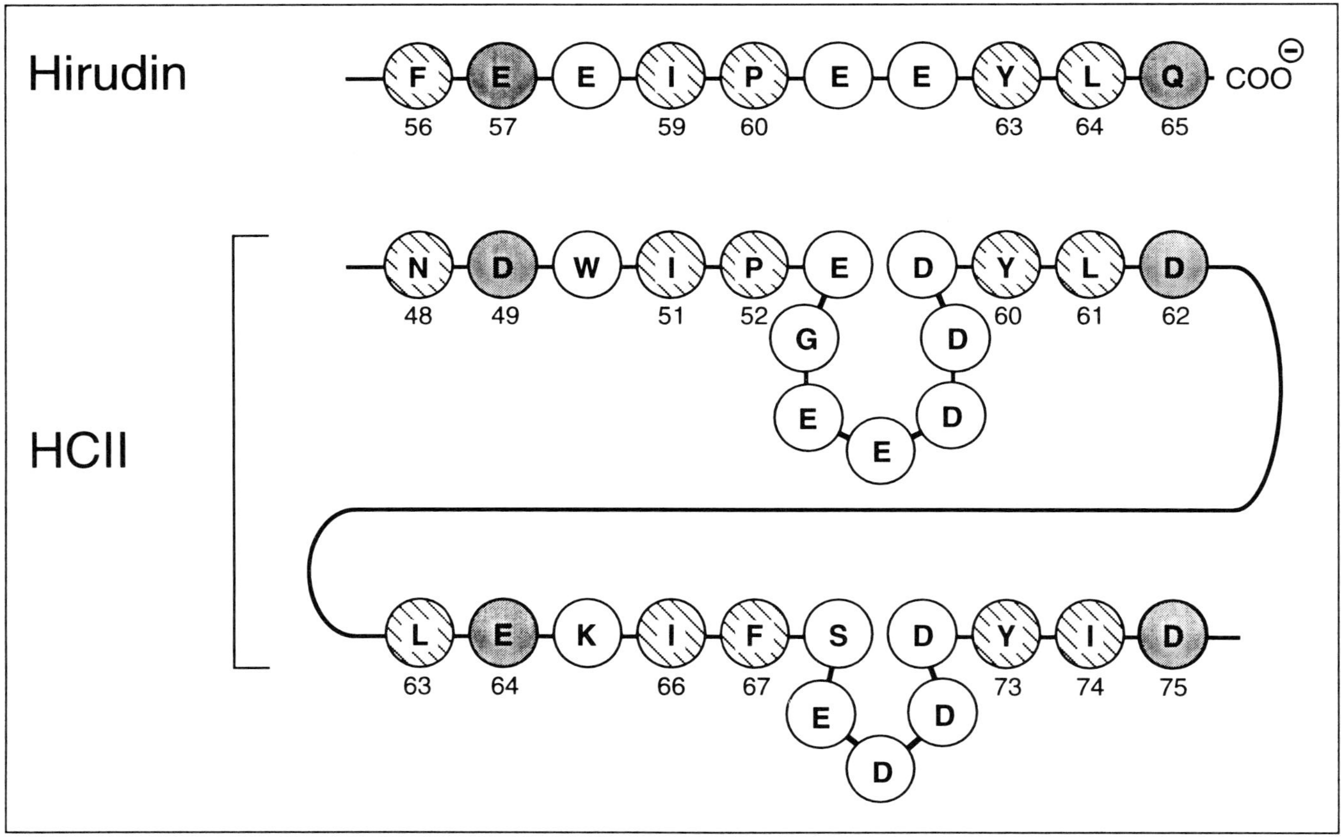

Fig. 4.11. Homology of the N-terminal acidic domain of heparin cofactor II to the carboxy-terminal domain of hirudin The sequence of the carboxy-terminal tail of hirudin, residues 56-65, is aligned with the tandem repeated hirudin-like sequences in the N-terminal region of heparin cofactor II, residues 48-75. Amino acid residues in hirudin which form contacts with anion binding exosite I of thrombin in the X-ray crystal structure of the complex are indicated by cross-hatched symbols for hydrophobic contacts and shaded symbols for ionic contacts. The homologous residues in heparin cofactor II are similarly shaded. Adapted with permission from Journal of Biological Chemistry.[185] Copyright 1991 by American Society for Biochemistry and Molecular Biology. Figure kindly provided by Dr. Douglas Tollefsen.

complex has been investigated by mutagenesis of residues in the heparin binding region of thrombin, which is also referred to as anion binding exosite II.[99] Such mutations decrease the binding of both heparin and dermatan sulfate to thrombin, consistent with these glycosaminoglycans binding to a common site on the proteinase. Nevertheless, such mutations have no effects on dermatan sulfate acceleration of

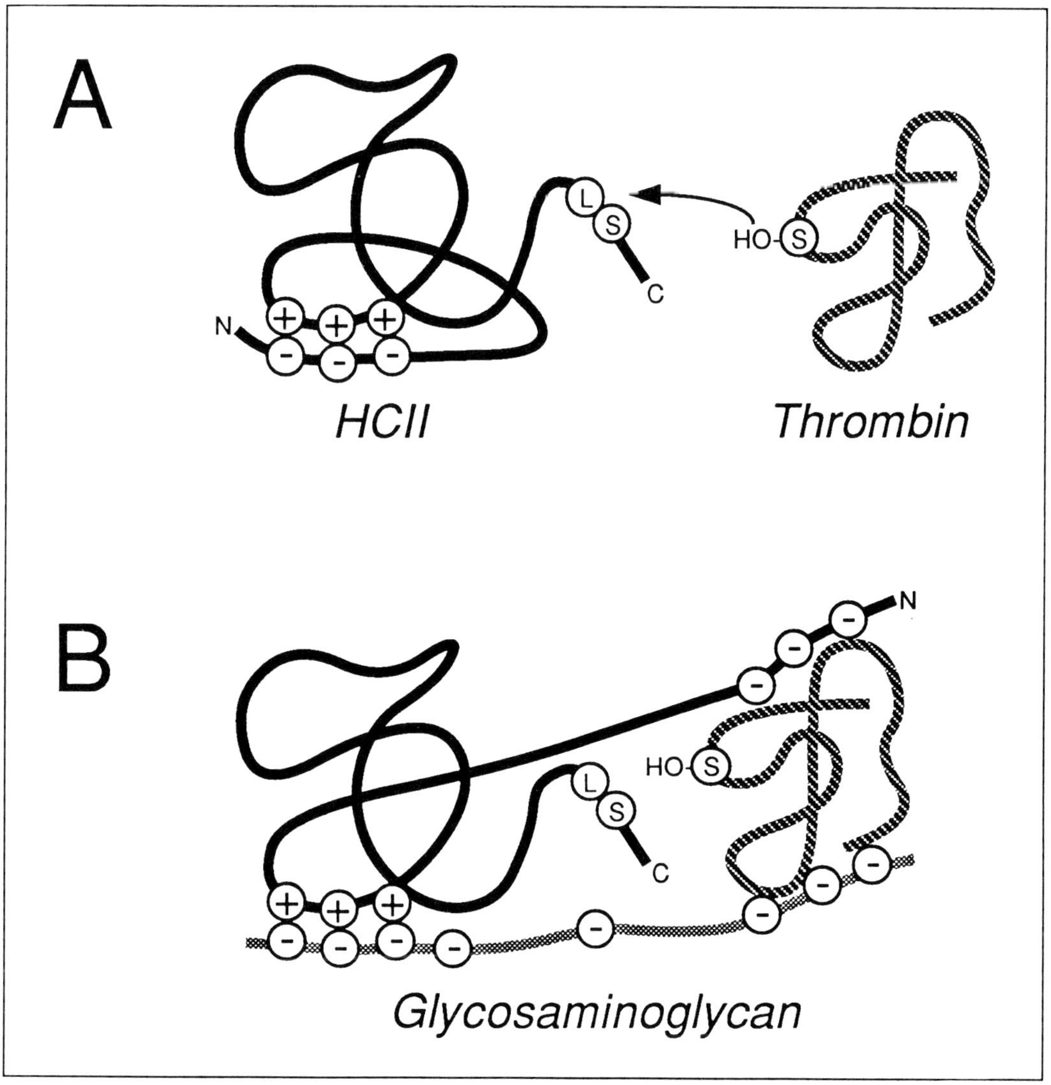

Fig. 4.12. Mechanism of glycosaminoglycan activation of heparin cofactor II. A depicts the reaction of thrombin with heparin cofactor II in the absence of glycosaminoglycan. The N-terminal acidic hirudin-like recognition sequence for thrombin is masked in this case by interaction with the basic glycosaminoglycan binding region in helix D so that it cannot assist in recognition of the Leu-Ser reactive bond of the inhibitor by thrombin. B depicts the reaction in the presence of glycosaminoglycan. In this case, glycosaminoglycan binding to the D helix region of the inhibitor disrupts the intramolecular interaction with the acidic N-terminal tail, allowing the acidic region to promote thrombin recognition by interacting with thrombin anion binding exosite I. Heparin additionally promotes this recognition through a bridging interaction with heparin binding exosite II of the proteinase. Only the conformational activation mechanism appears to be involved in dermatan sulfate activation. Adapted with permission from the Journal of Biological Chemistry.[185] Copyright 1991 by the American Society for Biochemistry and Molecular Biology. Figure kindly provided by Dr. Douglas Tollefsen.

heparin cofactor II inhibition of these variant proteinases and produce modest reductions (maximally 7-fold) in heparin accelerated reactions, implying that a bridging mechanism does not significantly contribute to the glycosaminoglycan rate enhancements. This conclusion is consistent with the observation that thrombin inhibition by the variant heparin cofactor II that lacks the hirudin-like acidic sequences in the amino-terminus is not accelerated by dermatan sulfate and is accelerated only 35-fold by heparin.[185] The predominant allosteric mechanism of glycosaminoglycan activation of heparin cofactor II suggests that the reduction in glycosaminoglycan rate enhancements seen at high glycosaminoglycan concentrations is not due to a bridging mechanism but rather may reflect the blocking of exosite I by glycosaminoglycan.[99] Evidence has been provided for such weak glycosaminoglycan interactions with thrombin exosite I.[191]

The physiological role of heparin cofactor II is not clear. Individuals with thrombotic disease are known who have deficiencies in the inhibitor. However, a similar frequency of heparin cofactor II deficiency in normal individuals has been documented, which suggests that deficiencies in this inhibitor per se may not be related to the development of thrombosis.[192,193] The inhibitor cannot substitute for antithrombin, since individuals with antithrombin deficiencies and normal levels of heparin cofactor II are still at risk for thrombosis.[194] The inhibitor can regulate thrombin activity in vivo since dermatan sulfate and other polyanionic agents are effective antithrombotic agents which function by activating heparin cofactor II,[195-197] and patients with thrombotic disease show elevated levels of heparin cofactor II-thrombin complexes.[157] The acceleration of intravascular thrombin inhibition by heparin cofactor II has not been demonstrated as it has with antithrombin.[198,199] Intravascular heparan sulfate or thrombomodulin-associated glycosaminoglycans therefore do not appear to activate heparin cofactor II. This serpin may thus function extravascularly in response to blood vessel injury. In keeping with this possibility, dermatan sulfate proteoglycans on fibroblasts and smooth muscle cells, or isolated from tissues, accelerate heparin cofactor II inhibition of thrombin.[200,201] Interestingly, a dermatan sulfate proteoglycan appears in the maternal and fetal circulation during pregnancy that is active in accelerating heparin cofactor II inhibition of thrombin, suggesting a possible role for the serpin in protecting the mother and fetus from thrombosis during pregnancy.[202] Glycosaminoglycan binding proteins have been shown to modulate glycosaminoglycan activation of the inhibitor as with antithrombin.[157] The observation that neutrophil proteinases cleave amino-terminal peptides, which can then act as potent chemoattractants from the inhibitor, suggests that such peptides may play a role in recruiting leukocytes to sites of vessel injury, thereby implicating heparin cofactor II in the inflammatory response to injury.[170,203,204]

3. PROTEIN C INHIBITOR

Protein C inhibitor (PCI) is a heparin-activatable inhibitor of the anticoagulant proteinase, activated protein C (Fig. 4.1). This serpin has a molecular weight of 57 kDa of which 43.8 kDa is protein and the remainder carbohydrate.[205] It circulates at low levels (2-5 μg/ml or ~0.1 μM) in blood plasma and 10-fold lower levels in urine,[205-207] but is found at much higher levels in seminal plasma (~3 μM).[208,209] Cloning and sequencing of the gene has indicated that the inhibitor consists of a 387 residue polypeptide chain with a 19 residue signal peptide.[210] The protein contains a single free cysteine and five potential asparagine glycosylation sites. The pI range is 4.5-6.[211] The gene is 11.5 kilobases long from the putative transcription initiation site to the polyadenylation site and is organized into 5 exons and 4 introns.[212] It is localized on chromosome 14q32.1, close to the genes for the serpins α_1-proteinase inhibitor, α_1-antichymotrypsin and corticosteroid binding globulin, suggesting a common evolutionary ancestor of these serpins.[213] A recombinant inhibitor has been expressed in baculovirus.[214]

The reactive site of PCI is Arg 354-Ser 355, compatible with the substrate specificity of its target proteinases.[210] Both P1 Arg and P2 Phe but not P1' Ser residues influence the inhibitor specificity for these proteinases.[214,215] In addition to inhibiting activated protein C, other target proteinases are urokinase- and tissue-type plasminogen activators,[206,207,216] acrosin,[217] prostate-specific antigen,[208] factor XIa and plasma kallikrein,[218] and thrombin.[173,219] Surprisingly, chymotrypsin is inactivated faster than most target proteinases,[220] suggesting that PCI may have overlapping reactive sites, as has been shown for α_2-antiplasmin.[221] In the absence of heparin, thrombin bound to thrombomodulin is inhibited the fastest (2.4 x 10^6 $M^{-1}s^{-1}$),[219] with free thrombin, plasma kallikrein and acrosin inhibited at 10- to 100-fold slower rates (Table 4.4). In the presence of optimal heparin, acrosin is the favored target since its inhibition rate constant shows the largest heparin acceleration (100-fold).[217] Of the blood plasma proteinases inactivated, activated protein C, thrombin, factor XIa and plasma kallikrein are inhibited the fastest in the presence of heparin, with rate constants between 10^4 and 10^5 $M^{-1}s^{-1}$. Plasma kallikrein inhibition is not accelerated by heparin, possibly due to a requirement for high molecular weight kininogen to bridge the interaction of enzyme with polysaccharide.[151] Tissue kallikrein is inhibited with a reasonable rate constant in the absence of heparin, but in the presence of polysaccharide the inhibition is blocked.[222]

PCI is activated by heparin and heparan sulfate, but not by dermatan sulfate.[220] Other polyanions are also activators. PCI is eluted from heparin-agarose at ~0.5 M NaCl, suggesting that PCI binds heparin with a lower affinity than antithrombin but with a higher affinity than heparin cofactor II.[173] The site for heparin binding appears to be distinct from that in other heparin-activatable serpins, since the conserved basic residues in helix D of this class of serpins are not conserved in PCI (Figs. 4.4 and 4.5). Based on calculations of the electrostatic surface potential of a structural model of PCI which was developed from the crystal structure of α_1-pro-

Table 4.4 Second order rate constants for protein C inhibitor-proteinase reactions measured in the absence and presence of heparin at the indicated temperatures (pH 7.4-7.8)

		Rate constant ($M^{-1}s^{-1}$)			
Target proteinase	**Temp (°C)**	**– Heparin**	**+ Heparin**	**Enhancement**	**Ref.**
Acrosin	37	2.4 x 10^5	5.6 x 10^7	230	217
Activated protein C	23	4.8 x 10^2	2.5 x 10^4	52	220
Activated protein C	37	8.0 x 10^2	1.5 x 10^4	18	228
Chymotrypsin	23	1.1 x 10^6	3.3 x 10^6	3	220
Factor Xa	23	3.2 x 10^3	3.5 x 10^3		220
Factor XIa	37	9.4 x 10^3	9.1 x 10^4	10	218
Plasma kallikrein	37	1.1 x 10^5	1.0 x 10^5		218
tPA (2 chain)	37	< 10^3	n.d.		207
Thrombin	23	2.0 x 10^4	1.9 x 10^5	10	220
Thrombin	37	3.3 x 10^4	2.0 x 10^5	6	228
Thrombin-TM[a]	23	2.4 x 10^6	n.d.[b]		219
Tissue kallikrein	37	2.3 x 10^4	Not inhibited		222
Trypsin	23	n.d.	1.7 x 10^4		214
uPA	37	8 x 10^3	9 x 10^4	11	207

a, TM, thrombomodulin; b, n.d., not determined.

teinase inhibitor, two adjacent α-helices, A^+ (preceding helix A) and H, were proposed to form a positively charged groove for heparin binding similar to the two helix heparin binding motif proposed for the platelet factor four-heparin interaction.[223] In agreement with this localization, a monoclonal antibody to helix A^+ blocks heparin binding and activation of the inhibitor.[223] Recent mutagenesis studies have confirmed the involvement of helix H in heparin binding by showing that deletion of the H-helix in a recombinant PCI fusion protein, as well as mutations of basic residues in the H helix of a full-length recombinant PCI, decrease heparin binding and activation of proteinase inhibition.[224] Heparin does not appear to act catalytically in its acceleration of PCI reactions as it does in antithrombin reactions, since PCI-proteinase complexes appear to bind heparin with an affinity similar to that of PCI.[220] Elastase cleavage of PCI in the reactive center also does not affect heparin binding affinity, implying that the serpin conformational change induced by proteinase binding or reactive center loop cleavage does not affect the heparin binding site.[220]

PCI is thought to be important for regulating activated protein C in plasma, since activated protein C-PCI complexes are found in the plasma of patients with thrombotic disease[225,226] or are formed when the proteinase is added to plasma[205] or injected in vivo.[227] Heparin or heparan sulfate may be important for the in vivo regulation since the inhibition rate constant is slow in the absence of polysaccharide.[220,228] The recent finding that thrombin bound to thrombomodulin is also an important PCI target is in keeping with a role for PCI in regulating the protein C anticoagulant pathway.[219] Urokinase has also been shown to be a target of PCI in blood plasma.[216] PCI appears to play important roles in regulating proteinases in seminal plasma. The inhibitor is present in high levels in seminal plasma and forms complexes with the tissue kallikrein-like proteinase, prostate-specific antigen, that is secreted by the prostate gland.[208,209] However, the molar levels of prostate-specific antigen (50 μM) greatly exceed those of PCI, suggesting that PCI regulates other proteinases. Endogenous PCI complexes with both urokinase- and tissue-type plasminogen activators have been demonstrated in seminal plasma, and such complexes also form when these proteinases are added to seminal plasma, with heparin stimulating their levels.[206] However, based on the rapid rate constant measured for the inhibition of the sperm proteinase acrosin, its substantial enhancement by heparin and the presence of heparin-like glycosaminoglycans at the site of acrosin action, i.e., in the female reproductive tract, acrosin has been proposed to be the primary target of PCI.[217] In the seminal fluid, PCI has been suggested to be a scavenger of acrosin released from damaged sperm, whereas in the female reproductive tract, PCI may regulate the binding of sperm to the zona pellucida matrix of the egg via acrosin and thereby the consequent invasion of the matrix required to fertilize the egg. PCI has also been recently shown to be expressed in kidney tubules and endogenous PCI-urokinase complexes have been demonstrated in urine or kidney tubule cells, consistent with these complexes originating from the kidney tubule.[229] Based on such observations and the expression of heparin-like glycosaminoglycans on epithelial kidney cells, PCI has been proposed to regulate urokinase activity in the kidney tubule and urine and thereby promote an appropriate pericellular function of the enzyme in the tubules.

4. PROTEASE NEXIN 1

Protease nexin 1 (PN1) is a 42-43 kDa glycoprotein inhibitor of thrombin which contains 6% carbohydrate.[230,231] In contrast to the localization of other heparin-activatable serpins, protease nexin 1 does not circulate in blood plasma. Instead it is secreted by a number of cell types, including fibroblasts,[232] platelets,[233] and glial cells.[234] Cloning and sequencing of cDNAs indicate that the fibroblast and glial-derived human inhibitors are identical.[234,235] Alternative splicing results in two forms of the protein which differ at position 310, which is Arg in the α-form and Thr-Gly in the β-form, resulting in polypeptide chains of 378 or 379 amino acid residues.[235] Both forms are synthesized with a 19 amino acid signal peptide and exhibit inhibitory activity. The α-form of rat PN1 has 379 residues and 84% sequence homology with the human protein.[236] There are three putative N-linked glycosylation sites and three cysteine residues in human PN1, all in the free reduced state.[237] The protein has a pI of 7.5-7.8.[230] The gene has been localized to chromosome 2q33-35.[238] Recombinant protease nexin 1 has been expressed in yeast,[239] baculovirus[240] and mammalian cells.[235]

The reactive center P1-P1' residues are Arg 345-Ser 346, compatible with the substrate specificity of the inhibitor's target proteinases.[235,236] The main target proteinases appear to be thrombin, urokinase and plasmin which are inhibited with second order rate constants ranging from 10^5 to 10^6 $M^{-1}s^{-1}$ in the absence of heparin.[231,232] The thrombin inhibition rate constant is enhanced about 300-fold in the presence of optimal levels of heparin.[231,241] tPA, factor Xa and γ-nerve growth factor are also inhibited at appreciable rates, whereas leukocyte elastase and cathepsin G are not detectably inactivated (Table 4.5).[231,237] Rapid kinetic studies have indicated that PN1 inhibition of thrombin is a two-step process, similar to that shown for the antithrombin-thrombin reaction,[93] with the formation of an initial encounter complex with K_D 3.4 μM followed by the conversion to a stable complex with a first order rate constant of 6 s^{-1}.[242]

In addition to heparin, heparan sulfate and chondroitin sulfates A and C are effective glycosaminoglycan activators of the PN1-thrombin reaction. Dermatan sulfate, keratan sulfate and hyaluronic acid are not activators.[243] The synthetic polyanion, dextran sulfate, also accelerates the reaction to an extent equal to that of heparin, implying that no sequence-specificity is involved in the activating effect.[243] The lack of a sequence-specific binding site in heparin for PN1 is in keeping with the elution of commercial heparin from immobilized PN1 as a single broad band between 0.1-0.6 M NaCl rather than as discrete fractions, and with the similar accelerating effects of heparins that elute at low and high salt concentrations from the

Table 4.5 Second order rate constants for protease nexin 1-proteinase reactions measured in the absence and presence of heparin at the indicated temperatures. (pH 7.8 - 8.2)

Target proteinase	Temp (°C)	Rate constant ($M^{-1}s^{-1}$) – Heparin	Rate constant ($M^{-1}s^{-1}$) + Heparin	Enhancement	Ref.
Acrosin	37	4.8×10^3	2.4×10^6	500	217
Activated protein C	37	5.2×10^3	2.3×10^5	44	228
Factor Xa	23	5.1×10^3	3.5×10^5	70	237
Factor Xa	37	7.3×10^3	–	–	231
γ-NGF[a]	37	4.0×10^3	–	–	231
Plasmin	23	1×10^5	1×10^5	–	237
Plasmin	37	1.3×10^5	–	–	231
tPA (1 chain)	37	1.5×10^3	–	–	231
tPA (2 chain)	37	3.0×10^4	–	–	231
Thrombin	37	1.4×10^6	$0.7\text{-}1.2 \times 10^9$	500-900	244
Trypsin	23	2.2×10^6	1.0×10^7	4.5	237
Trypsin	37	4.2×10^6	–	–	231
uPA	37	1.5×10^5	–	–	231

a, γ-NGF, γ-nerve growth factor

immobilized inhibitor.[244] The dependence of the heparin accelerating activity on polysaccharide concentration is bell-shaped, consistent with a ternary complex bridging mechanism for the rate-enhancing effect.[241] However, further data are needed to validate this possibility, given that a similar bell-shaped dependence of the heparin effect on polysaccharide concentration for the heparin cofactor II-thrombin reaction is not due to a bridging mechanism.[99] The optimal heparin accelerated inhibition rate constant of 1 x 10^9 $M^{-1}s^{-1}$ is the highest reported for a serpin reaction. This optimum is reached at heparin concentrations of about 20 nM, analogous to the optimum polysaccharide level for acceleration of the antithrombin-thrombin reaction.[244] However, the binding of heparin to PN1 does not appear to be as tight as it is to antithrombin, as judged from the lower salt concentration of 0.55-0.6 M required to elute PN1 relative to antithrombin from heparin-agarose.[231] Comparison of heparin-accelerated reactions of PN1 with α- and γ-thrombins have suggested that anion binding exosite I of thrombin does not play a significant role in heparin rate enhancement.[241] This supports a heparin accelerating mechanism similar to that of antithrombin but different from that of heparin cofactor II. Binding of thrombin or cleavage in the reactive center loop by nontarget proteinases does not alter the affinity of PN1 for heparin as it does with antithrombin, suggesting that the heparin accelerating effect is not catalytic.[237] The basic residues conserved in the D-helix region of most heparin-activatable serpins are highly conserved in PN1, consistent with this region being the site for heparin binding (Figs. 4.4 and 4.5). Site-directed mutagenesis studies have confirmed such a localization.[240] Thus, mutation to Glu of the seven Lys residues in the D-helix and the loop extending beyond this helix, i.e., residues 71-86 homologous to residues 121-136 of antithrombin, abolishes the heparin-accelerating effect on the PN1-thrombin reaction without affecting the unaccelerated reaction. A marked reduction of the heparin acceleration to ~4-fold also results from mutation of just Lys 83, 84, and 86.

The sites of synthesis of PN1 have indicated that the inhibitor functions predominantly to regulate the extravascular wound healing functions of thrombin. Secretion of PN1 by fibroblasts, smooth muscle cells and platelets is thought to regulate the procoagulant, chemotactic, mitogenic and cell differentiation actions of thrombin at the site of blood vessel injury which promote wound healing.[232,233,245] Extracellular matrix degradation by urokinase in the wound healing process may also be regulated by fibroblast PN1.[246] PN1 regulation of these proteinases is likely to occur in the vicinity of the cells that secrete the inhibitor, since PN1-proteinase complexes bind to such cells, resulting in their internalization and degradation.[247] Both the urokinase receptor and the α_2-macroglobulin receptor have been implicated in the binding and internalization of PN1-proteinase complexes.[248]

A role for PN1 in the regulation of extravascular thrombin is in keeping with the ability of heparan sulfate glycosaminoglycans on the fibroblast surface or in the extracellular matrix to accelerate the inhibition of thrombin by PN1, but not by antithrombin.[243,249] Chondroitin sulfate may also contribute to this acceleration. Binding of PN1 to fibroblast surface glycosaminoglycans or to collagen type IV in the extracellular matrix also blocks PN1 inhibition of urokinase and plasmin without affecting thrombin inhibition, suggesting that thrombin is the primary PN1 target on the cell surface and in the extracellular matrix.[250,251] That PN1 is involved in regulating proteinases which promote wound healing is further supported by the upregulation of PN1 synthesis in fibroblasts and in neuroblastoma cells by the injury-related factors interleukin 1, tumor necrosis factor-α, transforming growth factor-β and platelet-derived growth factor, and downregulation by the anti-inflammatory glucocorticoid dexamethasone.[252,253]

PN1 secreted by glial and certain other

neuronal cells promotes the outgrowth of neurite processes on neuronal cells in culture by inhibiting the neurite retraction activity of thrombin[254] and also inhibits the mitogenic effects of thrombin on astrocytes,[255,256] suggesting roles for the inhibitor in neuronal development.[257] The ability of PN1 to rescue motoneurons from injury-induced or natural programmed cell death also suggests an important role in regulating the wound healing functions of thrombin in the nervous system.[258] The reduction of PN1 protein in the brains of patients with Alzheimer's disease who suffer progressive neuronal loss supports such a role.[259] PN1 may also function to regulate the procoagulant activity of tumorigenic cells[260] as well as proteinases of the seminal fluid.[261,262]

5. PLASMINOGEN ACTIVATOR INHIBITOR-1

Plasminogen activator inhibitor-1 (PAI-1) is a 52 kDa molecular weight glycoprotein inhibitor of urokinase-type and tissue-type plasminogen activators.[263-265] It is synthesized by endothelial, platelet and hepatocyte cells as well as a number of other cell types consistent with the multiple physiologic functions and loci of action of the plasminogen activator targets of the inhibitor.[266] In plasma it circulates at rather low concentrations (~20 ng/ml) i.e., in the nanomolar range, but its levels may be enhanced 10-fold at sites of vessel injury by release of the inhibitor from activated platelets.[267-269] DNA sequencing has shown that the protein is 379 amino acids long and is synthesized with a 23 amino acid signal peptide, consistent with it being a secreted protein.[270-273] The protein molecular weight is 43 to 45 kilodaltons with the remainder carbohydrate. There are three potential N-linked glycosylation sites but no cysteines. The pI of the protein is 4.5-5.0.[265] The gene has been localized to chromosome 7q[270] and recombinant PAI-1 has been expressed in *E. coli*.[274-277]

The reactive center P1-P1' residues are Arg 346-Met 347.[270-273] The inhibitor activity is sensitive to oxidation of one or more methionine residues, similar to the behavior of α_1-proteinase inhibitor.[278] In addition to urokinase and tPA, thrombin,[276] activated protein C,[279-281] plasmin, factor XIIa, factor XIa, and plasma kallikrein[282,283] are inhibited at significant rates (see Tables 4.6 and 8.2). Heparin produces an ~90-fold enhancement of the rate of thrombin inhibition, but does not affect the rate at which other proteinases are inactivated.[282] Vitronectin also substantially increases the thrombin inhibition rate constant (~200-fold), whereas the rates of inhibition of tPA, urokinase and other proteinases are modestly enhanced by these effectors, suggesting that heparin and vitronectin modify the target proteinase specificity of PAI-1.[276,282,284] However, vitronectin does accelerate 1,000-fold the inhibition of tPA by R346M PAI-1 variants which have low activity toward the enzyme.[285] Random or site-directed mutagenesis of the reactive center P3, P2, P1 and P1' residues has demonstrated a requirement for a basic residue, either Arg or Lys, in the P1 site for inhibition of urokinase, but not for tPA which is inhibited by variant PAI-1s with P1 substitutions of Asn, Gln, His, Ser, Thr, Leu, Met, Phe, Tyr and Trp.[277,286-288] By contrast, there is a considerable tolerance for a number of amino acids in the P1' site for inhibition of uPA and tPA when the P1 residue is Arg or Lys. Thrombin shows a different specificity profile with a number of substitutions possible at P1 and P1' when inhibitory activity is measured in the presence of heparin. Second sites of interaction thus appear to be important for the inhibition of tPA and thrombin.[288] The different specificities of each proteinase for PAI-1 result from insertion loops in the active site region of the catalytic chains[289-291] as well as interaction determinants in the noncatalytic chains.[285,292]

A unique aspect of the inhibitor is its tendency to spontaneously inactivate or convert to a latent form[293] in which the reactive center loop has completely inserted as a central strand into β-sheet A.[294] The half-life of this process is about 1 hour at 37°C, pH 7.4.[295-297] Denaturation of latent PAI-1 with SDS, guanidine or other denaturants followed by renaturation partially restores activity.[293] Negatively charged phospholipid vesicles have also been reported to reactivate latent PAI-1.[298] Vitronectin stabilizes the active form of the inhibitor in which the reactive center loop is presumably exposed on the protein surface.[299] Reaction with plasminogen activators induces dissociation of the PAI-1-enzyme complexes from vitronectin. Replacement of the reactive center loop with that of other serpins or with a serpin consensus sequence does not affect the active to latent transition or inhibitory function, suggesting that the instability of the active PAI-1 conformation does not result from the reactive center sequence.[295] Disruption of the R30-E350 salt bridge observed in the crystal structure of latent PAI-1[296] or substitution of Glu at P12, P10, P8, P6 and P4 residues which insert into β-sheet A during the conversion to the latent inhibitor[297] increases the half-life of the native state up to 8-fold, possibly by destabilizing the latent state. Random mutagenesis studies followed by screening for variants which retain activity after prolonged 37°C incubation has allowed identification of variant inhibitors with up to 100-fold increased half-lives. Such studies have indicated that less favorable interactions of hydrophobic residues that stabilize the internal packing of the protein are mostly responsible for the decreased stability of PAI-1 relative to other serpins.[300] Vitronectin binds with a K_D in the range 0.3-190 nM.[301,302] The vitronectin binding domain has been localized to a binding surface distinct from the heparin binding site that bridges α-helices C and E to β-strand 1A and involves contacts with Gln 55, Phe 109, Met 110, Leu 116, and Gln 123.[303] The region between amino acids 110 and 145 and a smaller sequence encompassing residues 115-130 within this region have also been implicated in vitronectin binding, from studies using monoclonal antibodies, PAI-1 peptides or PAI-1/PAI-2 chimeras to map this site.[304,305] The importance of uncharged residues in the binding suggests a nonionic interaction, consistent with the stability of vitronectin complexes in high salt. The interaction of vitronectin with β-sheet A has been

Table 4.6 Second order rate constants for PAI-1-proteinase reactions measured in the absence and presence of heparin and vitronectin at the indicated temperatures (pH 7.4-8.0)

Target proteinase	Temp (°C)	Second order rate constant ($M^{-1}s^{-1}$) – Heparin	+ Heparin	+ Vitronectin	Ref.
Activated protein C	23	1.1×10^{4} [a]	–	–	281
Factor Xa	37	$< 10^{1}$	$< 10^{1}$	$< 10^{1}$	282
Factor XIa	37	2.1×10^{5}	–	–	283
Factor XIIa	37	6.7×10^{3}	6.7×10^{3}	7.4×10^{3}	282
Plasma kallikrein	37	6.0×10^{4}	–	–	283
Plasmin	37	2.7×10^{4}	2.7×10^{4}	1.1×10^{5}	282
tPA (1 chain)	23	6.8×10^{6}	–	–	295
tPA (2 chain)	23	2.3×10^{7}	–	–	295
tPA (2 chain)	37	2.6×10^{7}	2.6×10^{7}	3.1×10^{7}	282
Thrombin	25	6.1×10^{2}	–	1.6×10^{5}	284
Thrombin	37	1.1×10^{3}	1.0×10^{5}	2.2×10^{5}	282
uPA	23	1.1×10^{7}	–	–	295
uPA	37	4.8×10^{6}	4.8×10^{6}	3.7×10^{7}	282

a, pH 8.0; b, porcine platelet PAI-1.

proposed to stabilize the active conformation of PAI-1 by the need to disrupt this interaction to allow opening of β-sheet A and insertion of the reactive center loop. This proposal is consistent with the lower affinity binding of vitronectin to latent PAI 1.[303]

Besides heparin, heparan sulfate, dermatan sulfate and chondroitin 4-sulfate are active in accelerating the inhibition of thrombin by PAI-1, although at least 10-fold higher concentrations of these glycosaminoglycans are required for optimal stimulation.[306,307] Heparin appears to bind with an affinity about 50-fold greater than that of the other active glycosaminoglycans. The elution of PAI-1 from heparin-agarose at 0.4-0.5 M NaCl[306] indicates a heparin affinity comparable to those of heparin cofactor II and protein C inhibitor. Keratan sulfate, chondroitin 6-sulfate and hyaluronic acid do not accelerate thrombin inhibition, presumably due to their inability to bind to the inhibitor. The heparin accelerating effect shows a bell-shaped dependence on polysaccharide concentration that is optimal at 0.1-1 μg/ml of unfractionated polysaccharide under non-physiological conditions of pH and ionic strength and at 25 μg/ml under more physiological conditions.[308] Moreover, heparin chains greater than 14 saccharides in length are required to accelerate the PAI-1 thrombin reaction. This behavior is consistent with, but not proof of, a ternary complex bridging mechanism for the acceleration.[307] The cluster of basic residues typically found in helix D of the heparin-activatable serpins is conserved in PAI-1 (Fig. 4.5). Mutagenesis of these residues resulted in decreased heparin binding and acceleration of thrombin inhibition, consistent with helix D being the site on PAI-1 for heparin binding.[309] In addition to accelerating PAI-1 inhibition of thrombin, heparin also stimulates PAI-1 inactivation by thrombin, presumably by a substrate-like cleavage in the reactive center loop.[308]

PAI-1 is the predominant inhibitor of urokinase- and tissue-type plasminogen activators which function in fibrinolysis (Fig. 4.1) and tissue remodeling.[266,310,311] The inhibitor plays an essential role in regulating fibrinolysis since PAI-1 deficiency is associated with bleeding problems[312] and elevated levels of PAI-1 correlate with an increased risk of thrombosis.[313] Consistent with this role, disruption of the PAI-1 gene in mice produces a mild hyperfibrinolytic state and a greater resistance to venous thrombosis without affecting hemostasis.[310,311] PAI-1 binding to fibrin is essential for it to regulate plasminogen activation on the clot surface and thereby regulate clot lysis.[314-316] A monoclonal antibody that inhibits PAI-1 activity stimulated tPA-dependent fibrinolysis in animal models of thrombosis, suggesting that antagonists of fibrin-bound PAI-1 may be effective therapeutic agents in patients showing resistance to thrombolytic therapy.[317] Thrombin bound to the extracellular matrix is resistant to antithrombin inhibition,[318] but shows increased susceptibility to PAI-1 or protease nexin 1 inactivation, suggesting that thrombin may be an additional important PAI-1 target in the extravascular compartment. That PAI-1-vitronectin complexes in endothelial cell matrix promotes exclusive complexing of added thrombin with PAI-1 supports this view.[319-321] Plasma vitronectin may additionally bind PAI-1 released from endothelial cells or platelets at the site of a thrombus and promote the formation of PAI-1-thrombin complexes. The resulting depletion of PAI-1 could explain in part the known ability of thrombin to stimulate fibrinolysis. Anticoagulant thrombin bound to thrombomodulin is protected from PAI-1 inactivation, consistent with PAI-1 interactions with procoagulant thrombin providing an important link between coagulation and fibrinolysis.[319] PAI-1 regulation of fibrinolysis may also involve a feedback mechanism in which plasmin cleavage of vitronectin releases bound PAI-1, thereby promoting PAI-1 inactivation of plasminogen activators.[322,323] In addition to regulating fibrinolysis, PAI-1 is thought to regulate the extracellular matrix turnover functions of plasminogen activators involved in normal tissue remodeling processes and pathological tumor cell invasion.[324] The regulation of PAI-1 synthesis by certain cytokines supports its role in controlling the physiological functions of plasminogen activators.[325]

6. COMPARISONS BETWEEN ACTIVATION MECHANISMS

Comparison of the mechanisms by which the heparin-dependent serpins are activated by glycosaminoglycans suggests a number of common as well as disparate features. One common feature is the presence of a site on the inhibitor for binding heparin and activating the inhibitor. This site appears to involve helix for all the inhibitors except for protein C inhibitor, with the site in antithrombin additionally involving helix A. In protein C inhibitor, helix H instead seems to be the primary site for heparin binding (Fig. 4.4). Given this major difference in heparin binding site localization, it is likely that the H helix site evolved independently of the more prevalent D helix site and that the latter site had a common evolutionary origin.

The interaction of heparin with this group of serpins involves a specific polysaccharide sequence only in the case of antithrombin. No evidence for sequence-specific binding of heparin exists for the other serpins in this group. One consequence of the sequence-specificity of heparin binding to antithrombin is that it results in a high-affinity interaction. Lower concentrations of polysaccharide are therefore required to activate antithrombin compared with those required for most other heparin-activatable serpins. The high-affinity interaction derives from an induced complementary fit of the specific polysaccharide sequence with the inhibitor, which involves both electrostatic and non-electrostatic interactions. The induced fit binding mechanism requires a conformational change in antithrombin, which is transmitted to the reactive center loop, leading to inhibitor activation. The high-affinity interaction is lost following proteinase binding due to reactive center loop insertion into β-sheet A. This allows heparin to act as a catalyst and minimizes heparin-promoted inhibitor inactivation by reaction as a substrate. The reduction of heparin binding affinity by a proteinase-induced conformational change appears to be unique to antithrombin, since other heparin-activatable serpins do not show such changes in heparin affinity subsequent to their reaction with proteinase.

Interestingly, protease nexin 1 is activated by heparin concentrations similar to those for antithrombin, suggesting a high-affinity interaction is involved with this serpin as well. However, this may reflect a larger number of basic residues in helix D of this serpin, which can make more electrostatic interactions with heparin (Fig. 4.5), since there is no evidence for a sequence-specific site in heparin for this serpin or the induction of a serpin conformational change as a result of heparin binding. For heparin cofactor II, protein C inhibitor and PAI-1, the probable electrostatic interaction of heparin with these inhibitors is of low or moderate-affinity, which is reflected in the higher concentrations of polysaccharide needed to effect substantial rate enhancement. Heparin activation of these latter inhibitors thus appears to resemble the activation of antithrombin by heparin molecules lacking a sequence-specific antithrombin binding site, which also appears to be mediated mostly by electrostatic interactions. Whereas heparin and the related glycosaminoglycan, heparan sulfate, are both capable of activating this group of serpins, heparin cofactor II, and to a lesser extent PAI-1 are also activated by another glycosaminoglycan, dermatan sulfate. An overlapping but distinct site in helix D possibly involving a sequence-specific site, appears to be involved in the heparin cofactor II activation. Nevertheless, dermatan sulfate levels comparable to those of heparin are required for this activation, possibly reflecting the lower extent of sulfation of dermatan sulfate relative to heparin. In the case of PAI-1, dermatan sulfate activation is weaker than heparin and requires higher levels of glycosaminoglycan.

A ternary complex bridging mechanism in which heparin promotes the inhibitor-proteinase reaction by the binding of both proteins to a single polysaccharide chain in a ternary complex appears to explain the heparin activating effect on the reactions of antithrombin, protease nexin 1, PAI-1 and protein C inhibitor with thrombin and possibly other proteinases as well. Heparin enhancement of the heparin cofactor II-thrombin reaction is also partly due to such a bridging mechanism. In keeping with such a mechanism, a heparin binding site on thrombin known as exosite II has been identified and its role in heparin bridging antithrombin and to a lesser extent also heparin cofactor II with thrombin established. The other known effector of PAI-1 reactions with thrombin, the extracellular matrix protein vitronectin, also appears to accelerate PAI-1-thrombin reactions by such a bridging mechanism. The use of both the serpin D-helix and H-helix for heparin binding presents certain problems for the proper alignment of thrombin with the inhibitor when the two proteins are bound to the same heparin chain in a ternary complex. Thus, if the heparin binding exosite II on thrombin is used for bridging of thrombin with all heparin-activatable serpins, the different sites of heparin binding on the serpin would be expected to orient the two proteins differently with respect to each other on the heparin chain. Thus a less favorable orientation of the proteinase active site with the inhibitor reactive center may account for the smaller heparin rate enhancements of protein C inhibitor reactions relative to those of other heparin-activatable serpins.

Contrasting this bridging mechanism, heparin can also activate antithrombin and heparin cofactor II by an allosteric mechanism in which conformational changes in the serpin make it more reactive with a target proteinase. In the case of antithrombin, the allosteric mechanism predominates with factor Xa as the target and involves conformational changes in the reactive center loop which are transmitted by heparin binding. The nature of these changes, which appear to specifically enhance factor Xa recognition and not thrombin recognition, have yet to be clarified. With heparin cofactor II, heparin binding exposes a unique amino terminal binding site instead, which contains the determinants for recognition of the inhibitor by thrombin anion binding exosite I. Since the exposure of this site involves the disruption of an intramolecular interaction between the acidic amino-terminal region and the helix D glycosaminoglycan binding site, both heparin and dermatan sulfate activate the inhibitor by the same allosteric mechanism. One difference between dermatan sulfate and heparin is that the activation is all allosteric with dermatan sulfate and involves no bridging component whereas bridging makes a small additional contribution to the heparin inhibition rate enhancement. Whereas no evidence exists for an allosteric mechanism of heparin activation of protease nexin-1, protein C inhibitor or PAI-1 reactions, further data are needed before such a mechanism can be ruled out.

It is thus clear that heparin and other glycosaminoglycans may activate serpin-proteinase reactions by enhancing the recognition of serpins by their target proteinases either through protein-protein or protein-polysaccharide interactions. In the former case, heparin may either induce an optimal conformation of the serpin reactive center loop or may expose unique proteinase recognition sites in the inhibitor outside of the reactive center loop. In the latter case, heparin itself may provide additional recognition sites for proteinase in the serpin-heparin complex.

REFERENCES

1. Leon M, Aiach M, Coezy E, et al. Antithrombin III synthesis in rat liver parenchymal cells. Thromb Res 1983; 30: 369-375.
2. Conard J, Brosstad F, Lie Larsen M, et al. Molar antithrombin concentration in normal human plasma. Haemostasis 1983; 13:363-368.
3. Collen D, Schetz J, DeCock F, et al. Metabolism of antithrombin III (Heparin cofactor) in man: effects of venous thrombosis and of heparin administration. Eur J Clin Invest 1977; 7:27-35.
4. Nordenman B, Nyström C and Björk I. The size and shape of human and bovine antithrombin III. Eur J Biochem 1977; 78:195-203.
5. Petersen TE, Dudek-Wojciechowska G, Sottrup-Jensen L and Magnusson S. Primary structure of antithrombin III (heparin cofactor). Partial homology between α_1-antitrypsin and antithrombin III. In: Collen D, Wiman B, Verstraete M, eds. The Physiological Inhibitors of Blood Coagulation and Fibrinolysis. Amsterdam: Elsevier/North-Holland, 1979:43-54.
6. Bock SC, Wion KL, Vehar GA, et al. Cloning and expression of the cDNA for human antithrombin III. Nucleic Acids Res 1982; 10:8113-8125.
7. Prochownik EV, Markham AF and Orkin SH. Isolation of a cDNA clone for human antithrombin III. J Biol Chem 1983; 258:8389-8394.
8. Chandra T, Stackhouse R, Kidd VJ, et al. Isolation and sequence characterization of a cDNA clone of human antithrombin III. Proc Natl Acad Sci USA 1983; 80: 1845-1848.
9. Mejdoub H, Le Ret M, Boulanger Y, et al. The complete amino acid sequence of bovine antithrombin (ATIII). J Prot Chem 1991; 10:205-212.
10. Niessen RWLM, Sturk A, Hordijk PL, et al. Sequence characterization of a sheep cDNA for antithrombin III. Biochim Biophys Acta 1992; 1171:207-210.
11. Sheffield WP, Brothers AB, Wells MJ, et al. Molecular cloning and expression of rabbit antithrombin III. Blood 1992; 79: 2330-2339.
12. Wu JK, Sheffield WP and Blajchman MA. Molecular cloning and cell-free expression of mouse antithrombin III. Thromb Haemostas 1992; 68:291-296.
13. Franzén L-E, Svensson S and Larm O. Structural studies on the carbohydrate portion of human antithrombin III. J Biol Chem 1980; 255:5090-5093.
14. Mizuochi T, Fujii J, Kurachi K, et al. Structural studies of the carbohydrate moiety of human antithrombin III. Arch Biochem Biophys 1980; 203:458-465.
15. Carlson TH and Atencio AC. Isolation and partial characterization of two distinct types of antithrombin III from rabbit. Thromb Res 1982; 27:23-34.
16. Peterson CB and Blackburn MN. Isolation and characterization of an antithrombin III variant with reduced carbohydrate content and enhanced heparin binding. J Biol Chem 1985; 260:610-615.
17. Brennan SO, George PM and Jordan RE. Physiological variant of antithrombin III lacks carbohydrate sidechain at Asn 135. FEBS Lett 1987; 219:431-436.
18. Picard V, Erdsal-Badju E and Bock SC. Partial glycosylation of antithrombin III asparagine-135 is caused by the serine in the third position of its N-glycosylation consensus sequence and is responsible for production of the β-antithrombin III isoform with enhanced heparin affinity. Biochemistry 1995; 34:8433-8440.
19. Borsodi AD and Bradshaw RA. Isolation of antithrombin III from normal and α_1-antitrypsin-deficient human plasma. Thromb Haemostas 1977; 38:475-485.
20. Kao FT, Morse HG, Law ML, et al. Genetic mapping of the structural gene for antithrombin III to human chromosome 1. Hum Genet 1984; 67:34-36.
21. Bock SC, Harris JF, Balazs I, et al. Assignment of the human antithrombin III structural gene to chromosome 1q23-25. Cytogenet Cell Genet 1985; 39:67-69.
22. Prochownik EV, Bock SC and Orkin SH. Intron structure of the human antithrombin III gene differs from that of other members of the serine protease inhibitor superfamily. J Biol Chem 1985; 260:9608-9612.

23. Bock SC, Marrinan JA and Radziejewska E. Antithrombin III Utah: proline-407 to leucine mutation in a highly conserved region near the inhibitor reactive site [published erratum appears in Biochemistry 1989 Apr 18;28(8):3628]. Biochemistry 1988; 27: 6171-6178.
24. Olds RJ, Lane DA, Chowdhury V, et al. Complete nucleotide sequence of the antithrombin gene: evidence for homologous recombination causing thrombophilia. Biochemistry 1993; 32:4216-4224.
25. Austin RC, Rachubinski RA, Fernandez-Rachubinski F, et al. Expression in a cell-free system of normal and variant forms of human antithrombin III. Ability to bind heparin and react with α-thrombin. Blood 1990; 76:1521-1529.
26. Bröker M, Ragg H and Karges HE. Expression of human antithrombin III in *Saccharomyces cerevisiae* and *Saccharomyces pombe*. Biochem Biophys Res Commun 1987; 908:203-213.
27. Dingermann T, Troidi E, Bröker M, et al. Expression of human antithrombin III in the cellular slime mould *Dictyostelium discoideum*. Appl Microbiol Biotechnol 1991; 35:496-503.
28. Gillespie LS, Hillesland KK and Knauer DJ. Expression of biologically active human antithrombin III by recombinant baculovirus in *Spodoptera frugiperda* cells. J Biol Chem 1991;266:3995-4001.
29. Stephens AW, Siddiqui A and Hirs CHW. Expression of functionally active human antithrombin III. Proc Natl Acad Sci USA 1987; 84:3886-3890.
30. Wasley LC, Atha DH, Bauer KA, et al. Expression and characterization of human antithrombin III synthesized in mammalian cells [published erratum appears in J Biol Chem 1988 May 5;263(13):6459]. J Biol Chem 1987; 262:14766-14772.
31. Zettlmeißl G, Conrad HS, Nimtz M, et al. Characterization of recombinant human antithrombin III synthesized in Chinese hamster ovary cells. J Biol Chem 1989; 264:21153-21159.
32. Björk I, Ylinenjärvi K, Olson ST, et al. Decreased affinity of recombinant antithrombin for heparin due to increased glycosylation. Biochem J 1992; 286: 793-800.
33. Fan B, Crews BC, Turko IV, et al. Heterogeneity of recombinant human antithrombin III expressed in baby hamster kidney cells. Effect of glycosylation differences on heparin binding and structure. J Biol Chem 1993; 268:17588-17596.
34. Rosenberg RD. Chemistry of the hemostatic mechanism and its relationship to the action of heparin. Fed Proc 1977; 36:10-18.
35. Danielsson Å and Björk I. Mechanism of inactivation of trypsin by antithrombin. Biochem J 1982; 207:21-28.
36. Olson ST, Stephens AW, Hirs CHW, et al. Kinetic characterization of the proteinase binding defect in a reactive site variant of the serpin, antithrombin. Role of the P1' residue in transition-state stabilization of antithrombin-proteinase complex formation. J Biol Chem 1995; 270:9717-9724.
37. Björk I, Jackson CM, Jörnvall H, et al. The active site of antithrombin. Release of the same proteolytically cleaved form of the inhibitor from complexes with factor IXa, factor Xa, and thrombin. J Biol Chem 1982; 257:2406-2411.
38. Erdjument H, Lane DA, Panico M, et al. Single amino acid substitutions in the reactive site of antithrombin leading to thrombosis. Congenital substitution of arginine 393 to cysteine in antithrombin Northwick Park and to histidine in antithrombin Glasgow. J Biol Chem 1988; 263:5589-5593.
39. Lane DA, Erdjument H, Thompson E, et al. A novel amino acid substitution in the reactive site of a congenital variant antithrombin. Antithrombin Pescara, Arg^{393} to Pro caused by a CGT to CCT mutation. J Biol Chem 1989; 264:10200-10204.
40. Stephens AW, Thalley BS and Hirs CHW. Antithrombin III Denver, a reactive site variant. J Biol Chem 1987; 262:1044-1048.
41. Stephens AW, Siddiqui A and Hirs CHW. Site-directed mutagenesis of the reactive center (Serine 394) of antithrombin III. J Biol Chem 1988; 263:15849-15852.
42. Theunissen HJM, Dijkema R, Grootenhuis PDJ, et al. Dissociation of heparin-dependent thrombin and factor Xa inhibitory

activities of antithrombin-III by mutations in the reactive site. J Biol Chem 1993; 268:9035-9040.
43. Björk I and Lindahl U. Mechanism of the anticoagulant action of heparin. Mol Cell Biochem 1982; 48:161-182.
44. Lindahl U, Bäckström G, Thunberg L, et al. Evidence for a 3-O-sulfated D-glucosamine residue in the antithrombin-binding sequence of heparin. Proc Natl Acad Sci USA 1980; 77:6551-6555.
45. Casu B, Oreste P, Torri G, et al. The structure of heparin oligosaccharide fragments with high anti-(factor Xa) activity containing the minimal antithrombin III-binding sequence. Chemical and ^{13}C nuclear magnetic resonance studies. Biochem J 1981; 197:599-609.
46. Thunberg L, Bäckström G and Lindahl U. Further characterization of the antithrombin-binding sequence in heparin. Carbohydr Res 1982; 100:393-410.
47. Choay J, Petitou M, Lormeau JC, et al. Structure-activity relationship in heparin: A synthetic pentasaccharide with high-affinity for antithrombin III and eliciting high anti-factor Xa activity. Biochem Biophys Res Commun 1983; 116:492-499.
48. Atha DH, Stephens AW and Rosenberg RD. Evaluation of critical groups required for the binding of heparin to antithrombin. Proc Natl Acad Sci USA 1984; 81:1030-1034.
49. Oscarsson LG, Pejler G and Lindahl U. Location of the antithrombin-binding sequence in the heparin chain. J Biol Chem 1989; 264:296-304.
50. Danielsson Å and Björk I. Binding to antithrombin of heparin fractions with different molecular weights. Biochem J 1981; 193:427-433.
51. Jordan RE, Favreau LV, Braswell EH, et al. Heparin with two binding sites for antithrombin or platelet factor 4. J Biol Chem 1982; 257:400-406.
52. Nesheim ME, Blackburn MN, Lawler CM, et al. Dependence of antithrombin III and thrombin binding stoichiometries and catalytic activity on the molecular weight of affinity-purified heparin. J Biol Chem 1986; 261:3214-3221.
53. Nordenman B and Björk I. Binding of low-affinity and high-affinity heparin to antithrombin. Ultraviolet difference spectroscopy and circular dichroism studies. Biochemistry 1978; 17:3339-3344.
54. Nordenman B, Danielsson Å and Björk I. The binding of low-affinity and high-affinity heparin to antithrombin. Fluorescence studies. Eur J Biochem 1978; 90:1-6.
55. Jordan R, Beeler D and Rosenberg RD. Fractionation of low molecular weight heparin species and their interaction with antithrombin. J Biol Chem 1979; 254:2902-2913.
56. Höök M, Björk I, Hopwood J, et al. Anticoagulant activity of heparin: Separation of high-activity and low-activity heparin species by affinity chromatography on immobilized antithrombin. FEBS Lett 1976; 66:90-93.
57. Andersson LO, Barrowcliffe TW, Holmer E, et al. Anticoagulant properties of heparin fractionated by affinity chromatography on matrix-bound antithrombin III and by gel filtration. Thromb Res 1976; 9:575-583.
58. Lam LH, Silbert JE and Rosenberg RD. The separation of active and inactive forms of heparin. Biochem Biophys Res Commun 1976; 69:570-577.
59. Marcum JA and Rosenberg RD. Anticoagulantly active heparin-like molecules from vascular tissue. Biochemistry 1984; 23:1730-1737.
60. Tollefsen DM, Pestka CA and Monafo WJ. Activation of heparin cofactor II by dermatan sulfate. J Biol Chem 1983; 258:6713-6716.
61. Marcum JA, Atha DH, Fritze LMS, et al. Cloned bovine aortic endothelial cells synthesize anticoagulantly active heparan sulfate proteoglycan. J Biol Chem 1986; 261:7507-7517.
62. Olson ST, Srinivasan KR, Björk I, et al. Binding of high affinity heparin to antithrombin III. Stopped flow kinetic studies of the binding interaction. J Biol Chem 1981; 256:11073-11079.
63. Olson ST, Björk I, Sheffer R, et al. Role of the antithrombin-binding pentasaccharide in heparin acceleration of antithrombin-proteinase reactions. Resolution of the anti-

thrombin conformational change contribution to heparin rate enhancement. J Biol Chem 1992; 267:12528-12538.
64. Nordenman B and Björk I. Influence of ionic strength and pH on the interaction between high-affinity heparin and antithrombin. Biochim Biophys Acta 1981; 672:227-238.
65. Gettins P. Antithrombin III and its interaction with heparin. Comparison of the human, bovine, and porcine proteins by ^1H NMR spectroscopy. Biochemistry 1987; 26:1391-1398.
66. Gettins P and Wooten EW. On the domain structure of antithrombin III. Tentative localization of the heparin binding region using ^1H NMR spectroscopy. Biochemistry 1987; 26:4403-4408.
67. Horne AP and Gettins P. ^1H NMR spectroscopic studies on the interactions between human plasma antithrombin III and defined low molecular weight heparin fragments. Biochemistry 1992; 31:2286-2294.
68. Olson ST and Shore JD. Binding of high affinity heparin to antithrombin III. Characterization of the protein fluorescence enhancement. J Biol Chem 1981; 256: 11065-11072.
69. Chang J-Y. Binding of heparin to human antithrombin III activates selective modification at lysine 236. Lys 107, lys 125, and lys 136 are situated within the heparin binding site of antithrombin III. J Biol Chem 1989; 264:3111-3115.
70. Kress LF and Catanese JJ. Identification of the cleavage sites resulting from enzymatic inactivation of human antithrombin III by *Crotalus adamanteus* proteinase II in the presence and absence of heparin. Biochemistry 1981; 20:7432-7438.
71. Gettins PGW, Fan B, Crews BC, et al. Transmission of conformational change from the heparin binding site to the reactive center of antithrombin. Biochemistry 1993; 32:8385-8389.
72. Dawes J, James K and Lane DA. Conformational change in antithrombin induced by heparin, probed with a monoclonal antibody against the 1C/4B region. Biochemistry 1994; 33:4375-4383.
73. Huber R and Carrell RW. Implications of the three-dimensional structure of α_1-antitrypsin for structure and function of serpins. Biochemistry 1989; 28:8951-8966.
74. Carrell RW, Stein PE, Fermi G, et al. Biological implications of a 3Å structure of dimeric antithrombin. Structure 1994; 2:257-270
75. Olson ST and Björk I. Regulation of thrombin activity by antithrombin and heparin. Sem Thromb Hemostas 1994; 20:373-409.
76. Villanueva GB. Predictions of the secondary structure of antithrombin III and the location of the heparin binding site. J Biol Chem 1984; 259:2531 2536.
77. Fan B, Turko IV and Gettins PGW. Lysine-heparin interactions in antithrombin. Properties of K125M and K290M, K294M, K297M variants. Biochemistry 1994; 33:14156-14161.
78. Grootenhuis PDJ and van Boeckel CAA. Constructing a molecular model of the interaction between antithrombin III and a potent heparin analogue. J Am Chem Soc 1991; 113:2743-2747.
79. van Boeckel CAA, Grootenhuis PDJ and Visser A. A mechanism for heparin-induced potentiation of antithrombin III. Nature Struct Biol 1994; 1:423-425.
80. Schreuder HA, de Boer B, Dijkema R, et al. The intact and cleaved human antithrombin III complex as a model for serpin-proteinase interactions. Nature Struct Biol 1994; 1:48-54.
81. Björk I and Fish WW. Production *in vitro* and properties of a modified form of bovine antithrombin, cleaved at the active site by thrombin. J Biol Chem 1982; 257:9487-9493.
82. Fish WW, Danielsson Å, Nordling K, et al. Denaturation behavior of antithrombin in guanidinium chloride. Irreversibility of unfolding caused by aggregation. Biochemistry 1985; 24:1510-1517.
83. Carrell RW, Evans DL and Stein PE. Mobile reactive centre of serpins and the control of thrombosis. Nature 1991; 353: 576-578.
84. Erdsal-Badju E, Lu A, Picard V et al. A mechanism for heparin activation of antithrombin III based on identification of the heparin binding site. 1995; (submitted for publication).

85. Lane DA, Denton J, Flynn AM, et al. Anticoagulant activities of heparin oligosaccharides and their neutralization by platelet factor 4. Biochem J 1984; 218:725-732.
86. Danielsson Å, Raub E, Lindahl U, et al. Role of ternary complexes, in which heparin binds both antithrombin and proteinase, in the acceleration of the reactions between antithrombin and thrombin or factor Xa. J Biol Chem 1986; 261:15467-15473.
87. Laurent TC, Tengblad A, Thunberg L, et al. The molecular weight dependence of the anti-coagulant activity of heparin. Biochem J 1978; 175:691-701.
88. Grootenhuis PDJ, Westerduin P, Meuleman D, et al. Rational design of synthetic heparin analogues with tailor-made coagulation inhibitory activity. Nature Struct Biol 1995; 2:736-739.
89. Jordan RE, Oosta GM, Gardner WT, et al. The kinetics of hemostatic enzyme-antithrombin interactions in the presence of low molecular weight heparin. J Biol Chem 1980; 255:10081-10090.
90. Olson ST. Transient kinetics of heparin-catalyzed protease inactivation by antithrombin III. Linkage of protease-inhibitor-heparin interactions in the reaction with thrombin. J Biol Chem 1988; 263: 1698-1708.
91. Griffith MJ. Kinetics of the heparin-enhanced antithrombin III/thrombin reaction. Evidence for a template model for the mechanism of action of heparin. J Biol Chem 1982; 257:7360-7365.
92. Nesheim ME. A simple rate law that describes the kinetics of the heparin-catalyzed reaction between antithrombin III and thrombin. J Biol Chem 1983; 258: 14708-14717.
93. Olson ST and Shore JD. Demonstration of a two-step mechanism for inhibition of α-thrombin by antithrombin III and identification of the step affected by heparin. J Biol Chem 1982; 257:14891-14895.
94. Olson ST and Björk I. Predominant contribution of surface approximation to the mechanism of heparin acceleration of the antithrombin-thrombin reaction. Elucidation from salt concentration effects. J Biol Chem 1991; 266:6353-6364.
95. Craig PA, Olson ST and Shore JD. Transient kinetics of heparin-catalyzed protease inactivation by antithrombin III. Characterization of assembly, product formation, and heparin dissociation steps in the factor Xa reaction. J Biol Chem 1989; 264: 5452-5461.
96. Pomerantz MW and Owen WG. A catalytic role for heparin. Evidence for a ternary complex of heparin cofactor, thrombin and heparin. Biochim Biophys Acta 1978; 535:66-77.
97. Machovich R, Staub M and Patthy L. Decreased heparin sensitivity of cyclohexanedione-modified thrombin. Eur J Biochem 1978; 83:473-477.
98. Gan Z-R, Li Y, Chen Z, et al. Identification of basic amino acid residues in thrombin essential for heparin-catalyzed inactivation by antithrombin III. J Biol Chem 1994; 269:1301-1305.
99. Sheehan JP, Tollefsen DM and Sadler JE. Heparin cofactor II is regulated allosterically and not primarily by template effects. Studies with mutant thrombins and glycosaminoglycans. J Biol Chem 1994; 269:32747-32751.
100. Owen BA and Owen WG. Interaction of factor Xa with heparin does not contribute to the inhibition of factor Xa by antithrombin III-heparin. Biochemistry 1990; 29:9412-9417.
101. Streusand VJ, Björk I, Gettins PGW, et al. Mechanism of acceleration of antithrombin-proteinase reactions by low affinity heparin. Role of the antithrombin binding pentasaccharide in heparin rate enhancements. J Biol Chem 1995; 270:9043-9051.
102. Hopkins PCR, Crowther DC, Carrell RW, et al. Development of a novel recombinant serpin with potential antithrombotic properties. J Biol Chem 1995; 270: 11866-11871.
103. Björk I and Nordenman B. Acceleration of the reaction between thrombin and antithrombin III by non-stoichiometric amounts of heparin. Eur J Biochem 1976; 68: 507-511.
104. Carlström AS, Liedén K and Björk I. Decreased binding of heparin to antithrombin following the interaction between anti-

thrombin and thrombin. Thromb Res 1977; 11:785-797.

105. Olson ST and Shore JD. Transient kinetics of heparin-catalyzed protease inactivation by antithrombin III. The reaction step limiting heparin turnover in thrombin neutralization. J Biol Chem 1986; 261: 13151-13159.
106. Björk I, Ylinenjärvi K, Olson ST, et al. Conversion of antithrombin from an inhibitor of thrombin to a substrate with reduced heparin affinity and enhanced conformational stability by binding of a tetradecapeptide corresponding to the P1 to P14 region of the putative reactive bond loop of the inhibitor. J Biol Chem 1992; 267: 1976-1982.
107. Fish WW, Orre K and Björk I. The production of an inactive form of antithrombin through limited proteolysis by thrombin. FEBS Lett 1979; 98:103-106.
108. Olson ST. Heparin and ionic strength-dependent conversion of antithrombin III from an inhibitor to a substrate of α-thrombin. J Biol Chem 1985; 260:10153-10160.
109. Egeberg O. Inherited antithrombin deficiency causing thrombophilia. Thromb Diath Haemorrh 1965; 13:516-530.
110. Abildgaard U. Antithrombin and related inhibitors of blood coagulation. Recent Adv Blood Coag 1981; 3:151-173.
111. Vogel CN, Kingdon HS and Lundblad RL. Correlation of in vivo and in vitro inhibition of thrombin by plasma inhibitors. J Lab Clin Med 1979; 93:661-673.
112. Jesty J. The kinetics of inhibition of α-thrombin in human plasma. J Biol Chem 1986; 261:10313-10318.
113. Fuchs HE and Pizzo SV. Regulation of factor Xa in vitro in human and mouse plasma and in vivo in mouse. Role of the endothelium and plasma proteinase inhibitors. J Clin Invest 1983; 72:2041-2049.
114. Gitel SN, Medina VM and Wessler S. Inhibition of human activated factor X by antithrombin III and α_1-proteinase inhibitor in human plasma. J Biol Chem 1984; 259:6890-6895.
115. Fuchs HE, Trapp HG, Griffith MJ, et al. Regulation of Factor IX_a in vitro in human and mouse plasma and *in vivo* in the mouse. Role of the endothelium and the plasma proteinase inhibitors. J Clin Invest 1984; 73:1696-1703.
116. Lawson JH, Butenas S, Ribarik N, et al. Complex-dependent inhibition of factor VIIa by antithrombin and heparin. J Biol Chem 1993; 268:767-770.
117. Olson ST, Sheffer R and Francis AM. High molecular weight kininogen potentiates the heparin-accelerated inhibition of plasma kallikrein by antithrombin: Role for antithrombin in the regulation of kallikrein. Biochemistry 1993; 32:12136-12147.
118. Olson ST, Sheffer R and Shore JD. Parallel procoagulant and anticoagulant pathways for high molecular weight kininogen coagulant function. Agents Act Supp 1992; 38: 241-248.
119. Lane DA, Ireland H, Olds RJ, et al. Antithrombin III: A database of mutations. Thromb Haemostas 1991; 66:657-661.
120. Lane DA, Olds RJ, Boisclair M, et al. Antithrombin III mutation database: First update. Thromb Haemostas 1993; 70: 361-369.
121. Lane DA, Olds RJ, Conrad J, et al. Pleiotropic effects of antithrombin strand 1C substitution mutations. J Clin Invest 1992; 90:2422-2433.
122. Jacobsson KG and Lindahl U. Attempted determination of endogenous heparin in blood. Thromb Haemostas 1979; 42:84.
123. de Agostini A, Watkins SC, Slayter HS, et al. Localization of anticoagulantly active heparan sulfate proteoglycans in vascular endothelium: Antithrombin binding on cultured endothelial cells and perfused rat aorta. J Cell Biol 1990; 111:1293-1304.
124. Felsch JS and Owen WG. Endogenous antithrombin associated with microvascular endothelium. Quantitative analysis in perfused rat hearts. Biochemistry 1994; 33:818-822.
125. Marcum JA, McKenney JB and Rosenberg RD. Acceleration of thrombin-antithrombin complex formation in rat hindquarters via heparin-like molecules bound to the endothelium. J Clin Invest 1984; 74:341-350.
126. Lollar P, MacIntosh SC and Owen WG. Reaction of antithrombin III with thrombin bound to the vascular endothelium.

Analysis in a recirculating perfused rabbit heart preparation. J Biol Chem 1984; 259:4335-4338.

127. Bourin MC, Boffa MC, Björk I, et al. Functional domains of rabbit thrombomodulin. Proc Natl Acad Sci USA 1986; 83: 5924-5928.

128. Preissner KT, Delvos U and Müller-Berghaus G. Binding of thrombin to thrombomodulin accelerates inhibition of the enzyme by antithrombin III. Evidence for a heparin-independent mechanism. Biochemistry 1987; 26:2521-2528.

129. Bourin MC, Ohlin AK, Lane DA, et al. Relationship between anticoagulant activities and polyanionic properties of rabbit thrombomodulin. J Biol Chem 1988; 263:8044-8052.

130. Bourin MC, Lundgren-Åkerlund E and Lindahl U. Isolation and characterization of the glycosaminoglycan component of rabbit thrombomodulin proteoglycan. J Biol Chem 1990; 265:15424-15431.

131. Preissner KT, Koyama T, Müller D, et al. Domain structure of the endothelial cell receptor thrombomodulin as deduced from modulation of its anticoagulant functions. J Biol Chem 1990; 265:4915-4922.

132. Jakubowski HV, Kline MD and Owen WG. The effect of bovine thrombomodulin on the specificity of bovine thrombin. J Biol Chem 1986; 261:3876-3882.

133. Bourin MC. Effect of rabbit thrombomodulin in thrombin inhibition by antithrombin in the presence of heparin. Thromb Res 1989; 54:27-39.

134. Miletich JP, Jackson CM and Majerus PW. Properties of the factor Xa binding site on human platelets. J Biol Chem 1978; 253:6908-6916.

135. Stern DM, Nawroth PP, Handley D, et al. An endothelial cell-dependent pathway of coagulation. Proc Natl Acad Sci USA 1985; 82:2523-2527.

136. Lindhout T, Baruch D, Schoen P, et al. Thrombin generation and inactivation in the presence of antithrombin III and heparin. Biochemistry 1986; 25:5962-5969.

137. Schoen P and Lindhout T. The *in situ* inhibition of prothrombinase-formed human α-thrombin and meizothrombin (desF1) by antithrombin III and heparin. J Biol Chem 1987; 262:11268-11274.

138. Walker FJ and Esmon CT. The effect of prothrombin fragment 2 on the inhibition of thrombin by antithrombin III. J Biol Chem 1979; 254:5618-5622.

139. Arni RK, Padmanabhan K, Padmanabhan KP, et al. Structures of the noncovalent complexes of human and bovine prothrombin fragment 2 with human PPACK-thrombin. Biochemistry 1993; 32:4727-4737.

140. Marciniak E. Factor Xa inactivation by antithrombin III: Evidence for biological stabilization of factor Xa by factor V-phospholipid complex. Br J Haematol 1973; 24:391-400.

141. Barrowcliffe TW, Havercroft SJ, Kemball-Cook G, et al. The effect of Ca^{2+}, phospholipid, and factor V on the anti-(factor Xa) activity of heparin and its high-affinity oligosaccharides. Biochem J 1987; 243:31-37.

142. Schoen P, Lindhout T, Willems G, et al. Antithrombin III-dependent anti-prothrombinase activity of heparin and heparin fragments. J Biol Chem 1989; 264: 10002-10007.

143. Naski MC and Shafer JA. α-Thrombin-catalyzed hydrolysis of fibrin I: Alternative binding modes and the accessibility of the active site in fibrin I-bound α-thrombin. J Biol Chem 1990; 265:1401-1407.

144. Hogg PJ and Jackson CM. Fibrin monomer protects thrombin from inactivation by heparin-antithrombin III: Implications for heparin efficacy. Proc Natl Acad Sci USA 1989; 86:3619-3623.

145. Hogg PJ and Jackson CM. Heparin promotes the binding of thrombin to fibrin polymer. Quantitative characterization of a thrombin-fibrin polymer-heparin ternary complex. J Biol Chem 1990; 265:241-247.

146. Eisenberg PR, Siegel JE, Abendschein DR, et al. Importance of Factor-Xa in determining the procoagulant activity of whole-blood clots. J Clin Invest 1993; 91:1877-1883.

147. Lane DA. Heparin binding and neutralizing proteins. In: Lane DA, Lindahl U, eds. Heparin. Chemical and Biological Properties. Clinical Applications. London: Edward Arnold, 1989:363-392.

148. Barrowcliffe TW, Merton RE, Havercroft

SJ, et al. Low-affinity heparin potentiates the action of high-affinity heparin oligosaccharides. Thromb Res 1984; 34:125-133.
149. Young E, Prins M, Levine MN, et al. Heparin binding to plasma proteins, an important mechanism for heparin resistance. Thromb Haemostas 1992; 67:639-643.
150. de Boer HC, de Groot PG, Bouma BN, et al. Ternary vitronectin-thrombin-antithrombin III complexes in human plasma. Detection and mode of association. J Biol Chem 1993; 268:1279-1283.
151. Olson ST, Francis AM, Sheffer R, et al. Parallel mechanisms of high molecular weight kininogen action as a cofactor in kallikrein inactivation and prekallikrein activation reactions. Biochemistry 1993; 32:12148-12159.
152. Jochum M, Lander S, Heimburger N, et al. Effect of human granulocytic elastase on isolated human antithrombin III. Hoppe-Seyler's Z Physiol Chem 1981; 362: 103-112.
153. Jordan RE, Kilpatrick J and Nelson RM. Heparin promotes the inactivation of antithrombin by neutrophil elastase. Science 1987; 237:777-779.
154. Jordan RE, Nelson RM, Kilpatrick J, et al. Inactivation of human antithrombin by neutrophil elastase. Kinetics of the heparin-dependent reaction. J Biol Chem 1989; 264:10493-10500.
155. Mast AE, Enghild JJ, Nagase H, et al. Kinetics and physiologic relevance of the interaction of α_1-proteinase inhibitor, α_1-antichymotrypsin, and antithrombin III by matrix metalloproteinases-1 (tissue collagenase), -2 (72kDa gelatinase/type IV collagenase), and -3 (stromelysin). J Biol Chem 1991; 266:15810-15816.
156. Tollefsen DM, Majerus DW and Blank MK. Heparin cofactor II. Purification and properties of a heparin-dependent inhibitor of thrombin in human plasma. J Biol Chem 1982; 257:2162-2169.
157. Tollefsen DM and Pestka CA. Modulation of heparin cofactor II activity by histidine-rich glycoprotein and platelet factor 4. J Clin Invest 1985; 75:496-501.
158. Ragg H. A new member of the plasma protease inhibitor gene family. Nucl Acids Res 1986; 14:1073-1088.
159. Inhorn RC and Tollefsen DM. Isolation and characterization of a partial cDNA clone for heparin cofactor II. Biochem Biophys Res Commun 1986; 137:431-436.
160. Blinder MA, Marasa JC, Reynolds CH, et al. Heparin cofactor II: cDNA sequence, chromosome localization, restriction fragment length polymorphism, and expression in *Escherichia coli*. Biochemistry 1988; 27:752-759.
161. Church FC, Meade JB and Pratt CW. Structure-function relationships in heparin cofactor II: Spectral analysis of aromatic residues and absence of a role for sulfhydryl groups in thrombin inhibition. Arch Biochem Biophys 1987; 259:331-340.
162. Hortin G, Tollefsen DM and Strauss AW. Identification of two sites of sulfation of human heparin cofactor II. J Biol Chem 1986; 261:15827-15830.
163. Herzog R, Lutz S, Blin N, et al. Complete nucleotide sequence of the gene for human heparin cofactor II and mapping to chromosome band 22q11. Biochemistry 1991; 30:1350-1357.
164. Ragg H and Preibisch G. Structure and expression of the gene coding for the human serpin hLS2. J Biol Chem 1988; 263:12129-12134.
165. Ragg H, Ulshöfer T and Gerewitz J. On the activation of human leuserpin-2, a thrombin inhibitor, by glycosaminoglycans. J Biol Chem 1990; 265:5211-5218.
166. Blinder MA, Andersson TR, Abildgaard U, et al. Heparin cofactor II_{Oslo}. Mutation of Arg-189 to His decreases the affinity for dermatan sulfate. J Biol Chem 1989; 264:5128-5133.
167. Griffith MJ, Noyes CM, Tyndall JA, et al. Structural evidence for leucine at the reactive site of heparin cofactor II. Biochemistry 1985; 24:6777-6782.
168. Morii M and Travis J. Amino acid sequence at the reactive site of human α_1-antichymotrypsin. J Biol Chem 1983; 258: 12749-12752.
169. Church FC, Noyes CM and Griffith MJ. Inhibition of chymotrypsin by heparin cofactor II. Proc Natl Acad Sci USA 1985; 82:6431-6434.

170. Pratt CW, Tobin RB and Church FC. Interaction of heparin cofactor II with neutrophil elastase and cathepsin G. J Biol Chem 1990; 265:6092-6097.
171. Parker KA and Tollefsen DM. The protease specificity of heparin cofactor II. Inhibition of thrombin generated during coagulation. J Biol Chem 1985; 260:3501-3505.
172. Derechin VM, Blinder MA and Tollefsen DM. Substitution of arginine for Leu^{444} in the reactive site of heparin cofactor II enhances the rate of thrombin inhibition. J Biol Chem 1990; 265:5623-5628.
173. Pratt CW, Whinna HC and Church FC. A comparison of three heparin-binding serine proteinase inhibitors. J Biol Chem 1992; 267:8795-8801.
174. Scully MF, Ellis V and Kakkar VV. Pentosan polyphosphate: Activation of heparin cofactor II or antithrombin III according to molecular weight fractionation. Thromb Res 1986; 41:489-499.
175. Beguin SS, Dol F and Hemker HC. Influence of lactobionic acid on the kinetics of thrombin in human plasma. Sem Thromb Hemostas 1991; 17, Supp. 1:126-128.
176. Church FC, Meade JB, Treanor RE, et al. Antithrombin activity of fucoidan. The interaction with heparin cofactor II, antithrombin III, and thrombin. J Biol Chem 1989; 264:3618-3623.
177. Yamagishi R, Niwa M, Kondo S, et al. Purification and biological property of heparin cofactor II: activation of heparin cofactor II and antithrombin III by dextran sulfate and various glycosaminoglycans. Thromb Res 1984; 36:633-642.
178. Church FC, Pratt CW, Treanor RE, et al. Antithrombin action of phosvitin and other phosphate-containing polyanions is mediated by heparin cofactor II. FEBS Lett 1988; 237:26-30.
179. Church FC, Treanor RE, Sherill GB, et al. Carboxylate polyanions accelerate inhibition of thrombin by heparin cofactor II. Biochem Biophys Res Commun 1987; 148:362-368.
180. Hurst RE, Poon MC and Griffith MJ. Structure-activity relationships of heparin. Independence of heparin charge density and antithrombin-binding domains in thrombin inhibition by antithrombin and heparin cofactor II. J Clin Invest 1983; 72: 1042-1045.
181. Petitou M, Lormeau J-C, Perly B, et al. Is there a unique sequence in heparin for interaction with heparin cofactor II? Structural and biological studies of heparin-derived oligosaccharides. J Biol Chem 1988; 263:8685-8690.
182. Tollefsen DM, Sugimori T and Maimone MM. Effect of low molecular weight heparin preparations on the inhibition of thrombin by heparin cofactor II. Sem Thromb Hemostas 1990; 16 Suppl:66-70.
183. Maimone MM and Tollefsen DM. Structure of a dermatan sulfate hexasaccharide that binds to heparin cofactor II with high affinity. J Biol Chem 1990; 265:18263-18271.
184. Scully MF, Ellis V, Seno N, et al. Effect of oversulfated chondroitin and dermatan sulfate upon thrombin and factor Xa inactivation by antithrombin III or heparin cofactor II. Biochem J 1988; 254:547-551.
185. Van Deerlin VMD and Tollefsen DM. The N-terminal acidic domain of heparin cofactor II mediates the inhibition of α-thrombin in the presence of glycosaminoglycans. J Biol Chem 1991; 266:20223-20231.
186. Whinna HC, Blinder MA, Szewczyk M, et al. Role of lysine 173 in heparin binding to heparin cofactor II. J Biol Chem 1991; 266:8129-8135.
187. Blinder MA and Tollefsen DM. Site-directed mutagenesis of arginine 103 and lysine 185 in the proposed glycosaminoglycan binding site of heparin cofactor II. J Biol Chem 1990; 265:286-291.
188. Ragg H, Ulshöfer T and Gerewitz J. Glycosaminoglycan-mediated leuserpin-2/thrombin interaction. Structure-function relationships. J Biol Chem 1990; 265: 22386-22391.
189. Rogers SJ, Pratt CW, Whinna HC, et al. Role of thrombin exosites in inhibition by heparin cofactor II. J Biol Chem 1992; 267:3613-3617.
190. Sheehan JP, Wu Q, Tollefsen DM, et al. Mutagenesis of thrombin selectively modulates inhibition by serpins heparin cofactor II and antithrombin III. Interaction with the anion-binding exosite determines heparin cofactor II specificity. J Biol Chem

1993; 268:3639-3645.

191. Olson ST, Halvorson HR and Björk I. Quantitative characterization of the thrombin-heparin interaction. Discrimination between specific and nonspecific binding models. J Biol Chem 1991; 266:6342-6352.
192. Andersson TR, Larsen ML, Handeland GF, et al. Heparin cofactor II activity in plasma: application of an automated assay method to the study of a normal adult population. Scand J Haematol 1986; 36:96-102.
193. Bertina RM, van der Linden IK, Engesser L, et al. Hereditary heparin cofactor II deficiency and the risk of development of thrombosis. Thromb Haemostas 1987; 57:196-200.
194. Griffith MJ, Carraway T, White GC, et al. Heparin cofactor activities in a family with hereditary antithrombin III deficiency: evidence for a second heparin cofactor in human plasma. Blood 1983; 61:111-118.
195. Buchanan MR, Boneu B, Ofosu F, et al. The relative importance of thrombin inhibition and factor Xa inhibition to the antithrombotic effects of heparin. Blood 1985; 65:198.
196. Fernandez F, van Ryn J, Ofosu FA, et al. The haemorrhagic and antithrombotic effects of dermatan sulphate. Br J Haematol 1986; 64:309-317.
197. Carrie D, Caranobe C, Gabaig AM, et al. Effects of heparin, dermatan sulfate and of their association on the inhibition of venous thrombosis growth in the rabbit. Thromb Haemostas 1992; 68:637-641.
198. MacIntosh S and Owen WG. Regulation of the clearance and inhibition of intravascular thrombin. Bull Sanofi Thromb Res Found 1987; 1:8-18.
199. Pasche B, Swedenborg J, Frebelius S, et al. Heparin cofactor II significance for the inhibition of thrombin at the injured vessel wall. Thromb Res 1991; 62:409-419.
200. McGuire EA and Tollefsen DM. Activation of heparin cofactor II by fibroblasts and vascular smooth muscle cells. J Biol Chem 1987; 262:169-175.
201. Whinna HC, Choi HU, Rosenberg LC, et al. Interaction of heparin cofactor II with biglycan and decorin. J Biol Chem 1993; 268:3920-3924.
202. Andrew M, Mitchell L, Berry L, et al. An anticoagulant dermatan sulfate proteoglycan circulates in the pregnant woman and her fetus. J Clin Invest 1992; 89:321-326.
203. Hoffman M, Pratt CW, Brown RL, et al. Heparin cofactor II-proteinase reaction products exhibit neutrophil chemoattractant activity. Blood 1989; 73:1682-1685.
204. Church FC, Pratt CW and Hoffman M. Leukocyte chemoattractant peptides from the serpin heparin cofactor II. J Biol Chem 1991; 266:704-709.
205. Suzuki K, Nishioka J and Hashimoto S. Protein C inhibitor purification from human plasma and characterization. J Biol Chem 1983; 258:163-168.
206. España F, Estellés A, Fernández PJ, et al. Evidence for the regulation of urokinase and tissue type plasminogen activators by the serpin, protein C inhibitor, in semen and blood plasma. Thromb Haemostas 1993; 70:989-994.
207. Stump DC, Thienpont M and Collen D. Purification and characterization of a novel inhibitor of urokinase from human urine. Quantitation and preliminary characterization in plasma. J Biol Chem 1986; 261:12759-12766.
208. España F, Gilabert J, Estelles A, et al. Functionally active protein C inhibitor/plasminogen activator inhibitor-3 (PCI/PAI-3) is secreted in seminal vesicles, occurs at high concentrations in human seminal plasma and complexes with prostate-specific antigen. Thromb Res 1991; 64:309-320.
209. Laurell M, Christensson A, Abrahamsson P-A, et al. Protein C inhibitor in human body fluids. Seminal plasma is rich in inhibitor antigen deriving from cells throughout the male reproductive system. J Clin Invest 1992; 89:1094-1101.
210. Suzuki K, Deyashiki Y, Nishioka J, et al. Characterization of a cDNA for human protein C inhibitor. A new member of the plasma serine protease inhibitor superfamily. J Biol Chem 1987; 262:611-616.
211. Suzuki K. Protein C inhibitor. Methods Enzymol 1993; 222:385-399.
212. Meijers JCM and Chung DW. Organization of the gene coding for human protein C inhibitor (plasminogen activator inhibi-

tor-3). Assignment of the gene to chromosome 14. J Biol Chem 1991; 266: 15028-15034.
213. Billingsley GD, Walter MA, Hammond GL, et al. Physical mapping of four serpin genes: α_1-antitrypsin, α_1-antichymotrypsin, corticosteroid-binding globulin, and protein C inhibitor, within a 280-kb region on chromosome 14q32.1. Am J Hum Genet 1993; 52:343-353.
214. Phillips JE, Cooper ST, Potter EE, et al. Mutagenesis of recombinant protein C inhibitor reactive site residues alters target proteinase specificity. J Biol Chem 1994; 269:16696-16700.
215. Cooper ST and Church FC. Reactive site mutants of recombinant protein C inhibitor. Biochim Biophys Acta 1995; 1246: 29-33.
216. Geiger M, Huber K, Wojta J, et al. Complex formation between urokinase and plasma protein C inhibitor in vitro and in vivo. Blood 1989; 74:722-728.
217. Hermans JM, Jones R and Stone SR. Rapid inhibition of the sperm protease acrosin by protein C inhibitor. Biochemistry 1994; 33:5440-5444.
218. Meijers JCM, Kanters DHAJ, Vlooswijk RAA, et al. Inactivation of human plasma kallikrein and factor XIa by protein C inhibitor. Biochemistry 1988; 27:4231-4237.
219. Rezaie AR, Cooper ST, Church FC, et al. Protein C inhibitor is a potent inhibitor of the thrombin-thrombomodulin complex. J Biol Chem 1995; 270:(In Press).
220. Pratt CW and Church FC. Heparin binding to protein C inhibitor. J Biol Chem 1992; 267:8789-8794.
221. Potempa J, Shieh B-H and Travis J. Alpha-2-antiplasmin: A serpin with two separate but overlapping reactive sites. Science 1988; 241:699-670.
222. Ecke S, Geiger M, Resch I, et al. Inhibition of tissue kallikrein by protein C inhibitor. Evidence for identity of protein C inhibitor with the kallikrein binding protein. J Biol Chem 1992; 267:7049-7052.
223. Kuhn LA, Griffin JH, Fisher CL, et al. Elucidating the structural chemistry of glycosaminoglycan recognition by protein C inhibitor. Proc Natl Acad Sci USA 1990; 87:8506-8510.
224. Shirk RA, Elisen MGLM, Meijers JCM, et al. Role of the H helix in heparin binding to protein C inhibitor. J Biol Chem 1994; 269:28690-28695.
225. España F, Vicente V, Tabenero D, et al. Determination of plasma protein C inhibitor and of two activated protein C-inhibitor complexes in normals and in patients with intravascular coagulation and thrombotic disease. Thromb Res 1990; 59:593-608.
226. Minamikawa K, Wada H, Wakita Y, et al. Increased activated protein C-protein C inhibitor complex levels in patients with pulmonary embolism. Thromb Haemostas 1994; 71:192-194.
227. Heeb MJ, España F and Griffin JH. Inhibition and complexation of activated protein C by two major inhibitors in plasma. Blood 1989; 73:446-454.
228. Hermans JM and Stone SR. Interaction of activated protein C with serpins. Biochem J 1993; 295:239-245.
229. Radtke KP, Fernandez JA, Greengard JS, et al. Protein C inhibitor is expressed in tubular cells of human kidney. J Clin Invest 1994; 94:2117-2124.
230. Scott RW and Baker JB. Purification of human protease nexin. J Biol Chem 1983; 258:10439-10444.
231. Scott RW, Bergman BL, Bajpai A, et al. Protease nexin. Properties and a modified purification procedure. J Biol Chem 1985; 260:7029-7034.
232. Baker JB, Low DA, Simmer RL, et al. Protease-nexin: a cellular component that links thrombin and plasminogen activator and mediates their binding to cells. Cell 1980; 21:37-45.
233. Gronke RS, Bergman BL and Baker JB. Thrombin interaction with platelets. Influence of a platelet protease nexin. J Biol Chem 1987; 262:3030-3036.
234. Gloor S, Odink K, Guenther J, et al. A glia-derived neurite promoting factor with protease inhibitory activity belongs to the protease nexins. Cell 1986; 47:687-693.
235. McGrogan M, Kennedy J, Li MP, et al. Molecular cloning and expression of two forms of human protease nexin 1. Biotechnology 1988; 6:172-177.

236. Sommer J, Gloor SM, Rovelli GF, et al. cDNA sequence coding for a rat glia-derived nexin and its homology to members of the serpin superfamily. Biochemistry 1987; 26:6407-6410.

237. Evans DL, McGrogan M, Scott RW, et al. Protease specificity and heparin binding and activation of recombinant protease nexin I. J Biol Chem 1991; 266:22307-22312.

238. Carter RE, Cerosaletti KM, Burkin DJ, et al. The gene for the serpin thrombin inhibitor (PI7), protease nexin I, is located on human chromosome 2q33-q35 and on syntenic regions in the mouse and sheep genomes. Genomics 1995; 27:196-199.

239. Sommer J, Meyhack B, Rovelli G, et al. Synthesis of glia-derived nexin in yeast. Gene 1989; 85:453-459.

240. Stone SR, Brown-Luedi ML, Rovelli G, et al. Localization of the heparin-binding site of glia-derived nexin/protease nexin-1 by site-directed mutagenesis. Biochemistry 1994; 33:7731-7735.

241. Wallace A, Rovelli G, Hofsteenge J, et al. Effect of heparin on the glia-derived-nexin-thrombin interaction. Biochem J 1989; 257:191-196.

242. Stone SR and Hermans JM. Inhibitory mechanism of serpins. Interaction of thrombin with antithrombin and protease nexin 1. Biochemistry 1995; 34:5164-5172.

243. Farrell DH and Cunningham DD. Glycosaminoglycans on fibroblasts accelerate thrombin inhibition by protease nexin-1. Biochem J 1987; 245:543-550.

244. Rovelli G, Stone SR, Guidolin A, et al. Characterization of the heparin-binding site of glia-derived nexin/protease nexin-1. Biochemistry 1992; 31:3542-3549.

245. Low DA, Scott RW, Baker JB, et al. Cells regulate their mitogenic response to thrombin through release of protease nexin. Nature 1982; 298:476-478.

246. Bergman BL, Scott RW, Bajpai A, et al. Inhibition of tumor-cell-mediated extracellular matrix destruction by a fibroblast proteinase inhibitor, protease nexin I. Proc Natl Acad Sci USA 1986; 83:996-1000.

247. Low DA, Baker JB, Koonce WC, et al. Released protease-nexin regulates cellular binding, internalization, and degradation of serine proteases. Proc Natl Acad Sci USA 1981; 78:2340-2344.

248. Conese M, Olson D and Blasi F. Protease nexin-1-urokinase complexes are internalized and degraded through a mechanism that requires both urokinase receptor and α_2-macroglobulin receptor. J Biol Chem 1994; 269:17886-17892.

249. Farrell DH and Cunningham DD. Human fibroblasts accelerate the inhibition of thrombin by protease nexin. Proc Natl Acad Sci USA 1986; 83:6858-6862.

250. Wagner SL, Lau AL and Cunningham DD. Binding of protease nexin-1 to the fibroblast surface alters its target protease specificity. J Biol Chem 1989; 264:611-615.

251. Donovan FM, Vaughan PJ and Cunningham DD. Regulation of protease nexin-1 target protease specificity by collagen type IV. J Biol Chem 1994; 269:17199-17205.

252. Guttridge DC, Lau AL and Cunningham DD. Protease nexin-1, a thrombin inhibitor, is regulated by interleukin-1 and dexamethasone in normal human fibroblasts. J Biol Chem 1993; 268:18966-18974.

253. Vaughn PJ and Cunningham DD. Regulation of protease nexin-1 synthesis and secretion in cultured brain cells by injury-related factors. J Biol Chem 1993; 268: 3720-3727.

254. Guenther J, Nick H and Monard D. A glia-derived neurite-promoting factor with protease inhibitory activity. EMBO J 1985; 4:1963-1966.

255. Loret C, Sensebrenner M and Labourdette G. Differential phenotypic expression induced in cultured rat astroblasts by acidic fibroblast growth factor, epidermal growth factor and thrombin. J Biol Chem 1989; 264:8319-8327.

256. Cavanaugh K, Gurwitz D, Cunningham DD, et al. Reciprocal modulation of astrocyte stellation by thrombin and protease nexin-1. J Neurochem 1990; 54:1735-1743.

257. Monard D. Cell-derived proteases and protease inhibitors as regulators of neurite outgrowth. Trends Neurosci 1988; 11:541-544.

258. Houenou LJ, Turner PL, Li L, et al. A serine protease inhibitor, protease nexin I, rescues motoneurons from naturally occurring and axotomy-induced cell death. Proc Natl Acad

Sci USA 1995; 92:895-899.
259. Wagner SL, Geddes JW, Corman CW, et al. Protease nexin-1, an antithrombin with neurite outgrowth activity, is reduced in Alzheimer disease. Proc Natl Acad Sci USA 1989; 86:8284-8288.
260. Kazama Y, Komiyama Y and Kisiel W. Tissue factor pathway inhibitor and protease nexin-1 are major factor Xa binding proteins on the HepG2 cell surface. Blood 1993; 81:676-682.
261. Vassalli J-D, Huarte J, Bosco D, et al. Protease-nexin I as an androgen-dependent secretory product of the murine seminal vesicle. EMBO J 1993; 12:1871-1878.
262. Hermans JM, Monard D, Jones R, et al. Inhibition of acrosin by serpins. A suicide substrate mechanism. Biochemistry 1995; 34:3678-3685.
263. Coleman PL, Barovski PA and Gelehrter TD. The dexamethasone-induced inhibitor of fibrinolytic activity in hepatoma cells. J Biol Chem 1982; 257:4260-4264.
264. Loskutoff DJ, van Mourik JA, Erickson LA, et al. Detection of an unusually stable fibrinolytic inhibitor produced in bovine endothelial cells. Proc Natl Acad Sci USA 1983; 80:2956-2960.
265. van Mourik JA, Lawrence DA and Loskutoff DJ. Purification of an inhibitor of plasminogen activator (antiactivator) synthesized by endothelial cells. J Biol Chem 1984; 259:14914-14921.
266. Schleef RR and Loskutoff DJ. Fibrinolytic system of vascular endothelial cells. Role of plasminogen activator inhibitors. Haemostasis 1988; 18:328-341.
267. Erickson LA, Ginsburg MH and Loskutoff DJ. Detection and partial characterization of an inhibitor of plasminogen activator in human platelets. J Clin Invest 1984; 74:1465-1472.
268. Booth NA, Simpson AJ, Croll A, et al. Plasminogen activator inhibitor (PAI-1) in plasma and platelets. Br J Haematol 1988; 70:327-333.
269. Juhan-Vague I, Moerman B, DeCock F, et al. Plasma levels of a specific inhibitor of tissue-type plasminogen activator (and urokinase) in normal and pathological conditions. Thromb Res 1984; 33:523-530.
270. Ginsburg D, Zeheb R, Yang AY, et al. cDNA cloning of human plasminogen activator-inhibitor from endothelial cells. J Clin Invest 1986; 78:1673-1680.
271. Ny T, Sawdey M, Lawrence D, et al. Cloning and sequence of a cDNA coding for the human β-migrating endothelial-cell-type plasminogen activator inhibitor. Proc Natl Acad Sci USA 1986; 83:6776-6780.
272. Pannekoek H, Veerman H, Lambers H, et al. Endothelial plasminogen activator inhibitor (PAI): a new member of the serpin gene family. EMBO J 1986; 5:2539-2544.
273. Andreasen PA, Riccio A, Welinder KG, et al. Plasminogen activator inhibitor type-1: reactive center and amino-terminal heterogeneity determined by protein and cDNA sequencing. FEBS Lett 1986; 209:213-218.
274. Alessi MC, Declerck PJ, De Mol M, et al. Purification and characterization of natural and recombinant human plasminogen activator inhibitor 1 (PAI 1). Eur J Biochem 1988; 175:531-540.
275. Lawrence DA, Strandberg L, GrundströmT, et al. Purification of active human plasminogen activator inhibitor-1 from *Escherichia coli*. Comparison with natural and recombinant forms purified from eucaryotic cells. Eur J Biochem 1989; 186:523-533.
276. Ehrlich HJ, Gebbink RK, Keijer J, et al. Alteration of serpin specificity by a protein cofactor. Vitronectin endows plasminogen activator inhibitor 1 with thrombin inhibitory properties. J Biol Chem 1990; 265: 13029-13035.
277. Shubeita HE, Cottey TL, Franke AE, et al. Mutational and immunochemical analysis of plasminogen activator inhibitor 1. J Biol Chem 1990; 265:18379-18385.
278. Lawrence DA and Loskutoff DJ. Inactivation of plasminogen activator inhibitor by oxidants. Biochemistry 1986; 25: 6351-6355.
279. Sakata Y, Curriden S, Lawrence D, et al. Activated protein C stimulates the fibrinolytic activity of cultured endothelial cells and decreases antiactivator activity. Proc Natl Acad Sci USA 1985; 82:1121-1125.
280. van Hinsberg VWM, Bertina RM, van Wijngaarden A, et al. Activated protein C decreases plasminogen-activator-inhibitor

activity in endothelial cell conditioned medium. Blood 1985; 65:444-451.

281. Fay WP and Owen WG. Platelet plasminogen activator inhibitor: Purification and characterization of interaction with plasminogen activators and activated protein C. Biochemistry 1989; 28:5773-5778.
282. Keijer J, Linders M, Wegman JJ, et al. On the target specificity of plasminogen activator inhibitor 1: Role of heparin, vitronectin, and the reactive site. Blood 1991; 78: 1254-1261.
283. Berrettini M, Schleef RR, España F, et al. Interaction of type 1 plasminogen activator inhibitor with the enzymes of the contact activation system. J Biol Chem 1989; 264:11738-11743.
284. Naski MC, Lawrence DA, Mosher DF, et al. Kinetics of inactivation of alpha-thrombin by plasminogen activator inhibitor-1—comparison of the effects of native and urea-treated forms of vitronectin. J Biol Chem 1993; 268:12367-12372.
285. Keijer J, Ehrlich HJ, Linders M, et al. Vitronectin governs the interaction between plasminogen activator inhibitor-1 and tissue-type plasminogen activator. J Biol Chem 1991; 266:10700-10707.
286. York JD, Li P and Gardell SJ. Combinatorial mutagenesis of the reactive site region in plasminogen activator inhibitor 1. J Biol Chem 1991; 266:8495-8500.
287. Sherman PM, Lawrence DA, Yang AY, et al. Saturation mutagenesis of the plasminogen activator inhibitor-1 reactive center. J Biol Chem 1992; 267:7588-7595.
288. Sherman PM, Lawrence DA, Verhamme IM, et al. Identification of tissue-type plasminogen activator-specific plasminogen activator inhibitor-1 mutants. Evidence that second sites of interaction contribute to target specificity. J Biol Chem 1995; 270:9301-9306.
289. Madison EL, Goldsmith EJ, Gerard RD, et al. Amino acid residues that affect interaction of tissue-type plasminogen activator with plasminogen activator inhibitor 1. Proc Natl Acad Sci USA 1990; 87:3530-3533.
290. Adams DS, Griffin LA, Nachajko WR, et al. A synthetic DNA encoding a modified human urokinase resistant to inhibition by serum plasminogen activator inhibitor. J Biol Chem 1991; 266:8476-8482.
291. Horrevoets AJG, Tans G, Smilde AE, et al. Thrombin-variable region 1 (VR1) evidence for the dominant contribution of VR1 of serine proteases to their interaction with plasminogen activator. J Biol Chem 1993; 268:779-782.
292. De Serrano VS and Castellino FJ. Structural determinants of the noncatalytic chain of tissue-type plasminogen activator that modulate its association rate with plasminogen activator inhibitor 1. J Biol Chem 1990; 265:10473-10478.
293. Hekman CM and Loskutoff DJ. Endothelial cells produce a latent inhibitor of plasminogen activators that can be activated by denaturants. J Biol Chem 1985; 260:11581-11587.
294. Mottonen J, Strand A, Symersky J, et al. Structural basis of latency in plasminogen activator inhibitor 1. Nature 1992; 355: 270-273.
295. Lawrence DA, Strandberg L, Ericson J, et al. Structure-function studies of the SERPIN plasminogen activator inhibitor type 1. Analysis of chimeric strained loop mutants. J Biol Chem 1990; 265:20293-20301.
296. Lawrence DA, Olson ST, Palaniappan S, et al. Engineering plasminogen activator inhibitor 1 mutants with increased functional stability. Biochemistry 1994; 33: 3643-3648.
297. Tucker HM, Mottonen J, Goldsmith EJ, et al. Engineering of plasminogen activator inhibitor-1 to reduce the rate of latency transition. Nature Struct Biol 1995; 2:442-445.
298. Lambers JWJ, Cammenga M, Konig BW, et al. Activation of human endothelial cell-type plasminogen activator inhibitor (PAI-1) by negatively charged phospholipids. J Biol Chem 1987; 262:17492-17496.
299. Declerck PJ, De Mol M, Alessi MC, et al. Purification and characterization of a plasminogen activator inhibitor 1 binding protein from human plasma. Identification as a multimeric form of S protein (vitronectin). J Biol Chem 1988; 263:15454-15461.
300. Berkenpas MB, Lawrence DA and Ginsburg D. Molecular evolution of plasminogen ac-

tivator inhibitor-1 functional stability. EMBO J 1995; 14:2969-2977.
301. Seiffert D and Loskutoff DJ. Evidence that type-1 plasminogen activator inhibitor binds to the somatomedin B domain of vitronectin. J Biol Chem 1991; 266:2824-2830.
302. Salonen EM, Vaheri A, Pollanen J, et al. Interaction of plasminogen activator inhibitor (PAI-1) with vitronectin. J Biol Chem 1989; 264:6339-6343.
303. Lawrence DA, Berkenpas MB, Palaniappan S, et al. Localization of vitronectin binding domain in plasminogen activator inhibitor-1. J Biol Chem 1994; 269:15223-15228.
304. Van Meijer M, Gebbink RK, Preissner KT, et al. Determination of the vitronectin binding site on plasminogen activator inhibitor 1 (PAI-1). FEBS Lett 1994; 352:342-346.
305. Padmanabhan J and Sane DC. Localization of a vitronectin binding region of plasminogen activator inhibitor-1. Thromb Haemost 1995; 73:829-834.
306. Ehrlich HJ, Keijer J, Preissner KT, et al. Functional interaction of plasminogen activator inhibitor type 1 (PAI-1) and heparin. Biochemistry 1991; 30:1021-1028.
307. Gebbink RK, Reynolds CH, Tollefsen DM, et al. Specific glycosaminoglycans support inhibition of thrombin by plasminogen activator inhibitor 1. Biochemistry 1993; 32:1675-1680.
308. Patston PA and Schapira M. Low-affinity heparin stimulates the inactivation of plasminogen activator inhibitor-1 by thrombin. Blood 1994; 84:1164-1172.
309. Ehrlich HJ, Gebbink RK, Keiijer J, et al. Elucidation of structural requirements on plasminogen activator inhibitor-1 for binding to heparin. J Biol Chem 1992; 267: 11606-11611.
310. Carmeliet P, Kieckens L, Schoonjans L, et al. Plasminogen activator inhibitor 1 gene-deficient mice. I. Generation by homologous recombination and characterization. J Clin Invest 1993; 92:2746-2755.
311. Carmeliet P, Stassen JM, Schoonjans L, et al. Plasminogen activator inhibitor-1 gene-deficient mice. II. Effects on hemostasis, thrombosis, amd thrombolysis. J Clin Invest 1993; 92:2756-2760.
312. Fay WP, Shapiro AD, Shih JL, et al. Complete deficiency of plasminogen activator inhibitor type 1 due to frame shift mutation. N Engl J Med 1992; 327:1729-1733.
313. Hamsten A, Wiman B, de Faire U, et al. Increased plasma levels of a rapid inhibitor of tissue plasminogen activator in young survivors of myocardial infarction. N Engl J Med 1985; 313:1557-1563.
314. Wagner OF, deVries C, Hohmann C, et al. Interaction between plasminogen-activator inhibitor type-1 (PAI-1) bound to fibrin and either tissue-type plasminogen-activator (t-PA) or urokinase-type plasminogen activator (u-PA)—binding of t-PA/PAI-1 complexes to fibrin mediated by both the finger and the kringle-2 domains of t-PA. J Clin Invest 1989; 84:647-655.
315. Reilly CF and Hutzelmann JE. Plasminogen activator inhibitor-1 binds to fibrin and inhibits tissue-type plasminogen activator-mediated fibrin dissolution. J Biol Chem 1992; 267:17128-17135.
316. Stringer HAR and Pannekoek H. The significance of fibrin binding by plasminogen activator inhibitor 1 for the mechanism of tissue-type plasminogen activator-mediated fibrinolysis. J Biol Chem 1995; 270: 11205-11208.
317. Levi M, Biemond BJ, van Zonneveld AJ, et al. Inhibition of plasminogen activator inhibitor-1 results in promotion of endogenous thrombolysis and inhibition of thrombus extension in models of experimental thrombosis. Circulation 1992; 85:305-312.
318. Bar-Shavit R, Eldor A and Vlodavsky I. Binding of thrombin to subendothelial extracellular matrix. Protection and expression of functional properties. J Clin Invest 1989; 84:1096-1104.
319. Ehrlich HJ, Gebbink RK, Preissner KT, et al. Thrombin neutralizes plasminogen activator inhibitor 1 (PAI-1) that is complexed with vitronectin in the endothelial cell matrix. J Cell Biol 1991; 115:1773-1781.
320. Mimuro J and Loskutoff DJ. Binding of type 1 plasminogen activator inhibitor to the extracellular matrix of cultured bovine endothelial cells. J Biol Chem 1989; 264:5058-5063.

321. Owensby DA, Morton PA, Wun TC, et al. Binding of plasminogen activator inhibitor type-1 to extracellular matrix of Hep G2 cells. Evidence that the binding protein is vitronectin. J Biol Chem 1991; 266: 4334-4340.
322. Chain D, Kreizman T, Shapira H, et al. Plasmin cleavage of vitronectin. Identification of the site and consequent attenuation in binding plasminogen activator inhibitor-1. FEBS Lett 1991; 285:251-256.
323. Kost C, Stüber W, Ehrlich HJ, et al. Mapping of binding sites for heparin, plasminogen activator-1, and plasminogen to vitronectin's heparin-binding region reveals a novel vitronectin-binding feedback mechanism for the control of plasmin formation. J Biol Chem 1992; 267:12098-12105.
324. Cajot JJ, Bamat J, Bergonzelli E, et al. Plasminogen activator inhibitor type 1 is a potent natural inhibitor of extracellular matrix degradation by fibrosarcoma and colon carcinoma cells. Proc Natl Acad Sci USA 1990; 87:6939-6943.
325. Sawdey M, Podor TJ and Loskutoff DJ. Regulation of type 1 plasminogen activator gene expression in cultured bovine aortic endothelial cells. Induction by transforming growth factor-β, lipopolysaccharide, and tumor necrosis factor-α. J Biol Chem 1989; 264:10396-10401.
326. Pixley RA, Schapira M and Colman RW. The regulation of human factor XIIa by plasma proteinase inhibitors. J Biol Chem 1985; 260:1723-1729.
327. Olson ST and Björk I. Mechanism of action of heparin and heparin-like antithrombotics. Persp Drug Discovery Design 1994; 1:479-501.
328. Kjellén L and Lindahl U. Proteoglycans: Structures and interactions. Annu Rev Biochem 1991; 60:443-475.
329. Pratt CW and Church FC. General features of the heparin-binding serpins antithrombin, heparin cofactor-II and protein-C inhibitor. Blood Coag Fibrinolysis 1993; 4:479-490.
330. Bode W, Mayr I, Baumann U, et al. The refined 1.9Å crystal structure of human α-thrombin: Interaction with D-Phe-Pro-Arg chloromethylketone and significance of the Tyr-Pro-Pro-Trp insertion segment. EMBO J 1989; 8:3467-3475.
331. Mulloy B, Forster MJ, Jones C, et al. NMR and molecular-modelling studies of the solution conformation of heparin. Biochem J 1993; 293:849-858.

CHAPTER 5

Other Inhibitory Serpins

The previous chapter dealt with the properties of the glycosaminoglycan-activatable serpins from human plasma, namely, antithrombin, heparin cofactor II, activated protein C inhibitor, protease nexin 1 and PAI-1. This chapter deals with the other inhibitory serpins found in human plasma (α_1-proteinase inhibitor, α_1-antichymotrypsin, α_2-antiplasmin, C1-inhibitor, PAI-2, kallistatin), as well as those found in rodents, intracellular locations, insects, schistosomes, plants and viruses. Some of these serpins have been mentioned in previous chapters, particularly in regard to structure and mechanism. This chapter provides an introduction to the extensive literature available on these serpins and references are made to both review articles and original papers. A number of the serpins described have been cloned and recombinant proteins produced. A listing of recombinant serpins (both wild type and variant) and a synopsis of the major results can be found in Patston and Gettins.[1] The biosynthesis and clinical significance of selected serpins are dealt with in Chapters 7 and 8 respectively.

1. α_1-PROTEINASE INHIBITOR

α_1-Proteinase inhibitor is also known as α_1-antitrypsin because it was originally identified as an inhibitor of pancreatic trypsin.[2] This is a misnomer given that α_1-proteinase inhibitor can inhibit a number of serine proteinases, although the most important physiological target for α_1-proteinase inhibitor is neutrophil elastase.[3] This is emphasized by the fact that deficiency of α_1-proteinase inhibitor can result in emphysema (see Chapter 8). The second order rate constant for the inhibition of neutrophil elastase is 6.5×10^7 $M^{-1}s^{-1}$,[3] although this can be reduced to 4.6×10^4 $M^{-1}s^{-1}$ in the presence of heparin[4] (see Table 8.2 for rate constants with other serine proteinases). Recently it has been shown that α_1-proteinase inhibitor can inhibit serine proteinases of the structurally-distinct subtilisin class.[5] Human α_1-proteinase inhibitor is the most abundant plasma serpin and usually circulates in plasma at a concentration of between 1.0 and 2.0 mg/ml.

The molecular weight of α_1-proteinase inhibitor on SDS-PAGE is 52 kDa. It has 394 amino acids with the P1 residue being Met 358. This methionine is susceptible to oxidation, which reduces the activity of the inhibitor (see Chapter 8). The amino acid numbering of

α_1-proteinase inhibitor is the convention for numbering of all serpins. Thus the number of a residue in a particular serpin is compared back to the equivalent residue in α_1-proteinase inhibitor, in a manner analogous to the serine proteinase numbering being based on the residue numbering in chymotrypsin. The P1 residue of methionine determines the specificity for elastase. Changing the methionine to arginine alters the specificity to thrombin as is described for the α_1-proteinase inhibitor-Pittsburgh mutant in Chapter 8. α_1-Proteinase inhibitor was the first serpin structure to be determined by X-ray crystallography (see Chapter 2).[6] It has also been shown that latent and polymeric forms of the protein can be made.[7-9] Polymer formation is associated with the pathologic Z variant of α_1-proteinase inhibitor (see Chapter 8). The structure of the latent form is presumed to be similar to that of latent PAI-1. The structure of the polymers is not known but the two models for the structure involve insertion of the reactive center loop into either β-sheet A or β-sheet C. There are three Asn-linked carbohydrates at positions 46, 83 and 247. The structure of the carbohydates may vary such that the biantennary oligosaccharides at positions 46 and 83 can be replaced by triantennary oligosaccharides. There is a single cysteine residue at position 232 which can form disulfides with glutathione or free cysteine. The redox state of this thiol has been suggested to regulate the inhibitory activity[10] and α_1-proteinase inhibitor has been shown to bind to extracellular matrix components perhaps via this thiol.[11] The cDNA has been sequenced[12,13] and the gene localized to chromosome 14q32.1.[14] This locus also contains the genes for the serpins α_1-antichymotrypsin, corticosteroid binding globulin, protein C inhibitor and an α_1-proteinase inhibitor-like pseudogene.[15,16] The α_1-proteinase inhibitor gene has four introns, which are in the same positions as the introns in the angiotensinogen and α_1-antichymotrypsin genes.[17,18] A feature of α_1-proteinase inhibitor is that the reactive center loop is highly susceptible to inactivating proteolysis by non-target enzymes. In particular a neutrophil metalloproteinase, tissue collagenase (MMP-1), type IV collagenase (MMP-2) and stromelysin (MMP-3) can all inactivate α_1-proteinase inhibitor, reactions which might have significance in inflammation.[19-24] Further information on the properties of α_1-proteinase inhibitor can be found in references 25-28.

The human protein is the best studied however α_1-proteinase inhibitor has also been described in baboon,[12] monkey,[29] sheep,[30] goat,[31] horse,[32-34] donkey,[35] pig,[36] dog,[37] rabbit,[38] chinchilla,[39] opossum,[40] rat,[41,42] mouse,[43] guinea pig,[44] mink,[45] mongolian gerbil,[46] chipmunk,[47] wallaby,[48] fat tailed insectivore *Smithopsis crassicaudata,*[49] eastern barred bandicoot,[50] and carp.[51,52] A detailed comparison of mammalian α_1-proteinase inhibitors can be found in Patterson[53] and further details of rodent α_1-proteinase inhibitors are given below.

Human α_1-proteinase inhibitor has been shown to be highly polymorphic, with over 75 different alleles identified.[26-28,54] Although they are often the cause of α_1-proteinase inhibitor deficiency, some of the alleles represent silent DNA variations. Some aspects of α_1-proteinase inhibitor deficiency are described in Chapter 8. Before the advent of DNA sequence analysis the routine method to screen for α_1-proteinase inhibitor allelic variants was by isoelectric focusing. The nomenclature for α_1-proteinase inhibitor variants was standardized at a meeting in Rouen in 1978.[55] The lettering for the variants was originally based on mobility in acid starch electrophoretic gels, namely F (fast), M (medium), S (slow) and Z (the most cathodal).[56] The most common normal allele found in 70% of North American Caucasians is termed M1. Other normal alleles which vary by one or two base changes respectively are M2 and M3, which make up 10% each of this population (see Table 8.1). There are null alleles where there is no detectable protein and a large number of alleles which lead to dysfunctional protein. Another important feature of α_1-proteinase inhibitor is that it is an acute phase reactant. Thus, during an inflammatory response the plasma levels can increase 3- or 4-fold. Further information on α_1-proteinase inhibitor biosynthesis can be found in Chapter 7.

2. α_1-ANTICHYMOTRYPSIN

As with α_1-proteinase inhibitor, α_1-antichymotrypsin has been misnamed. The true target for this serpin is likely to be neutrophil cathepsin G, which is inhibited with a rate constant of 5.1 x 10^7 $M^{-1}s^{-1}$.[3] The physiological function is not known however, although the plasma concentration (usually 0.5 mg/mL) increases 4-fold within hours of an inflammatory stimulus.[57] α_1-Antichymotrypsin is also implicated in the pathogenesis of Alzheimer's disease (discussed in Chapter 8). In addition, α_1-antichymotrypsin-proteinase complexes have some unique bioactive properties (discussed in Chapter 7). The cDNA for α_1-antichymotrypsin has been sequenced[58,59] and the structure of the gene determined and shown to be homologous to that of α_1-proteinase inhibitor.[17] α_1-Antichymotrypsin has 398 amino acids. It has 45% sequence homology to α_1-proteinase inhibitor and 29% homology to ovalbumin. The molecular weight on SDS-PAGE is 65 kDa. There are four probable carbohydrate attachment sites at Asn 70, 104, 163 and 247. There is one cysteine residue at position 238.[59,60] The P1 residue is Leu 358, which is consistent with the specificity for chymotrypsin-like enzymes. The α_1-antichymotrypsin gene maps to chromosome 14 and is closely linked to the genes for α_1-proteinase inhibitor, protein C inhibitor and cortisol binding globulin.[15] The gene has four introns and is similar to the gene for α_1-proteinase inhibitor and angiotensinogen. α_1-Antichymotrypsin also has a DNA binding site (see Chapter 2). The cDNA of bovine α_1-antichymotrypsin has been cloned and shown to exist in a number of isoforms.[61] Two serpins related to α_1-antichymotrypsin have been isolated from bovine plasma and milk.[62] Two serpins have also been purified from goat plasma and shown by N-terminal sequencing to be homologous to α_1-antichymotrypsin. However, one was a trypsin inhibitor and one was an elastase inhibitor.[63] The trypsin inhibitor was named contrapsin which also describes the α_1-antichymotrypsin-like serpins in rodent plasma (see below). A porcine α_1-antichymotrypsin has also been identified.[64] An α_1-antichymotrypsin-like serpin has been found in bovine adrenal medulla and pituitary which inhibits the prohormone thiol protease, an enzyme responsible for the processing of enkephalin in chromaffin granules.[65] This α_1-antichymotrypsin-like serpin cross-reacted with antisera to human plasma α_1-antichymotrypsin. Human plasma α_1-antichymotrypsin also inhibited this enzyme, thereby making it another member of the serpin family that can inhibit cysteine proteinases. Further study has shown that there are specific bovine adrenal medulla and pituitary isoforms of α_1-antichymotrypsin which, in contrast to the human plasma protein, have arginine at P1.[66] Moreover, these isoforms are different from the bovine liver isoforms of α_1-antichymotrypsin that have been previously identified.[61]

3. α_2-ANTIPLASMIN

α_2-Antiplasmin (α_2-plasmin inhibitor) is the primary inhibitor of plasmin, the enzyme responsible for the degradation of fibrin. The second order rate constant for the inhibition of plasmin is 3.8 x 10^7 $M^{-1}s^{-1}$.[67] α_2-Antiplasmin consists of 452 residues, as determined from protein and DNA sequencing, with the P1-P1' bond being Arg 364-Met 365.[68-72] The gene has 10 exons and 9 introns distributed within 16 kilobases and is located on chromosome 17p13.[71,73] There are four potential asparagine-linked glycosylation sites at positions 87, 256, 270 and 277. There are two disulfides connecting cysteines 64-104 and 31-113.[69,70] A form of α_2-antiplasmin with 12 additional residues at the amino terminus has also been isolated from plasma, and probably represents the full length protein when synthesized.[72,74,75] It has also been shown that chymotrypsin uses Met 365-Ser 366 as the P1-P1' bond and it was proposed that an alternative P1-P1' could also occur in other serpins.[75,76] The molecular weight is 67 kDa by SDS-PAGE and the plasma concentration is 70 μg/ml. Details on purification and assay of human α_2-antiplasmin can be found in Aoki et al.[77] Rat[78] and bovine[79] α_2-antiplasmin have also been identified.

α_2-Antiplasmin has been used to show that serpin proteinase complexes can be reversible. In the case of complexes with chymotrypsin and trypsin[80] this may indeed be the case (see Chapter 3). However, studies with plasmin[81,82] have failed to consider that cleaved inactive serpin and active enzyme can be generated either through the substrate pathway (k_3 in Fig. 3.2) or by slow turnover of stable complex (k_5).

Two important features of α_2-antiplasmin are the plasminogen binding site and the fibrin crosslinking site. The plasminogen binding site consists of a series of lysine residues located within a 26 amino acid region in the carboxyl terminal end of α_2-antiplasmin, which bind to lysine binding sites in the kringle domains of plasminogen. Synthetic peptides corresponding to various parts of the carboxyl terminal domain of α_2-antiplasmin inhibit the reaction between plasmin and α_2-antiplasmin.[83,84] α_2-Antiplasmin binds to plasminogen at the same sites as fibrin. Consequently α_2-antiplasmin is a competitive inhibitor of plasminogen binding to fibrin, and thereby delays fibrinolysis.[85,86] The converse is also true, as plasmin bound to fibrin or to cell surfaces has a reduced rate of inhibition by α_2-antiplasmin.[87-89] It has also been suggested that lipoprotein (a), which shares considerable homology with plasminogen, can modulate the inhibition of plasmin by α_2-antiplasmin.[90] Further evidence that the interaction between α_2-antiplasmin and the lysine binding site on plasmin is partly responsible for the rapid reaction between the two proteins is given by the finding that blocking this interaction with 6-aminohexanoic acid markedly reduce the inhibition rate.[67,81]

In addition to the C-terminal domain responsible for binding to plasminogen, α_2-antiplasmin has a binding site for fibrin. When blood clots, about 25% of the α_2-antiplasmin present in the plasma environment is cross-linked to the Aα-chain of fibrin. This reaction is catalyzed by the activated form of factor XIII.[91-94] The cross-linking site is on a glutamine, the second residue in the amino terminus of α_2-antiplasmin.[95] This mechanism prevents the degradation of the fibrin by plasmin and thereby plays an important role in clot stabilization.

A form of α_2-antiplasmin has been detected in plasma which does not have the plasminogen binding site due to partial degradation of the carboxy terminal domain. This form is a less efficient inhibitor of plasmin than the native form and represents about 30% of α_2-antiplasmin in plasma.[96,97]

4. C1-INHIBITOR

C1-inhibitor was named as a result of being the only known physiological inhibitor of the activated C1r and C1s serine proteinase subunits of the C1 component of the classical pathway of complement (Fig. 5.1). The rate constants for inhibition of C1r and C1s have been reported to be 2.8 x 10^3 and 1.2 x 10^4 $M^{-1}s^{-1}$ respectively,[98] which are relatively slow compared to other serpin-proteinase reactions but may reflect the limited substrate specificity of these proteinases. Using different methodology a rate for the inhibition of C1s of about 5 x 10^4 $M^{-1}s^{-1}$ has also been reported.[99] C1-inhibitor is also an important inhibitor of plasma kallikrein[100-103] and factor XIIa[104,105] which are inhibited with rate constants of 1.7 x 10^4 $M^{-1}s^{-1}$ and 3.1 x 10^3 $M^{-1}s^{-1}$ respectively (see also Table 8.2). C1-inhibitor can prevent the spontaneous activation of C1, presumably by inhibiting any C1r and C1s that forms in the absence of activator.[106,107] C1-inhibitor can also reverse the inhibition of immune precipitation caused by C1.[108] A

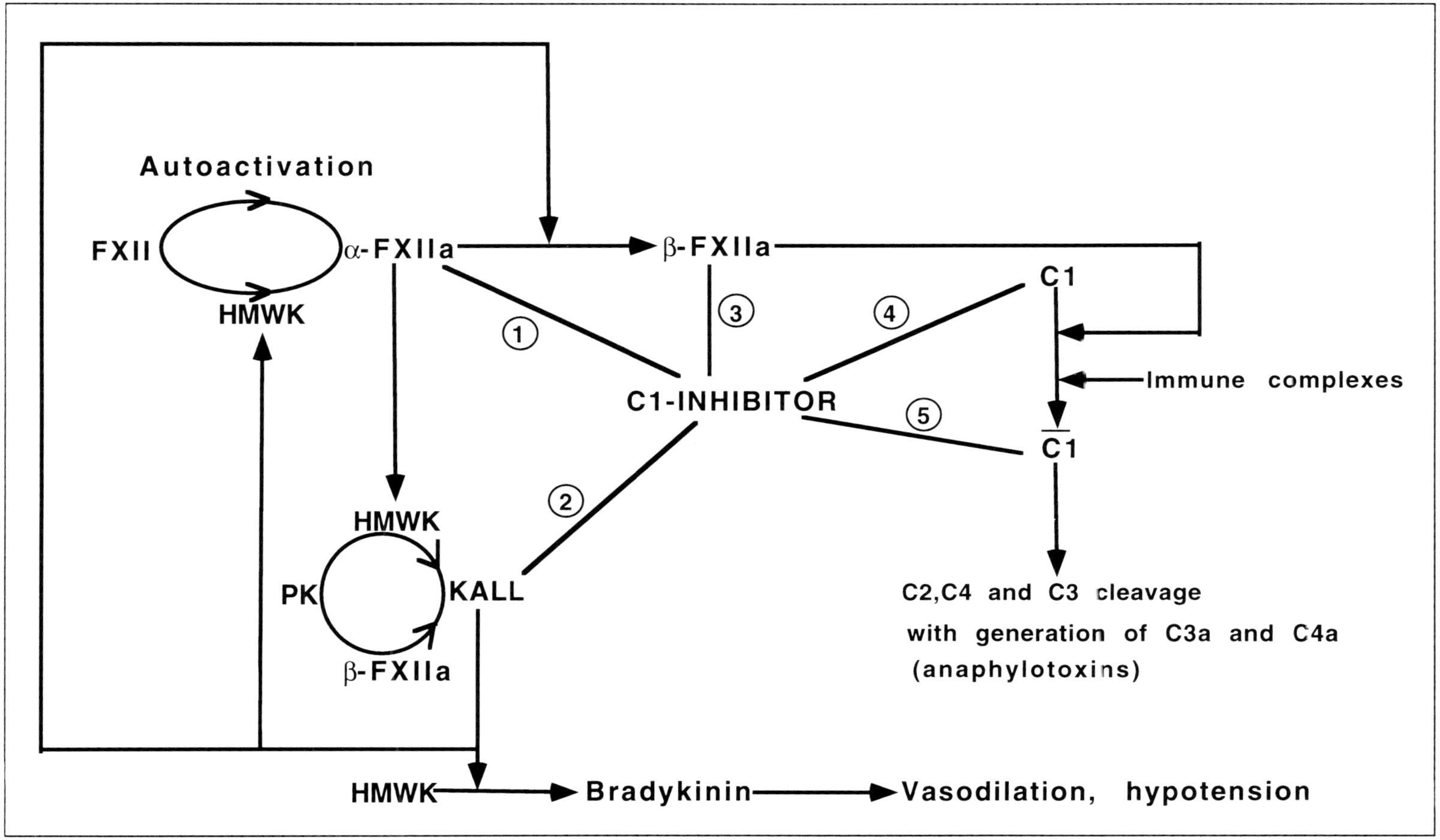

Fig. 5.1. The pathways controlled by C1-inhibitor. C1-inhibitor inactivates α-FXIIa (1), KALL (2), β-FXIIa (3), and the activated C1r and C1s subunits of C1 (5). C1 inhibitor also prevents the spontaneous activation of C1 in the absence of activator (4). Further details can be found in Kaplan et al,[254] De La Cadena et al[255] and Winkelstein and Colten.[256] Abbreviations: FXII, factor XII; α-FXIIa, activated factor XII; β-FXIIa, activated factor XII fragment; HMWK, high molecular weight kininogen; PK, plasma prekallikrein; KALL, activated plasma kallikrein; C1, macromolecular complement component C1 complex ($C1q$-$C1r_2$-$C1s_2$); $\overline{C1}$, activated C1.

number of reports have shown that heparin can enhance the activity of C1-inhibitor towards C1[98,99,109,110] although one report has shown that heparin can also cause an increase in the cleavage of C1-inhibitor by C1s.[111]

C1-inhibitor was first purified from serum by Pensky et al,[112] and subsequently a number of methods for purification from plasma have been described.[113-119] The reactive site P1-P1' bond was identified by protein chemistry as Arg-Thr,[120] and subsequently shown to correspond to residues 444-445 from the sequence of the cDNA and genomic DNA.[121-125] The gene for C1-inhibitor is on chromosome 11p11.2-q13, is 17 kb long and has eight exons.[121,122,125,126] The intron positions are not found in other serpins. The gene is unusual in containing 17 *Alu* repeat elements within the introns. An alternative P1 residue of Ala 443 has been identified for the reaction with chymotrypsin,[127] but it is not known if this also applies to other proteinases. A comprehensive review of all aspects of C1-inhibitor can be found in Davis.[128] In addition to the human protein, C1-inhibitor has been identified in guinea pig and rat.[129,130]

The molecular weight on SDS-PAGE of C1-inhibitor is about 105 kDa, but based on the amino acid sequence (478 residues) it would be expected to be 53 kDa.[121] There are at least 17 carbohydrate groups attached, 6 of which are N-linked at Asn 3, 47, 59, 216, 231 and 330. The remainder are O-linked. Ser 42 and Thr 26, 66, 70 and 74 are definitely glycosylated and Thr 49 and 61 probably are. The consensus glycosylation sequence Glx-Pro-Thr-Thr (or minor variations thereof) occurs seven times between residues 63 and 97, and although only three of these sites have definitively been shown to be occupied the others probably are.[121] Due to the heavily glycosylated nature of the protein the pI is 2.7.[113] Characterization of the carbohydrate composition and neutron scattering analysis have shown that the true molecular weight is closer to 76 kDa[131] and that higher values from electrophoresis are due to artifacts resulting from the glycosylation. Electron microscopy showed that C1-inhibitor is a two domain protein with a 'head and tail' like structure.[132] This asymmetry accounts for the hydrodynamic properties of the protein.[133] There are two disulfides connecting cysteine 101-406 and 108-183.[121] A unique feature of C1-inhibitor is that homology with other serpins does not begin until about residue 120.[121] Starting from the amino terminus are 50 residues which have no homology with any known protein and have no known function.[120,134] Residues 50-100 contain seven consensus O-linked glycosylation sites and consequently share homology to mucins. Extensive glycosylation within this region accounts for the elongated shape of this domain. A deletion mutant lacking residues 1-98 appears to have unaltered inhibitory function,[135] and desialylation of the protein was similarly without any obvious effects on activity.[136] It has been proposed that the amino terminal region blocks the interaction of C1r and C1s with C1q when these proteinases are in complex with C1-inhibitor, thereby reducing their affinity for C1q.[131,133] Alternatively, this domain may be important in mediating interactions of C1-inhibitor with the native C1 complex.[137,138]

Some naturally occurring mutants of C1-inhibitor (Ala 436 to Thr, Val 451 to Met, Pro 476 to Ser and Phe 455 to Ser) have been reported to undergo spontaneous polymerization, accounting for the C1-inhibitor deficiency in these individuals.[139,140] Polymerization can also readily occur as a result of heating or mild denaturation of the normal plasma protein.[141] The mechanism of polymerization is not known, but by analogy to the Z-variant of α_1-proteinase inhibitor and the dimeric crystal structures of antithrombin it is possible that it involves the interaction of the reactive center loop of one molecule with β-sheet A or C of another.

5. PLASMINOGEN ACTIVATOR INHIBITOR-2 (PAI-2)

Plasminogen activator inhibitor-2 (PAI-2) was originally isolated from human placenta and shown to be an inhibitor of tPA and uPA but to be distinct from PAI-1.[142,143] PAI-2 is elevated in pregnancy during the last trimester.[144] Expression in monocytes and U-937 cells has also been shown.[145-148] A number of groups have determined the sequence of the cDNA[147-150] and the structure of the gene.[151,152] The mRNA is 2.0 kb and encodes a protein of 415 amino acids, which has a deduced molecular weight of 46 kDa and P1-P1' residues of Arg-Thr (residues 380-381). There are also three potential N-glycosylation sites and 6 cysteines at positions 5, 79, 145, 161, 406 and 414. In spite of functional similarity to PAI-1 the gene is most closely related to that of chicken ovalbumin, with 6 of the 7 PAI-2 introns being in identical positions. The gene is on chromosome 18q21.2-22, adjacent to the genes for the serpins squamous cell carcinoma antigens 1 and 2 and maspin.[153] An interesting feature of both PAI-2 and ovalbumin is that they are secreted without a typical propeptide leader sequence. As this process is not entirely efficient, PAI-2 can also accumulate as an intracellular serpin as a product of the same mRNA.[154,155] Efficiency of secretion can be enhanced by point mutations which increase the hydrophobicity of two regions near the amino terminus.[156] Even though PAI-2 is elevated in pregnancy, the physiological role of this serpin is not known. There is, however, much evidence implicating the plasminogen activator/plasmin system in processes such as embryonic development, tissue remodeling and angiogenesis.[157] The finding that PAI-2 can be crosslinked to trophoblast membranes by tissue transglutaminase or activated factor XIII has been proposed to localize PAI-2 to sites of tissue damage.[158]

6. KALLISTATIN

Kallistatin is an inhibitor of tissue kallikrein, a serine proteinase distinct from plasma kallikrein of the contact system. Tissue kallikrein cleaves low molecular weight kininogen to release vasoactive kinins and probably plays a role in blood pressure regulation. Although α_1-proteinase inhibitor and activated protein C inhibitor have been reported to inhibit tissue kallikrein[159,160] another inhibitor, previously known as human kallikrein-binding protein,[161] has been shown to be a serpin and named kallistatin.[162] Kallistatin has 401 amino acids, and is 45% homologous to human α_1-proteinase inhibitor, α_1-antichymotrypsin and activated protein C inhibitor. Cysteines are present at positions 5 and 237. A new name for kallistatin is human proteinase inhibitor-4, the gene of which is 9816 base pairs long and has five exons and four introns. The structure is similar to genes for α_1-proteinase inhibitor, α_1-antichymotrypsin and protein C inhibitor and is localized to the same locus as these genes, namely 14q31-q32.1.[14] The 5' flanking region of the gene has CAAT and TATA boxes, AP-1 and AP-2 binding sites, a GC rich region and a cAMP response element.[14] Kallistatin has the unusual P1-P1' residues of Phe-Ser. Kallistatin inhibits tissue kallikrein with a relatively slow rate constant of 2.6×10^4 $M^{-1}s^{-1}$.[163] This reaction can be prevented by the addition of heparin. The slow reaction with tissue kallikrein is partially explained by the fact that this proteinase prefers to cleave at Arg or Lys residues. The plasma concentration is only 13.5 μg/ml, and so the role of this serpin in the prevention of hypertension remains to be determined. However, in the spontaneously hypertensive rat, levels of rat kallistatin are reduced.[164] In these rats the expression of kallistatin mRNA in the kidney was significantly reduced.[165] Rat kallistatin, also known as Spi-2.3 (see Rodent serpins, below) inhibits a number of rat tissue kallikreins in vitro.[166] It shares 45% homology with the human protein.[162]

7. SQUAMOUS CELL CARCINOMA ANTIGEN

Squamous cell carcinoma antigen (SCCA) was originally isolated from a squamous cell carcinoma of uterine tissue and was named TA-4.[167] Levels of this serpin are used in the diagnosis of this carcinoma. The cDNA of this protein has been cloned and expressed in *E. coli* and found to code for a serpin of 390 amino acids with a putative P1-P1' bond of Ser-Ser.[168] The overall homology to PAI-2 is 43%. Subsequently it was found that there are two closely linked SCCA genes with sequence homology of 92%, the second gene product having a Leu-Ser active site.[153] This finding of two closely related SCCA gene products explained the observations that purified SCCA showed heterogeneity on isoelectric focusing, with both neutral (pI 6.3-6.6) and acidic fractions (pI 5.9-6.2). The neutral form is detected in the cytoplasm of normal and of some malignant cells whereas the acidic form is expressed primarily in malignant cells and is also the major form in the plasma of cancer patients. SCCA-1, which was the form originally cloned, is the neutral form and SCCA-2 is the acidic form. The gene structures indicate that the SCCAs have homology to ovalbumin and map to chromosome 18q21.3, along with the genes for PAI-2 and the tumor suppressor serpin maspin. Recombinant SCCA-1 was shown to inhibit chymotrypsin based on the ability to form a stable complex on SDS-PAGE.[169] In contrast a study using SCCA purified from patient sera indicated that it did not inhibit serine proteinases, but could inhibit the cysteine proteinases papain and cathepsin L.[170] The purification involved affinity chromatography on a column of immobilized antibodies to a high molecular weight cysteine proteinase inhibitor called psoriastatin. The exact relationship of this SCCA to SCCA-1 and -2 is not clear as yet.

8. RODENT SERPINS

A polymorphic genetic locus for serpins, called the *Spi* locus, has been identified in rats and mice. The *Spi-1* genes are those most closely related to human α_1-proteinase inhibitor whereas the *Spi-2* genes are those most closely related to human α_1-antichymotrypsin. This has complicated the nomenclature and the comparisons with human serpins. In humans there are probably single genes coding for α_1-proteinase inhibitor and α_1-antichymotrypsin, whereas in rats there may be 4 or more closely related genes at each *Spi* locus and even more in mice.[171] The mouse is the best studied of the rodent systems. The mouse serpin contrapsin, which is a trypsin inhibitor, shows about 60% homology to human α_1-antichymotrypsin overall, but in the reactive center region there is only 10% homology.[172,173] This serpin was found to be part of the *Spi-2* gene locus and mapped to chromosome 12.[174] The murine *Spi-2* locus appears to contain as many as 10 closely linked genes, which show the most divergence in the reactive center region.[171] This great divergence in the reactive center of orthologous serpins suggested that there was gene duplication and accelerated evolution (or positive selection) of rodent serpins, perhaps due to the diversity of exogenous proteinases encountered in infectious and parasitic organisms.[175] Similar conclusions can be drawn from a study comparing five murine *Spi-1* locus (α_1-proteinase inhibitor-like) genes.[176]

Another murine serpin has been identified and designated Spi-3.[177] The gene was found by probing a mouse library with the cDNA for human proteinase inhibitor-6 (see section 9, below). The mouse serpin has 76% homology with the human serpin, has an Arg-Cys reactive center and could form complex with thrombin. The gene has similarity to members of the ovalbumin gene family (including PAI-2) but lacks one intron, indicating that it should be in a separate gene family. The gene is on chromosome 13 whereas the gene for mouse PAI-2 is on chromosome 1.

The rat is less well defined genetically than the mouse, but similar ideas have emerged concerning serpin diversity. As with the mouse, the nomenclature is confusing. A serpin from rat has been cloned that is 70% homologous to human α_1-proteinase inhibitor[42] and is likely to be the major serpin produced by the *Spi-1* locus.

This is probably equivalent to the rat serpin-I purified from rat serum. Rat serpin-II is likely to be another product of the *Spi-1* locus.[178] There appear to be three main genes in the *Spi-2* locus[179] (Table 5.1). The hormonal regulation of these serpins has been studied in some detail.[179-183]

The guinea pig also appears to have contrapsin and at least two α_1-proteinase inhibitor-like serpins, although this system is not as well defined.[44,184-186] In Asian chipmunks a serpin which shows homology to rat α1-proteinase inhibitor has been shown to be elevated during hibernation.[47]

9. INTRACELLULAR SERPINS

Although serpins were originally considered only to regulate extracellular proteolysis, one of the original members of the family, ovalbumin, does not have a typical pro-peptide signal sequence (see Chapter 6). There are now a number of inhibitory serpins within mammalian cells that have been identified, in addition to PAI-2 (described above). The most extensively studied of these is the equine leukocyte elastase inhibitor (ELEI), which has a molecular weight of 35.2 kDa and is composed of 379 residues. It is not glycosylated. There is one cysteine at position 214. This serpin is present in the cytosol of horse leukoctyes. It inhibits porcine pancreatic and human neutrophil elastase with second order rate constants of greater than $10^7\ M^{-1}s^{-1}$. ELEI also rapidly inhibited cathepsin G.[187] The primary structure of ELEI has been determined and the P1-P1' bond found to be Ala-Met (residues 344-345). ELEI is 48% homologous to PAI-2 compared to only 30% homologous to α_1-proteinase inhibitor with which it has functional homology, which may reflect that intracellular serpins can be grouped together in the same evolutionary subgroup. Another intriguing finding is that ELEI copurified with thymosin β_4, a protein which is involved with cytoskeletal rearrangement.[188] ELEI has also been crystallized in the cleaved form[189] and the structure shown to be essentially the same as those of other cleaved inhibitory serpins (see Chapter 2). ELEI has also been cloned and the cDNA expressed in *E. coli*.[190] Similar intracellular serpins have been identified in human and porcine leukocytes.[191-193] The human protein has a molecular weight of 42 kDa and consists of 379 amino acids. The reactive center bond is an unusual Cys-Met (residues 344-345). Although it shares nucleotide and amino acid homology with PAI-2, the gene maps to chromosome 6p24,[194] in contrast to the PAI-2 gene which is on chromosome 18.[153] Recombinant protein has been produced in a baculovirus system and has been shown to have inhibitory activity against neutrophil cathepsin G and proteinase-3.[195]

A novel 38 kDa serpin, which has thrombin inhibitory activity and was originally called placental thrombin inhibitor, has been purified from human placenta and the cytosol of K562 leukemic cells.[196] A newer designation for this serpin is human proteinase inhibitor 6.[177] This serpin has been cloned and the P1-P1' bond identified as Arg-Cys (residues 341-342). The protein has 376 amino acids. As well as the cysteine at position 342, there are

Table 5.1. Nomenclature of rat Spi-2 locus serpins

Spi-2 serpin	Alternative names	Ref.
Spi-2.1		175
	λ Spi-1	180
	SPI-1	181
	Cpi-23	250
Spi-2.2		175
	SPI-3	181
Spi-2.3		175
	λ Spi-2	180
	Rat kallistatin or rat kallikrein binding protein (RKBP)	251
	Cpi-21	250
	Rat growth hormone regulated protein	252
	Rat thyroid hormone regulated protein	253
	SPI-2	181

cysteines at positions 36, 100, 110, 350 and 370. This serpin is most closely related to ovalbumin, PAI-2 and the leukocyte elastase inhibitors. The cDNA has also been expressed in COS-1 cells. Northern blotting reveals that mRNA for this serpin is present in most tissues but most strongly in cardiac and skeletal muscles.[197] The gene has been localized to chromosome 6p25,[198] close to the gene for human monocyte elastase inhibitor.[194] This serpin has been cloned independently by others[199] who have also identified a similar serpin in a monkey kidney epithelial cell line.[200]

An intracellular serpin has been purified from hemocytes of the horseshoe crab and called *Limulus* intracellular coagulation inhibitor or LICI. LICI was purified and was shown to consist of 394 amino acids with an Arg-Ser reactive center from the sequence of the cloned cDNA. LICI inhibited the *Limulus* lipopolysaccharide-sensitive factor C with a second order rate constant of 2.5 x 10^6 $M^{-1}s^{-1}$. It also inhibited thrombin, salivary kallikrein, plasmin and trypsin but not *Limulus* clotting enzyme, *Limulus* factor B, factor Xa, factor XIa, t PA, uPA, elastase or chymotrypsin. LICI shows most homology to PAI-2 (40%) and human monocyte elastase inhibitor (39%) in spite of the evolutionary distance between humans and *Limulus*.[201] A second *Limulus* serpin has been purified and cloned and called LICI2. This serpin, which has a Lys-Ser reactive center, has 386 amino acids and inhibits Limulus clotting enzyme with a rate of 4.3 x 10^5 $M^{-1}s^{-1}$ in addition to inhibiting plasmin, thrombin and tPA. LICI2 is 42% homologous to LICI and 36% homologous to PAI-2.[202]

10. INSECT SERPINS

A cDNA encoding a serpin was isolated from the larval fat body of a lepidopteran insect, the tobacco hornworm *Manduca sexta*. The predicted molecular weight of the protein was 43.5 kDa and the predicted P1 residue was alanine, giving rise to its name of Ala-serpin. A protein of 47 kDa with identical amino terminal sequence to the predicted protein has been isolated and found to inhibit porcine pancreatic elastase and bovine chymotrypsin.[203] Expression of this serpin has been proposed to play a role in antennae morphogenesis and olfactory neuron axon outgrowth.[204] Subsequently other serpins were identified that can inhibit chymotrypsin and trypsin.[205] By use of antisera to these three serpins, cDNA libraries from fat body and hemolymph of *Manduca sexta* were screened and 38 clones were identified. The differences in these clones were in the region coded by exon 9 of which there are at least 12 variants. Exon 9 codes for the reactive center region. Thus, by use of multiple exon 9 variants, the potential exists for production of serpins with diverse specificity from one gene. The structure of the *Manduca sexta* gene is somewhat homologous to that of human and rat PAI-1 and the provocative suggestion has been made that such exon diversity may also occur with these genes.[206] Analysis of the Ala-serpin mRNA shows that it is most abundant in the fat larval body and less so in hematocytes. Also, the mRNA is most abundant in the 4th and 5th instar larvae and decreases after molts. The hormone, 20-hydroxyecdysone is a negative regulator of expression.[207]

Serpins have also been identified in the silkworm, *Bombyx mori*. Two of these serpins inhibit chymotrypsin and have been named antichymotrypsin I and II.[208,209] The name antichymotrypsin is based on inhibition studies and does not imply a close phylogenetic relationship with human α_1-antichymotrypsin. A serpin with anti-trypsin-like activity has also been identified.[210] The true target enzymes of these serpins are not known.

A protein found only in the adult male accessory gland of *Drosophila melanogaster*, known as accessory gland peptide 76A, is a serpin. After mating the levels in the gland drop but the serpin is now found in the female uterus, where it decreases over the next six hours.[211]

11. PARASITE SERPINS

A serpin has been identified in the human parasite *Schistosoma haematobium*. The encoded protein has 409 amino acids and a putative P1 residue of Phe. The native protein is expressed as a 58 kDa glycoprotein on the surface of the worms.[212] Northern blotting of mRNA from *Schistosoma mansoni* with the *Schistosoma haematobium* serpin cDNA showed a weak hybridization band suggesting the presence of related serpins in both species. Individuals infected with *Schistosoma haematobium* generate antibodies to the serpin, but these antibodies are not cross reactive to sera from individuals with *Schistosoma mansoni* infection.[213] Although the function of these serpins is not known, it is possible that they help the parasite to evade the host's immune defense mechanisms. In this regard it should be noted that crude lysates of *Schistosoma mansoni* inhibit the activity of factor XIIa.[214,215] A 56 kDa inhibitor of *Schistosoma mansoni* 28 kDa protease and elastase has also been identified in this organism, although it was not proven to be a serpin.[216]

A serpin has been found in the filarial worm *Brugia malayi*, the causative agent of elephantiasis. A 1,287 base pair cDNA which coded for a serpin was identified by reverse transcriptase PCR. Transcription began between 8 and 9 days of larval development and continued through adult and microfiliarial stages. Western blotting revealed the protein to be 44 kDa, although a 75 kDa form was found secreted by the infective and fourth stage larval forms, which might represent a complex with a target enzyme. The predicted P1-P1' residues are Arg-Met.[217]

12. PLANT SERPINS

The only plants in which serpins have been identified are barley and wheat. A protein from the barley endosperm called barley protein Z_4, which has a molecular weight of 43 kDa, has been shown to be a serpin.[218,219] Other serpins called protein Z_7 and protein Zx have also been identified.[220,221] No proteinase inhibitory activity has been reported for Z_4 or Zx, but some data indicates that Z_7 can inhibit chymotrypsin.[220] A number of related serpins have been detected in wheat grain, one of which was able to form an SDS stable complex with chymotrypsin.[222] The role of these serpins is not known but it is likely that they provide protection from insect or microbial proteinases.

13. VIRAL SERPINS

A gene in the cowpox virus was shown to code for a protein with homology to the serpins and was proposed to account for the hemorrhage in lesions caused by the virus, perhaps through thrombin inhibition.[223] The protein has a molecular mass of 38 kDa and the gene was originally called the 38K gene.[224] The unusual reactive center residues of Asp-Cys, the cytoplasmic location and early nature of this gene product (within the first five hours of viral replication), and the appearance of hemorrhage after three days indicated that it was unlikely to be an inhibitor of thrombin or participate in the hemorrhage. It was shown that this protein inhibited an unusual cysteine proteinase called interleukin-1β converting enzyme (or ICE), which is required for the production of mature interleukin-1β. This is the best characterized demonstration that a serpin can inhibit a proteinase from another mechanistic class.[225] This serpin is now called cytokine response modifier A or CrmA. The association rate constant for the inhibition of ICE is $1.7 \times 10^7\ M^{-1}s^{-1}$. The complex formed is not stable to SDS but has a K_i of 4×10^{-12} M. The CrmA protein shows the typical inhibitory serpin characteristics of being in a metastable conformation.[226]

The inhibition of ICE by CrmA has generated a lot of interest as ICE has been implicated as a mediator of apoptosis. A number of studies have shown that overexpression of ICE (or the homologous *ced-3* and *Nedd-2* genes of *C. elegans* and mouse respectively) can cause cells to undergo apoptosis and that apoptosis can be prevented by expression of CrmA.[227,228] In addition, expression of CrmA can prevent apoptosis induced by a variety of stimuli.[229-234] Further information on the role of ICE in apoptosis can be found in Kumar.[235] It is now known that there are a number of ICE-like cysteine proteinases which might participate in apoptosis and which might be inhibited by CrmA.[236-239] CrmA has also been shown to inhibit the serine proteinase granzyme B with a second order rate constant of 2.9 x 10^5 $M^{-1}s^{-1}$.[240] Granzyme B can induce apoptosis in target cells indicating that it might have some substrates in common with ICE. Although ICE and granzyme B are from different mechanistic classes they both have a preference for Asp as the P1 residue, explaining in part the ability of CrmA to inhibit both proteinases. The virus has found an efficient way around the host anti-viral mechanisms by using CrmA to inhibit both the inflammatory response and the death of cells containing the replicating virus.

In addition to CrmA from cowpox virus, serpins have been reported in a number of other pox viruses. In the vaccinia virus, three serpins have been identified.[241-243] One of these (SPI-3) controls the ability of infected cells to fuse.[244] One serpin has been found in the fowl pox virus[245] and one in myxoma virus designated SERP 1.[246] SERP 1 has been shown to interfere with the host inflammatory response[247] and in vitro inhibits plasmin, tPA, uPA and C1s.[248] A comprehensive review of Poxvirus serpins can be found in Turner et al.[249]

References

1. Patston PA and Gettins PGW. A database of recombinant wild-type and mutant serpins. Thromb Haemost 1994; 72: 166-179.
2. Schultze HE, Heide K and Haupt H. α_1-Antitrypsin aus Humanserum. Klin Wochenschr 1962; 40:427-429.
3. Beatty K, Bieth J and Travis J. Kinetics of association of serine proteinases with native and oxidized α-1-proteinase inhibitor and α-1-antichymotrypsin. J Biol Chem 1980; 255:3931-3934.
4. Frommherz KJ, Faller B and Bieth JG. Heparin strongly decreases the rate of inhibition of neutrophil elastase by α_1-proteinase inhibitor. J Biol Chem 1991; 266: 15356-15362.
5. Komiyama T and Salvesen G. Interaction of subtilisins with serpins. J Cell Biochem 1995; Suppl 19b:247(Abstract).
6. Löbermann H, Tokuoka R, Deisenhofer J, et al. Human α_1-proteinase inhibitor. Crystal structure analysis of two crystal modifications and preliminary analysis of the implications for function. J Mol Biol 1984; 177:731-757.
7. Lomas DA, Elliott PR, Chang W-SW, et al. Preparation and characterization of latent α_1-antitrypsin. J Biol Chem 1995; 270:5282-5288.
8. Mast AE, Enghild JJ and Salvesen GS. Conformation of the reactive site loop of α_1-proteinase inhibitor probed by limited proteolysis. Biochemistry 1992; 31: 2720-2728.
9. Lomas DA, Evans DL, Stone SR, et al. Effect of the Z mutation on the physical and inhibitory properties of α_1-antitrypsin. Biochemistry 1993; 32:500-508.
10. Tyagi SC and Simon SR. Role of disulfide exchange in α_1-protease inhibitor. Biochemistry 1992; 31:10584-10590.
11. Rinehart AR, Mallya S and Simon SR. Human α_1-proteinase inhibitor binds to extracellular matrix *in vitro*. Am J Resp Cell Mol Biol 1993; 9:666-679.
12. Kurachi K, Chandra T, Degen SJF, et al. Cloning and sequence of cDNA coding for α_1-antitrypsin. Proc Natl Acad Sci USA 1981; 78:6826-6830.
13. Long GL, Chandra T, Woo SLC, et al. Complete sequence of the cDNA for human α_1-antitrypsin and the S variant. Biochemistry 1984; 23:4824-4837.
14. Chai KX, Ward DC, Chao J, et al. Molecular cloning, sequence analysis, and chromosomal localization of the human proteinase inhibitor 4 (kallistatin) gene. Genomics 1994; 23:370-378.
15. Billingsley GD, Walter MA, Hammond GL, et al. Physical mapping of four serpin genes: α_1-Antitrypsin, α_1-antichymotrypsin, corti-

costeroid-binding globulin, and protein C inhibitor, within a 280-kb region on chromosome 14q32.1. Am J Hum Genet 1993; 52:334-353.
16. Byth BC, Billingsley GD and Cox DW. Physical and genetic mapping of the serpin gene cluster at 14q32.1: Allelic association and a unique haplotype associated with α_1-antitrypsin deficiency. Am J Hum Genet 1994; 55:126-133.
17. Bao J, Sifers RN, Kidd VJ, et al. Molecular evolution of serpins: Homologous structure of the human α_1-antichymotrypsin and α_1-antitrypsin genes. Biochemistry 1987; 26:7755-7759.
18. Tanaka T, Hiroaki O and Nakanishi S. Common structural organization of the angiotensinogen and the α1-antitrypsin genes. J Biol Chem 1984; 259:8063-8065.
19. Desrochers PE, Mookhtiar K, Van Wart HE, et al. Proteolytic inactivation of α_1-proteinase inhibitor and α_1-antichymotrypsin by oxidatively activated human neutophil metalloproteinases. J Biol Chem 1992; 267:5005-5012.
20. Vissers MCM, George PM, Bathurst IC, et al. Cleavage and inactivation of α_1-antitrypsin by metalloproteinases released from neutrophils. J Clin Invest 1988; 82:706-711.
21. Desrochers PE, Jeffrey JJ and Weiss SJ. Interstitial collagenase (matrix metalloproteinase-1) expresses serpinase activity. J Clin Invest 1991; 87:2258-2265.
22. Mast AE, Enghild JJ, Nagase H, et al. Kinetics and physiologic relevance of the interaction of α_1-proteinase inhibitor, α_1-antichymotrypsin, and antithrombin III by matrix metalloproteinases-1 (tissue collagenase), -2 (72kDa gelatinase/type IV collagenase), and -3 (stromelysin). J Biol Chem 1991; 266:15810-15816.
23. Winyard PG, Zhang Z, Chadwick K, et al. Proteolytic inactivation of human α_1-antitrypsin by human stromelysin. FEBS Lett 1991; 279:91-94.
24. Pei D, Majmudar G and Weiss SJ. Hydrolytic inactivation of a breast carcinoma cell-derived serpin by human stromelysin-3. J Biol Chem 1994; 269:25849-25855.
25. Travis J and Salvesen G. Human plasma proteinase inhibitors. Annu Rev Biochem 1983; 52:655-709.
26. Carrell RW, Jeppsson J-O, Laurell C-B, et al. Structure and variation of human α_1-antitrypsin. Nature 1982; 298:329-334.
27. Perlmutter DH and Pierce JA. The α_1-antitrypsin gene and emphysema. Am J Physiol 1989; 257:L147-L162.
28. Cox DW. α_1-Antitrypsin deficiency. In: Scriver CR, Beaudet AL, Sly WS, Valle D, eds. The Metabolic Basis of Inherited Disease. New York: McGraw Hill, 1989; 2409-2437.
29. Berninger RW and Mathis RK. Isolation and characterization of α-1-antitrypsin from Rhesus-monkey serum. Biochem J 1986; 159:95-104.
30. Mistry R, Snashall PD, Totty N, et al. Isolation and characterization of sheep α-$_1$-proteinase inhibitor. Biochem J 1991; 273: 1071-1082.
31. Salahuddin P. Isolation and characterization of alpha-1-proteinase inhibitor from goat plasma. Biochem Int 1991; 24:321-338.
32. Pellegrini A and VonFellenberg R. Fractionation and partial characterization of α_1-proteinase isoinhibitors of horse. Biochim Biophys Acta 1980; 616:351-361.
33. Patterson SD, Bell K and Shaw DC. The equine major plasma serpin multigene family: Partial characterization including sequence of the reactive-site regions. Biochem Genet 1991; 29:477-499.
34. Potempa J, Wunderlich JK and Travis J. Comparative properties of three functionally different but structurally related serpin variants from horse plasma. Biochem J 1991; 274:465-471.
35. Braend M and Romagnoli A. Variation of acidic prealbumins in the donkey (*Equus asinus*). Anim Blood Grps Biochem Genet 1980; 11:77-80.
36. Gahne B and Juneja RK. Extensive genetic polymorphism of four plasma α-protease inhibitors in pigs and evidence for tight linkage between the structure loci of these inhibitors. Anim Genet 1986; 17:135-157.
37. Abrams WR, Kimbel P and Weinbaum G. Purification and characterization of canine α_1-antiproteinase. Biochemistry 1978; 17:3556-3561.

38. Koj A and Regeoczi E. Effect of experimental inflammation on the synthesis and distribution of antithrombin III and α_1-antitrypsin in rabbits. Br J Exp Path 1978; 59:473-481.
39. Diven WF, Vietmeier B, Hempel J, et al. Purification and N-terminal characterization of *Chinchilla villidera* α-1-antitrypsin. Comp Biochem Physiol 1990; 95B:39-44.
40. Catanese JJ and Kress LF. Opossum serum α_1-proteinase inhibitor: Purification, linear sequence, and resistance to inactivation by rattlesnake venom metalloproteinases. Biochemistry 1993; 32:509-515.
41. Schwarzenberg SJ, Sharp HL, Berry SA, et al. Hormonal regulation of serum alpha$_1$-antitrypsin and hepatic alpha$_1$-antitrypsin mRNA in rats. Biochem Biophys Res Commun 1987; 147:936-941.
42. Chao S, Chai KX, Chao L, et al. Molecular cloning and primary structure of rat α_1-antitrypsin. Biochemistry 1990; 29: 323-329.
43. Lamontagne L, Gauldie J and Koj A. Ontogeny and tissue distribution of alpha-1-antitrypsin in the mouse. Biochim Biophys Acta 1981; 662:15-21.
44. Kobayashi S and Nagasawa S. Protease inhibitors of guinea pig serum. I. Isolation of two functionally different trypsin inhibitors from guinea pig serum. Biochim Biophys Acta 1974; 342:372-381.
45. Junueja RK, Lundin LG and Gahne B. Genetic polymorphism of an α-1-protease inhibitor in mink plasma. Hereditas 1981; 94:249-252.
46. Goto K, Suzuki Y, Yoshida K, et al. Plasma α-1-antiproteinase from the Mongolian gerbil, *Meriones unguiculatus*: Isolation, partial characterization, sequencing of cDNA, and implications for molecular evolution. J Biochem (Tokyo) 1994; 116:582-588.
47. Kondo N and Kondo J. Identification of novel blood proteins specific for mammalian hibernation. J Biol Chem 1992; 267:473-478.
48. Patterson SD and Bell K. The carbohydrate side chains of the major plasma serpins of horse and wallaby: Analyses of enzymatic and chemically treated (including "Smith degradation") protein blots by lectin binding. Biochem Int 1990; 20:429-436.
49. Bennett JH, Hayman DL and Hope RM. Novel sex differences in linkage values and meiotic chromosome behaviour in a marsupial. Nature 1986; 323:59-60.
50. Sherwin WB. Postranslational modification of α_1-antitrypsin (protease inhibitor) and alkaline phosphatase in the marsupial, *Peramales gunni*. Biochem Genet 1990; 28:111-115.
51. Huang C-J, Chen C-C, Chen H-J, et al. A protease inhibitor of the serpin family is a major protein in carp perimeningeal fluid: I. Protein purification and characterization. J Neurochem 1995; 64:1715-1720.
52. Huang CJ, Lee MS, Huang FL, et al. A protease inhibitor of the serpin family is a major protein in carp perimeningeal fluid. 2. cDNA cloning, sequence analysis, and *Escherichia coli* expression. J Neurochem 1995; 64:1721-1727.
53. Patterson SD. Mammalian α_1-antitrypsins: Comparative biochemistry and genetics of the major plasma serpin. Comp Biochem Physiol [B] 1991; 100:439-454.
54. Crystal RG. α_1-Antitrypsin deficiency, emphysema, and liver disease. J Clin Invest 1990; 85:1343-1352.
55. Cox DW, Johnson AM and Fagerhol MK. Report of nomenclature meeting for alpha-1-antitrypsin. Hum Genet 1980; 53: 429-433.
56. Fagerhol MK. Serum Pi types in Norwegians. Acta Pathol Microbiol Scand 1967; 70:421.
57. Berninger RW. Alpha-1-antichymotrypsin. J Med 1985; 16:101-127.
58. Chandra T, Stackhouse R, Kidd VJ, et al. Sequence homology between human α_1-antichymotrypsin, α_1-antitrypsin and antithrombin III. Biochemistry 1983; 266: 5055-5061.
59. Rubin H, Wang ZM, Nickbarg EB, et al. Cloning, expression, purification, and biological activity of recombinant native and variant human α_1-antichymotrypsin. J Biol Chem 1990; 265:1199-1207.
60. Baumann U, Huber R, Bode W, et al. Crystal structure of cleaved α_1-antichymotrypsin at 2.7Å resolution and its comparison with other serpins. J Mol Biol

1991; 218:595-606.
61. Hwang S-R, Kohn AB and Hook VYH. Molecular cloning reveals isoforms of bovine α_1-antichymotrypsin. Proc Natl Acad Sci USA 1994; 91:9579-9583.
62. Christensen S and Sottrup-Jensen L. Characterization of two serpins from bovine plasma and milk. Biochem J 1994; 303: 383-390.
63. Potempa J, Enghild JJ and Travis J. The primary elastase inhibitor (elastasin) and trypsin inhibitor (contrapsin) in the goat are serpins related to human α_1-anti-chymotrypsin. Biochem J 1995; 306:191-197.
64. Buchman TG, Cabin DE, Vickers S, et al. Molecular biology of circulatory shock. Part II. Expression of four groups of hepatic genes is enhanced after resusitation from cardiogenic shock. Surgery 1990; 108:559-566.
65. Hook VYH, Purviance RT, Azaryan AV, et al. Purification and characterization of α_1-antichymotrypsin-like protease inhibitor that regulates thiol protease involved in enkephalin precursor processing. J Biol Chem 1993; 268:20570-20577.
66. Hwang S-R, Kohn AB and Hook VYH. Unique reactive site domains of neuroendocrine isoforms of alpha-1-antichymotrypsin from bovine adrenal medulla and pituitary revealed by molecular cloning. FEBS Lett 1995; 368:471-476.
67. Wiman B and Collen D. On the kinetics of the reaction between human antiplasmin and plasmin. Eur J Biochem 1978; 84:573-578.
68. Wiman B and Collen D. On the mechanism of the reaction between human α_2-antiplasmin and plasmin. J Biol Chem 1979; 254:9291-9297.
69. Holmes WE, Nelles L, Lijnen HR, et al. Primary structure of human α_2-antiplasmin, a serine protease inhibitor (serpin). J Biol Chem 1987; 262:1659-1664.
70. Lijnen HR, Holmes WR, Van Hoef B, et al. Amino-acid sequence of human α_2-antiplasmin. Eur J Biochem 1987; 166: 565-574.
71. Hirosawa S, Nakamura Y, Miura O, et al. Organization of the human α_2-plasmin inhibitor gene. Proc Natl Acad Sci USA 1988; 85:6836-6840.
72. Sumi Y, Ichikawa Y, Nakamura Y, et al. Expression and characterization of pro α_2-plasmin inhibitor. J Biochem 1989; 106:703-707.
73. Kato A, Hirosawa S, Toyota S, et al. Localization of the human α_2-plasmin inhibitor gene (PLI) to 17p13. Cytogenet Cell Genet 1993; 62:190-192.
74. Bangert K, Johnsen AH, Christensen U, et al. Different N-terminal forms of α_2-plasmin inhibitor in human plasma. Biochem J 1993; 291:623-625.
75. Enghild JJ, Valnickova Z, Thøgerson IB, et al. An examination of the inhibitory mechanism of serpins by analysing the interaction of trypsin and chymotrypsin with α_2-antiplasmin. Biochem J 1993; 291:933-938.
76. Potempa J, Shieh B-H and Travis J. Alpha-2-antiplasmin: A serpin with two separate but overlapping reactive sites. Science 1988; 241:699-670.
77. Aoki N, Sumi Y, Miura O, et al. Human α-2-plasmin inhibitor. Methods Enzymol 1993; 223:185-197.
78. Högstorp H and Carlin G. Studies on biosynthesis of α-2-antiplasmin in rat liver cells. Thromb Res 1988; 49:307-318.
79. Christensen S, Berglund L and Sottrup-Jensen L. Primary structure of bovine α_2-antiplasmin. FEBS Lett 1994; 343:223-228.
80. Shieh B-H, Potempa J and Travis J. The use of α_2-antiplasmin as a model for the demonstration of complex reversibility in serpins. J Biol Chem 1989; 264: 13420-13423.
81. Longstaff C and Gafney PJ. Serpin-serine protease binding kinetics: α_2-antiplasmin as a model inhibitor. Biochemistry 1991; 30:979-986.
82. Longstaff C and Gaffney PJ. Studies on the mechanism of binding of serpins and serine proteases. Blood Coag Fibrinolysis 1992; 3:89-97.
83. Sugiyama N, Sasaki T, Iwamoto M, et al. Binding site of α_2-plasmin inhibitor to plasminogen. Biochim Biophys Acta 1988; 952:1-7.
84. Hortin GL, Trimpe BL and Fok KF. Plasmin's peptide-binding specificity: Characterization of ligand sites in α_2-antiplasmin. Thromb Res 1989; 54:621-632.
85. Aoki N, Moroi M and Tachiya K. Effects of

α_1-plasmin inhibitor on fibrin clot lysis: Its comparison with α_2-macroglobulin. Thromb Haemostas 1978; 39:22-31.
86. Rákóczi I, Wiman B and Collen D. On the biological significance of the specific interaction between fibrin, plasminogen and antiplasmin. Biochim Biophys Acta 1978; 540:295-300.
87. Anonick PK and Gonias SL. Soluble fibrin preparations inhibit the reaction of plasmin with alpha-2-macroglobulin. Biochem J 1991; 275:53-59.
88. Hall SW, Humphries JE and Gonias SL. Inhibition of cell surface receptor-bound plasmin by α_1-antiplasmin and α_2-macroglobulin. J Biol Chem 1991; 266: 12329-12336.
89. Christensen U and Clemmensen I. Kinetic properties of the primary inhibitor of plasmin from human plasma. Biochem J 1977; 163:389-391.
90. Edelberg JM and Pizzo SV. Lipoprotein (a) promotes plasmin inhibition by α_2-antiplasmin. Biochem J 1992; 286:79-84.
91. Sakata Y and Aoki N. Cross-linking of α_2-plasmin inhibitor to fibrin by fibrin-stabilizing factor. J Clin Invest 1980; 65:290-297.
92. Sakata Y and Aoki N. Significance of cross-linking of α_2-plasmin inhibitor to fibrin in inhibition of fibrinolysis and in hemostasis. J Clin Invest 1982; 69:536-542.
93. Tamaki T and Aoki N. Cross-linking of α_2-plasmin inhibitor and fibronectin to fibrin by fibrin-stabilizing factor. Biochim Biophys Acta 1981; 661:280-286.
94. Tamaki T and Aoki N. Cross-linking of α_2-plasmin inhibitor catalyzed by activated fibrin-stabilizing factor. J Biol Chem 1982; 257:14767-14772.
95. Ichinose A, Tanaka T and Aoki N. Factor XII-mediated cross-linking of NH2-terminal peptide of α_2-plasmin inhibitor to fibrin. FEBS Lett 1981; 153:369-371.
96. Clemmensen I, Thorsen S, Müllertz S, et al. Properties of three different molecular forms of the α_2-plasmin inhibitor. Eur J Biochem 1981; 120:105-112.
97. Kluft C and Los N. Demonstration of two forms of α_2-antiplasmin in human plasma by modified crossed immunoelectrophoresis. Thromb Res 1981; 21:65-71.
98. Sim RB, Arlaud G and Colomb M. Kinetics of reaction of human C1-inhibitor with the human complement system proteases C1r and C1s. Biochim Biophys Acta 1980; 612:433-449.
99. Lennick M, Brew SA and Ingham K. Kinetics of interaction of C1 inhibitor with complement C1s. Biochemistry 1986; 25:3890-3898.
100. Schapira M, deAgostini A and Colman RW. Contribution of plasma protease inhibitors to the inactivation of kallikrein in plasma. J Clin Invest 1982; 69:462-468.
101. van der Graaf F, Koedam JA, Griffin JH, et al. Interaction of human plasma kallikrein and its light chain with C1 inhibitor. Biochemistry 1983; 22:4860-4866.
102. van der Graaf F, Koedam JA and Bouma BN. Inactivation of kallikrein in plasma. J Clin Invest 1983; 71:149-158.
103. Harpel PC, Lewin MC and Kaplan AP. Distribution of plasma kallikrein between C1 inactivator and α_2-macroglobulin-kallikrein complexes. J Biol Chem 1985; 260:4257-4263.
104. de Agostini A, Lijnen HR, Pixley RA, et al. Inactivation of factor XII active fragment in normal plasma. Predominant role of C1-inhibitor. J Clin Invest 1984; 73:1542-1549.
105. Pixley RA, Schapira M and Colman RW. The regulation of human factor XIIa by plasma proteinase inhibitors. J Biol Chem 1985; 260:1723-1729.
106. Ziccardi RJ. A new role for C1-inhibitor in homeostasis: control of activation of the first component of complement. J Immunol 1982; 128:2505-2508.
107. Tseng Y, Poon PH, Závodszky P, et al. Spontaneous activation of serum C1 in vitro. Role of C1 inhibitor. J Immunol 1991; 147:1884-1890.
108. Schifferli JA, Steiger G and Schapira M. The role of C1, C1-inactivator and C4 in modulating immune precipitation. Clin Exp Immunol 1985; 60:605-612.
109. Rent R, Myhrman R, Fiedel BA, et al. Potentiation of C1-esterase inhibitor activity by heparin. Clin Exp Immunol 1976; 23:264-271.

110. Caughman GB, Boackle RJ and Vesely J. A postulated mechanism for heparin's potentiation of C1 Inhibitor function. Mol Immunol 1982; 287:295-290.
111. Weiss V and Engel J. Heparin-stimulated modification of C1-inhibitor by subcomponent C1s of human complement. Hoppe-Seyler's Z Physiol Chem 1983; 364: 295-301.
112. Pensky J, Levy LR and Lepow IH. Partial purification of a serum inhibitor of C'1-esterase. J Biol Chem 1961; 236:1674-1679.
113. Haupt H, Heimburger N, Kranz T, et al. Ein Beitrag zur Isolierung und Charakterisierung des C1-Inactivators aus Humanplasma. Eur J Biochem 1970; 17:254-261.
114. Harrison RA. Human C1 Inhibitor: Improvised isolation and preliminary structural characterization. Biochemistry 1983; 22:5001-5007.
115. Alsenz J and Loos M. Simplified methods for the purification, quantitation, and functional estimation of human complement C1-inhibitor (C1-INH) with a monoclonal anti-C1-INH antibody. J Immunol Methods 1987; 96:107-114.
116. Prograis LJ Jr, Hammer CH, Katusha K, et al. Purification of C1-inhibitor. A new approach for the isolation of this biologically important plasma protease inhibitor. J Immunol Methods 1987; 99:113-122.
117. Schapira M, de Agostini A and Colman RW. C1-inhibitor: The predominant inhibitor of plasma kallikrein. Methods Enzymol 1988; 163:179-185.
118. Donaldson VH and Falconieri MW. A modification of an affinity procedure for purification of human C1 inhibitor that provides a homogeneous stable preparation. J Immunol Methods 1993; 157:101-104.
119. Davis III AE, Aulak KS, Zahedi K, et al. C1 inhibitor. Methods Enzymol 1993; 223:97-120.
120. Salvesen GS, Catanese JJ, Kress LF, et al. Primary structure of the reactive site of human C1-inhibitor. J Biol Chem 1985; 260:2432-2436.
121. Bock SC, Skriver K, Nielsen E, et al. Human C1 inhibitor: Primary structure, cDNA cloning, and chromosomal localization. Biochemistry 1986; 25:4292-4301.
122. Davis AE, Whitehead AS, Harrison RA, et al. Human inhibitor of the first component of complement, C1: Characterization of cDNA clones and localization of the gene to chromosome 11. Proc Natl Acad Sci USA 1986; 3161:3165-3160.
123. Que B and Petra PH. Isolation and analysis of a cDNA coding for human C1 inhibitor. Biochem Biophys Res Commun 1986; 137:620-625.
124. Tosi M, Duponchel C, Bourgarel P, et al. Molecular cloning of C1 inhibitor: sequence homologies with α_1 antitrypsin and other members of the serpin superfamily. Gene 1986; 42:265-272.
125. Carter PE, Dunbar BA and Fothergill JE. Genomic and cDNA cloning of the human C1 inhibitor. Intron-exon junctions and comparison with other serpins. Eur J Biochem 1988; 163:169-160.
126. Carter PE, Duponchel C, Tosi M, et al. Complete nucleotide sequence of the gene for human C1 inhibitor with an unusually high density of *Alu* repeats. Eur J Biochem 1991; 197:301-308.
127. Aulak KS, Davis AE III., Donaldson VH, et al. Chymotrypsin inhibitory activity of normal C1-inhibitor and a P1 Arg to His mutant: Evidence for the presence of overlapping reactive centers. Prot Sci 1993; 2:727-732.
128. Davis AE. C1 inhibitor and hereditary angioneurotic edema. Annu Rev Immunol 1988; 6:595-628.
129. Loos M, Wolf HU and Opferkuch W. The C1 inactivator from guinea pig serum-III. Characterization and kinetic data of the reaction between the inactivator and EAC1 and EAC14. Immunochemistry 1972; 9:451-459.
130. Schwögler S, Odenthal M, Knittel T, et al. Fat storing cells of the rat liver synthesize and secrete C1-esterase inhibitor; Modulation by cytokines. Hepatology 1992; 16:794-802.
131. Perkins SJ, Smith KF, Amatayakul S, et al. Two-domain structure of the native and reactive center cleaved forms of C1 inhibitor of human complement by neutron scattering. J Mol Biol 1990; 214:751-763.
132. Odermatt E, Berger H and Sano Y. Size

and shape of human C1-inhibitor. FEBS Lett 1981; 131:283-285.
133. Perkins SJ. Hydrodynamic data show that C1 inhibitor of complement forms compact complexes with C1r and C1s. FEBS Lett 1990; 271:89-92.
134. Patston PA, Qi M, Schifferli JA, et al. The effect of cleavage by a *Crotalus atrox* α-proteinase fraction on the properties of C1 inhibitor. Toxicon 1995; 33:53-61.
135. Coutinho M, Aulak KS and Davis AE III. Functional analysis of the serpin domain of C1 inhibitor. J Immunol 1994; 153: 3648-3654.
136. Schoenberger OL. Characterization of carbohydrate chains of C1-inhibitor and of desialylated C1-inhibitor. FEBS Lett 1992; 314:430-434.
137. Tenner AJ and Frank MM. Activator-bound C1 is less susceptible to inactivation by C1 inhibition than is fluid-phase C1. J Immunol 1986; 137:625-630.
138. Ziccardi RJ. Demonstration of the interaction of native C1 with monomeric immunoglobulin and C1 inhibitor. J Immunol 1985; 134:2559-2563.
139. Aulak KS, Eldering E, Hack CE, et al. A hinge region mutation in C1-inhibitor (Ala^{436}→Thr) results in nonsubstrate-like behavior and in polymerization of the molecule. J Biol Chem 1993; 268:18088-18094.
140. Eldering E, Verpy E, Roem D, et al. COOH-terminal substitutions in the serpin C1 inhibitor that cause loop overinsertion and subsequent multimerization. J Biol Chem 1995; 270:2579-2587.
141. Patston PA, Hauert J, Michaud M, et al. Formation and properties of C1-inhibitor polymers. FEBS Lett 1995; 368:401-404.
142. Åstedt B, Lecander I, Brodin T, et al. Purification of a specific placental plasminogen activator inhibitor by monoclonal antibody and its complex formation with plasminogen activator. Thromb Haemostas 1985; 53:122-125.
143. Wun TC and Reich E. An inhibitor of plasminogen activator from human placenta. Purification and characterization. J Biol Chem 1987; 262:3646-3653.
144. Lecander I and Åstedt B. Isolation of a new specific plasminogen activator inhibitor from pregnancy plasma. Br J Haematol 1986; 62:221-228.
145. Kruithof EKO, Vassalli J, Mattaliano RJ, et al. Purification and characterization of a plasminogen activator inhibitor from the histiocytic lymphoma cell line U-937. J Biol Chem 1986; 261:11207-11213.
146. Wohlwend A, Belin D and Vassalli J. Plasminogen activator-specific inhibitors produced by human monocytes/macrophages. J Exp Med 1987; 165:320-339.
147. Schleuning W, Medcalf RL, Hession C, et al. Plasminogen activator inhibitor 2: Regulation of gene transcription during phorbol ester-mediated differentiation of U-937 human histiocytic lymphoma cells. Mol Cell Biol 1987; 7:4564-4567.
148. Webb AC, Collins KL, Snyder SE, et al. Human monocyte Arg-serpin cDNA: sequence, chromosomal assignment and homology to plasminogen activator inhibitor. J Exp Med 1987; 166:77-94.
149. Ye RD, Wun T and Sadler JE. cDNA cloning and expression in *Escherichia coli* of a plasminogen activator from human placenta. J Biol Chem 1987; 262:3718-3725.
150. Antalis TM, Clark MA, Barnes T, et al. Cloning and expression of a cDNA coding for a human monocyte-derived plasminogen activator inhibitor. Proc Natl Acad Sci USA 1988; 85:985-989.
151. Ye RD, Ahern SM, Le Beau MM, et al. Structure of the gene for human plasminogen activator inhibitor-2. The nearest mammalian homologue of chicken ovalbumin. J Biol Chem 1989; 264:5495-5502.
152. Samia JA, Alexander SJ, Horton KW, et al. Chromosomal organization and localization of the human urokinase inhibitor gene: Perfect structural conservation with ovalbumin. Genomics 1990; 6:159-167.
153. Schneider SS, Schick C, Fish KE, et al. A serine proteinase inhibitor locus at 18q21.3 contains a tandem duplication of the human squamous cell carcinoma antigen gene. Proc Natl Acad Sci USA 1995; 92: 3147-3151.
154. Belin D, Wohlwend A, Schleunung W-D, et al. Facultative polypeptide translocation allows a single mRNA to encode the secreted and cytosolic forms of plasminogen

activator inhibitor 2. EMBO J 1989; 8: 3287-3294.

155. Mikus P, Urano T, Liljeström P, et al. Plasminogen-activator inhibitor type 2 (PAI-2) is a spontaneously polymerizing SERPIN. Biochemical characterization of the recombinant intracellular and extracellular forms. Eur J Biochem 1993; 218:1071-1082.

156. von Heijne G, Liljestruom P, Mikus P, et al. The efficiency of the uncleaved secretion signal in the plasminogen activator inhibitor type-2 protein can be enhanced by point mutations that increase its hydrophobicity. J Biol Chem 1991; 266:15240 15243.

157. Vassalli JD, Sappino AP and Belin D. The plasminogen activator/plasmin system. J Clin Invest 1991; 88:1067-1072.

158. Jensen PH, Lorand L, Ebbesen P, et al. Type-2 plasminogen-activator inhibitor is a substrate for trophoblast transglutaminase and Factor-XIII(a)—transglutaminase-catalyzed cross-linking to cellular and extracellular structures. Eur J Biochem 1993; 214:141-146.

159. Geiger R, Stuckstedte U, Clausnitzer B, et al. Progressive inhibition of human glandular (urinary) kallikrein by human serum and identification of the progressive anti-kallikrein as α_1-antitrypsin (α_1-proteinase inhibitor). Hoppe-Seyler's Z Physiol Chem 1981; 362:317-325.

160. Ecke S, Geiger M, Resch I, et al. Inhibition of tissue kallikrein by protein C inhibitor. Evidence for identity of protein C inhibitor with the kallikrein binding protein. J Biol Chem 1992; 267:7048-7052.

161. Chen LM, Chao L, Mayfield RY, et al. Differential interactions of human kallikrein-binding protein and α_1-antitrypsin with human tissue kallikrein. Biochem J 1990; 267:79-84.

162. Chai KX, Chen LM, Chao J, et al. Kallistatin—a novel human serine proteinase inhibitor—molecular cloning, tissue distribution, and expression in *Escherichia coli*. J Biol Chem 1993; 268:24498-24505.

163. Zhou GX, Chao L and Chao J. Kallistatin: A novel human tissue kallikrein inhibitor. Purification, characterization, and reactive center sequence. J Biol Chem 1992; 267:25873-25880.

164. Chao J, Chai KX, Chen LM, et al. Tissue kallikrein-binding protein is a serpin. I. Purification, characterization, and distribution in normotensive and spontaneously hypertensive rats. J Biol Chem 1990; 265:16394-16401.

165. Yang T, Terada Y, Nonoguchi H, et al. Distribution of kallikrein-binding protein mRNA in kidneys and difference between SHR and WKY rats. Am J Physiol 1994; 267:F325-F330.

166. Serveau C, Moreau T, Zhou GX, et al. Inhibition of rat tissue kallikrein gene family members by rat kallikrein-binding protein and α 1-protease inhibitor. FEBS Lett 1992; 309:405-408.

167. Kato H and Torigoe T. Radioimmunoassay for tumor antigen of human cervical squamous cell carcinoma. Cancer 1977; 40: 1621-1628.

168. Suminami Y, Kishi F, Sekiguchi K, et al. Squamous cell carcinoma antigen is a new member of the serine protease inhibitors. Biochem Biophys Res Commun 1991; 181:51-58.

169. Nawata S, Tsunaga N, Numa F, et al. Serine protease inhibitor activity of recombinant squamous cell carcinoma antigen towards chymotrypsin, as demonstrated by sodium dodecyl sulfate polyacrylamide gel electrophoresis. Electrophoresis 1995; 16: 1027-1030.

170. Takeda A, Yamamoto T, Nakamura Y, et al. Squamous cell carcinoma antigen is a potent inhibitor of cysteine proteinase cathepsin L. FEBS Lett 1995; 359:78-80.

171. Inglis JD and Hill RE. The murine Spi-2 proteinase inhibitor locus: a multigene family with a hypervariable reactive site domain. EMBO J 1991; 10:255-261.

172. Hill RE, Shaw PH, Boyd PA, et al. Plasma protease inhibitors in mouse and man: divergence within the reactive center regions. Nature 1984; 311:175-177.

173. Suzuki Y, Yamamoto K and Sinohara H. Molecular cloning and sequence analysis of full-length cDNA coding for mouse contrapsin. J Biochem 1990; 108:344-346.

174. Hill RE, Shaw PH, Barth RK, et al. A genetic locus closely linked to a protease inhibitor gene complex controls the level of

multiple RNA transcripts. Mol Cell Biol 1985; 5:2114-2122.

175. Hill RE and Hastie ND. Accelerated evolution in the reactive centre regions of serine protease inhibitors. Nature 1987; 326: 96-99.
176. Borriello F and Krauter K. Multiple murine α_1-protease inhibitor genes show unusual evolutionary divergence. Proc Natl Acad Sci USA 1991; 88:9417-9421.
177. Sun J, Rose JB and Bird P. Gene structure, chromosomal localization, and expression of the murine homologue of human proteinase inhibitor 6 (PI-6) suggests divergence of PI-6 from the ovalbumin serpins. J Biol Chem 1995; 270:16089-16096.
178. Pirie-Shepherd SR, Miller HRP and Ryle A. Differential inhibition of rat mast cell proteinase I and II by members of the α-1-proteinase inhibitor family of serine proteinase inhibitors. J Biol Chem 1991; 266:17314-17319.
179. Schwarzenberg SJ, Yoon JB, Sharp HL, et al. Homologous rat hepatic protease inhibitor genes show divergent functional responses to inflammation. Am J Physiol 1989; 256:C413-C419.
180. Yoon Y-B, Towle HC and Seelig S. Growth hormone induces two mRNA species of the serine protease inhibitor gene family in rat liver. J Biol Chem 1987; 262:4284-4289.
181. Pages G, Rouayrenc JF, LeCam G, et al. Molecular characterization of three rat-liver serine protease inhibitors affected by inflammation and hypophysectomy. Protein and mRNA analysis and cDNA cloning. Eur J Biochem 1990; 190:385-391.
182. Paquerau L, Vilarem MJ, Rossi V, et al. Regulation of two rat-serine protease inhibitor gene promoters by somatotropin and glucocorticoids. Study with intact hepatocytes and cell-free systems. Eur J Biochem 1992; 209:1053-1061.
183. Berry SA, Bergad PL and Bundy MV. Expression of growth hormone-responsive serpin messenger RNAs in perinatal rat liver. Am J Physiol 1993; 264:E973-E980.
184. Suzuki Y, Yoshida K, Ichimiya T, et al. Trypsin inhibitors in guinea pig plasma: Isolation and characterization of contrapsin and two isoforms of α-1-antiproteinase and acute phase response of four major trypsin inhibitors. J Biochem 1990; 107:173-179.
185. Suzuki Y, Yoshida K, Honda E, et al. Molecular cloning and sequence analysis of cDNAs coding for guinea pig α_1-antiproteinases S and F and contrapsin. J Biol Chem 1991; 266:928-932.
186. Imamura T and Kambara T. The major plasma kallikrein inhibitor of guinea pig plasma. Biochim Biophys Acta 1992; 1118:261-266.
187. Potempa J, Dubin A, Watorek W, et al. An elastase inhibitor from equine leukocyte cytosol belongs to the serpin superfamily. J Biol Chem 1988; 263:7364-7369.
188. Dubin A, Travis J, Enghild JJ, et al. Equine leukocyte elastase inhibitor. Primary structure and identification as a thymosin-binding protein. J Biol Chem 1992; 267: 6576-6583.
189. Baumann U, Bode W, Huber R, et al. Crystal structure of cleaved equine leucocyte elastase inhibitor determined at 1.95Å resolution. J Mol Biol 1992; 226:1207-1218.
190. Kordula T, Dubin A, Schooltink H, et al. Molecular cloning and expression of an intracellular serpin - an elastase inhibitor from horse leucocytes. Biochem J 1993; 293: 187-193.
191. Remold-O'Donnell E, Nixon JC and Rose RM. Elastase inhibitor. Characterization of the human elastase inhibitor associated with monocytes, macrophages and neutrophils. J Exp Med 1989; 169:1071-1086.
192. Remold-O'Donnell E, Chin J, Alberts M. Sequence and molecular characterization of human monocyte/neutrophil elastase inhibitor. Proc Natl Acad Sci USA 1992; 89:5635-5639.
193. Teschauer WF, Mentele R and Sommerhoff CP. Primary structure of a porcine leukocyte serpin. Eur J Biochem 1993; 217: 519-526.
194. Evans E, Cooley J and Remold-O'Donnell E. Characterization and chromosomal localization of $ELANH_2$, the gene encoding human monocyte/neutrophil elastase inhibitor. Genomics 1995; 28:235-240.
195. Sugimori T, Cooley J, Hoidal JR, et al. Inhibitory properties of recombinant human monocyte/neutrophil elastase inhibitor. Am

J Resp Cell Mol Biol 1995; 13:314-322.

196. Coughlin PB, Tetaz T and Salem HH. Identification and purification of a novel serine proteinase inhibitor. J Biol Chem 1993; 268:9541-9547.
197. Coughlin P, Sun JR, Cerruti L, et al. Cloning and molecular characterization of a human intracellular serine proteinase inhibitor. Proc Natl Acad Sci USA 1993; 90:9417-9421.
198. Coughlin P, Nicholl J, Sun H, et al. Chromosomal mapping of the human proteinase inhibitor-6 (PI6) gene to 6p25 by fluorescence in situ hybridization. Genomics 1995, 26:431-433.
199. Morgenstern KA, Sprecher C, Holth L, et al. Complementary DNA cloning and kinetic characterization of a novel intracellular serine proteinase inhibitor: Mechanism of action with trypsin and factor Xa as model proteinases. Biochemistry 1994; 33:3432-3441.
200. Morgenstern KA, Henzel WJ, Baker JB, et al. Isolation and characterization of an intracellular serine proteinase inhibitor from a monkey kidney epithelial cell line. J Biol Chem 1993; 268:21560-21568.
201. Miura Y, Kawabata S and Iwanaga S. A *Limulus* intracellular coagulation inhibitor with characteristics of the serpin superfamily. Purification, characterization and cDNA cloning. J Biol Chem 1994; 269:542-547.
202. Miura Y, Kawabata S, Wakamiya Y, et al. A limulus intracellular coagulation inhibitor type 2. Purification, characterization, cDNA cloning and tissue localization. J Biol Chem 1995; 270:558-565.
203. Kanost MR, Prasad SV and Wells MA. Primary structure of a member of the serpin superfamily of proteinase inhibitors from an insect, *Manduca sexta*. J Biol Chem 1989; 264:965-972.
204. Hanneman E and Kanost MR. Differential alaserpin expression during development of the antennae in the tobacco hawkmoth, *Manduca sexta*. Arch Insect Biochem Physiol 1992; 19:39-52.
205. Kanost MR. Isolation and characterization of 4 serine proteinase inhibitors (serpins) from hemolymph of *Manduca sexta*. Insect Biochem 1990; 20:141-147.
206. Jiang H, Wang Y and Kanost MR. Mutually exclusive exon use and reactive center diversity in insect serpins. J Biol Chem 1994; 269:55-58.
207. Kanost MR, Prasad SV, Huang YL, et al. Regulation of serpin gene-1 in *Manduca sexta*. Insect Biochem Mol Biol 1995; 25:285-291.
208. Sasaki T. Patchwork-structure serpins from silkworm (*Bombyx mori*) larval hemolymph. Eur J Biochem 1991; 202:255-261.
209. Narumi H, Hishida T, Sasaki T, et al. Molecular cloning of silkworm (*Bombyx mori*) antichymotrypsin. A new member of the serpin superfamily of proteins from insects. Eur J Biochem 1993; 214:181-187.
210. Takagi H, Narumi H, Nakamura K, et al. Amino acid sequence of silkworm (*Bombyx mori*) hemolymph antitrypsin deduced from its cDNA nucleotide sequence: Confirmation of its homology with serpins. J Biochem 1990; 108:372-378.
211. Coleman S, Drahn B, Petersen G, et al. A *Drosophila* male accessory gland protein that is a member of the serpin superfamily of protease inhibitors is transferred to the female during mating. Insect Biochem Mol Biol 1995; 25:203-207.
212. Blanton RE, Licate LS and Aman RA. Characterization of a native and recombinant *Schistosoma haematobium* serine protease inhibitor gene product. Mol Biochem Parasitol 1994; 63:1-11.
213. Li Z, King CL, Ogundipe JO, et al. Preferential recognition by human IgE and IgG4 of a species-specific *Schistosoma haemotobium* serine protease inhibitor. J Inf Dis 1995; 171:416-422.
214. Tsang VCW, Hubbard WJ and Damian RT. Coagulation factor XIIa (activated Hageman factor) inhibitor from adult *Schistosoma mansoni*. Am J Med Trop Med Hyg 1977; 26:243-247.
215. Foster CB, Flanigan TP, DeStigter KK, et al. Inhibition of the activation of Hageman factor (factor XII) by extracts of *Schistosoma mansoni*. J Lab Clin Med 1992; 120: 735-739.
216. Ghendler Y, Arnon R and Fishelson Z. *Schistosoma mansoni*: Isolation and characterization of Smpi56, a novel serine proteinase

inhibitor. Exp Parasit 1994; 78:121-131.

217. Yenbutr P and Scott AL. Molecular cloning of a serine protease inhibitor from *Brugia malayi*. Infect Immunol 1995; 63: 1745-1753.

218. Hejgaard J, Rasmussen SK, Brandt A, et al. Sequence homology between barley endosperm protein Z and protease inhibitors of the α_1-antitrypsin family. FEBS Lett 1985; 180:89-94.

219. Brandt A, Svendsen I and Hejgaard J. A plant serpin gene. Structure, organization and expresion of the gene encoding barley protein Z_4. Eur J Biochem 1990; 194: 499-505.

220. Lundgard R and Svensson B. A 39kD barley seed protein of the serpin superfamily inhibits α-chymotrypsin. Carlsberg Res Commun 1990; 54:173-180.

221. Rasmussen SK. A gene coding for a new plant serpin. Biochim Biophys Acta 1993; 1172:151-154.

222. Rosenkrands I, Hejgaard J, Rasmussen SK, et al. Serpins from wheat grain. FEBS Lett 1994; 343:75-80.

223. Pickup DJ, Ink BS, Hu W, et al. Hemorrhage in lesions caused by cowpox virus is induced by a viral protein that is related to plasma protein inhibitors of serine proteases. Proc Natl Acad Sci USA 1986; 83: 7698-7702.

224. Palumbo GJ, Pickup DJ, Fredrickson TN, et al. Inhibition of an inflammatory response is mediated by a 38-kDA protein of cowpox virus. Virology 1989; 171:262-273.

225. Ray CA, Black RA, Kronheim SR, et al. Viral inhibition of inflammation. Cowpox virus encodes an inhibitor of the interleukin-1β converting enzyme. Cell 1992; 69: 597-604.

226. Komiyama T, Ray CA, Pickup DJ, et al. Inhibition of interleukin-1β converting enzyme by the cowpox virus serpin CrmA. An example of cross-class inhibition. J Biol Chem 1994; 269:19331-19337.

227. Miura M, Zhu H, Rotello R, et al. Induction of apoptosis in fibroblasts by IL-1β-converting enzyme, a mammalian homolog of the *C. elegans* cell death gene *ced-3*. Cell 1993; 75:653-660.

228. Kumar S, Kinoshita M, Noda M, et al. Induction of apoptosis by the mouse *Nedd2* gene, which encodes a protein similar to the product of the *Caenorhabdtis elegans* cell death gene *ced-3* and the mammalian IL-1 β-converting enzyme. Genes and Development 1994; 8:1613-1626.

229. Gagliardini V, Fernandez PA, Lee RKK, et al. Prevention of vertebrate neuronal death by the *crmA* gene. Science 1994; 263: 836-828.

230. Boudreau N, Sympson CJ, Werb Z, et al. Suppression of ICE and apoptosis in mammary epithelial cells by extracellular matrix. Science 1995; 267:891-893.

231. Tewari M and Dixit VM. Fas- and tumor necrosis factor-induced apoptosis is inhibited by the poxvirus *crmA* gene product. J Biol Chem 1995; 270:3255-3260.

232. Enari M, Hug H and Nagata S. Involvement of ICE-like protease in Fas-mediated apoptosis. Nature 1995; 375:78-81.

233. Los M, Van de Craen M, Penning LC, et al. Requirement of an ICE/CED-3 protease for Fas/APO-1-mediated apoptosis. Nature 1995; 375:81-83.

234. Tewari M, Beidler DR and Dixit VM. CrmA-inhibitable cleavage of the 70-kDa protein component of the U1 small nuclear ribonucleoprotein during Fas- and tumor necrosis factor induced apoptosis. J Biol Chem 1995; 270:18738-18741.

235. Kumar S. ICE-like proteases in apoptosis. Trends Biochem Sci 1995; 20:198-202.

236. Martin SJ and Green DR. Protease activation during apoptosis. Death by a thousand cuts? Cell 1995; 82:349-352.

237. Munday NA, Vaillancourt JP, Ali A, et al. Molecular cloning and pro apoptotic activity of ICE(rel)II and ICE(rel)III, members of the ICE/CED-3 family of cysteine proteinases. J Biol Chem 1995; 270: 15870-15876.

238. Wang XD, Pai JT, Weidenfeld EA, et al. Purification of an interleukin-1-β converting enzyme-related cysteine proteinase that cleaves sterol regulatory element-binding protein between the Leu zipper transmembrane domains. J Biol Chem 1995; 270: 18044-18056.

239. Tewari M, Quan LT, O'Rourke K, et al. Yama/CPP32β, a mammalian homolog of CED-3, is a CrmA-inhibitable protease that cleaves the death substrate poly(ADP-ribose) polymerase. Cell 1995; 81:801-809.
240. Quan LT, Caputo A, Bleackley RC, et al. Granzyme B is inhibited by the cowpox virus serpin cytokine response modifier A. J Biol Chem 1995; 270:10377-10379.
241. Boursnell MEG, Foulds IJ, Campbell JI, et al. Non-essential genes in the vaccinia virus Hind III fragment: a gene related to serine protease inhibitors and a gene related to the 37K vaccinia virus major envelope antigen. J Gen Virol 1988; 69:2995-3003.
242. Kotwal GJ and Moss B. Vaccinia virus encodes two proteins that are structurally related to members of the plasma serine protease inhibitor superfamily. J Virol 1989; 63:600-606.
243. Smith G, Howard ST and Chan YS. Vaccina virus encodes a family of genes with homology to serine proteinase inhibitors. J Gen Virol 1989; 70:2333-2343.
244. Turner PC and Moyer RW. An orthopoxvirus serpin-like gene controls the ability of infected cells to fuse. J Virol 1992; 66:2076-2085.
245. Tomley F, Binns M, Campbell J, et al. Sequence analysis of an 11.2 kilobase, near terminus, BamHI fragment of fowlpox virus. J Gen Virol 1988; 69:1025-1040.
246. Upton C, Macen JL, Wishart DS, et al. Myxoma virus and malignant rabbit firoma virus encode a serpin-like protein important for virus virulence. Virology 1990; 179:618-631.
247. Macen JL, Upton C, Nation N, et al. SERP1, a serine proteinase inhibitor encoded by myxoma virus is a secreted glycoprotein that interferes with inflammation. Virology 1993; 195:348-363.
248. Lomas DA, Evans DL, Upton C, et al. Inhibition of plasmin, urokinase, tissue plasminogen activator and C1s by a myxoma virus proteinase inhibitor. J Biol Chem 1993; 268:516-521.
249. Turner PC, Musy PY and Moyer RW. Poxvirus serpins. In: McFadden G, ed. Viroceptors, Virokines and Related Immune Modulators Encoded by DNA Viruses. Austin: R.G. Landes Co., 1994:67-88.
250. Okhubo K, Ogata S, Misumi Y, et al. Molecular cloning and characterizaton of rat contrapsin-like protease inhibitor and related proteins. J Biochem 1991; 109:385-391.
251. Chai KX, Ma JX, Murray SR, et al. Molecular cloning and analysis of rat kallikrein-binding protein gene. J Biol Chem 1991; 16029:16036-16030.
252. Le Cam A, Pages G, Auberger P, et al. Study of a growth hormone-regulated protein secreted by rat hepatocytes: cDNA cloning, anti-protease activity and regulation of its synthesis by various hormones. EMBO J 1987; 6:1225-1232.
253. Tecce MF, Dozin B, Magnuson MA, et al. Transcriptional regulation by thyroid hormone of an mRNA homologous to a protease inhibitor. Biochemistry 1986; 25: 5831-5834.
254. Kaplan AP and Silverberg M. The coagulation-kinin pathway of human plasma. Blood 1987; 70:1-15.
255. De La Cadena RA, Wachtfogel YT and Colman RW. Contact activation pathway: Inflammation and coagulation. In: Colman RW, Hirsh J, Marder VJ, Salzman EW, eds. Hemostasis and Thrombosis; Basic Principles and Clinical Practice. Philadelphia: J.B. Lippincott Company, 1994; 219-240.
256. Winkelstein JA and Colten HR. Genetically determined disorders of the complement system. In: Scriver CR, Beaudet AL, Sly WS, Valle D, eds. The Metabolic Basis of Inherited Disease New York: McGraw Hill, 1989; 2711-2737.

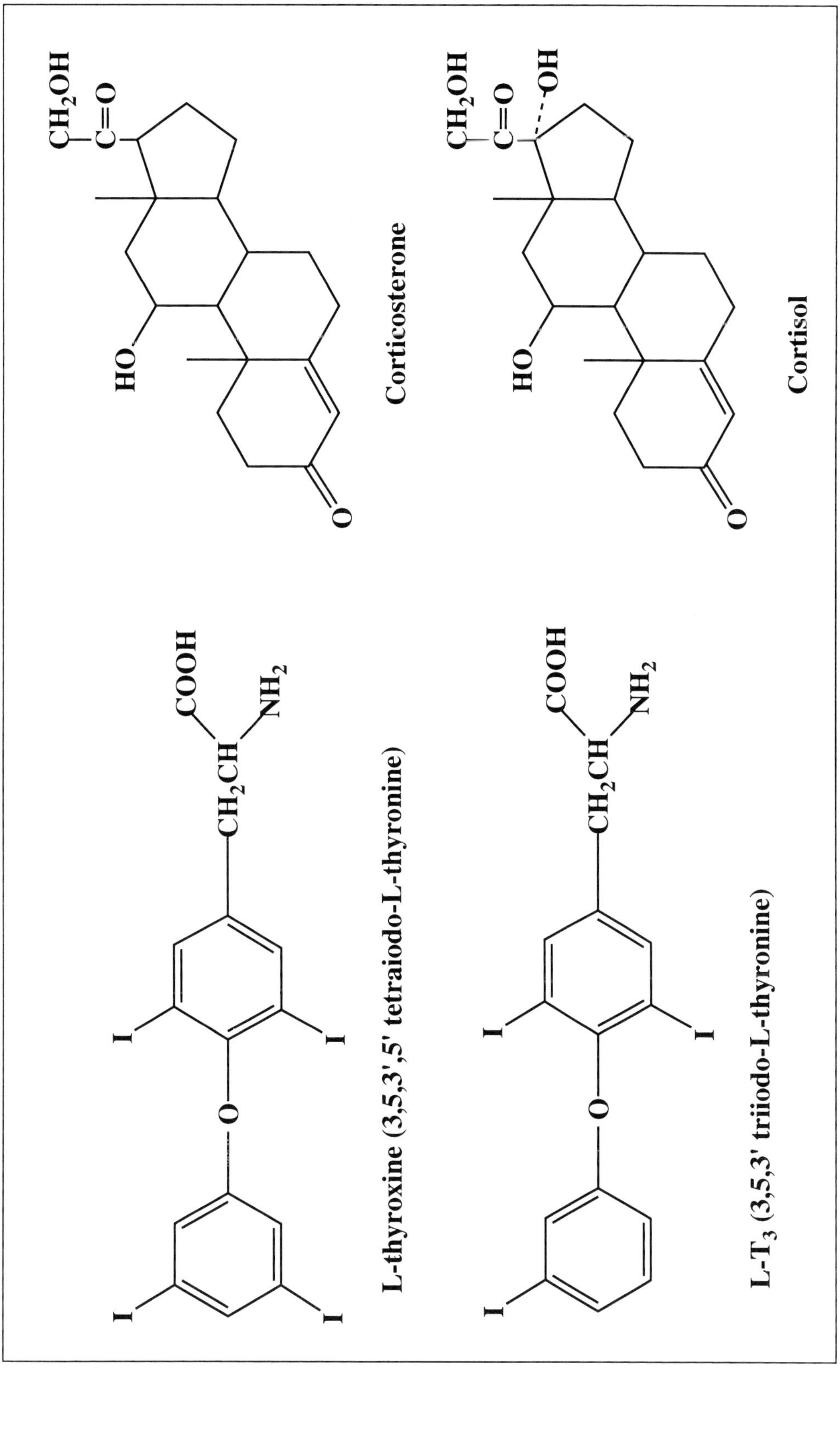

Fig. 6.2. Structures of thyroid and steroid hormones transported by TBG and CBG respectively.

thyroxine and triiodothyronine respectively.[32,33] Both ligands show a strong bell shaped pH-dependence for binding, with an optimum between pH 6.8 and 7.7.[32] Although there are no structures of either TBG or TBG-thyroxine complex, there have been two similar proposals made for the location and nature of the thyroxine binding site of TBG. Huber and Carrell proposed that the thyroxine binding site is located in a pocket formed between β-sheets B and C.[34] Although access to such a pocket is restricted in α_1-proteinase inhibitor by tryptophan 238, access in TBG might be possible as a result of substitution of the much smaller threonine for tryptophan at this position. In support of such a proposal are limited chemical cross-linking data which show that lysine 256 (α_1-PI numbering) at the end of strand 3 of β-sheet B, and thus at the entrance to the proposed pocket, can be cross-linked to the hormone.[35] Also in other proteins that bind small hydrophobic ligands, including the retinol binding proteins,[36] fatty acid binding proteins[37] and transthyretin, which binds both retinal and thyroxine,[38] the ligand binding site is within the walls of a β-barrel analogous to that formed by the B and C sheets of serpins. An attempt has been made to model the thyroxine-TBG interaction by docking thyroxine with an α-PI-derived, energy-minimized model of TBG[39] and using the prediction of Huber and Carrell for the location of the thyroxine binding site. The TBG model has a cavity ~11Å deep and 17Å across, which could accommodate the thyroxine molecule, which is 6Å x 12Å. Although the proposed binding site is characterized as hydrophobic, there are a number of accessible charged side chains that may be important in interactions with the phenol, carboxyl and amino groups of the hormone.

Human TBG is a 395 residue secreted glycoprotein that is biosynthesized as a 415 residue precursor with a 20 residue hydrophobic leader sequence.[40] It is synthesized in the liver as a relatively rare translation product of total liver mRNA.[41] The gene is localized to the long arm of the X chromosome (Xq21-22)[42-44] and is composed of 5 exons, with similar size, exon/intron organization and phase to those of α_1-PI, α_1-antichymotrypsin and CBG.[45] A site for hepatocyte nuclear factor 1 (HNF-1) has been shown to be essential for transcription of the TBG gene.[45] There are 4 potential glycosylation sites at positions 16 (NAT), 79 (NLT), 145 (NIS) and 391 (NPT)[40] that are likely to be completely utilized, based on chemical analysis.[46] The carbohydrate represents 14.6% of the total mass of 54 kDa[46] and is composed of biantennary complex carbohydrate chains containing no fucose.[47] Microheterogeneity results in four major and five minor bands by isoelectric focusing[48,49] with pIs of the major bands between 4.25 and 4.55.[49] The rat TBG gene has been cloned and codes for a protein 3 residues longer at the N-terminus than the human protein, with which it shares 76% homology.[50] The 5' untranslated region of the gene has an HNF-1 binding site similar to that of the human TBG gene.[50] Human TBG has been expressed in reticulocyte lysate and *Xenopus* oocytes[51] and in the mammalian cell lines HepG2, CV-1 and CHO.[45]

It has been reported that cleavage of TBG in the reactive center loop region by neutrophil elastase does not lead to an alteration in thyroxine affinity although it does result in stability changes consistent with expansion of β-sheet A as a result of insertion of the cleaved reactive center loop as an additional strand.[52] The lack of change in thyroxine affinity is compatible with the proposed location of the thyroxine binding site between the β-sheets B and C, i.e., not involving β-sheet A and therefore not very sensitive to large changes in β-sheet A. It might be predicted that if a latent form of TBG could be prepared, analogous to the latent form of PAI-1, it would have reduced thyroxine affinity as a result of the partial disruption of the C sheet that occurs in addition to the A sheet expansion. The ability of the cleaved reactive center loop to insert into β-sheet A is as expected from the P14-P10 sequence (Fig. 6.1). The absence of inhibitory properties may be more the result of Gln-Pro at P1-P1', leading to preferential cleavage at other sites within the reactive center and consequent failure to disrupt the substrate mechanism and thereby trap the proteinase in complex.

Several naturally-occurring variants of TBG have been reported (Table 6.1), some of which have alterations in thyroxine affinity and/or in levels of secretion, from normal to completely absent. In vitro expression of $TBG^{PRO-227}$, an inherited abnormality, showed that this defect results in impaired secretion.[51] Inherited abnormalities have been reviewed in 1989 by Refetoff.[53] The reductions in affinity for thyroid hormones are relatively small and represent changes in ΔG of between 0.4 and 1.3 kcal mol^{-1} (3.3-11% of the total binding energy). Some of these changes may affect the thyroxine binding site directly, whereas the variant TBG-Gary, which results in creation of a new N-linked glycosylation site at residue 96,[54] may result in hindered access to the binding site as a result of additional surface carbohydrate. There is a parallel in antithrombin, in which removal of carbohydrate at residue 135 increases the affinity for the ligand heparin without altering the structure of the protein,[55] whereas the presence of a new glycosylated carbohydrate attachment site in a natural variant (antithrombin Rouen III Ile7→Asn) results in reduced heparin affinity.[56] A possibly more important practical consequence of the TBG-Gary defect is a large reduction in the level of secreted TBG, with concomitant increase in intracellular degradation.[54] Individuals with TBG deficiency are euthyroid and show no endocrine abnormalities.[43]

(II) Corticosteroid Binding Globulin

Corticosteroid binding globulin (CBG), also known as transcortin, is responsible for carrying more than 90% of the circulating glucocorticoids cortisol and corticosterone (Fig. 6.2) in humans[57] and rats.[58] A protein with the properties of CBG appears to be ubiquitous in the plasma of vertebrates[59] and may also be present in lower organisms such as yeast.[60] Although a diminished level of CBG in humans appears not to be an uncommon genetic trait, there has been no substantiated instance of a homozygous defect, indicating that a certain minimum plasma level must be present for development and that a complete absence of CBG may be lethal in utero.[61] The high proportion of corticosteroid bound to CBG in vivo results from the high affinity and high specificity of CBG for these particular steroids. Although

Table 6.1. Natural variants of TBG

TBG Variant	Mutation	[TBG]	Ref
TBG-C	None (normal wild-type)	Normal	53
TBG-Poly	L283F	Normal	53
TBG-Chicago	Not determined	Normal	154
TBG-S	D171N	Reduced	53
TBG-San Diego	S23T, L283F	Reduced affinity	155,156
TBG-A	A191T, L283F	Reduced affinity	157
TBG-Québec	L283F, H331T	Reduced	158
TBG-Montreal	A113P	Reduced	159,160
TBG-Gary	I96Q	Reduced	54
TBG-PDJ	P363L	Reduced	161
TBG-CDJ	Codon 352 nucleotide deletion	Complete deficiency	162
TBG-CD1	Unknown plus L283F	Complete deficiency	53
TBG-CD5	L227P, L283F	Complete deficiency	53
TBG-CD6	Truncation from frame-shift	Complete deficiency	163

both progesterone and testosterone can also bind to CBG, the much lower affinity of these interactions makes it unlikely that they are important ligands, except perhaps in reproductive tissue where their concentrations may be much greater than those of the corticosteroids.

As with ovalbumin and TBG, CBG had been known for many years before being recognized as a member of the serpin superfamily from the determination of its primary structure, deduced from the human cDNA sequence.[62] Human CBG is one of the smaller serpins, containing only 383 amino acids with a polypeptide mass of 42,646 Daltons.[63] Rabbit CBG is the same size,[64] but rat CBG is 9 residues shorter.[63] Rat CBG lacks five residues at the N-terminus and has three internal deletions of one or two amino acids. These three serpins share about 60-70% identity,[65] with the putative glucocorticoid-binding region having one of the highest degrees of sequence similarity (see below). CBG from squirrel monkey, a New World monkey, is unusual in appearing to be a non-covalent heterodimer that has reduced corticosteroid affinity and resulting high levels of free cortisol in serum.[66] This may be a dimer analogous to those found in the crystals of human antithrombin examined by X-ray diffraction[67,68] and composed of one latent or cleaved molecule that had lost normal activity and one molecule that was still active. If the squirrel monkey CBG is such a dimer, the reduced corticocosteroid affinity could be understood in terms of the two structures present and their different corticosteroid affinities (see below). CBGs are glycoproteins, with 6 potential asparagine glycosylation sites in human and rat CBGs and 4 in rabbit CBG. Methylation analysis of human CBG indicates that five of the six sites are used on average,[69] resulting in 25% of the mass being carbohydrate. Only two of the potential glycosylation sites are conserved between all three species, one at position 74 (83 α_1-PI numbering) and one at position 238 (249 α_1-PI numbering). Both of these conserved sites are in regions of high sequence similarity. Furthermore, the 238 site is adjacent to the putative glucocorticoid binding region. Somewhat surprisingly, one of the potential glycosylation sites in human CBG, which appears to be at least partly utilized,[70] is located at position P3 of the reactive center region, only two residues removed from the P1-P1' neutrophil elastase cleavage site.[52] It has been reported that glycosylation at this position does not affect the ability of the P1-P1' bond to be cleaved by neutrophil elastase.[70] Nevertheless, it has been shown that glycosylation is necessary for activity[71] and that processing is site specific.[72] In particular, it has been shown that position 238 must be glycosylated for corticosteroid binding activity,[73] but that carbohydrate at this position is only needed during protein folding, suggesting that there may be a specific sugar-polypeptide interaction required for correct formation of the corticosteroid binding site during folding.[72]

The proposed binding site for the thyroid hormones in TBG in the cavity of the β-barrel formed by the β-sheets B and C (discussed above in section 6.2.I) is also the proposed binding site for the corticosteroids in CBG. The dimensions of the pocket are appropriate to accommodate the ligands in the orientation indicated by an electron paramagnetic mapping study, which showed that the steroid ring is deeply buried with the D ring and the side chain pointing towards the protein surface.[74] As with TBG, localization of the ligand binding site of CBG was based on a combination of the appropriate size and hydrophobic nature of the pocket and the results of chemical modification and mutagenesis experiments. Thus hydrophobic cluster analysis, which is a method developed to identify three-dimensional hydrophobic regions in proteins and to permit structural comparison of proteins even when sequence identity is very low, has shown that the region between residues 241 and 252 aligns with the hydrophobic region of cytochromes P450c17 and P450c21 that constitutes the steroid binding site in these proteins.[75] This region in CBG also shows high sequence conservation between species. Affinity-labeling of one of the two cysteine residues in human CBG by 6-bromo-progesterone has implicated a cysteine as being in or near the corticosteroid binding site.[76] Mutagenesis of one of the cysteines, Cys 228 (237 α_1-PI numbering), to alanine or serine resulted in no alteration in steroid binding properties, suggesting either that it is Cys 60 that is in the steroid binding site or that Cys 228 is in the binding site but is not a critical residue for high affinity binding.[77] However, given the location of Cys 228 close to the putative glucocorticoid binding pocket and of Cys 60 (69 α_1-PI numbering) well away from this region, it seems more likely that Cys 228 is the residue that is affinity labeled. This is further supported by the primary structure of both rat and rabbit CBGs, which show that Cys 228 is not only conserved, but is the only cysteine present in these CBGs.[63,64] Cys 228 is located in a highly conserved region of the primary structure, in contrast to the region surrounding the residue equivalent to Cys 60 of human CBG.

In keeping with this localization of the steroid binding site is a recent study of the effect of tryptophan substitutions on the steroid binding properties of CBG.[78] Tryptophan had been previously implicated in high affinity binding of steroids from fluorescence quenching[79,80] and UV difference spectroscopy measurements[81] as well as by affinity labeling of tryptophan.[82] Replacement of tryptophan 371 (380 α_1-PI numbering) by phenylalanine resulted in a large decrease in affinity for cortisol, such that binding of [^{3}H]-cortisol was not detectable. In contrast, replacement of any of the other three tryptophans by phenylalanine caused a reduction in binding energy ($\Delta\Delta G$) of at most 1.4 kcal mol^{-1} out of a total binding energy of 12.3 kcal mol^{-1}. Tryptophan 371 has also been identified as the major site of photoaffinity labeling of CBG by Δ^6 derivatives of corticosteroids.[83] This tryptophan is located between strands 4 and 5 of β-sheet B and is thus close to the putative corticosteroid binding pocket. Another mutation that has been characterized involves a Leu→His substitution at position 93 in human CBG.[84] The mutation resulted in reduced affinity for cortisol, corresponding to a $\Delta\Delta G$ for binding of 0.7 kcal mol^{-1}. This is again a relatively small reduction in overall binding affinity, but one that may have practical consequences for corticosteroid uptake and delivery.

In a manner similar to TBG, reaction of CBG with neutrophil elastase results in cleavage that appears to be specific to the reactive center loop region and produces a conformation with increased thermal stability consistent with loop insertion into β-sheet A. In contrast to TBG, however, such alteration in structure results in a tenfold reduction in K_a for the hormone ligand.[52] It has been proposed that such a reduction in K_a could have great practical consequences in vivo and be directly related to the anti-inflammatory role of corticosteroids.[85] Thus, at the surface of activated neutrophils present at sites of inflammation, elastase secreted by the neutrophil might react with CBG, cleaving the reactive center loop and thereby reducing the affinity for corticosteroids.[86] The released corticosteroids would then be delivered at an appropriate inflammatory site. Depending on the concentrations of CBG and free corticosteroid, such a modest change in K_a could result in a large percentage increase in free steroid concentration. For example consider a CBG concentration of 100 nM and a total cortisol concentration of 10 nM. In the high affinity state (K_a~2.6 x 10^8 M^{-1}) the free cortisol concentration would be about 0.4 nM, whereas in the low affinity state (K_a~2.6 x 10^7 M^{-1}) it would increase to about 3 nM, a seven-fold increase. The reaction between CBG and neutrophil elastase at the surface of neutrophils is likely to be further enhanced by the inactivating effect that neutrophil-secreted superoxide radical has on the primary plasma inhibitor of neutrophil elastase, α_1-proteinase inhibitor.

Such a role for CBG in delivering glucocorticoids specifically to sites of inflammation is in keeping with the close genetic relationship between CBG and α_1-proteinase inhibitor. These serpins not only show high structural similarity at the protein level,[75] but have very similar organization of their genes,[87] both of which are located in the same region of chromosome 14 (q31-32.1), together with another serpin proposed to be involved in regulation of inflammation, α_1-antichymotrypsin.[88] This suggests that the origin of CBG was in gene duplication and subsequent evolution from being an inhibitor of elastase to a substrate, with development of a binding site for the anti-inflammatory glucocorticoids that is modulated by elastase in a way that results in release of the ligand at sites of inflammation.[70] The primary structure of the reactive center of CBG shows differences from that of α_1-proteinase inhibitor that might further facilitate this function. Thus the P1 residue, which is methionine in α_1-proteinase inhibitor and which is susceptible to oxidation with resultant decrease in the rate of reaction with elastase, is a threonine in human CBG and a valine in rabbit CBG. The rate of cleavage of CBG by elastase is thus not likely to be adversely affected by the high levels of superoxide radicals released by neutrophils. Somewhat puzzling in this regard is the reactive center sequence of rat CBG, which appears to have arginine at P1 and is thus not likely to be cleaved at this site by elastase. However, since insertion of the reactive center loop can occur even for cleavages at sites other than P1-P1',[89] rat CBG may undergo conformational change and consequent release of bound corticosteroid as a result of reaction with elastase at a different site within the reactive center region.

4. ANGIOTENSINOGEN

Human angiotensinogen is a 452 residue plasma serpin glycoprotein that serves as the high molecular weight precursor of the potent vasoconstrictor angiotensin II (Fig. 6.3). It is synthesized as a 476 residue pre-protein with a 25 residue hydrophobic leader sequence.[90] The organization of the 12 kb angiotensinogen gene into 5 exons and 4 introns[91] parallels that of human α_1-PI and α_1-antichymotrypsin.[92] The extra length of exon 2, which codes for the N-terminal region that generates angiotensin II (see below), compared with α_1-PI and α_1-antichymotrypsin may indicate later evolution to provide its biological function as a non-inhibitory serpin that is the progenitor of angiotensin II. Expression of the gene is cell-type specific.[91] Regulation involves interaction between a 24 bp core fragment in the 3' untranslated region of the gene and an unidentified enhancer, named hAEF-1 (human angiotensinogen enhancing factor 1).[93]

There are four potential glycosylation sites in human angiotensinogen at positions 14 (<u>N</u>ES), 137 (<u>N</u>CT), 271 (<u>N</u>ST) and 295 (<u>N</u>FS).[90] The protein sequence of human preangiotensinogen shows about 60-65% identity with those of rat,[94] sheep[95] and mouse[96] angiotensinogens. A major difference between human and other mammalian angiotensinogens is in one of the sites of glycosylation. The glycosylation site in the N-terminal region is at residue 23 in rat and mouse angiotensinogens and not present in sheep angiotensinogen, compared with residue 14 in the human protein. Since formation of angiotensin II involves initial cleavage by the aspartate proteinase renin (EC 3.4.23.15) of the N-terminal decapeptide of angiotensinogen to give angiotensin I, this glycosylation difference may account for the species specificity of human angiotensinogen for cleavage only by human or primate renin, whereas other mammalian angiotensinogens are cleaved by most mammalian renins.[97] This is reasonable since the seven residues that flank the cleavage site are the primary determinants of renin specificity.[98] Amino acid substitutions in this region can result in potent renin inhibitors.[99] Angiotensin I is further processed by angiotensin converting enzyme (EC 3.4.15.1) by removal of the C-terminal dipeptide to give the biologically active octapeptide angiotensin II. A high molecular weight form of angiotensinogen has been reported (332 kDa)

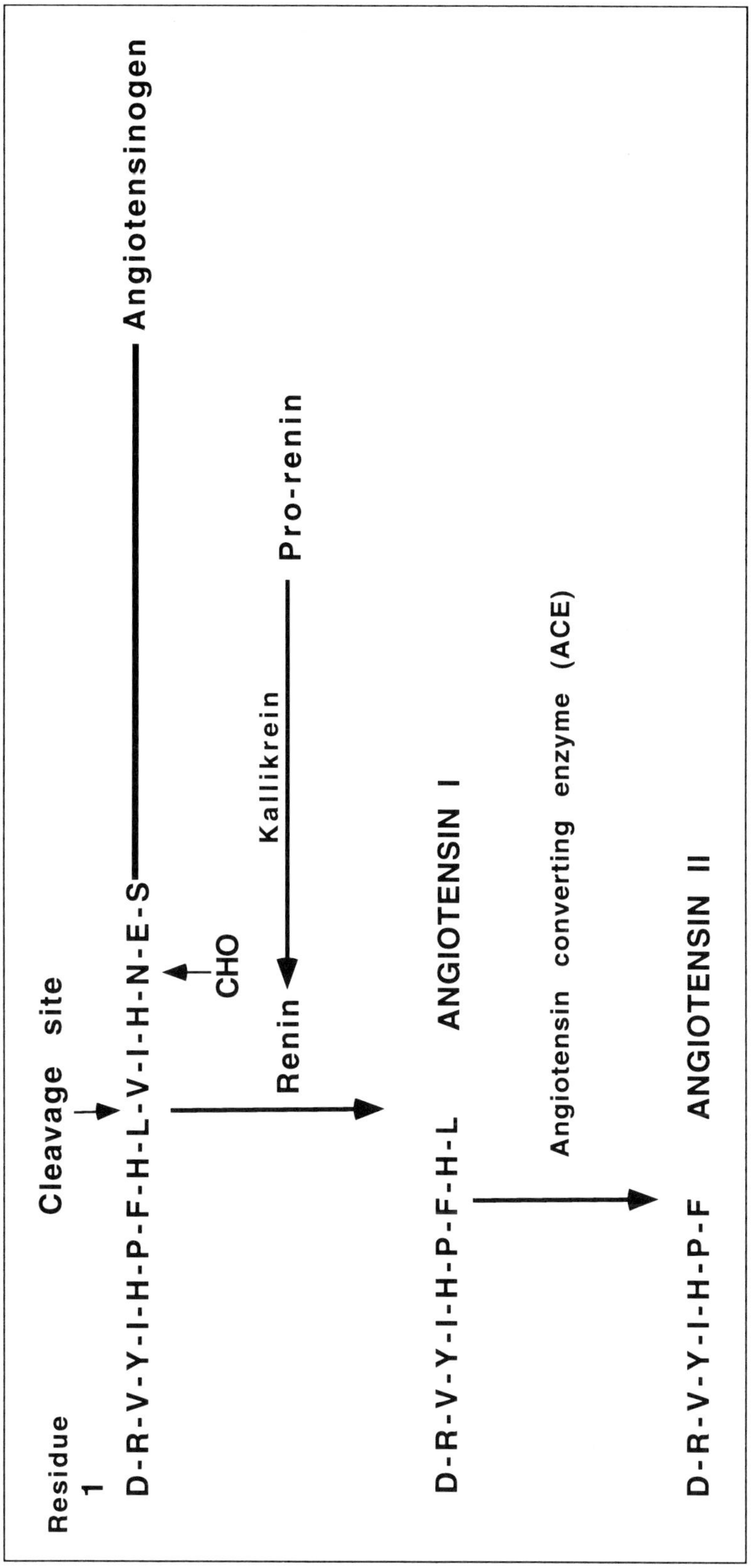

Fig. 6.3. Scheme for the conversion of angiotensinogen into angiotensin II.

that is present at much higher levels in amniotic fluid and the plasma of pregnant women than in plasma from non-pregnant women and of men.[100] Using monoclonal antibodies against pro-eosinophil major basic protein, a high molecular weight pregnancy-associated complex between the pro-eosinophil basic protein, complement component C3dg and angiotensinogen was demonstrated.[101] This was a disulfide cross-linked complex that probably represents the additional high molecular weight form of angiotensinogen present during pregnancy. The affinity of the high molecular weight form for renin is higher than for angiotensinogen alone[102] and may confer an advantage during pregnancy in the kinetics of production of angiotensin peptides.

Since the angiotensins are the physiologically important products of angiotensinogen, these are the parts of the serpin that have been most studied. In contrast, the properties of the remainder of the serpin have been little examined. This is in part due to the low concentration of angiotensinogen in plasma and thus poor availability for biochemical and biophysical characterization. Expression of functional recombinant angiotensinogen has been reported in *E. coli*[103] and in CHO cells,[104] which may result in greater availability of this serpin for biophysical characterization.

Although cleavage by renin in the N-terminal region is the most important proteolytic reaction with angiotensinogen, it has been shown that V8 proteinase cleaves angiotensinogen specifically in the reactive center loop region, at the P10-P9 bond,[105] suggesting that the reactive center loop has the exposed conformation typical of serpins. In contrast to TBG and CBG this cleavage does not result in increased stability,[105,106] implying that expansion of β-sheet A to accommodate the N-terminal portion of the cleaved reactive center region does not occur. Whereas this failure to loop-insert may be due to the shortness of the remaining portion of the reactive center region following V8 proteolysis, it is more likely to be due to the presence of unfavorable residues at both P14 and P12. P12 is proline in both rat and human angiotensinogens[90,94] and P14 is a charged residue (arginine in human and glutamate in rat and sheep) rather than the serine, threonine, or valine found in most loop-insertable serpins. Both in the nature of the residue at P14 and in the failure to loop insert following reactive center loop cleavage, angiotensinogen thus behaves like ovalbumin rather than like CBG or TBG.

Most interest in angiotensinogen arises from its potential role in hypertension. A genetic linkage has been shown between angiotensinogen and essential hypertension, with angiotensinogen variants being associated with the disease and with differences in plasma concentrations.[107,108] An M235T variant that occurs with high frequency in the Japanese population is correlated with essential hypertension[109] and with increased risk of coronary heart disease.[110] An in vivo model, using mice containing 0, 1, 2, 3, and 4 copies of the murine angiotensinogen gene, showed a strong, but not linear, correspondence of angiotensinogen level with gene copy number, with 145% of the normal level being expressed with four copies.[111] All mice were normal at birth, but those mice with no angiotensinogen gene died before weaning. An increase in blood pressure of about 8 mm Hg per gene copy was found despite the presence of normal compensatory mechanisms of blood pressure regulation.[111] In a transgenic mouse model, the presence of human angiotensinogen mRNA was demonstrated in liver, kidney, heart, adrenal gland, ovary, brain and both white and brown adipose tissue.[112] The plasma levels of angiotensinogen were 150-fold higher than in humans, though with normal blood pressure, perhaps through the inability of mouse renin to process the N-terminus. Administration of a single bolus of human renin resulted in an increase in blood pressure of 30 mm Hg within 2 min,[112] supporting a role for angiotensinogen levels resulting in hypertension. A mutation within the N-terminal region (L10F) has been found in a patient with preeclampsia,[113] an illness associated with pregnancy when there is increased demand on the renin-angiotensin system. This mutation results in an alteration of the kinetics of the renin-angiotensinogen interaction, with 10-fold reduction in K_m, from 1.1 to 0.09 μM, leading to an overall increase of 2-fold in k_{cat}/K_m.

5. PIGMENT EPITHELIUM-DERIVED FACTOR

A recently discovered non-inhibitory serpin has been identified independently by two groups, based on two different biological properties and has thus been given two names; PEDF (for pigment epithelium-derived factor) and EPC-1 (for early population doubling level cDNA-1). This serpin was first identified as a protein secreted by cultured human pigment epithelial cells from the fetal retina and was found to induce a neuronal phenotype in cultured human retinoblastoma cells when added in nanogram amounts.[114,115] It has now been shown to be expressed at high levels in both fetal and adult monkey retinal pigment epithelial cells.[116] In response to PEDF, retinoblastoma cells extend long dendrite-like structures which form an elaborate meshwork (Fig. 6.4). Levels of neuron-specific enolase and neurofilament 200 kDa protein, both indicators of neuronal maturation, are also elevated in response to PEDF treatment. The neurotrophic effects of PEDF are not, however, confined to neurons of the eye. Thus PEDF has been shown to be a potent survival factor for an abundant mammalian brain neuronal cell subtype, cerebellar granule cells, but not to affect mitosis or neurite outgrowth in these cells.[117]

The later isolation of cDNAs for several genes that are differentially-expressed in young versus senescent population doubling level (PDL) human WI-38, fetal lung-derived, fibroblast-like cells, led to the identification of one gene that is expressed at levels ≥ 100-fold higher in young versus senescent cells.[118] This cDNA was named EPC-1 for early PDL cDNA-1.[118] The encoded protein is identical to PEDF. It was shown that EPC-1 mRNA and the corresponding secreted protein are abundantly present in early PDL WI-38 cells that have been serum-starved or density-arrested in their growth, but that the levels of both mRNA and protein become negligible both in exponentially growing early PDL WI-38 cells and in late PDL WI-38 cells under all growth conditions.[118] Thus entry into the G_0 state appears to induce expression of this serpin. Possible involvement of PEDF in cellular differentiation and senescence has also been shown from studies on cultured retinal pigment epithelial cells. Levels of PEDF and of the mRNA transcript decrease with cell passage, such that they are no longer detectable by the 10th passage, at which time the cells have changed their morphology from epithelial-like to fibroblast-like.[116] It has also been shown that the EPC-1 transcript is restricted to specific cell types, being present in fibroblasts, vascular smooth muscle cells, T lymphocytes and retinal pigmented epithelial cells, but absent from vascular endothelial cells, glial cells, and peritoneal mesothelial cells.[119] In addition it was demonstrated that related DNAs are present in all mammalian species examined (human, monkey, rat, mouse, dog, cow, and rabbit) as well as in chicken, but absent, by Southern blot analysis, from yeast.[119]

As deduced from the cDNA-derived amino acid sequence, human PEDF is synthesized as a 418 amino acid protein containing a 17 residue signal sequence. There is one possible glycosylation site at position 285 (<u>N</u>LT), which may bear carbohydrate, given that the mobility of PEDF observed on polyacrylamide gels is lower than would be expected for a non-glycosylated protein of this size.[115] There are three cysteine residues, though it is not known whether any of these is involved in a disulfide bond. It has been proposed that retinal pigment epithelial cells might secrete PEDF into the interphotoreceptor matrix during fetal development and there modulate retinal neurite outgrowth and maturation.[120] In support of this PEDF has been shown to be present in the interphotoreceptor matrix of bovine eyes[121] and to have the same neurotrophic activity on retinoblastoma cells as had previously been shown for the crude extract from the interphotoreceptor matrix. Purified bovine PEDF has been shown to be N-glycosylated by the effect that N-glycosidase F treatment has on electrophoretic mobility and to carry about 5% carbohydrate, consistent with one site of glycosylation.

PEDF has been expressed in *E. coli* as a truncated form (Asp 44-Pro418) which is associated predominantly with inclusion bodies. Biologically active PEDF was obtained from these inclusion bodies by denaturation in 5 M guanidine HCl and dilution-renaturation. Thus neither glyco-

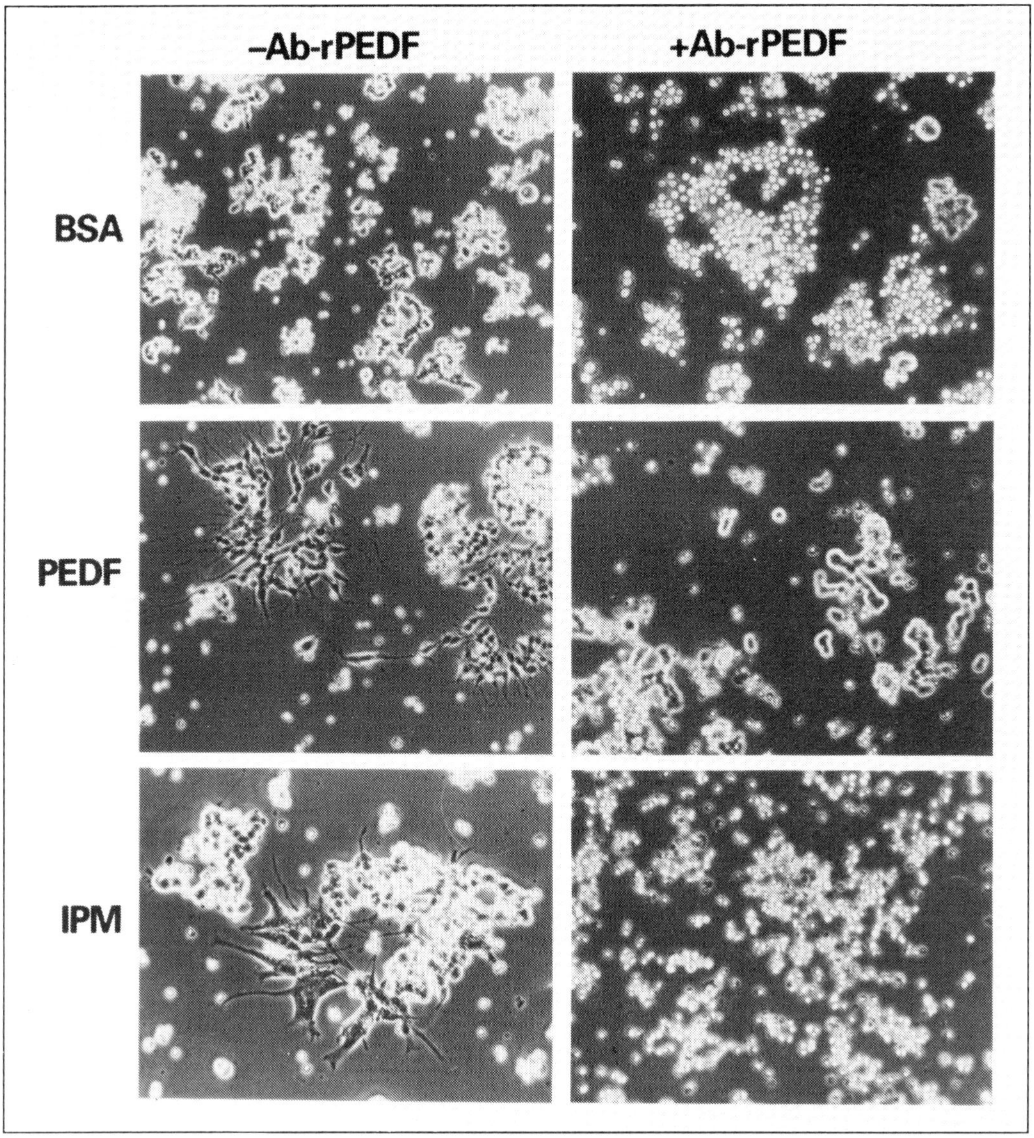

Fig. 6.4. Biological effect of PEDF on retinoblastoma Y79 cell morphology. In the absence of anti-PEDF antibody (left panels), neurite outgrowth is apparent upon treatment of cells with PEDF or interphotoreceptor matrix (IPM) but not upon treatment with bovine serum albumin (BSA). In the presence of anti-PEDF antibody (right panels), the biological activity of PEDF is abrogated. Reprinted with permission from Protein Expression and Purification.*[121] Copyright 1995 by Academic Press Inc. Figure kindly provided by Dr. Patricia Becerra.*

sylation nor the amino terminal 27 residues (after processing) are required for biological activity. However, a minimum concentration of 2M guanidine HCl was needed to keep the protein in solution, suggesting that the recombinant protein does not adopt its native conformation. Despite this the recombinant protein still retained biological activity by promoting neurite outgrowth in Y-79 cells,[122] suggesting that the biologically active region may reside within a single portion of the primary structure and not require the proper three-dimensional fold for activity. This is further supported by the presence of neurite-promoting activity in both reactive center loop cleaved PEDF and in recombinant fragments with larger C-terminal truncations.[123]

Neither recombinant human PEDF[122] nor bovine eye PEDF[121] showed proteinase inhibitory activity against a spectrum of proteinases, including chymotrypsin, trypsin, endoproteinase Lys-C, subtilisin, and elastase. The reason for this was not clear for the recombinant human PEDF, since it was necessary to maintain 0.5 M urea to keep the protein in solution, thereby possibly promoting formation of a polymerized or latent form. However, the inability of the bovine protein to form SDS-stable complexes with these proteinases strongly suggests the absence of inhibitory activity for PEDF. The proteinases give rise to limited cleavages within the reactive center region, but the resultant cleaved PEDF species show no enhanced thermal stability, indicating that the reactive center loop does not undergo sheet insertion.[123,124] In keeping with this, the primary structure of the reactive center region appears not to be favorable towards facile loop insertion. There are significant differences between the reactive center loop sequences of PEDF and of a number of inhibitory serpins (Fig. 6.1). In particular there is a proline at P10 that may reduce the rate of loop insertion and thereby result in predominantly substrate reaction of the serpin. A Thr→Pro substitution at P12 in a recombinant α_1-proteinase inhibitor variant resulted in conversion of the serpin from an inhibitor to a predominant (though not exclusive) substrate.[125]

6. UTERINE MILK PROTEINS

Two closely-related and very abundant serpins are present in the uterine secretions of the sheep during pregnancy and also as a result of chronic administration of progesterone. These serpins, which have been termed "uterine milk proteins" or UTMP, represent the dominant progesterone-induced proteins of uterine mik, with as much as 15 g being isolated from one uterine horn of a pregnant ewe.[126] The two sheep proteins are indistinguishable by peptide mapping,[127] but differ slightly in electrophoretic mobility, suggesting a single polypeptide with some difference in post-translational modification, though both are reported to carry the mannose-6-phosphate recognition marker.[128] They were identified as members of the serpin family from the cDNA-derived primary structure.[129] Analogous proteins, although less well characterized than those from the sheep, have been identified in uterine secretions from the pig[130] and the cow.[131]

Although no proteinase inhibitory properties have been found for UTMP,[129] a number of other properties appear to be associated with them and may account both for the acute phase, progesterone-induced, nature of their biosynthesis and their abundance in uterine fluid. Thus uterine milk proteins have been shown to inhibit lymphocyte proliferation in pregnant ewes[132,133] and to be responsible, through their abundance, for the majority of such suppression caused by uterine fluids even though a much less abundant uterine fluid protein, megasuppressin, has a higher anti-proliferative specific activity.[132] UTMP have also been shown to suppress phytohemagglutinin-stimulated lymphocyte blastogenesis.[134] Another potentially important activity is as an inhibitor of natural killer cells. At physiological concentrations UTMP have been shown to inhibit the cytotoxic activities of cultured natural killer cells from both sheep and mice.[135] Finally it has been shown that UTMP bind to immunoglobulins, but in a class-specific manner, such that preference is for IgA, but with binding also to IgM in a salt concentration dependent manner.[136] Although the significance of these activities has not been determined, it is likely that they are involved in successfully bringing the pregnancy to term.

The primary structure deduced from the cDNA of sheep UTMP corresponds to a mature protein of 404 residues with a 25 residue hydrophobic signal sequence. There are two possible N-glycosylation sites. The reported sequence alignment with five other serpin sequences (ovalbumin, human antithrombin, human protein C inhibitor, human α_1-antichymotrypsin and bovine α_1-proteinase inhibitor) places the potential P1-P1' scissile bond between Ala 365 and Asn 366 and consequently makes the P14 residue Asp 353.[129] An alternative alignment is consistent with Leu 350 being the P14 residue (Fig. 6.1). In either case the P14 residue is not one found in inhibitory serpins and in the case of a negatively charged residue has only been found for the non-loop-inserting serpin rat angiotensinogen. The unusual composition and sequence of the reactive center region compared with other inhibitory serpins, together with unfavorable candidate P14 residues, make it unlikely that UTMP are inhibitors of any proteinases or even that they undergo proteinase-induced conformational change. In this regard, in their relative abundance, and in their induction by progesterone during fetal development, they have considerable similarity with hen ovalbumin. It would be interesting to see if these similarities extend further to common physiological roles.

7. *XENOPUS* SERPIN

One of four mRNAs that become highly abundant as a result of estrogen treatment of *Xenopus* frogs has been shown to encode a 45 kDa protein of the serpin family.[137] The protein has been named Ep-45. An independently identified *Xenopus* protein, named pNiXa because of its affinity for nickel, is present both in *Xenopus* oocytes and developing embryos and is identical to Ep-45.[138] The *Xenopus* genome contains one copy of the *Ep-45* gene. The mRNA encodes a 436 residue protein containing a 16 residue hydrophobic leader sequence, expected for a secreted protein. There are three potential N-linked glycosylation sites, at positions 244 (NPT), 260 (NVT) and 289 (NDT). Sequence alignment with other serpins leads to a prediction of the P1-P1' residues being Thr-Lys, a combination not found in any other serpin, and most closely similar to mouse contrapsin (Arg-Lys).[139] Although the P11 residue (valine) is also unusual, the threonine at P14, alanine at P12 and glycine at P10 are not incompatible with this being an inhibitory serpin. It may, however, have more in common with other steroid hormone-induced serpins such as UTMP and ovalbumin and serve a role as a non-inhibitory, but nevertheless important protein during egg maturation and embryogenesis.

8. MASPIN

By use of subtractive hybridization and differential display techniques a number of potential tumor suppressor genes have been identified. One of these, termed maspin, is a serpin of 375 residues which shares 43% homology with equine leucocyte elastase inhibitor and 31% homology with PAI-2 and has a putative Arg-Ile reactive center bond.[140] The maspin gene is located on chromosome 18 at q21.3, together with plasminogen activator inhibitor-2, and two forms of squamous cell carcinoma antigen (SCCA1 and SCCA2).[18] A single 3.0 kb maspin mRNA is present in normal mammary epithelial cells and results in protein expression. The mRNA is missing in most mammary tumor cell lines. Transfection of MDA-MB-435 mammary carcinoma cells with the maspin gene abrogated their ability to induce tumors and metastasize in nude mice. In addition, immunostaining revealed that maspin antigen was reduced in advanced tumors. Recombinant maspin has been produced and shown to inhibit the invasion of breast tumor cells across basement membrane. This activity was abolished if the reactive center was cleaved by trypsin, suggesting that inhibition of a proteinase could be essential for this activity.[141] However, it has recently been shown that maspin does not undergo loop insertion following cleavage within the reactive center loop,[142] making it very unlikely that it is capable of functioning as a proteinase inhibitor. The failure to undergo loop insertion is not surprising given the unusual and probably unfavorable hinge region sequence of the reactive center loop.[143] Thus the P12 residue is charged (aspartate) and the common Ala-Ala-Ala-Ala motif between P12 and P9 is replaced by Asp-Gly-Gly-Asp.

9. COLLAGEN BINDING PROTEIN

A novel family of intracellular serpins are those which have been independently identified as J6, HSP47, gp46, colligin, and collagen binding protein. These proteins all appear to be the same protein, allowing for minor variations in sequence due to species differences.[144] Although these proteins are the same they will be discussed separately. J6 was identified as a retinoic acid inducible gene in mouse F9 teratocarcinoma cells, and was shown to be a serpin. The predicted P1-P1' residues were Arg-Ser. The time course for the induction of J6 expression was similar to that of laminin B1 and collagen (IV) $\alpha 1$.[145] The gene for J6 has been characterized and shown to have similarity to the gene for α_1-proteinase inhibitor rather than ovalbumin as might be expected for an intracellular serpin.[146] Also, a potential heat shock element was found, suggesting that the gene might be heat inducible. Further analysis revealed that the gene contains a GAGATAG sequence which is repeated four times. This sequence binds to a retinoic acid induced protein which could regulate the expression of J6.[147] Subsequently, the retinoic acid-inducible GATA-binding protein, GATA-4, has been shown to bind to the consensus motif in the retinoic acid-sensitive region of the J6 promoter.[148] Expression of HSP47 was also studied from F9 cells (and thus is identical to J6) and this gene was shown to be inducible by retinoic acid and heat shock. Expression of collagen paralleled that of HSP47. Sequencing of the cDNA revealed the homology to serpins and showed that the C-terminal end of the protein contained a putative endoplasmic reticulum retention signal.[149] The gene for HSP47 has been studied and shown to contain a TATA box, four Sp1-binding sites and one AP-1 binding site, as well as a heat shock response element.[150] An HSP47 from chicken embryo fibroblasts has also been identified and shown to bind collagen, to have an endoplasmic reticulum retention signal, to be a heat shock protein and to be a serpin.[151] Colligin, which is also called collagen binding protein or gp46, is present in the endoplasmic reticulum of rat L6 myoblasts. It was shown to be a heat shock protein, to bind to collagen and to be a serpin.[152] Recombinant colligin has been produced and the interaction with procollagen 1 studied.[153] The function of this family of serpins is not known. They are heat shock proteins and might function as molecular chaperones in the synthesis of collagen. It is not known whether they have proteinase inhibitory activity and whether such activity is involved in regulating collagen biosynthesis. However, their primary structure in the reactive center loop region makes it likely that they are not capable of the rapid loop insertion required to make them inhibitory serpins (Fig. 6.1). Thus the presence of phenylalanine at P12 and glutamine at P10 make this quite different from the inhibitory serpins, which commonly contain alanine at both of these positions. In addition a proline at P13 rather than the usual glutamate is unlikely to promote loop insertion.

References

1. Hunt LT and Dayhoff MO. A surprising new protein superfamily containing ovalbumin, antithrombin III and α_1-proteinase inhibitor. Biochem Biophys Res Commun 1980; 95:864-871.
2. Stein PE, Leslie AGW, Finch JT, et al. Crystal structure of uncleaved ovalbumin at 1.95Å resolution. J Mol Biol 1991; 221:941-959.
3. Warner RC. Egg proteins. In: Neurath H, Bailey K, eds. The Proteins. New York: Academic Press, 1954:435-485.
4. Hofmeister F. Über die Darstellung von krystallisirtem Eieralbumin und die Krystallisirbarkeit colloider Stoffe. Z Physiol Chem 1892; 14:165-172.
5. McReynolds L, O'Malley BW, Nisbet AD, et al. Sequence of chicken ovalbumin mRNA. Nature 1978; 273:723-728.
6. Nisbet AD, Saundry RH, Moir AJG, et al. The complete amino acid sequence of hen ovalbumin. Eur J Biochem 1981; 115: 335-345.
7. O'Malley BW and Means AR. Female steroid hormones and target cell nuclei. Science 1974; 182:610-620.
8. Cunningham LW, Clouse RW and Ford JD. Heterogeneity of the carbohydrate moiety of crystalline ovalbumin. Biochim Biophys Acta 1963; 78:379-381.
9. Tai T, Yamashita K, Ito S, et al. Structure of the carbohydrate moiety of ovalbumin glycopeptide III and the difference in specificity of endo-β-N-acetylglucosaminidases C11 and H. J Biol Chem 1977; 252: 6687-6694.
10. Atkinson PH, Grey A, Carver JP, et al. Demonstration of heterogeneity of chick ovalbumin glycopeptides using 360MHz proton magnetic resonance spectroscopy. Biochemistry 1981; 20:3979-3986.
11. Ishihara H, Takahashi N, Ito J, et al. Either high-mannose-type or hybrid-type oligosaccharide is linked to the same asparagine residue in ovalbumin. Biochim Biophys Acta 1981; 669:216-221.
12. Mucha J, Klaudiny J, Klaudinyová V, et al. The sequence of Japanese quail ovalbumin cDNA. Nucl Acids Res 1990; 18:5553.
13. Cann JR. Electrophoretic analysis of ovalbumin. J Am Chem Soc 1949; 71:907-909.
14. Robinson A, Meredith C and Austen BM. Isolation and properties of the signal region from ovalbumin. FEBS Lett 1986; 203: 243-246.
15. Narita K and Ishii J. N-terminal sequence in ovalbumin. J Biochem 1962; 52:367-373.
16. Goux WJ and Venkatasubramanian PN. Metal ion binding properties of hen ovalbumin and S-ovalbumin: Characterization of the metal ion binding site by ^{31}P NMR and water proton relaxation rate enhancements. Biochemistry 1986; 25:84-94.
17. Heilig R, Muraskowsky R, Kloepfer C, et al. The ovalbumin gene family: complete sequence and structure of the Y gene. Nucl Acids Res 1982; 10:4363-4382.
18. Schneider SS, Schick C, Fish KE, et al. A serine proteinase inhibitor locus at 18q21.3 contains a tandem duplication of the human squamous cell carcinoma antigen gene. Proc Natl Acad Sci USA 1995; 92: 3147-3151.
19. Ye RD, Ahern SM, Le Beau MM, et al. Structure of the gene for human plasminogen activator inhibitor-2. The nearest mammalian homologue of chicken ovalbumin. J Biol Chem 1989; 264:5495-5502.
20. Evans E, Cooley J and Remold-O'Donnell E. Characterization and chromosomal localization of $ELANH_2$, the gene encoding human monocyte/neutrophil elastase inhibitor. Genomics 1995; 28:235-240.
21. Remold-O'Donnell E. The ovalbumin family of serpin proteins. FEBS Lett 1993; 315:105-108.
22. Ottesen M and Wollenberger A. Stepwise degradation of the peptides liberated in the transformation of ovalbumin to plakalbumin. Compt rend Lab Carlsberg Sér chim 1956; 28:463-475.
23. Ottesen M. The transformation of ovalbumin into plakalbumin. Compt rend Lab Carlsberg Sér chim 1958; 30:211-270.
24. Wright HT. Ovalbumin is an elastase substrate. J Biol Chem 1984; 259: 14335-14336.
25. Carrell RW and Owen MC. Plakalbumin, α_1-antitrypsin, antithrombin and the mechanism of inflammatory thrombosis. Nature 1985; 317:730-732.

26. Gettins P. Absence of large-scale conformational change upon limited proteolysis of ovalbumin, the prototypic serpin. J Biol Chem 1989; 264:3781-3785.
27. Bruch M, Weiss V and Engel J. Plasma serine proteinase inhibitors (serpins) exhibit major conformational changes and a large increase in conformational stability upon cleavage at their reactive sites. J Biol Chem 1988; 263:16626-16630.
28. Wright HT, Qian HX and Huber R. Crystal structure of plakalbumin, a proteolytically nicked form of ovalbumin. Its relationship to the structure of cleaved α-1-proteinase inhibitor. J Mol Biol 1990; 213:513-528.
29. Huntington JA, Patston PA and Gettins PGW. S-ovalbumin, an ovalbumin conformer with properties analogous to those of loop inserted serpins. Prot Sci 1995; 4:613-621.
30. Smith MB and Back JF. Studies on ovalbumin II. The formation and properties of S-ovalbumin, a more stable form of ovalbumin. Aust J Biol Sci 1965; 18:365-377.
31. Royal A, Garapin A, Cami B, et al. The ovalbumin gene region: common features in the organization of three genes expressed in chicken oviduct under hormonal control. Nature 1979; 279:125-132.
32. Korcek L and Tabachnik M. Thyroxine-protein interactions. Interaction of thyroxine and triiodothyronine with human thyroxine-binding globulin. J Biol Chem 1976; 251:3558-3562.
33. Bartalena L. Recent achievements in studies on thyroid hormone-binding proteins. Endocrine Reviews 1990; 11:47-64.
34. Huber R and Carrell RW. Implications of the three-dimensional structure of α_1-antitrypsin for structure and function of serpins. Biochemistry 1989; 28:8951-8966.
35. Tabachnik M and Perret V. Specific labeling of the thyroxine binding site in the thyroxine-binding globulin. Determination of the amino acid composition of a labeled peptide fragment isolated from a proteolytic digest of the derivatized protein. Biochem Int 1987; 15:409-417.
36. Newcomer ME, Jones TA, Åqvist J, et al. The three-dimensional structure of retinol-binding protein. EMBO J 1984; 3:1451-4.
37. Sacchettini JC and Gordon JI. Rat intestinal fatty acid binding protein. A model system for analyzing the forces that can bind fatty acids to proteins. J Biol Chem 1993; 268:18399-18402.
38. Noy N, Slosberg E and Scarlata S. Interactions of retinal with binding proteins: studies with retinol-binding protein and transthyretin. Biochemistry 1992; 31: 11118-11124.
39. Jarvis JA, Munro SLA and Craik DJ. Homology model of thyroxine binding globulin and elucidation of the thyroid binding site. Prot Eng 1992; 5:61-67.
40. Flink IL, Bailey TJ, Gustafson TA, et al. Complete amino acid sequence of human thyroxine-binding globulin deduced from cloned DNA: Close homology to the serine antiproteases. Proc Natl Acad Sci USA 1986; 83:7708-7712.
41. Bartalena L, Tera JR and Robbins J. Characterization of nascent and secreted thyroxine-binding globulin in cultured hepatic (HepG2) cells. J Biol Chem 1984; 259: 13605-13609.
42. Refetoff S, Robin NI and Alper CA. Study of four new kindreds with inherited thyroxine-binding globulin abnormalities. Possible mutations of a single gene locus. J Clin Invest 1972; 51:848-867.
43. Burr WA, Ramsden DB and Hoffenburg R. A study of 19 kindreds with inherited increase or decrease of thyroxine-binding globulin. Q J Med 1980; 49:295-313.
44. Trent JM, Flink IL, Morkin E, et al. Localization of the human thyroxine-binding globulin gene to the long arm of the X chromosome (Xq21-22). Am J Hum Genet 1987; 41:428-435.
45. Hayashi Y, Mori Y, Janssen OE, et al. Human thyroxine-binding globulin gene—complete sequence and transcriptional regulation. Mol Endocrinol 1993; 7:1049-1060.
46. Zinn AB, Marshall JS and Carlson DM. Preparation of glycopeptides and oligosaccharides from thyroxine-binding globulin. J Biol Chem 1978; 253:6761-6767.
47. Zinn AB, Marshall JS and Carlson DM. Carbohydrate structures of thyroxine-binding globulin and their effects on hepatocyte

membrane binding. J Biol Chem 1978; 253:6768-6773.

48. Marshall JS, Pensky J and Williams S. Studies on human thyroxine-binding globulin. VII. Isoelectric focussing evidence for microheterogeneity of thyroxine-binding globulin. Arch Biochem Biophys 1973; 156:456-462.
49. Gärtner R, Henze R, Horn K, et al. Thyroxine-binding globulin: investigation of microheterogeneity. J Clin Endocrinol Metab 1981; 52:657-664.
50. Tani Y, Mori Y, Miura Y, et al. Molecular cloning of the rat thyroxine-binding globulin gene and analysis of its promoter activity. Endocrinology 1994; 135:2731-2736.
51. Janssen OE and Refetoff S. *In vitro* expression of thyroxine-binding globulin (TBG) variants. Impaired secretion of $TBG^{Pro\,227}$ but not $TBG^{Pro\,113}$. J Biol Chem 1992; 267: 13998-14004.
52. Pemberton PA, Stein PE, Pepys MB, et al. Hormone binding globulins undergo serpin conformational change in inflammation. Nature 1988; 336:257-258.
53. Refetoff S. Inherited thyroxine-binding globulin abnormalities in man. Endocrine Reviews 1989; 10:275-293.
54. Kambe F, Seo H, Mori Y, et al. An additional carbohydrate chain in the variant thyroxine binding globulin-Gary ($TBG^{Asn\text{-}96}$) impairs its secretion. Mol Endocrinol 1992; 6:443-449.
55. Peterson CB and Blackburn MN. Isolation and characterization of an antithrombin III variant with reduced carbohydrate content and enhanced heparin binding. J Biol Chem 1985; 260:610-615.
56. Brennan SO, Borg T-Y, George PM, et al. New carbohydrate site in mutant antithrombin (7Ile→Asn) with decreased heparin affinity. FEBS Lett 1988; 237:118-122.
57. Siiteri PK, Murai JT, Hammond GL, et al. The serum transport of steroid hormones. Recent Prog Horm Res 1982; 38:457-503.
58. Henning SJ. Plasma concentrations of total and free corticosterone during development in the rat. Am J Physiol 1978; 235:E451.
59. Seal US and Doe RP. Corticosteroid-binding globulin: species distribution and small-scale purification. Endocrinology 1963; 73:371-376.
60. Loose DS and Feldman D. Characterization of a unique corticosterone-binding protein in *Candida albicans*. J Biol Chem 1982; 257:4925-4930.
61. Rosner W, Darmstadt RA and Toppel S. A simple precipitin test for corticosteroid-binding globulin: use in a search for the absence of corticosteroid-binding globulin in 10,000 subjects. J Clin Endocrinol Metab 1973; 37:983-985.
62. Hammond GL, Smith CL, Goping IS, et al. Primary structure of human corticosteroid binding globulin, deduced from hepatic and pulmonary cDNAs, exhibits homology with serine protease inhibitors. Proc Natl Acad Sci USA 1987; 84:5153-5157.
63. Smith CL and Hammond GL. Rat corticosteroid binding globulin: primary structure and messenger ribonucleic acid levels in the liver under different physiological conditions. Mol Endocrinol 1989; 3:420-426.
64. Seralini G-E, Smith CL and Hammond GL. Rabbit corticosteroid-binding globulin: Primary structure and biosynthesis during pregnancy. Mol Endocrinol 1990; 4: 1166-1172.
65. Hammond GL, Smith CL and Underhill DA. Molecular studies of corticosteroid binding globulin structure, biosynthesis and function. J Steroid Biochem Molec Biol 1991; 40:755-762.
66. Kuhn RW, VestWeber C and Siiteri PK. Purification and properties of squirrel monkey (*Saimiri sciureus*) corticosteroid binding globulin. Biochemistry 1988; 27: 2579-2586.
67. Schreuder HA, de Boer B, Dijkema R, et al. The intact and cleaved human antithrombin III complex as a model for serpin-proteinase interactions. Nature Struct Biol 1994; 1:48-54.
68. Carrell RW, Stein PE, Fermi G, et al. Biological implications of a 3Å structure of dimeric antithrombin. Structure 1994; 2:257-270.
69. Akhrem AA, Avvakumov GV, Akhrem LV, et al. Structural organization of the carbohydrate moiety of human transcortin as determined by methylation analysis of the whole glycoprotein. Biochim Biophys Acta

1982; 714:177-180.

70. Hammond GL, Smith CL, Paterson NAM, et al. A role for corticosteroid-binding globulin in delivery of cortisol to activated neutrophils. J Clin Endocrinol Metab 1990; 71:34-39.
71. Ghose-Dastidar J, Ross JBA and Green R. Expression of biologically active human corticosteroid binding globulin by insect cells: Acquisition of function requires glycosylation and transport. Proc Natl Acad Sci USA 1991; 88:6408-6412.
72. Avvakumov GV and Hammond GL. Glycosylation of human corticosteroid binding globulin. Differential processing and significance of carbohydrate chains at individual sites. Biochemistry 1994; 33:5759-5765.
73. Avvakumov GV, Warmels-Rodenhiser S and Hammond GL. Glycosylation of human corticosteroid-binding globulin at asparagine 238 is necessary for steroid binding. J Biol Chem 1993; 268:862-866.
74. Defaye G, Basset M, Monnier N, et al. Electron spin resonance study of human transcortin thiol groups and binding site topography. Biochim Biophys Acta 1980; 623:280-294.
75. Mornon JP, Bissery V, Gaboriaud C, et al. Hydrophobic cluster analysis (HCA) of the hormone-binding domain of receptor proteins. J Steroid Biochem Molec Biol 1989; 34:355-361.
76. Khan MS and Rosner W. Investigation of the binding site of human corticosteroid-binding globulin by affinity labeling. Demonstration of a cysteinyl residue in the binding site. J Biol Chem 1977; 252:1895-1900.
77. Ghose-Dastidar J, Green R and Ross JBA. Identification of the cysteine in the steroid-binding site of human corticosteroid binding globulin by site-directed mutagenesis and site-specific chemical modification. J Steroid Biochem Molec Biol 1994; 48: 139-144.
78. Avvakumov GV and Hammond GL. Substitutions of tryptophan residues in human corticosteroid-binding globulin: Impact on steroid binding and glycosylation. J Steroid Biochem Molec Biol 1994; 49:191-194.
79. Stroupe SD, Harding GB, Forsthoefel MW, et al. Kinetic and equilibrium studies on steroid interaction with human corticosteroid-binding globulin. Biochemistry 1978; 17:171-182.
80. Akhrem AA, Avvakumov GV, Kukushkina II, et al. Tryptophan in the steroid-binding site of transcortin. Bioorg Khim 1978; 4:421-423.
81. Akhrem AA, Sviridov OV and Strel'chyonok OA. Specific interactions in transcortin-steroid complex. Bioorg Khim 1979; 5: 711-720.
82. Akhrem AA, Sviridov OV, Strel'chyonok OA, et al. Study of steroid-binding site in human transcortin using tryptophan-specific reagents. Bioorg Khim 1981; 7:662-669.
83. Grenot C, Blachère T, De Ravel MR, et al. Identification of Trp-371 as the main site of specific photoaffinity labeling of corticosteroid binding globulin using Δ^6 derivatives of cortisol, corticosterone, and progesterone as unsubstituted photoreagents. Biochemistry 1994; 33:8969-8981.
84. Smith CL, Power SGA and Hammond GL. A Leu→His substitution at residue 93 in human corticosteroid binding globulin results in reduced affinity for cortisol. J Steroid Biochem Molec Biol 1992; 42:671-676.
85. Hammond GL. Molecular properties of corticosteroid binding globulin and the sex-steroid binding proteins. Endocrine Reviews 1990; 11:65-79.
86. Hammond GL, Smith CL, Underhill CM, et al. Interaction between corticosteroid binding globulin and activated leukocytes *in vitro*. Biochem Biophys Res Commun 1990; 172:172-177.
87. Underhill DA and Hammond GL. Organization of the human corticosteroid binding globulin gene and analysis of its 5' flanking region. Mol Endocrinol 1989; 3:1448-1454.
88. Seralini G-E, Bérubé D, Gagné R, et al. The human corticosteroid binding globulin gene is located on chromosome 14 q31-q32.1 near two other serine protease inhibitor genes. Hum Genet 1990; 86:73-75.
89. Gettins P and Harten B. Properties of thrombin- and elastase-modified human antithrombin III. Biochemistry 1988; 27:3634-3639.
90. Kageyama R, Ohkubo H and Nakanishi S. Primary structure of human preangio-

tensinogen deduced from the cloned cDNA sequence. Biochemistry 1984; 23: 3603-3609.

91. Fukamizu A, Takahashi S, Seo MS, et al. Structure and expression of the human angiotensinogen gene. Identification of a unique and highly active promoter. J Biol Chem 1990; 265:7576-7582.
92. Tanaka T, Hiroaki O and Nakanishi S. Common structural organization of the angiotensinogen and the α1-antitrypsin genes. J Biol Chem 1984; 259:8063-8065.
93. Nibu Y, Takahashi S, Tanimoto K, et al. Identification of cell type-dependent enhancer core element located in the 3' downstream region of the human angiotensinogen gene. J Biol Chem 1994; 269:28598-28605.
94. Ohkubo H, Kageyama R, Ujihara M, et al. Cloning and sequence analysis of cDNA for rat angiotensinogen. Proc Natl Acad Sci USA 1983; 80:2196-2200.
95. Nagase M, Suzuki F, Fukamizu A, et al. Sequencing and expression of sheep angiotensinogen cDNA. Biosci Biotechnol Biochem 1994; 58:1884-1885.
96. Clouston WM, Evans BA, Haralambidis J, et al. Molecular cloning of the mouse angiotensinogen gene. Genomics 1988; 2:240-248.
97. Braun-Menendez E, Fasciolo JC, Leloir LF et al. Renal Hypertension, Springfield: Thomas, 1946:134.
98. Skeggs LT, Lentz KE, Kahn JR, et al. Kinetics of the reaction of renin with nine synthetic substrates. J Exp Med 1968; 128:13-34.
99. Poulsen K, Haber E and Burton J. Specificity of human renin. Studies with peptide inhibitors. Biochim Biophys Acta 1976; 452:533-537.
100. Tewksbury DA and Tryon ES. Immunochemical comparison of high molecular weight angiotensinogen from amniotic fluid, plasma of men, and plasma of pregnant women. Am J Hypertension 1989; 2: 411-413.
101. Oxvig C, Haaning J, Kristensen L, et al. Identification of angiotensinogen and complement C3dg as novel proteins binding the proform of eosinophil major basic protein in human pregnancy serum and plasma. J Biol Chem 1995; 270: 13645-13651.
102. Sato Y, Hiwada K and Kokubu T. Physicochemical characteristics of human high molecular weight angiotensinogen. Life Sciences 1985; 37:371-377.
103. Kunapoli SP, Prasad GL and Kumar A. Expression of human angiotensinogen cDNA in *Escherichia coli*. J Biol Chem 1987; 262:7672-7675.
104. Hatae T, Takimoto E, Kukamizu A, et al. Expression and purification of human angiotensinogen in Chinese hamster ovary cells. Biochim Biophys Acta 1992; 1121: 335-338.
105. Mast AE, Enghild JJ, Pizzo SV, et al. Analysis of the plasma elimination kinetics and conformational stabilities of native, proteinase-complexed, and reactive site cleaved serpins: Comparison of α_1-proteinase inhibitor, α_1-antichymotrypsin, antithrombin III, α_2-antiplasmin, angiotensinogen, and ovalbumin. Biochemistry 1991; 30:1723-1730.
106. Stein PE, Tewkesbury DA and Carrell RW. Ovalbumin and angiotensinogen lack serpin S-R conformational change. Biochem J 1989; 262:103-107.
107. Jeunemaitre X, Soubrier F, Kotelevtsev YV, et al. Molecular basis of human hypertension: role of angiotensinogen. Cell 1992; 71:169-180.
108. Caulfield M, Lavender P, Newell-Price J, et al. Linkage of the angiotensinogen gene locus to human essential hypertension in African Caribbeans. J Clin Invest 1995; 96:687-692.
109. Hata A, Namikawa C, Sasaki M, et al. Angiotensinogen as a risk factor for essential hypertension in Japan. J Clin Invest 1994; 93:1285-1287.
110. Katsuya T, Koike G, Yee TW, et al. Association of angiotensinogen gene T235 variant with increased risk of coronary heart disease. Lancet 1995; 345:1600-1603.
111. Kim H-S, Krege JH, Kluckman KD, et al. Genetic control of blood pressure and the angiotensinogen locus. Proc Natl Acad Sci USA 1995; 92:2735-2739.
112. Yang G, Merrill DC, Thompson MW, et al. Functional expression of the human angiotensinogen gene in transgenic mice. J

Biol Chem 1994; 269:32497-32502.

113. Inoue I, Rohrwasser A, Helin C, et al. A mutation of angiotensinogen in a patient with preeclampsia leads to altered kinetics of the renin-angiotensin system. J Biol Chem 1995; 270:11430-11436.
114. Tombran-Tink J and Johnson L. Neuronal differentiation of retinoblastoma cells induced by medium conditioned by human RPE cells. Invest Ophthalmol Vis Sci 1989; 30:1700-1707.
115. Tombran-Tink J, Chader GJ and Johnson LV. PEDF: A pigment epithelium-derived factor with potent neuronal differentiative activity. Exp Eye Res 1991; 53:411-414.
116. Tombran-Tink J, Shivaram SM, Chader GJ, et al. Expression, secretion, and age-related downregulation of pigment epithelium-derived factor, a serpin with neurotrophic activity. J Neurosci 1995; 15:4992-5003.
117. Tanikawa T, Becerra SP, Chader GJ, et al. Pigment epithelium-derived factor (PEDF) is a survival factor for cerebellar granule cells in culture. J Neurochem 1995; 64:2509-2517.
118. Pignolo RJ, Cristofalo VJ and Rotenberg MO. Senescent WI-38 cells fail to express *EPC-1*, a gene induced in young cells upon entry into the G_0 state. J Biol Chem 1993; 268:8949-8957.
119. Pignolo RJ, Rotenberg MO and Cristofalo VJ. Analysis of *EPC-1* growth state-dependent expression, specificity, and conservation of related sequences. J Cell Physiol 1995; 162:110-118.
120. Steele FR, Chader GJ, Johnson LV, et al. Pigment epithelium-derived factor: Neurotrophic activity and identification as a member of the serine protease inhibitor gene family. Proc Natl Acad Sci USA 1993; 90:1526-1530.
121. Wu Y-Q, Notario V, Chader GJ, et al. Identification of pigment epithelium-derived factor in the interphotoreceptor matrix of bovine eyes. Prot Exp Purifn 1995; 6:447-456.
122. Becerra SP, Palmer I, Kumar A, et al. Overexpression of fetal human epithelium-derived factor in *Escherichia coli*. A functionally active neurotrophic factor. J Biol Chem 1993; 268:23148-23156.
123. Becerra SP, Sagasti A, Spinella P, et al. Pigment epithelium-derived factor (PEDF) behaves like a non-inhibitory serpin: Neurotrophic activity does not require the serpin reactive loop. J Biol Chem 1995; 270:(In Press).
124. Becerra SP. PEDF behaves as a non-inhibitory serpin. FASEB J 1995; 9:A1439. (Abstract).
125. Hopkins PCR, Carrell RW and Stone SR. Effects of mutations in the hinge region of serpins. Biochemistry 1993; 32:7650-7657.
126. Moffatt RJ, Bazer FW, Roberts RM, et al. Secretory function of the ovine uterus: effects of gestation and steroid replacement therapy. J Animal Sci 1987; 65:1400-1410.
127. Moffatt RJ, Bazer FW, Hansen PJ, et al. Purification, secretion and immunocytochemical localization of the uterine milk proteins, major progesterone-induced proteins in uterine secretions of the sheep. Biol Reprod 1987; 36:419-430.
128. Hansen PJ, Ing NH, Moffatt RJ, et al. Biochemical characterization and biosynthesis of the uterine milk proteins of the pregnant sheep uterus. Biol Reprod 1987; 36:405-418.
129. Ing NH and Roberts RM. The major progesterone-modulated proteins secreted into the sheep uterus are members of the serpin superfamily of serine protease inhibitors. J Biol Chem 1989; 264:3372-3379.
130. Malathy PV, Imakawa K, Simmen RCM, et al. Molecular cloning of the uteroferrin-associated protein, a major progesterone-induced serpin secreted by the porcine uterus, and the expression of its mRNA during pregnancy. Mol Endocrinol 1990; 4:428-440.
131. Leslie MV and Hansen PJ. Progesterone-regulated secretion of the serpin-like proteins of the ovine and bovine uterus. Steroids 1991; 56:589-597.
132. Skopets B and Hansen PJ. Identification of the predominant proteins in uterine fluids of unilaterally pregnant ewes that inhibit lymphocyte proliferation. Biol Reprod 1993; 49:997-1007.
133. Stephenson DC, Hansen PJ, Newton GR, et al. Inhibition of lymphocyte proliferation by uterine fluid from pregnant ewes. Biol Reprod 1989; 41:1063-1075.

134. Segerson EC, Moffat RJ, Bazer FW, et al. Suppression of phytohemagglutinin-stimulated lymphocyte blastogenesis by ovine uterine milk proteins. Biol Reprod 1984; 30:1175-1186.
135. Liu W-J and Hansen PJ. Effect of the progesterone-induced serpin-like proteins of the sheep endometrium on natural-killer cell activity in sheep and mice. Biol Reprod 1993; 56:1008-1014.
136. Hansen PJ and Newton GR. Binding of immunoglobulins to the major progesterone-induced proteins of the sheep uterus. Arch Biochem Biophys 1988; 260:208-217.
137. Holland LJ, Suksang C, Wall AA, et al. A major estrogen-regulated protein secreted from the liver of *Xenopus laevis* is a member of the serpin superfamily. Nucleotide sequence of cDNA and hormonal induction of mRNA. J Biol Chem 1992; 267: 7053-7059.
138. Beck BL, Henjum DC, Antonijczuk K, et al. *pNiXa*, a Ni^{2+}-binding protein in *Xenopus* oocytes and embryos, shows identity to *EP45*, an estrogen-regulated hepatic serpin. Res Commun Chem Pathol Pharmacol 1992; 77:1-14.
139. Hill RE and Hastie ND. Accelerated evolution in the reactive centre regions of serine protease inhibitors. Nature 1987; 326: 96-99.
140. Zou Z, Anisowicz A, Hendrix MJC, et al. Maspin, a serpin with tumor-supressing activity in human mammary epithelial cells. Science 1994; 263:526-529.
141. Sheng S, Pemberton PA and Sager R. Production, purification, and characterization of recombinant maspin proteins. J Biol Chem 1994; 269:30988-30993.
142. Pemberton PA, Wong DT, Gibson HL, et al. The tumor suppressor maspin does not undergo the stressed to relaxed transition or inhibit trypsin-like serine proteases. Evidence that maspin is not a protease inhibitory serpin. J Biol Chem 1995; 270: 15832-15837.
143. Hopkins PCR and Whisstock J. Function of maspin. Science 1994; 265:1893-1894.
144. Clarke EP and Sanwal BD. Cloning of human collagen-binding protein, and its homology with rat gp46, chick hsp47 and mouse J6 proteins. Biochim Biophys Acta 1992; 1129:246-248.
145. Wang SY and Gudas LJ. A retinoic acid inducible mRNA from F9 teratocarcinoma cells encodes a novel protease inhibitor homologue. J Biol Chem 1990; 265:15818-15822.
146. Wang SY. Structure of the gene and its retinoic acid-regulatory region for murine J6 serpin. An F9 teratocarcinoma cell retinoic acid-inducible protein. J Biol Chem 1992; 267:15362-15366.
147. Wang SY. A retinoic acid-inducible GATA-binding protein binds to the regulatory region of J6 serpin gene. J Biol Chem 1994; 269:607-613.
148. Bielinska M and Wilson DB. Regulation of J6 gene expression by transcription factor GATA-4. Biochem J 1995; 307:183-189.
149. Takechi H, Hirayoshi K, Nakai A, et al. Molecular cloning of a mouse 47-kDa heat shock protein (HSP47), a collagen-binding stress protein, and its expression during the differentiation of F9 teratocarcinoma cells. Eur J Biochem 1992; 206:323-329.
150. Hosokawa N, Takechi H, Yokota S, et al. Structure of the gene encoding the mouse 47-kDa heat-shock protein (HSP47). Gene 1993; 126:187-193.
151. Hirayoshi K, Kudo H, Takechi H, et al. HSP47: a tissue specific, transformation-sensitive, collagen-binding heat shock protein of chicken embryo fibroblasts. Mol Cell Biol 1991; 11:4036-4044.
152. Clarke EP, Cates GA, Ball EH, et al. A collagen-binding protein in the endoplasmic reticulum of myoblasts exhibits relationship with serine protease inhibitors. J Biol Chem 1991; 266:17230-17235.
153. Jain N, Brickenden A, Lorimer I, et al. Interaction of procollagen I and other collagens with colligin. Biochem J 1994; 304:61-68.
154. Takamatsu J and Refetoff S. Inherited heat-stable variant thyroxine-binding globulin (TBG-Chicago). J Clin Endocrinol Metab 1986; 63:1140-1144.
155. Sarne DH, Refetoff S, Nelson JC, et al. A new inherited abnormality of thyroxine-binding globulin (TBG-San Diego) with

decreased affinity for thyroxine and triiodothyronine. J Clin Endocrinol Metab 1989; 68:114-119.
156. Bertenshaw R, Sarne D, Tornari J, et al. Sequencing of the variant thyroxine binding globulin (TBG)-San Diego reveals two nucleotide substitutions. Biochim Biophys Acta 1992; 1139:307-310.
157. Takeda K, Mori Y, Sobieszczyk S, et al. Sequence of the variant thyroxine-binding globulin of Australian aborigines. Only one of two amino acid replacements is responsible for its altered properties. J Clin Invest 1989; 83:1344-1348
158. Bertenshaw R, Takeda K and Refetoff S. Sequencing of the variant thyroxine-binding globulin (TBG)-Quebec reveals two nucleotide substitutions. Am J Hum Genet 1991; 48:741-744.
159. Takamatsu J, Refetoff S, Charbonneau M, et al. Two new inherited defects of the thyroxine-binding globulin (TBG) molecule presenting as partial TBG deficiency. J Clin Invest 1987; 79:833-840.
160. Janssen OE, Takeda K and Refetoff S. Sequence of the variant thyroxine binding globulin (TBG) in a Montreal family with partial TBG deficiency. Hum Genet 1991; 87:119-122.
161. Miura Y, Mori Y, Yamamori I, et al. Sequence of a variant thyroxine-binding globulin (TBG) in a family with partial TBG deficiency in Japanese (TBG-PDJ). Endocrine J 1993; 40:127-132.
162. Yamamori I, Mori Y, Miura Y, et al. Gene screening of 23 Japanese families with complete thyroxine-binding globulin deficiency: Identification of a nucleotide deletion at codon 352 as a common cause. Endocrine J 1993; 40:563-569.
163. Li P, Janssen OE, Takeda K, et al. Complete thyroxine-binding globulin (TBG) deficiency caused by a single nucleotide deletion in the TBG gene. Metabolism 1991; 40:1231-1234.
164. Bock SC, Wion KL, Vehar GA, et al. Cloning and expression of the cDNA for human antithrombin III. Nucl Acids Res 1982; 10:8113-8125.
165. Chandra T, Stackhouse R, Kidd VJ, et al. Sequence homology between human α_1-antichymotrypsin, α_1-antitrypsin and antithrombin III. Biochemistry 1983; 266:5055-5061.
166. Long GL, Chandra T, Woo SLC, et al. Complete sequence of the cDNA for human α_1-antitrypsin and the S variant. Biochemistry 1984; 23:4824-4837.
167. Ye RD, Wun T and Sadler JE. cDNA cloning and expression in *Escherichia coli* of a plasminogen activator from human placenta. J Biol Chem 1987; 262:3718-3725.

CHAPTER 7

Serpin Biosynthesis, Clearance and Receptors

This chapter deals with serpin biosynthesis and its regulation, the clearance of the serpin-proteinase complexes and receptors for the complexes. Defective biosynthesis is involved in some of the inherited serpin deficiencies and specific examples are given in Chapter 8. Understanding the regulation of serpin biosynthesis could ultimately lead to better treatments for serpin deficiency states. Receptors for serpin-proteinase complexes and their biological function represent, as will be indicated, a controversial area where a consensus has yet to emerge.

1. SERPIN BIOSYNTHESIS

As is found with many plasma proteins, the liver is likely to be the major site of biosynthesis of circulating serpins. However, many extrahepatic tissues produce serpins presumably for the regulation of localized proteolytic pathways. The regulation of the biosynthesis of many of the plasma serpins by hormones and cytokines has been extensively studied, in many cases using cell lines. It is beyond the scope of this article to discuss all of these papers in detail, consequently only a few specific examples will be given. Decreased serpin synthesis leading to pathogenetic states is discussed in Chapter 8. The plasma concentration of a number of serpins is given in Table 1.1.

α_1-Proteinase inhibitor is a major plasma protein, but is also produced in a number of extrahepatic locations, for example by monocytes and neutrophils,[1] cornea,[2] kidney and intestine.[3,4] The finding that the gene is under the control of tissue specific promoters[5] and transcriptional initiation sites[6] suggests that there are important regulatory mechanisms yet to be elucidated. The endoplasmic reticulum transmembrane protein calnexin, which is a type of molecular chaperone, has been shown to associate transiently with newly expressed α_1-proteinase inhibitor and α_1-antichymotrypsin in the endoplasmic reticulum of HepG2 cells.[7] Calnexin appears to bind to incompletely folded glycoproteins and is involved in the regulated release of these proteins from the endoplasmic reticulum. The possibility that α_1-proteinase inhibitor synthesis can be regulated by α_1-proteinase inhibitor-proteinase complexes binding to a cell surface receptor is discussed below. Both α_1-proteinase inhibitor and α_1-antichymotrypsin are acute phase reactants. In response to an inflammatory stimulus or infection the plasma concentrations can increase up to 4-fold.[8,9] The possibility that there might be co-ordinate regulation of α_1-proteinase inhibitor and of other genes at the 14q32.1 locus during inflammation has been discussed in Chapter 1.

In addition to localization in plasma, protein C inhibitor antigen has been identified in seminal fluid and in a variety of cells in the male reproductive tract. mRNA for the inhibitor has also been identified in prostate, testes, seminal vesicles and epidydimal glands.[10,11] This serpin is an inhibitor of the sperm proteinase acrosin with a second order inhibition rate constant of 2.4 x 10^5 $M^{-1}s^{-1}$.[12] Heparin can increase this rate constant 230-fold. Protein C inhibitor has also been found in complexes with tPA and urokinase in seminal fluid.[13] Protein C inhibitor is also expressed in tubular cells of human kidney[14] (see also Chapter 4).

The main site of antithrombin biosynthesis is the liver[15,16] but biosynthesis can also occur in the kidney and endothelial cells.[17-19] Similarly, in the case of C1-inhibitor, liver is the main source of the circulating protein,[20] as has been verified by use of HepG2 cells,[21] although synthesis at extrahepatic sites such as endothelial cells,[22] synovium,[23,24] fibroblasts,[25] monocytes[26] and brain glial cells[27,28] has also been demonstrated. Gamma interferon is the major regulator of C1-inhibitor biosynthesis in most cell types.[29-31]

Protease nexin 1, which is important in regulating proteolytic pathways in the brain, is produced by both astrocytes and glial cells.[32-34] In the case of glial cells, synthesis is increased by cytokines such as interleukin-1, tumor necrosis factor-α and transforming growth factor-β. Protease nexin 1 is also produced by HepG2 cells[35] and fibroblasts.[36]

The most interesting plasma serpin from the context of biosynthesis is PAI-1. This is present in plasma at low concentrations (0.4 nM).[37,38] A number of cell types express PAI-1, although a physiologically important pool of PAI-1 is present in the α-granule of platelets.[39-42] mRNA for PAI-1 has been shown to be present in both platelets and megakaryocytes.[43] A model system for the processing of PAI-1 into platelet storage granules has been constructed by using a mouse pituitary cell line transfected with human PAI-1 cDNA.[44] Platelet PAI-1 is likely to have an important role in regulating both endogenous and therapeutic thrombolysis.[45-48]

PAI-1 is also involved in the regulation of other processes such as angiogenesis and ovulation[49,50] and in the rat PAI-1 is an immediate early response gene in regenerating liver.[51] Among the agonists shown to stimulate PAI-1 synthesis by various cell types are glucocorticoids,[52-55] tumor necrosis factor,[56] lipopolysaccharide,[57,58] transforming growth factor-β,[59-63] hepatocyte growth factor,[64] epidermal growth factor,[65] insulin[66] and low density lipoproteins.[67,68] Further details on the regulation of PAI-1 expression by various agonists can be found in review articles.[69-72]

2. CLEARANCE OF SERPIN-PROTEINASE COMPLEXES

In man native serpins generally have a half life in the circulation of over 24 hours whereas serpin-proteinase complexes have a half life of a few hours,[73,74] indicating that specific mechanisms exist to remove the complexes from the circulation. In a series of experiments which compared the clearance of ^{125}I labeled human serpins and their proteinase complexes from the circulation of experimental animals, it was shown that the complexes were cleared from the circulation about 10-fold faster than the unreacted serpin. When injected into mice, rats or rabbits, native human serpins have a half life of 5 or 6 hours and the complexes have a half life of 10-30 minutes. Competition studies showed that radiolabeled antithrombin-thrombin complex and α_1-proteinase inhibitor-trypsin complex appeared to be removed by the same pathway,[75] whereas α_2-antiplasmin-plasmin was

antithrombin, α_2-macroglobulin-methylamine, asialoorosomucoid and mannose-BSA did not compete.[118] Further studies showed that uptake and degradation of antithrombin and α_1-proteinase inhibitor complexes are likely to be mediated via the same receptor.[119] This receptor was proposed to be the one responsible for the circulatory clearance of these complexes in rat in vivo (see above), and has obvious similarity to the SEC receptor. A receptor for the uptake and degradation of the antithrombin-thrombin complex was identified on bovine corneal endothelial cells.[120] Cyanogen bromide degradation of iodinated antithrombin indicated that the cellular binding domain was located within residues 253-314 of antithrombin. This region includes β-sheet strands 6C and 5A and is located to the amino terminal side of the reactive center. Thus, this is distinct from the putative binding site of the SEC receptor which is on the carboxy-terminal side of the reactive center and is associated with β-sheet strand 4B. The uptake and degradation of antithrombin-thrombin complex via a receptor-mediated process by the monocyte-like U937 cell line has also been reported. This receptor is probably not the SEC receptor, as the SEC receptor binding peptide did not compete.[121] The thrombin-antithrombin complex binds to vitronectin in plasma and this has been suggested to be involved in clearance of this complex from circulation, perhaps via conformational changes on vitronectin and a vitronectin receptor.[122] A receptor which appears to be distinct from the SEC receptor and that binds to the cathepsin G-α_1-antichymotrypsin complex has been shown on mouse spinal cord astrocytes.[123] Similarly, it is not known if a receptor for tPA-PAI-1 complexes detected on hepatocytes is LRP.[124] Complexes of protease nexin 1 with thrombin and urokinase bind to fibroblasts and astrocytes with subsequent uptake and degradation. Binding occurs through the serpin moiety and is inhibitable by heparin.[33,125,126] Again, the relationship of this receptor to SEC or LRP is not known.

6. BIOLOGICAL ACTIVITY OF CLEAVED AND COMPLEXED SERPINS

The possibility that cleaved or complexed serpins might stimulate their own biosynthesis has been mentioned above. Numerous reports also indicate that cleaved or complexed serpins might also have other novel biological activities not present on the native protein. An example of a bioactive peptide released from a serpin is that of angiotensin II, but it should be stressed that in this case the peptide is released from the amino terminus of the angiotensinogen (see Chapter 6). The findings that cleaved α_1-proteinase inhibitor can stimulate neutrophil chemotaxis have been discussed above in the context that the SEC receptor can mediate this activity. Chemotactic peptides have also been shown to arise from the proteolysis of heparin cofactor II by cathepsin G and neutrophil elastase.[127] Similar results were obtained when heparin cofactor II was incubated with stimulated neutrophils and the supernatant from these incubations used for chemotaxis assays.[128] The thrombin-heparin cofactor II complex is not chemotactic, indicating that cleavage within the reactive center of the serpin was not responsible for generation of the activity in contrast to the situation with α_1-proteinase inhibitor.[129] The chemotactic activity was shown to reside in a peptide corresponding to residues Asp 49-Tyr 60 of heparin cofactor II. It was suggested that proteolysis of heparin cofactor II to release bioactive peptides could be an important inflammatory mechanism.[130]

Another serpin which has been shown to have bioregulatory properties is α_1-antichymotrypsin. The α_1-antichymotrypsin-cathepsin G complex stimulated interleukin-6 production by human lung fibroblasts. Conditioned medium from these cells could stimulate HepG2 cells to produce α_1-antichymotrypsin and other acute-phase response proteins.[131] α_1-Antichymotrypsin cleaved in the reactive center by trypsin or *Staphylococcus aureus* metalloproteinase was chemotactic for neutrophils,[132] suggesting a mechanism similar to that observed for cleaved α_1-proteinase inhibitor. A novel function for unreacted α_1-antichymotrypsin has also been identified. When neutrophils are incubated with the potent chemotactic agent fMet-Leu-Phe a respiratory burst takes place associated with the assembly of an active NADPH oxidase and the generation of superoxide radical anions. This activity was inhibitable by native α_1-antichymotrypsin (at concentrations as low as 2 μM) but not by native α_1-proteinase inhibitor. The α_1-antichymotrypsin-chymotrypsin complex was inhibitory as were recombinant non-glycosylated α_1-antichymotrypsin P1 variants, indicating that a region distinct from the reactive center was responsible for the inhibition.[133] Further analysis using an isolated cell-free NADPH oxidase preparation showed that native α_1-antichymotrypsin was not inhibitory but α_1-antichymotrypsin-proteinase complexes were, indicating that in the earlier study the native α_1-antichymotrypsin had reacted with endogenous proteinases in order to mediate the effect.[134] The α_1-antichymotrypsin-proteinase complexes appear to inhibit the initial activation of the oxidase rather than inhibiting the active enzyme. Moreover, the α_1-antichymotrypsin-proteinase complexes did not inhibit agonist-stimulated calcium mobilization or protein kinase C activity.[135]

References

1. du Bois RM, Bernaudin JF, Paako P, et al. Human neutrophils express the α1-antitrypsin gene and produce α1-antitrypsin. Blood 1991; 77:2724-2730.
2. Twining SS, Fukuchi T, Yue BY, et al. Corneal syntheis of α_1-proteinase inhibitor (α_1-antitrypsin). Invest Ophthalmol Vis Sci 1994; 35:458-462.
3. Kelsey GD, Povey S, Bygrave AE, et al. Species and tissue specific expression of human α_1-antitrypsin in transgenic mice. Genes & Development 1987; 1:161-171.
4. Molmenti EP, Perlmutter DH and Rubin DC. Cell-Specific expression of alpha(1)-antitrypsin in human intestinal epithelium. J Clin Invest 1993; 92:2022-2034.
5. Perlino E, Cortese R and Ciliberto G. The human α_1-antitrypsin gene is transcribed from two different promoters in macrophages and hepatocytes. EMBO J 1987; 6:2767-2771.
6. Hafeez W, Ciliberto G and Perlmutter DH. Constitutive and modulated expression of the human α_1-antitrypsin gene. Different transcriptional initiation sites used in three different cell types. J Clin Invest 1992; 89:1214-1222.
7. Ou WJ, Cameron PH, Thomas DY, et al. Association of folding intermediates of glycoproteins with calnexin during protein maturation. Nature 1993; 364:771-776.
8. Koj A and Regeoczi E. Effect of experimental inflammation on the synthesis and distribution of antithrombin III and α_1-antitrypsin in rabbits. Br J Exp Path 1978; 59:473-481.
9. Sehgal PB, Grieninger G and Tosato G. Regulation of the acute phase and immune responses: Interleuklin 6. Ann N Y Acad Sci 1989; 557:1-583.
10. Laurell M, Christensson A, Abrahamsson PA, et al. Protein C inhibitor in human body fluids. Seminal plasma is rich in inhibitor antigen deriving from cells throughout the male reproductive system. J Clin Invest 1992; 89:1094-1101.
11. Zheng X, Geiger M, Ecke S, et al. Inhibition of acrosin by protein C inhibitor and localization of protein C inhibitor to spermatozoa. Am J Physiol 1994; 267:C466-C472.
12. Hermans JM, Jones R and Stone SR. Rapid inhibition of the sperm protease acrosin by protein C inhibitor. Biochemistry 1994; 33:5440-5444.
13. España F, Estellés A, Fernández PJ, et al. Evidence for the regulation of urokinase and tissue type plasminogen activators by the serpin, protein C inhibitor, in semen and blood plasma. Thromb Haemostas 1993; 70:989-994.
14. Radtke KP, Fernandez JA, Greengard JS, et al. Protein C inhibitor is expressed in tubular cells of human kidney. J Clin Invest 1994; 94:2117-2124.

15. Owens MR and Miller LL. Net biosynthesis of antithrombin III by the isolated rat liver perfused for 12-24 hours: Compared with rat fibrinogen and α2(acute phase) globulin, antithrombin III is not an acute phase protein. Biochim Biophys Acta 1980; 62:30-39.
16. Fair DS and Bahnak BR. Human hepatoma cells secrete single chain factor X, prothrombin and antithrombin III. Blood 1984; 64:194-204.
17. Chan TK and Chan V. Antithrombin III, the major modulator of intravascular coagulation is synthesized by human epithelial cells. Thromb Haemostas 1981; 46: 504-506.
18. Lee AKY, Chan V and Chan TK. The identification and localization of antithrombin III in human tissues. Thromb Res 1979; 14:209-217.
19. Kouteva Y, Schapira M and Patston PA. The effect of sex and age on antithrombin biosynthesis in the rat. Thromb Res 1995; 78:521-529.
20. Johnson AM, Alper CA, Rosen FS, et al. C1-inhibitor: Evidence for decreased hepatic synthesis in hereditary angioneurotic edema. Science 1971; 173:553-554.
21. Morris KM, Aden DP, Knowles BB, et al. Complement biosynthesis by the human hepatoma derived cell line Hep G2. J Clin Invest 1982; 70:906-913.
22. Schmaier AH, Murray SC, Heda GD, et al. Synthesis and expression of C1 inhibitor by human umbilical vein endothelial cells. J Biol Chem 1989; 264:18173-18179.
23. Moffat GJ, Lappin D, Birnie GD, et al. Complement biosynthesis in human synovial tissue. Clin Exp Immunol 1989; 78:54-60.
24. Breitner S, Störkel S, Reichel W, et al. Complement components C1q, C1r/C1s, and C1 INH in rheumatoid arthritis. Correlation of in situ hybridization and northern blot results with function and protein concentration in synovium and primary cell cultures. Arthritis & Rheumatism 1995; 38:492-498.
25. Katz Y and Strunk RC. Synthesis and regulation of C1 inhibitor in human skin fibroblasts. J Immunol 1989; 142:2041-2045.
26. Bensa JC, Reboul A and Colomb MG. Biosynthesis *in vitro* of complement subcomponents C1q, C1s and C1 inhibitor by resting and stimulated human monocytes. Biochem J 1983; 216:385-392.
27. Walker DG, Yasuhara O, Patston PA, et al. Complement C1 inhibitor is produced by brain tissue and is cleaved in Alzheimer disease. Brain Res 1995; 675:75-82.
28. Gasque P, Ischenko A, Legoendoc J, et al. Expression of the complement classical pathway by human glioma in culture. A model for complement expression by nerve cells. J Biol Chem 1993; 268:25068-25074.
29. Heda GD, Mardente S, Weiner L, et al. Interferon gamma increases in vitro and in vivo expression of C1 inhibitor. Blood 1990; 75:2401-2407.
30. Lappin DF, Guc D, Hill A, et al. Effect of γ-interferon on complement gene expression in different cell types. Biochem J 1992; 281:437-442.
31. Zahedi K, Prada AE and Davis AE III. Transcriptional regulation of the C1 inhibitor gene by γ-interferon. J Biol Chem 1994; 269:9669-9674.
32. Guenther J, Nick H and Monard D. A glia-derived neurite-promoting factor with protease inhibitory activity. EMBO J 1985; 4:1963-1966.
33. Rosenblatt DE, Cotman CW, Nieto-Sampredro M, et al. Identification of a protease inhibitor produced by astrocytes that is structurally and functionally homologous to human protease nexin I. Brain Res 1987; 415:40-48.
34. Vaughan PJ and Cunningham DD. Regulation of protease nexin-1 synthesis and secretion in cultures brain cells by injury-related factors. J Biol Chem 1993; 268: 3720-3727.
35. Kazama Y, Komiyama Y and Kisiel W. Tissue factor pathway inhibitor and protease nexin-1 are major factor Xa binding proteins on the HepG2 cell surface. Blood 1993; 81:676-682.
36. Farrell DH, Wagner SL, Yuan RH, et al. Localization of protease nexin-I on the fibroblast extracellular matrix. J Cell Physiol 1988; 134:179-188.

37. Booth NA, Simpson AJ, Croll A, et al. Plasminogen activator inhibitor (PAI-1) in plasma and platelets. Br J Haematol 1988; 70:327-333.
38. Declerck PJ, Alessi MC, Verstreken I, et al. Measurement of plasminogen activator inhibitor 1 in biological fluids with a murine monoclonal antibody-based enzyme-linked immunosorbent assay. Blood 1988; 71: 220-225.
39. Erickson LA, Hekman CM and Loskutoff DJ. The primary plasminogen-activator inhibitors in endothelial cells, platelets, serum, and plasma are immunologically related. Proc Natl Acad Sci USA 1985; 82:8710-8714.
40. Kruithof EKO, Tran-Thang C and Bachman FW. Studies on the release of a plasminogen activator inhibitor by human platelets. Thromb Haemostas 1986; 55:201-205.
41. Fay WP and Owen WG. Platelet plasminogen activator inhibitor: Purification and characterization of interaction with plasminogen activators and activated protein C. Biochemistry 1989; 28:5773-5778.
42. Lang IM, Marsh JJ, Moser KM, et al. Presence of active and latent type 1 plasminogen activator inhibitor associated with porcine platelets. Blood 1992; 80:2269-2274.
43. Konkle BA, Schick PK, He X, et al. Plasminogen activator inhibitor-1 mRNA is expressed in platelets and megakaryocytes and the megakaryoblastic cell line CHRF-288. Arteriosclerosis and Thrombosis 1993; 13:669-674.
44. Gombau L and Schleef RR. Processing of type 1 plasminogen activator inhibitor (PAI-1) into the regulated secretory pathway. J Biol Chem 1994; 269:3875-3880.
45. Potter van Loon BJ, Rijken DC, Brommer EJP, et al. The amount of plasminogen, tissue-type plasminogen and plasminogen activator inhibitor type 1 in human thrombi and the relation to ex-vivo lysis. Thromb Haemostas 1992; 67:101-105.
46. Braaten JV, Handt JV, Jerome WG, et al. Regulation of fibrinolysis by platelet-released plasminogen activator inhibitor 1; Light scattering and ultrastructural examination of lysis of a model platelet-fibrin thrombus. Blood 1993; 81:1290-1299.
47. Torr-Brown SR and Sobel BE. Attenuation of thrombolysis by release of plasminogen activator inhibitor type-1 from platelets. Thromb Res 1993; 72:413-421.
48. Serizawa K, Urano T, Kozima Y, et al. The potential role of platelet PAI-1 in t-PA mediated clot lysis of platelet rich plasma. Thromb Res 1993; 71:289-300.
49. Vassalli JD, Sappino AP and Belin D. The plasminogen activator/plasmin system. J Clin Invest 1991; 88:1067-1072.
50. Peng XR, Hsueh AJW and Ny T. Transient and cell-specific expression of tissue-type plasminogen activator and plasminogen activator inhibitor type-1 results in controlled and directed proteolysis during gonadotrophin-induced ovulation. Eur J Biochem 1993; 214:147-156.
51. Thornton AJ, Bruzdzinski CJ, Raper SE, et al. Plasminogen activator inhibitor-1 is an immediate early response gene in regenerating rat liver. Cancer Res 1994; 54: 1337-1343.
52. Medcalf R, van den Berg E and Schleuning WD. Glucocorticoid-modulated gene expression of tissue- and urinary-type plasminogen activator and plasminogen activator inhibitor 1 and 2. J Cell Biol 1988; 106:971-978.
53. Lund LR, Georg B, Nielsen LS, et al. Plasminogen activator inhibitor type 1: cell-specific and differentiation-induced expression and regulation in human cell lines, as determined by enzyme-linked immunosorbent assay. Mol Cell Endocrin 1988; 60:43-53.
54. Konkle BA, Schuster SJ, Kelly MD, et al. Plasminogen activator inhibitor-1 messenger RNA expression is induced in rat hepatocytes in vivo by dexamethasone. Blood 1992; 79:2636-2642.
55. Bruzdzinski CJ, Johnson MR, Goble CA, et al. Mechanism of glucocorticoid induction of the rat plasminogen activator inhibitor-1 gene in HTC rat hepatoma cells: Identification of *cis*-acting regulatory elements. Mol Endocrinol 1993; 7:1169-1177.
56. van Hinsbergh VWM, Kooistra T, van den Berg EA, et al. Tumor necrosis factor increases the production of plasminogen activator inhibitor in human endothelial cells

in vitro and in rats in vivo. Blood 1988; 72:1467-1473.

57. Sawdey M, Podor TJ and Loskutoff DJ. Regulation of type 1 plasminogen activator gene expression in cultured bovine aortic endothelial cells. Induction by transforming growth factor-β, lipopolysaccharide, and tumor necrosis factor-α. J Biol Chem 1989; 264:10396-10401.
58. Sawdey MS and Loskutoff DJ. Regulation of murine type 1 plasminogen activator inhibitor gene expression in vivo. Tissue specificity and induction by lipopolysaccharide, tumor necrosis factor-α, and transforming growth factor-β. J Clin Invest 1991; 88:1346-1353.
59. Westerhausen DR, Hopkins WE and Billadello JJ. Multiple transforming growth factor-β-inducible elements regulate expression of the plasminogen activator inhibitor type-1 gene in HepG2 cells. J Biol Chem 1991; 266:1092-1100.
60. Lund LR, Riccio A, Andreasen PA, et al. Transforming growth factor-β is a strong and fast acting regulator of the level of type-1 plasminogen activator inhibitor mRNA in WI-38 human lung fibroblasts. EMBO J 1987; 6:1281-1286.
61. Reilly CF and McFall RC. Platelet-derived growth factor and transforming growth factor-β regulate plasminogen activator inhibitor-1 synthesis in vascular smooth muscle cells. J Biol Chem 1991; 266:9419-9427.
62. Slivka SR and Loskutoff DJ. Platelets stimulate endothelial cells to synthesize type 1 plasminogen activator inhibitor. Evaluation of the role of transforming growth factor β. Blood 1991; 77:1013-1019.
63. Hopkins WE, Fujii S and Sobel BE. Synergistic induction of plasminogen activator inhibitor type-1 in HEP G2 cells by thrombin and transforming growth factor. Blood 1992; 79:75-81.
64. Wojta J, Nakamura T, Fabry A, et al. Hepatocyte growth factor stimulates expression of plasminogen activator inhibitor type 1 and tissue factor in Hep G2 cells. Blood 1994; 84:151-157.
65. Lucore CL, Fujii S, Wun TC, et al. Regulation of the expression of type 1 plasminogen activator inhibitor in Hep G2 cells by epidermal growth factor. J Biol Chem 1988; 263:15845-15848.
66. Grant PJ, Rüegg M and Medcalf RL. Basal expression and insulin-mediated induction of PAI-1 mRNA in Hep G2 cells. Fibrinolysis 1991; 5:81-86.
67. Latron Y, Chauton M, Anfosso F, et al. Stimulating effect of oxidized low density lipoproteins on plasminogen activator inhibitor-1 synthesis by endothelial cells. Arteriosclerosis and Thrombosis 1991; 11:1821-1829.
68. Tremoli E, Camera M, Maderna P, et al. Increased synthesis of plasminogen activator inhibitor-1 by cultured human endothelial cells exposed to native and modified LDLs: An LDL receptor-independent phenomenon. Arteriosclerosis and Thrombosis 1993; 13:338-346.
69. Loskutoff DJ, Sawdey M and Mimuro J. Type 1 plasminogen activator inhibitor. Prog Hemostas Thromb 1989; 9:87-115.
70. Andreasen PA, Georg B, Lund LR, et al. Plasminogen activator inhibitors: hormonally regulated serpins. Mol Cell Endocrin 1990; 68:1-19.
71. Grant PJ and Medcalf RL. Hormonal regulation of haemostasis and the molecular biology of the fibrinolytic system. Clin Sci 1990; 78:3-11.
72. Krishnamurti C and Alving BM. Plasminogen activator inhibitor type 1: Biochemistry and evidence for modulation of fibrinolysis in vivo. Sem Thromb Hemostas 1992; 18:67-80.
73. Ohlsson K and Laurell CB. The disappearance of enzyme-inhibitor complexes from the circulation of man. Clin Sci Mol Med 1976; 51:87-92.
74. Collen D and Wiman B. Turnover of α_2-antiplasmin and of the plasmin α_2-antiplasmin complex. In: Collen D, Wiman B, Verstraete M, eds. The Physiological Inhibitors of Coagulation and Fibrinolysis. Amsterdam: Elsevier/North Holland Biomedical Press, 1979:171-176.
75. Fuchs HE, Shifman MA and Pizzo SV. In vivo catabolism of α_1-proteinase inhibitor-trypsin, antithrombin III-thrombin and α_2-macroglobulin-methylamine. Biochim Biophys Acta 1982; 716:151-157.

76. Gonias SL, Fuchs HE and Pizzo SV. A unique pathway for the plasma elimination of α_2-antiplasmin-protease complexes in mice. Thromb Haemostas 1982; 48: 208-210.
77. Pratt CW, Church FC and Pizzo SV. *In vivo* catabolism of heparin cofactor II and its complex with thrombin: Evidence for a common receptor-mediated clearance pathway for three serine proteinase inhibitors. Arch Biochem Biophys 1988; 262:111-117.
78. Pizzo SV, Mast AE, Feldman SR, et al. In vivo catabolism of α_1-antichymotrypsin is mediated by the Serpin receptor which binds α_1-proteinase inhibitor, antithrombin III and heparin cofactor II. Biochim Biophys Acta 1988; 967:158-162.
79. Mast AE, Enghild JJ, Pizzo SV, et al. Analysis of the plasma elimination kinetics and conformational stabilities of native, proteinase-complexed, and reactive site cleaved serpins: Comparison of α_1-proteinase inhibitor, α_1-antichymotrypsin, antithrombin III, α_2-antiplasmin, angiotensinogen, and ovalbumin. Biochemistry 1991; 30:1723-1730.
80. Enghild JJ, Valnickova Z, Thogersen IB, et al. Complexes between serpins and inactive proteinases are not thermodynamically stable but are recognized by serpin receptors. J Biol Chem 1994; 269:20159-20166.
81. Xiong W, Tang CQ, Zhou GX, et al. In vivo catabolism of human kallikrein-binding protein and its complex with tissue kallikrein. J Lab Clin Med 1992; 119: 514-521.
82. de Smet BJGL, de Boer J-P, Agterberg J, et al. Clearance of human native, proteinase-complexed, and proteolytically inactivated C1-inhibitor in rats. Blood 1993; 81:56-61.
83. Patston PA, Qi M, Schifferli JA, et al. The effect of cleavage by a *Crotalus atrox* α-proteinase fraction on the properties of C1 inhibitor. Toxicon 1995; 33:53-61.
84. Minta JO. The role of sialic acid in the functional activity and the hepatic clearance of C1-INH. J Immunol 1981; 126:245-249.
85. Perlmutter DH, Travis J and Punsal PI. Elastase regulates the synthesis of its inhibitor, α_1-proteinase inhibitor, and exaggerates the defect in homozygous PiZZ α_1PI deficiency. J Clin Invest 1988; 81: 1774-1780.
86. Perlmutter DH and Punsal PI. Distinct and additive effects of elastase and endotoxin on expression of α_1-proteinase inhibitor in mononuclear phagocytes. J Biol Chem 1988; 263:16499-16503.
87. Barbey-Morel C and Perlmutter DH. Effect of Pseudomonas elastase on human mononuclear phagocyte α_1-antitrypsin expression. Pediatr Res 1991; 29:130-140.
88. Banda MJ, Rice AG, Griffin GL, et al. α_1-Proteinase inhibitor is a neutrophil chemoattractant after proteolytic inactivation by macrophage elastase. J Biol Chem 1988; 263:4481-4484.
89. Banda MJ, Rice AG, Griffin GL, et al. The inhibitory complex of human α_1-proteinase inhibitor and human leukocyte elastase is a neutrophil chemoattractant. J Exp Med 1988; 167:1608-1615.
90. Perlmutter DH, Glover GI, Rivetna M, et al. Identification of a serpin-enzyme complex receptor on human hepatoma cells and human monocytes. Proc Natl Acad Sci USA 1990; 87:3753-3757.
91. Joslin G, Wittwer A, Adams S, et al. Cross-competition for binding of α_1-antitrypsin (α_1AT)-elastase complexes to the serpin-enzyme complex receptor by other serpin-enzyme complexes and by proteolytically modified α_1AT. J Biol Chem 1993; 268:1886-1893.
92. Hoffman M, Fuchs HE and Pizzo SV. The macrophage-mediated regulation of hepatocyte synthesis of antithrombin III and α_1-proteinase inhibitor. Thromb Res 1986; 41:707-715.
93. Patston PA, Medcalf RL, Kourteva Y, et al. C1-inhibitor-serine proteinase complexes and the biosynthesis of C1-inhibitor by Hep G2 and U937 cells. Blood 1993; 82: 3371-3379.
94. Perlmutter DH, Joslin G, Nelson P, et al. Endocytosis and degradation of α_1-antitrypsin-protease complexes is mediated by the serpin-enzyme complex (SEC) receptor. J Biol Chem 1990; 265:16713-16716.
95. Joslin G, Fallon RJ, Bullock J, et al. The SEC receptor recognizes a pentapeptide neodomain of α_1-antitrypsin-protease complexes. J Biol Chem 1991; 266:

11282-11288.
96. Schulze AJ, Huber R, Bode W, et al. Structural aspects of serpin inhibition. FEBS Lett 1994; 344:117-124.
97. Joslin G, Krause JE, Hershey AD, et al. Amyloid-β peptide, substance P, and bombesin bind to the serpin-enzyme complex receptor. J Biol Chem 1991; 266: 21897-21902.
98. Joslin G, Griffin GL, August AM, et al. The serpin-enzyme complex (SEC) receptor mediates the neutrophil chemotactic effect of α_1-antitrypsin-elastase complexes and amyloid-β peptide. J Clin Invest 1992; 90:1150-1154.
99. Khalil Z, Sanderson K, Isberg P, et al. β $A_4$25-35 modulates substance P effect on rat skin microvasculature in aged rats: pharmacological manipulation using SEC-receptor ligands. Brain Res 1994; 651:227-235.
100. Cubellis MV, Andreasen PA, Ragno P, et al. Accessibility of receptor bound urokinase to type-1 plasminogen activator inhibitor. Proc Natl Acad Sci USA 1989; 86:4828-4832.
101. Ellis V, Wun TC, Behrendt N, et al. Inhibition of receptor-bound urokinase by plasminogen-activator inhibitors. J Biol Chem 1990; 265:9904-9908.
102. Olson D, Pöllännen J, Høyer-Hansen X, et al. Internalization of the urokinase-plasminogen activator inhibitor type-1 complex is mediated by the urokinase receptor. J Biol Chem 1992; 267:9129-9133.
103. Nykjær A, Petersen CM, Møller B, et al. Purified α_2-macroglobulin receptor/LDL receptor-related protein binds urokinase plasminogen activator inhibitor type-1 complex. Evidence that the α_2-macroglobulin receptor mediates cellular degradation of urokinase receptor-bound complexes. J Biol Chem 1992; 267:14543-14546.
104. Herz J, Clouthier DE and Hammer RE. LDL-receptor-related protein internalizes and degrades uPA-PAI-1 complexes and is essential for embryo implantation. Cell 1992; 71:411-421.
105. Andreasen PA, Sottrup-Jensen L, Kjøller L, et al. Receptor mediated endocytosis of plasminogen activators and activator/inhibitor complexes. FEBS Lett 1994; 338:239-245.
106. Bu G, Warshawsky I and Schwartz AL. Cellular receptors for the plasminogen activators. Blood 1994; 83:3427-3436.
107. Moestrup SK, Nielsen S, Andreasen P, et al. Epithelial glycoprotein-330 mediates endocytosis of plasminogen activator-plasminogen activator inhibitor type-1 complexes. J Biol Chem 1993; 268:16564-16570.
108. Kounnas MZ, Chappell DA, Strickland DK, et al. Glycoprotein-330, a member of the low density lipoprotein receptor family, binds lipoprotein lipase in vitro. J Biol Chem 1993, 268:14176-14181.
109. Heegard CW, Simonsen ACW, Oka K, et al. Very low density lipoprotein receptor binds and mediates endocytosis of urokinase-type plasminogen activator-type 1 plasminogen activator inhibitor complex. J Biol Chem 1995; 270:20855-20861.
110. Morton PA, Owensby DA, Wun TC, et al. Identification of determinants involved in binding of tissue-type plasminogen activator inhibitor-type 1 complexes to HepG2 cells. J Biol Chem 1990; 265:14093-14099.
111. Underhill DM, Owensby DA, Morton PA, et al. Endocytosis and lysosomal delivery of tissue plasminogen activator-inhibitor 1 complexes in HepG2 cells. Blood 1992; 80:2746-2754.
112. Orth K, Madison EL, Gething M-J, et al. Complexes of tissue-type plasminogen activator and its serpin inhibitor plasminogen activator inhibitor type 1 are internalized by means of the low density lipoprotein-related protein/α_2-macroglobulin receptor. Proc Natl Acad Sci USA 1992; 89:7422-7426.
113. Bu G, Maksymovitch EA and Schwartz AL. Receptor-mediated endocytosis of tissue-type plasminogen activator by low density lipoprotein receptor-related protein on human hepatoma Hep G2 cells. J Biol Chem 1993; 268:13002-13009.
114. Grimsley PG, Normyle JF, Brandt RA, et al. Urokinase binding and catabolism by HepG2 cells is plasminogen activator inhibitor-1 dependent, analogous to interactions of tissue-type plasminogen activators with these cells. Thromb Res 1995; 79:353-361.
115. Simon DI, Xu H and Vaughan DE. Cathe-

psin D-like aspartyl protease activity mediates the degradation of tissue-type plasminogen activator/plasminogen activator inhibitor-1 complexes in human monocytes. Biochim Biophys Acta 1995; 1268:143-151.

116. Conese M, Olson D and Blasi F. Protease nexin-1-urokinase complexes are internalized and degraded through a mechanism that requires both urokinase receptor and α_2-macroglobulin receptor. J Biol Chem 1994; 269:17886-17892.

117. Poller W, Willnow TE, Hilpert J, et al. Differential recognition of α_1-antitrypsin-elastase and α_1-antichymotrypsin-cathepsin G complexes by the low density lipoprotein receptor-related protein. J Biol Chem 1995; 270:2841-2845.

118. Fuchs HE, Shifman MA, Michalopoulous G, et al. Hepatocyte receptors for antithrombin III-proteinase complexes. J Cell Biochem 1984; 24:197-206.

119. Fuchs HE, Michalopoulous GK and Pizzo SV. Hepatocyte uptake of α_1-proteinase inhibitor-trypsin complexes in vitro: evidence for a shared uptake mechanism for proteinase complexes of α_1-proteinase inhibitor and antithrombin III. J Cell Biochem 1984; 25:231-243.

120. Knoller S and Savion N. Characterization of the cellular binding domain and the effects of monoclonal antibodies and thrombin inhibitors on the binding and internalization of the antithrombin III-thrombin complex by cultured cells. Eur J Biochem 1991; 195:801-806.

121. Takeya H, Hamada T, Kume M, et al. Receptor-mediated endocytosis of thrombin-antithrombin III complex by the human monocytoid cell line U937. Biochem Biophys Res Commun 1994; 200:1334-1340.

122. de Boer HC, de Groot PG, Bouma BN, et al. Ternary vitronectin-thrombin-antithrombin III complexes in human plasma. Detection and mode of association. J Biol Chem 1993; 268:1279-1283.

123. Chen M, Conn KJ and Festoff BW. A receptor for cathepsin G:alpha(1)-antichymotrypsin complexes on mouse spinal cord astrocytes. Neurology 1993; 43: 1223-1227.

124. Wing LR, Bennett B and Booth N. The receptor for tissue plasminogen activator (t-PA) in complex with its inhibitor, PAI-1, on human hepatocytes. FEBS Lett 1991; 278:95-97.

125. Low DA, Baker JB, Koonce WC, et al. Released protease-nexin regulates cellular binding, internalization, and degradation of serine proteases. Proc Natl Acad Sci USA 1981; 78:2340-2344.

126. Van Nostrand WE, McKay LD, Baker JB, et al. Functional and structural similarities between protease nexin I and C1 inhibitor. J Biol Chem 1988; 263:3979-3983.

127. Hoffman M, Pratt CW, Brown RL, et al. Heparin cofactor II-proteinase reaction products exhibit neutrophil chemoattractant activity. Blood 1989; 73:1680-1685.

128. Corbin LW, Church FC and Hoffman M. Production of chemotactic peptides by neutrophil degradation of heparin cofactor II. Thromb Res 1990; 57:77-85.

129. Hoffman M, Pratt CW, Corbin LW, et al. Characteristics of the chemotactic activity of heparin cofactor II proteolysis products. J Leuk Biol 1990; 48:156-162.

130. Church FC, Pratt CW and Hoffman M. Leukocyte chemoattractant peptides from the serpin heparin cofactor II. J Biol Chem 1991; 266:704-709.

131. Kurdowska A and Travis J. Acute phase protein stimulation by α_1-antichymotrypsin-cathepsin G complexes. Evidence for the involvement of interleukin-6. J Biol Chem 1990; 265:21023-21026.

132. Potempa J, Fedak D, Dubin A, et al. Proteolytic inactivation of α-1-antichymotrypsin. Sites of cleavage and generation of chemotactic activity. J Biol Chem 1991; 266:21482-21487.

133. Kilpatrick L, Johnson JL, Nickbarg EB, et al. Inhibition of human neutrophil superoxide generation by α1-antichymotrypsin. J Immunol 1991; 146:2388-2393.

134. Schuster MG, Enriquez PM, Curran P, et al. Regulation of neutrophil superoxide by antichymotrypsin-chymotrypsin complexes. J Biol Chem 1992; 267:5056-5059.

135. Kilpatrick L, McCawley L, Nachiappan V, et al. α-1-Antichymotrypsin inhibits the

α_1-proteinase inhibitor. α_1-Proteinase inhibitor isolated from the bronchoalveolar lavage fluid of smokers (with normal plasma α_1-proteinase inhibitor activity) had significantly reduced elastase inhibitory activity compared with the α_1 proteinase inhibitor isolated from the lavage fluid of non-smokers.[25] In addition to the direct action of cigarette smoke on α_1-proteinase inhibitor, oxidants spontaneously released from the alveolar macrophages of smokers can inactivate α_1-proteinase inhibitor.[26]

As well as the α_1-proteinase inhibitor variants responsible for emphysema, a very important variant in terms of the lessons learned from it concerning serpin specificity is α_1-proteinase inhibitor Pittsburgh. This variant was first described in a patient with a fatal bleeding disorder.[27] The mutation was identified as a Met to Arg change at the P1 position (residue 358).[28] Studies with both purified and recombinant proteins showed that this changed the specificity from elastase-like enzymes to enzymes with specificity for Arg at P1 such as thrombin, which is inhibited with a second order rate constant of 3×10^5 $M^{-1}s^{-1}$.[29-33] This was believed to account for the bleeding diathesis. In addition, however, this variant could inhibit plasma kallikrein, β-factor XIIa and activated protein C in vitro[29,32,34] (Table 8.2). Another patient with this same disorder has been identified but in this case the patient was asymptomatic, although reduced levels of protein C and elevated levels of proalbumin were present.[35,36] The α_1-proteinase inhibitor Pittsburgh variant has provided the basis for further variants of α_1-proteinase inhibitor[31,37-40] and provided the impetus for site directed mutagenesis at P1 in a number of serpins (see reference 41 and Chapter 3). A further variant of α_1-proteinase inhibitor Pittsburgh was constructed with an additional Pro to Ala change at position 357 (P2), thereby producing an α_1-proteinase inhibitor variant with the same P2 and P1 residues as C1-inhibitor and PAI-1.[37,38] Although this changed the second order inhibition rate constants

Table 8.2. Second order rate constants for the inhibition of serine proteinases by α_1-proteinase inhibitor, α_1-proteinase inhibitor-Arg358, α_1-proteinase inhibitor-Ala357,Arg358, C1 inhibitor and PAI-1

Serpin	Second order rate constants ($M^{-1}s^{-1}$) for indicated proteinase								
	Elastase	Thrombin	Kallikrein	Factor XIIa	Protein C	C1s	tPA	Urokinase	Plasmin
α_1-Proteinase inhibitor (Met 358)	6.5×10^{7} [a]	48 [a]	< 10 [b]	< 10 [c]	11 [d]	< 10 [e]	< 10 [e]	<10 [e]	< 10 [e]
α_1-Proteinase inhibitor-Arg 358	2.2×10^{3} [f]	3.1×10^{5} [f]	6.9×10^{4} [g]	3.5×10^{4} [g]	4.9×10^{4} [d]	< 10 [e]	137 [e]	8.9×10^{3} [e]	2.5×10^{6} [e]
α_1-Proteinase inhibitor-Ala 357, Arg 358	not determined	7.3×10^{5} [h]	3.6×10^{5} [h]	9.2×10^{3} [h]	2.7×10^{4} [d]	310 [e]	590 [e]	1.4×10^{4} [e]	1×10^{5} [e]
C1 inhibitor	0 [i]	< 10 [j]	1.7×10^{4} [b]	3.1×10^{3} [c]	< 1 [k]	5×10^{4} [l]	< 10 [m]	< 10 [e]	1×10^{4} [e]
PAI-1	0 [n]	610 [o]	6×10^{4} [p]	1.6×10^{4} [p]	1.1×10^{4} [q]	< 10 [e]	1.8×10^{7} [r]	1.8×10^{7} [r]	7×10^{5} [s]

References: a, 24; b, 153; c, 154; d, 34; e, 38; f, 30; g, 32; h, 37; i, 155; j, 29; k, 156; l, 157; m, 158; n, 159; o, 160; p, 161; q, 162; r, 163; s, 164.

somewhat, the changes were not predictable a priori, and the double mutant had different specificity from both C1-inhibitor and PAI-1 (Table 8.2).

2. ANTITHROMBIN DEFICIENCY

The clinical, biochemical and genetic aspects of inherited antithrombin deficiency have been dealt with extensively in a series of review articles[42-46] and so details will not be described here. A database of known antithrombin mutations has also been published.[47] Occurrence of hereditary antithrombin deficiency can be as much as 1 in 500 of the general population[48-50] and up to 1 in 20 for patients with a history of thrombosis.[51] The use of antithrombin concentrates for the treatment of inherited antithrombin deficiency and of acquired deficiency, such as in disseminated intravascular coagulation has been reviewed.[52,53]

Type I antithrombin deficiency is a quantitative deficiency with only 50% activity and antigen being present due to synthesis from only one allele. Type II deficient individuals have 50-100% activity but normal antigenic levels due to the production of variant inactive protein from one allele. Examples of Type I and II antithrombin deficiencies are given in Table 8.3. Type II variants can be subdivided into three classes. Type IIa are molecules with mutations within the reactive center loop, which have reduced (or absent) abilities to react with thrombin and factor Xa or can react but exhibit substrate properties. Type IIb are those which have mutations within the heparin binding site and so have reduced heparin affinity and cannot be stimulated, although they do have normal basal inhibitory activity. Type IIc are the pleiotropic mutants which have reduced heparin affinity as well as reduced inhibitory activity. In this latter group the mutations cluster around the P9'-P16' region, which is part of β strands 1C and 4B. This suggests that this region might provide an important link between the heparin binding site and the reactive center loop.[54]

3. C1-INHIBITOR DEFICIENCY

In a classic paper, Osler[55] described the symptoms of a condition called hereditary angioedema (HAE) which is characterized by recurrent attacks of swellings in the face, larynx, intestine and extremities, which in the case of laryngeal edema can be life threatening. It was also noted that the predisposition to these attacks was transmitted in an autosomal dominant manner, which was subsequently confirmed.[56] Much later it was determined that a deficiency in complement inhibition was involved[57] and Donaldson and Evans[58] showed that HAE was caused by deficiency of C1-inhibitor. The facial effects are pronounced.[59]

Type I deficiency involves synthesis from only one allele and is the common form of C1-inhibitor deficiency (85% of cases). Defects in both the gene and the mRNA for C1-inhibitor have been documented.[60-63] An unusual feature of the type I deficiency is that the circulating levels of functional inhibitor are 5-30% of normal rather than the expected 50%. It has been shown that transinhibition of the normal allele can occur, whereby the mutant mRNA/protein can cause reduced translation of the wild type mRNA.[64]

Type II C1-inhibitor deficiency has normal antigenic levels of protein but 50% activity. The 50% inactive protein is due to a variety of mutations, most of which cluster in the reactive center region thereby preventing interaction with proteinase or altering loop insertion and complex formation.[65-70] The susceptibility of the C1-inhibitor gene to mutation seems to result from clusters of *Alu* repeats as well as a CpG dinucleotide in the codon for the P1 residue.[71-73] There is also a form of acquired C1-inhibitor deficiency in which the patient has autoantibodies that impair the function. In these patients the C1-inhibitor is usually present in a cleaved form, as is often also the case with type II variants.[74-78]

An unresolved issue is the mechanism of the pathogenesis of the HAE attacks. Given that C1-inhibitor regulates the

QUESTIONNAIRE

Receive a FREE BOOK of your choice

Please help us out—Just answer the questions below, then select the book of your choice from the list on the back and return this card.

R.G. Landes Company publishes five book series: *Medical Intelligence Unit, Molecular Biology Intelligence Unit, Neuroscience Intelligence Unit, Tissue Engineering Intelligence Unit* and *Biotechnology Intelligence Unit.* We also publish comprehensive, shorter than book-length reports on well-circumscribed topics in molecular biology and medicine. The authors of our books and reports are acknowledged leaders in their fields and the topics are unique. Almost without exception, there are no other comprehensive publications on these topics.

Our goal is to publish material in important and rapidly changing areas of bioscience for sophisticated scientists. To achieve this goal, we have accelerated our publishing program to conform to the fast pace in which information grows in bioscience. Most of our books and reports are published within 90 to 120 days of receipt of the manuscript.

Please circle your response to the questions below.

1. We would like to sell our *books* to scientists and students at a deep discount. But we can only do this as part of a prepaid subscription program. The retail price range for our books is $59-$99. Would you pay $196 to select four *books* per year from any of our Intelligence Units–$49 per book–as part of a prepaid program?

   Yes **No**

2. We would like to sell our *reports* to scientists and students at a deep discount. But we can only do this as part of a prepaid subscription program. The retail price range for our reports is $39-$59. Would you pay $145 to select five *reports* per year–$29 per report–as part of a prepaid program?

   Yes **No**

3. Would you pay $39–the retail price range of our books is $59-$99–to receive any single book in our Intelligence Units if it is spiral bound, but in every other way identical to the more expensive hardcover version?

   Yes **No**

To receive your free book, please fill out the shipping information below, select your free book choice from the list on the back of this survey and mail this card to:

R.G. Landes Company, 909 S. Pine Street, Georgetown, Texas 78626 U.S.A.

Your Name ______________________________

Address ______________________________

City ____________________ State/Province: ____________________

Country: ____________________ Postal Code: ____________________

My computer type is Macintosh______ ; IBM-compatible ______ ; Other ______

Do you own ____ or plan to purchase ___ a CD-ROM drive?

Available Free Titles

Please check three titles in order of preference.
Your request will be filled based on availability. Thank you.

- ❐ Water Channels
  Alan Verkman,
  University of California-San Francisco
- ❐ The Na,K-ATPase:
  Structure-Function Relationship
  J.-D. Horisberger, University of Lausanne
- ❐ Intrathymic Development of T Cells
  J. Nikolic-Zugic,
  Memorial Sloan-Kettering Cancer Center
- ❐ Cyclic GMP
  Thomas Lincoln, University of Alabama
- ❐ Primordial VRM System and the Evolution of Vertebrate Immunity
  John Stewart, Institut Pasteur-Paris
- ❐ Thyroid Hormone Regulation of Gene Expression
  Graham R. Williams, University of Birmingham
- ❐ Mechanisms of Immunological Self Tolerance
  Guido Kroemer, CNRS Génétique Moléculaire et Biologie du Développement-Villejuif
- ❐ The Costimulatory Pathway for T Cell Responses
  Yang Liu, New York University
- ❐ Molecular Genetics of Drosophila Oogenesis
  Paul F. Lasko, McGill University
- ❐ Mechanism of Steroid Hormone Regulation of Gene Transcription
  M.-J. Tsai & Bert W. O'Malley, Baylor University
- ❐ Liver Gene Expression
  François Tronche & Moshe Yaniv,
  Institut Pasteur-Paris
- ❐ RNA Polymerase III Transcription
  R.J. White, University of Cambridge
- ❐ src Family of Tyrosine Kinases in Leukocytes
  Tomas Mustelin, La Jolla Institute
- ❐ MHC Antigens and NK Cells
  Rafael Solana & Jose Peña,
  University of Córdoba
- ❐ Kinetic Modeling of Gene Expression
  James L. Hargrove, University of Georgia
- ❐ PCR and the Analysis of the T Cell Receptor Repertoire
  Jorge Oksenberg, Michael Panzara & Lawrence Steinman, Stanford University
- ❐ Myointimal Hyperplasia
  Philip Dobrin, Loyola University
- ❐ Transgenic Mice as an In Vivo Model of Self-Reactivity
  David Ferrick & Lisa DiMolfetto-Landon, University of California-Davis and Pamela Ohashi, Ontario Cancer Institute
- ❐ Cytogenetics of Bone and Soft Tissue Tumors
  Avery A. Sandberg, Genetrix & Julia A. Bridge,
  University of Nebraska
- ❐ The Th1-Th2 Paradigm and Transplantation
  Robin Lowry, Emory University
- ❐ Phagocyte Production and Function Following Thermal Injury
  Verlyn Peterson & Daniel R. Ambruso,
  University of Colorado
- ❐ Human T Lymphocyte Activation Deficiencies
  José Regueiro, Carlos Rodríguez-Gallego and Antonio Arnaiz-Villena,
  Hospital 12 de Octubre-Madrid
- ❐ Monoclonal Antibody in Detection and Treatment of Colon Cancer
  Edward W. Martin, Jr., Ohio State University
- ❐ Enteric Physiology of the Transplanted Intestine
  Michael Sarr & Nadey S. Hakim, Mayo Clinic
- ❐ Artificial Chordae in Mitral Valve Surgery
  Claudio Zussa, S. Maria dei Battuti Hospital-Treviso
- ❐ Injury and Tumor Implantation
  Satya Murthy & Edward Scanlon,
  Northwestern University
- ❐ Support of the Acutely Failing Liver
  A.A. Demetriou, Cedars-Sinai
- ❐ Reactive Metabolites of Oxygen and Nitrogen in Biology and Medicine
  Matthew Grisham, Louisiana State-Shreveport
- ❐ Biology of Lung Cancer
  Adi Gazdar & Paul Carbone,
  Southwestern Medical Center
- ❐ Quantitative Measurement of Venous Incompetence
  Paul S. van Bemmelen, Southern Illinois University and John J. Bergan, Scripps Memorial Hospital
- ❐ Adhesion Molecules in Organ Transplants
  Gustav Steinhoff, University of Kiel
- ❐ Purging in Bone Marrow Transplantation
  Subhash C. Gulati,
  Memorial Sloan-Kettering Cancer Center
- ❐ Trauma 2000: Strategies for the New Millennium
  David J. Dries & Richard L. Gamelli,
  Loyola University

activity of the C1 component of complement and that complement activation during HAE attacks has been documented (evidence reviewed in Davis[79]) it is possible that a complement derived peptide (such as a C2-derived kinin) is involved. However, C1-inhibitor also regulates the contact system (see Fig. 5.1) and evidence has been presented that bradykinin, generated by kallikrein cleavage of high and low molecular weight kininogens, is an important mediator.[80] Pre-kallikrein activation and consumption of high molecular weight kininogen in hereditary angioedema have been shown.[81] The finding that heparin enhances the inactivation of kallikrein in the presence of high molecular weight kininogen[82] supports the idea that heparin could be used in the management of hereditary angioedema. Plasmin activation also seems to play a role in HAE attacks.[83] Plasmin can cleave C2 and C3 to release

Table 8.3. Examples of naturally occurring Type II antithrombin variants

Name	Mutation	Effect of mutation
Type IIa (reactive center mutants)		
Hamilton	Ala 382→Thr (P12)	Substrate for proteinases
Charleville	Ala 384→Pro (P10)	Substrate for proteinases
Northwick Park	Arg 393→Cys (P1)	Disulfide linked to albumin, no interaction with target proteinases
Glasgow	Arg 393→His (P1)	Increased heparin affinity, no interaction with target proteinases
Pescara	Arg 393→Pro (P1)	Same as Glasgow variant
Denver	Ser 394→Leu (P1')	Reduced rates of proteinase inhibition
Type IIb (heparin binding site mutants)		
Rouen III	Ile 7→Asn	Reduced heparin affinity. Potential new glycosylation site at position 7
Rouen IV	Arg 24→Cys	Reduced heparin affinity due to loss of basic residue
Toyama	Arg 47→Cys	Same as Rouen IV variant
Rouen II	Arg 47→Ser	Same as Rouen IV variant
Budapest 3	Leu 99→Phe	Reduced heparin affinity due to proximity of residue 99 to heparin binding site
Southport	Leu 99→Val	Same as Budapest 3 variant
Nagasaki	Ser 116→Pro	Reduced heparin affinity due to proximity of residue 116 to heparin binding site
Geneva	Arg 129→Gln	Reduced heparin affinity due to loss of critical arginine
Type IIc (pleiotropic mutants)		
Rosny	Phe 402→Cys (P6')	Decreased activity, heparin affinity and plasma concentration
Oslo	Ala 404→Thr (P8')	Same as Rosny variant
La Rochelle	Asn 405→Lys (P9')	Same as Rosny variant
Kyoto	Arg 406→Thr (P10')	Decreased activity and concentration
Budapest 5	Pro 407→Thr (P11')	Decreased activity and heparin affinity
Utah	Pro 407→Leu (P11')	Same as Rosny variant
Budapest	Pro 429→Leu	Same as Rosny variant

Further details and references to original articles can be found in references 1, 45 and 47.

bioactive peptides as well as activating C1, and indeed the antifibrinolytic agent EACA is successfully used to treat HAE. There are comprehensive reviews of biochemical and genetic aspects of C1-inhibitor deficiency and HAE.[79,84,85] Clinical aspects of HAE and treatment regimens are discussed in Orfan and Kolski[86] and Elnicki.[87]

4. α_2-ANTIPLASMIN DEFICIENCY

Deficiency of this serpin is somewhat rare but a growing number of cases are being reported (Griffin et al[88] and references therein). Homozygous deficiency is associated with serious bleeding, whereas heterozygous deficiency is less severe. Acquired deficiencies are also known.[89] Some variant α_2-antiplasmin molecules have been studied in detail. The α_2-antiplasmin-Okinawa variant has a deletion of Glu 137 which leads to a block of intracellular transport from the endoplasmic reticulum to the Golgi body.[90] The α_2-antiplasmin-Enschede variant has an alanine insertion in the reactive center region which converts it from an inhibitor to a substrate of plasmin.[91-93]

5. THE ROLE OF PAI-1 IN VASCULAR DISEASE

PAI-1 deficiency is somewhat rare but seems to be associated with bleeding episodes particularly after trauma or surgery.[94-97] More problematic seems the case of an excess of PAI-1. The findings that the plasma concentration of PAI-1 is only 0.4 nM and that expression of the gene is subject to extensive regulation indicates how detrimental high levels of PAI-1 can be. Elevated PAI-1 has been suggested to be a risk factor for myocardial infarction.[98-100] Moreover, high levels of PAI-1 and increased expression have been found to be associated with the development of thrombi and atherosclerotic plaques.[101-106] In support of these data, transgenic mice expressing the human PAI-1 gene develop venous occlusions.[107] Elevated PAI-1 has also been suggested to be responsible for the fibrin deposition in hemolytic uremic syndrome.[108]

6. SERPINS AND ALZHEIMER'S DISEASE

There is a growing body of evidence that proteolytic pathways are essential for neural processes such as synaptic plasticity and long term potentiation.[109-113] Also, serpins have been implicated in neurological diseases such as Alzheimer's disease.[114] PAI-2, C1-inhibitor and antithrombin have all been found in amyloid plaques.[115-117] Protease nexin 1, which can regulate neurite outgrowth (see Chapter 4), is reduced in Alzheimer's disease tissue, although this may reflect increased complexing with target proteinases.[118] The most important serpin, however, is likely to be α_1-antichymotrypsin. This serpin is found in the amyloid plaques[119,120] and is expressed by astrocytes from Alzheimer's disease brains,[121] although serum α_1-antichymotrypsin does not seem to be a reliable diagnostic marker for Alzheimer's disease.[122] α_1-Antichymotrypsin can modulate the ability of the amyloid β-peptide to form filaments.[123-125]

Another risk factor for Alzheimer's disease is the presence of certain apolipoprotein E polymorphic variants.[126] The *APOE*4* allele is a strong susceptibility marker. However, genetic linkage studies indicate that the risk is greatest when a certain polymorphic variant of α_1-antichymotrypsin is also present.[127] This allelic variant of α_1-antichymotrypsin has an Ala to Thr mutation in the signal peptide (residue -15) and was previously shown to be present in healthy individuals with no abnormality in plasma α_1-antichymotrypsin levels.[128]

7. SERPINS AND CANCER

Two serpins implicated in cancer are maspin and squamous cell carcinoma antigen. The acidic variant of squamous cell carcinoma antigen (SCCA2) is expressed primarily by malignant cells and is the major form in the plasma of cancer patients. The gene for this serpin is present at chromosome 18q21.3 in a locus which also contains the genes for SCCA1, PAI-2 and the tumor suppressor serpin maspin.

This gene cluster is closely linked to the gene for Bcl-2 and the immunoglobulin heavy chain locus, translocations between which are involved in many B-cell lymphomas, suggesting that translocations of the serpin cluster genes could also be involved in cancer.[129] It is also possible that rearrangements of the serpin gene cluster at 14q32.1 are involved in some types of cancer.[130] In regard to PAI-2 there is an extensive literature on the role of plasminogen activators and their inhibitors in cancer and some comprehensive reviews have been written.[131-133] In general, elevated levels of uPA and the uPA receptor and low levels of PAI-1 are associated with malignancy.[134-138] However, the situation with PAI-2 is less clear. Over-expression of PAI-2 in a metastatic melanoma cell line producing uPA reduced the ability of these cells to cause tumors in mice,[139] but elevated PAI-1 and PAI-2 have been reported to be associated with lymphoma.[140] Cells expressing PAI-2 are protected from apoptosis induced by a variety of stimuli, suggesting that PAI-2 might inhibit a proteinase involved in the apoptotic pathway[141,142] (see Chapter 5, section 13 for more information on serpins and apoptosis).

8. SERPINS AS TARGETS FOR THERAPEUTIC INTERVENTION

The most important serpin that is used as a target for therapuetic intervention is antithrombin through the use of the anticoagulant drug heparin. The mechanism of activation of antithrombin by heparin has been described in Chapter 4. This is the only example of a drug which specifically acts via a serpin. Replacement therapy is available for some serpin deficiencies, particularly antithrombin deficiency. C1-inhibitor has been used to prevent pulmonary dysfunction induced by endotoxin in dogs[143] suggesting a possible therapeutic role. The use of recombinant variants of α_1-proteinase inhibitor to treat endotoxic shock is not as promising as once hoped,[144,145] although an α_1-proteinase inhibitor variant with specificity for the processing protease furin could inhibit processing of HIV gp 160, a step essential to viral infectivity.[39] However, the future for serpins as therapeutic tools probably lies, at least in the near term, in gene therapy for serpin deficient states. The serpin that has been studied in this regard is α_1-proteinase inhibitor. A number of reasons can be given for the use of this serpin: i) As many as 10% of people of European origin are heterozygous for a defective α_1-proteinase inhibitor gene (usually S or Z variants); ii) emphysema, although disabling and life threatening, is chronic rather than acute; iii) the lung is a relatively easy organ to target through the vasculature or more directly through the airways; iv) α_1-proteinase inhibitor has been studied very extensively at both the biochemical and genetic levels; and v) recombinant α_1-proteinase inhibitor is comparatively easy to produce in vitro, suggesting that in vivo expression in transfected animals might be possible.

Many of the papers in which the α_1-proteinase inhibitor gene has been successfully transfected into a variety of eukaryotic cell types are listed in Patston and Gettins.[41] Mouse fibroblasts transfected with the gene via a retroviral vector expressed α_1-proteinase inhibitor up to 4 weeks after transplantation into a mouse.[146] Hepatocytes from mice transgenic for human α_1-proteinase inhibitor could be transplanted into a non-transgenic mouse and still express the gene after six months.[147] Similarly, retrovirally transfected dog hepatocytes could be transplanted and express α_1-proteinase inhibitor.[148] Another approach has been to target the lung directly, since this is where the tissue damage in emphysema occurs. Thus, adenovirus-mediated gene transfer has been used to target lung epithelium of rats in vitro and in vivo[149] and of humans in vitro.[150] Another approach to target the lung directly is to use aerosol administration of liposomes containing the expression plasmid through the airways, a technique that has been shown to be feasible in rabbits.[151]

References

1. Stein PE and Carrell RW. What do dysfunctional serpins tell us about molecular mobility and disease? Nature Struct Biol 1995; 2:96-113.
2. Cox DW. α_1-Antitrypsin deficiency. In: Scriver CR, Beaudet AL, Sly WS, Valle D, eds. The Metabolic Basis of Inherited Disease. New York: McGraw Hill, 1989: 2409-2437.
3. Perlmutter DH and Pierce JA. The α_1-antitrypsin gene and emphysema. Am J Physiol 1989; 257:L147-L162.
4. Snider GL. Emphysema: the first two centuries and beyond. A historical overview, with suggestions for future research: Part 1. Am Rev Respir Dis 1992; 146:1334-1244.
5. Snider GL. Emphysema: the first two centuries and beyond. A historical overview, with suggestions for future research: Part 2. Am Rev Respir Dis 1992; 146:1615-1622.
6. Evans MD and Pryor WA. Cigarette smoking, emphysema, and damage to α_1-proteinase inhibitor. Am J Physiol 1994; 266: L593-L611.
7. Ogushi F, Fells GA, Hubbard RC, et al. Z-type α1-antitrypsin is less competent than M1-type α1-antitrypsin as an inhibitor of neutrophil elastase. J Clin Invest 1987; 80:1366-1374.
8. Lomas DA, Evans DL, Stone SR, et al. Effect of the Z mutation on the physical and inhibitory properties of α_1-antitrypsin. Biochemistry 1993; 32:500-508.
9. Brantly M, Courtney M and Crystal RG. Repair of the secretion defect in the Z form of α1-antitrypsin by addition of a second mutation. Science 1988; 242:1700-1702.
10. Sifers RN, Hardick CP and Woo SLC. Disruption of the 290-342 salt bridge is not responsible for the secretory defect of the PiZ α_1-antitrypsin variant. J Biol Chem 1989; 264:2997-3001.
11. McCracken AA, Kruse KB and Brown JL. Molecular basis for defective secretion of the Z variant of human alpha-1-proteinase inhibitor: Secretion of variants having altered potential for salt bridge formation between amino acids 290 and 342. Mol Cell Biol 1989; 9:1406-1414.
12. McCracken AA, Kruse KB, Valentine J, et al. Construction and expression of α_1-proteinase inhibitor mutants and the effect of these mutations on secretion of the variant inhibitors. J Biol Chem 1991; 266:7578-7582.
13. Lomas DA, Evans DL, Finch JT, et al. The mechanism of Z α_1-antitrypsin accumulation in the liver. Nature 1992; 357:605-607.
14. Sifers RN. Defective protein folding as a cause of disease. Nature Struct Biol 1995; 2:355-357.
15. Yu M-H, Lee KN and Kim J. The Z type variation of human α_1-antitrypsin causes a protein folding defect. Nature Struct Biol 1995; 2:363-367.
16. Holmes MD, Brantly ML and Crystal RG. Molecular basis of the heterogeneity among the P-family of alpha-1-antitrypsin alleles. Am Rev Respir Dis 1990; 142:1185-1192.
17. Curiel DT, Chytil A, Courtney M, et al. Serum α1-antitrypsin deficiency associated with the common S-type ($Glu^{264}\rightarrow Val$) mutation results from intracellular degradation. J Biol Chem 1989; 264:10477-10486.
18. Lomas DA, Finch JT, Seyama K, et al. α_1-Antitrypsin S_{iiyama} ($Ser^{53}\rightarrow Phe$). Further evidence for intracellular loop-sheet polymerization. J Biol Chem 1993; 268:15333-15335.
19. Holmes MD, Brantly ML, Fells GA, et al. α1-Antitrypsin $W_{bethesda}$: Molecular basis of an unusual α1-antitrypsin deficiency variant. Biochem Biophys Res Commun 1990; 170:1013-1020.
20. Curiel DT, Holmes MD, Okayama H, et al. Molecular basis of the liver and lung disease associated with the α1-antitrypsin allele M_{malton}. J Biol Chem 1989; 264: 13938-13945.
21. Curiel DT, Vogelmeier C, Hubbard RC, et al. Molecular basis of α1-antitrypsin deficiency and emphysema associated with the α1-antitrypsin $M_{mineral\ springs}$ allele. Mol Cell Biol 1990; 10:47-56.
22. Lomas DA, Elliott PR, Sidhar SK, et al. α_1-Antitrypsin M_{malton} (Phe^{52}-deleted) forms loop-sheet polymers *in vivo.* Evidence for the C sheet mechanism of polymerization. J Biol Chem 1995; 270:16864-16870.
23. Sifers RN, Brashears-Macatee S, Kidd VJ,

et al. A frameshift mutation results in a truncated α_1-antitrypsin that is retained within the rough endoplasmic reticulum. J Biol Chem 1988; 263:7330-7335.

24. Beatty K, Bieth J and Travis J. Kinetics of association of serine proteinases with native and oxidized α-1-proteinase inhibitor and α-1-antichymotrypsin. J Biol Chem 1980; 255:3931-3934.
25. Ogushi F, Hubbard RC, Vogelmeier C, et al. Risk factors for emphysema. Cigarette smoking is associated with a reduction in the association rate constant of lung α_1-antitrypsin for neutrophil elastase. J Clin Invest 1991; 87:1060-1065.
26. Hubbard RC, Ogushi F, Fells GA, et al. Oxidants spontaneously released by alveolar macrophages of cigarette smokers can inactivate the active site of α1-antitrypsin, rendering it ineffective as an inhibitor of neutrophil elastase. J Clin Invest 1987; 80:1289-1295.
27. Lewis JH, Iammarino RM, Spero JA, et al. Antithrombin Pittsburgh: an α_1-antitrypsin variant causing hemorrhagic disease. Blood 1987; 51:129-137.
28. Owen MC, Brennan SO, Lewis JH, et al. Mutation of antitrypsin to antithrombin. α_1-Antitrypsin Pittsburgh (358 Met Arg), a fatal bleeding disorder. N Engl J Med 1983; 309:694-698.
29. Scott CF, Carrell RW, Glaser CB, et al. Alpha-1-antitrypsin-Pittsburgh. A potent inhibitor of human plasma factor XIa, kallikrein, and factor XII. J Clin Invest 1986; 77:631-634.
30. Travis J, Matheson NR, George PM, et al. Kinetic studies on the interaction of α_1-proteinase inhibitor (Pittsburgh) with trypsin-like serine proteinases. Biol Chem Hoppe-Seyler 1986; 367:853-859.
31. Jallat S, Carvallo D, Tessier LH, et al. Altered specificities of genetically engineered α_1 antitrypsin variants. Prot Eng 1986; 1:29-35.
32. Schapira M, Ramus M-A, Jallat S, et al. Recombinant α-antitrypsin Pittsburgh (Met^{358}-Arg) is a potent inhibitor of plasma kallikrein and activated factor XII fragment. J Clin Invest 1986; 77:635-637.
33. George PM, Pemberton P, Bathurst IC, et al. Characterization of antithrombins produced by active site mutagenesis of human α_1-antitrypsin expressed in yeast. Blood 1989; 73:490-496.
34. Heeb MJ, Bischoff R, Courtney M, et al. Inhibition of activated protein C by recombinant α_1-antitrypsin variants with substitution of arginine or leucine for methionine. J Biol Chem 1990; 265:2365-2369.
35. Vidaud D, Emmerich J, Alhenc-Gelas M, et al. Met 358 to Arg of alpha-1-antitrypsin associated with protein C deficiency in a patient with mild bleeding tendency. J Clin Invest 1992; 89:1537-1543.
36. Brennan SO, Sheat JM and Aiach M. Repair of the secretion defect in the Z form of α1-antitrypsin Pittsburgh. Clin Chim Acta 1993; 214:123-128.
37. Schapira M, Ramus M-A, Waeber B, et al. Protection by recombinant α_1-antitrypsin Ala^{357} Arg 358 against arterial hypotension induced by factor XII fragment. J Clin Invest 1987; 80:582-585.
38. Patston PA, Roodi N, Schifferli JA, et al. Reactivity of α_1-antitrypsin mutants against proteolytic enzymes of the kallikrein-kinin, complement, and fibrinolytic systems. J Biol Chem 1990; 265:10786-10791.
39. Anderson ED, Thomas L, Hayflick JS, et al. Inhibition of HIV-1 gp160-dependent membrane fusion by a furin-directed α_1-antitrypsin variant. J Biol Chem 1993; 268:24887-24891.
40. Hopkins PCR, Crowther DC, Carrell RW, et al. Development of a novel recombinant serpin with potential antithrombotic properties. J Biol Chem 1995; 270:11866-71.
41. Patston PA and Gettins PGW. A database of recombinant wild-type and mutant serpins. Thromb Haemost 1994; 72: 166-179.
42. Tollefsen DM. Antithrombin deficiency. In: Scriver CR, Beaudet AL, Sly WS,Valle D, eds. The Metabolic Basis of Inherited Disease. New York: McGraw-Hill, 1989:2207-2218.
43. Hathaway WE. Clinical aspects of antithrombin III deficiency. Sem Hematol 1991; 28:19-23.
44. Blajchman MA, Austin RC, Fernandez-Rachubinski F, et al. Molecular basis of

inherited human antithrombin deficiency. Blood 1992; 80:2159-2171.

45. Olds RJ, Lane DA, Mille B, et al. Antithrombin, the principal inhibitor of thrombin. Sem Thromb Hemostas 1994; 20: 353-372.
46. Perry DJ. Antithrombin and its inherited disorders. Blood Rev 1994; 8:37-55.
47. Lane DA, Olds RJ, Boisclair M, et al. Antithrombin III mutation database: First update. Thromb Haemostas 1993; 70: 361-369.
48. Tait RC, Walker ID, Islam SIAM, et al. Influence of demographic factors on antithrombin III activity in a healthy population. Br J Haematol 1993; 84:476-480.
49. Tait RC, Walker ID, Perry DJ, et al. Prevalence of antithrombin deficiency in the healthy population. Br J Haematol 1994; 87:106-112.
50. Wells PS, Blajchman MA, Henderson P, et al. Prevalence of antithrombin deficiency in healthy blood donors: a cross sectional study. Am J Hematol 1994; 45:321-324.
51. Harper PL, Luddington RJ, Daly M, et al. The incidence of dysfunctional antithrombin variants: four cases in 210 patients with thromboembolic disease. Br J Haematol 1991; 77:360-364.
52. Menache D. Antithrombin III concentrates. Hematology/Oncology Clinics of North America 1992; 6:1115-1120.
53. Bick RL. Disseminated intravascular coagulation. Hematology/Oncology Clinics of North America 1992; 6:1259-1285.
54. Lane DA, Olds RJ, Conrad J, et al. Pleiotropic effects of antithrombin strand 1C substitution mutations. J Clin Invest 1992; 90:2422-2433.
55. Osler W. Hereditary angioneurotic edema. Am J Med Sci 1888; 95:362-367.
56. Crowder JR and Crowder TR. Five generations of angioneurotic edema. Arch Int Med 1917; 20:840-852.
57. Landerman NS. Hereditary angioneurotic edema. Arch Int Med 1962; 33:316-329.
58. Donaldson VH and Evans RR. A biochemical abnormality in hereditary angioneurotic edema. Absence of serum inhibitor C1-esterase. Am J Med 1963; 35:37-44.
59. Hammerschmidt DE and Donaldson V. Hereditary angioedema. J Lab Clin Med 1993; 121:148.
60. Stoppa-Lyonnet D, Tosi M, Laurent J, et al. Altered C1 inhibitor genes in type I hereditary angioedema. N Engl J Med 1987; 317:1-6.
61. Cicardi M, Igarashi T, Kim MS, et al. Restriction fragment length polymorphism of the C1 inhibitor gene in hereditary angioneurotic edema. J Clin Invest 1987; 80:1640-1643.
62. Ariga T, Igarashi T, Ramesh N, et al. Type I C1 inhibitor deficiency with a small messenger RNA resulting from deletion of one exon. J Clin Invest 1989; 83:1888-1893.
63. Frangi D, Cicardi M, Sica A, et al. Nonsense mutations affect C1 inhibitor messenger RNA levels in patients with type I angioneurotic edema. J Clin Invest 1991; 88:755-759.
64. Kramer J, Rosen FS, Colten HR, et al. Transinhibition of C1 inhibitor synthesis in type I hereditary angioneurotic edema. J Clin Invest 1993; 91:1258-1262.
65. Aulak KS, Pemberton PA, Rosen FS, et al. Dysfunctional C1-inhibitor (At), isolated from a type II hereditary angio-oedema plasma, contains a P1' reactive centre ($Arg^{444}\rightarrow$His) mutation. Biochem J 1988; 253:615-618.
66. Levy NJ, Ramesh N, Cicardi M, et al. Type II hereditary angioneurotic edema that may result from a single nucleotide change in the codon for alanine-436 in the C1 inhibitor gene. Proc Natl Acad Sci USA 1990; 87:265-268.
67. Skriver K, Wikoff WR, Patston PA, et al. Substrate properties of C1 inhibitor Ma (alanine$\rightarrow$glutamic acid). Genetic and structural evidence suggesting that the P12 region contains critical determinants of serine protease inhibitor inhibitor/substrate status. J Biol Chem 1991; 266:9216-9221.
68. Aulak KS, Eldering E, Hack CE, et al. A hinge region mutation in C1-inhibitor ($Ala^{436}\rightarrow$Thr) results in nonsubstrate-like behavior and in polymerization of the molecule. J Biol Chem 1993; 268:18088-18094.
69. Verpy E, Couture-Tosi E, Eldering E, et al. Crucial residues in the carboxy-terminal end of C1 inhibitor revealed by pathogenic

mutants impaired in secretion or function. J Clin Invest 1995; 95:350-359.
70. Eldering E, Verpy E, Roem D, et al. COOH-terminal substitutions in the serpin C1 inhibitor that cause loop overinsertion and subsequent multimerization. J Biol Chem 1995; 270:2579-2587.
71. Skriver K, Radziejewska E, Silbermann JA, et al. CpG mutations in the reactive site of human C1 inhibitor. J Biol Chem 1989; 264:3066-3071.
72. Stoppa-Lyonnet D, Carter PE, Meo T, et al. Clusters of intragenic *Alu* repeats predispose the human C1 inhibitor locus to deleterious rearrangements. Proc Natl Acad Sci USA 1990; 87:1551-1555.
73. Stoppa-Lyonnet D, Duponchel C, Meo T, et al. Recombinational biases in the rearranged C1-inhibitor genes of hereditary angioedema patients. Am J Hum Genet 1991; 49:1055-1062.
74. Jackson J, Sim RB, Whaley K, et al. Autoantibody facilitated cleavage of C1-inhibitor in autoimmune angioedema. J Clin Invest 1989; 83:698-707.
75. Alsenz J, Lambris JD, Bork K, et al. Acquired C1 inhibitor (C1-INH) deficiency type II. Replacement therapy with C1-INH and analysis of patients' C1-INH and anti-C1-INH autoantibodies. J Clin Invest 1989; 83:1794-1799.
76. Zuraw BL and Altman LC. Acute consumption of C1 inhibitor in a patient with acquired C1-inhibitor deficiency syndrome. J Allergy Clin Immunol 1991; 88:908-918.
77. Donaldson VH, Bernstein DI, Wagner CJ, et al. Angioneurotic edema with acquired C1 inhibitor deficiency and autoantibody to C1 inhibitor: Response to plasmapheresis and cytotoxic therapy. J Lab Clin Med 1992; 119:397-406.
78. Mandle R, Baron C, Ruox E, et al. Acquired C1 inhibitor deficiency as a result of an autoantibody to the reactive center region of C1 inhibitor. J Immunol 1994; 152:4680-4689.
79. Davis AE. C1 inhibitor and hereditary angioneurotic edema. Annu Rev Immunol 1988; 6:595-628.
80. Shoemaker LR, Schurman SJ, Donaldson VH, et al. Hereditary angioneurotic oedema: Characterization of plasma kinin and vascular permeability-enhancing activities. Clin Exp Immunol 1994; 95:22-28.
81. Schapira M, Silver LD, Scott CF, et al. Prekallikrein activation and high-molecular weight kininogen consumption in hereditary angioedema. N Engl J Med 1983; 308:1050-1054.
82. Olson ST, Sheffer R and Francis AM. High molecular weight kininogen potentiates the heparin-accelerated inhibition of plasma kallikrein by antithrombin: Role for antithrombin in the regulation of kallikrein. Biochemistry 1993; 32:12136-12147.
83. Cugno M, Hack CE, de Boer JP, et al. Generation of plasmin during acute attacks of hereditary angioedema. J Lab Clin Med 1993; 121:38-43.
84. Davis III AE, Aulak KS, Zahedi K, et al. C1 inhibitor. Methods Enzymol 1993; 223:97-120.
85. Donaldson VH and Bissler JJ. C1 inhibitors and their genes. An update. J Lab Clin Med 1992; 119:330-333.
86. Orfan NA and Kolski GB. Angioedema and C1 inhibitor deficiency. Annals Allergy 1992; 69:167-172.
87. Elnicki DM. Hereditary angioedema. Southern Medical J 1992; 85:1084-1090.
88. Griffin GC, Mammen EF, Sokol RJ, et al. α_2-Antiplasmin deficiency. An overlooked cause of hemorrhage. Am J Ped Heme/Onc 1993; 15:328-330.
89. Okajima K, Kohno I, Soe G, et al. Direct evidence for systemic fibrinogenolysis in patients with acquired α_2-plasmin inhibitor deficiency. Am J Hematol 1994; 45:16-24.
90. Miura O, Sugahara Y and Aoki N. Hereditary α_2-plasmin inhibitor deficiency caused by a transport-deficient mutation (α_2-PI-Okinawa). Deletion of Glu^{137} by a trinucleotide deletion blocks intracellular transport. J Biol Chem 1989; 264:18213-18219.
91. Kluft C, Nieuwenhuis HK, Rijken DC, et al. α_2-Antiplasmin Enschede: Dysfunctional α_2-antiplasmin molecule associated with an autosomal hemorrhagic disorder. J Clin Invest 1987; 80:1391-1400.
92. Holmes WE, Lijnen HR, Nelles L, et al. α_2-Antiplasmin Enschede: Alanine insertion and abolition of plasmin inhibitory activity.

Science 1987; 238:209-211.

93. Rijken DC, Groeneveld E, Kluft C, et al. α_2-Antiplasmin Enschede is not an inhibitor, but a substrate, of plasmin. Biochem J 1988; 255:609-615.
94. Schleef RR, Higgins DL, Pillemer E, et al. Bleeding diathesis due to decreased functional activity of type 1 plasminogen activator inhibitor. J Clin Invest 1989; 83:1747-1752.
95. Diéval J, Nguyen G, Delobel J, et al. A lifelong bleeding disorder associated with a deficiency of plasminogen activator inhibitor type 1. Blood 1991; 77:528-532.
96. Fay WP, Shapiro AD, Shih JL, et al. Complete deficiency of plasminogen activator inhibitor type 1 due to frame shift mutation. N Engl J Med 1992; 327:1729-1733.
97. Lee MH, Vosburgh E, Anderson K, et al. Deficiency of plasma plasminogen activator inhibitor 1 results in hyperfibrinolytic bleeding. Blood 1993; 81:2357-2362.
98. Hamsten A, Wiman B, de Faire U, et al. Increased plasma levels of a rapid inhibitor of tissue plasminogen activator in young survivors of myocardial infarction. N Engl J Med 1985; 313:1557-1563.
99. Hamsten A, de Faire U, Walldius G, et al. Plasminogen activator inhibitor in plasma. Risk factor for recurrent myocardial infarction. Lancet 1987; 549:3-9.
100. Cortellaro M, Cofrancesco E, Boschetti C, et al. Increased fibrin turnover and high PAI-1 activity as predictors of ischemic events in atherosclerotic patients. A case controlled study. Arteriosclerosis and Thrombosis 1993; 13:1412-1417.
101. Sawa H, Fujii S and Sobel BE. Augmented arterial wall expression of type-1 plasminogen activator inhibitor induced by thrombosis. Arteriosclerosis and Thrombosis 1992; 12:1507-1515.
102. Schneiderman J, Sawdey MS, Keeton MR, et al. Increased type 1 plasminogen activator inhibitor gene expression in atherosclerotic human arteries. Proc Natl Acad Sci USA 1992; 89:6998-7002.
103. Lupu F, Bergonzelli GE, Heim DA, et al. Localization and production of plasminogen activator inhibitor-1 in human healthy and atherosclerotic arteries. Arteriosclerosis and Thrombosis 1993; 13:1090-1100.
104. Lang IM, Marsh JJ, Olman MA, et al. Expression of type I plasminogen activator inhibitor in chronic pulmonary thromboemboli. Circulation 1994; 89:2715-2721.
105. Chomiki N, Henry M, Alessi MC, et al. Plasminogen activator inhibitor-1 expression in human liver or atherosclerotic vessel walls. Thromb Haemostas 1994; 72:44-53.
106. Padró T, Emmeis JJ, Stems M, et al. Quantification of plasminogen activators and their inhibitors in the aortic vessel wall in relation to the presence and severity of atherosclerotic disease. Arteriosclerosis Thromb Vasc Biol 1995; 15:893-902.
107. Erickson LA, Fici GJ, Lund JE, et al. Development of venous occlusions in mice transgenic for the plasminogen activator inhibitor 1 gene. Nature 1990; 346:74-76.
108. Bergstein JM, Riley M and Bang NU. Role of plasminogen-activator inhibitor type 1 in the pathogenesis and outcome of the hemolytic uremic syndrome. N Engl J Med 1992; 327:755-759.
109. Hantaï D, Rao JS and Festoff BW. Serine proteases and serpins: their possible roles in the motor system. Rev Neurol 1988; 144:680-687.
110. Festoff BW. Serine proteases and their serpin inhibitors in the nervous system; regulation in development and in degeneration and malignant disease, New York: Plenum Press, 1990.
111. Qian Z, Gilbert ME, Colico MA, et al. Tissue-plasminogen activator is induced as an immediate-early gene during seizure, kindling and long-term potentiation. Nature 1993; 361:453-457.
112. Sappino AP, Madani R, Huarte J, et al. Extracellular proteolysis in the adult murine brain. J Clin Invest 1993; 92:679-685.
113. Smirnova IV, Ho GJ, Fenton JW II, et al. Extravascular proteolysis and the nervous system: Serine protease/serpin balance. Sem Thromb Hemostas 1994; 20:426-432.
114. Ho GJ, Smirnova IV, Akaabourne M, et al. Serine proteases and their serpin inhibitors in Alzheimer's disease. Biomed Pharmacother 1994; 48:296-304.
115. Akiyama H, Ikeda K, Kondo H, et al. Microglia express the type 2 plasminogen

activator inhibitor in the brain of control subjects and patients with Alzheimer's disease. Neurosci Lett 1993; 164:233-235.

116. Walker DG, Yasuhara O, Patston PA, et al. Complement C1 inhibitor is produced by brain tissue and is cleaved in Alzheimer disease. Brain Res 1995; 675:75-82.

117. Kalaria RN, Golde T, Kroon SN, et al. Serine protease inhibitor antithrombin-III and its messenger RNA in the pathogenesis of Alzheimer's disease. Am J Pathol 1993; 143:886-893.

118. Wagner SL, Geddes JW, Cotman CW, et al. Protease nexin-1, an antithrombin with neurite outgrowth activity, is reduced in Alzheimer disease. Proc Natl Acad Sci USA 1989; 86:8284-8288.

119. Abraham CR, Selkoe DJ and Potter H. Immunochemical identification of the serine protease inhibitor α_1-antichymotrypsin in the brain amyloid deposits of Alzheimer's disease. Cell 1988; 52:487-501.

120. Abraham CR, Shirahama T and Potter H. α1-Antichymotrypsin is associated solely with amyloid deposits containing the β-protein: amyloid and cell localization of α1-antichymotrypsin. Neurobiol Aging 1990; 11:123-129.

121. Pasternack JM, Abraham CR, Van Dyke BJ, et al. Astrocytes in Alzheimer's disease gray matter express α_1-antichymotrypsin mRNA. Am J Pathol 1989; 135:827-834.

122. Hinds TR, Kukull WA, Van Belle G, et al. Relationship between serum α_1-antichymotrypsin and Alzheimer's disease. Neurobiol Aging 1994; 15:21-27.

123. Ma J, Yee A, Brewer HB Jr et al. Amyloid-associated proteins α_1-antichymotrypsin and apolipoprotein E promote assembly of Alzheimer β-protein into filaments. Nature 1994; 372:92-94.

124. Eriksson S, Janciauskiene S and Lannfelt L. α_1-Antichymotrypsin regulates Alzheimer β-amyloid peptide fibril formation. Proc Natl Acad Sci USA 1995; 92:2313-2317.

125. Fraser PE, Nguyen JT, Mclachlan DR, et al. Alpha-1-antichymotrypsin binding to Alzheimer A-beta peptides is sequence specific and induces fibril disaggregation in vitro. J Neurochem 1993; 61:298-305.

126. Kamboh MI. Apolipoprotein E polymorphism and susceptibility to Alzheimer's disease. Hum Biol 1995; 67:195-215.

127. Kamboh MI, Sanghera DK, Ferrell RE, et al. *APOE*4*-associated Alzheimer's disease risk is modified by α1-antichymotrypsin polymorphism. Nature Genetics 1995; 10:486-488.

128. Faber JP, Poller W, Olek K, et al. The molecular basis of alpha(1)-antichymotrypsin deficiency in a heterozygote with liver and lung disease. J Hepatol 1993; 18:313-321.

129. Schneider SS, Schick C, Fish KE, et al. A serine proteinase inhibitor locus at 18q21.3 contains a tandem duplication of the human squamous cell carcinoma antigen gene. Proc Natl Acad Sci USA 1995; 92: 3147-3151.

130. Byth BC, Billingsley GD and Cox DW. Physical and genetic mapping of the serpin gene cluster at 14q32.1: Allelic association and a unique haplotype associated with α_1-antitrypsin deficiency. Am J Hum Genet 1994; 55:126-133.

131. Danø K, Andreasen PA, Grøndahl-Hansen J, et al. Plasminogen activators, tissue degradation and cancer. Adv Cancer Res 1985; 44:139-226.

132. Kwaan HC. The plasminogen-plasmin system in malignancy. Cancer Metastasis Rev 1992; 11:291-311.

133. Pappot H, Gardsvoll H, Romer J, et al. Plasminogen activator inhibitor type 1 in cancer. Therapeutic and prognostic implications. Biol Chem Hoppe-Seyler 1995; 376:259-267.

134. Landau BJ, Kwaan HC, Verrusio EN, et al. Elevated levels of urokinase-type plasminogen activator and plasminogen inhibitor type-1 in malignant human brain tumors. Cancer Res 1994; 54:1105-1108.

135. Nagayama M, Sato A, Hayakawa H, et al. Plasminogen activators and their inhibitors in non-small cell lung cancer. Low content of type-2 plasminogen activator inhibitor associated with tumor dissemination. Cancer 1994; 73:1398-1405.

136. Kobayashi H, Fujishiro S and Terao T. Impact of urokinase-type plasminogen activator and its inhibitor type 1 on prognosis in cervical cancer of the uterus. Cancer Res 1994; 54:6539-6548.

137. Chambers SK, Gertz RE Jr, Ivins CM, et al. The significance of urokinase-type plasminogen activator, its inhibitors, and its receptor in ascites of patients with epithelial ovarian cancer. Cancer 1995; 75: 1627-1633.
138. Rao JS, Rayford A, Morantz RA, et al. Increased levels of plasminogen activator inhibitor-1 (PAI-1) in human brain tumors. J Neuro-Oncology 1993; 17:215-221.
139. Mueller BM, Yu YB and Laug WE. Overexpression of plasminogen activator inhibitor 2 in human melanoma cells inhibits spontaneous metastasis in *scid/scid* mice. Proc Natl Acad Sci USA 1995; 92:205-209.
140. Semeraro N, Montemurro P, Giordano P, et al. Increased mononuclear cell tissue factor and type-2 plasminogen activator inhibitor and reduced plasma fibrinolytic capacity in children with lymphoma. Thromb Haemostas 1994; 72:54-57.
141. Kumar S and Baglioni C. Protection from TNF-mediated cytolysis by overexpression of plasminogen activator inhibitor type-2. J Biol Chem 1991; 266:20960-20964.
142. Gan H, Newman GW and Remold H. Plasminogen activator inhibitor type-2 prevents programmed cell death of human macrophages infected with *Mycobacterium avium*, serovar 4. J Immunol 1995; 155:1304-1315.
143. Guerro R, Velasco F, Rodriguez M, et al. Endotoxin-induced pulmonary dysfunction is prevented by C1-esterase inhibitor. J Clin Invest 1993; 91:2754-2760.
144. Colman RW, Flores DN, De La Cadena RA, et al. Recombinant α-1-antitrypsin Pittsburgh attenuates experimental gram-negative septicemia. Am J Physiol 1988; 130:418-426.
145. Harper PL, Taylor FB, Courtney M et al. Antitrypsin Pittsburgh in a septic shock model. Thromb Haemostas 1993; 69:729 (Abstract).
146. Garver RI Jr, Chytil A, Courtney M et al. Clonal gene therapy: Tranplanted mouse fibroblast clones express human α1-antitrypsin gene in vivo. Science 1987; 762-764.
147. Ponder KP, Gupta S, Leland F, et al. Mouse hepatocytes migrate to liver parenchyma and function indefinitely after intrasplenic transplantation. Proc Natl Acad Sci USA 1991; 88:1217-1221.
148. Kay MA, Baley P, Rothenberg S, et al. Expression of human α_1-antitrypsin in dogs after autologous transplantation of retroviral hepatocytes. Proc Natl Acad Sci USA 1992; 89:89-93.
149. Rosenfeld MA, Ronald G and Crystal RG. Gene therapy for pulmonary diseases. Pathol Biol 1993; 41:677-680.
150. Lemarchand P, Jaffe HA, Danel C, et al. Adenovirus-mediated transfer of a recombinant human α_1-antitrypsin cDNA to human endothelial cells. Proc Natl Acad Sci USA 1992; 89:6482-6486.
151. Canonico AE, Conary JT, Meyrick BO, et al. Aerosol and intravenous transfection of human α1-antitrypsin gene to lungs of rabbits. Am J Resp Cell Mol Biol 1994; 10:24-29.
152. Crystal RG, Brantly ML, Hubbard RC, et al. The alpha-1-antitrypsin gene and its mutations. Clinical consequences and strategies for therapy. Chest 1989; 95:196-208.
153. Schapira M, Scott CF and Colman RW. Protection of human plasma kallikrein from inactivation by C1-inhibitor and other protease inhibitors: The role of high molecular weight kininogen. Biochemistry 1981; 20:2738-2743.
154. de Agostini A, Lijnen HR, Pixley RA, et al. Inactivation of factor XII active fragment in normal plasma. Predominant role of C1-inhibitor. J Clin Invest 1984; 1542:1549-1540.
155. Pemberton PA, Harrison RA, Lachmann PJ, et al. The structural basis for neutrophil inactivation of C1 inhibitor. Biochem J 1989; 258:193-198.
156. Hermans JM and Stone SR. Interaction of activated protein C with serpins. Biochem J 1993; 295:239-245.
157. Lennick M, Brew SA and Ingham K. Kinetics of interaction of C1 inhibitor with complement C1s. Biochemistry 1986; 25:3890-3898.
158. Rånby M, Bergsdorf N and Nilsson T. Enzymatic properties of one- and two-chain forms of tissue plasminogen activator. Thromb Res 1982; 27:175-183.

159. Abe H, Matsuda E and Binder BR. Effect of type-1 plasminogen activator inhibitor on human leukocyte elastase. Thromb Res 1994; 73:361-369.
160. Naski MC, Lawrence DA, Mosher DF, et al. Kinetics of inactivation of alpha-thrombin by plasminogen activator inhibitor-1—comparison of the effects of native and urea-treated forms of vitronectin. J Biol Chem 1993; 268:12367-12372.
161. Berrettini M, Schleef RR, España F, et al. Interaction of type 1 plasminogen activator inhibitor with the enzymes of the contact activation system. J Biol Chem 1989; 264:11738-11743.
162. Fay WP and Owen WG. Platelet plasminogen activator inhibitor: Purification and characterization of interaction with plasminogen activators and activated protein C. Biochemistry 1989; 28:5773-5778.
163. Thorsen S, Philips M, Selmer J, et al. Kinetics of inhibition of tissue-type and urokinase-type plasminogen activator by plasminogen activator inhibitor type 1 and type 2. Eur J Biochem 1988; 175:55-59.
164. Hekman CM and Loskutoff DJ. Bovine plasminogen activator inhibitor 1: Specificity determinationa and comparison of the active, latent and guanidine-activated forms. Biochemistry 1988; 27:2911-2918.

CHAPTER 9

Unanswered Questions and Directions for Future Research

In this book we have attempted to give a comprehensive overview of the mechanism and functions of serpins. The rapid growth of serpin research, as indicated by the increase in serpin literature and the expanding list of serpins identified, attests to the interest this family of proteins has attracted from many different biological realms. From the preceding chapters it should be clear that there are a number of important questions that remain to be answered in both serpin biology and chemistry. The following is a summary of areas in which we anticipate that there will be significant advances in the future:

1. CRYSTAL STRUCTURE OF SERPIN-PROTEINASE COMPLEXES

The structure of a complex has yet to be determined. This information is clearly essential to understanding serpin mechanism and will resolve the issues of whether in the stable complex the serpin is uncleaved and in a tetrahedral geometry with the proteinase active serine hydroxyl or whether the serpin is cleaved and in the acyl-enzyme state. If the reactive center loop is uncleaved, it is important to know how far into β-sheet A the loop has inserted. Similarly if the reactive center loop is cleaved, has the loop inserted only a few residues or is it fully inserted such that the serpin in the complex is similar to the cleaved non-complexed forms? Clearly the more structures that are determined the more detailed the picture that will emerge of the factors that determine the mechanism of formation of the stable complex. At present it is not clear how the complex is stabilized and what prevents it from undergoing a normal substrate reaction. In addition it will be possible to determine the role of interactions between the serpin and the proteinase other than those involving the reactive center residues which might stabilize the complex. As with the structures of the non-complexed serpins, which have shown a remarkable diversity, it is entirely possible that different complexes will also show considerable variations in structure.

2. THE MECHANISM OF ACTIVATION OF ANTITHROMBIN BY HEPARIN

This is an area under intense investigation that will be greatly enhanced with the crystal structure of an antithrombin-heparin complex as well as by further mutagenesis studies. This should enable the changes in hydrogen bonds and salt bridges which lead to activation to be identified. It will be essential to determine how the binding of heparin alters the conformation of the D-helix, the reactive center loop and the P1 residue, β-sheets A and C, and the insertion of the P14 region. Any mechanism to account for the ability of heparin to activate antithrombin must also account for the requirement of a conformational change for factor Xa inhibition but not for thrombin inhibition. Studies on the mechanism of glycosaminoglycan activation of antithrombin have implications for the development of future generations of antithrombotic drugs with less deleterious side effects, just as low molecular weight heparin is rapidly being recognized as the next generation to succeed conventional heparin.[1,2]

3. ADDITIONAL STRUCTURES OF NATIVE SERPINS

The X-ray structures of intact serpins determined so far (ovalbumin, latent PAI-1, antithrombin and a variant α_1-antichymotrypsin) are all so different that there is a need for additional structures of both inhibitory and non-inhibitory serpins. Moreover, other biophysical techniques such as NMR, FT-IR, CD and neutron scattering will aid in understanding the differences in serpin properties in terms of differences in structure. Some additional serpin structures which it would be useful to determine are CBG-cortisol, TBG-thyroxine, PAI-1-vitronectin and α_1-antichymotrypsin-DNA. In addition, it is of interest to determine the structure of non-inhibitory serpins such as maspin, PEDF and UTMP, since these proteins must have domains that interact with other proteins to mediate their biological effects. The structure of these domains could provide insight into the functions of these proteins. An example of such serpin-protein interaction is the high molecular weight complex of angiotensinogen with pro-eosinophil basic protein and complement component C3dg that occurs during pregnancy.[3]

4. CLASS SPECIFICITY

Although originally named because of their ability to inhibit serine proteinases, it is now known that some serpins (CrmA, squamous cell carcinoma antigen and α_1-antichymotrypsin) can inhibit cysteine proteinases. Thus, the distinction between mechanistic class of enzyme and inhibition by serpins is less clear. It will be of great interest to compare the inhibition mechanism for serpins with the two classes of proteinases. It can be envisaged that kinetic approaches with wild type and variant serpins and proteinases as well as crystallographic studies will be used. Also, to extend this, it will be interesting to determine if aspartic proteinases or metalloproteinases can also be inhibited by serpins. There is clearly a profound biological significance to such studies, particularly if the inhibition of cysteine (and other) proteinases by serpins is more widespread than known at present. This might have a particular impact on the intracellular serpins, the functions of which are not known. The fact that the viral serpin CrmA inhibits interleukin-1β converting enzyme and can modulate apoptosis raises the intriguing possibility that intracellular serpins might have similar functions. With the exception of the inhibition of cathepsin L by SCCA, nothing is known about whether or not serpins can influence the proteolytic activity of proteosomes or lysosomal proteinases.

5. PROTEIN FOLDING AND FUNCTION

The ability of serpins to fold in a metastable conformation, as well as the conformational lability of serpins in general, is of great interest to the field of protein folding. How the serpin is able to fold into the metastable state is not known, nor is the involvement of molecular chaperones in this process. In this regard the ability of a few proteins to fold into metastable states with the ability to adopt alternative conformations has been recognized.[4-6] The ability of the reactive center loop to adopt either α-helical or β-sheet conformations has implications for both prions and the amyloid-β peptide in Alzheimer's disease, since both proteins undergo similar helix-to-sheet changes as part of the pathological process.[7-10] Indeed, due to the presence of a high number of alanines and glycines in the reactive center region of serpins, and in the cleavage sites of viral fusion peptides and prions, it has been suggested that such residues may be present in regions that exhibit structural similarity and the ability to be cleaved.[11] Similarly, the HIV glycoprotein gp120 has a region that undergoes a sheet-to-helix transition depending on the conditions.[12] It is remarkable that serpins have been adapted to have a variety of functions in addition to proteinase inhibition. A better understanding of the protein architecture is essential for fully understanding how the serpins can have such diversity of function (proteinase inhibition, hormone carrier, cell regulation) within one basic protein structure.

6. EVOLUTIONARY RELATIONSHIPS

The striking conservation in the sequences, probable mechanism, and structures of serpins from organisms as diverse as viruses, barley, insects and mammals clearly shows how remarkable the serpins are. Studies at both the protein (amino acid sequence) and genetic level (nucleotide sequence and intron/exon structure) have much to tell us about the phylogenetic relationships between organisms as well as the evolution of different serpins within a particular organism. At present the mammalian serpins make up the largest group, particularly with species variants of the same serpin, but undoubtedly serpins will be found in more species.

7. INVOLVEMENT OF PEDF, MASPIN AND OTHER SERPINS IN DIVERSE BIOLOGICAL PROCESSES

Both PEDF and maspin have some unique properties not related to proteinase inhibition. The same is true for the uterine milk proteins. The role of these proteins is not yet understood, nor is the relevance of the serpin structure to their function. Nothing is known about the interactions these serpins have with other proteins and how any signaling pathways may operate. In this regard, it is clear that we have much to learn about serpin receptors and biological functions of cleaved and complexed serpins. Similarly, the role of serpins in intracellular processes, such as apoptosis and cell growth and differentiation is likely to be vigorously investigated. Thus it is expected that we can add to the information on bio- and immuno-regulatory properties of serpins.

8. HUMAN DISEASE AND BIOTECHNOLOGY

The advances in molecular genetics and the increased understanding of the biochemistry of serpins have done much to enhance our understanding of the role of serpins in human disease and will continue to do so. This area is likely to expand both diagnostically and therapeutically, not only for serpin-deficient states but also as the role of serpins in the pathogenesis of disease caused by viruses and parasites is elucidated. It is quite possible that similar advances in veterinary and agricultural science will emerge. The idea that transgenic plants containing a serpin gene whose product can specifically provide protection against an unwanted insect or other pest can be developed is quite feasible.

9. THE IN VIVO REGULATION OF PROTEOLYTIC PATHWAYS

Complex biochemical pathways, whether they involve intracellular metabolic interconversions of low molecular weight organic compounds, intracellular signal transduction pathways or extracellular proteolytic cascades are usually depicted as a linear scheme of events, largely for simplicity in describing the reactions involved. The challenge is to understand how these various pathways are regulated. The full understanding of how, for example, blood coagulation is regulated will have to not only focus on the proteinase activation steps, the concentrations of substrates and reactants and their K_M values, the role of platelets and other cells, the flow conditions and so on, but also how the properties of antithrombin, heparin cofactor II and protein C inhibitor can be integrated into the scheme, on more than just a qualitative level. Similar logic applies to fibrinolysis. Any model for the regulation of proteolytic pathways needs to consider the unique properties of serpins and individual details of the interaction between a serpin and a specific proteinase as well as the influence of ligand binding on activity. It can be envisaged that serpins keep basal levels of proteinase activation under control, but when the generation of proteinase exceeds the rate of inhibition then propagation of the cascade via the next set of substrates and amplification via feedback loops can occur. The complexity of modeling biological processes is clearly illustrated in some recent articles.[13-18]

10. CONCLUDING REMARKS

The preceding eight chapters of this book have attempted to summarize much of what is known about a number of proteins from a wide range of sources in the context that they are serpins and therefore share much structural and functional homology. Although the similarities between serpins from organisms as evolutionarily distant as viruses, barley, *Drosophila*, chicken and man is quite remarkable, each serpin also has its own unique properties. The preceding sections of this chapter have highlighted some areas in which we are confident that progress will be made. Without doubt other exciting new areas of research will emerge as more serpins, from more species, with yet more unusual and interesting properties are discovered. We hope that, in addition to being a survey of what is known about serpins, this book will also provide an impetus for future discoveries in the area of serpin research.

REFERENCES

1. Hull RD and Pineo GF. Therapeutic use of low molecular weight heparins: Knowledge to date and their application to therapy. Sem Thromb Hemostas 1994; 20:339-344.
2. Hirsh J and Levine MN. Low molecular weight heparin. Blood 1994; 79:1-8.
3. Oxvig C, Haaning J, Kristensen L, et al. Identification of angiotensinogen and complement C3dg as novel proteins binding the proform of eosinophil major basic protein in human pregnancy serum and plasma. J Biol Chem 1995; 270: 13645-13651.
4. Baker D and Agard DA. Kinetics versus thermodynamics in protein folding. Biochemistry 1994; 33:7505-7509.
5. Turk B, Dolenc I, Zerovnik E, et al. Human cathepsin B is a metastable enzyme stabilized by specific ionic interactions associated with the active site. Biochemistry 1994; 33:14800-14806.
6. Murray AJ, Lewis SJ, Barclay AN, et al. One sequence, two folds: A metastable structure of CD2. Proc Natl Acad Sci USA 1995; 92:7337-7341.
7. Pan K-M, Baldwin M, Nguyen J, et al. Conversion of α-helices into β-sheets in the formation of the scrapie prion proteins. Proc Natl Acad Sci USA 1993; 90:10962-10966.
8. Talfous J, Marcinowski KJ, Klopman G, et al. Solution structure of residues 1-28 of the amyloid β-peptide. Biochemistry 1994; 33:7788-7796.

9. Lee JP, Stimson ER, Ghilardi JR, et al. ^{1}H NMR of β amyloid peptide congeners in water solution. Conformational changes correlate with plaque competence. Biochemistry 1995; 34:5191-5200.
10. Nguyen J, Baldwin MA, Cohen FE, et al. Prion protein peptides induce α-helix to β-sheet conformational transitions. Biochemistry 1995; 34:4186-4192.
11. Callebout I, Tasso A, Brasseur R, et al. Common prevalence of alanine and glycine in mobile reactive center loops of serpins and viral fusion peptides: Do prions possess a fusion peptide? J Computer-Aided Mol Des 1994; 8:175-191.
12. Reed J and Kinzel V. Primary structure elements responsible for the conformational switch in the envelope glycoprotein gp120 from human immunodeficiency virus type 1: LPCR is a motif governing folding. Proc Natl Acad Sci USA 1993; 90:6761-6765.
13. Hemker HC and Kessels H. Feedback mechanisms in coagulation. Haemostasis 1991; 21:189-196.
14. Fell DA. Metabolic control analysis: A survey of its theoretical and experimental development. Biochem J 1992; 286:313-330.
15. Jesty J, Beltrami E and Willems G. Mathematical analysis of a proteolytic positive-feedback loop: Dependence of lag time enzyme yields on the initial conditions and kinetic parameters. Biochemistry 1993; 32:6266-6274.
16. Jones KC and Mann KG. A model for the tissue factor pathway to thrombin. II. A mathematical simulation. J Biol Chem 1994; 269:23367-23373.
17. Bray D. Protein molecules as computational elements in living cells. Nature 1995; 376:307-312.
18. McAdams HH and Shapiro L. Circuit simulation of genetic networks. Science 1995; 269:650-656.

INDEX

Page numbers in italics denote figures (f) or tables (t).

A

α-lytic proteinase, 35, *35f*
α_1-antichymotrypsin, 34, 113, 167
  3-D structure, 15, 20, 23
    reactive center, 46
  biological activity, 167-168
  characterization, 113
  DNA binding, 29, *28f*
  in Alzheimer's disease, 113, 182
  variants, *39t*
α_1-antitrypsin. *See* α_1-proteinase inhibitor.
α_1-proteinase inhibitor, 15-16, 45, 111-113, 168
  3-D structure, 15-16, 19-20
  alleles, 112, 177-180, *178-179t*
    Pittsburgh, 38, 112, *178t*, 179
    Z variant, 112, *178t*
  characterization, 111-112
  expression, 161-162
  in pulmonary emphysema, 177-179
  loop insertion, 20, 44
  therapeutic, 183
  variants, *39t, 44t, 179t*
α_2-antiplasmin, 34, 113-114
  characterization, 113-114
  deficiency, 182
  Enschede, 182
  fibrin crosslinking site, 114
  plasminogen binding site, 114
α_2-macroglobulin (α_2M), 34, 165
α_2-macroglobulin/ low density lipoprotein related protein receptor (LRP). *See* Serpin-proteinase complexes/receptors.
Accessory gland peptide 76A, 120
Acrosin, *43t, 68t, 85t, 87t*
Activated protein C, *68t*, 80, *85t, 87t, 90t, 179t*
Ala-serpin, 120
Alzheimer's disease, 89, 113, 182, 195
Angioedema. *See* C1-inhibitor/in HAE.
Angiotensinogen, 144-147
  angiotensin II generation, 144-146, *145f*
  characterization, 144
  expression, 146
  in hypertension, 146
  in pregnancy, 144, 146
Anticoagulant, 65-92, 121, 180-183, 196
Antithrombin, 2, 34, 65-79
  3-D structure, 15, 17, 20-21, 24-26, *24-25f*
    reactive center, 46-47
  characterization, 65-67, *67t*
  deficiency, 180, *181t*, 183
  expression, 162
  heparin activation, 67-68, *68t*, 73-79, *74-75f, 77f*
  heparin binding, 24, 26-28, 287, 43, 46, 68-72, *73-76f*, 74-79, 141, 180
  inhibition of coagulation cascade proteinases, 67, *68t*, 78-79
  loop insertion, 21-23, *22f*, 44-45
  neoepitope, 45
  thrombin binding, 74-76, *75-76f*
  variants, *39t, 42t, 44t, 181t*
    pleitropic, 78, 180
    Rouen, 141
Apoptosis, 121-122

B

Barley protein Z_4, 121
Barley protein Z_7, 121
Blood coagulation pathway, *66f*
Bombyx mori, 120
Branched pathway. See Inhibition mechanism/suicide substrate inhibition.
Brugia malayi, 121

C

C1-inhibitor, 2, 114-116, *115f*
  characterization, 116
  C1s inhibition, 114-116, *115f, 179t*
  glycosylation, 116
  in hereditary angioedema (HAE), *3t*, 180-182
  kallikrein inhibition, 43, 114, *115f*
  neoepitope, 45
  variants, *39t, 44t*
C1s, *39t*, 114, *179t*
Cancer, 91, 117-118, 147, 150, 182-183
Catalytic triad. *See* Serpin structure/reactive center; Inhibition mechanism.
Cathepsin G, *3t*, 80, 113, 167
Chemotaxis, 88, 167-168
Chondroitin sulfate, 80, 87-88
Chromosomal location, 4-10, *9t*, 67, 80, 89, 117-118. *See also* Serpin superfamily/gene clusters.
Chymase, *39t, 43t*
Chymotrypsin, *39t*, 80, *80t, 85t*, 113
Circular dichroism, 70, 137
Collagen binding protein, 151
Colligin. *See* Collagen binding protein.
Complement activation pathway, *115f*
Corticosteroid binding globulin (CBG), 45, 141-144
  characterization, 142
  hormone binding, 26-29, *138f*, 141-143
CrmA (Cytokine response modifier A), 121-122
  characterization, 121
  inhibition of interkeukin-1β converting enzyme (ICE), 121-122

D

Dermatan sulfate, 80, *80t*, 85, 87, 92-93
Disulfide bonds, *67t*, 87, 113
DNA. See α_1-antichymotrypsin/DNA binding; Serpin structure/binding sites.

E

Elastase, *3t, 36t, 80t*, 111, 143-144, 163, *179t*
Emphysema, 177, *178t*, 179
Ep-45. *See Xenopus* serpin.
EPC-1 (Early population doubling level cDNA-1). *See* Pigment epithelium-derived factor (PEDF).
Equine leucocyte elastase inhibitor (ELEI), 15, *18f*, 119
Evolution. *See* Serpin gene structure.
Extracellular matrix, 88, 91

F

Factor IXa, *68t*, 78, 80
Factor VIII, *68t*, 78
Factor Xa, *68t*, 73-80, *85t, 87t, 90t*
Factor XIa, *68t*, 78, 80, *85t, 90t*
Factor XIIa, *68t*, 80, *90t*, 114, *179t*
Fibrinolysis, 65, 91, 196
Fluorescence
  of endogenous tryptophan, 70, 143
  of covalently-bound fluorophores, 45, 70
Free energy of interconversion, 34-35, *35f*

G

γ-NGF, *87t*
Gene clusters. *See* Serpin superfamily.
Gene duplication, 8, 118
Gene families, *9t*
Gene structure. *See* Serpin gene structure.
Gene therapy, 183
Gene Y protein, 1, 137
  characterization, 137
gp46. *See* Collagen binding protein.

H

Hemorrhage, *3t*, 179, 182
Heparan sulfate, 85, 87
Heparin activation, 2, 65-93, *67t, 80t, 85t, 87t, 90t*, 194. *See also individual serpins*; Inhibition mechanism; Serpin gene structure/binding sites.
  bridging mechanism, 73-76
  conformational activation mechanism, 70
  heparin pentasaccharide, *28f*, 69, *70f*
  heparin structure heterogeneity, 67, *69f, 77f*, 80-81
  high-affinity heparin, 68-69
  low-affinity heparin, 74-75
  regulation of proteinases of blood coagulation, 65, *66f*
Heparin cofactor II, 2, 38, 79-84, *82f*
  characterization, *67t*, 79-80
  chemotactic peptides, 84, 167
  glycosaminoglycan activation, 80-84, *80t, 83f*
  glycosaminoglycan binding, 81, *83f*
  in pregnancy, 84
  variants, *39t*, 81
Heparin pentasaccharide, 26, *28f, 70f*
Hereditary angioedema (HAE). *See* C1-inhibitor.
High molecular weight kininogen, *68t*, 78
Hinge region. *See* Serpin structure/reactive center.
Hormone-binding serpins. *See* Serpin superfamily.
HSP47. *See* Collagen binding protein.
Human proteinase inhibitor-4, *See* Kallistatin.
Human proteinase inhibitor-6, 1, 119-120
  characterization, 119
Hypertension, *3t*, 146
Hypotension, *3t*

I

Inhibition mechanism, 33-55, 194. *See also* Serpin-proteinase complexes; Serpin conformation.
  catalytic triad involvement, 37-38, 47-48
  heparin acceleration, 73-78, 80-89, *83f*, 91-93
  non-serpin inhibitors, 39-40
  protease specificity, 38-39, *39t*
  rate constants, 41-45, *42t*, 54, *54t, 68t, 77f*, 80, *80t*, 85, *85t*, 87, *87t*, 90, *90t*, 111, 113-114, 121, 147, *179t*
  stoichiometry of inhibition (SI), 41-43, *42-44t*, 45, *54t*
  substrate pathway, 47, *47f*
    complex stabilization, 48-51
  suicide substrate inhibition, 40-55, *49f*
    branched pathway mechanism, 40-44, *40f*, 54
Insect serpins, 120
Interkeukin-1β converting enzyme (ICE), 121-122
Intracellular serpins. *See* Serpin superfamily.

J

J6. *See* Collagen binding protein.

K

Kallikrein, 43, 85, 117, 181
  characterization, 117
Kallistatin, 8, 117
  rodent, 117, *119t*
Kazal inhibitors, 33, *34t, 36-37t*
Kinetics of proteinase inhibition. *See* Inhibition mechanism.
Kunitz inhibitors, *34t, 36-37t*, 37
  variants, *39t*

L

Leuserpin. *See* Heparin cofactor II.
Limulus intracellular coagulation inhibitor (LICI), 120
  characterization, 120
Loop insertion. *See* Serpin conformation/conformational change.

M

Manduca sexta, 120
Maspin, 150-151
  tumor inhibition, 150
Metastable state. *See* Serpin conformation.
Monocyte/neutrophil elastase inhibitor, 1

N

Neurite outgrowth, 89, 147-149, *148f*
NMR, 23, 38, 48, 70, 136-137
Non-inhibitory serpins, 135-151

O

Ov-serpins. *See* Serpin superfamily.
Ovalbumin, 1-2, 135-137
  3-D structure, 15-17, *17f*, 19, 23
    reactive center, 45-46
  characterization, 135-136
  cleavage, 136
  loop insertion, 136-137
  phosphorylation, 136
  S-ovalbumin, 137

P

PI-6. *See* Human placental thrombin inhibitor.
Parasite serpins, 120-121
Pigment epithelium-derived factor (PEDF), 147-149
  characterization, 147-149
  effect on cell differentiation, 147, *148f*
  expression, 147
Placental thrombin inhibitor. *See* Human proteinase inhibitor-6.
Plant serpins, 121
Plasma serpins. *See* Serpin superfamily.
Plasmin, *68t*, 80, *87t*, *90t*, 113, 122, *179t*
Plasminogen activator inhibitor-1 (PAI-1), 2
  3-D structure, 15, *21f*
  characterization, 89
  expression, 162
    elevated, 182
  glycosaminoglycan activation, 89, 91, *90t*
  heparin binding, 90-91
  in fibrinolysis, 91
  in tissue remodeling, 92
  in vascular disease, 182
  latent conformation, 20-21, *21f*
  loop insertion, 20, 45
  variants, *39t*, *44t*, 89-90
  vitronectin activation, 89-91, *90t*
  vitronectin binding, 29, 89-90
Plasminogen activator inhibitor-2 (PAI-2), 1, 3, 116-117
  characterization, 117
pNiXa. *See Xenopus* serpin.
Prions, 195
Protease nexin I (PN1), 43
  characterization, *67t*, 86-87
  expression, 162
  glycosaminoglycan activation, 87-88, *87t*
  in the nervous system, 89
  in wound healing, 88-89
Protein C inhibitor (PCI), 2, 24, 26
  characterization, *67t*, 84
  heparin and heparan sulfate activation, 85-86, *85t*, 162
  heparin binding, 24, 26-28, 85-86
  in seminal fluid, 86, 162
  variants, *39t*
Proteinase inhibitors, non-serpin, 2, 33-40, 48
  families, *34t*
  reactive center, 35-37, *37t*
  structures, *36t*

R

Reactive center. *See* Serpin structure; Proteinase inhibitors, non-serpin.
Rodent serpins, 118-119

S

Schistosoma haematobium, 120
Schistosoma mansoni, 121
SEC receptor. *See* Serpin-proteinase complexes/receptors.
Serpin biosynthesis, 161-162, 167
Serpin concentration, *3t*, *67t*
Serpin conformation
  conformational change, 2, *17f*, *22f*, *24-25f*, 46, 70, 72
    β-sheet expansion, 45, 72, 140
    loop insertion, 3, 20-22, 42-45, 47, 49-51, 136, 140
  latent conformation, 20-21, *21-22f*, 24, *24-25f*, 46
  metastable state, 2, 8, 23, 34-35, 51-52, 194-195
  X-ray structures, 15-28, *16t*, 37, 46, 52, 72, 136, 194
Serpin deficiencies, *3t*, 78, 141, 146, 177-182. *See also individual serpins.*
Serpin gene structure, 4, *5f*, *7f*, *9t*. *See also individual serpins.*
  evolution, 8, *9-10f*
Serpin hyperexpression, *3t*, 182
Serpin receptors. *See* Serpin-proteinase complexes/receptors.
Serpin regulation, 161-168. *See also individual serpins*; Heparin activation; Serpin biosynthesis; Serpin-proteinase complexes/clearance.

Serpin structure, 2, *67t*. *See also* Serpin conformation.
  binding sites, 27-28
    DNA, 29, *28f*
    heparin (and other glycosaminoglycans), 24, 26-29, *28f*, 65-93, *71-73f*, 81, 85-86, 90
    hormone, 26-27, 140, 142-143
    vitronectin, 29, 90
  C-terminal extension, 2
  core domain, 2-3, 17
  disulfides, 3
  glycosylation, 2, *5f*, *19t*
  N-terminal extension, 2-3, 24, 48, *82f*, 116, 144, 146
  phosphorylation, 2
  reactive center, 3, *6f*, 8, *10f*, 19-22, *22f*, *26f*, 46-47, *138f*, 149-150
    catalytic triad, 37-38, *47f*, 48, 50
    hinge region, 3, *44t*, 151
    scissile bond, 3, 47, *138f*
  secondary structure, 8, 17-28, *17-18f*, 52-54, *52-53f*
  signal sequence, 1, 3
  thermodynamic stability, 20, 23, 34-35, 136
Serpin superfamily, 1
  α1-PI family, 4
  gene clusters, 8, *9t*, 10, 112, 118-120, 144, 150, 182-183
  hormone-binding serpins, 26-27, 137-144
  intracellular serpins, 117, 119-120
  non-inhibitory serpins, 135-151
  ov-serpins, 1, 3, 8, 136
  plasma serpins, 2, *3t*, 65, 111
  species distribution, 2, *4t*, 8, 112-114, 117-122, 142, 195
Serpin-proteinase complexes, 2, 33, 38-45, *43t*, 47-54, *47f*, *76f*, 137,140
  clearance, 162-163
  cysteine proteinases, 51
  receptors, 163-167
    α_2-macroglobulin/low density lipoprotein related protein receptor (LRP), 165, *166f*
    serpin-enzyme complex (SEC) receptor, 163-165, *164f*
    urokinase receptor, 165
  stable complex, 48-51
  structure, 193-194
Serpin-tetradecapeptide binary complex, 42, 44-45
Signal sequence. *See* Serpin structure.
Soybean trypsin inhibitors (Bowman Birk), *34t*
Spi serpins, 118-119, *119t*
Squamous cell carcinoma antigen (SCCA), 1, 117-118
  characterization, 118
Squash seed inhibitors, 33, *34t*
Stability. *See* Serpin structure/thermodynamic stability.
Stoichiometry of inhibition. *See* Inhibition mechanism.
Streptomyces subtilisin inhibitors, *34t*, *37t*
  variants, *39t*
Substrate reaction, 76-78
Subtilisin, *39t*, 136
Suicide substrate inhibition. *See* Inhibition mechanism.

T

Thrombin, 74-81, 88-89, 91
Thrombosis, *3t*, 68, 78-79, 84, 91, 180
Thyroxine binding globulin (TBG), 45, 137-141
  characterization, 140
  hormone binding, 26-27, 137, *139f*, 140
  loop insertion, 140
  variants, 141, *141t*
Tissue factor, *68t*
Tissue-type plasminogen activator (tPA), 39, 80, 89, *85t*, *87t*, *90t*, 122, *179t*
Transcortin. *See* Corticosteroid binding protein (CBG).

U

Urokinase receptor. *See* Serpin-proteinase complexes/receptors.
Urokinase-type plasminogen activator (uPA), 80, *85t*, 86, *87t*, *90t*, 122, *179t*, 183
Uterine milk proteins (UTMP), 149-150
  characterization, 150

V

Viral serpins, 121-122
Vitronectin, 89-91

X

X-ray structures, 15-28, *16t*, 35-37, *36t*, 46, 52, 72, 136, 194
  molecular replacement, 16, 23-25
Xenopus serpin, 150

Z

Z variant. See α_1-proteinase inhibitor/alleles.

Neuroscience Intelligence Unit

Available and Upcoming Titles

- ❐ Neurodegenerative Diseases and Mitochondrial Metabolism
  M. Flint Beal, Harvard University
- ❐ Molecular and Cellular Mechanisms of Neostriatum
  Marjorie A. Ariano and D. James Surmeier, Chicago Medical School
- ❐ Ca^{2+} Regulation By Ca^{2+}-Binding Proteins in Neurodegenerative Disorders
  Claus W. Heizmann and Katharina Braun, University of Zurich, Federal Institute for Neurobiology, Magdeburg
- ❐ Measuring Movement and Locomotion: From Invertebrates to Humans
  Klaus-Peter Ossenkopp, Martin Kavaliers and Paul Sanberg, University of Western Ontario and University of South Florida
- ❐ Triple Repeats in Inherited Neurologic Disease
  Henry Epstein, University of Texas-Houston
- ❐ Cholecystokinin and Anxiety
  Jacques Bradwejn, McGill University
- ❐ Neurofilament Structure and Function
  Gerry Shaw, University of Florida
- ❐ Molecular and Functional Biology of Neurotropic Factors
  Karoly Nikolics, Genentech
- ❐ Prion-related Encephalopathies: Molecular Mechanisms
  Gianluigi Forloni, Istituto di Ricerche Farmacologiche "Mario Negri"-Milan
- ❐ Neurotoxins and Ion Channels
  Alan Harvey, A.J. Anderson and E.G. Rowan, University of Strathclyde
- ❐ Analysis and Modeling of the Mammalian Cortex
  Malcolm P. Young, University of Oxford
- ❐ Free Radical Metabolism and Brain Dysfunction
  Irène Ceballos-Picot, Hôpital Necker-Paris
- ❐ Molecular Mechanisms of the Action of Benzodiazepines
  Adam Doble and Ian L. Martin, Rhône-Poulenc Rorer and University of Alberta
- ❐ Neurodevelopmental Hypothesis of Schizophrenia
  John L. Waddington and Peter Buckley, Royal College of Surgeons-Ireland
- ❐ Synaptic Plasticity in the Retina
  H.J. Wagner, Mustafa Djamgoz and Reto Weiler, University of Tübingen
- ❐ Non-classical Properties of Acetylcholine
  Margaret Appleyard, Royal Free Hospital-London
- ❐ Molecular Mechanisms of Segmental Patterning in the Vertebrate Nervous System
  David G. Wilkinson, National Institute of Medical Research-UK
- ❐ Molecular Character of Memory in the Prefrontal Cortex
  Fraser Wilson, Yale University

T. Moritani
S. Ekholm
P.-L. Westesson

Diffusion-Weighted MR Imaging of the Brain

With 661 Figures and 11 Tables

Springer

ISBN 3-540-03439-0
Springer-Verlag Berlin Heidelberg New York

Library of Congress Cataloging-in-Publication Data
Moritani, T. (Toshio) Diffusion-weighted MR imaging of the brain/T. Moritani, S. Ekholm, P.-L. Westesson. p.; cm. Includes bibliographical references and index. ISBN 3-540-03439-0 (alk. paper) 1. Brain–Magnetic resonance imaging. 2. Brain–Diseases–Diagnosis. I. Ekholm, S. (Sven) II. Westesson, Per-Lennart III. Title. [DNLM: 1. Brain Diseases–diagnosis. 2. Diffusion Magnetic Resonance Imaging. 3. Brain–pathology. WL 141 M862 2004] RC386.6.M34M67 2004 616.8'047548–dc21

Springer-Verlag is a part of Springer Science+Business Media
springeronline.com

Printed in Germany

Cover design: E. Kirchner, Heidelberg
Product management and layout: B. Wieland, Heidelberg
Reproduction and typesetting: AM-production, Wiesloch
Printing and bookbinding: Stürtz AG, Würzburg

21/3150 – 5 4 3 2 1 0
Printed on acid-free paper

Preface

This book is the result of many years of clinical and academic interest in diffusion-weighted MR (DW) imaging of the brain. Researchers and clinicians at the University of Rochester started to collect DW images of a spectrum of abnormalities affecting the brain immediately after this technique became available. Several case series with clinical and radiographic correlations have been presented at the annual meetings of the American Society of Neuroradiology and the Radiological Society of North America via posters and scientific reports. Over time it became quite clear that we had a collection of DW images representing the majority of conditions that affect the brain and we felt a need to put them all together under one cover.

MR imaging has evolved dramatically since its introduction into clinical work in the mid-1980s. Looking back, there are several major steps that took MR imaging of the central nervous system to the next level. One of the first steps was the introduction of the clinical usefulness of contrast agents. Other steps were the development of fat suppression techniques, fast spin echo imaging, and, more recently, the development of a clinically useful DW imaging technique. DW imaging has revolutionized the imaging diagnosis of acute infarction in the brain. It is, however, quite clear from the series of cases shown in this book that DW imaging is useful for many other conditions. The time it takes to obtain a DW image is so short that in many institutions it is now being used as a routine part of any MR imaging of the brain.

The initial chapters on principles of DW imaging, normal DW appearance, and pitfalls and artifacts provide the bases for understanding DW imaging. This technique is complex and is associated with many pitfalls and artifacts. The following chapter on brain edema provides the basis for understanding the pathophysiology of signal alterations in DW images related to various pathological conditions. The images are correlated to corresponding neuropathologic slides and aid the understanding of the DW imaging representation of various types of brain edema. Chapters 5–13 cover DW imaging characteristics of different pathologic conditions and in Chap. 14 (pediatrics) we have collected DW images of pediatric conditions.

The book is organized according to major disease categories. This brings structure to the book, but is not optimal for the clinician sitting in front of a set of images and wondering what they might represent. For that reason we have a summary chapter entitled "How to Use This Book" (Chap. 15), which is organized from the opposite perspective. Thus, in Chap. 15 we have started with DW images and grouped them according to imaging characteristics. In each table we have listed differential diagnoses for each specific set of DW imaging characteristics and added thumbnail images with references to the corresponding chapters. The clinician can go directly to Chap. 15, determine the signal on the DW imaging, combine it with the T2 and ADC signal characteristics, and get a list of the conditions that match these imaging characteristics. The thumbnail images, the reference to corresponding chapter and knowledge about the patient's clinical presentation should allow the clinician to formulate a relatively narrow differential diagnosis for most clinical conditions. We think that this "reversed" chapter will make the book very useful for everyday work with DW imaging of the brain.

We are grateful for many pathological slides and fruitful discussions with Barbara Germin, MD, Department of Pathology, University of Rochester. We acknowledge the case contribution from the Department of Radiology, Showa University, Japan, collected during the primary author's time at Showa University. We would also like to thank Masahiro Ida, MD, Department of Radiology, Ebara Municipal Hospital, Japan; Minoru Morikawa, MD, Department of Radiology, Nagasaki University, Japan; R. Nuri Sener, MD, Department of Radiology, Ege University Hospital, Turkey; and Ryutarou Ukisu, MD, Department of Radiology, Showa University, Japan, all of whom contributed case studies. Our deepest gratitude goes to Ms Margaret Kowaluk and Ms Theresa Kubera, Medical Graphic Designers, Department of Radiology, University of Rochester, and Ms Belinda De Libero for her secretary work. We also wish to thank Yuji Numaguchi, MD, PhD, Department of Radiology, University of Rochester and St. Luke's Hospital, Japan, who gave us encouragement and support.

We want to thank the editorial staff at Springer-Verlag, without whose guidance, skills and knowledgeable advice this book would not have become a reality. We would also like to thank our colleagues, fellows and coworkers at the University of Rochester. Finally, but not least, we thank our families for giving us the time to complete this project.

It is our hope that our readers will find this book on "Diffusion-Weighted Imaging of the Brain" instructional and clinically useful.

October 2003

Toshio Moritani
Sven Ekholm
Per-Lennart Westesson

Authors

Toshio Moritani, MD, PhD
Assistant Research Professor
Division of Diagnostic
and Interventional Neuroradiology
Department of Radiology
University of Rochester
School of Medicine and Dentistry
Rochester, New York, USA

Sven Ekholm, MD, PhD
Professor of Radiology and Director of Research
Division of Diagnostic
and Interventional Neuroradiology
Department of Radiology
University of Rochester
School of Medicine and Dentistry
Rochester, New York

Professor of Radiology
University of Gothenburg
Gothenburg, Sweden

Per-Lennart Westesson, MD, PhD, DDS
Professor of Radiology
and Director of Division of Diagnostic
and Interventional Neuroradiology
Department of Radiology and
Professor of Clinical Dentistry
University of Rochester
School of Medicine and Dentistry
Rochester, New York, USA

Professor of Oral Diagnostic Sciences
State University of New York at Buffalo
Buffalo, New York, USA

Associate Professor of Oral Radiology
University of Lund
Lund, Sweden

Akio Hiwatashi, MD
Assistant Research Professor
Division of Diagnostic
and Interventional Neuroradiology
Department of Radiology
University of Rochester
School of Medicine and Dentistry
Rochester, New York, USA

Ramon R. de Guzman, MD
Fellow, Division of Diagnostic
and Interventional Neuroradiology
Department of Radiology
University of Rochester Medical Center
Rochester, New York, USA

Fellow, Philippine College of Radiology
Philippines

Jianhui Zhong, PhD
Associate Professor of Radiology,
Biomedical Engineering, and Physics
Director of MR Imaging Research
Department of Radiology
University of Rochester
School of Medicine and Dentistry
Rochester, New York, USA

Associate Professor of Radiology
Yale University School of Medicine
New Haven, Connecticut, USA

Contents

1 Basics of Diffusion Measurements by MRI . . . 1

1.1 Diffusion Imaging in MR . . . 1
1.2 Diffusion Imaging of the Brain . . . 1
1.3 Magnetic Resonance Principles of Diffusion Imaging . . . 1
1.4 Apparent Diffusion Coefficient . . . 2
1.5 Diffusion Represents a Molecular Event . . . 3
1.6 Requirements in Clinical Diffusion Imaging . . . 4
1.7 Setting the b-Value in Clinical DW Imaging . . . 4
1.8 Future Trends in Clinical DW Imaging . . . 5
References . . . 5

2 Diffusion-Weighted Imaging of the Normal Brain . . . 7

2.1 Introduction . . . 7
2.2 Adult Brain . . . 7
2.2.1 Low Signal in Basal Ganglia . . . 7
2.2.2 Diffusion-Weighted Imaging of Gray and White Matter . . . 7
2.2.3 Choroid Plexus . . . 7
2.3 Pediatric Brain . . . 9
2.3.1 Diffusion-Weighted Imaging and ADC of the Pediatric Brain . . . 9
2.4 Conclusion . . . 10
References . . . 10

3 Pitfalls and Artifacts of DW Imaging . . . 11

3.1 Introduction . . . 11
3.2 Influence of ADC and T2 on the DW Appearance . . . 11
3.2.1 Concepts . . . 11
3.2.2 Apparent Diffusion Coefficient Maps . . . 11
3.2.3 Exponential Images . . . 11
3.3 Clinical Conditions . . . 12
3.3.1 T2 Shine-through . . . 12
3.3.2 T2 Washout . . . 15
3.3.3 T2 Blackout . . . 16
3.4 Artifacts . . . 18
3.4.1 Eddy Current Artifacts . . . 18
3.4.2 Susceptibility Artifacts . . . 18
3.4.3 N/2 Ghosting Artifact (Nyquist Ghost) . . . 20
3.4.4 Chemical Shift . . . 20
3.4.5 Motion Artifacts . . . 20
3.5 Conclusion . . . 23
References . . . 23

4 Brain Edema . . . 25

4.1 Characterization and Classification of Brain Edema . . . 25
4.2 Definition and Classification of Cytotoxic Edema . . . 25
4.3 Pathophysiology of Cytotoxic Edema . . . 25
4.3.1 Energy Failure . . . 25
4.3.2 Excitotoxic Brain Injury . . . 27
4.4 Diffusion-Weighted Imaging and Cytotoxic Edema . . . 28
4.4.1 Conditions that Cause Cytotoxic Edema, and Reversibility . . . 29
4.5 Vasogenic or Interstitial Edema . . . 35
4.5.1 Conditions that Cause Vasogenic Edema . . . 36

4.6 Conclusion . . . 37
4.6.1 Cytotoxic or Cellular Edema . . . 37
4.6.2 Vasogenic Edema . . . 37
References . . . 37

5 Infarction . . . 39

5.1 Clinical Significance and Therapeutic Considerations for Brain Infarcts . . . 39
5.1.1 Stroke Mimickers . . . 39
5.1.2 Diffusion-Weighted Imaging. . . . 39
5.2 Diffusion-Weighted Imaging and Pathophysiology of Cerebral Infarction . . 39
5.3 Apparent Diffusion Coefficient . . . 40
5.3.1 Explanation for Restricted Diffusion. . . . 40
5.4 Time Course of Infarction . . . 40
5.4.1 Hyperacute (<6 Hours) . . . 40
5.4.2 Acute (6 Hours to 3 Days) . . . 42
5.4.3 Subacute (3 Days to 3 Weeks) . . . 43
5.4.4 Chronic (3 Weeks to 3 Months) . . 44
5.5 Diffusion-Weighted Imaging and ADC Characteristics of Gray and White Matter Ischemia . . . 45
5.5.1 Relative ADC . . . 45
5.6 Reversibility and Treatment . . . 45
5.7 Watershed Infarction. . . . 47
5.8 Perfusion Versus Diffusion Imaging . . . 48
5.9 Venous Infarction . . . 48
5.9.1 Predisposing Factors . . . 48
5.9.2 Pathophysiology . . . 48
5.10 Small Vessel Infarcts . . . 50
5.11 Brain Stem and Cerebellar Infarcts . . . 51
5.12 Corpus Callosum Infarcts . . . 52
5.13 Hemorrhagic Infarcts . . . 53
References . . . 54

6 Intracranial Hemorrhage . . . 55

6.1 Introduction . . . 55
6.2 Intraparenchymal Hemorrhages: Appearance and Evolution. . . . 55
6.2.1 Hyperacute Hematoma . . . 55
6.2.2 Acute Hematoma . . . 59
6.2.3 Early Subacute Hematoma . . . 59
6.2.4 Late Subacute Hematomas . . . 62
6.2.5 Chronic Hematomas . . . 62
6.3 Subarachnoid Hemorrhage . . . 62
6.4 Subdural and Epidural Hemorrhages . . . 64
6.5 Intraventricular Hemorrhage . . . 65
6.6 Intra-tumoral Hemorrhage . . . 66
6.7 Hemorrhage Related to Vascular Malformation . . . 67
6.8 Hemorrhage Related to Trauma . . . 68
6.9 Conclusions. . . . 68
References . . . 69

7 Vasculopathy and Vasculitis . . . 73

7.1 Definition . . . 73
7.2 Clinical Presentation . . . 73
7.3 Treatment . . . 73
7.4 Vasculitis of the CNS. . . . 73
7.4.1 Characterization of CNS Vasculitis. . . . 73
7.4.2 Primary Angitis of the Central Nervous System . . . 74
7.4.3 Giant Cell (Temporal) Arteritis . . 75
7.4.4 Takayasu's Arteritis (Aortitis Syndrome) . . . 78
7.4.5 Polyarteritis Nodosa . . . 80
7.4.6 Churg–Strauss Disease . . . 80
7.4.7 Other Small Vessel Vasculitis. . . . 80
7.4.8 Collagen Vascular Diseases . . . 80
7.4.9 Infectious Vasculitis . . . 82
7.4.10 Drug-Induced Vasculitis, Including Illicit Drugs . . . 83
7.5 Vasculopathy of the CNS . . . 84
7.5.1 Systemic Lupus Erythematosus . . 84
7.5.2 Moyamoya Disease . . . 86
7.5.3 Sickle Cell Disease . . . 86
7.5.4 Posterior Reversible Encephalopathy Syndrome . . . 86
7.5.5 Hypertensive Encephalopathy . . . 87
7.5.6 Preeclampsia/Eclampsia . . . 88
7.5.7 Immunosuppressive Drug-Induced Vasculopathy . . . 89
7.5.8 Uremic Encephalopathy and Hemolytic Uremic Syndrome . . . 91
7.5.9 Thrombotic Thrombocytopenic Purpura . . . 92
7.5.10 Cerebral Amyloid Angiopathy . . . 92
7.6 Conclusion . . . 92
References . . . 93

8 Epilepsy . . . 95

8.1 Definition . . . 95
8.2 Classification . . . 95
8.3 Mechanisms and Pathophysiology of Epilepsy . . . 95

8.4 Magnetic Resonance Imaging of Epilepsy 96
8.4.1 Diffusion-Weighted Imaging in Epilepsy 97
8.4.2 Magnetic Resonance Signal Alterations in Epilepsy 98
8.4.3 Ictal Stage to Periictal Stage . . . 99
8.4.4 Status Epilepticus 100
8.4.5 Cytotoxic Edema in Status Epilepticus. 101
8.4.6 Other Imaging Techniques for Epilepsy. 103
8.5 Hemiconvulsion–Hemiplegia Epilepsy Syndrome 103
8.6 Focal Lesion in the Splenium of the Corpus Callosum in Epileptic Patients. 103
8.7 Conclusion. 103
References. 105

9 Demyelinating and Degenerative Disease. 107

9.1 Demyelinating Disease 107
9.1.1 Multiple Sclerosis 107
9.1.2 Acute Disseminated Encephalomyelitis 112
9.1.3 Progressive Multifocal Leukoencephalopathy 112
9.2 Degenerative Disease 113
9.2.1 Wallerian or Transneuronal Degeneration. 113
9.2.2 Creutzfeldt–Jakob Disease 114
9.2.3 Amyotrophic Lateral Sclerosis . . 116
9.3 Conclusion. 116
References. 117

10 Toxic and Metabolic Disease 119

10.1 Toxic Disease 119
10.1.1 Chemotherapy-Induced Leukoencephalopathy 119
10.1.2 Heroin-Induced Spongiform Leukoencephalopathy 119
10.1.3 Cocaine, Phencyclidine Hydrochloride, Amphetamines and Related Catecholaminergics . 119
10.1.4 Central Pontine Myelinolysis and Extrapontine Myelinolysis . . 122
10.1.5 Wernicke Encephalopathy 123
10.1.6 Marchiafava–Bignami Disease . . 124
10.2 Metabolic Disease 125
10.2.1 Mitochondrial Encephalopathy . 125
10.2.2 Phenylketonuria 126
10.2.3 Other Metabolic Diseases and Leukodystrophies. 126
References. 129

11 Infectious Diseases 131

11.1 Overview of Brain Infections. 131
11.2 Bacterial Brain Abscess 131
11.3 Septic Emboli 132
11.4 Brain Abscess Caused by Unusual Bacteria 133
11.4.1 Differential Diagnosis 133
11.5 Bacterial Abscess in the Extra-Axial Space 137
11.5.1 Differential Diagnosis 137
11.6 Bacterial Vasculitis 141
11.7 Toxoplasmosis 141
11.7.1 Differential Diagnosis 143
11.8 Disseminated Aspergillosis 145
11.9 Herpes Encephalitis 145
11.10 Human Immunodeficiency Virus Infection 146
References. 147

12 Trauma 149

12.1 Introduction. 149
12.2 Diffuse Axonal Injury. 149
12.2.1 Location 149
12.2.2 Computed Tomography and MR Imaging. 154
12.2.3 Diffusion-Weighted Imaging . . . 154
12.3 Brain Contusion. 154
12.3.1 Location 154
12.3.2 Computed Tomography and MR Imaging. 156
12.3.3 Diffusion-Weighted Imaging Findings 156
12.4 Hemorrhage Related to Trauma 156
12.4.1 Computed Tomography and MR Imaging. 157
12.4.2 Diffusion-Weighted Imaging . . . 157
12.5 Vascular Injuries 159
References. 160

13 Brain Neoplasms 161

13.1 Introduction 161
13.2 Gliomas 161
13.2.1 High-Grade Tumors 161
13.2.2 Peritumoral Infiltration 168
13.2.3 Treatment Response 168
13.3 Epidermoid Tumors and Arachnoid Cysts 169
13.4 Primitive Neuroectodermal Tumors. . . . 171
13.5 Meningiomas 172
13.6 Malignant Lymphomas 174
13.7 Craniopharyngiomas 175
13.8 Metastases 176
13.9 Conclusion. 178
References. 178

14 Pediatrics 181

14.1 Water Content of the Pediatric Brain . . . 181
14.2 Normal Structures 181
14.3 Anisotropy. 182
14.4 Infarction and Ischemia 182
14.4.1 Moyamoya Disease 184
14.4.2 Sickle Cell Disease. 184
14.4.3 Hypoxic Ischemic Encephalopathy 184
14.5 Trauma 186
14.5.1 Battered Child Syndrome 186
14.5.2 Diffuse Axonal Injury and Brain Contusion 189
14.6 Encephalopathies 189
14.6.1 Mitochondrial Encephalopathy . 189
14.6.2 Acute Necrotizing Encephalopathy 190
14.6.3 Hypertensive Encephalopathy . . 190
14.7 Infections 191
14.7.1 Encephalitis 191
14.7.2 Brain Abscess 193
14.8 Brain Tumor 194
14.9 Dysmyelination and Demyelination. . . . 195
14.9.1 Pelizaeus–Merzbacher Disease . . 195
14.9.2 Vanishing White Matter Disease . 196
14.9.3 Metabolic or Toxic Leukoencephalopathies 196
14.9.4 Multiple Sclerosis 197
14.9.5 Osmotic Myelinolysis 197
14.10 Conclusion. 199
References. 199

15 How to Use This Book 201

Table 1 Differential diagnoses for lesions with a high diffusion signal associated with low ADC and iso intense T2 signal. 202
Table 2 Differential diagnoses for lesions with a high diffusion signal associated with iso-high ADC and a high intense T2 signal 203
Table 3 Differential diagnoses for lesions with a high diffusion signal associated with a low ADC and high intense T2 signal 208
Table 4 Differential diagnoses for lesions with an iso diffusion signal associated with a high ADC and high intense T2 signal. 218
Table 5 Differential diagnoses for lesions with a low diffusion signal associated with a high ADC and high intense T2 signal. 219
Table 6 Differential diagnoses for lesions with a low diffusion signal associated with a high ADC and iso intense T2 signal. 221
Table 7 Differential diagnoses for lesions with artifacts 222

Subject Index 225

Basics of Diffusion Measurements by MRI

In collaboration with J. Zhong

Diffusion occurs as a result of the constant movement of water molecules. Water makes up 60–80% of our body weight. The heat associated with our body temperature energizes the water molecules, causing them to "jerk" around randomly. This phenomenon is called "Brownian motion" after the scientist who first described it [1, 2]. It can be demonstrated by adding a few drops of ink to a still bucket of water. Initially, the ink will be concentrated in a very small volume, but it will quickly spread out (diffuse) and mix with the rest of the water. The speed of this process of diffusion gives physicists a measure of the property of water. Similarly, if we could put some "magic ink" into the brain tissue and follow its progress, we would gain knowledge about the brain tissue itself, as well as the kind of changes that may occur in the brain when it is affected by various disease processes.

1.1 Diffusion Imaging in MR

In the measurement of diffusion by MR, the "magic ink" is created by the magnetic field gradients [3]. When the patient enters the large tunnel of a static magnetic field, nuclear spins (small magnets inside each proton nucleus) are lined up along the direction of the big magnet. Magnetic field gradients of certain duration will then add a smaller magnetic field to spins located in different regions within the tissue. This is similar to marking the spins with "magic ink". By applying another gradient pulse at a later time, information is obtained about how much the spins have spread (diffused) during this time. This is analogous to comparing two snapshots, one taken at the moment when the ink is dropped into the water and one taken later, to obtain information about how the ink has spread in the water. However, the analogy of ink in water and what happens in the brain stops here.

1.2 Diffusion Imaging of the Brain

The brain is complex and full of fibrous, globular and other structures and membranes, which may or may not allow water to move freely. Because water spins will run into constituents of cells of different concentrations in different cellular compartments, they will spread at different rates when labeled with the "magic ink". In addition, they will not behave in the same way when they are moving in different directions. As described below, the former is measured as the diffusion rate, diffusion coefficient, or simply diffusivity, depending on the unit used, and the latter is more formally described as diffusion anisotropy, with a variety of parameters defined [4–7].

1.3 Magnetic Resonance Principles of Diffusion Imaging

In order to perform diffusion studies, one needs to apply field gradients in addition to the radiofrequency and gradient pulses used for conventional MR imaging. A simplified version of the most commonly used pulse-gradient spin-echo pulse sequence for diffusion imaging is shown in Fig. 1.1. During the time evolution (TE), a pair of field gradients is used to perform "diffusion-encoding." Each gradient in this gradient pair will last a time δ, with strength G (usually in units of mT/m), and the pair is separated by a time Δ. A more formal analysis will tell us that the intensity of the signal will depend on all these parameters, given by

$$S=S_0\exp(-b\,ADC), \qquad (1)$$

where ADC is the apparent diffusion coefficient, and b is the gradient factor, sometimes simply called the b-factor. S_0 is the signal intensity obtained when no diffusion gradients are used. The diffusion coefficient

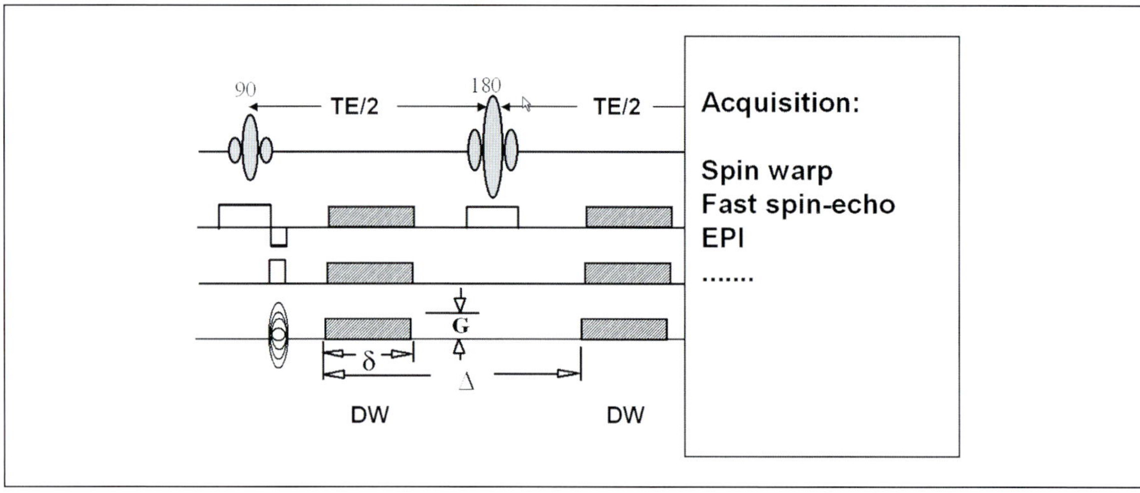

Figure 1.1

A typical pulse sequence for diffusion imaging. The *shaded areas* represent field gradient pulses. *DW* diffusion weighted, *TE* time evolution

calculated in this way is usually called "apparent" because it is often an average measure of much more complicated processes in the tissues, as discussed below. The b-factor is related to the gradient parameters δ, G and Δ (Fig. 1.1), usually in the form $b \propto (\delta G)^2 (\Delta - \delta / 3)$, and is set by the experimenter. The formula for the b-factor tells us that we can increase diffusion weighting (DW) by increasing either gradient timing, δ or Δ, or gradient strength, G.

Equation 1.1 suggests that there is a reduction in the measured signal intensity when DW is applied, $b \neq 0$, which can be understood with some simple reasoning. As the diffusing spins are moving inside the field gradient (an additional magnetic field varying in intensity from location to location), each spin is affected differently by the field. The alignment of the spins with each other is thus destroyed. Since the measured signal is a summation of tiny signals from all individual spins, the misalignment, or "dephasing", caused by the gradient pulses results in a drop in signal intensity; the longer the diffusion distance, the lower the signal (more dephasing; Fig. 1.2).

1.4 Apparent Diffusion Coefficient

From Eq. 1.1 it can be seen that when a fixed DW b-factor is used, tissues with a higher ADC value produce a lower signal intensity. Since brain cerebrospinal fluid (CSF) contains water that can move around freely, its ADC value is much higher than that of other brain tissues (either gray matter or white matter), which contain many more cellular structures and constituents. Consequently, in a DW image one typically sees dark CSF space (pronounced dephasing) and brighter tissue signals (less dephasing). It is also clear from Eq. 1.1 that if we collect a series of DW images with different b-values, we can calculate according to the expression for every pixel and obtain a parametric map of ADC values. The result is sometimes referred to as an ADC map. The calculated ADC map would have image pixel intensities reflecting the strength of diffusion in the pixels. Regions of CSF will therefore have higher intensity than other brain tissues – a reversal of DW images. There are several reasons why it is sometimes desirable to calculate an ADC map instead of just using DW images. One is the so-called T2-shine-through effect, which will be discussed in a later chapter. It can also be noted that S_0 in Eq. 1.1 is equal to the signal when no DW is used.

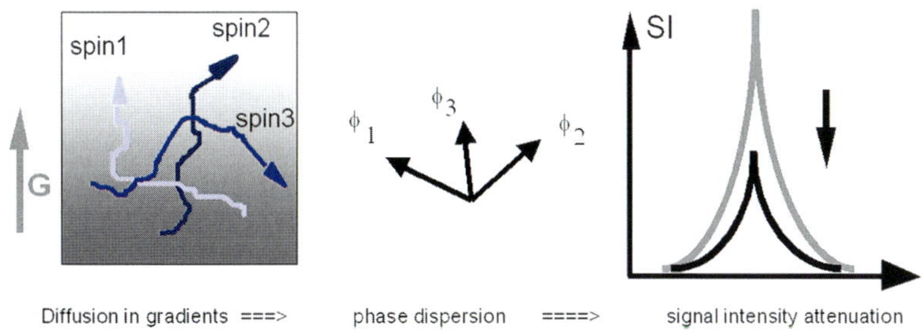

Figure 1.2

Effects of gradients on diffusing spins. Diffusion in the gradient causes dephasing of spins and therefore reduction in the measured signal intensity

Figure 1.1 suggests that this is actually the same as would be obtained from a simple spin-echo sequence. In most clinical scanners, a long TE time (tens of milliseconds) is needed to accommodate the diffusion pulses, so S_0 is often T2-weighted.

1.5 Diffusion Represents a Molecular Event

Even though an image pixel size in the order of millimeters is used in most clinical MR imaging, the information provided by diffusion imaging reflects cellular or molecular events in much smaller scales. This is because the molecular diffusion process is highly modulated by these events. It can be shown that water spins diffuse about tens of micrometers during a typical MR imaging measurement time, which coincides with the dimension of typical cellular structures. If spins experience minimal obstruction from cellular structures during this time (such as for spins in the CSF space), the measured diffusion is "free" and "isotropic", and ADC is just the intrinsic molecular diffusion coefficient. On the other hand, when diffusing spins run into cellular constituents and membranes, the value of ADC will be reduced when compared with the value in free space. What happens at the cellular level is represented schematically in Fig. 1.3. For patients with neurological abnormalities that change the water distribution in various cellular compartments, or change the ability of water to pass through cell membranes, the measured ADC values will also be altered [4–7]. Therefore the MR diffusion measurement offers a unique opportunity to obtain information about morphology otherwise inaccessible to conventional MR imaging methods. A wide range of pathological conditions can be explored with water diffusion measurements, as described later in this book. The measured ADC may also vary depending on the duration of the diffusion process, the direction in which diffusion is measured, and other factors. For diffusion in an anisotropic environment (such as in brain white matter, where bundles of axons with myelin layers wrapped around them make diffusion along the bundle much easier than across the bundle), diffusion becomes more complicated and a complete description of the process relies on what is called tensor analysis [8, 9].

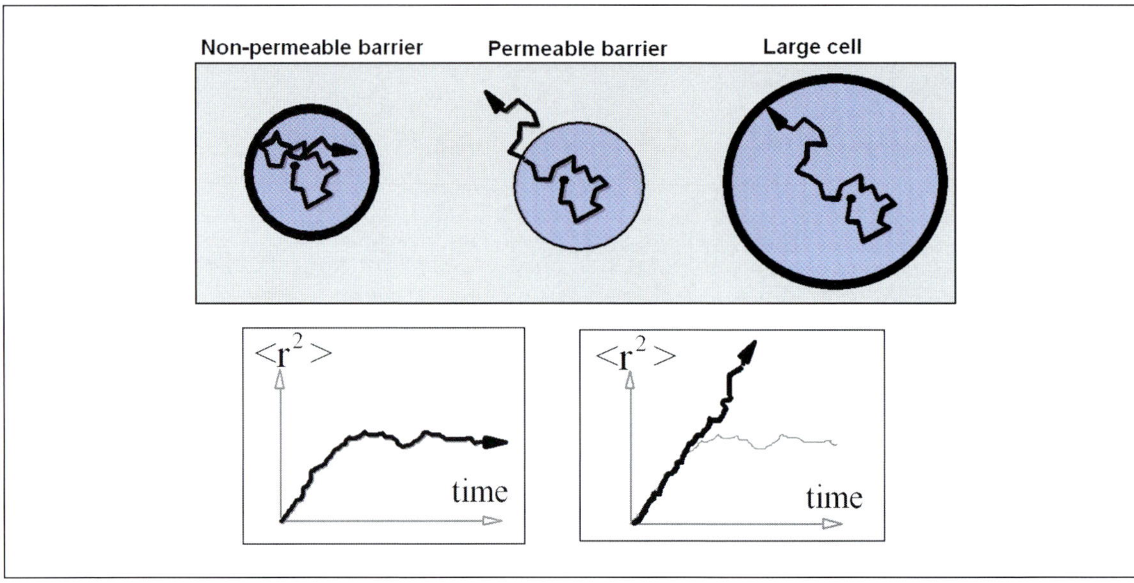

Figure 3.1

When water spins are diffusing among cellular structures, depending on the mean displacement (<r^2>) during the measurement time and the size of the cellular structure, their behavior can be quite different. Water inside a non-permeable cell (*top* and *bottom left*) experiences restriction or hindrance to diffusion. Diffusion barrier effects are minimal for water inside a permeable cell, or in a cell that is much larger than the mean displacement during diffusion (*top middle* and *right*, and *bottom right*)

1.6 Requirements in Clinical Diffusion Imaging

In a clinical environment, certain requirements are imposed for diffusion studies. A reasonable imaging time is often limited to several minutes for each type of measurement (T1 W, T2 W, diffusion, and others). Multiple slices (15–20) are required to cover most of the brain. A good spatial resolution (~5–8 mm thick, 1–3 mm in-plane) is required. A reasonably short TE (120 ms) to reduce T2 decay, and an adequate diffusion sensitivity (ADC ~0.2–1×10^{-3} mm^2/s for brain tissues) are also needed. However, most essential is the almost complete elimination of sensitivity to subject motion during scanning. The best compromise so far in most clinical practices of diffusion imaging is the use of the multi-directional (x,y,z), 2 b-factor (b=0, and b~1/ADC) single-shot echo-planar imaging technique. Sometimes fluid attenuation with inversion recovery (FLAIR) is used to eliminate signal in the highly diffusive CSF space. Separation from relaxation effects is achieved with calculation of ADC instead of just using DW, and elimination of anisotropic diffusion is achieved by averaging the diffusion measurements from three orthogonal directions.

1.7 Setting the b-Value in Clinical DW Imaging

In a clinical setting it is advisable to maintain the same b-value for all examinations, making it easier to learn to interpret these images and become aware of the appearance of findings in various disease processes. The studies and discussions presented in this book are limited to DW imaging using b-factors of 0 and 1,000. An upper b-factor around 1,000 has been available for most clinical scanners until now and DW imaging at these standard values has been shown to be a sensitive tool in detecting and delineating restricted diffusion, e.g. in acute ischemic lesions of the brain. However, this technique has become clinically important in many other disease processes, which will be discussed in this book.

1.8 Future Trends in Clinical DW Imaging

Newer DW imaging techniques are using even higher b-values: 8.000 and more. Some recent articles that explore the use of higher b-values imply that they will simplify clinical diffusion imaging [10]. The increased b-values may free up routine DW imaging from its most pressing problem, "T2 shine-through". At high b-values more attention will be focused on the actual physiological basis of restricted and facilitated diffusion. Clearly, much of the advantage of increased b-values may lie not with the diagnosis of lesions with restricted diffusion, especially acute infarcts, but with allowing a more complete understanding of other types of diseases.

The benefits of improved diffusion contrast at high b values come with the complication of prescription dependent measures of apparent diffusion. The ADC is conventionally derived from images taken at two different b-values. Because tissues are described by fast and slow components, the results of a two-point measurement will depend on the specific b-values chosen. If the lower b-value is set to 0 (a T2-weighted image) and the upper value is allowed to vary, the ADC will vary as a function of the upper value. Specifically, one would expect the measured ADC to decrease as the upper b-value increases.

Another area of DW imaging that will evolve over the next few years is diffusion tensor imaging, which is becoming available in many modern clinical scanners. When these and other techniques become more accessible and technically more mature, they may provide more specific measurements.

References

1. Brown R (1928) A brief account of microscopical observations made in the months of June, July, and August, 1827, on the particles contained in the pollen of plants; and on the general existence of active molecules in organic and inorganic bodies. Phil Mag 4:161–173
2. Doob JL (1942) The Brownian movement and stochastic equations. Ann Math 43:351–369
3. Stejeskal EO, Tanner JE (1965) Spin diffusion measurements: spin echoes in the presence of time-dependent field gradient. J Chem Phys 42:288–292.
4. Le Bihan D, Breton E, Lallemand D, Grenier P, Cabanis E, Laval-Jeantet M (1986) MR imaging of intravoxel incoherent motions: application to diffusion and perfusion in neurologic disorders. Radiology Nov; 161(2):401–407
5. Moseley ME, Kucharczyk J, Mintorovitch J (1990) Diffusion-weighted MR imaging of acute stroke: correlation with T2-weighted and magnetic susceptibility-enhanced MR imaging in cats. AJNR Am J Neuroradiol 11:423
6. Moonen CT, Pekar J, Vleeschouwer de MHM, Gelderen van P, Zijl van PCM, DesPres D (1991) Restricted and anisotropic displacement of water in healthy cat brain and in stroke studied by NMR diffusion imaging. Magn Reson Med 19:327
7. Zhong J, Petroff OAC, Prichard JW, Gore JC (1993) Changes in water diffusion and relaxation properties of rat cerebrum during status epilepticus. Magn Reson Med 30:241
8. Basser PJ, Mattiello J, Le Bihan D (1994) Estimation of the effective self-diffusion tensor from the NMR spin-echo. J Magn Reson 103:247–254
9. Le Bihan D, Van Zijl P (2002) From the diffusion coefficient to the diffusion tensor. NMR Biomed Nov-Dec; 15(7–8):431–434
10. Clark CA, Le Bihan D (2000) Water diffusion compartmentation and anisotropy at high b values in the human brain. Magn Reson Med Dec; 44(6):852–859

Diffusion-Weighted Imaging of the Normal Brain

In collaboration with A. Hiwatashi and J. Zhong

2.1 Introduction

Diffusion-weighted (DW) images are usually obtained in three orthogonal orientations using spin-echo type single-shot DW echo-planar imaging with b-values around 0 and 1,000 s/mm^2. These three planes are combined into isotropic DW images, and apparent diffusion coefficient (ADC) maps are calculated on a pixel-by-pixel basis (Fig. 2.1). To avoid misinterpretations, it is important to recognize the normal findings on DW images and ADC maps.

2.2 Adult Brain

2.2.1 Low Signal in Basal Ganglia

Isotropic DW imaging in adult brain often shows low signal intensity in the basal ganglia (Fig. 2.1). This low signal is caused by normal iron deposition. The hypointensity on DW images of these areas is essentially related to T2 contrast, which is also shown on b_0 images. ADC maps usually show the areas as iso-intense, but it can be hyper- or hypointense depending on the paramagnetic susceptibility artifact of iron deposition.

2.2.2 Diffusion-Weighted Imaging of Gray and White Matter

Gray matter on DW images is generally hyperintense when compared with white matter. ADC of gray ($0.76\pm0.13\times10^{-3}$ mm^2/s) and white matter ($0.77\pm0.18\times10^{-3}$ mm^2/s) are, however, identical in the adult brain [1]. There are several reports about ADC increasing with age [2–7], but this increase is minimal and has been observed in all parts of the brain. It is usually more apparent in the white matter and lentiform nucleus than in the rest of the brain. Focal areas of DW hyperintensities are often seen in posterior limbs of internal capsule, corticospinal tracts, medial lemniscus and the decussation of the superior cerebellar peduncles (Fig. 2.1). These DW hyperintensities are caused by T2 contrast and represent normal findings without clinical significance. ADC maps are usually isointense in these areas.

2.2.3 Choroid Plexus

The choroid plexus occasionally shows prominent hyperintensity on DW imaging associated with mild elevation of ADC. In these situations the ADC is often higher than in white matter, but lower than in cerebrospinal fluid. The high DW signal is believed to represent gelatinous cystic changes of the choroid plexus, which can occur with age (Fig. 2.2).

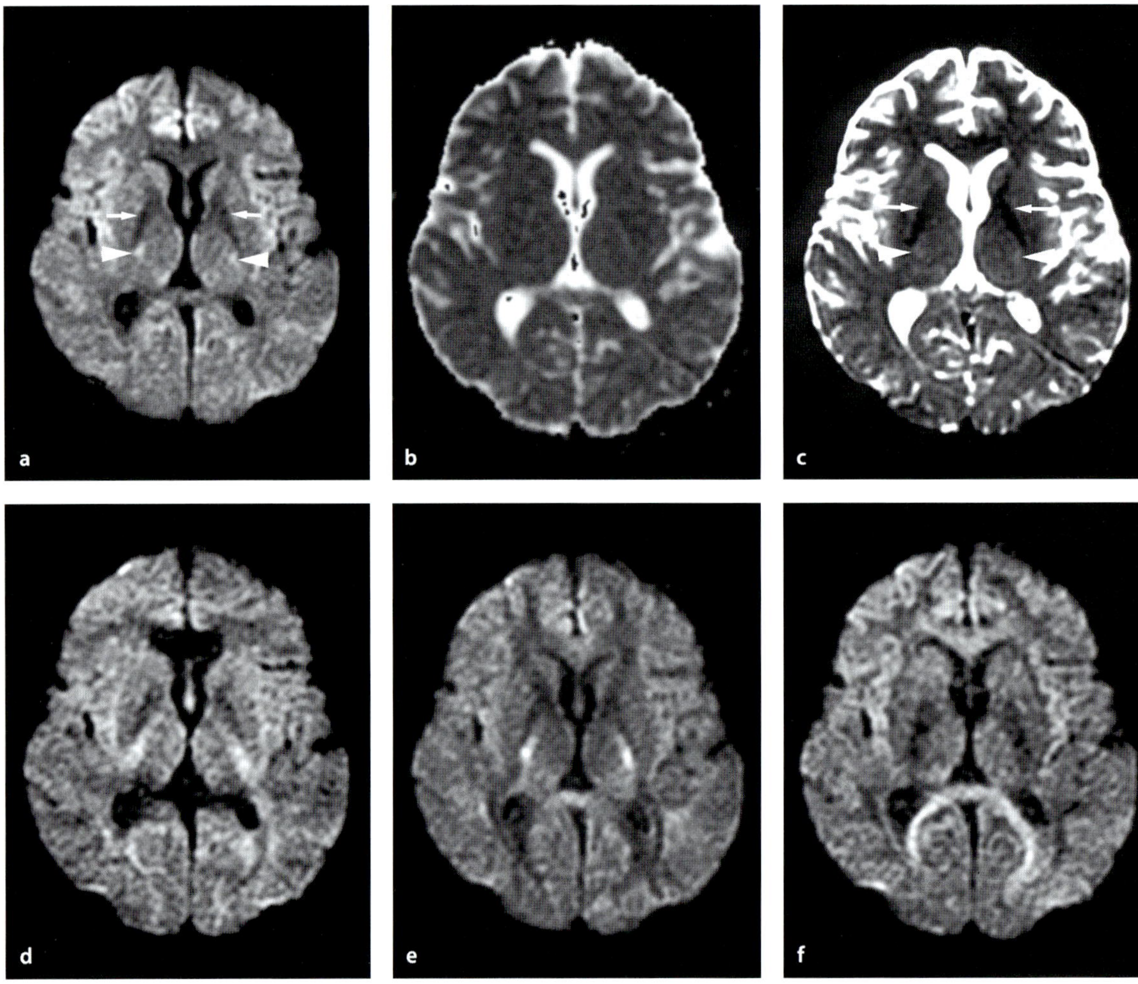

Figure 2.1 a–f

Normal adult brain of a 40-year-old male without neurological deficits. **a** Isotropic DW image is obtained by combining b_0 image and three orthogonal unidirectional images (*x, y, z* axis). The bilateral globi pallidi have low signal on DW image as a result of physiological iron deposition (*arrows*). Corticospinal tracts have mildly high signal on DW image (*arrowheads*). Gray matter shows mildly high signal compared to white matter. These signal changes on isotropic DW imaging are normal and are caused by T2 contrast. **b** ADC map shows homogeneous ADC values in globi pallidi, corticospinal tracts, gray and white matter. **c** b_0 image shows low signal in globi pallidi (*arrows*), high signal in corticospinal tracts (*arrowheads*), and the gray–white matter contrast. **d–f** Diffusion weighting is applied in *x* axis (**d**), *y* axis (**e**), and *z* axis (**f**)

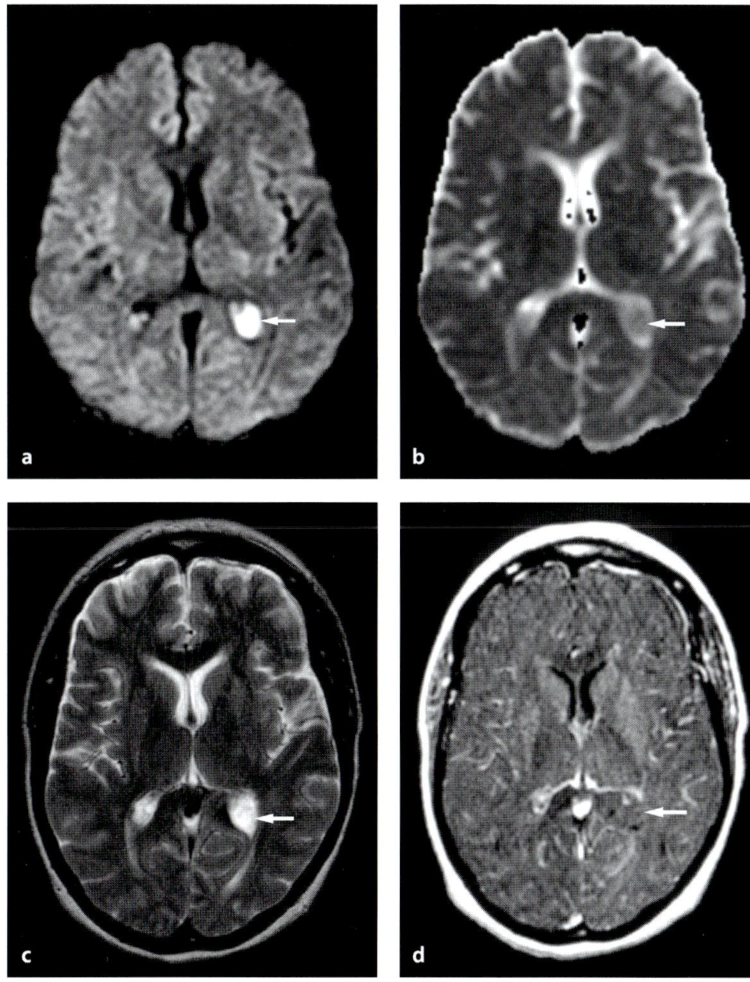

Figure 2.2 a–d

Cystic changes in the choroid plexus. **a** DW image shows hyperintensity in cystic changes of the left choroid plexus (*arrow*). **b** ADC values of the cystic changes are lower than those of the CSF, which may represent viscous gelatinous materials, but higher than those of brain parenchyma (*arrow*). **c** T2-weighted image shows the cystic changes as hyperintensity (*arrow*). **d** Gadolinium-enhanced T1-weighted image with magnetization transfer contrast reveals no enhancement in it (*arrow*)

2.3 Pediatric Brain

2.3.1 Diffusion-Weighted Imaging and ADC of the Pediatric Brain

The normal brain of neonates and infants has significantly higher ADC values than the adult brain [8–13] (Fig. 2.3). ADC in neonates and infants varies markedly within different areas of the brain and is higher in white matter (1.13×10^{-3} mm^2/s) than in gray matter (1.02×10^{-3} mm^2/s) [13]. ADC at birth is higher in subcortical white matter (1.88×10^{-3} mm^2/s) than in both the anterior (1.30×10^{-3} mm^2/s) and posterior limbs of the internal capsule (1.09×10^{-3} mm^2/s). It is also higher in cortex and the caudate nucleus (1.34×10^{-3} mm^2/s) than in the thalamus and the lentiform nucleus (1.20×10^{-3} mm^2/s) [13]. With the exception of the cerebrospinal fluid (CSF), there is a trend of decreasing ADC with increasing maturation in most areas of the pediatric brain. These ADC changes seem to reflect a combination of different factors, including a reduction of overall water content, cellular maturation and white matter myelination. In neonates and infants, ischemia is usually global and can therefore resemble the normal image with elevated DW signal and decreased ADC. White matter diseases can also be mimicked by the normal, age-related appearance of DW imaging and ADC. Out of necessity, the ADC values will therefore have to be age related for a correct interpretation of the DW images of the pediatric brains.

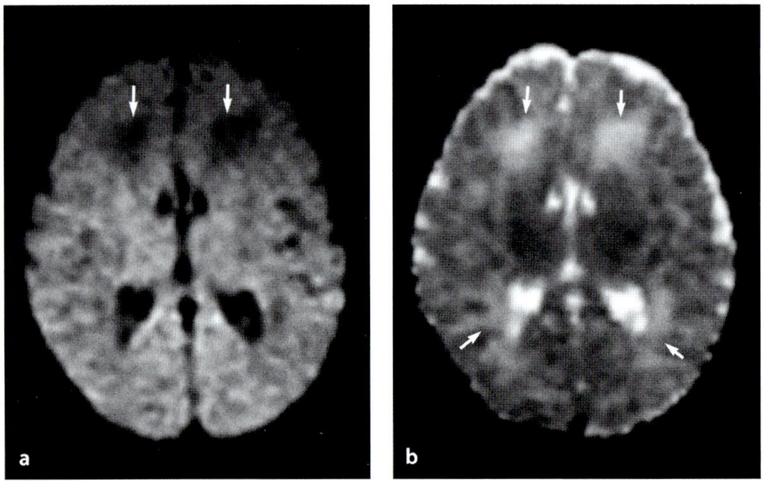

Figure 2.3 a, b

Normal neonatal brain. a The appearance of the pediatric brain on DW images varies with age. In neonates it is normal to have low DW signal intensities in the frontal deep white matter (*arrows*). b ADC values of the corresponding areas are high in neonatal brain, especially in the white matter (*arrows*). These ADC changes seem to reflect a combination of factors, including a reduction of overall water content, cellular maturation, and white matter myelination

2.4 Conclusion

Good knowledge of the DW appearance of the normal adult and pediatric brain and variations is necessary to avoid misinterpretation. In children it is also important to match the findings with those of normal children of the same age.

References

1. Yoshiura T, Wu O, Sorensen AG (1999) Advanced MR techniques: diffusion MR imaging, perfusion MR imaging, and spectroscopy. Neuroimaging Clin N Am 9:439–453
2. Chun T, Filippi CG, Zimmerman RD, Ulug AM (2000) Diffusion changes in the aging human brain. Am J Neuroradiol 21:1078–1083
3. Engelter ST, Provenzale JM, Petrella JR, DeLong DM, MacFall JR (2000) The effect of aging on the apparent diffusion coefficient of normal-appearing white matter. Am J Roentgenol 175:425–430
4. Helenius J, Soinne L, Perkio J (2002) Diffusion-weighted MR imaging in normal human brains in various age groups. Am J Neuroradiol 23:194–199
5. Gideon P, Thomsen C, Henriksen O (1994) Increased self-diffusion of brain water in normal aging. J Magn Reson Imaging 4:185–188
6. Nusbaum AO, Tang CY, Buchsbaum MS, Wei TC, Atlas SW (2001) Regional and global changes in cerebral diffusion with normal aging. Am J Neuroradiol 22:136–142
7. Abe O, Aoki S, Hayashi N, et al. (2002) Normal aging in the central nervous system: quantitative MR diffusion-tensor analysis. Neurobiol Aging 23:433–441
8. Sakuma H, Nomura Y, Takeda K, et al. (1991) Adult and neonatal human brain: diffusional anisotropy and myelination with diffusion-weighted MR imaging. Radiology 180:229– 233
9. Morriss MC, Zimmerman RA, Bilaniuk LT, Hunter JV, Haselgrove JC (1999) Changes in brain water diffusion during childhood. Neuroradiology 41:929–934
10. Tanner SF, Ramenghi LA, Ridgway JP, et al. (2000) Quantitative comparison of intrabrain diffusion in adults and preterm and term neonates and infants. Am J Roentgenol 174:1643–1649
11. Neil JJ, Shiran SI, McKinstry RC, et al. (1998) Normal brain in human newborns: apparent diffusion coefficient and diffusion anisotropy measured by using diffusion tensor MR imaging. Radiology 209:57–66
12. Engelbrecht V, Scherer A, Rassek M, Witsack HJ, Modder U (2002) Diffusion-weighted MR imaging in the brain in children: findings in the normal brain and in the brain with white matter diseases. Radiology 222:410–418
13. Forbes KP, Pipe JG, Bird CR (2002) Changes in brain water diffusion during the 1st year of life. Radiology 222:405–409

Pitfalls and Artifacts of DW Imaging

In collaboration with A. Hiwatashi and J. Zhong

3.1 Introduction

There are many inherent artifacts and pitfalls in diffusion-weighted (DW) imaging of the brain that are important to recognize to avoid misinterpretations.

3.2 Influence of ADC and T2 on the DW Appearance

Diffusion-weighted images are inherently T2 weighted and changes in T2 signal characteristics will thus influence the appearance of DW images independent of tissue diffusibility [1–16]. The effect of T2 prolongation, so-called "T2 shine-through", is well known. Less well known is the balance between apparent diffusion coefficient (ADC) and T2, sometimes called T2 washout. Also the effect of T2 shortening, or T2 blackout, and magnetic susceptibility effects will influence the DW appearance in many situations. This chapter will illustrate and discuss the effects of T2 and ADC on DW images.

3.2.1 Concepts

The signal intensity (SI) on DW images is influenced by T2, ADC, the b factor, the spin density (SD) and the echo time (TE), and is calculated as follows:

$$SI=SI_{b=0}e^{-bADC}$$

However,

$$SI_{b=0}=kSD(1-e^{-TR/T1})e^{-TE/T2}$$

For TR >>T1

$$SI=kSDe^{-TE/T2}e^{-bADC}$$

where k is a constant, TR is repetition time, and $SI_{b=0}$ is the signal intensity on the spin-echo echo-planar image (b_0 image) [1, 2, 5, 7, 8, 10, 12, 16].

To evaluate the tissue T2 and ADC, we should pay attention to the images discussed below as well as isotropic DW images and b_0 images [3–5, 7, 8, 10, 11, 13–16].

3.2.2 Apparent Diffusion Coefficient Maps

To evaluate the diffusibility, ADC is calculated as:

$$ADC=-ln\,(SI/SI_{b=0})/b$$

Subsequently, increased ADC causes decreased SI on DW images, and decreased ADC causes increased SI on DW images [3–5, 7, 10, 15, 16].

3.2.3 Exponential Images

To remove the T2-weighted contrast, the DW image can be divided by the b_0 image to create an "exponential image" [4, 7, 10, 15].

The signal intensity (SI_{eDWI}) on the exponential image is calculated as:

$$SI_{eDWI}=SI/SI_{b=0}=e^{-bADC}$$

Therefore, this image can eliminate the effect of T2. Contrary to ADC maps, hyperintensity on exponential DW images means decreased ADC, and hypointensity means increased ADC.

3.3 Clinical Conditions

3.3.1 T2 Shine-through

This is a well-known phenomenon that causes hyperintensity on DW images by means of T2 prolongation [3–5, 7, 8, 10, 11, 15, 16]. If ADC is decreased at the same time, this can result in an accentuation of the hyperintensity on DW images (Figs. 3.1, 3.2 and 3.3).

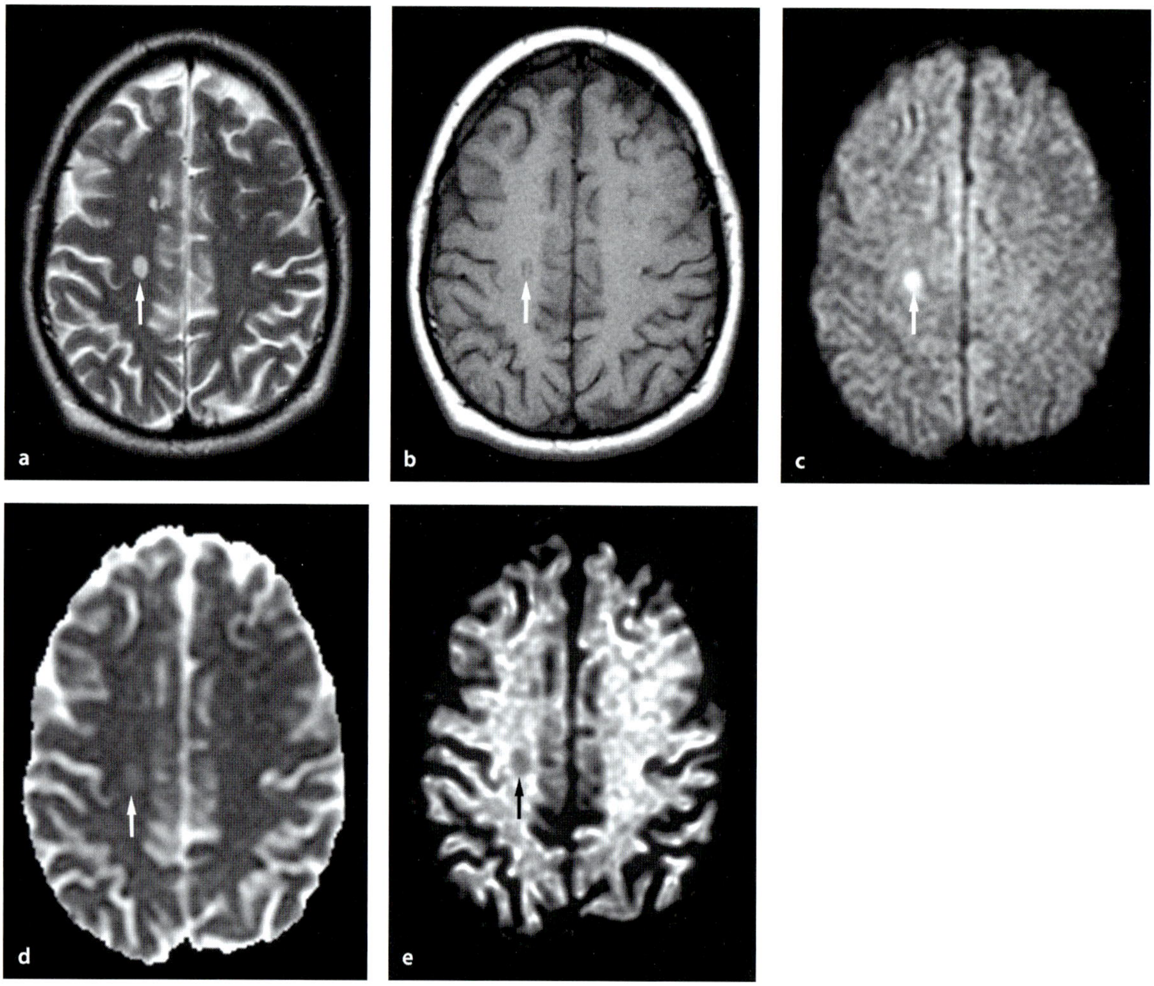

Figure 3.1 a–e

T2 shine-through in a 35-year-old female with multiple sclerosis and weakness of the lower extremities. **a** T2-weighted image shows several hyperintense lesions, with the largest one in the right frontal lobe (*arrow*). **b** On T1-weighted image the lesion was hypointense (*arrow*) and did not enhance with contrast (not shown). **c** On DW image the lesion is hyperintense (*arrow*). **d** ADC map also shows hyperintensity in the lesion (1.2×10^{-3} mm^2/s; *arrow*). **e** Exponential image eliminates the T2 effect and shows the lesion to be hypointense (*arrow*). This confirms that the hyperintensity on DW image is due to a T2 shine-through

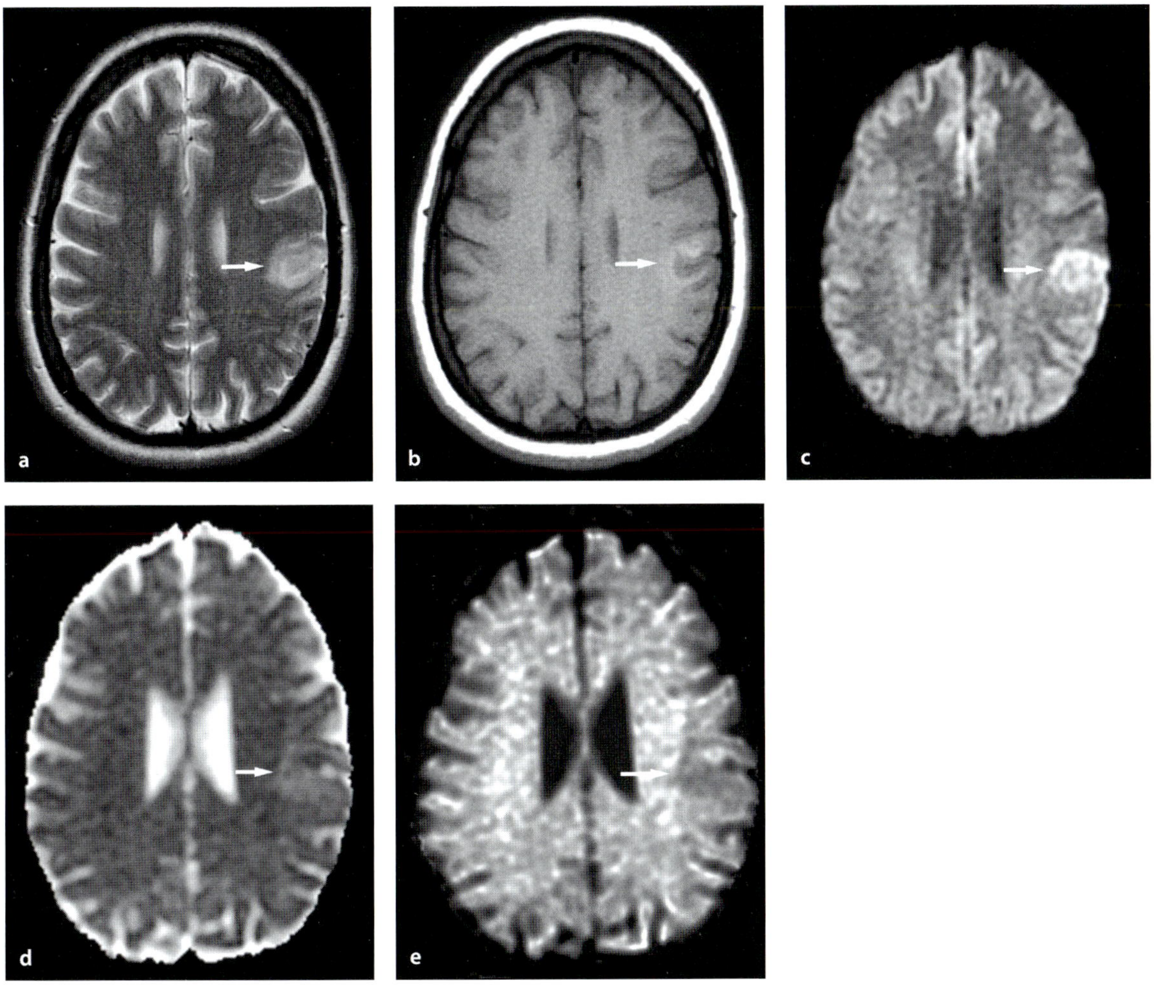

Figure 3.2 a–e

T2 shine-through in a 45-year-old female with seizures caused by anaplastic astrocytoma. **a** T2-weighted image shows a hyperintense lesion in the left frontal lobe (*arrow*). **b** On T1-weighted image the lesion is hypointense with peripheral hyperintense area (*arrow*). The lesion did not enhance with contrast (not shown). **c** DW image shows hyperintensity (*arrow*). **d** ADC map also shows hyperintensity in the lesion ($0.98–1.35\times10^{-3}$ mm^2/s; *arrow*). **e** Exponential image eliminates the T2 effect and shows the lesion to be hypointense (*arrow*). This confirms that the hyperintensity on DW image is due to a T2 shine-through

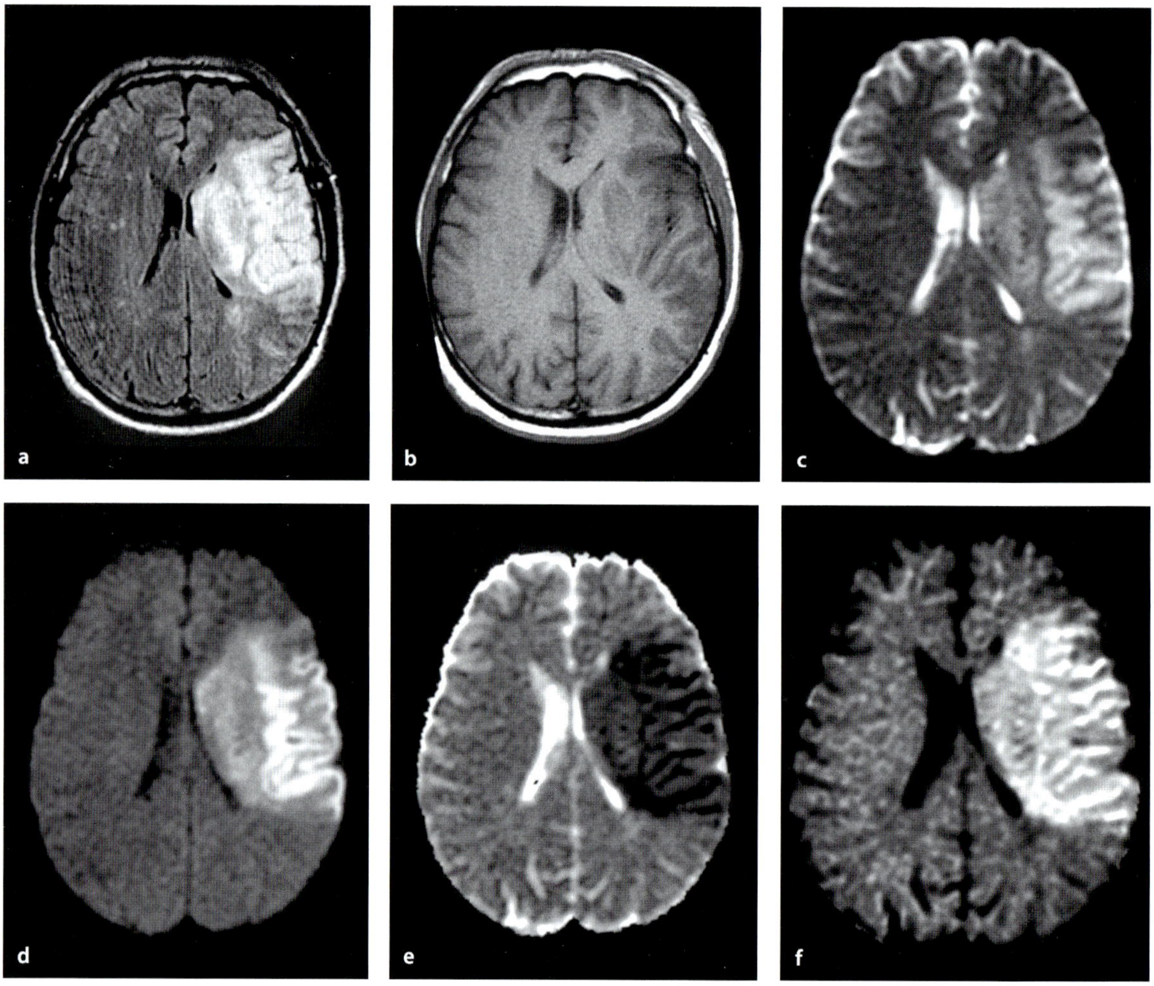

Figure 3.3 a–f

T2 shine-through and restricted diffusion in a 56-year-old male with right-sided weakness due to acute infarction. MR imaging obtained 24 hours after the onset of symptoms. **a** FLAIR image shows a hyperintense lesion in the left middle cerebral artery territory. **b** On T1-weighted image the lesion is hypointense. **c** On T2-weighted image (b_0) the lesion is hyperintense. **d** DW image also shows hyperintensity in the lesion. **e** ADC map shows hypointensity in the lesion ($0.27–0.45 \times 10^{-3}$ mm^2/s). **f** On the exponential image, which eliminates the T2 effect, the lesion remains hyperintense. This confirms that the DW hyperintensity is due to both restricted diffusion and T2 prolongation

3.3.2 T2 Washout

This implies that isointensity on DW images is the result of a balance between hyperintensity on T2-weighted images and increased ADC [13, 14, 16]. This is often seen in vasogenic edema, where the combination of increased ADC and hyperintensity on T2-weighted images will result in isointensity on DW images (Fig. 3.4).

To the best of our knowledge there have been no systematic reports on pathological conditions with isointensity on DW images, caused by a balance of hypointensity on T2-weighted images and decreased ADC.

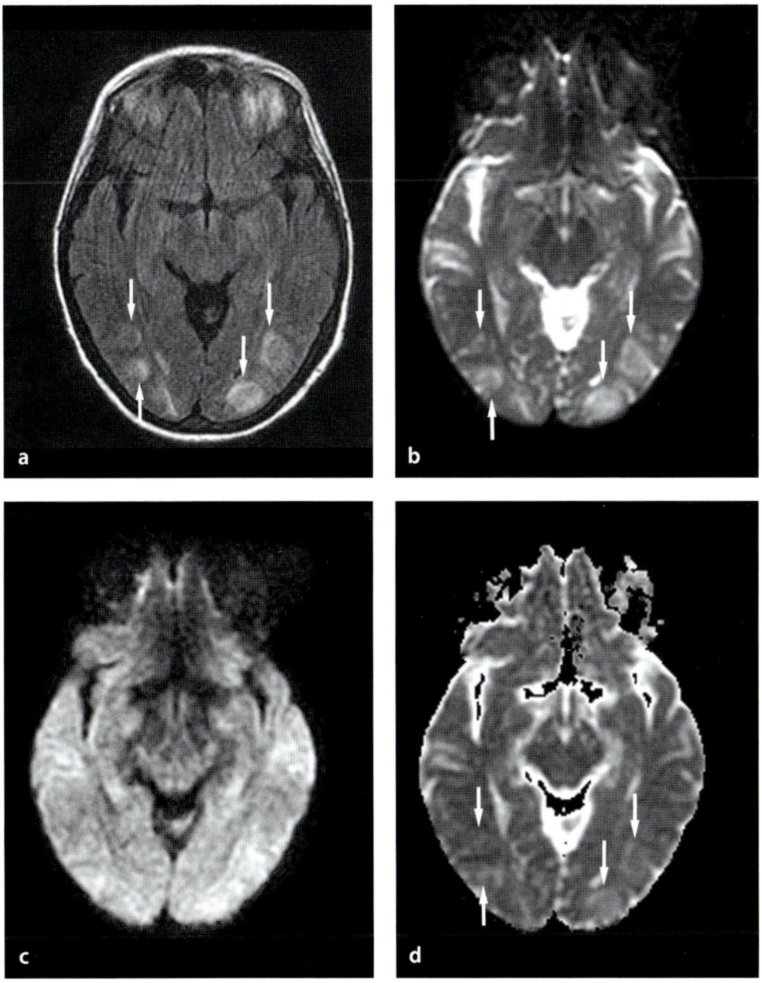

Figure 3.4 a–d

T2 washout in a 45-year-old female with hypertension, seizures and posterior reversible encephalopathy syndrome. **a** FLAIR image shows hyperintense lesions in the bilateral occipital lobe (*arrows*). **b** T2-weighted image (b_0) also shows hyperintensity of the lesions (*arrows*). **c** DW image shows mild hyperintensity in the lesions. **d** ADC map shows hyperintensity of the lesions (1.18–1.38×10^{-3} mm^2/s; *arrows*). With the strong T2 prolongation one would expect more hyperintensity on the DW image, but the T2 shine-through effect is reduced by the hyperintensity on the ADC, resulting in a balance between increased diffusibility and hyperintensity on the T2-weighted image (T2 wash-out)

3.3.3 T2 Blackout

This indicates hypointensity on DW images caused by hypointensity on T2-weighted images and is typically seen in some hematomas [9,16]. Paramagnetic susceptibility artifacts may occur in this situation (Figs. 3.5 and 3.6).

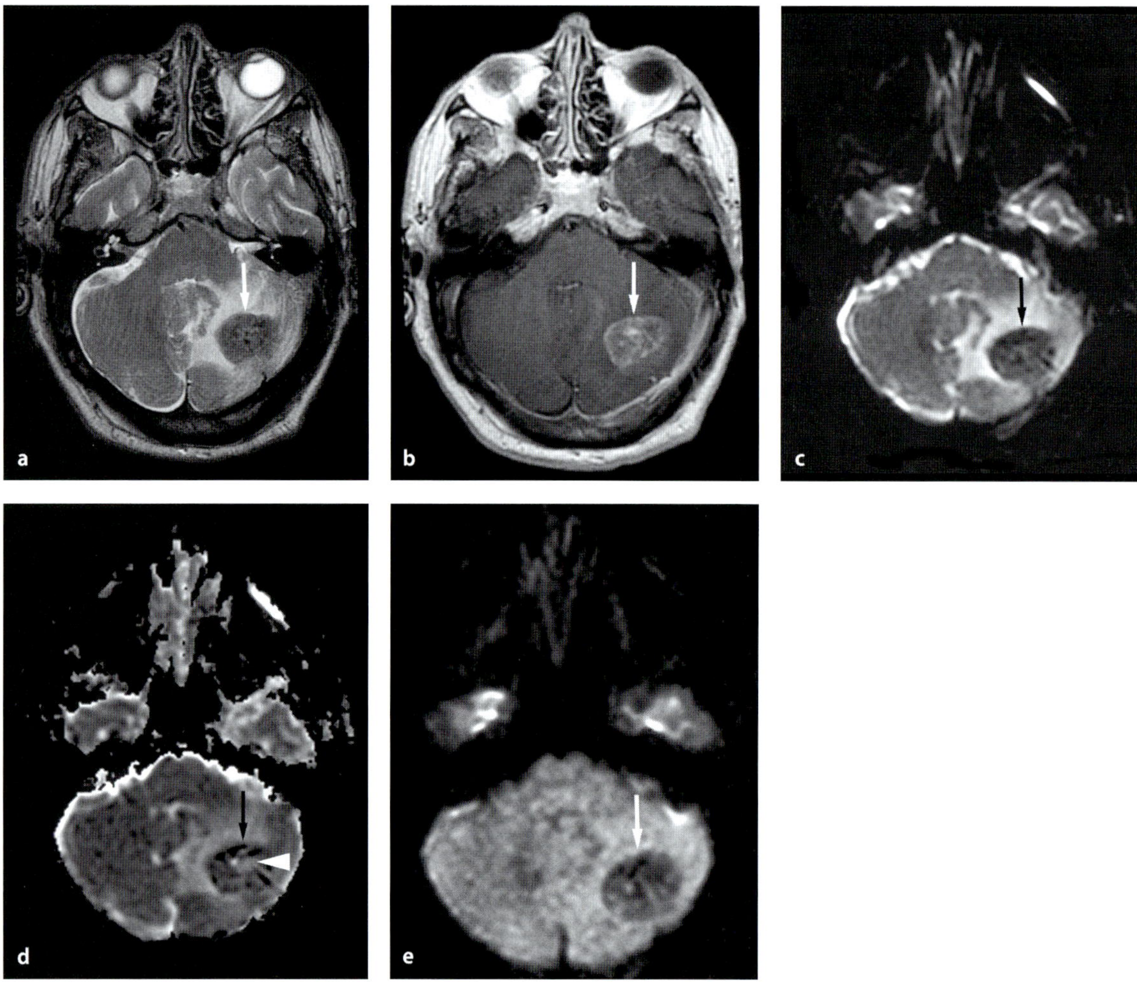

Figure 3.5 a–e

T2 blackout in lung cancer metastasis in a 62-year-old male with adenocarcinoma of the lung. **a** T2-weighted image shows a hypointense mass (*arrow*) with surrounding edema in the left cerebellar hemisphere. **b** Gadolinium-enhanced T1-weighted image shows heterogeneous enhancement of the mass (*arrow*). **c** T2-weighted image (b_0) also shows hypointensity in the lesion with surrounding hyperintense edema (*arrow*). **d** ADC map shows central hyperintensity ($1.63–2.35\times10^{-3}$ mm^2/s; *arrowhead*) and peripheral hypointensity ($1.13–1.38\times10^{-3}$ mm^2/s; *arrow*) of the mass. There is also hyperintensity of the surrounding tissue, consistent with vasogenic edema. **e** DW image shows heterogeneous hypointensity of the mass (*arrow*) and isointensity of the surrounding edema. The DW hypointensity of the mass (*arrow*) is due to the increased diffusibility and hypointensity on T2-weighted image. The isointensity in the surrounding edema is due to the balance between the increased diffusibility and hyperintensity on T2-weighted image (T2 washout)

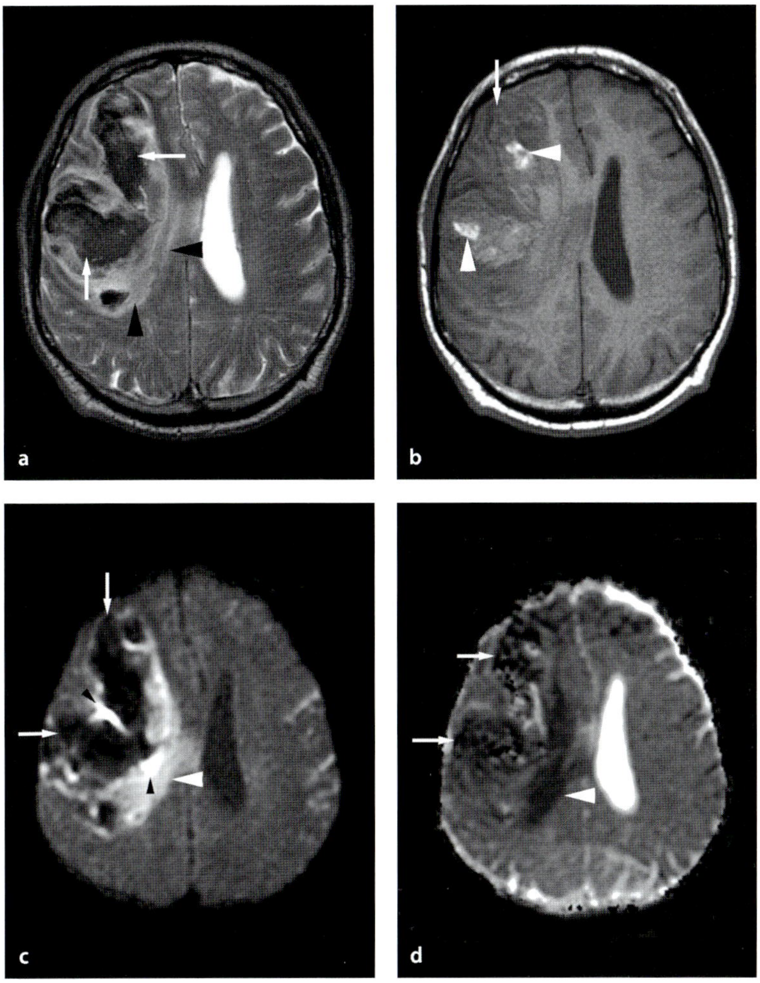

Figure 3.6 a–d

T2 blackout from susceptibility artifacts in acute hemorrhage (deoxyhemoglobin and intracellular met-hemoglobin) in a 74-year-old male with left-sided weakness. MR imaging was obtained 24 hours after the onset of symptoms. **a** T2-weighted image shows hypointense lesions in the right frontoparietal lobes (*arrows* deoxyhemoglobin and intracellular met-hemoglobin) with areas of surrounding hyperintensity consistent with edema (*arrowheads*). **b** T1-weighted image shows the heterogeneous lesion with hypointensity (*arrow* deoxyhemoglobin) and hyperintensity (*arrowheads* intracellular met-hemoglobin). **c** DW image shows hypointensity (*arrows* deoxy-hemoglobin and intracellular met-hemoglobin) and hyperintensity in region of edema (*arrowhead*). The surrounding hyperintense rims (*small arrowheads*) are due to magnetic susceptibility artifacts. **d** ADC could not be calculated accurately in the T2 "dark" hematoma due to magnetic susceptibility artifacts (*arrows*). The surrounding areas of hypointensity (*arrowhead*) probably correspond to cytotoxic edema surrounding the hematoma. This example shows how T2 hypointensity from susceptibility effects can produce a complex appearance in and around cerebral hemorrhage

3.4 Artifacts

Numerous artifacts can be generated during acquisition of DW images. There are five main artifacts of single-shot DW echo-planar imaging:

1. Eddy current artifacts due to echo-planar imaging phase-encode and readout gradients, and motion-probing gradient pulses for diffusion weighting
2. Susceptibility artifacts
3. N/2 ghosting artifacts
4. Chemical shift artifacts
5. Motion artifacts

We will discuss each artifact separately.

3.4.1 Eddy Current Artifacts

Eddy currents are electrical currents induced in a conductor by a changing magnetic field. Eddy currents can occur in patients and in the MR scanner itself, including cables or wires, gradient coils, cryoshields and radiofrequency shields [17]. Eddy currents are particularly severe when gradients are turned on and off quickly, as in echo-planar imaging pulse sequences. Gradient waveforms are distorted due to eddy currents, which results in image artifacts, including spatial blurring and misregistration. In single-shot DW echo-planar imaging, eddy currents are due to both echo-planar imaging gradients and motion-probing gradients, which lead to image distortions (Fig. 3.7). Correction of image distortion is essential to calculate ADC values and especially to quantify anisotropy with diffusion tensor imaging. **Correction methods:** (1) correction of distortion by using post-processing [18–21], (2) pre-emphasis or pre-compensation, purposely distorting the gradient-driving currents [22, 23], (3) shielded gradients, redesigning the magnet to incorporate shielding coils between the gradient coils and main windings [24].

3.4.2 Susceptibility Artifacts

Single-shot echo-planar imaging is sensitive to susceptibility artifacts, especially frequency and phase errors due to paramagnetic susceptibility effects. These artifacts are seen near the skull base, especially near the air in the sinus and mastoid (Fig. 3.8). Susceptibility artifacts are more severe along the phase-encoding direction and phase encoding should thus be along the anterior–posterior direction for axial DW images. Coronal and sagittal DW images are helpful in detecting lesions in certain locations, such as the hippocampus and brain stem, and to identify susceptibility artifacts (Fig. 3.9). Increased matrix size leads to elongation of readout time, which causes even larger image distortions.
Correction methods: (1) multi-shot echo-planar imaging (to reduce the readout time, to enable high-resolution scan) [25, 26], (2) line scan [27,28], (3) single-shot fast spin echo (SSFSE) [29, 30], (4) periodically rotated overlapping parallel lines with enhanced reconstruction (PROPELLER) [31, 32], (5) sensitive encoding (SENSE)/array spatial and sensitivity-encoding technique (ASSET), undersampling of k-space enables effective band width and shortens readout time, providing thin section and high-resolution matrix [33].

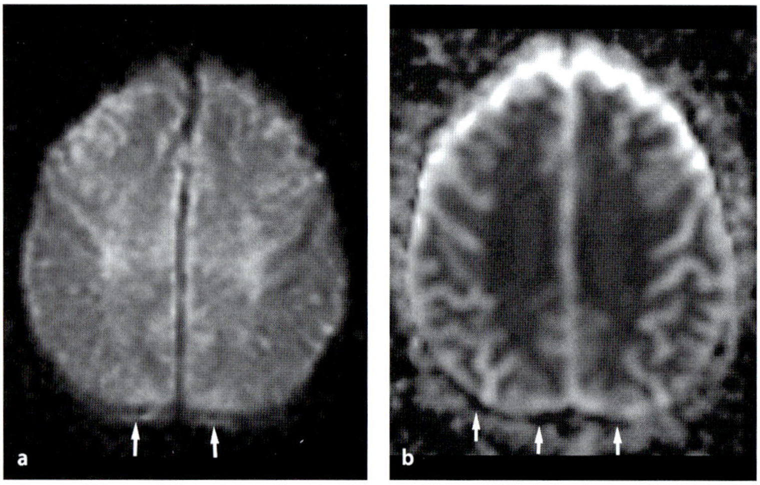

Figure 3.7 a,b

Misregistration due to eddy current artifact. **a, b** Misregistration artifact is noted in the occipital regions (*arrows*) on DW image (**a**) and the ADC map (**b**). Gradient waveforms are distorted due to eddy currents, which results in this misregistration

Figure 3.8

Susceptibility artifact. Susceptibility artifacts are seen near the air content of the mastoids (*arrows*). This is generally prominent in echo-planar sequences

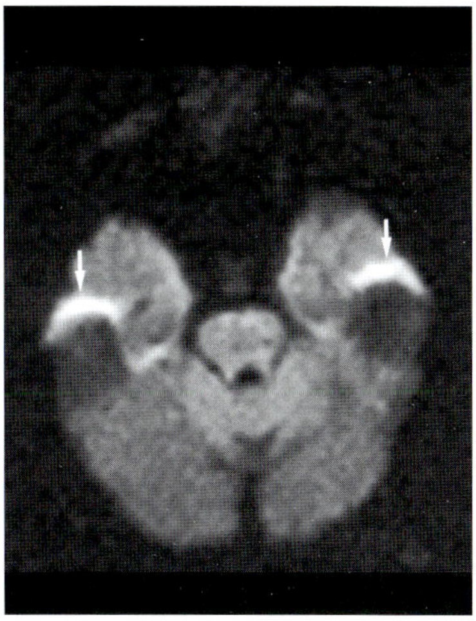

Figure 3.9 a–d

Susceptibility artifact in coronal and sagittal plane DW images. Coronal DW image (**a**) and the ADC map (**b**) are used to evaluate the hippocampus, but susceptibility artifacts distort the image near the mastoids. Sagittal DW image (**c**) and the ADC map (**d**) show a pontine infarct as hyperintense with decreased ADC (*arrowhead*). Susceptibility artifacts are caused by air in the ethmoid and sphenoid sinuses (*arrows*)

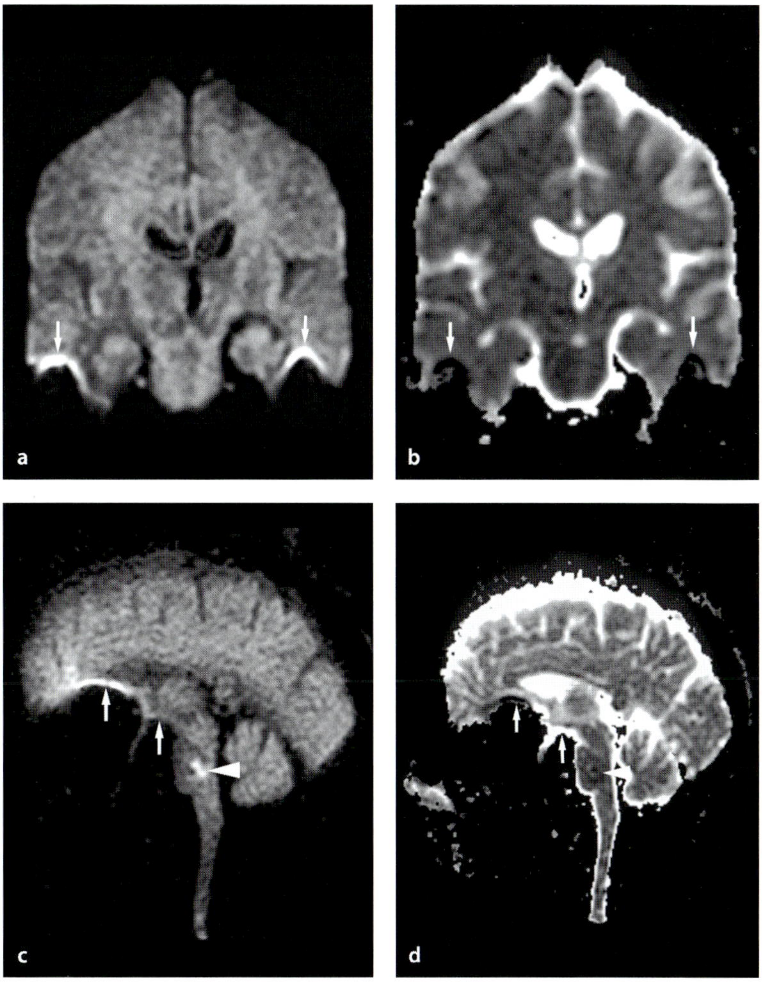

3.4.3 N/2 Ghosting Artifact (Nyquist Ghost)

N/2 ghosting artifact occurs when there are differences between the even and odd lines of the k-space. Phase error is due to hardware imperfections (eddy currents, imperfect timing of even and odd echo, imperfect gradients, and magnetic field inhomogeneity), which can be produced by on-off switching during readout gradients. The ghosts in this artifact are always shifted by half of the field of view in the phase-encoding direction (Fig. 3.10). This ghost can produce severe artifacts when ADC maps are calculated. **Correction methods:** (1) reduce eddy currents, (2) adjust gradients and magnetic field homogeneity, (3) high b-value, (4) fluid-attenuated inverse-recovery (FLAIR) DW imaging (reduce cerebrospinal fluid signal) [33, 34].

3.4.4 Chemical Shift

In echo-planar DW imaging, chemical shift artifacts due to the different resonance frequencies in water and fat are produced along the phase-encoding direction, while they are along the frequency-encoding direction in conventional spin-echo type MR imaging. This artifact is more severe in echo-planar imaging than in conventional spin-echo type MR imaging. Effective fat suppression techniques, such as the chemical shift selective (CHESS) method and the spectral selective radiofrequency excitation method are necessary.
Correction methods: appropriate fat suppression techniques.

3.4.5 Motion Artifacts

The sources of motion artifacts include gross head motion, respiratory motions, cardiac-related pulsations and patient bed vibration due to gradient pulses. Single-shot DW echo-planar imaging has relatively low sensitivity to patient motion, because each image is acquired in about 100–300 ms and the total acquisition time is less than 40 s. If one of the *x*, *y*, *z* or b_0 images is corrupted by motion artifacts during a scan, or if patient head motion occurs between scans, the isotopic DW images and the ADC maps will have these artifacts (Figs. 3.11 and 3.12). In those cases, unidirectional and b_0 images from the raw data of DW imaging can be free from the motion artifacts and remain diagnostically useful. Long (tens of ms) gradient pulses to reach sufficient diffusion weighting often increase sensitivity to motion.
Correction methods: (1) For a fixed b-factor, use high-gradient amplitude but reduce gradient pulse duration to minimize the sensitivity to motion, (2) post-processing to correct for phase error (Navigator method) [35–37], (3) elimination of phase-encode step (line scan method, projection reconstruction), (4) minimize time for phase error accumulation (single-shot echo-planar imaging, hybrid method with multishot echo-planar imaging), (5) SSFSE, (6) PROPELLER [38], (7) SENSE.

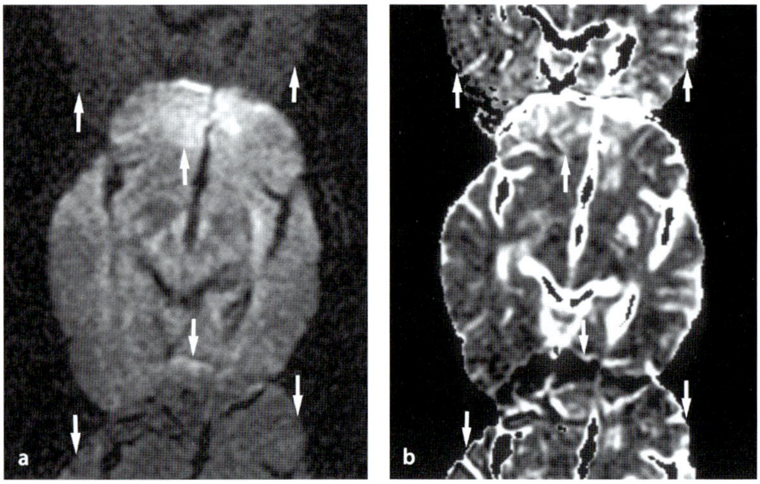

Figure 3.10 a,b

N/2 ghosting artifact. **a** DW image shows N/2 ghosting artifacts (*arrows*), which are always shifted by half of the field of view in the phase-encoding direction. **b** On the ADC map severe N/2 ghosting artifacts are also seen (*arrows*)

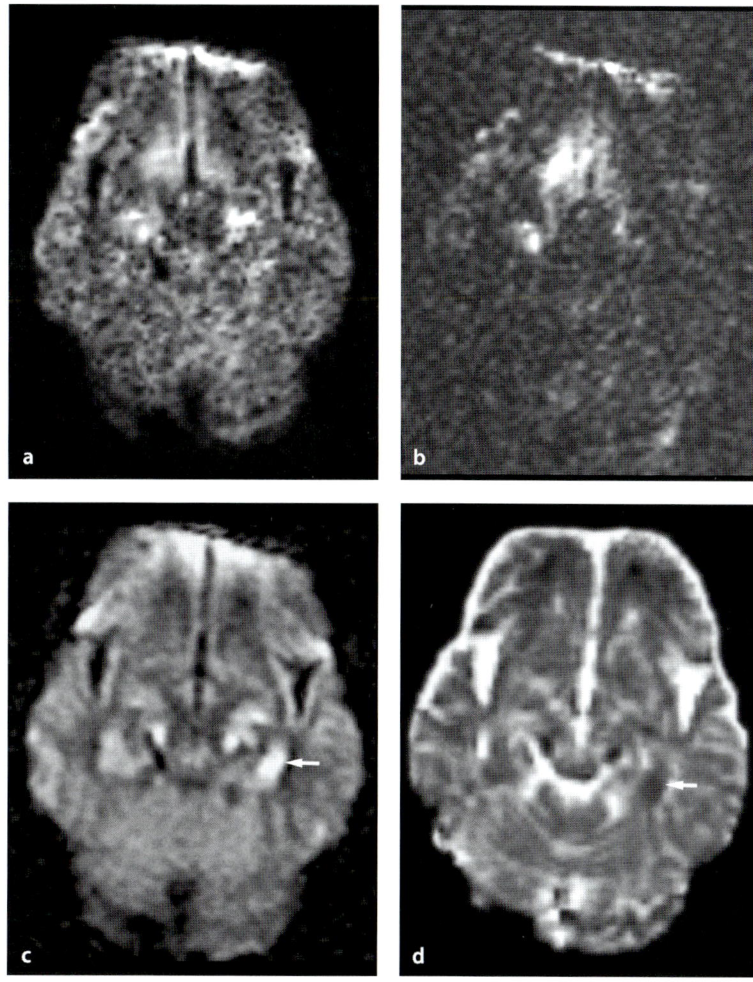

Figure 3.11 a–d

Motion artifacts due to head motion during the scan of a patient with status epilepticus. **a** It is difficult to evaluate the DW image because of severe motion artifacts. **b** In the raw data of the DW imaging, the *x* axis image is corrupted by head motion during the scan. **c** The *y* axis image is free from the artifacts. This image shows a hyperintense lesion in the left hippocampus (*arrow*). **d** ADC map of *y* axis image also shows decreased ADC of the lesion (transferred to a workstation for image processing, using a home-made code, which is based on the numerical computation software)

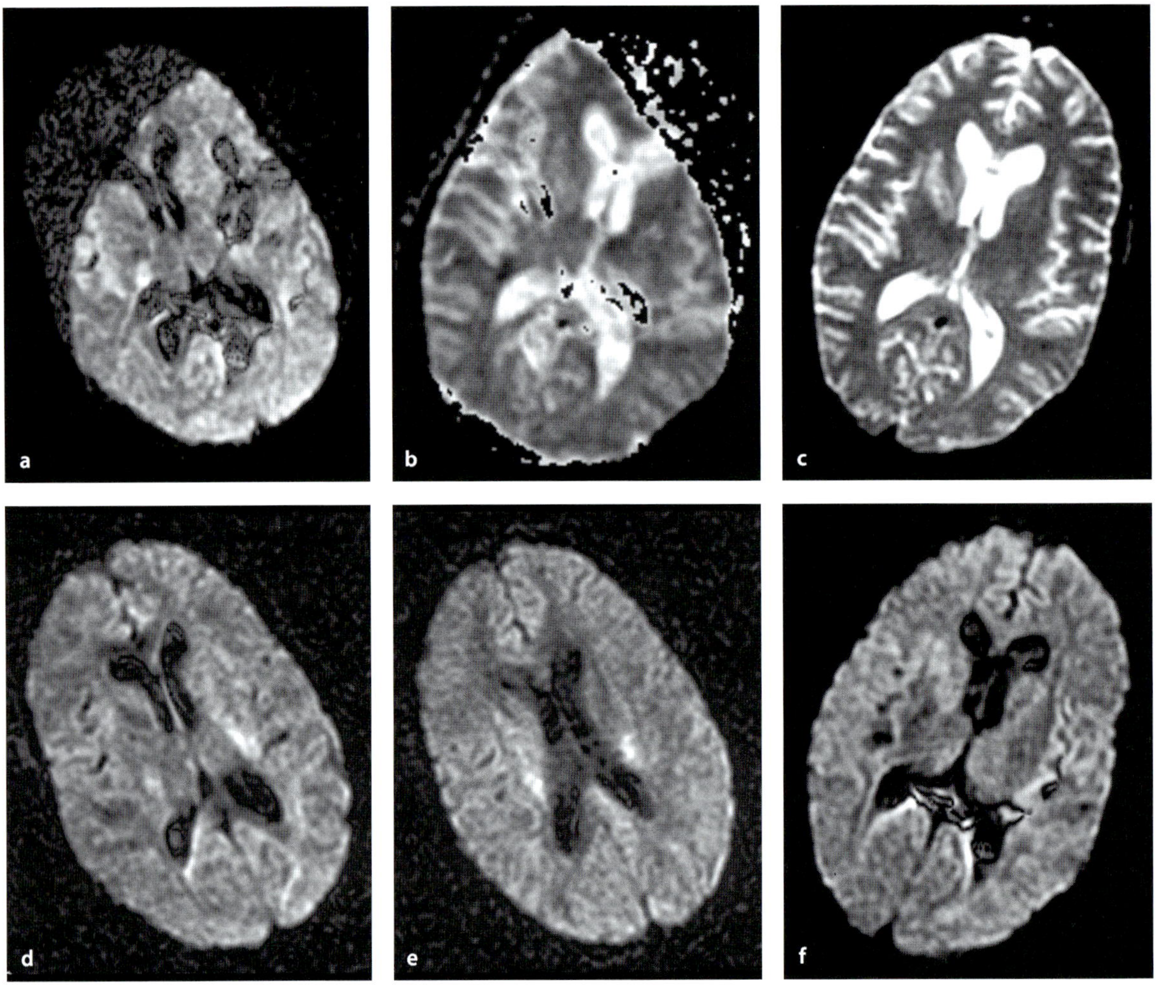

Figure 3.12 a–f

Motion artifacts due to head motion between the scans. Chronic infarcts in the right basal ganglia. **a** DW image has motion artifacts due to head motion between the scans. This image appears overlapping of b_0, and *x, y, z* axis images. **b** ADC map also shows severe motion artifacts. **c** b_0, **d** *x* axis, **e** *y* axis, **f** *z* axis. b_0 and unidirectional images are all free from the artifacts

3.5 Conclusion

Diffusion-weighted images are inherently T2 weighted and the interpretation of signal intensity on DW images requires a correlation between b_0 images, ADC maps and exponential images to uncover the underlying pathophysiologic condition. It is also important to understand a variety of artifacts to avoid misinterpreting the DW images. Understanding inherent artifacts and the way to reduce the artifacts on DW imaging will improve the quality and accuracy of DW imaging.

References

1. Stejkal EO, Tanner J (1965) Spin diffusion measurements: spin echoes in the presence of a time-dependent field gradient. J Chemical Phys 42:288–292
2. LeBihan D, Breton E, Lallemand D, Grenier P, Cabanis E, Laval-Jeantet (1986) MR imaging of intravoxel incoherent motions: application to diffusion and perfusion in neurologic disorders. Radiology 161:401–407
3. Warach S, Gaa J, Siewert B, Wielopolski P, Edelman RR (1995) Acute human stroke studied by whole brain echo planar diffusion-weighted magnetic resonance imaging. Ann Neurol 37:231–241
4. Provenzale JM, Engelter ST, Petrella JR, Smith JS, MacFall JR (1999) Use of MR exponential diffusion-weighted images to eradicate T2 "shine-through" effect. AJR Am J Roentgenol 172:537–539
5. Burdette JH, Elster AD, Ricci PE (1999) Acute cerebral infarction: quantification of spin-density and T2 shine-through phenomena on diffusion-weighted MR images. Radiology 212:333–339
6. Coley SC, Porter DA, Calamante F, Chong WK, Connelly A (1999) Quantitative MR diffusion mapping and cyclosporine-induced neurotoxicity. AJNR Am J Neuroradiol 20:1507–1510
7. Schaefer PW, Grant PE, Gonzalez RG (2000) Diffusion-weighted MR imaging of the brain. Radiology 217:331–345
8. Field A (2001) Diffusion and perfusion imaging. In: Elster AD, Burdette JH (eds). Questions and answers in magnetic resonance imaging. Mosby, St. Louis, Missouri pp 194–214
9. Maldjian JA, Listerud J, Moonis G, Siddiqi F (2001) Computing diffusion rates in T2-dark hematomas and areas of low T2 signal. AJNR Am J Neuroradiol 22:112–128
10. Engelter ST, Provenzale JM, Petrella JR, Alberts MJ, DeLong DM, MacFall JR (2001) Use of exponential diffusion imaging to determine the age of ischemic infarcts. J Neuroimaging 11:141–147
11. Chen S, Ikawa F, Kurisu K, Arita K, Takaba J, Kanou Y (2001) Quantitative MR evaluation of intracranial epidermoid tumors by fast fluid-attenuated inversion recovery imaging and echo-planar diffusion-weighted imaging. AJNR Am J Neuroradiol 22:1089–1096
12. Geijer B, Sundgren PC, Lindgren A, Brockstedt S, Stahlberg F, Holtas S (2001) The value of b required to avoid T2 shine-through from old lacunar infarcts in diffusion-weighted imaging. Neuroradiology 43:511–517
13. Casey S (2001) "T2 washout": an explanation for normal diffusion-weighted images despite abnormal apparent diffusion coefficient maps. AJNR Am J Neuroradiol 22:1450–1451
14. Provenzale JM, Petrella JR, Cruz LC Jr, Wong JC, Engelter S, Barboriak DP (2001) Quantitative assessment of diffusion abnormalities in posterior reversible encephalopathy syndrome. AJNR Am J Neuroradiol 22:1455–1461
15. Eastwood JD, Engelter ST, MacFall JF, Delong DM, Provenzale JM (2003) Quantitative assessment of the time course of infarct signal intensity on diffusion-weighted images. AJNR Am J Neuroradiol 24:680–687
16. Hiwatashi A, Kinoshita T, Moritani T, et al. (2003) Hypointensity on diffusion-weighted MRI related to T2 shortening and susceptibility effects. AJR Am J Roentgenol (in press)
17. Elster AD, Burdette JH (2001) Scanner hardware. In: Elster AD, Burdette JH (eds). Questions and answers in magnetic resonance imaging. Mosby, St. Louis, Missouri pp 54–71
18. Haselgrove JC, Moore JR (1996) Correction for distortion of echo-planar images used to calculate the apparent diffusion coefficient. Magn Reson Med 36:960–964
19. Horsfield MA (1999) Mapping eddy current induced fields for the correction of diffusion-weighted echo planar images. Magn Reson Imaging 17:1335–1345
20. Jezzard P, Barnett AS, Pierpaoli C (1998) Characterization of and correction for eddy current artifacts in echo planar diffusion imaging. Magn Reson Med 39:801–812
21. Calamante F, Porter DA, Gadian DG, Connelly A (1999) Correction for eddy current induced Bo shifts in diffusion-weighted echo-planar imaging. Magn Reson Med 41:95–102
22. Schmithorst VJ, Dardzinski BJ (2002) Automatic gradient preemphasis adjustment: a 15-minute journey to improved diffusion-weighted echo-planar imaging. Magn Reson Med 47:208–212
23. Papadakis NG, Martin KM, Pickard JD, Hall LD, Carpenter TA, Huang CL (2000) Gradient preemphasis calibration in diffusion-weighted echo-planar imaging. Magn Reson Med 44:616–624
24. Chapman BL (1999) Shielded gradients. And the general solution to the near field problem of electromagnet design. MAGMA 9:146–151
25. Robson MD, Anderson AW, Gore JC (1997) Diffusion-weighted multiple shot echo planar imaging of humans without navigation. Magn Reson Med 38:82–88
26. Brockstedt S, Moore JR, Thomsen C, Holtas S, Stahlberg F (2000) High-resolution diffusion imaging using phase-corrected segmented echo-planar imaging. Magn Reson Imaging 18:649–657
27. Maier SE, Gudbjartsson H, Patz S, et al. (1998) Line scan diffusion imaging: characterization in healthy subjects and stroke patients. AJR Am J Roentgenol 171:85–93
28. Robertson RL, Maier SE, Mulkern RV, Vajapayam S, Robson CD, Barnes PD (2000) MR line-scan diffusion imaging of the spinal cord in children. AJNR Am J Neuroradiol 21:1344–1348

29. Lovblad KO, Jakob PM, Chen Q, et al. (1998) Turbo spin-echo diffusion-weighted MR of ischemic stroke. Am J Neuroradiol 19:201–208
30. Alsop DC (1997) Phase insensitive preparation of single-shot RARE: application to diffusion imaging in humans. Magn Reson Med 38:527–533
31. Forbes KP, Pipe JG, Karis JP, Farthing V, Heiserman JE (2003) Brain imaging in the unsedated pediatric patient: comparison of periodically rotated overlapping parallel lines with enhanced reconstruction and single-shot fast spin-echo sequences. AJNR Am J Neuroradiol 24:794–798
32. Bammer R, Keeling SL, Augustin M, et al. (2001) Improved diffusion-weighted single-shot echo-planar imaging (EPI) in stroke using sensitivity encoding (SENSE). Magn Reson Med 46:548–554
33. Porter DA, Calamante F, Gadian DG, Connelly A (1999) The effect of residual Nyquist ghost in quantitative echo-planar diffusion imaging. Magn Reson Med 42:385–392
34. Bastin ME (2001) On the use of the FLAIR technique to improve the correction of eddy current induced artefacts in MR diffusion tensor imaging. Magn Reson Imaging 19:937–950
35. Anderson AW, Gore JC (1994) Analysis and correction of motion artifacts in diffusion weighted imaging. Magn Reson Med 32:379–387
36. Ordidge RJ, Helpern JA, Qing ZX, Knight RA, Nagesh V (1994) Correction of motional artifacts in diffusion-weighted MR images using navigator echoes. Magn Reson Imaging 12:455–460
37. Dietrich O, Heiland S, Benner T, Sartor K (2000) Reducing motion artefacts in diffusion-weighted MRI of the brain: efficacy of navigator echo correction and pulse triggering. Neuroradiology 42:85–91
38. Pipe JG (1999) Motion correction with PROPELLER MRI: application to head motion and free-breathing cardiac imaging. Magn Reson Med 42:963–969

Brain Edema

4.1 Characterization and Classification of Brain Edema

Brain edema is defined as accumulation of excess fluid in cells or in the extracellular space. Brain edema can be classified as cytotoxic (cellular), vasogenic [1] or interstitial. Cytotoxic and vasogenic edema usually coexist in pathological conditions such as infarction, hypoxic ischemic encephalopathy, trauma, or multiple sclerosis. The edema may primarily be either vasogenic or cytotoxic, but as the process evolves over time, the injury leads to a combination of cellular swelling and vascular damage. Interstitial edema occurs with hydrocephalus, water intoxication, or plasma hyposmolarity.

Conventional MR imaging does not always allow distinction between the different forms of edema. However, diffusion-weighted (DW) imaging, which is based on the microscopic movement of water molecules in brain tissue, can differentiate cytotoxic edema from vasogenic and interstitial edema [2].

4.2 Definition and Classification of Cytotoxic Edema

Cytotoxic or cellular edema is an abnormal uptake of fluid in the cytoplasm due to abnormal cellular osmoregulation. This kind of edema may accompany various processes that damage cells, such as ischemia, trauma, toxic metabolic disease, demyelination, and even the early phase of degeneration. Classification of the involved cell types may explain the pathophysiology and different prognosis of these conditions.

In normal brain tissues, the gray and white matters are mainly composed of neurons, glial cells, axons, and myelin sheaths (Fig. 4.1). In the gray matter, cytotoxic edema occurs mainly in neurons and glial cells (Fig. 4.2). In the white matter, however, cytotoxic edema occurs in glial cells, axons (axonal swelling) (Fig. 4.3) and myelin sheaths (intramyelinic edema) (Fig. 4.4) [1].

4.3 Pathophysiology of Cytotoxic Edema

4.3.1 Energy Failure

In ischemia or hypoxia, cytotoxic edema is mainly caused by energy failure [3]. The insult initiates substrate depletion, which leads to a decrease in intracellular ATP used for oxidative phosphorylation, and a failure of the sodium–potassium pump. This will cause an influx of sodium and calcium into the cells, subsequently increasing the osmotic gradient and the transport of water into the cells, resulting in cellular swelling. Moreover, in an attempt to produce ATP, the cells switch from oxidative phosphorylation to anaerobic glycolysis, resulting in intracellular lactate. This will further increase the osmotic gradient across the cell membrane, which exacerbates the cytotoxic edema.

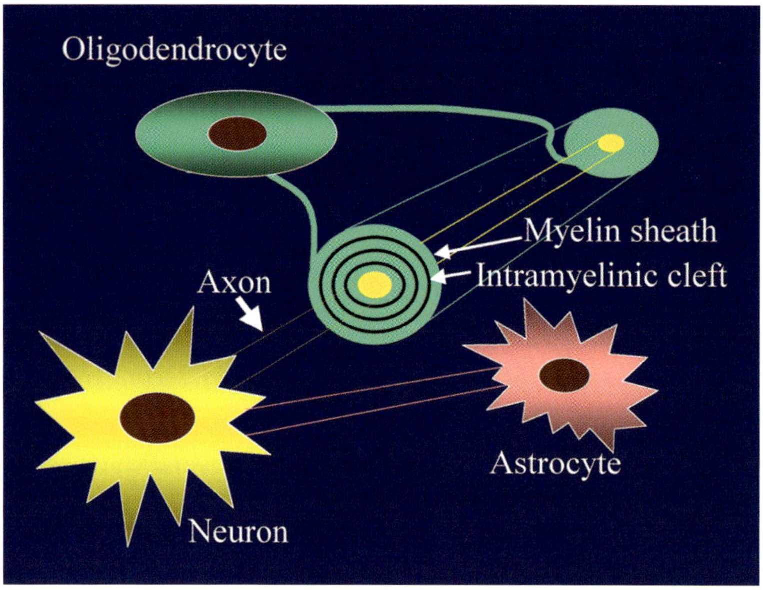

Figure 4.1

Fig. 4.1. Normal brain tissue is mainly composed of neurons, glial cells (astrocytes or oligodendrocytes), axons and myelin sheaths surrounded by an extracellular space

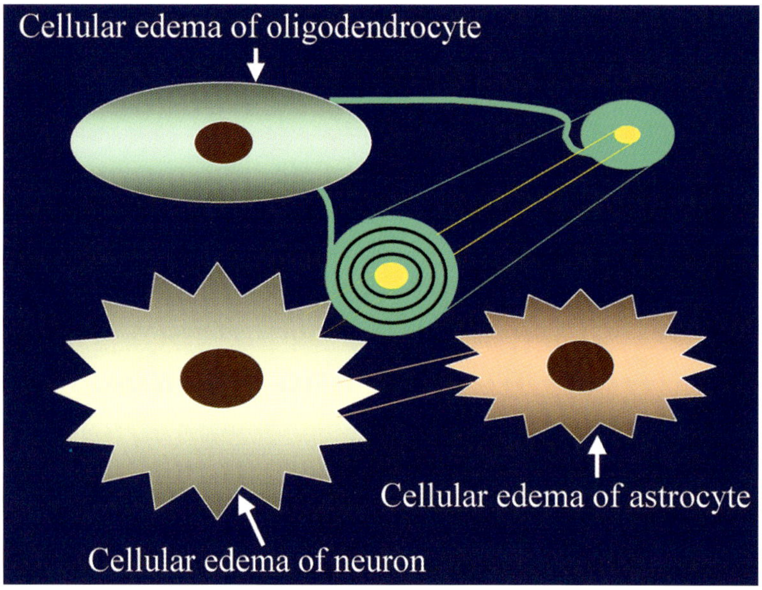

Figure 4.2

Cytotoxic edema occurs in neurons and glial cells (astrocytes and oligodendrocytes). These cells are vulnerable to ischemia. As cells increase in size, there is a shift of water from extracellular to intracellular compartments, which can occur without a net gain in water (compared with Fig. 4.1). Cytotoxic edema results in increased intracellular space and decreased extracellular space, which may cause a decrease in ADC

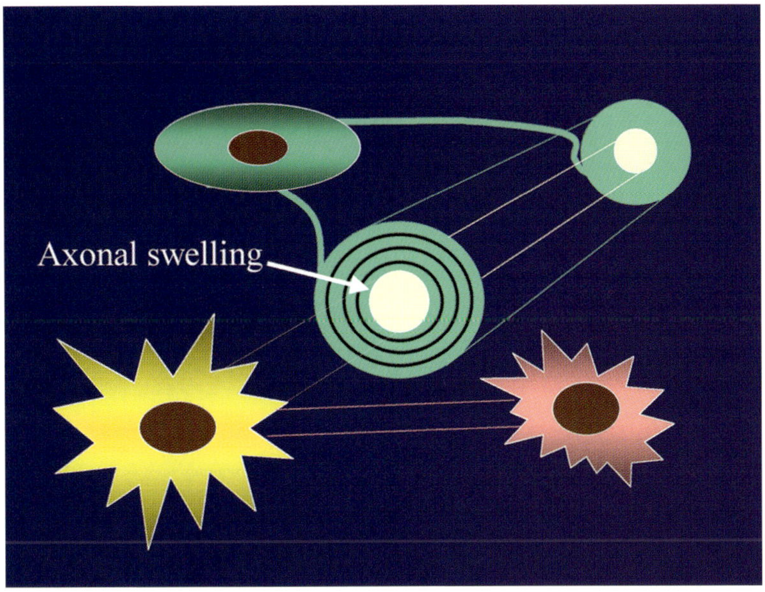

Figure 4.3

Cytotoxic edema can occur in axons

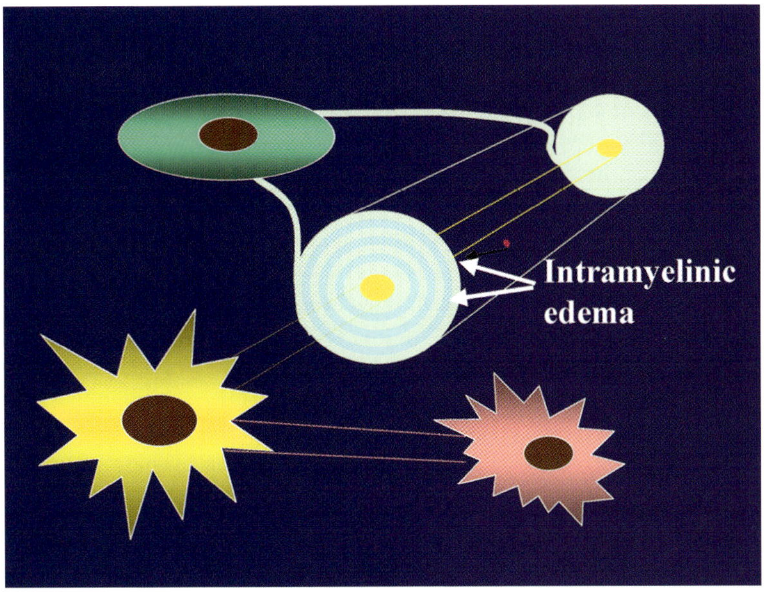

Figure 4.4

Cytotoxic edema can occur in myelin sheaths in which edema is found in either the myelin sheath itself or in the intramyelinic cleft

4.3.2 Excitotoxic Brain Injury

Energy failure is not the only mechanism responsible for the cytotoxic edema [3]. Membrane transporters can be triggered or inhibited by a range of excitatory neurotransmitters, such as glutamate and aspartate, but also other agents such as cytokines and free radicals [4]. Any cell, including neuron, glia, axon and myelin sheath can be a target of these toxic substances; however, reactive astrocytes play a significant role in cellular and tissue repair by detoxifying various noxious substances (such as glutamate, free radical, ammonia and metals). Neuropathologic examination shows that the acutely reactive astrocytes have swollen cytoplasm and neutrophil, consistent with cytotoxic edema [5].

High glutamate in the synaptic and extracellular space is one of the important mechanisms associated

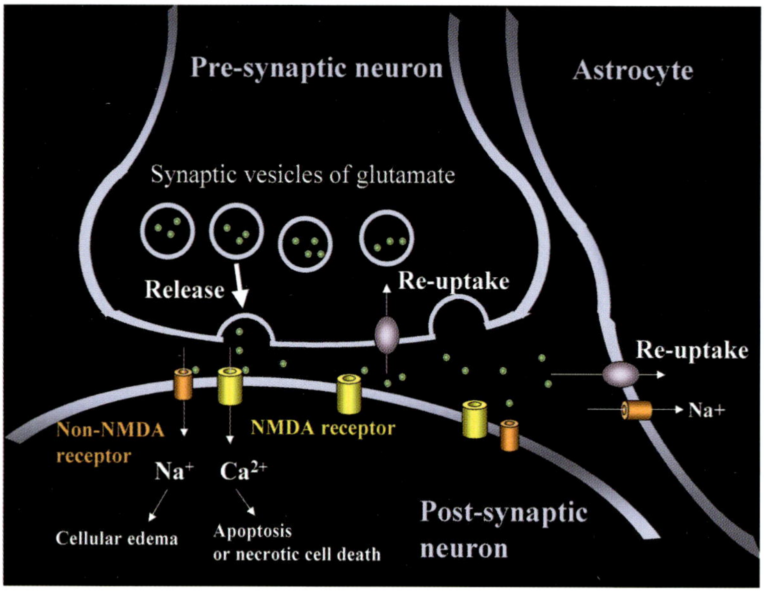

Figure 4.5 Excitotoxic mechanisms in the neuron and astrocyte. In the neuron, glutamate is released from the pre-synaptic terminal into the synaptic cleft. The glutamate binding to NMDA receptors allows entry of Ca^{2+} into the post-synaptic neuron, which can result in necrotic cell death or apoptosis. The glutamate binding to non-NMDA receptors allows entry of Na^{+} into the post-synaptic neuron, resulting in cytotoxic edema of the neuron. Re-uptake of extracellular glutamate takes place at the pre-synaptic terminals and in adjacent astrocytes. Similar mechanisms also cause cytotoxic edema in the astrocyte

with cytotoxic edema of various diseases, including infarction, hypoxic ischemic encephalopathy, status epilepticus, and traumatic brain injury, such as diffuse axonal injury, contusion and shaken baby syndrome [6, 7]. Increased extracellular glutamate is a direct cause of excitotoxic brain injury. In acute excitotoxic injury, increased extracellular glutamate results from an increased release/ leakage of glutamate or a decreased re-uptake (Fig. 4.5). Neuronal glutamate is released from the pre-synaptic terminal into the synaptic cleft. The glutamate binding to N-methyl-D-aspartate (NMDA) receptors allows entry of Ca^{2+} into the post-synaptic neuron, which results in necrotic cell death or apoptosis. The glutamate binding to non-NMDA receptors allows entry of Na^{+} into the post-synaptic neuron, resulting in cytotoxic edema. Apoptosis is defined as a programmed cell death and is histologically characterized by fragmentation of DNA in the nucleus of the cell. Re-uptake of extracellular glutamate takes place at the pre-synaptic terminals and in adjacent glial cells, which may cause cytotoxic edema (acute phase of reactive astrocytosis).

Whatever the cause, cytotoxic edema can result in necrosis or delayed neuronal death, or degeneration similar to apoptosis with various amounts of reactive gliosis. Whether necrosis or apoptosis ensues, it may neurochemically depend on the levels of adenosine triphosphate or cytosol calcium ions that trigger protease and lipase production [8].

4.4 Diffusion-Weighted Imaging and Cytotoxic Edema

Cytotoxic edema characteristically shows hyperintensity on DW images associated with decreased apparent diffusion coefficient (ADC). The precise mechanisms underlying the reduction in ADC are unknown. The most common explanation is a shift of extracellular water to the intracellular space. However, the observed 40% reduction of ADC cannot be explained by an increase in intracellular water alone, even if all extracellular fluid went intracellular [9]. There must be a reduction in diffusivity of water molecules in the intracellular space which may be explained by the large number of intracellular organelles, that may act as obstacles for diffusion. A decrease in intracellular ADC could also be due to a decrease in the energy-dependent intracellular circulation or an increase in cytoplasmic viscosity from a swelling of intracellular organelles [10].

Tumors, hemorrhages, abscesses and coagulative necrosis also result in a decrease in ADC. The mechanisms underlying the reduction in ADC in those lesions are also unknown, but can be related to hypercellularity or hyperviscosity of the pathological tissue [11, 12].

4.4.1 Conditions that Cause Cytotoxic Edema, and Reversibility

Cytotoxic edema of neuron and glial cells may accompany infarction [13–16], hypoxic ischemic encephalopathy [17, 18], traumatic brain injury [19, 20], status epilepticus [6, 21, 22], encephalitis [23] and Creutzfeldt–Jakob disease [24, 25].

Neurons and glial cells are the cells most vulnerable to ischemia and hypoxia, but if the ischemia is severe, myelin sheaths and axons may also be involved [3]. These differences among cell types for cytotoxic edema can explain the different time courses of DW abnormalities between gray and white matter in cerebral infarction and hypoxic ischemic encephalopathy. In arterial infarction, the area of cytotoxic edema

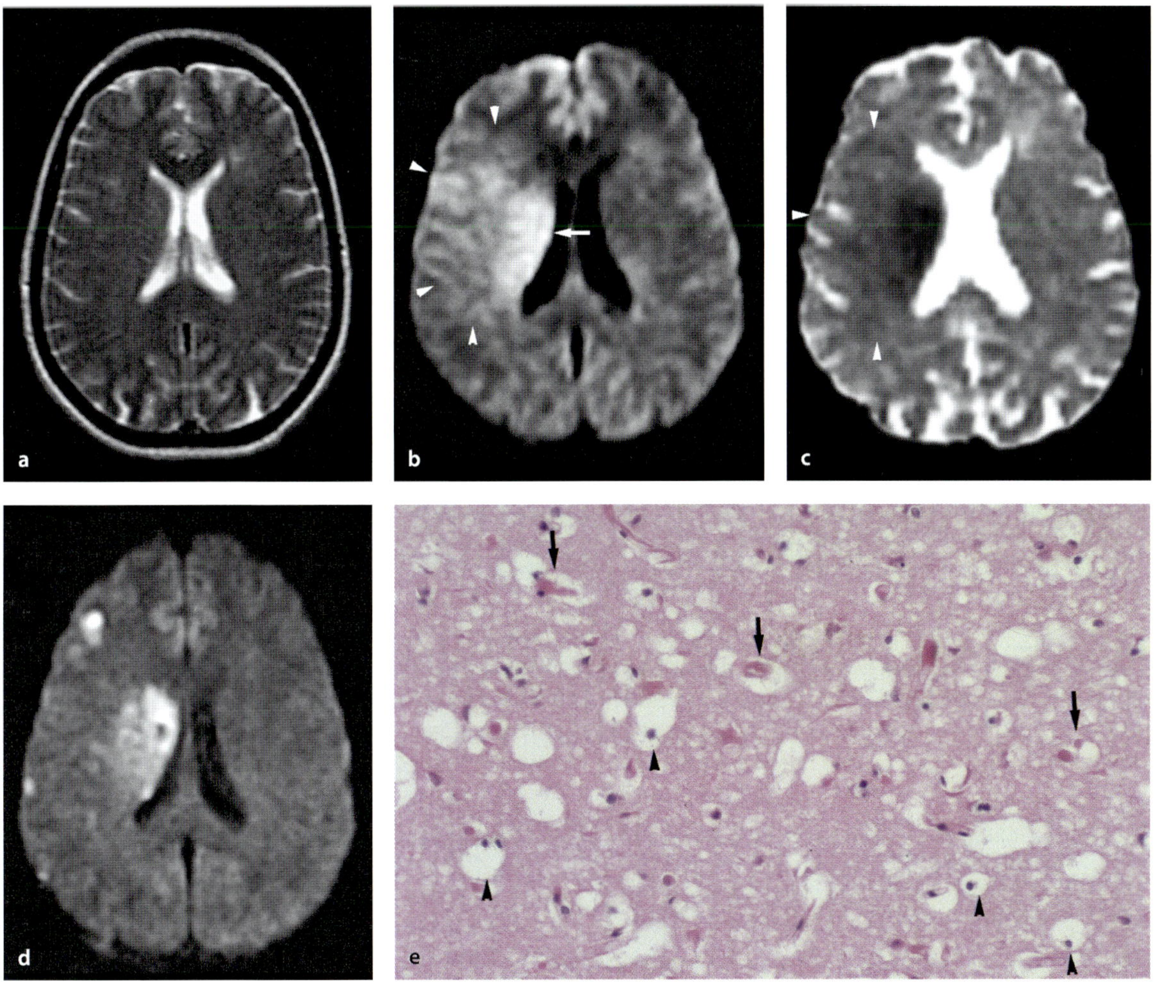

Figure 4.6 a–e

Hyperacute cerebral infarction (3 h after onset) in a 39-year-old woman with decreased consciousness. Her neurologic functions improved after intra-arterial thrombolytic therapy. **a** T2-weighted image appears normal. **b** DW image shows a hyperintense lesion in the right corona radiata (*arrow*) and a slightly hyperintense lesion in the right middle cerebral artery (MCA) territory, which may correspond to ischemic penumbra (*arrowheads*). **c** ADC map shows a definite decrease in ADC in the corona radiata and slightly decreased ADC in the right MCA territory (*arrowheads*). **d** On DW image after fibrinolytic therapy (3 days after onset), the hyperintense lesion in the cortical area is largely resolved, with remaining small, peripheral infarcts. Early cytotoxic edema with slightly decreased ADC does not always result in infarction. **e** Another case. Pathological specimen of cytotoxic edema in the cortex in an acute stroke shows swelling of neurons (*arrows*) and glial cells (*arrowheads*) (hematoxylin–eosin stain, original magnification ×200). (From [36])

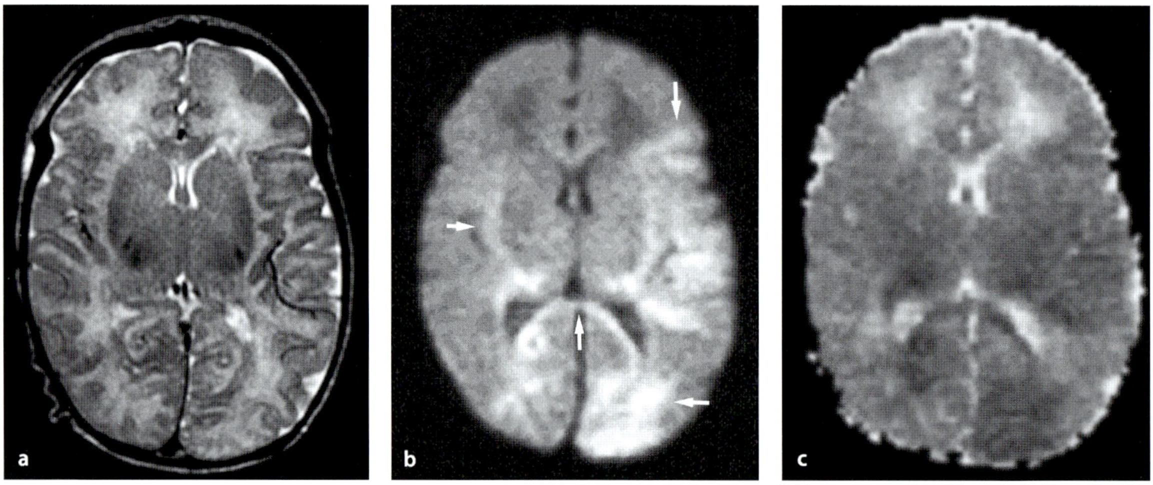

Figure 4.7 a–d

A 2-day-old term girl with hypoxic ischemic encephalopathy due to perinatal hypoxia–ischemia event. **a** T2-weighted image appears normal. **b** DW image shows hyperintense lesions in the temporo-occipital gray and white matter including the corpus callosum (*arrows*). Low intensity in bilateral frontal deep white matter (*arrowheads*) is a normal finding in a patient of this age. **c** ADC map shows these lesions as decreased ADC representing cytotoxic edema. Increased ADC in the frontal deep white matter is also a normal finding in a patient of this age. These ischemic lesions are more clearly seen on DW imaging than on the ADC map because DW imaging depicts subtle T2 contrast abnormalities (T2 shine-through effect) in addition to the contrast of diffusion restriction of these lesions

on DW imaging seems to be irreversibly damaged tissue, resulting in coagulative or liquefactive necrosis. However, mild decreased ADC in the ischemic penumbra can be reversible after intra-arterial or intravenous fibrinolytic therapy (Fig. 4.6). In transient ischemic attacks and venous infarctions, an initially abnormal signal on DW imaging has occasionally been reversed, partially or completely, on follow-up MR images. Hypoxic ischemic encephalopathy (Fig. 4.7) and traumatic brain injury are usually related to irreversible brain damage.

In status epilepticus, cytotoxic edema is often reversed, partially or completely, but may result in selective necrosis, brain atrophy or gliosis (Fig. 4.8). A cytotoxic edema of reactive astrocytes in the acute phase can be responsible for the reversible signal abnormalities [6].

In Creutzfeldt–Jakob disease, the area of cytotoxic edema will eventually develop into prominent brain atrophy (Fig. 4.9). Axonal swelling can also accompany diffuse axonal injury (Fig. 4.10) and the early phase of wallerian degeneration (Fig. 4.11) [26].

Intramyelinic edema may accompany the acute phase of multiple sclerosis (Fig. 4.12) [27, 28], toxic or metabolic leukoencephalopathy (Fig. 4.13) [29–31] and osmotic myelinolysis [32]. Partially or completely reversible lesions are also seen in these diseases. The explanation for this reversibility may be an intramyelinic edema where the edema is often primarily located in the intramyelinic cleft [1].

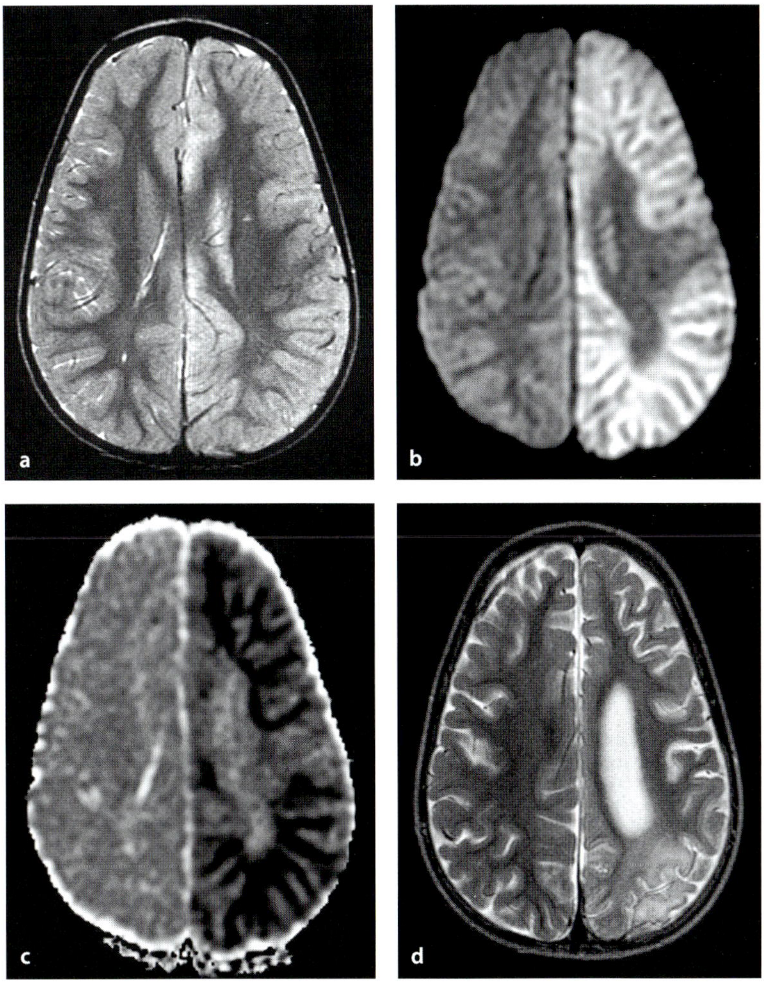

Figure 4.8 a–d

Status epilepticus in a 2-year-old girl 24 hours after onset. **a** T2-weighted image shows diffuse cortical hyperintensity in the entire left hemisphere cortex. **b** DW image shows diffuse hyperintensity mainly in the gray matter of the left hemisphere. **c** ADC map shows decreased ADC of these lesions. **d** Diffuse brain atrophy and hyperintense lesions in the left hemisphere are seen on a 5-month follow-up T2-weighted image

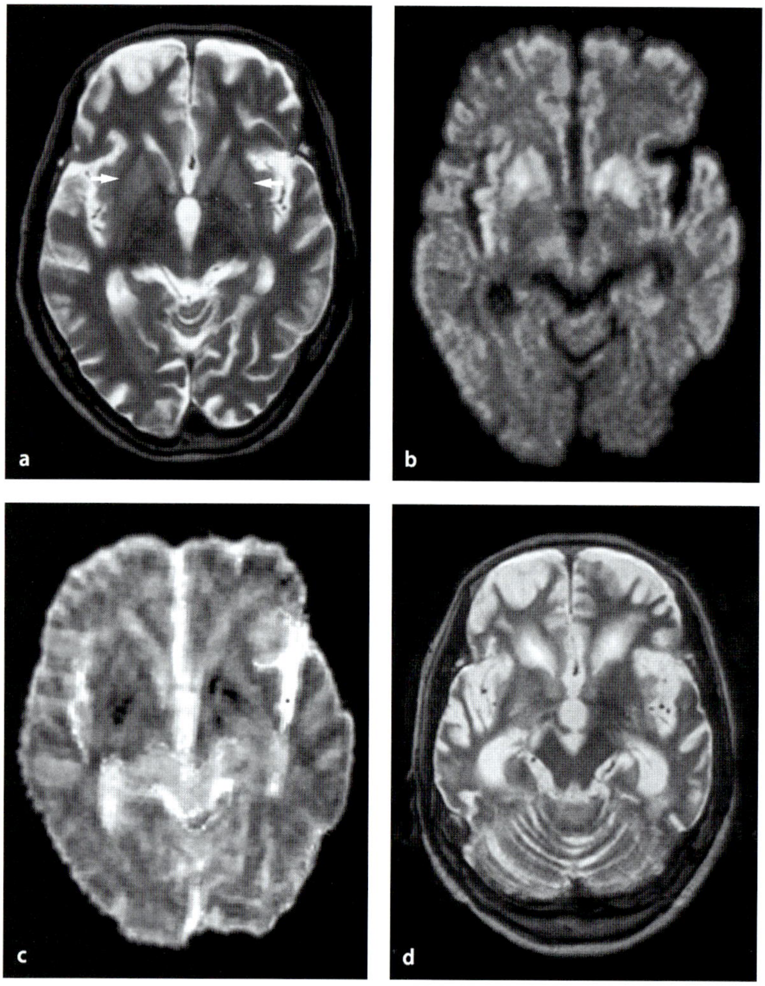

Figure 4.9 a–d

Creutzfeldt–Jakob disease in a 72-year-old woman with progressive dementia. **a** T2-weighted image demonstrates mildly increased signal bilaterally in the caudate nuclei and putamina (*arrows*). **b** DW image clearly demonstrates bilateral, symmetrical increase in signal intensity in the caudate nuclei and putamina. **c** ADC map shows these lesions as decreased ADC. **d** 4-month follow-up MRI shows prominent brain atrophy. (From [37])

Figure 4.10 a–c

Diffuse axonal injury in an 18-year-old female 48 h after motor vehicle accident. **a** T2-weighted image shows mildly hyperintense lesions in the corpus callosum and the white matter of bilateral frontal lobes (*arrows*). **b** DW image demonstrates diffuse axonal injury as high signal intensity, representing cytotoxic edema (*arrows*). **c** ADC map shows decreased ADC lesions in the anterior to posterior corpus callosum and the frontal deep white matter (*arrows*). (From [37])

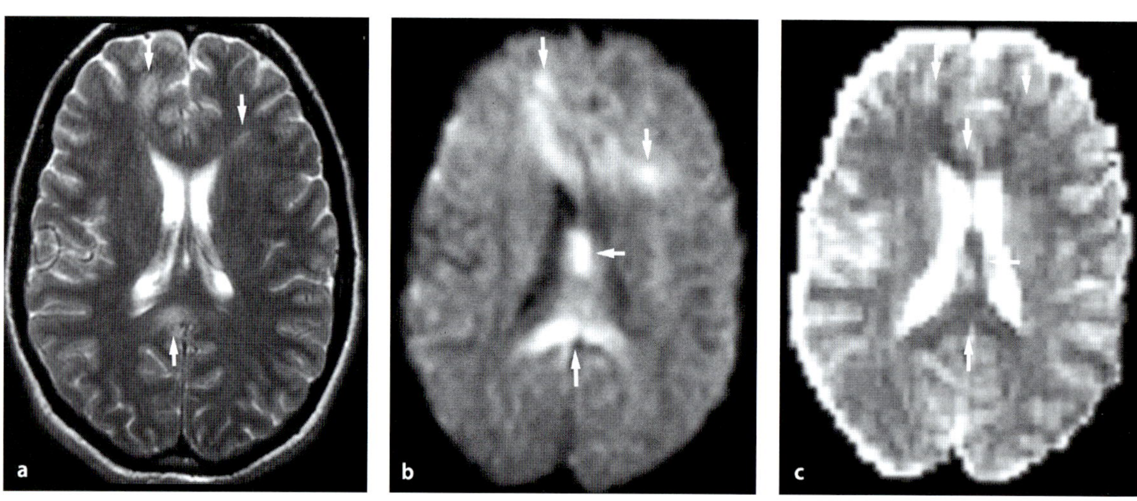

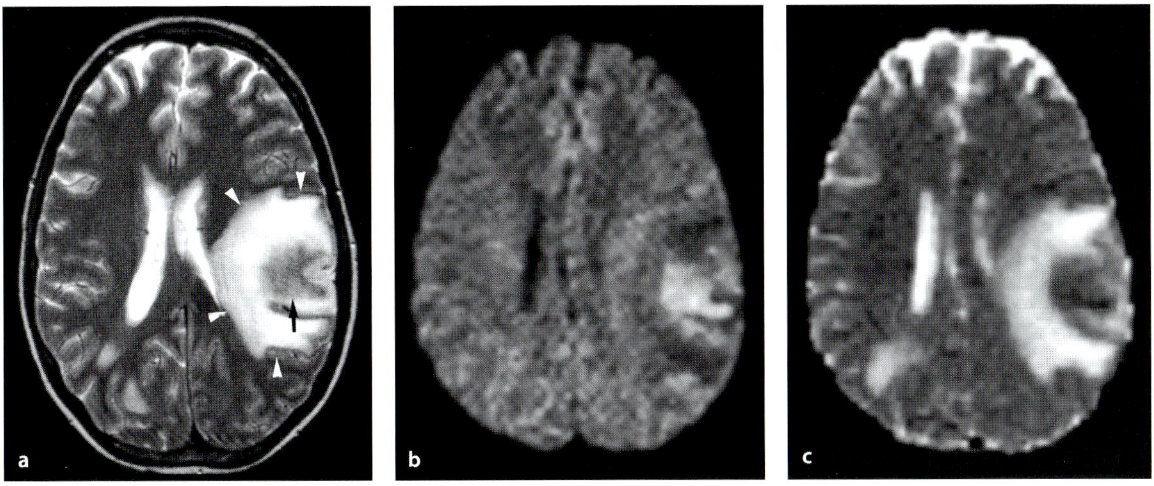

Figure 4.16 a–c

Cerebral toxoplasmosis and vasogenic edema in an 18-year-old woman with headache. **a** T2-weighted image shows central necrosis as slightly hyperintense (*arrow*) and peripheral vasogenic edema as very hyperintense in the left hemisphere (*arrowheads*). Multiple lesions of toxoplasmosis are also seen in the right occipital and left periventricular areas. **b** DW image reveals vasogenic edema as hypointense, while the central necrosis shows hyperintensity on DW image. **c** ADC map shows increased ADC from the vasogenic edema. Decreased ADC of the central necrosis is probably due to hyperviscosity of the coagulative necrosis

4.6 Conclusion

4.6.1 Cytotoxic or Cellular Edema

Cytotoxic or cellular edema is hyperintense on DW images and associated with decreased ADC. It can occur in neurons, glial cells, axons (axonal swelling) and myelin sheaths (intramyelinic edema). Cytotoxic edema may be present not only in infarction/ischemia and trauma, but also in status epilepticus, the acute phase of multiple sclerosis, toxic or metabolic leukoencephalopathy, osmotic myelinolysis, encephalitis, and presumably in the early phase of transneuronal or wallerian degeneration and Creutzfeldt–Jakob disease. The differential diagnosis for hyperintense DW images also includes tumor, abscess and hemorrhage, conditions that also may have decreased ADC. The decreased ADC in these latter conditions may be due to hypercellularity and/or hyperviscosity rather than the cytotoxic edema.

4.6.2 Vasogenic Edema

Vasogenic edema has a variable appearance on DW images, with increased ADC. It is reversible but occasionally associated with cytotoxic edema, which usually is not reversible. DW images and ADC maps are useful for understanding MR images of various diseases with cytotoxic and/or vasogenic edema. These images are more sensitive than conventional MRI to determine the extent of edema in both gray and white matter.

References

1. Milhorat TH (1992) Classification of the cerebral edemas with reference to hydrocephalus and pseudotumor cerebri. Child's Nerv Syst 8:301–306
2. Ebisu T, Naruse S, Horikawa Y, et al. (1993) Discrimination between different types of white matter edema with diffusion-weighted MR imaging. J Magn Res Imaging 3:863–868
3. Ironside JW, Pickard JD (2002) Raised intracranial pressure, oedema and hydrocephalus. In: Graham DI, Lantos PL (eds) Greenfield's neuropathology, 7th edn, pp 193–231

4. Lipton SA, Rosenberg PA (1994) Excitatory amino acids as a final common pathway for neurologic disorders. N Engl J Med 330:613–622
5. Chan S, et al. (1996) Reversible signal abnormalities in the hippocampus and neocortex after prolonged seizures. AJNR Am J Neuroradiol 17:1725–1731
6. Mark LP, Prost RW, Ulmer JL, et al. (2001) Pictorial review of glutamate excitotoxicity: fundamental concepts for neuroimaging. AJNR Am J Neuroradiol 22:1813–1824
7. Moritani T, Shrier D, Wang H, et al. (2002) Excitotoxic mechanism in pediatric brain. Neurographics Vol. 2, Article 1, http://foundation.asnr.org/neurographics/
8. Sharp FR, Swanson RA, Honkaniemi J, et al. (1998) Neurochemistry and molecular biology. In: Barnett HJM, Mohr JP, Stein BM, et al. (eds) Stroke pathophysiology, diagnosis, and management, pp 54–56
9. Duong TO, Ackerman JJH, Ying HS, et al. (1998) Evaluation of extra- and intracellular apparent diffusion in normal and globally ischemic rat brain via ^{19}F NMR. Magn Reson Med 40:1–13
10. van der Toorn A, Sykova EDRM, Vorisek I, et al. (1996) Dynamic changes in water ADC, energy metabolism, extracellular space volume, and tortuosity in neonatal rat brain during global ischemia. Magn Reson Med 36:52–56
11. Tien RD, Felsberg GJ, Friedman H, et al. (1993) MR imaging of high-grade cerebral gliomas: value of diffusion weighted echoplanar pulse sequence. AJR Am J Roentgenol 162:671–677
12. Desprechins B, Stadnik T, Koerts G, et al. (1999) Use of diffusion-weighted MR imaging in differential diagnosis between intracerebral necrotic tumors and cerebral abscesses. AJNR Am J Neuroradiol 20:1252–1257
13. Desmond PM, Lovell AC, Rawlinson AA, et al. (2001) The value of apparent diffusion coefficient maps in early cerebral ischemia. AJNR Am J Neuroradiol 22:1260–1267
14. Burdette JH, Ricci PE, Petitti N, et al. (1998) Cerebral infarction: Time course of signal intensity changes on diffusion-weighted MR images. AJR Am J Roentgenol 171: 791–795
15. Kamal AK, Segal AZ, Ulug AM (2002) Quantitative diffusion-weighted MR imaging in transient ischemic attacks. AJNR Am J Neuroradiol 23:1533–1538
16. Forbes KP, Pipe JG, Heiserman JE (2001) Evidence for cytotoxic edema in the pathogenesis of cerebral venous infarction. AJNR Am J Neuroradiol 22:450–455
17. Arbelaez A, Castillo M, Mukheri SK (1999) Diffusion-weighted MR imaging of global cerebral anoxia. AJNR Am J Neuroradiol 20:999–1007
18. Wolf RL, Zimmerman RA, Clancy R, et al. (2001) Quantitative apparent diffusion coefficient measurements in term neonates for early detection of hypoxic-ischemic brain injury: initial experience. Radiology 218:825–833
19. Barzo P, Marmarou A, Fatouros P, et al. (1997) Contribution of vasogenic and cellular edema to traumatic brain swelling measured by diffusion-weighted imaging. J Neurosurg 87:900–907
20. Liu AY, Maldjian JA, Bagley LJ, Sinson GP, et al. (1999) Traumatic brain injury: diffusion-weighted MR imaging findings. AJNR Am J Neuroradiol 20:1636–1641
21. Kim JA, Chung JI, Yoon PH, et al. (2001) Transient MR signal changes in patients with generalized tonicoclonic seizure or status epilepticus: periictal diffusion-weighted imaging. AJNR Am J Neuroradiol 22:1149–1160
22. Men S, Lee DH, Barron JR, et al. (2000) Selective neuronal necrosis associated with status epilepticus; MR findings. AJNR Am J Neuroradiol 21:1837–1840
23. Tsuchiya K, Katase S, Yoshino A, et al. (1999) Diffusion-weighted MR imaging of encephalitis. AJR Am J Roentgenol 173:1097–1099
24. Demaerel P, Baert AL, Vanopdenbosch L, et al. (1997) Diffusion-weighted magnetic resonance imaging in Creutzfeldt–Jakob disease. Lancet 349:847–848
25. Bahn MM, Parchi P. (1999) Abnormal diffusion-weighted magnetic resonance images in Creutzfeldt–Jakob disease. Arch Neurol 56:577–583
26. Castillo M, Mukheriji SK (1999) Early abnormalities related to postinfarction wallerian degeneration: evaluation with MR diffusion-weighted imaging. JCAT 23:1004–1007
27. Verity MA (1997) Toxic disorders. In: Graham DI, Lantos PL (eds) Greenfield's neuropathology, 6th edn, pp 755–811
28. Tievsky AL, Ptak T, Farkas J (1999) Investigation of apparent diffusion coefficient and diffusion tenser anisotropy in acute and chronic multiple sclerosis lesion. AJNR Am J Neuroradiol 20:1491–1499
29. Matsumoto S, Nishizawa S, Murakami S, et al. (1995) Carmofur-induced leukoencephalopathy: MRI. Neuroradiology 37:649–652
30. Phillips MD, McGraw P, Lowe MJ, et al. (2001) Diffusion-weighted imaging of white matter abnormalities in patients with phenylketonuria. AJNR Am J Neuroradiol 22:1583–1586
31. Sener RN (2002) Metachromatic leukodystrophy: diffusion MR imaging findings. AJNR Am J Neuroradiol 23:1424–1426
32. Cramer SC, Stegbauer KC, Schneider A, et al. (2001) Decreased diffusion in central pontine myelinolysis. AJNR Am J Neuroradiol 22:1476–1479
33. Gonatas NK, Zimmerman HM, Levine S (1963) Ultrastructure of inflammation with edema in the rat brain. Am J Pathol 42:455–469
34. Mukherjee P, McKinstry RC (2001) Reversible posterior leukoencephalopathy syndrome: evaluation with diffusion-tensor MR imaging. Radiology 219:756–765
35. Keller E, Flancke S, Urbach H, et al. (1999) Diffusion- and perfusion-weighted magnetic resonance imaging in deep cerebral venous thrombosis. Stroke 30:1144–1146
36. Moritani T (2002) Classification of brain edema. In Koredewakaru Diffusion MRI, Tokyo: Shujunsha pp 128–137
37. Moritani T, Shrier DA, (2000) Numaguchi Y, et al. Diffusion-weighted echo-planar MR imaging: clinical applications and pitfalls – a pictorial essay. Clin Imaging 24: 181–92.

Infarction

In collaboration with R. de Guzman

5.1 Clinical Significance and Therapeutic Considerations for Brain Infarcts

Stroke is the third leading cause of death in the USA, and cerebral infarction is the most common cause of disability among adult Americans. Until recently these patients were mainly imaged with computed tomography (CT) to establish if the cause of stroke was ischemic or hemorrhagic. Treatment was above all aimed to reduce the risk for further embolic events. In a few instances intra-arterial or intravenous thrombolysis has been instituted, but in most cases this is not feasible because of the narrow therapeutic window. Thrombolytic treatment has a risk of hemorrhagic complications, which is why it has become important to establish the potential benefit of thrombolysis for the individual patient. CT as well as conventional MR imaging have sensitivities below 50% with regard to detection of infarcts in the hyperacute stage, within 6 hours.

5.1.1 Stroke Mimickers

There is a long list of conditions that mimic the symptoms of an acute ischemic stroke. The most common ones include intracranial hemorrhage, migraines, seizures, functional and metabolic disorders, and also vasogenic edema syndromes. It is important to visualize and verify that an ischemic lesion is indeed the cause of the clinical symptoms before therapy is initiated, as these non-ischemic stroke mimickers should not be treated with thrombolysis and such therapy could actually be harmful. Moreover, in older patients it is not uncommon to detect older lesions with prolonged T2 that are indistinguishable from acute lesions using conventional MR imaging.

5.1.2 Diffusion-Weighted Imaging

In recent years, diffusion-weighted (DW) imaging has been proven as the most sensitive MR imaging technique to diagnose hyperacute cerebral infarction. The detection of acute ischemic lesions is based on alterations in motion of water molecules. It is a very sensitive technique, which is not significantly affected by patient motion. DW imaging of the brain can usually be accomplished in less than 2 minutes.

The ischemic event results in restricted diffusion of the affected tissue, which can be seen as early as 30 minutes after ictus. A few rare cases of false-negative DW imaging have been reported [1, 2]. These infarcts were seen on perfusion-weighted images and later on DW imaging [1, 2].

5.2 Diffusion-Weighted Imaging and Pathophysiology of Cerebral Infarction

The abnormal imaging finding of cerebral and cerebellar infarctions is an area of hyperintensity on DW imaging of the involved vascular territory. This hyperintensity is presumed to be caused by cytotoxic edema as a result of cessation of ATP production. Under normal circumstances, ATP maintains the Na^+/K^+ pump activity and other intracellular energy-related processes. When the Na^+/K^+ pump is not functioning properly, an inability to remove excess water from the cells develops, resulting in intracellular edema. The outcome of this on DW imaging is restriction of water diffusion, which results in a signal increase on DW imaging and a decrease in diffusion shown as a reduced apparent diffusion coefficient (ADC) [3]. These findings in acute stroke usually represent irreversible damage of brain tissue, or infarction [4].

5.3 Apparent Diffusion Coefficient

The ADC is used to determine whether the signal abnormality on DW images is caused by restricted diffusion or a T2 shine-through effect, as seen in subacute–chronic infarctions. ADC represents the degree of diffusibility of water molecules and aids in detecting subtle fluid changes in the hyperacute–acute stages of ischemic stroke. Reduced diffusion is seen as an area of low signal intensity on ADC maps.

5.3.1 Explanation for Restricted Diffusion

Several mechanisms have been proposed to explain the restricted diffusion in ischemia. These include cellular necrosis, shift of fluid from extra- to intracellular spaces causing a reduction in size and increase in tortuosity of the extracellular space, but there is also rather strong evidence that at least parts of these findings relate to a reduction in intracellular diffusion [5].

Regions of decreased or restricted diffusion are best seen on DW imaging, while ADC maps will verify the findings by eliminating the T2 shine-through effect as a cause of the increased signal intensity on DW imaging. DW imaging and ADC can also show changes in diffusion that vary for the different stages of a stroke [6], and they can possibly distinguish between multiple strokes over time versus a single, progressive stroke by determining the time course of a cerebral infarction.

Apparent diffusion coefficient values may also be of help in the future to assist in selecting patients with salvageable tissue within an ischemic penumbra for thrombolysis. Intermediate ADC values are noted in the ischemic penumbra, indicating tissue at risk of infarction [7]. An approach that is used more often to select patients who may benefit from thrombolysis is by comparing DW imaging and perfusion MR imaging to look for hypoperfused but not diffusion-restricted regions. The mismatch between DW imaging and perfusion demonstrates affected tissue that is still salvageable and not yet infarcted: the penumbra.

5.4 Time Course of Infarction

Infarctions may be classified as hyperacute (less than 6 hours from time of onset of symptoms), acute (6 hours to 3 days), subacute (3 days to 3 weeks), or chronic (3 weeks to 3 months), each having its characteristic signal abnormalities (Table 5.1).

5.4.1 Hyperacute (< 6 Hours)

One of the main clinical applications of DW imaging is to detect a hyperacute cerebral infarction. This information is critical, particularly in cases of territorial thromboembolic infarction, as thrombolytic therapy can be started within the golden period of 3 hours from onset of symptoms. Such treatment can result in early reperfusion and reduce the extension of the infarction [3, 8]. The DW signal intensity is increased during the hyperacute stage (<6 hours), with corresponding low ADC, manifested as a dark area on the ADC map (Fig. 5.1).

Table 5.1. Time course of thromboembolic infarction of the middle cerebral artery [7]

	< 6 hours	3 days	7 days	30 days
T2	Isointense	Bright	Bright	Bright
DW imaging	Bright	Very bright	Bright	Isointense
ADC	Dark	Very dark	Dark	Bright

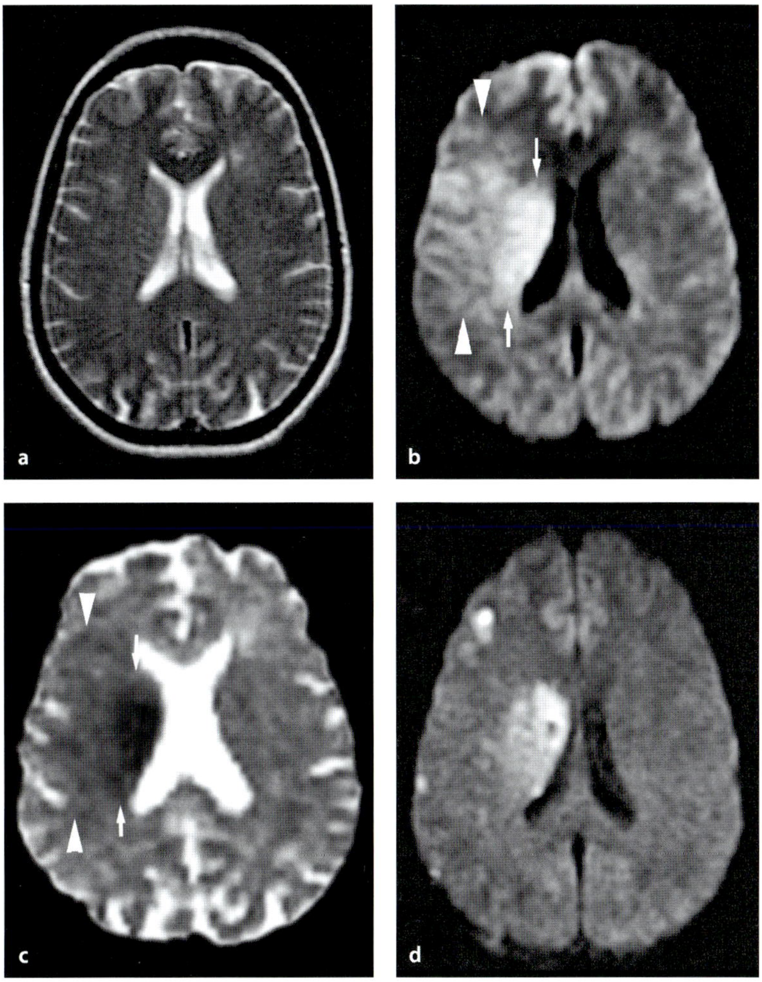

Figure 5.1 a–d

Hyperacute infarction (2 hours after onset) in a 39-year-old woman with decreased consciousness. The symptoms improved after intra-arterial fibrinolytic therapy. **a** T2-weighted image appears normal. **b** DW image shows a hyperintense lesion in the right corona radiata (*arrows*) and a slightly hyperintense lesion in the right middle cerebral artery (MCA) territory (*arrowheads*). **c** ADC map shows decreased ADC in the corona radiata (*arrows*) and slightly decreased ADC in the cortical area of the MCA territory (*arrowheads*). **d** On DW image after fibrinolytic therapy (3 days after onset), the hyperintense lesion in the cortical area mostly resolved with peripheral small infarcts. Early cytotoxic edema with mild decreased ADC does not always result in infarction after treatment

5.4.2 Acute (6 Hours to 3 Days)

Almost all acute (6 hours to 3 days) stroke patients examined within 24 hours of onset of symptoms show abnormal signal intensity on DW imaging [9]. At this stage the infarctions show a further increase in DW signal intensity and also a lower ADC than in the hyperacute stage (Fig. 5.2).

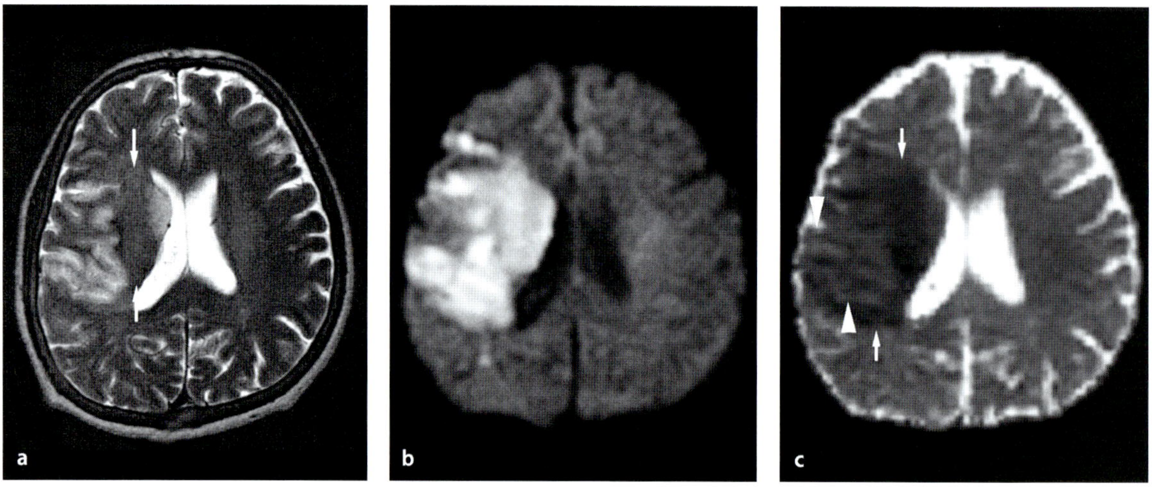

Figure 5.2 a–c

Acute infarction (24 hours after onset) in a 56-year-old man with left hemiparesis. **a** T2-weighted image shows hyperintense lesions preferentially involving the right posterior frontal cortex and the right caudate region, sparing the right corona radiata (*arrows*). This finding is consistent with a relatively greater involvement of gray matter in the early infarction. **b** DW image shows the entire right MCA territory as hyperintense. **c** Decreased ADC is seen in the right MCA territory (*arrows*). However, some cortical lesions seem to be isointense or have a slightly increased ADC (*arrowheads*). This may reflect relative vulnerability for brain tissue. Hyperintensity on DW image of these cortical lesions is due to a T2 shine-through effect. DW images and ADC maps are more sensitive than conventional MRI for showing both gray and white matter involvement. ADC maps precisely reflect diffusion restrictions of the lesion within the gray and white matter

5.4.3 Subacute (3 Days to 3 Weeks)

As the infarct continues to evolve into the subacute stage (3 days to 3 weeks), there is pseudo-normalization of the ADC, most likely attributed to a combination of (a) persistence of cytotoxic edema, and (b) development of vasogenic edema and cell membrane disruption, which results in increased amounts of extracellular water. The hyperintensity on DW imaging usually decreases within 1–2 weeks [10], but is still slightly hyperintense, while ADC is usually normalized within 10 days [11]. This time gap is thought to result from T2 shine-through effects on DW imaging in the late subacute infarction (Fig. 5.3).

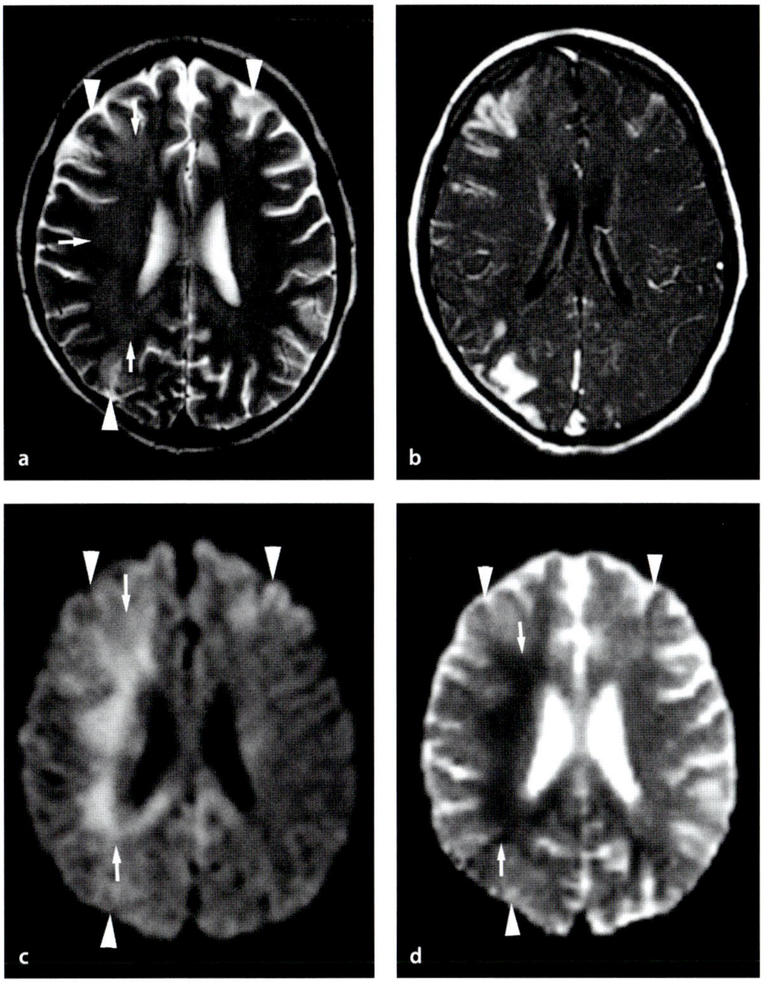

Figure 5.3 a–d

Subacute infarction (10 days after onset) in a 19-year-old woman with loss of consciousness due to cerebral embolism after cardiac surgery for endocarditis. **a** T2-weighted image shows hyperintense lesions in the gray (*arrowheads*) and white matter (*arrows*) in the right hemisphere and left frontal region. **b** Gadolinium T1-weighted image with magnetization transfer contrast shows gyral enhancement in the cortical lesions, representing subacute infarcts. **c** DW image also shows hyperintense lesions in the right deep white matter (*arrows*), and gray matter of both frontal and right parieto-occipital regions (*arrowheads*). **d** The ADC map shows decreased ADC in the right deep white matter lesion (*arrows*), and normal or slightly increased ADC in the gray matter lesions (*arrowheads*). The prolonged decreased ADC in the white matter may reflect edema of myelin sheaths or axons

5.4.4 Chronic (3 Weeks to 3 Months)

In the chronic stage (3 weeks to 3 months) of infarction, there is a more or less complete necrosis of the cells and at this stage there is an increase in ADC with a bright signal on the ADC map. On T2-weighted imaging the infarction is seen as a bright signal and this, in combination with the increased ADC, will result in a decrease in the signal on DW imaging; the infarction is isointense with surrounding tissue (Fig. 5.4).

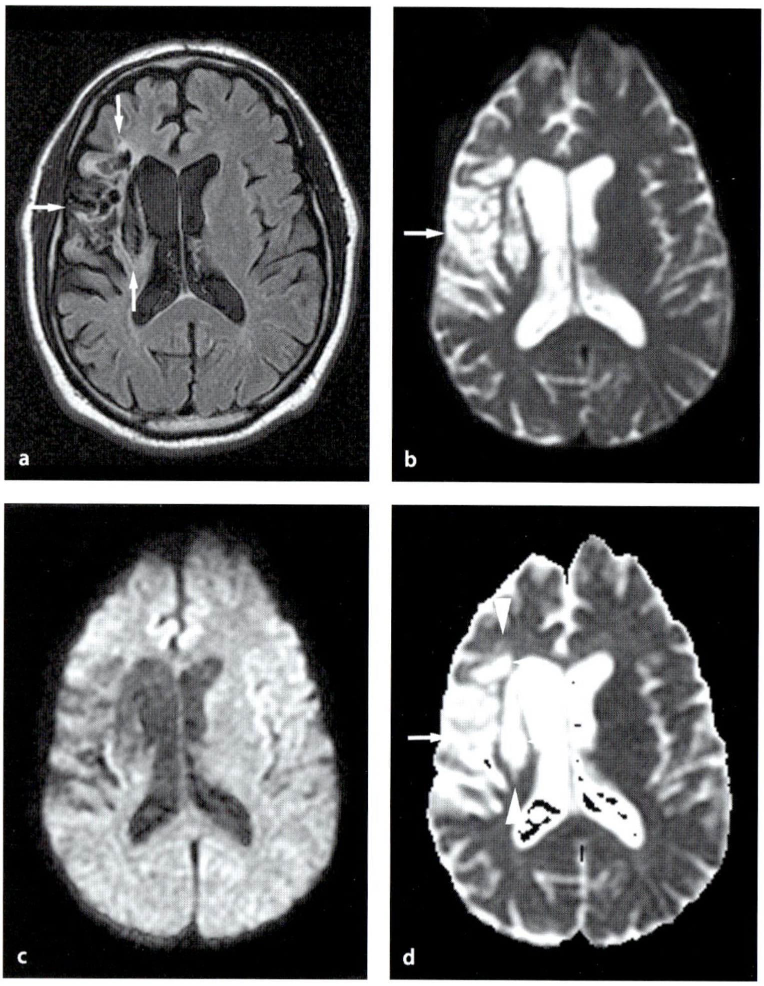

Figure 5.4 a–d

Chronic infarction (10 months after onset) in a 54-year-old man with numbness and weakness of the left lower extremity. **a** Fluid-attenuated inversion-recovery (FLAIR) image shows chronic infarction in the right MCA territory as a cystic lesion with low signal intensity (cystic necrosis) and peripheral mild hyperintensity (gliosis) with atrophy (*arrows*). **b** b_0 image shows the hyperintense cystic lesion (*arrow*). **c** DW image shows chronic infarction as hypointense cystic areas, and iso- or slightly hyperintense areas representing gliosis. **d** The ADC map shows marked increased ADC in the cystic necrosis (*arrow*), and slightly increased ADC in the gliotic periphery of the lesion (*arrowheads*)

5.5 Diffusion-Weighted Imaging and ADC Characteristics of Gray and White Matter Ischemia

Diffusion-weighted imaging and ADC maps are more sensitive than conventional MR imaging in demonstrating both gray and white matter ischemia (Figs. 5.1, 5.2 and 5.3). Changes in ADC values in acute infarctions seem to be different for gray matter and white matter. Thus, there is a more prominent decrease in ADC in white matter than in gray matter. This decrease also remains for a longer period than in gray matter (Figs. 5.2 and 5.3). One of the explanations for these phenomena is that necrosis may be completed earlier in gray matter infarctions than in white matter infarctions. Another explanation is that the prominent and prolonged decrease in ADC in white matter may reflect cytotoxic edema in different cell types, such as myelin sheaths, axons and glial cells [5, 6, 11].

5.5.1 Relative ADC

The time course of relative ADC is slightly different in gray matter when compared with the relative ADC in white matter [6]. Fiebach et al. observed a decrease in the relative ADC up to 3 days after the stroke and an increase in relative ADC from the third to the tenth day. The relative ADC increased slightly faster in gray matter than in white, which may be due to the variability between these two tissue types at any stage in the ischemic process, which leads to an altered diffusion. The observed diffusion contrast in gray and white matter could be caused by differences in the mismatch between blood supply and metabolic demand, the type and/or severity of the histopathologic response to ischemic injury (vulnerability) or mechanisms by which histopathologic changes lead to altered diffusion [12]. Regarding the histopathologic response, gray matter has traditionally been considered to be more vulnerable than white matter to early ischemia. More recent findings in experimental models of stroke have demonstrated that ischemic damage to white matter occurs earlier and with greater severity than previously appreciated [13]. However, if this is true for humans as well is to our knowledge, not yet established.

5.6 Reversibility and Treatment

Reversible ADC is rare but can be found in cases of transient ischemic attack in which imaging was performed within 4 hours, venous infarction, hemiplegic migraine and transient global amnesia. In these rare clinical settings, ischemia does not progress to complete necrosis but a minor subclinical, irreversible injury cannot be ruled out [5].

Clinically, the area of cytotoxic edema with bright DW signal seems to be irreversibly damaged resulting in permanent infarction. In early cerebral ischemia, mildly decreased ADC in the ischemic penumbra is indicative of viable tissue, but hypoperfused tissue at risk of infarction [14]. After intra-arterial or intravenous fibrinolytic therapy, or spontaneous lysis of a clot, abnormal signal in such areas is occasionally reversed, partially or completely (Figs. 5.4 and 5.5).

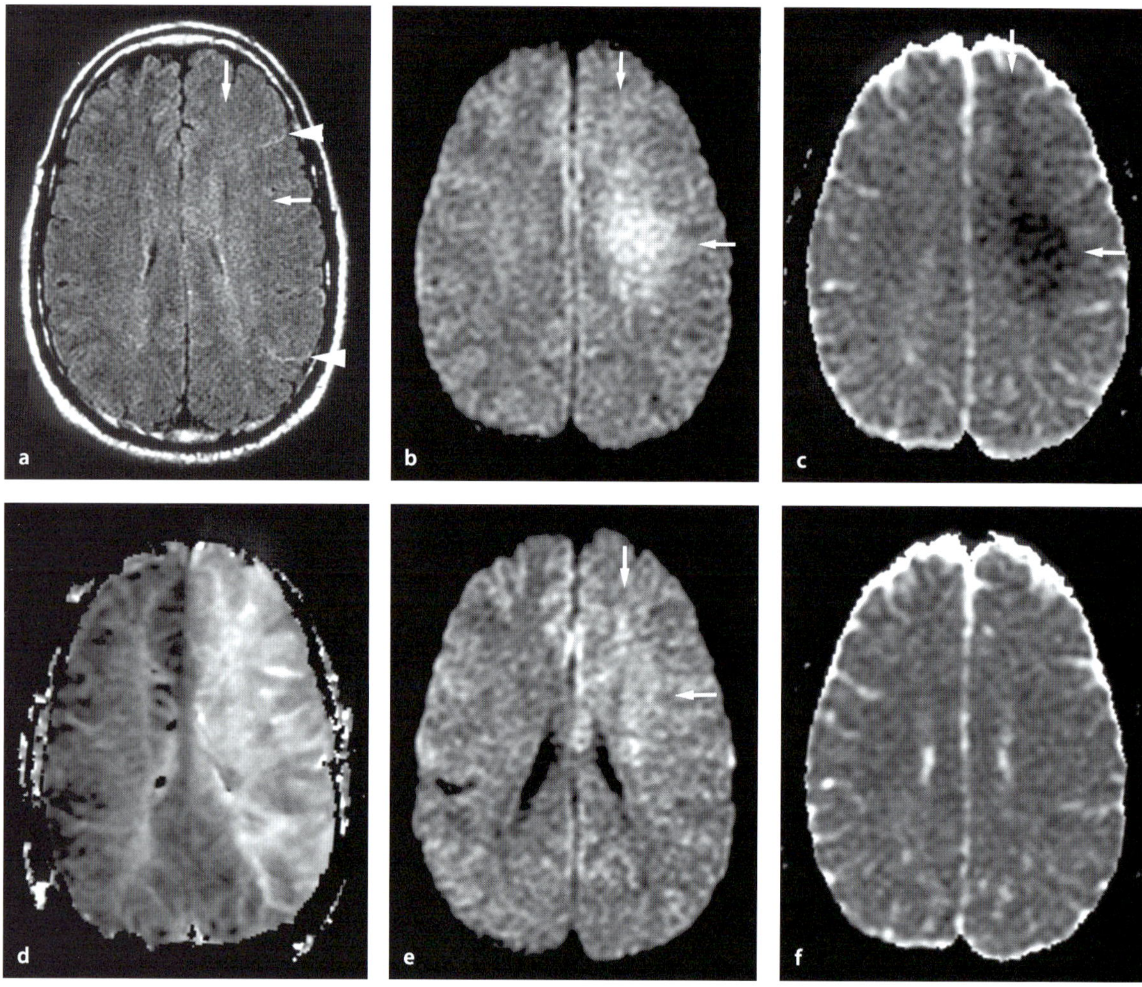

Figure 5.5 a–f

Reversible ischemia with cytotoxic edema (2 h from onset) in a 39-year-old man with left internal carotid artery dissection, presenting with right-sided weakness. **a** FLAIR image shows a subtle hyperintensity in the left frontoparietal white matter (*arrows*), and linear hyperintensity representing slow flow in the peripheral arteries (*arrowheads*). **b, c** DW image (**b**) shows a hyperintense lesion with decreased ADC (**c**) in the left frontoparietal white matter, representing cytotoxic edema (*arrows*). **d** Perfusion-weighted image shows increase in mean transit time of the entire left anterior and MCA territories. **e** Follow-up DW image 2 days later shows only a very subtle hyperintensity in the left frontal white matter (*arrows*). **f** ADC was normalized, which is in accordance with clinical improvement. Early ischemia with cytotoxic edema may have spontaneously resolved

5.7 Watershed Infarction

Watershed infarction may develop between two major vascular territories or within a single territory in the supraganglionic white matter, a border zone of the superficial and deep penetrating arterioles (Fig. 5.6). As mentioned above, thrombolytic therapy within the first 3 hours from acute onset of symptoms can be effective to limit the size of the infarct under those circumstances. This is, however, not the case in watershed infarctions, as the basic etiology for these lesions is a significant reduction in perfusion secondary to an overall decrease in cerebral blood flow with subsequent poor perfusion pressure distally [7].

There is a difference in the evolution time of ADC between watershed and thromboembolic infarction, the latter having an earlier normalization (Table 5.2). However, T2 signal intensity is the same for both types of infarction. The reason for this difference most likely lies in the different pathophysiologic features and cerebral perfusion of the two stroke subtypes. It is important to note that strokes with different pathogenetic, hemodynamic mechanisms may have different evolution in the ADC courses as well [7].

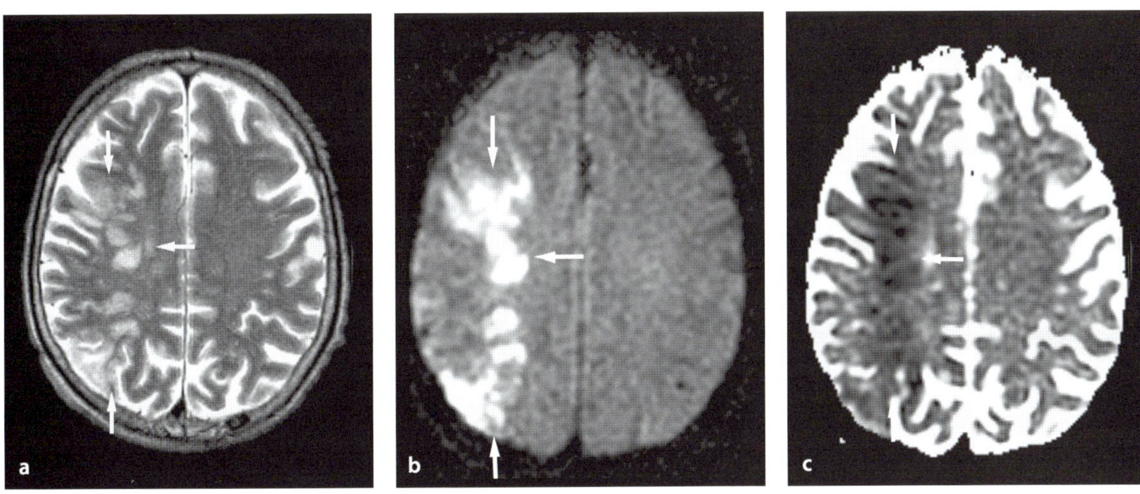

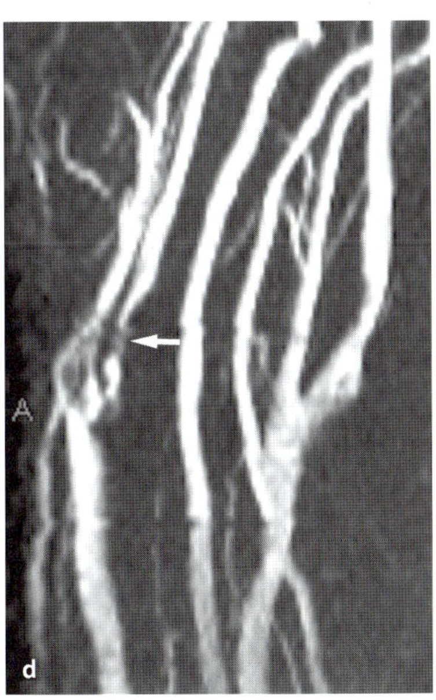

Figure 5.6 a–d

Watershed infarction. 66-year old man presented with stroke and seizure. **a** T2-weighted image shows multiple hyperintense lesions in the right frontoparietal area (arrows). **b, c** DW image shows these lesions as very hyperintense with decreased ADC, representing acute infarcts in the watershed area between anterior and middle cerebral arteries (arrows). **d** MRA shows bilateral stenoses of internal carotid, and middle and anterior cerebral arteries (arrows).

Table 5.2. Time course of watershed infarction of middle cerebral arterial territory [7]

	3 days	7 days	14 days	30 days
T2	Bright	Bright	Bright	Bright
DW imaging	Bright	Bright	Bright	Bright
ADC	Dark	Dark	Dark	Less dark

5.8 Perfusion Versus Diffusion Imaging

Perfusion MR imaging may be more sensitive than DW imaging in the detection of a hyperacute cerebral infarction, but it currently entails extensive post-processing to create interpretable perfusion maps (Figs. 5.5 and 5.6). Moreover, MR perfusion determines the degree of blood flow reduction at the level of the cerebral microvasculature, but it will not tell if a hypoperfused area represents an area of infarction or severe hypoperfusion. Perfusion MR can, however, be matched with the infarcted area on DW images and can demonstrate the area of hypoperfusion outside the infarction – the so-called penumbra. This is the area where neural tissue is at risk for infarction if perfusion is not re-established and ischemic penumbra is assumed to be salvageable by means of thrombolysis.

5.9 Venous Infarction

Cerebral venous sinus thrombosis accounts for only a small percentage of cerebral infarctions in general. Because of its non-specific presentation, cerebral venous sinus thrombosis can be difficult to diagnose. In about 50% of cases, cerebral venous sinus thrombosis results in cerebral venous infarction. This usually presents as a hemorrhagic infarction or focal edema in regions that are not typical for an arterial vascular distribution, usually occurring within the white matter or at the gray–white matter junction (Fig. 5.7).

5.9.1 Predisposing Factors

There are several predisposing factors for thrombus formation within the cerebral venous sinuses. These include pregnancy, infection, extrinsic compression or local invasion by tumor, dehydration, oral contraceptives, hypercoagulable state, trauma, drug abuse. It may also be idiopathic. Thrombus initially forms within the venous sinuses, eventually extending to the veins draining into the sinuses, leading to infarction.

5.9.2 Pathophysiology

The pathophysiological mechanisms that lead to cerebral venous infarction are unclear. It has been postulated that: (1) retrograde venous pressure may cause breakdown of the blood–brain barrier, with leakage of fluid (vasogenic edema) and hemorrhage into the extracellular space or, (2) retrograde venous pressure may cause a decrease in cerebral blood flow, causing tissue damage similar to that seen in arterial infarctions. Restricted water diffusion suggesting cytotoxic edema is found in patients with acute cerebral venous infarction [2]. However, in cases with only vasogenic edema, ADC is increased with variable DW signals (Fig. 5.7). Preferred imaging modalities when suspecting cerebral venous sinus thrombosis are conventional MR imaging combined with MR venography.

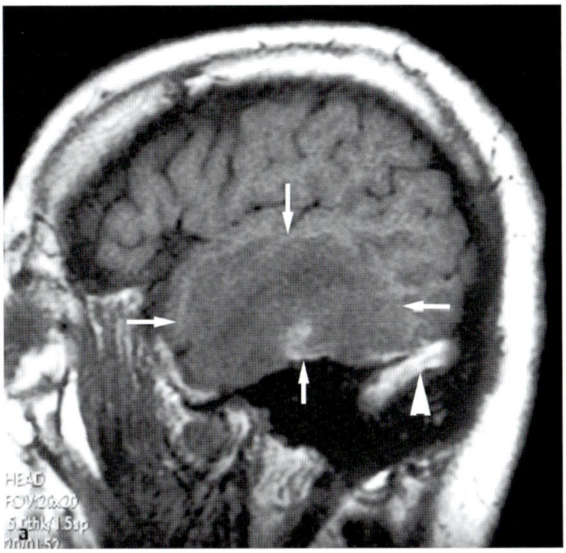

Figure 5.7 a–d

Venous infarction in a 57-year-old man with dysarthria. **a** Sagittal T1-weighted image shows a large area of hyperintensity in the left temporal lobe (arrows) with a small area (hemorrhage). The hyperintensity in the left transverse sinus represents sinus thrombosis (arrowhead). **b** T2-weighted image shows a hyperintense lesion in the left temporal lobe (*arrow*). **c** DW image reveals this lesion as mildly hyperintense. **d** ADC is increased, representing vasogenic edema. On DW image, the lesion is overlapped with diamagnetic susceptibility artifacts from air in the mastoid cells

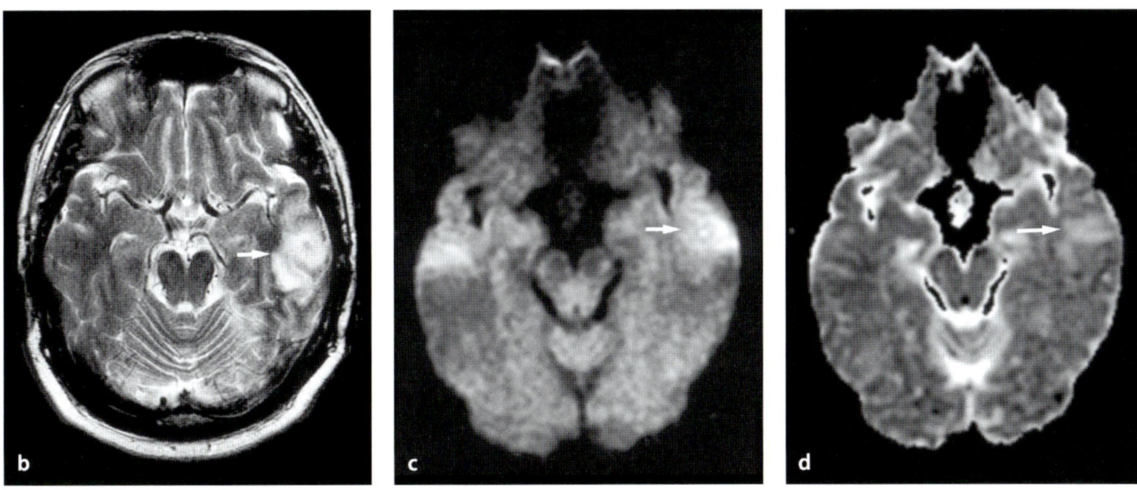

5.10 Small Vessel Infarcts

These are small infarcts measuring approximately 5–15 mm, usually seen in the basal ganglia, internal capsule, thalamus, pons and corona radiata. They account for about 20% of all infarctions and are secondary to an embolus, thrombus or atheromatous lesion within long, single, penetrating end arterioles.

These infarcts show increased signal in DW imaging with low ADC values (Fig. 5.8). However, unlike the usual time course of cerebral infarctions, they may show a prolonged increase in DW imaging signal and decrease in ADC values, sometimes seen beyond 60 days after onset of symptoms [15].

Differential diagnoses include widened perivascular spaces (Virchow–Robin spaces) and subependymal myelin pallor.

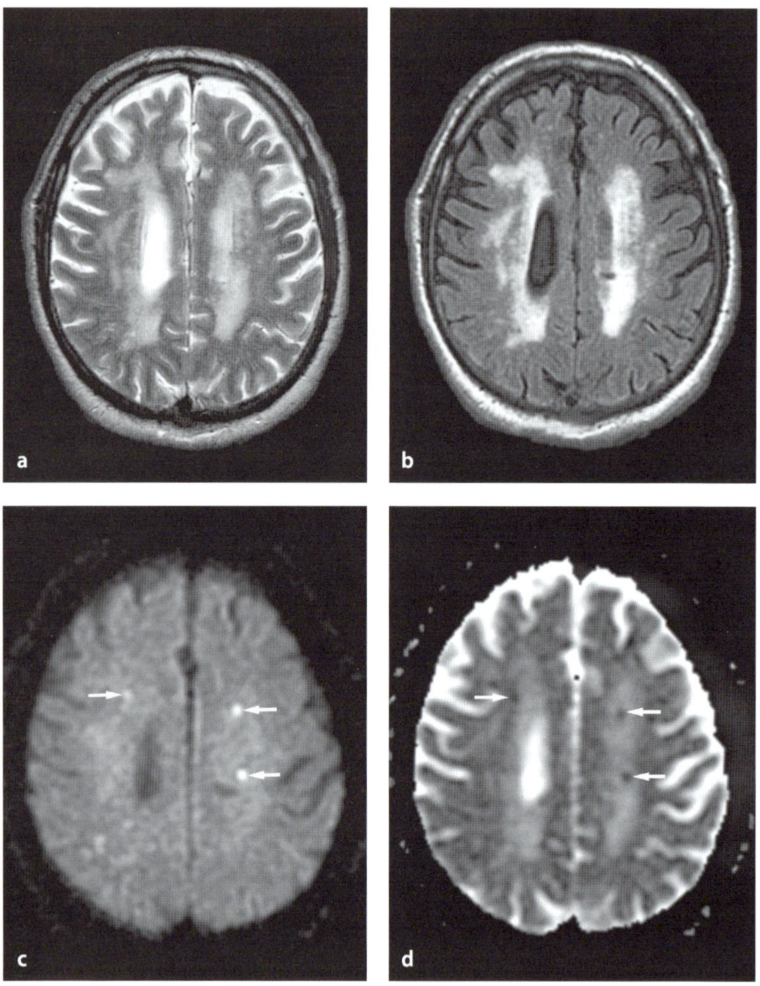

Figure 5.8 a–d

Small vessel infarcts in a 69-year-old man. **a, b** T2-weighted and FLAIR images show periventricular hyperintensities; however, it is difficult to detect acute small infarcts. **c** DW image clearly shows multiple hyperintensity spots in the white matter, representing the acute phase of small vessel infarcts. **d** ADC

5.11 Brain Stem and Cerebellar Infarcts

Cerebellar infarction is caused by occlusion of one of the major posterior circulation branches, which include the superior cerebellar, the anterior and posterior inferior cerebellar arteries, and the basilar artery. The posterior inferior cerebellar artery (PICA) supplies the postero-inferior portions of the cerebellum and is the most commonly obstructed cerebellar artery. The size of the infarct is important because a large infarct may cause a significant mass effect on the fourth ventricle and lead to hydrocephalus as well as brain stem compression. PICA infarctions can also result in the so-called lateral medullary (Wallenberg) syndrome, manifested by ipsilateral Horner's syndrome, ataxia, dysphagia, vertigo, nystagmus, hiccups and contralateral numbness, diminished pain and temperature sensation. The brain stem and cerebellar infarcts behave similar to cerebral infarcts on DW imaging and ADC maps (Fig. 5.9).

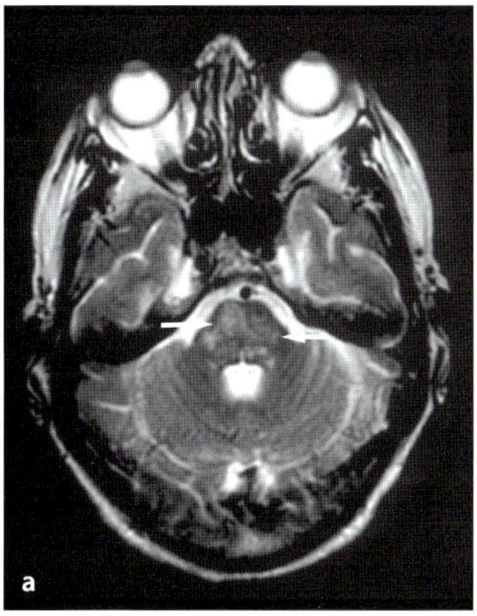

Figure 5.9 a–c

Brain stem infarction in an 83-year-old woman with slurred speech and gait difficulties. **a** T2-weighted image show hyperintense lesions in the pons (*arrows*). **b** DW image clearly shows a hyperintense lesion. **c** ADC is decreased in the right side of the pons, representing acute infarction (*arrow*)

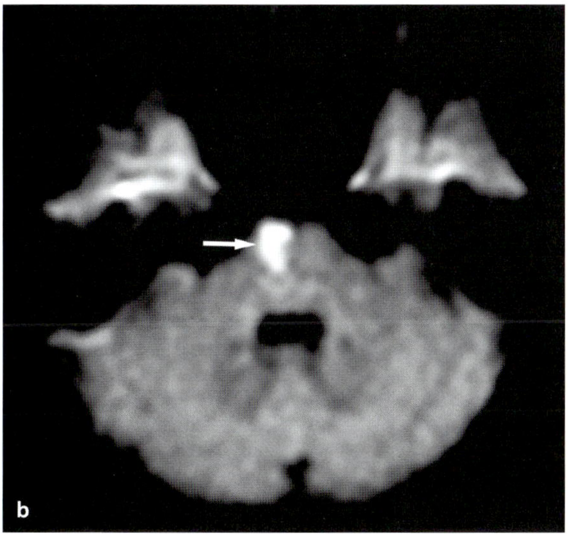

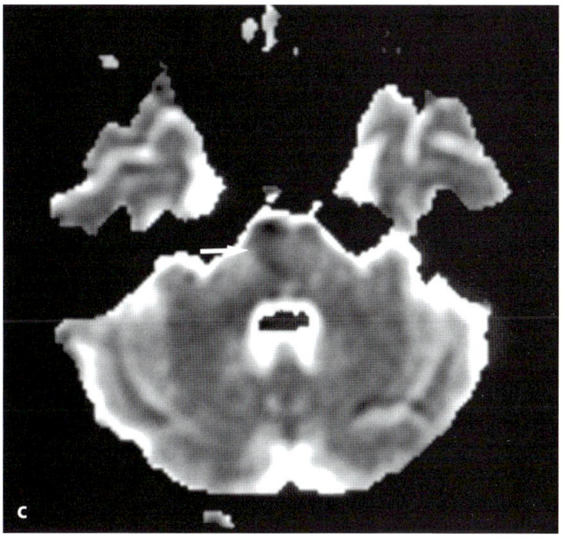

5.12 Corpus Callosum Infarcts

Isolated corpus callosum infarction due to pericallosal artery disease is rare, but can present as an alien hand syndrome. These patients fail to recognize the ownership of one hand when placed in certain positions or situations [16]. These lesions are readily detected by DW imaging and have signal characteristics similar to cerebral infarcts (Fig. 5.10).

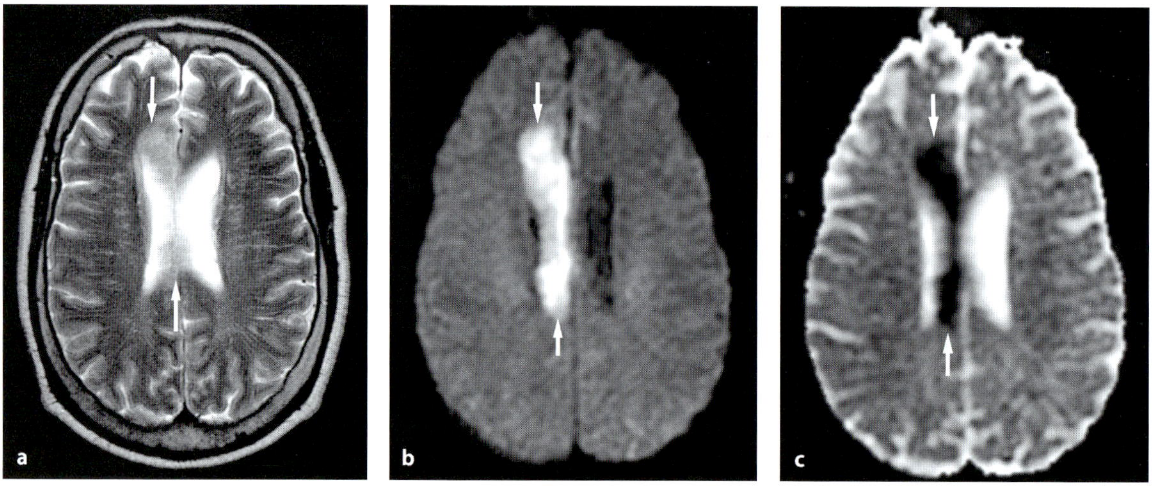

Figure 5.10 a–c

Corpus callosum infarction in a 64-year-old man with left-sided weakness. **a** T2-weighted image shows a hyperintense lesion in the anterior part and body of the corpus callosum extending into right frontal white matter (*arrows*). **b** DW image clearly shows this lesion as hyperintense. **c** ADC is decreased, representing acute infarction (*arrows*)

5.13 Hemorrhagic Infarcts

About 40–50% of all stroke patients develop hemorrhagic transformation of their infarcts (Fig. 5.11). This usually occurs during the first week after onset of symptoms. The cause may be a spontaneous lysis of an embolus, which took place at a time when endothelial cells of the vessel had also been damaged by the ischemia, thus resulting in a breakthrough hem orrhage into the infarcted region. The incidence of hemorrhage is increased with use of thrombolytic therapy, as well as in the presence of certain clinical conditions, such as hypertension, embolic etiology, use of anticoagulant therapy and increasing stroke severity.

Studies have shown that neuroimaging can predict which lesions are prone to progress into a hemorrhagic infarction [17]. Thus, ischemic lesions with a significantly greater percentage of low ADC values have a higher risk for hemorrhagic transformation than lesions with a smaller proportion of low ADC [17].

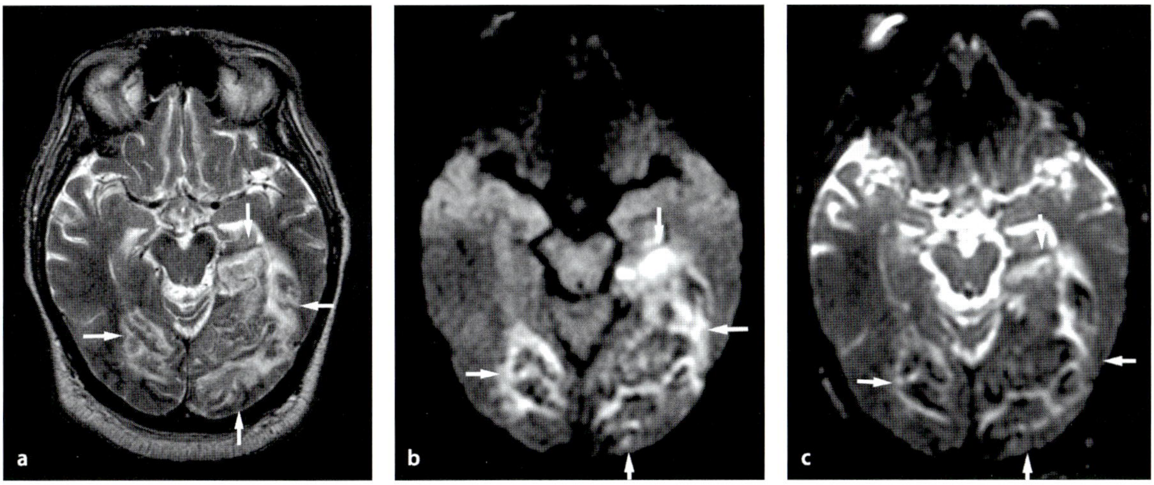

Figure 5.11 a–c

Hemorrhagic infarction in a 64-year-old man with mental status change. **a** T2-weighted image shows mixed hyper- and hypointense lesions in bilateral occipital lobes (*arrows*). **b** DW image shows these lesions as mixed hyper-, hypo- and isointense, representing acute hemorrhagic infarction (*arrows*). **c** ADC is decreased with areas of hyperintensities

References

1. Lefkowitz D, LaBenz M, Nudo Sr, et al. (1999) Hyperacute ischemic stroke missed by diffusion-weighted imaging. AJNR Am J Neuroradiol 20:1871–1875
2. Wang PYK, Barker PB, Wityk RJ, et al. (1999) Diffusion-negative stroke: a report of two cases. AJNR Am J Neuroradiol 20:1876–1880
3. Provenzale JM, Sorensen AG (1999) Diffusion-weighted MR imaging in acute stroke: theoretic considerations and clinical applications. AJR Am J Radiol 173:1459–1467
4. Keller E, Flacke S, Urbach H, Schild H. (1999) Diffusion- and perfusion-weighted magnetic resonance imaging in deep cerebral venous thrombosis. Stroke 30:1144–1146
5. Grant PE, He J, Halpern EF, Wu O, Schaefer PW, et al. (2001) Frequency and clinical context of decreased apparent diffusion coefficient reversal in the human brain. Radiology 221:43–50
6. Fiebach, JB, Jansen O, Schellinger PD, et al. (2002) Serial analysis of the apparent diffusion coefficient time course in human stroke. Neuroradiology 44:294–298
7. Desmond PM, Lovell AC, Rawlinson AA, et al. (2001) The value of apparent diffusion coefficient maps in early cerebral ischemia. AJNR Am J Neuroradiol 22:1260–1267
8. Huang II, Chen CY, Chung HW, et al. (2001) Time course of cerebral infarction in the middle cerebral arterial territory: deep watershed versus territorial subtypes on diffusion-weighted MR images. Radiology 221:35–42
9. Burdette JH, Ricci PE, Petitti N, et al. (1998) Cerebral infarction: time course of signal intensity changes on diffusion-weighted MR images. Am J Radiol 171:791–795
10. Mukherjee P, Bahn MM, McKinstry RC, et al. (2000) Differences between gray and white matter water diffusion in stroke: diffusion-tensor MR imaging in 12 patients. Radiology 215:211–220
11. Krueger K, Kugel H, Grond M, et al. (2000) Late resolution of diffusion-weighted MRI changes in a patient with prolonged reversible ischemic neurological deficit after thrombolytic therapy. Stroke 31:2715–2718
12. Fitzek C, Tintera J, Muller-Forell W, et al. (1998) Differentiation of recent and old cerebral infarcts by diffusion-weighted MRI. Neuroradiology 40:778–782
13. Pan Foni L, Garcia GH, Gutierreg JA (1996) Cerebral white matter is highly vulnerable to ischemia Strole 27: 1641–1647.
14. Kuroiwa T, Nagaoka T, Ueki M, et al. (1998) Different apparent diffusion coefficient: water content correlations of gray and white matter during early ischemia. Stroke 29:859–865
15. Geijer B, Lindgren A, Brockstedt S, et al. (2001) Persistent high-signal on diffusion-weighted MRI in the late stages of small cortical and lacunar ischemic lesions. Neuroradiology 43:115–122
16. Suwanwela NC, Leelacheavasit N (2002) Isolated corpus callosal infarction secondary to pericallosal artery disease presenting as alien hand syndrome. J Neurosurg Psychiatry 72:533–536
17. Tong DC, Adami A, Moseley ME, et al. (2000) Relationship between apparent diffusion coefficient and subsequent hemorrhagic transformation following acute ischemic stroke. Stroke 31:2378–2384

Intracranial Hemorrhage

In collaboration with A. Hiwatashi

6.1 Introduction

Intracranial hemorrhages are often characterized according to their location, such as intraparenchymal, subarachnoid, subdural, epidural and intraventricular hemorrhages. The etiology of these hemorrhages includes a variety of heterogeneous conditions, such as trauma, hypertension, infarction, infection, neoplasm, vascular malformations, vasculitis, vasculopathy, coagulopathy, and drugs. This chapter will describe diffusion-weighted (DW) imaging characteristics of intracranial hemorrhages in relation to their location and evolutionary stage.

6.2 Intraparenchymal Hemorrhages: Appearance and Evolution

The classic pattern for the temporal evolution of intracerebral hematomas on MRI at 1.5 T is well known [1–38]. However, determination of the age of a hemorrhage is often inaccurate because of variations between individual patients. The change in signal intensity over time depends on many factors, such as the oxygenation state of hemoglobin, the status of red blood cell membranes, hematocrit, proteins and clot formation [1–34, 37, 38]. Among these, the evolution of hemoglobin and the red cell membrane integrity are the most important [1–23, 25, 28, 30, 31, 33, 34]. The transition of oxy-hemoglobin to deoxy-hemoglobin and thereafter to met-hemoglobin depends primarily upon the oxygen tension in the vicinity of the lesion as well as inside the hematoma itself. In the hyperacute stage, oxy-hemoglobin will dominate initially, but transformation into deoxy-hemoglobin will soon take place and deoxy-hemoglobin will dominate after a few days – the acute hematoma [7, 11, 14, 19, 22, 23, 25, 28, 29, 33, 34]. After approximately one week, deoxy-hemoglobin will transform to met-hemoglobin [3, 4, 7, 10, 11, 14, 19, 22, 23, 28–30, 33, 34]. However, the rate of this oxidation to met-hemoglobin will depend on the oxygen tension in the tissue, which may further complicate the temporal pattern of expected signal changes. Initially, met-hemoglobin will be found within intact red blood cells (the early subacute stage), but when the red cell membranes start to rupture, met-hemoglobin will be found in the extracellular fluid space (the late subacute stage), which takes place about two weeks following hemorrhage [7, 11, 14, 19, 22, 23, 28, 29, 33, 34]. The final stage, the chronic stage, is the result of continuous phagocytation of the breakdown products of hemoglobin, ferritin and hemosiderin, which starts about one month following hemorrhage. The products, ferritin and hemosiderin, will remain within the phagocytic cells, which accumulate in the periphery of the hematoma, where they may remain for years, maybe indefinitely, as a marker of an old hemorrhage [6, 7, 11, 14, 17, 19–23, 27, 33, 34].

6.2.1 Hyperacute Hematoma

In the early stage of a hyperacute hematoma (Figs. 6.1 and 6.2), oxygenated hemoglobin within intact red blood cells is dominant. Oxy-hemoglobin is a diamagnetic substance and will, as such, generate an opposing magnetic field that reduces the applied magnetic field, as in most normal tissues in the body. Since there are no unpaired electrons in the iron of oxygenated hemoglobin, both longitudinal and transverse relaxation will be created by the so-called proton–proton, dipole–dipole interactions. At this stage, hematomas will have shorter relaxation times than water due to their protein content and will be slightly hypo- or iso-intense when compared with brain parenchyma on T1-weighted images. On T2-weighted images, oxy-hemoglobin will be seen as a slightly hyperintense region because of the high water content [7, 11, 14, 19, 22, 23, 25, 28, 29, 33, 34].

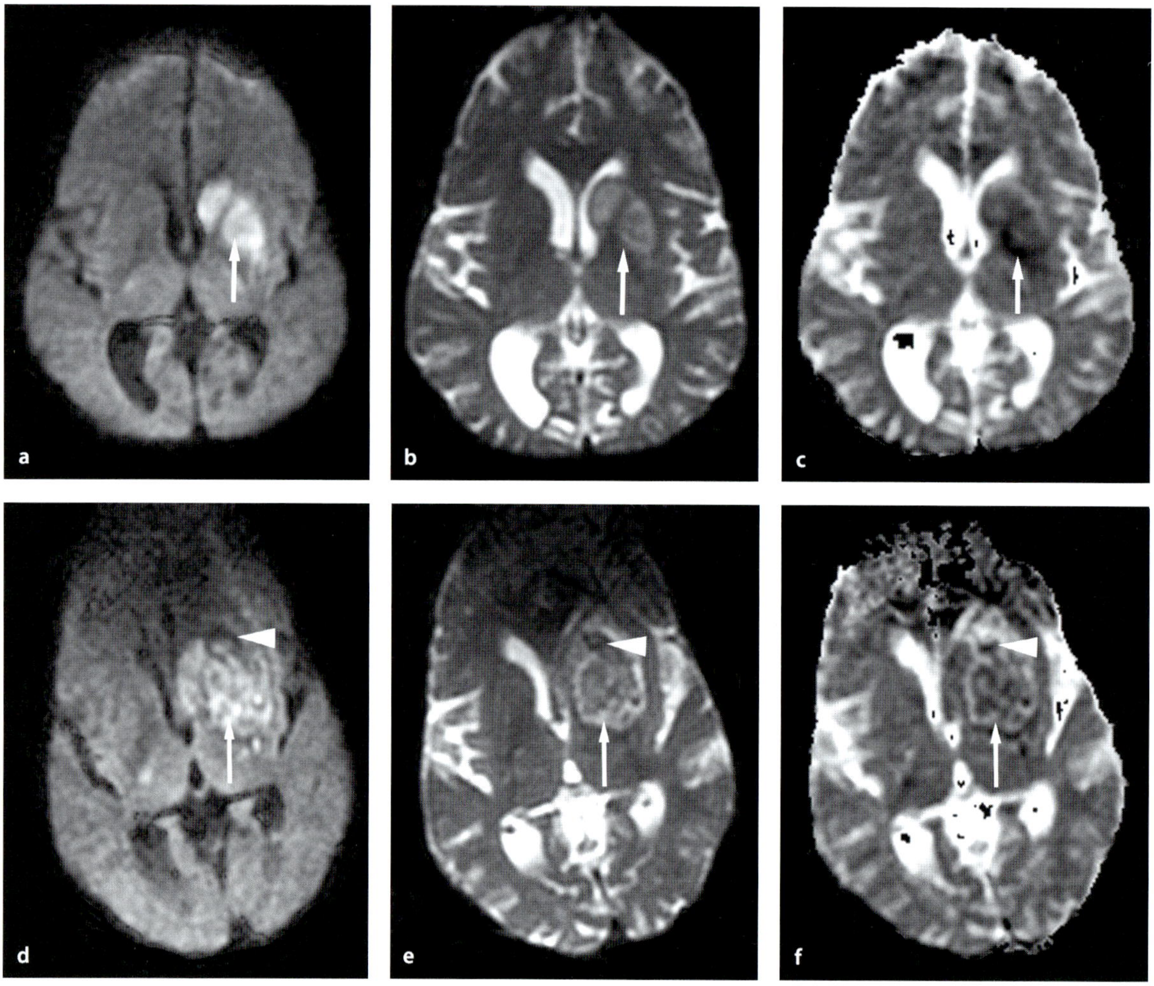

Figure 6.1 a–f

Acute infarction with hyperacute hemorrhage (oxy-hemoglobin/deoxy-hemoglobin). A 72-year-old woman suffered from a right-sided weakness. MR imaging 30 hours after the onset of symptoms (**a–c**) showed a lesion in the right basal ganglia (*arrow*) that is hyperintense on DW (**a**) and b_0 images (**b**). On the ADC map (**c**) the lesion is hypointense (*arrow*), indicating that this is a non-hemorrhagic acute infarction. During the end of the MR scanning, 30.5 hours after the onset of symptoms (**d–f**), the patient suddenly lost consciousness. The study was repeated, showing the increased size of the basal ganglia lesion. On the DW image (**d**) there is now a heterogeneous hyperintensity (*arrow*), consistent with oxy-hemoglobin. The b_0 image (**e**) also shows the lesion as heterogeneous, hyperintense (*arrow*), supporting oxy-hemoglobin. Note the new regions of hypointensity around the lesion (*arrowhead*), indicating deoxy-hemoglobin (**d–f**). The ADC map (**f**) shows the lesions as heterogeneously hypointense, consistent with oxy-hemoglobin (*arrow*). However, it is difficult to calculate the ADC due to the presence of paramagnetic deoxy-hemoglobin (*arrowhead*).

Results of DW imaging of hematomas at this stage have not been well characterized. In our experience, however, a hyperacute intraparenchymal hemorrhage is hyperintense on DW images, with a decreased apparent diffusion coefficient (ADC). This is in accordance with observations of other authors [28, 33]. The possible causes for the decreased ADC are shrinkage of extracellular space due to clot retraction, changes in the concentration of hemoglobin and a high viscosity [13, 19, 22–24, 28, 33–35].

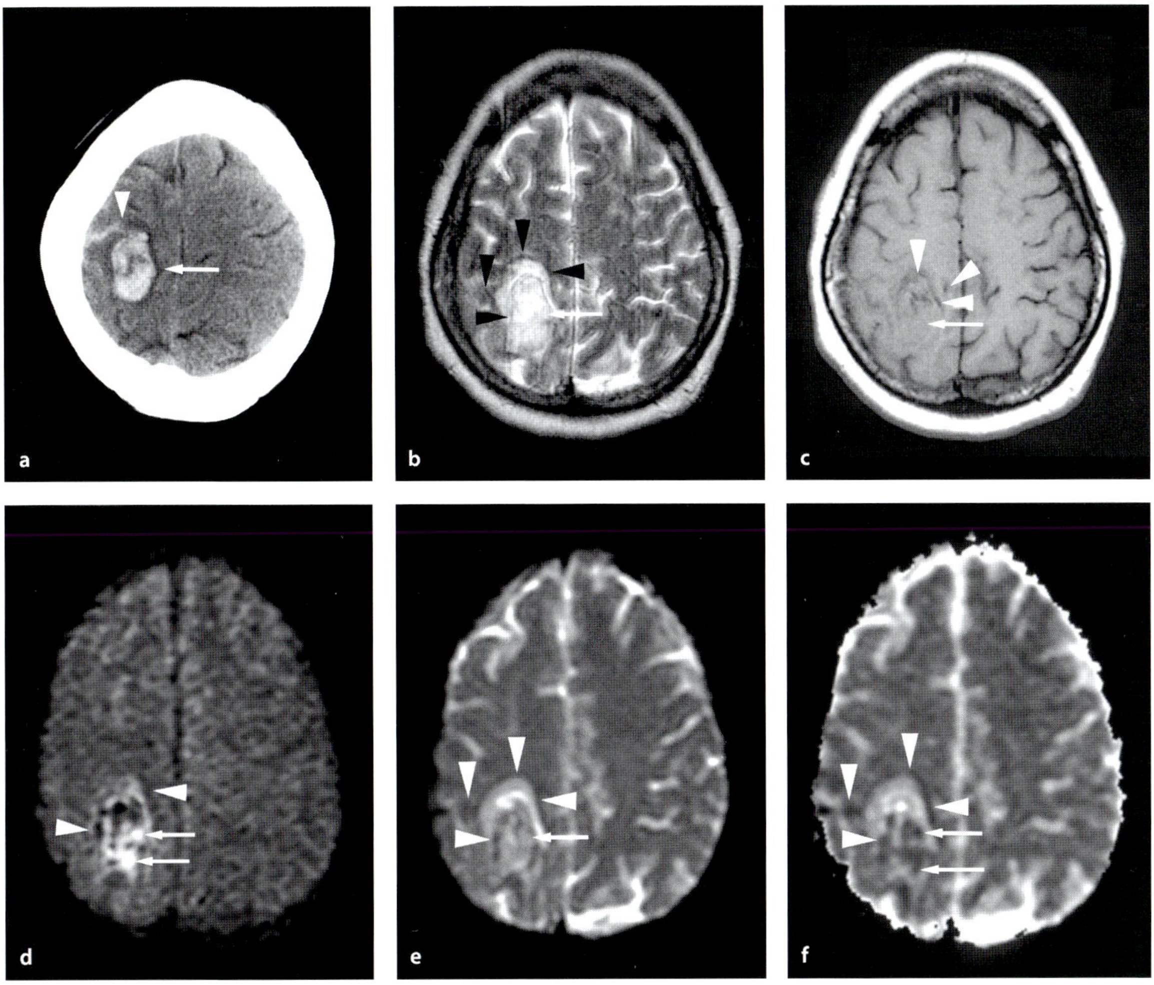

Figure 6.2 a–f

Hyperacute and chronic hemorrhage (oxy-hemoglobin/deoxy-hemoglobin, met-hemoglobin and hemosiderin/ferritin). A 40-year-old woman developed an acute left-sided weakness. A non-enhanced CT scan 6 hours after the onset of symptoms (a) shows a hyperdense lesion (*arrow*) in the right frontoparietal region associated with a small amount of subarachnoid hemorrhage (*arrowhead*). MR imaging 12 hours after the onset of symptoms (b–h) shows a heterogeneous lesion with areas of central hyperintensity on the T2-weighted image (b) (*arrow*; oxy-hemoglobin) with a peripheral hypointensity from the susceptibility influence of paramagnetic material (*arrowheads*; deoxy-hemoglobin). The T1-weighted image (c) shows the lesion as heterogeneous with areas of isointensity (*arrow*; oxy-hemoglobin) and hypointensity (*arrowheads*; deoxy-hemoglobin). Both the DW (d) and b_0 images (e) show the lesion as heterogeneous with areas of hyperintensity (*arrows*; oxy-hemoglobin) and hypointensity (*arrowheads*; deoxy-hemoglobin). The hypointensity is more prominent on these sequences than on the T2-weighted image (b). The ADC map (f) has a similar appearance of a heterogeneous lesion with areas of hypointensity (*arrows*; oxy-hemoglobin and *arrowheads*; deoxy-hemoglobin). As usual, when there is a strong susceptibility influence, it is difficult to calculate ADC.

▶

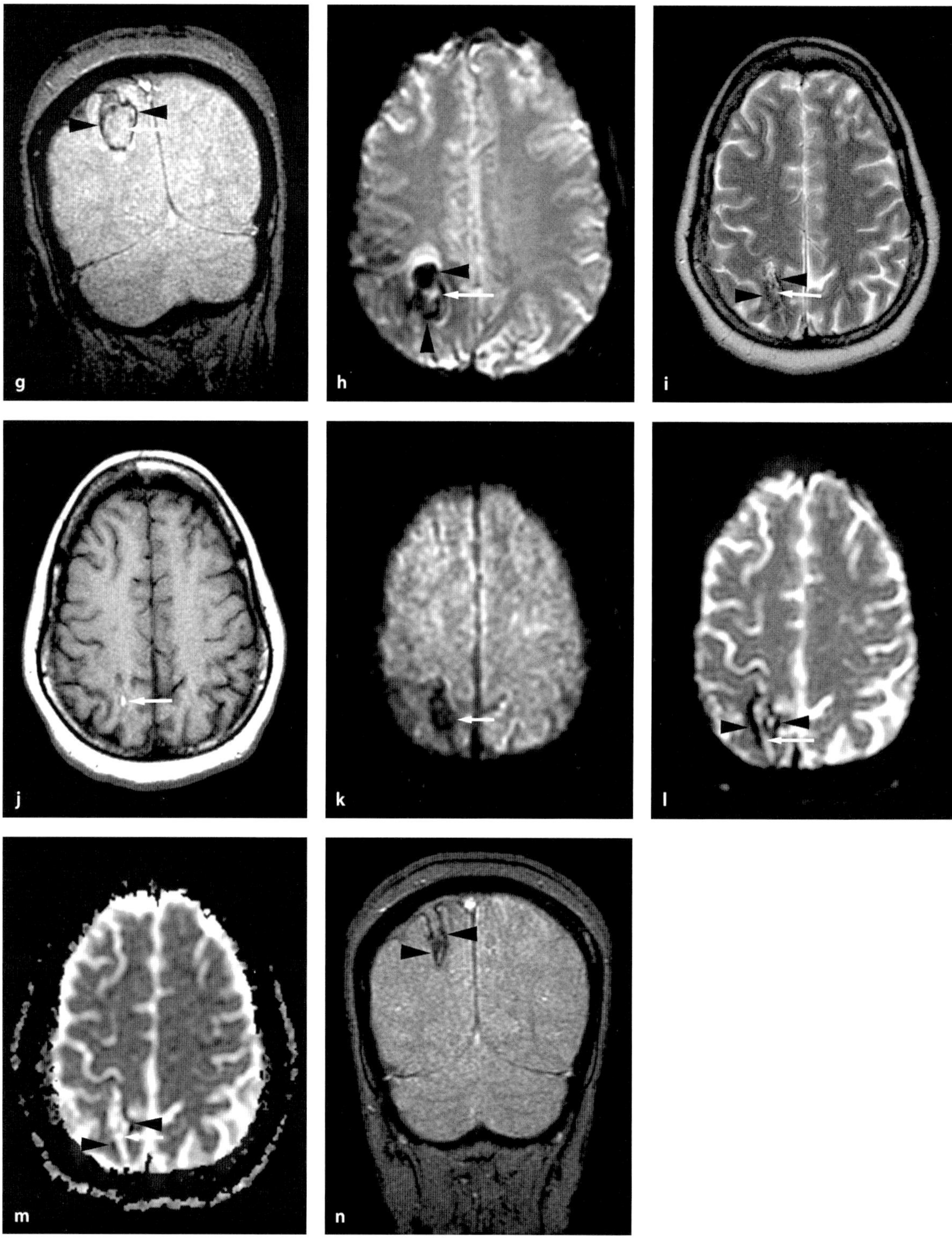
g
h
i
j
k
l
m
n

Figure 6.2 g–n

The coronal GRE image (**g**) and the GRE–echo-planar image (**h**) are more sensitive in depicting the susceptibility effect from deoxy-hemoglobin (*arrowheads*). The diamagnetic oxy-hemoglobin will, as expected, show an isointense signal (*arrow*), similar to the other sequences. Four months after the onset of symptoms (**i–n**), the patient had a new MR examination. On a T2-weighted image (**i**), the lesion is still heterogeneous, but there are now areas of hyperintensity (extracellular met-hemoglobin; *arrow*) and hypointensity (*arrowheads*; hemosiderin/ferritin). The T1-weighted image (**j**) shows the heterogeneous lesion with areas of hyperintensity (*arrow*; extracellular met-hemoglobin) and hypointensity (hemosiderin/ferritin). The DW image (**k**) show the lesion as hypointense overall and the same is seen on the b_0 image (**l**), but this "T2-weighted" image has a small hyperintense zone in the center of the hematoma, probably due to extracellular met-hemoglobin (*arrow*). The hypointensity (*arrowheads*) on the b_0 image (**l**) is more pronounced than on the conventional T2-weighted image (**i**), demonstrating the higher sensitivity to susceptibility with this imaging sequence. The susceptibility effect will make it difficult to calculate ADC accurately, but the ADC map (**m**) will demonstrate the susceptibility from hemosiderin/ferritin as peripheral hypointense areas (*arrowheads*). In the center there are hyperintense areas (*arrow*), probably containing extracellular met-hemoglobin. The hypointensity on the DW image (**k**) is probably a combined effect of T2 shortening, magnetic susceptibility and increased diffusibility. A marked peripheral hypointensity from hemosiderin/ferritin will also be seen when a GRE technique (**n**) is used (*arrowheads*)

◄

6.2.2 Acute Hematoma

Following hemorrhage there is normally a gradual oxygen desaturation of hemoglobin transforming oxy-hemoglobin into deoxy-hemoglobin (Fig. 6.3) [7, 11, 14, 19, 22, 23, 25, 28, 29, 33, 34]. The loss of oxygen will change the binding geometry of iron from a six-ligand system to a five-ligand system, leaving four unpaired electrons and making it paramagnetic. When exposed to a magnetic field, paramagnetic substances will enhance the applied field locally. This will influence image contrast by means of so-called magnetic susceptibility-induced relaxation, which only effects transverse relaxation (T2* effect). The susceptibility effects create local field inhomogeneities, with a rapid loss of transverse magnetization of protons within this region resulting in signal loss on T2-weighted images. Deoxy-hemoglobin will therefore demonstrate a marked hypointensity on T2-weighted images [1, 3, 7–9, 11–15, 19, 22–25, 30–34]. Besides the concentration of deoxy-hemoglobin, red blood cell concentration and/or clot formation may also contribute to this T2 hypointensity. Since this is an effect of magnetic susceptibility, gradient-echo (GRE) images will be more sensitive than DW- and T2-weighted images for the detection of acute, as well as chronic hematomas [19, 26, 27, 32]. The T1 relaxivity effect is related to dipole–dipole relaxation, in this case the dipoles of water and unpaired electrons of the paramagnetic center. This would normally result in a shortening of relaxation time; however, the molecular configuration of deoxy-hemoglobin will act as a shield for such a close approach of water molecules to the unpaired electrons of the paramagnetic, ferrous iron. The acute hematoma containing deoxy-hemoglobin will thus show an iso- to hypointense signal on T1-weighted images, similar to oxy-hemoglobin [5, 22, 23, 34].

Diffusion-weighted images of an acute hematoma show a marked hypointensity [28–31, 33], caused by the magnetic field inhomogeneity created by the paramagnetic deoxy-hemoglobin [6, 7, 14, 18, 22, 23, 25, 33]. Although the ADC has been reported to be decreased, accurate calculations are often difficult [28–30, 33, 36].

6.2.3 Early Subacute Hematoma

In the early stage of the subacute hematoma (Figs. 6.2 and 6.3), there is a decline in the energy state of the red blood cell and hemoglobin is oxidized to met-hemoglobin [3, 4, 7, 10, 11, 14, 19, 22, 23, 28–30, 33, 34]. In met-hemoglobin the iron is still bound to the heme moiety within the globin protein, but it is now in the ferric state with five unpaired electrons. This transformation normally starts in the periphery of the hemorrhage and gradually evolves to the center. In the transition to met-hemoglobin, conformational changes will take place in the molecule and water protons will now have access to the unpaired electrons of iron in met-hemoglobin, creating a proton–electron, dipole–dipole interaction. Dipolar relaxation enhancement will then take place, making met-hemoglobin appear hyperintense on T1-weighted images. Met-hemoglobin, as a paramagnetic substance, will induce magnetic susceptibility relaxation affecting the transverse relaxation (T2* effect), which results in a marked hypointensity on T2-weighted images [3, 6, 7, 10, 14, 22–24, 28–31, 33, 34].

On DW imaging, intracellular met-hemoglobin shows hypointensity due to these paramagnetic susceptibility effects and ADC measurements are not reliable due to the susceptibility effects [28–30, 33].

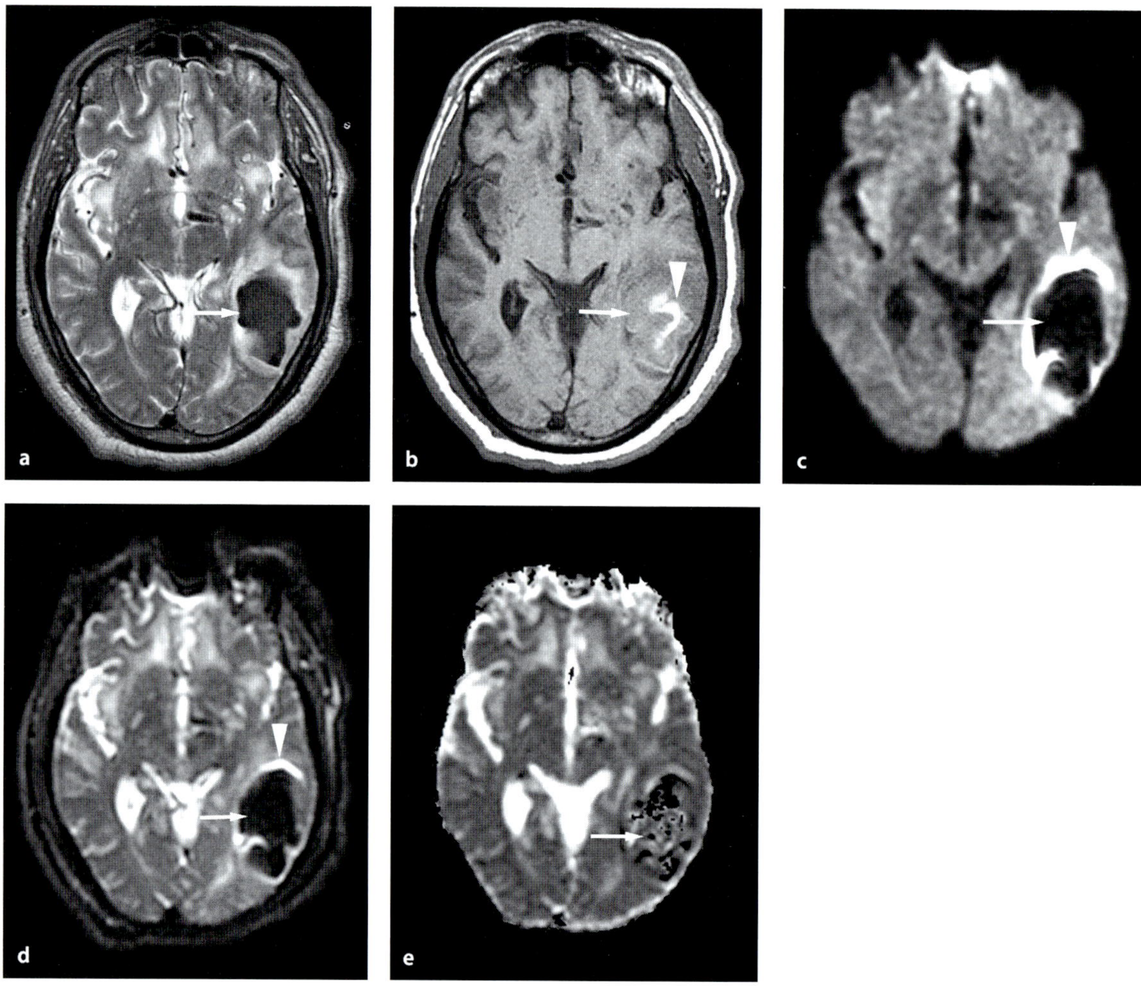

Figure 6.3 a–e

Acute to early subacute hemorrhage (deoxy-hemoglobin and intracellular met-hemoglobin). A 49-year-old man with headache and aphasia was referred for MR imaging 24 hours after the onset of symptoms (**a–e**). This study shows a left temporal lobe lesion that is hypointense on the T2-weighted image (**a**) (*arrow*; deoxy-hemoglobin and intracellular met-hemoglobin) with surrounding edema. On the T1-weighted image (**b**) the lesion is heterogeneous with areas of hypointensity (*arrow*; deoxyhemoglobin) and hyperintensity (*arrowhead*; intracellular met-hemoglobin). The DW image (**c**) demonstrates hypointensity (*arrow*; deoxy-hemoglobin and intracellular met-hemoglobin). The surrounding hyperintense rim (*arrowhead*) is a magnetic susceptibility artifact. This peripheral artifact is also seen around the hypointensity (*arrow*) created by deoxy-hemoglobin and intracellular met-hemoglobin on the b_0 image (**d**). ADC cannot be calculated accurately, which is easy to understand when looking at the extremely heterogeneous lesion depicted on the ADC map (**e**)

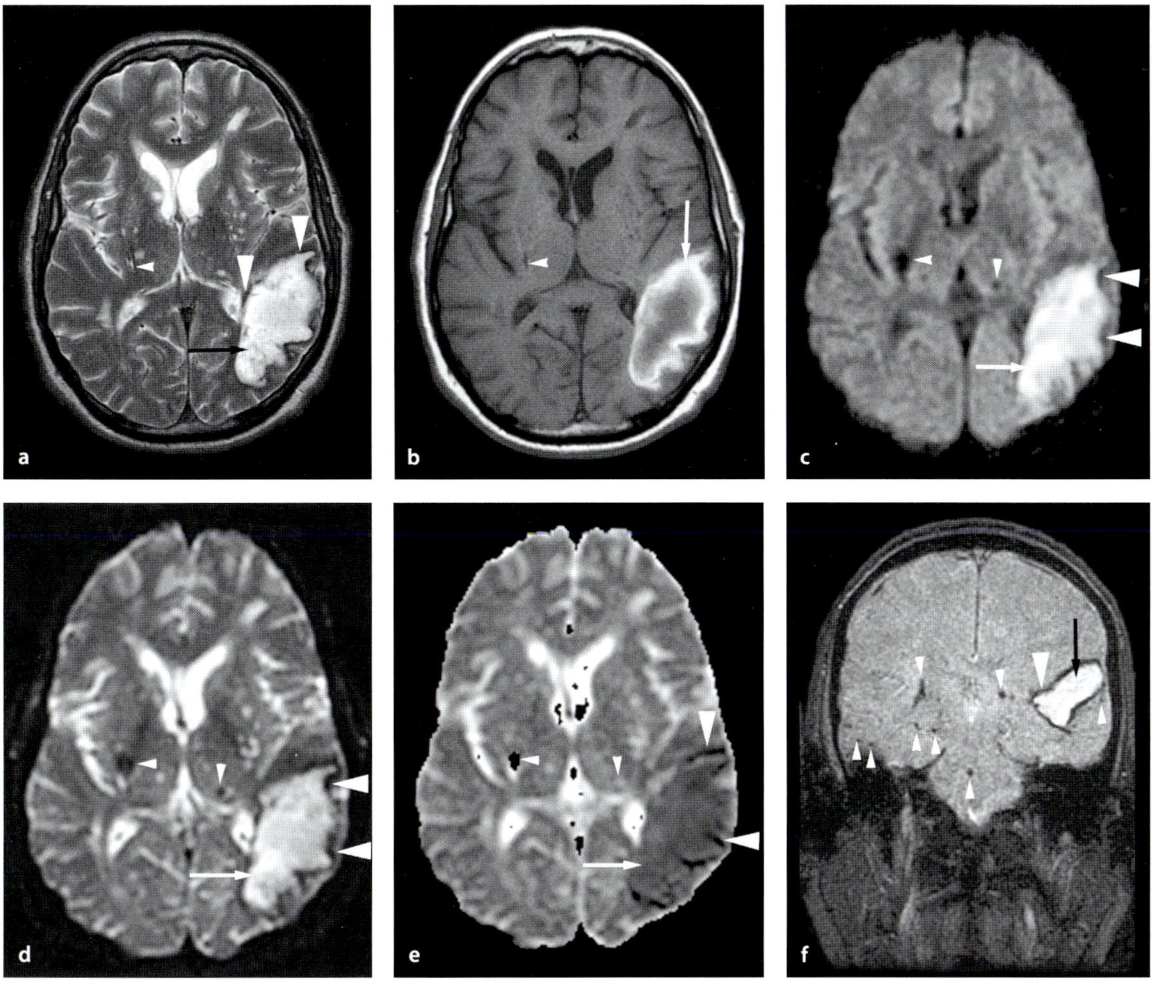

Figure 6.4 a–f

Late subacute-chronic hemorrhage (extracellular met-hemoglobin and hemosiderin/ferritin). A 52-year-old man with a history of chronic hypertension complained of headache and aphasia. MR examination 2 months after the onset of symptoms shows a left temporal lobe lesion that is hyperintense on the T2-weighted image (a) (*arrow*; extracellular met-hemoglobin) and surrounded by a hypointense rim (*arrowheads*; hemosiderin/ferritin). Another hypointense lesion is visualized in the right basal ganglia (*small arrowhead*; hemosiderin/ferritin), compatible with chronic hemorrhage secondary to hypertension. The T1-weighted image (b) shows the temporal lobe lesion as hyperintense (*arrow*; extracellular met-hemoglobin). The lesion in the right basal ganglia is also hypointense on this sequence (*small arrowhead*; hemosiderin/ferritin). On the DW image (c) the temporal lobe lesion is hyperintense (*arrow*; extracellular met-hemoglobin) with a hypointense rim (*arrowheads*; hemosiderin/ferritin). The basal ganglia lesion remains hypointense, but the signal void is more extensive than on the conventional T2-weighted image (a), since GRE does not compensate for signal loss due to local magnetic field inhomogeneities and is thus more sensitive than regular spin-echo imaging. This increased susceptibility sensitivity revealed a second, old hemorrhagic lesion in the left thalamus (small *arrowhead*; hemosiderin/ferritin). The b_0 image (d) also shows the hyperintense lesion (*arrow*; extracellular met-hemoglobin) with the hypointense rim (*arrowheads*; hemosiderin/ferritin) as well as the older lesions in the basal ganglia and thalamus (*small arrowheads*; hemosiderin/ferritin). On the ADC map (e) the temporal lesion is somewhat hypointense (*arrow*; extracellular met-hemoglobin) with a hypointense rim (*arrowheads*; hemosiderin/ferritin). The other lesions are also visualized (*small arrowheads*); however, ADC cannot be calculated. Finally, the coronal GRE image (f) shows the temporal lesion as hyperintense (*arrow*; extracellular met-hemoglobin) with a hypointense rim (*arrowhead*; hemosiderin/ferritin). The GRE sequence is the most sensitive and shows multiple small hypointense lesions (old hemorrhagic breakdown products) in the cerebral hemispheres and in the pons (*small arrowheads*; hemosiderin/ferritin)

6.2.4 Late Subacute Hematomas

The decline in energy state of the red blood cell will eventually damage the integrity of the red cell membrane, releasing the intracellular content to the extracellular fluid space (Figs. 6.2 and 6.4). Subsequently, there will be a dilution of the paramagnetic met-hemoglobin in extracellular fluid, reducing the susceptibility effect of met-hemoglobin [7, 11, 14, 19, 22, 23, 28, 29, 33, 34]. The signal intensity on T2-weighted images will thus relate to the water content, creating a hyperintense signal on T2-weighted images. Extracellular met-hemoglobin will, however, still have high signal intensity on T1-weighted images created by the same proton–electron, dipole–dipole relaxation as described in early subacute hematomas [5–7, 10, 14, 22, 23].

It has been reported that late subacute hematomas are hyperintense on DW imaging [28, 29]. The ADC value for late subacute hematoma is controversial. Ebisu et al. [24] reported decreased ADC in hematomas of hemorrhagic infarctions, whereas Atlas et al. reported ADC values higher than normal white matter in late subacute hematomas and suggested this was due to increased diffusibility [28]. Finally, Kang et al. reported decreased ADC in late subacute hematomas and thought this was due to high viscosity and cellularity [33].

6.2.5 Chronic Hematomas

Over time, met-hemoglobin will be resorbed or degraded and the effect on T1 enhancement will be reduced (Figs. 6.2 and 6.4). The high water content in the chronic stage will result in prolonged T1 as well as T2 relaxation. From the start of the hemorrhage there is a continuous phagocytation of heme proteins. Ferritin and hemosiderin, the final breakdown products of hemoglobin, will remain within the phagocytic cells, which accumulate in the periphery of the hematoma, where they may remain indefinitely as a marker of an old hemorrhage [6, 7, 11, 14, 17, 19–23, 27, 33, 34, 37, 38]. Ferritin and hemosiderin within these cells will have no access to water protons and thus there are no relaxivity effects. Magnetic susceptibility is the only factor influencing the signal, creating a marked signal loss on T2-weighted images. As mentioned earlier, the magnetic susceptibility effects are most prominent on T2*-weighted images [7, 17, 22, 23, 34].

Diffusion-weighted images are also hyperintense in chronic hematomas [29, 33]. The ADC value has been reported to be increased, but is often difficult to measure accurately due to magnetic susceptibility artifacts [28–30, 33, 36].

6.3 Subarachnoid Hemorrhage

Computed tomography (CT) is still essential in the diagnosis of acute subarachnoid hemorrhages (Fig. 6.5), as the sensitivity and usefulness of MR imaging is controversial [4, 5, 9, 34, 39–45]. Fluid-attenuated inversion-recovery (FLAIR) imaging has a high sensitivity for subarachnoid hemorrhage [46–49]. However, the specificity is low because there are several other causes for the appearance of subarachnoid hemorrhage on FLAIR imaging, such as high proton concentration, mass effect, vascular disease, contrast medium and use of specific intravenous anesthetic agents [50–53].

It is often difficult to detect subarachnoid hemorrhage on DW images [31, 54]. Lin et al. detected subarachnoid hemorrhage in two of four cases on GRE imaging, but it could not be detected on b_0 images [31]. Wiesmann et al. reported that proton density and FLAIR images could detect subarachnoid hemorrhage, but T2-weighted and DW images could not [54].

However, DW images may be useful to visualize parenchymal injuries secondary to subarachnoid hemorrhage. Ischemic changes, probably related to subarachnoid hemorrhage, have shown hyperintensity on DW images in both clinical and animal studies [55–60]. This finding depends on the timing of imaging and the severity of injury.

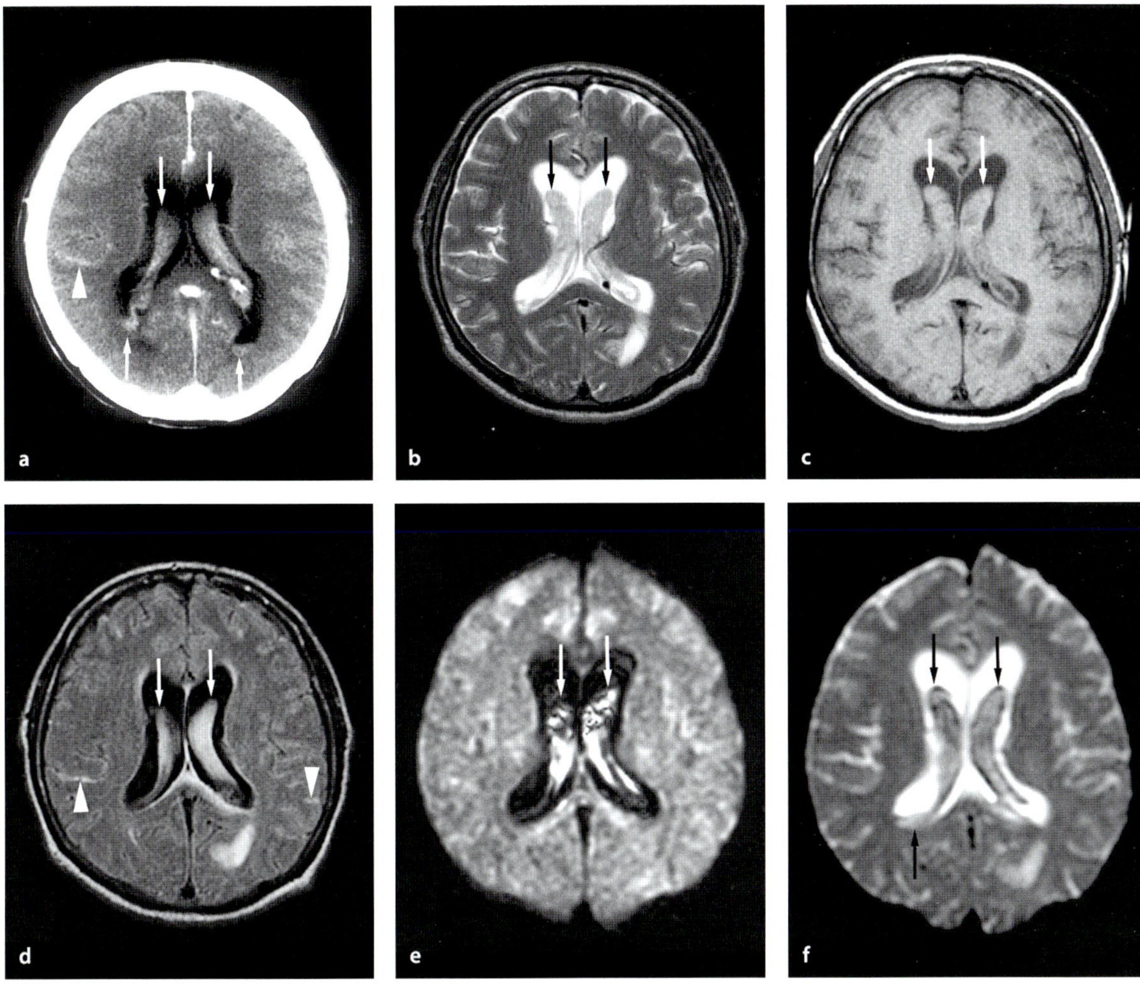

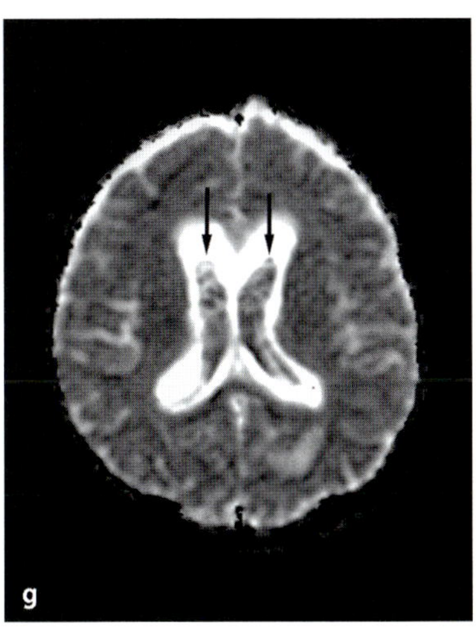

Figure 6.5 a–g

Subarachnoid and intraventricular hemorrhage due to arteriovenous malformation (intracellular met-hemoglobin). A 48-year-old female presented with acute onset of severe headache had a CT scan 24 hours after the onset of symptoms (**a**), which shows diffuse subarachnoid (*arrowhead*) and intraventricular hemorrhage (*arrows*). Forty-eight hours after the onset of symptoms (**b–g**) she had an MRI. On this examination the T2-weighted image (**b**) showed intraventricular hemorrhage (*arrows*), which is hypointense when compared with the cerebrospinal fluid. The diffuse subarachnoid hemorrhage cannot be visualized on the T2-weighted image. The T1-weighted image (**c**) shows the intraventricular hemorrhage (*arrows*) as hyperintense when compared with cerebrospinal fluid, but neither sequence can visualize the subarachnoid hemorrhage. The FLAIR image (**d**), however, shows both the subarachnoid (*arrowheads*) and intraventricular hemorrhage (*arrows*). The subarachnoid hemorrhage cannot be visualized on the DW image (**e**), but the intraventricular hemorrhage (*arrows*) can easily be seen as hyperintense when compared with cerebrospinal fluid and brain parenchyma. As expected, the subarachnoid hemorrhage cannot be demonstrated on either b_0 image (**f**) or ADC maps (**g**), but both sequences will depict the intraventricular hematoma as hypointense (*arrows*)

6.4 Subdural and Epidural Hemorrhages

Subdural and epidural hematomas (Fig. 6.6) are well demonstrated on T1- and T2-weighted images. In the acute and subacute stages, the signal intensity is often similar to intraparenchymal hematomas. In the chronic stage, MR images often show hyperintensity of the hematoma, which may be iso- to hypodense on CT [34, 61–67].

Diffusion-weighted imaging findings of subdural and epidural hematomas have not been well described. Lin et al. reported that all three lesions could be detected on b_0 images, but GRE images were better at detecting lesions [31]. The benefit of DW imaging is probably for the detection of underlying or associated parenchymal lesions [68, 69].

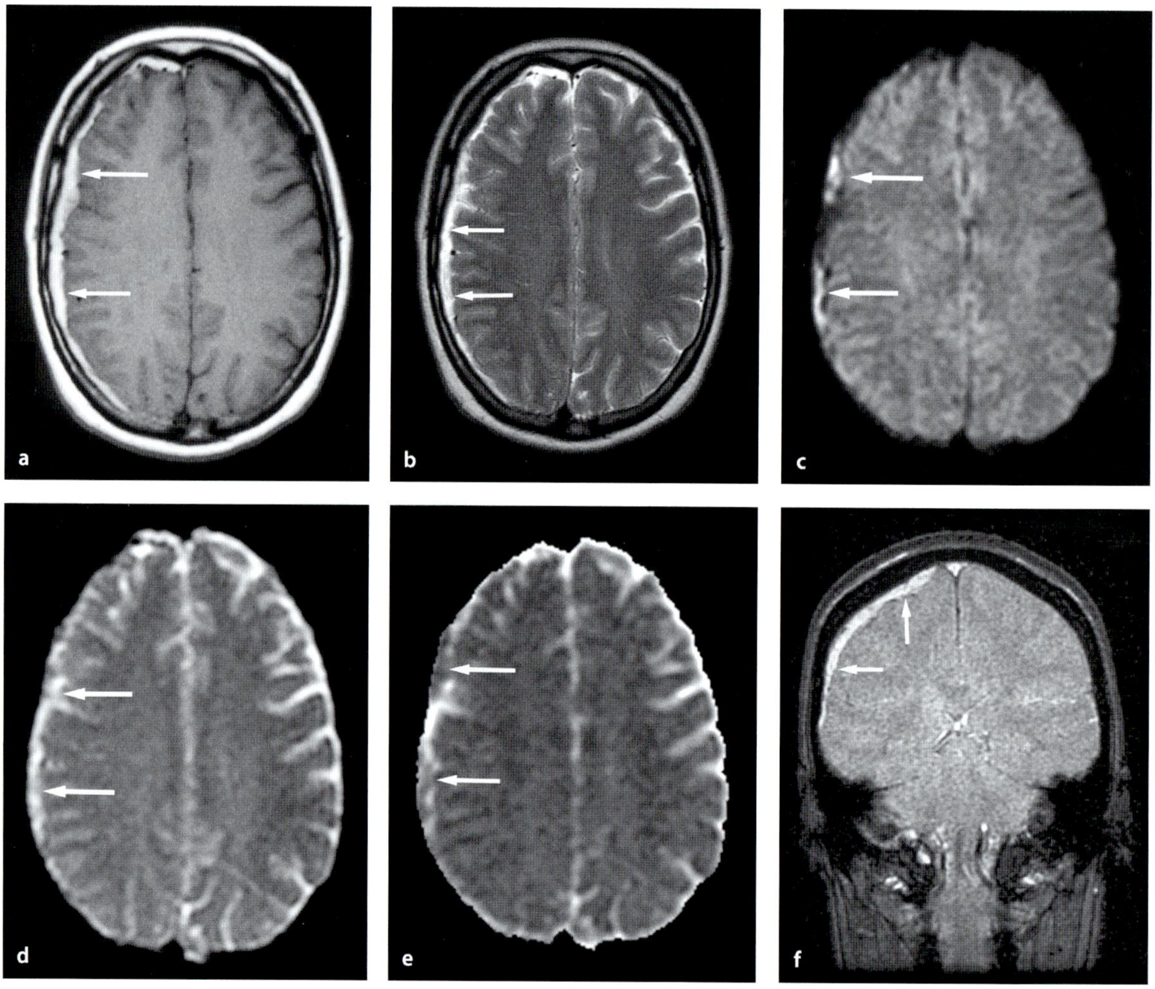

Figure 6.6 a–f

Subacute subdural hemorrhage (extracellular met-hemoglobin). A 49-year-old woman with headache following head trauma had an MR examination 2 weeks after the trauma, which shows right subdural hemorrhage as a hyperintense lesion (*arrows*) on the T1-weighted (a) and T2-weighted (b) images. The lesion was hyperintense on the DW (c) and the b_0 (d) images (*arrows*). On the ADC map (e) there are hypointense lesions (*arrows*), which correspond to the hyperintense lesions on the DW image (c). The coronal GRE image (f) shows the subdural hemorrhage to be hyperintense (*arrows*)

6.5 Intraventricular Hemorrhage

Intraventricular hemorrhages (Figs. 6.5 and 6.7) are well demonstrated on FLAIR, T1-, T2- and proton density-weighted images [34, 59, 70–72]. FLAIR has been reported to have the highest sensitivity for detection of intraventricular hematomas [71]. DW images can demonstrate intraventricular hemorrhages, but in general the GRE images have a higher sensitivity [31].

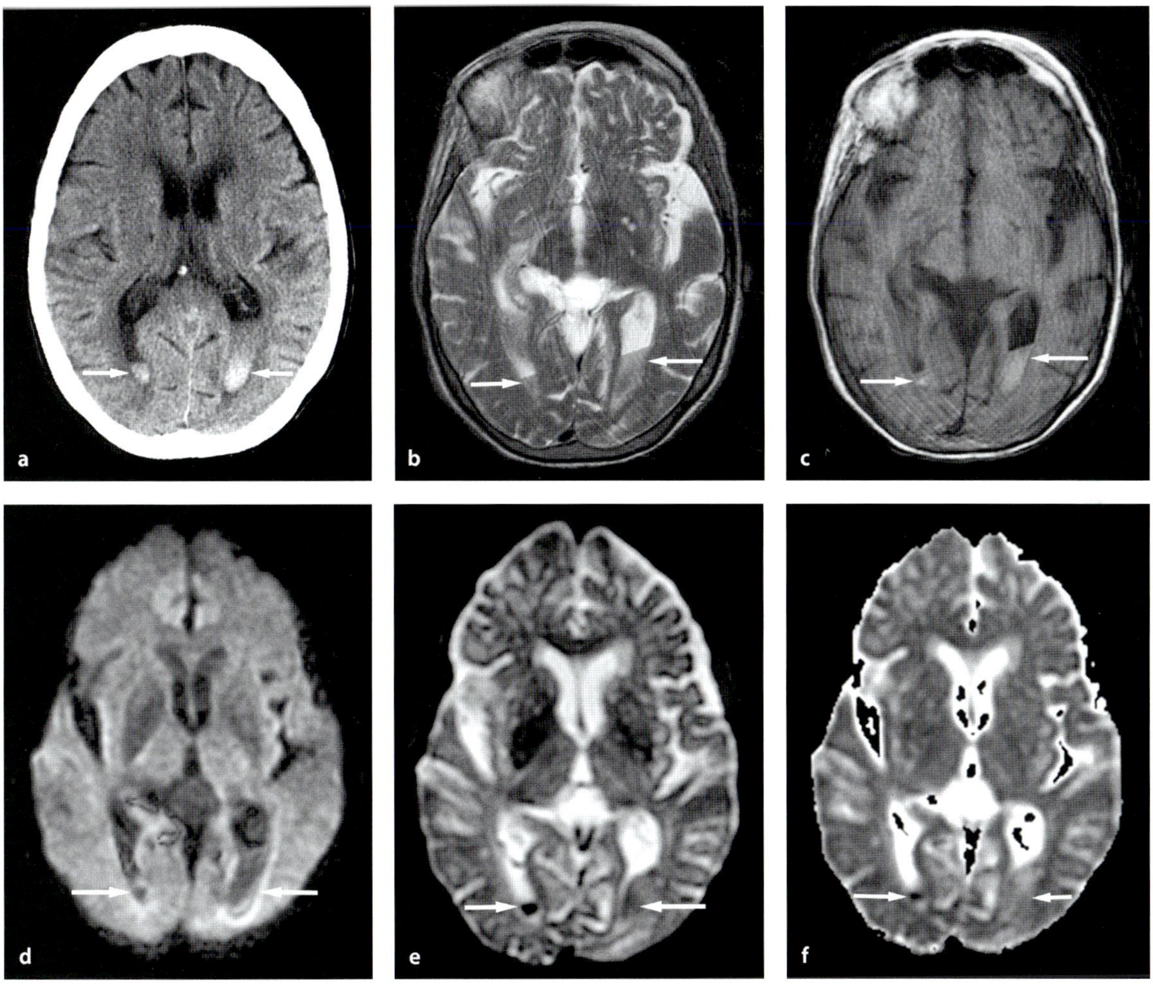

Figure 6.7 a–f

Intraventricular hemorrhage (intracellular met-hemoglobin). A 78-year-old woman with headache after thrombolytic therapy for femoral artery occlusion had a CT scans 6 hours after surgery (**a**). This shows intraventricular hemorrhage in the bilateral lateral ventricles (*arrows*). Three days after surgery (**b–f**), the T2-weighted image (**b**) shows hypointense lesions in the bilateral lateral ventricles with fluid–fluid levels (*arrows*) and the T1-weighted image (**c**) shows hyperintense lesions in the same distributions (*arrows*). The DW image (**d**) shows hypointense lesions (*arrows*) with surrounding hyperintensities. These are ascribed to magnetic susceptibility artifacts. The b_0 image (**e**) also shows hypointense lesions (*arrows*). These hypointensities are more prominent than on the T2-weighted images (**b**). The ADC map (**f**) shows hypointense lesions (*arrows*)

6.6 Intra-tumoral Hemorrhage

Many primary brain tumors and metastases can bleed [34, 73–78]. The signal intensity of intra-tumoral hemorrhage (Fig. 6.8) tends to be more complex and its evolution tends to be delayed when compared with non-neoplastic hemorrhages [34, 78]. DW and b_0 images are useful to detect the hemorrhage in tumors [31].

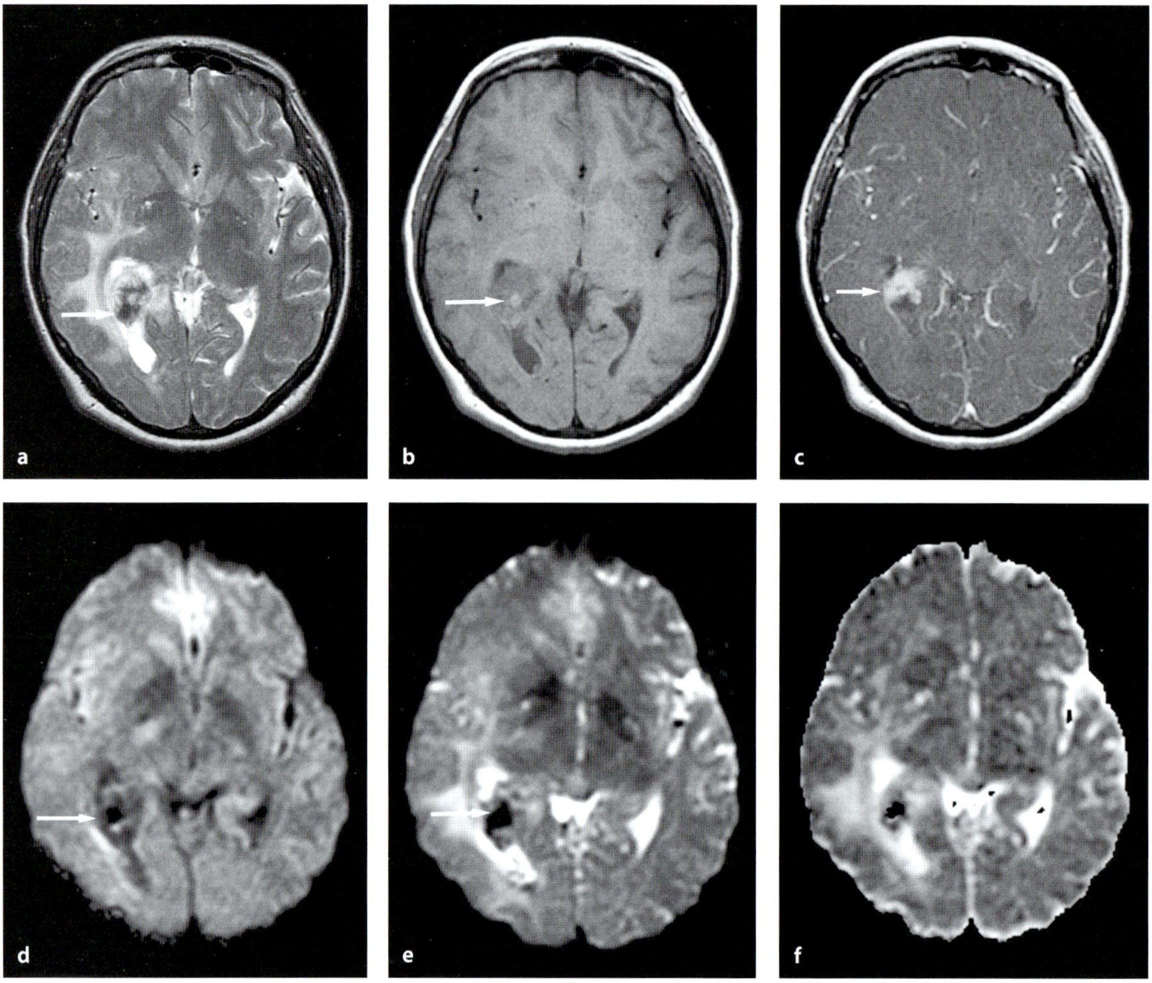

Figure 6.8 a–f

Hemorrhagic tumor. The T2-weighted MR image (**a**) in a 54-year-old woman with glioblastoma shows a mass lesion with heterogeneous intensity near the right lateral ventricle. The irregular hypointensities centrally in the lesion (*arrow*) indicate hemorrhage. The T1-weighted image (**b**) shows the heterogeneous hypointense to isointense mass, with a central area of higher signal intensity consistent with hemorrhage (*arrow*). The gadolinium-enhanced T1-weighted image (**c**) shows heterogeneous enhancement (*arrow*). On the DW image (**d**) the hemorrhage is heterogeneously hypointense (*arrow*). The b_0 image (**e**) shows the hemorrhage to be more hypointense (*arrow*) than on the T2-weighted image (**a**). The ADC (**f**) cannot be calculated due to magnetic susceptibility artifacts

6.7 Hemorrhage Related to Vascular Malformation

Vascular malformations can also cause intracranial hemorrhages (Fig. 6.9). Cavernous angioma is a vascular malformation that contains blood cavities surrounded by a single layer of endothelium [79–83]. MR imaging findings are well known and characterized as a central reticulated core with a peripheral rim of hypointensity due to the deposition of hemosiderin [79, 82, 83]. DW and b_0 images are useful for detecting hemorrhages related to vascular malformations.

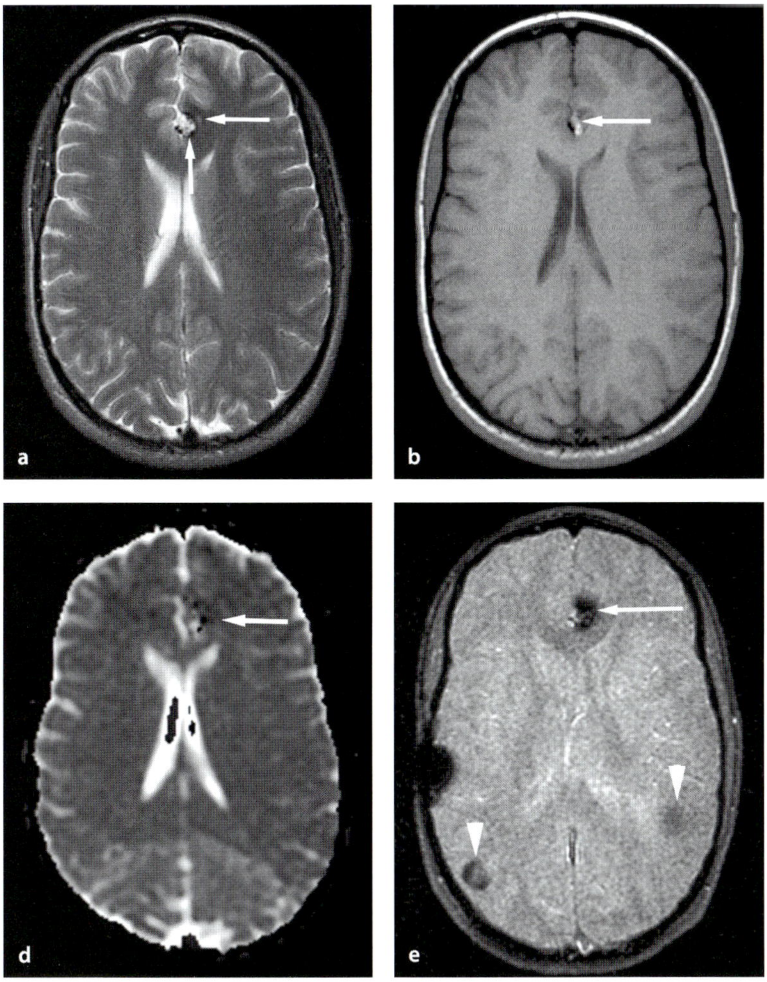

Figure 6.9 a–e

Multiple cavernous angiomas. The T2-weighted MR image (**a**) in a 30-year-old woman with seizures shows a hyperintense lesion in the left frontal lobe with a surrounding hypointense rim (*arrows*). This is a characteristic finding for a cavernous angioma. The lesion is hyperintense on the T1-weighted image (**b**) (*arrow*) and hypointense on the DW image (**c**) (*arrow*). The ADC map (**d**) shows heterogeneous intensity and the GRE image (**e**) shows marked hypointensity in the left frontal lobe (*arrow*). The hypointensities in the right and left temporo-occipital region (*arrowheads*) suggest multiple cavernous angiomas

6.8 Hemorrhage Related to Trauma

Trauma is one of the most common causes of intracranial hemorrhages in younger patients (Fig. 6.10). MR imaging is valuable in detecting intracranial injuries. DW and b_0 images and an ADC map can be more sensitive than conventional MR images to detect whether the abnormality includes diffuse axonal injury [84].

6.9 Conclusions

Diffusion-weighted imaging is often of limited value for diagnosis and staging of intracranial hemorrhages because accurate ADC measurements are only possible in the hyperacute stage, which contains diamagnetic oxy-hemoglobin, and in the late subacute phase, which contains extracellular met-hemoglobin, whose paramagnetic susceptibility artifacts are diminished by the dilution of extracellular fluid. CT and routine MR imaging continue to be the mainstay in diagnosing and characterizing intracranial hemorrhages. A thorough understanding of DW imaging characteristics is important, however, in order to avoid misinterpretations and inaccurate conclusions.

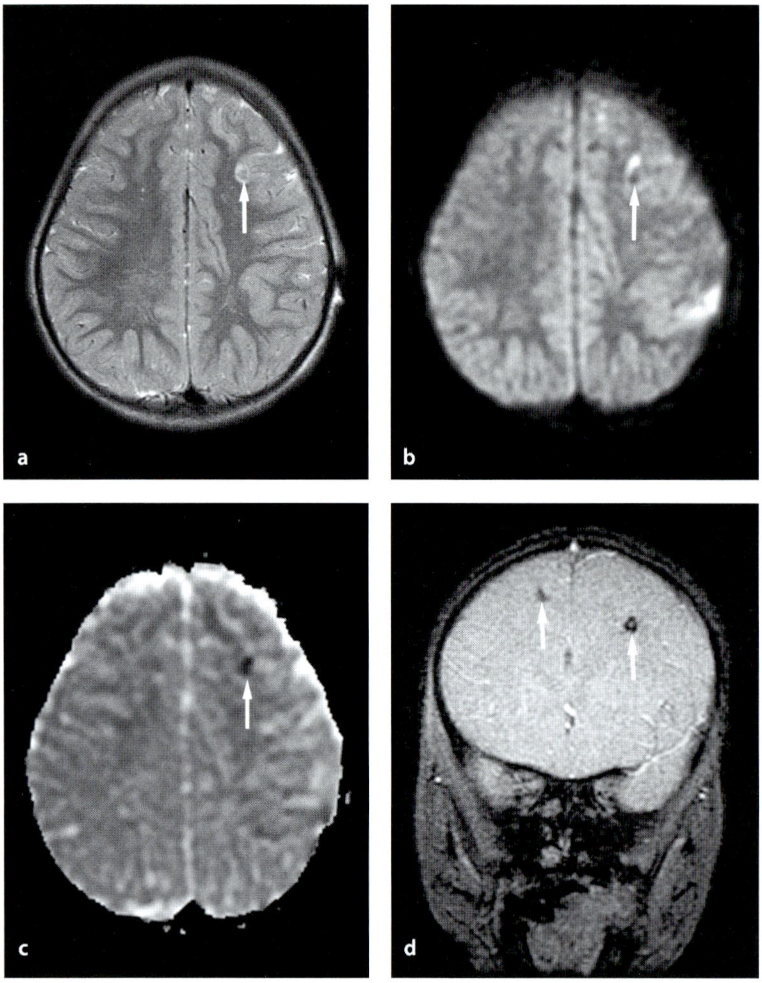

Figure 6.10 a–d

Trauma with hemorrhagic diffuse axonal injury. MR imaging of a 2-year-old girl 12 hours after a motor vehicle accident shows (**a**) a hyperintense lesion in the left frontal lobe on T2-weighted image (*arrow*). The corresponding area on the DW image (**b**) is hypointense with surrounding hyperintensity (*arrow*). The ADC map (**c**) shows hypointensity (*arrow*). The coronal GRE (**d**) image shows hypointense lesions in the bilateral frontal lobes (*arrows*)

References

1. Thulborn KR, Waterton JC, Matthews PM, Radda GK (1982) Oxygenation dependence of the transverse relaxation time of water protons in whole blood at high field. Biochim Biophys Acta 714:265–270
2. Fullerton GD, Potter JL, Dornbluth NC (1982) NMR relaxation of protons in tissues and other macromolecular water solutions. Magn Reson Imaging 1:209–226
3. Sipponen JT, Sepponen RE, Sivula A (1983) Nuclear magnetic resonance (NMR) imaging of intracerebral hemorrhage in the acute and resolving phases. J Comput Assist Tomogr 7:954–959
4. DeLaPaz RL, New PF, et al. (1984) NMR imaging of intracranial hemorrhage. J Comput Assist Tomogr 8:599–607
5. Bradley WG Jr, Schmidt PG (1985) Effect of methemoglobin formation on the MR appearance of subarachnoid hemorrhage. Radiology 156:99–103
6. Gomori JM, Grossman RI, Bilaniuk LT, Zimmerman RA, Goldberg HI (1985) High-field MR imaging of superficial siderosis of the central nervous system. J Comput Assist Tomogr 9:972–975
7. Gomori JM, Grossman RI, Goldberg HI, Zimmerman RA, Bilaniuk LT (1985) Intracranial hematomas: imaging by high-field MR. Radiology 157:87–93
8. Sipponen JT, Sepponen RE, Tanttu JI, Sivula A (1985) Intracranial hematomas studied by MR imaging at 0.17 and 0.02 T. J Comput Assist Tomogr 9:698–704
9. Grossman RI, Kemp SS, Ip CY, et al. (1986) Importance of oxygenation in the appearance of acute subarachnoid hemorrhage on high field magnetic resonance imaging. Acta Radiol Suppl 369:56–58
10. Di Chiro G, Brooks RA, Girton ME, et al. (1986) Sequential MR studies of intracerebral hematomas in monkeys. AJNR Am J Neuroradiol 7:193–199
11. Gomori JM, Grossman RI, Yu-Ip C, Asakura T (1987) NMR relaxation times of blood: dependence on field strength, oxidation state, and cell integrity. J Comput Assist Tomogr 11:684–690
12. Zimmerman RD, Heier LA, Snow RB, Liu DP, Kelly AB, Deck MD (1988) Acute intracranial hemorrhage: intensity changes on sequential MR scans at 0.5 T. AJR Am J Roentgenol 150:651–661
13. Hayman LA, Ford JJ, Taber KH, Saleem A, Round ME, Bryan RN (1988) T2 effect of hemoglobin concentration: assessment with in vitro MR spectroscopy. Radiology 168:489–491
14. Brooks RA, Di Chiro G, Patronas N (1989) MR imaging of cerebral hematomas at different field strengths: theory and applications. J Comput Assist Tomogr 13:194–206
15. Hayman LA, McArdle CB, Taber KH, et al. (1989) MR imaging of hyperacute intracranial hemorrhage in the cat. AJNR Am J Neuroradiol 10:681–686
16. Hayman LA, Taber KH, Ford JJ, et al. (1989) Effect of clot formation and retraction on spin-echo MR images of blood: an in vitro study. AJNR Am J Neuroradiol 10:1155–1158
17. Hardy PA, Kucharczyk W, Henkelman RM (1990) Cause of signal loss in MR images of old hemorrhagic lesions. Radiology 174:549–555
18. Bryant RG, Marill K, Blackmore C, Francis C (1990) Magnetic relaxation in blood and blood clots. Magn Reson Med 13:133–144
19. Clark RA, Watanabe AT, Bradley WG Jr, Roberts JD (1990) Acute hematomas: effects of deoxygenation, hematocrit, and fibrin-clot formation and retraction on T2 shortening. Radiology 175:201–206
20. Thulborn KR, Sorensen AG, Kowall NW, et al. (1990) The role of ferritin and hemosiderin in the MR appearance of cerebral hemorrhage: a histopathologic biochemical study in rats. AJNR Am J Neuroradiol 11:291–297
21. Bizzi A, Brooks RA, Brunetti A, et al. (1990) Role of iron and ferritin in MR imaging of the brain: a study in primates at different field strengths. Radiology 177:59–65
22. Bradley WG Jr (1993) MR appearance of hemorrhage in the brain. Radiology 189:15–26
23. Ekholm S (1996) Intracranial hemorrhages. Rivista di Neuroradiologia 9 (Suppl 1):17–21
24. Ebisu T, Tanaka C, Umeda M, et al. (1997) Hemorrhagic and nonhemorrhagic stroke: diagnosis with diffusion-weighted and T2-weighted echo-planar MR imaging. Radiology 203:823–828
25. Atlas SW, Thulborn KR (1998) MR detection of hyperacute parenchymal hemorrhage of the brain. AJNR Am J Neuroradiol 19:1471–1477
26. Liang L, Korogi Y, Sugahara T, et al. (1999) Detection of intracranial hemorrhage with susceptibility-weighted MR sequences. AJNR Am J Neuroradiol 20:1527–1534
27. Kinoshita T, Okudera T, Tamura H, Ogawa T, Hatazawa J (2000) Assessment of lacunar hemorrhage associated with hypertensive stroke by echo-planar gradient-echo T2*-weighted MRI. Stroke 31:1646–1650
28. Atlas SW, DuBois P, Singer MB, Lu D (2000) Diffusion measurements in intracranial hematomas: implications for MR imaging of acute stroke. AJNR Am J Neuroradiol 21:1190–1194
29. Schaefer PW, Grant PE, Gonzalez RG (2000) Diffusion-weighted MR imaging of the brain. Radiology 217:331–345
30. Maldjian JA, Listerud J, Moonis G, Siddiqi F (2001) Computing diffusion rates in T2-dark hematomas and areas of low T2 signal. AJNR Am J Neuroradiol 22:112–128
31. Lin DD, Filippi CG, Steever AB, Zimmerman RD (2001) Detection of intracranial hemorrhage: comparison between gradient-echo images and b(0) images obtained from diffusion-weighted echo-planar sequences. AJNR Am J Neuroradiol 22:1275–1281
32. Hermier M, Nighoghossian N, Derex L, et al. (2001) MRI of acute post-ischemic cerebral hemorrhage in stroke patients: diagnosis with T2*-weighted gradient-echo sequences. Neuroradiology 43:809–815
33. Kang BK, Na DG, Ryoo JW, Byun HS, Roh HG, Pyeun YS (2001) Diffusion-weighted MR imaging of intracerebral hemorrhage. Korean J Radiol 2:183–191
34. Atlas SW, Thulborn KR (2002) Intracranial hemorrhage. In: Magnetic resonance imaging of the brain and spine. Lippincott Williams & Wilkins, Philadelphia, pp 773–832
35. Latour LL, Svoboda K, Mitra PP, Sotak CH (1994) Time-dependent diffusion of water in a biological model system. Proc Natl Acad Sci USA 91:1229–1233
36. Does MD, Zhong J, Gore JC (1999) In vivo measurement of ADC change due to intravascular susceptibility variation. Magn Reson Med 41:236–240

37. Harrison PM, Fischbach FA, Hoy TG, Haggis GH (1967) Ferric oxyhydroxide core of ferritin. Nature 216:1188–1190
38. Munro HN, Linder MC (1978) Ferritin: structure, biosynthesis, and role in iron metabolism. Physiol Rev Apr;58(2):317–396
39. Chakeres DW, Bryan RN (1986) Acute subarachnoid hemorrhage: in vitro comparison of magnetic resonance and computed tomography. AJNR Am J Neuroradiol 7:223–228
40. Jenkins A, Hadley DM, Teasdale GM, Condon B, Macpherson P, Patterson J (1988) Magnetic resonance imaging of acute subarachnoid hemorrhage. J Neurosurg 68:731–736
41. Satoh S, Kadoya S (1988) Magnetic resonance imaging of subarachnoid hemorrhage. Neuroradiology 30:361–366
42. Atlas SW (1993) MR imaging is highly sensitive for acute subarachnoid hemorrhage ... not! Radiology 186:319–322
43. Ogawa T, Uemura K (1993) MR imaging is highly sensitive for acute subarachnoid hemorrhage ... not! Reply. Radiology 186:323
44. Ogawa T, Inugami A, Shimosegawa E, et al. (1993) Subarachnoid hemorrhage: evaluation with MR imaging. Radiology 186:345–351
45. Griffiths PD, Wilkinson ID, Mitchell P, et al. (2002) Multimodality MR imaging depiction of hemodynamic changes and cerebral ischemia in subarachnoid hemorrhage. AJNR Am J Neuroradiol 22:1690–1697
46. Noguchi K, Ogawa T, Inugami A, Toyoshima H, Okudera T, Uemura K (1994) MR of acute subarachnoid hemorrhage: a preliminary report of fluid-attenuated inversion-recovery pulse sequences. AJNR Am J Neuroradiol 15:1940–1943
47. Noguchi K, Ogawa T, Inugami A, et al. (1995) Acute subarachnoid hemorrhage: MR imaging with fluid-attenuated inversion recovery pulse sequences. Radiology 196:773–777
48. Noguchi K, Ogawa T, Seto H, et al. (1997) Subacute and chronic subarachnoid hemorrhage: diagnosis with fluid-attenuated inversion-recovery MR imaging. Radiology 203(1):257–262
49. Noguchi K, Seto H, Kamisaki Y, Tomizawa G, Toyoshima S, Watanabe N (2000) Comparison of fluid-attenuated inversion-recovery MR imaging with CT in a simulated model of acute subarachnoid hemorrhage. AJNR Am J Neuroradiol 21:923–927
50. Melhem ER, Jara H, Eustace S (1997) Fluid-attenuated inversion recovery MR imaging: identification of protein concentration thresholds for CSF hyperintensity. AJR Am J Roentgenol 169:859–862
51. Singer MB, Atlas SW, Drayer BP (1998) Subarachnoid space disease: diagnosis with fluid-attenuated inversion-recovery MR imaging and comparison with gadolinium-enhanced spin-echo MR imaging – blinded reader study. Radiology 208:417–422
52. Dechambre SD, Duprez T, Grandin CB, Lecouvet FE, Peeters A, Cosnard G (2000) High signal in cerebrospinal fluid mimicking subarachnoid haemorrhage on FLAIR following acute stroke and intravenous contrast medium. Neuroradiology 42:608–611
53. Taoka T, Yuh WT, White ML, Quets JP, Maley JE, Ueda T (2001) Sulcal hyperintensity on fluid-attenuated inversion recovery mr images in patients without apparent cerebrospinal fluid abnormality. AJR Am J Roentgenol 176:519–524
54. Wiesmann M, Mayer TE, Yousry I, Medele R, Hamann GF, Bruckmann H (2002) Detection of hyperacute subarachnoid hemorrhage of the brain by using magnetic resonance imaging. J Neurosurg 96:684–689
55. Busch E, Beaulieu C, de Crespigny A, Moseley ME (1998) Diffusion MR imaging during acute subarachnoid hemorrhage in rats. Stroke 29:2155–2161
56. Rordorf G, Koroshetz WJ, Copen WA, et al. (1999) Diffusion- and perfusion-weighted imaging in vasospasm after subarachnoid hemorrhage. Stroke 30:599–605
57. Domingo Z, Bradley JK, Blamire AM, Brindle K, Styles P, Rajagopalan B (2000) Diffusion weighted imaging and magnetic resonance spectroscopy in a low flow ischaemia model due to endothelin induced vasospasm. NMR Biomed 13:154–162
58. Condette-Auliac S, Bracard S, Anxionnat R, et al. (2001) Vasospasm after subarachnoid hemorrhage: interest in diffusion-weighted MR imaging. Stroke 32:1818–1824
59. Hadeishi H, Suzuki A, Yasui N, Hatazawa J, Shimosegawa E (2002) Diffusion-weighted magnetic resonance imaging in patients with subarachnoid hemorrhage. Neurosurgery 50:741–747
60. Leclerc X, Fichten A, Gauvrit JY, et al. (2002) Symptomatic vasospasm after subarachnoid haemorrhage: assessment of brain damage by diffusion and perfusion-weighted MRI and single-photon emission computed tomography. Neuroradiology 44:610–616
61. Fobben ES, Grossman RI, Atlas SW, Hackney DB, Goldberg HI, Zimmerman RA, Bilaniuk LT (1989) MR characteristics of subdural hematomas and hygromas at 1.5 T. AJR Am J Roentgenol 153:589–595
62. Ebisu T, Naruse S, Horikawa Y, Tanaka C, Higuchi T (1989) Nonacute subdural hematoma: fundamental interpretation of MR images based on biochemical and in vitro MR analysis. Radiology 171:449–453
63. Wilms G, Marchal G, Geusens E, et al. (1992) Isodense subdural haematomas on CT: MRI findings. Neuroradiology 34:497–499
64. Ashikaga R, Araki Y, Ishida O (1997) MRI of head injury using FLAIR. Neuroradiology 39:239–242
65. Williams VL, Hogg JP (2000) Magnetic resonance imaging of chronic subdural hematoma. Neurosurg Clin N Am Jul;11(3):491–498
66. Tsui EY, Fai Ma K, Cheung YK, Chan JH, Yuen MK (2000) Rapid spontaneous resolution and redistribution of acute subdural hematoma in a patient with chronic alcoholism: a case report. Eur J Radiol 36:53-57
67. Lee Y, Lee KS, Hwang DH, Lee IJ, Kim HB, Lee JY (2001) MR imaging of shaken baby syndrome manifested as chronic subdural hematoma. Korean J Radiol 2:171–174
68. Biousse V, Suh DY, Newman NJ, Davis PC, Mapstone T, Lambert SR (2002) Diffusion-weighted magnetic resonance imaging in Shaken Baby Syndrome. Am J Ophthalmol Feb;133(2):249–255
69. Mesiwala AH, Goodkin R (2002) Reversible ischemia detected by diffusion-weighted magnetic resonance imaging. Case illustration. J Neurosurg 97:230
70. Bakshi R, Kamran S, Kinkel PR, et al. (1999) MRI in cerebral intraventricular hemorrhage: analysis of 50 consecutive cases. Neuroradiology 41:401–409

versible vasogenic edema (Fig. 7.2) and can demonstrate multiphasic infarctions (Fig. 7.3). Prompt diagnosis is important, as primary angitis of the central nervous system is often fatal if not treated with aggressive immunosuppression [17].

7.4.3 Giant Cell (Temporal) Arteritis

The criteria of the American College of Rheumatology for the diagnosis of giant cell arteritis (Fig. 7.4) include at least three of the following: (1) age at disease onset >50 years, (2) new onset of headache, (3) claudication of jaw or tongue, (4) tenderness of the temporal artery on palpation or decreased pulsation, (5) erythrocyte sedimentation ratio >50 mm/h and

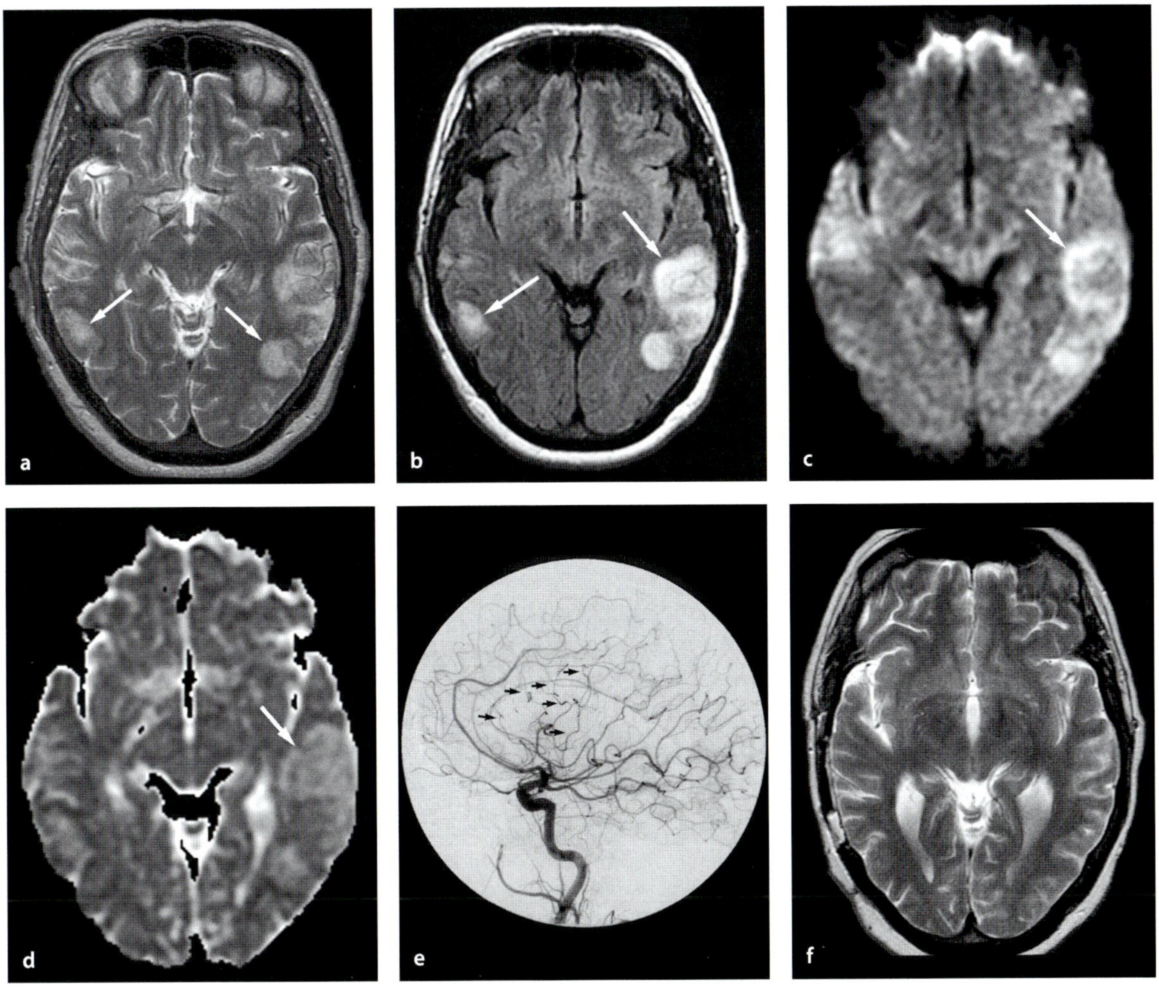

Figure 7.2 a–f

Primary angitis of central nervous system (proven by biopsy) in a 60-year-old woman with dizziness and speech difficulties. **a** T2-weighted image shows hyperintense lesions in the bilateral tempo-occipital cortices (*arrows*). **b** FLAIR image shows hyperintense lesions in the bilateral tempo-occipital cortices (*arrows*). **c** DW image shows slightly high signal in the lesions (*arrow*) with increased ADC (**d**), mainly representing vasogenic edema (*arrow*). **e** DSA shows multiple focal stenoses of distal branches of left middle cerebral arteries (*arrows*). **f** Two-month follow-up T2-weighted image shows no infarction in the bilateral tempo-occipital areas. (From [49])

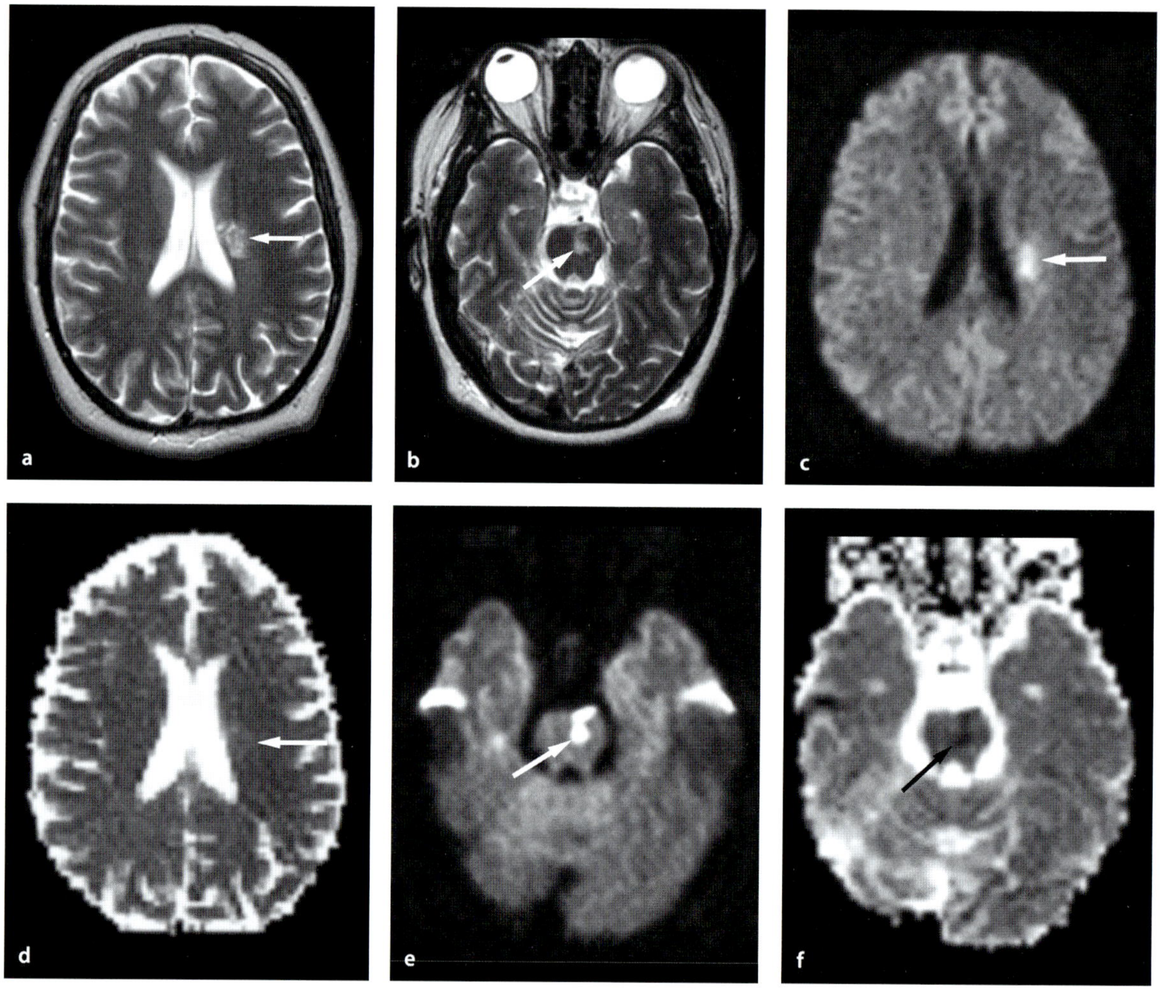

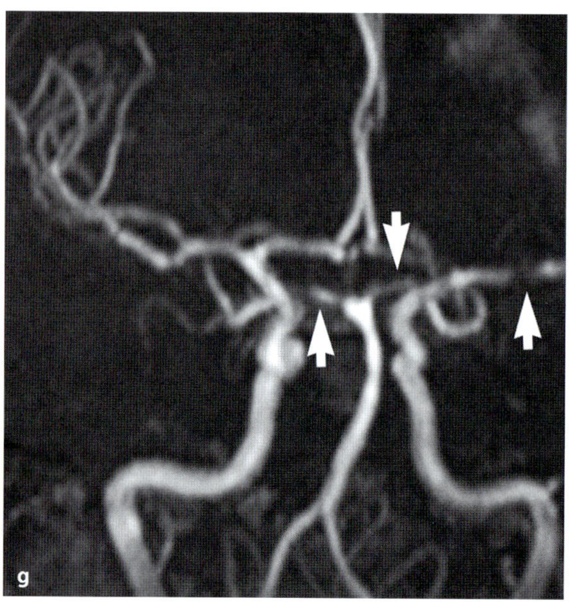

Figure 7.3 a–g

Primary angitis of central nervous system (proven by biopsy) in a 35-year-old woman with right hemiparesis. **a, b** T2-weighted image shows hyperintense lesions in the left internal capsule (*arrow* in **a**) and left side of the pons (*arrow* in **b**). **c** DW image shows slightly high signal in the left corona radiata, indicating subacute infarction. **a** T2-weighted image shows hyperintense lesions in the bilateral tempo-occipital cortices (*arrow*). **d** ADC map shows increased ADC in this lesion (*arrow*). **e** DW image simultaneously shows very high signal in the left side of the pons, indicating acute infarction (*arrow*). **f** ADC map shows decreased ADC in this lesion (*arrow*). **g** MR angiography shows stenosis in the left middle cerebral and posterior cerebral arteries (*arrows*). (From [49])

▶

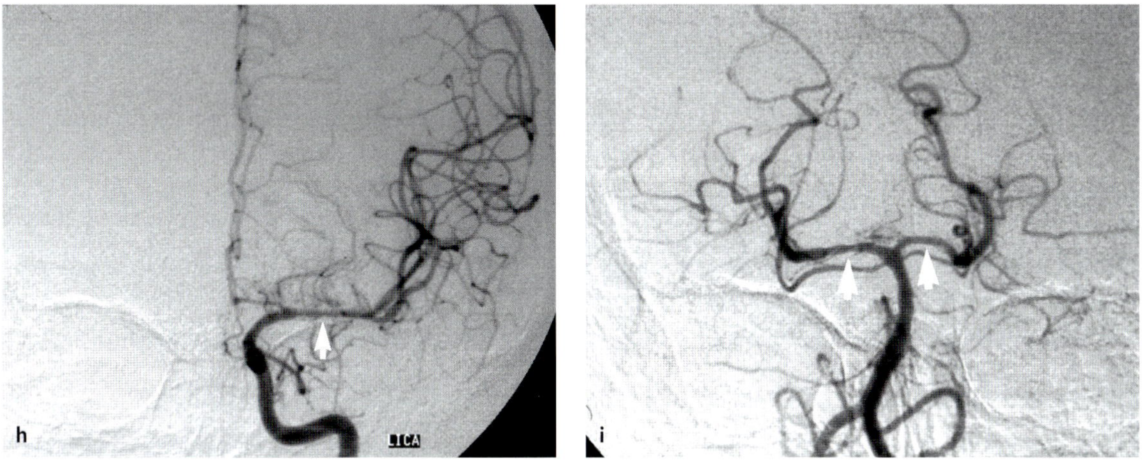

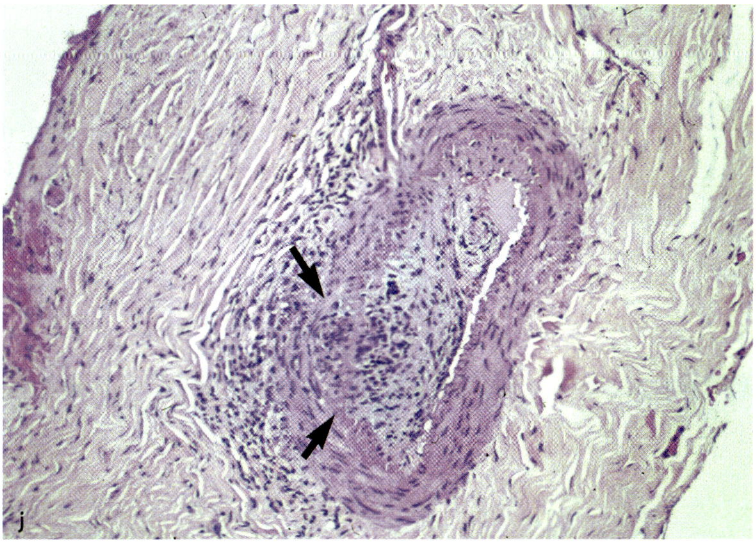

Figure 7.3 h–j

h, i Digital subtraction angiography (DSA) confirms the stenosis in the left middle cerebral (*arrow* in **h**) and posterior cerebral arteries (**i**) (*arrows* in **i**). **j** Pathological specimen by meningeal biopsy shows infiltration of the vessel walls with lymphocytes and multinucleated giant cells, with intramural granulomatous tissue formation (*arrows*). (From [49])

(6) temporal artery biopsy showing vasculitis with multinucleated giant cells.

Giant cell arteritis is probably a T cell-mediated vasculitis and it can affect medium to large arteries. The superficial temporal, vertebral and ophthalmic arteries are more commonly involved than the internal carotid arteries, while the intracranial arteries are rarely involved (Fig. 7.4) [18]. Abrupt and irreversible visual loss is the most dramatic complication of giant cell arteritis, while a TIA and stroke are rare (7%), but when present most often involve the vertebrobasilar territory. Steroids are effective, and giant cell arteritis is usually self-limited and rarely fatal.

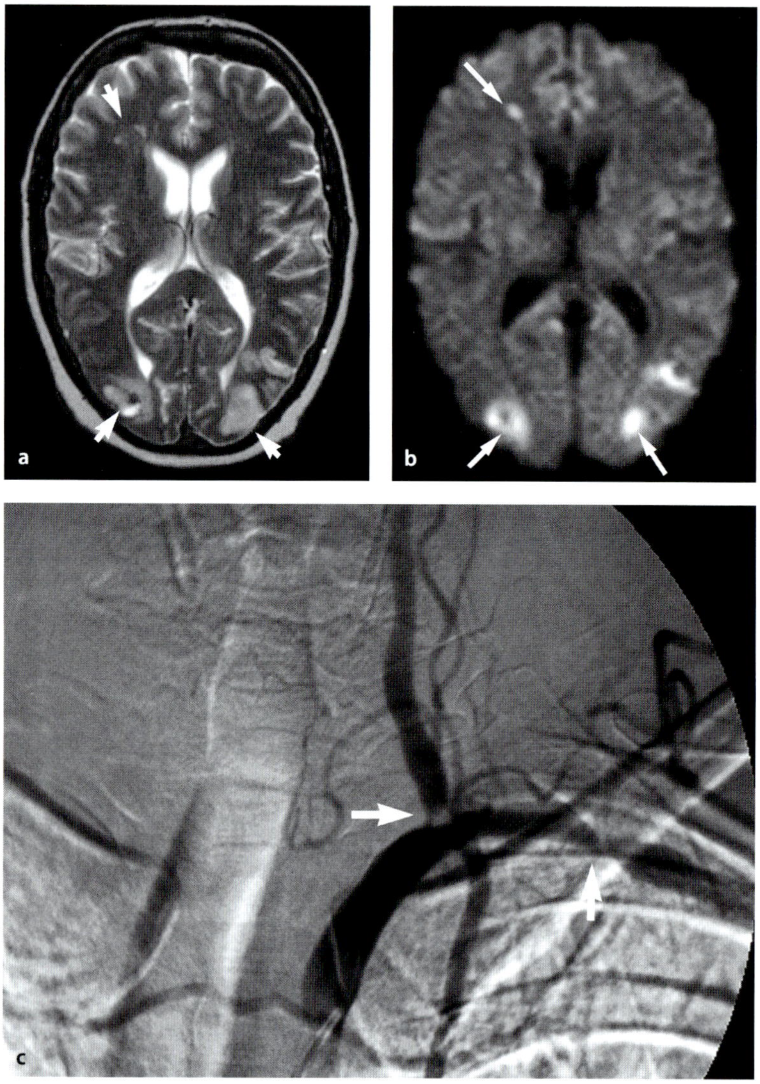

Figure 7.4 a–d

Giant cell arteritis (proven by biopsy) in a 48-year-old woman with visual loss. **a** T2-weighted image shows hyperintense lesions in the bilateral parieto-occipital cortices and right frontal deep white matter (*arrows*). **b** DW image shows these lesions as high signal intensity, representing acute infarcts (*arrows*). **c** DSA of the left subclavian artery shows stenoses of the left vertebral and subclavian arteries (*arrows*). **d** DSA of the left vertebral artery shows extensive multifocal arterial stenoses (*arrows*). (From [49])

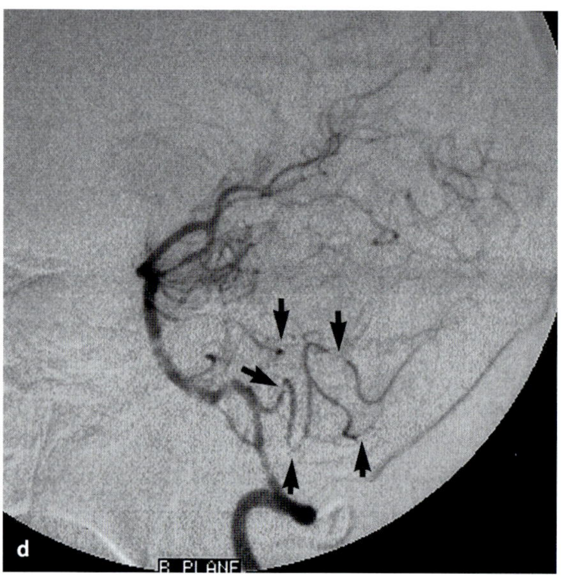

7.4.4 Takayasu's Arteritis (Aortitis Syndrome)

Takayasu's arteritis (Fig. 7.5) is a primary arteritis of unknown cause but probably also related to T cell-mediated inflammation. Takayasu's arteritis commonly affects large vessels including the aorta and its major branches to the arms and the head. It is more commonly seen in Asia and usually affects young women [19]. Pulseless upper extremities and hypertension are the common clues to suggest the diagnosis. Most patients are treated with steroids alone to reduce the inflammation. The prognosis is relatively good and 90% of patients are still alive after 10 years. TIA or stroke is rare but can occasionally occur in severe cases with significant stenosis of arteries supplying the CNS (Fig. 7.5).

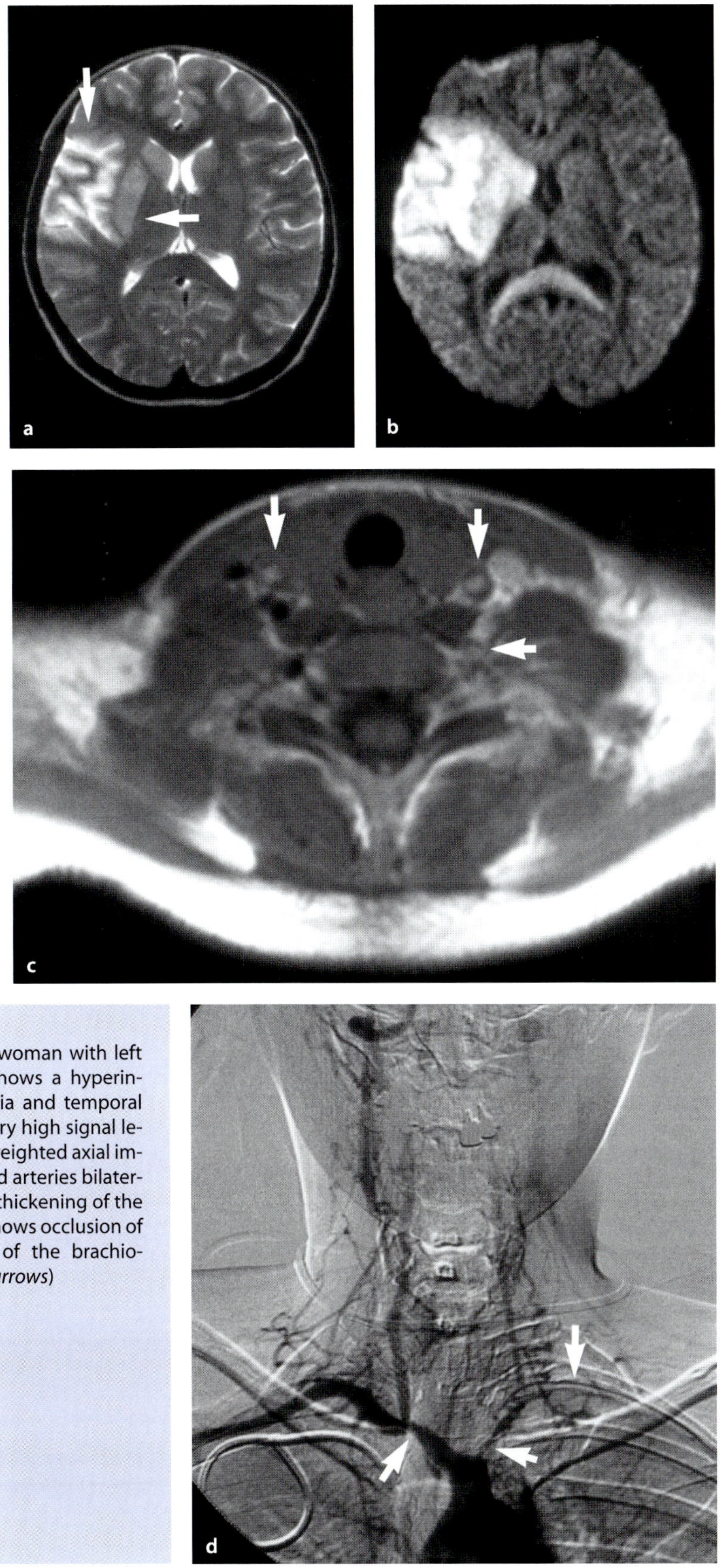

Figure 7.5 a–d

Takayasu's arteritis in an 18-year-old woman with left hemiparesis. **a** T2-weighted image shows a hyperintense lesion in the right basal ganglia and temporal lobe (*arrows*). **b** DW image shows a very high signal lesion, indicating an acute infarct. **c** T1-weighted axial image shows occlusion of internal carotid arteries bilaterally and the left vertebral artery, with thickening of the wall and thrombosis (*arrows*). **d** DSA shows occlusion of both carotid arteries, and stenosis of the brachiocephalic and left subclavian arteries (*arrows*)

7.4.5 Polyarteritis Nodosa

The criteria of the American College of Rheumatology for the diagnosis of polyarteritis nodosa include at least three of the following: (1) weight loss >4 kg, (2) livedo reticularis, (3) testicular pain or tenderness, (4) myalgias, weakness or leg tenderness, (5) mono- or polyneuropathy, (6) hypertension, (7) elevated blood creatinine or blood urea nitrogen, (8) hepatitis B antigen or antibodies in the serum, (9) aneurysm or occlusion of the visceral arteries and (10) granulocytes in small or medium-sized arteries on vessel wall biopsy. Neurologic abnormalities occur in 25–50% of cases. Ischemic stroke can result from vasculitis, severe hypertension or embolism secondary to cardiac involvement [20]. The treatment usually requires both cytotoxic agents and steroids.

7.4.6 Churg–Strauss Disease

This is an antineutrophil cytoplasmic autoantibody-mediated vasculitis, defined by at least four of the following: (1) asthma, (2) history of allergy, (3) eosinophilia (>10%), (4) mono- or polyneuropathy, (5) migratory or transitory pulmonary infiltrates and (6) sinusitis. A biopsy of affected organs, including small arteries, arterioles or venules shows extravascular eosinophils, which confirms the diagnosis. Neurological involvement occurs in 62% of cases, including stroke and intracerebral hemorrhage (Fig. 7.6) [21, 22]. Steroids usually stabilize this condition, but treatment with cyclophosphamide may be required. A normal angiogram does not exclude this form of vasculitis, as affected vessels are often smaller than the resolution of angiography.

7.4.7 Other Small Vessel Vasculitis

The incidence of CNS involvement in Wegener's granulomatosis varies from 11 to 44%, but stroke is a very rare complication [23]. Involvement of the CNS in other forms of small vessel vasculitis (microscopic polyangitis, Henoch–Schönlein purpura, essential cryoglobulinemia and hypersensitivity vasculitis) is rare, but Henoch–Schönlein purpura may show reversibility of lesions on MRI [7].

7.4.8 Collagen Vascular Diseases

Behçet's disease is a multisystem vasculitis of unknown origin. It is especially common in Middle Eastern and Mediterranean countries. CNS involvement has been described in 4–49% of cases [6]. The parenchymal distribution of lesions, especially at the mesodiencephalic junction (46%) supports small vessel vasculitis involving both the arterial and venous systems; mainly venules. The lesions are occasionally reversible on MRI, which mainly represents vasogenic edema, which is why DW imaging is useful in distinguishing them from infarction (Fig. 7.7). The treatment is usually a combination of cytotoxic agents and steroids. In other types of collagen diseases, such as scleroderma or rheumatoid arthritis, involvement of the CNS is very rare.

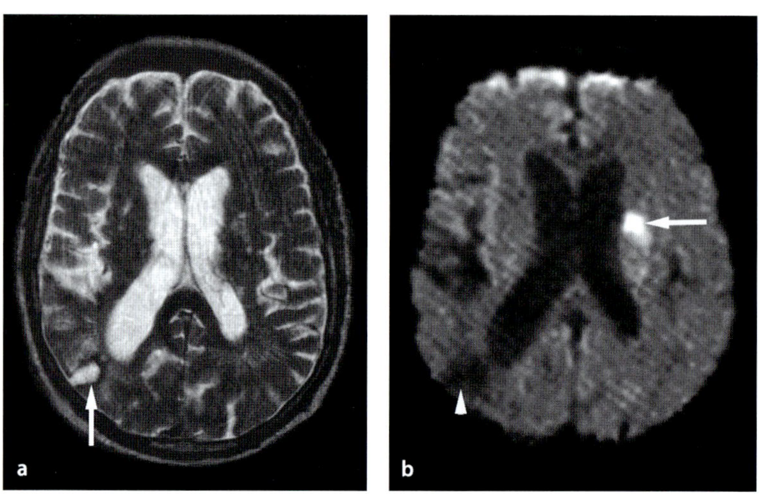

Figure 7.6 a,b

Churg–Strauss disease in a 65-year-old man with seizures. **A** T2-weighted image shows multiple hyperintense lesions in both corona radiata and right parieto-occipital area, the latter with a hemosiderin rim, representing old hemorrhage (*arrow*). **b** DW image shows a hyperintense lesion (*arrow*) in the left corona radiata, representing an acute infarct. An old hemorrhage shows hypointensity (*arrowhead*) on DW image. MR and DS angiography revealed no abnormalities (not shown). (From [49])

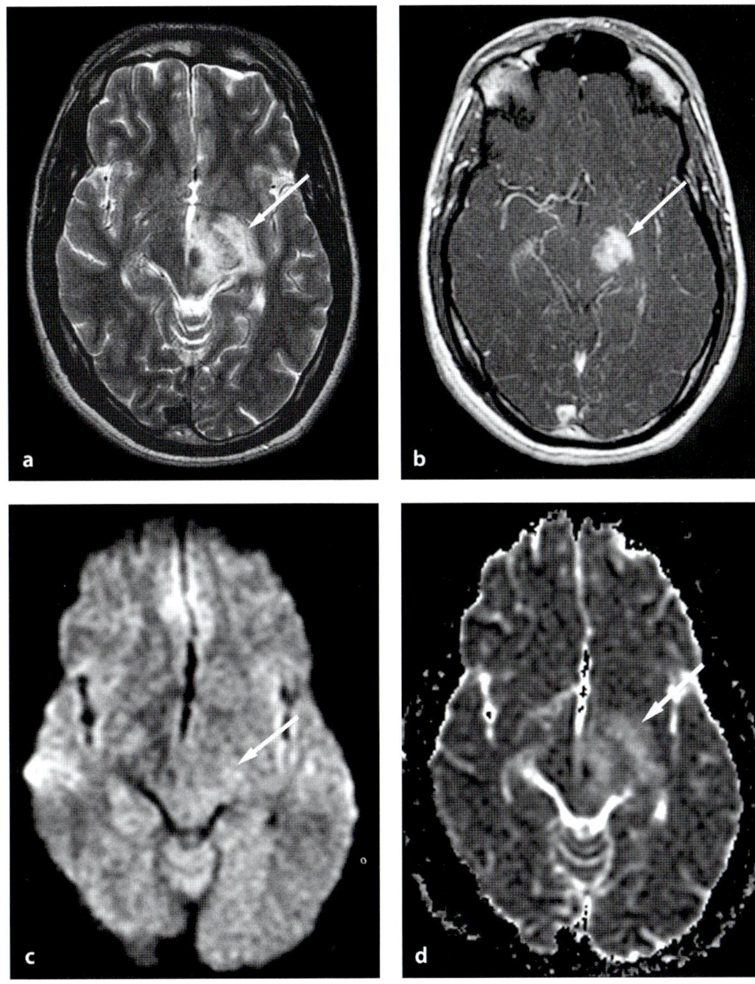

Figure 7.7 a–d

Neuro-Behçet's disease in a 24-year-old male. **a** T2-weighted image shows a hyperintense lesion in the left midbrain extending into the left temporal lobe, with enlargement of the left cerebral peduncle (*arrow*). **b** Gadolinium-enhanced T1-weighted image shows enhancement in this lesion (*arrow*). **c** DW image shows a hyperintense lesion with increased signal intensity in the left cerebral peduncle, probably representing vasogenic edema (*arrow*). **d** ADC map shows increased ADC in this lesion (*arrow*). (From [50])

7.4.9 Infectious Vasculitis

Infections can cause vasculitis both by direct invasion of the vessel walls and by an immune-mediated response to the pathogens. Bacterial, fungal and some viral vasculitis (e.g. herpes virus) cause a direct invasion of the vessel walls, usually resulting in infarction (Fig. 7.8) [2, 24]. Vasculitis with aseptic meningitis is probably related to an immunologic reaction, which can show reversible lesions. Aspergillus infiltrates and destroys the internal elastic lamina of major cerebral arteries, which results in infarction, abscess formation and hemorrhage [25] (Fig. 7.9). Infection of the infarcted tissue may be aggressive, and direct extension into the surrounding brain may progress quickly.

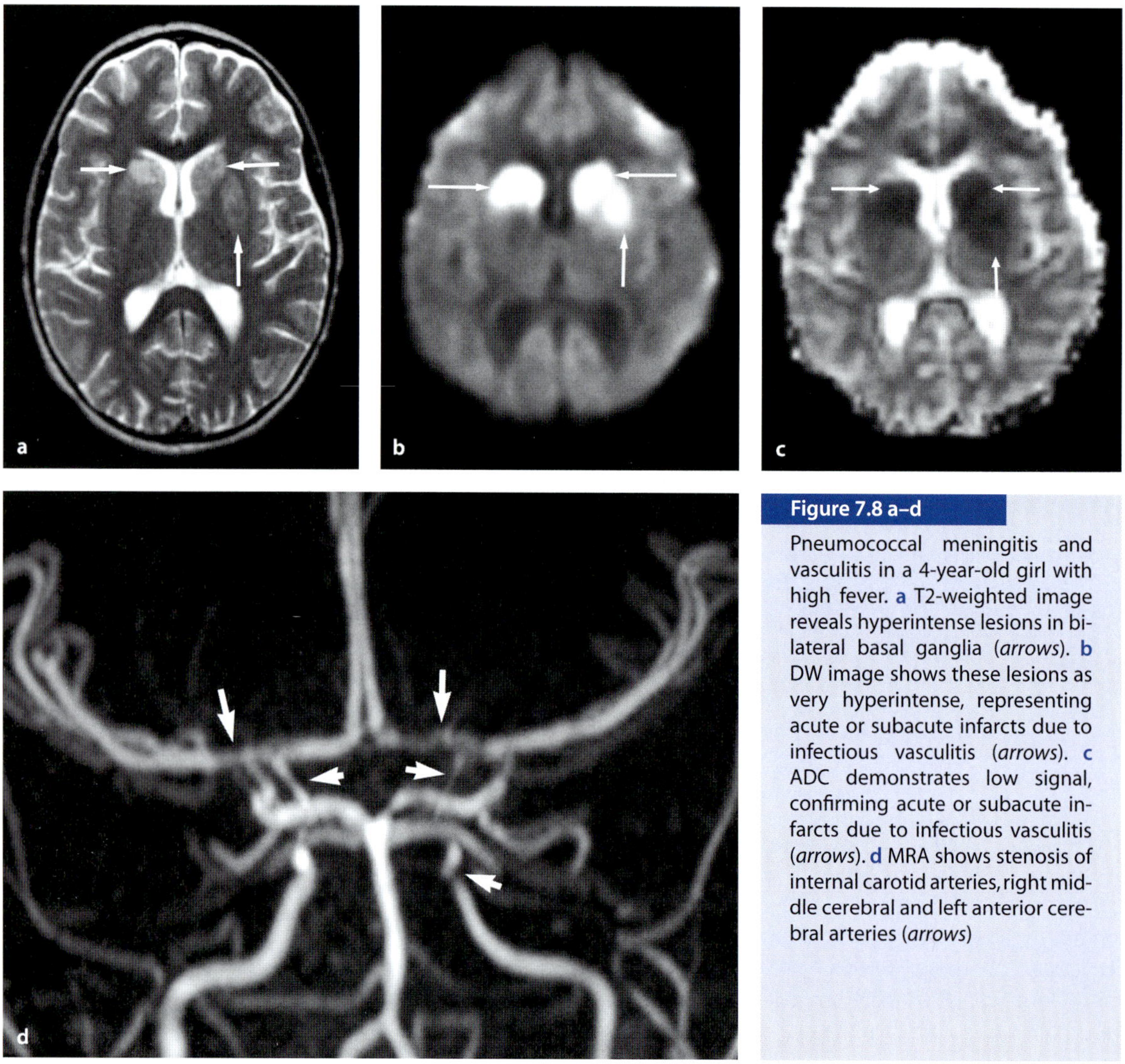

Figure 7.8 a–d

Pneumococcal meningitis and vasculitis in a 4-year-old girl with high fever. **a** T2-weighted image reveals hyperintense lesions in bilateral basal ganglia (*arrows*). **b** DW image shows these lesions as very hyperintense, representing acute or subacute infarcts due to infectious vasculitis (*arrows*). **c** ADC demonstrates low signal, confirming acute or subacute infarcts due to infectious vasculitis (*arrows*). **d** MRA shows stenosis of internal carotid arteries, right middle cerebral and left anterior cerebral arteries (*arrows*)

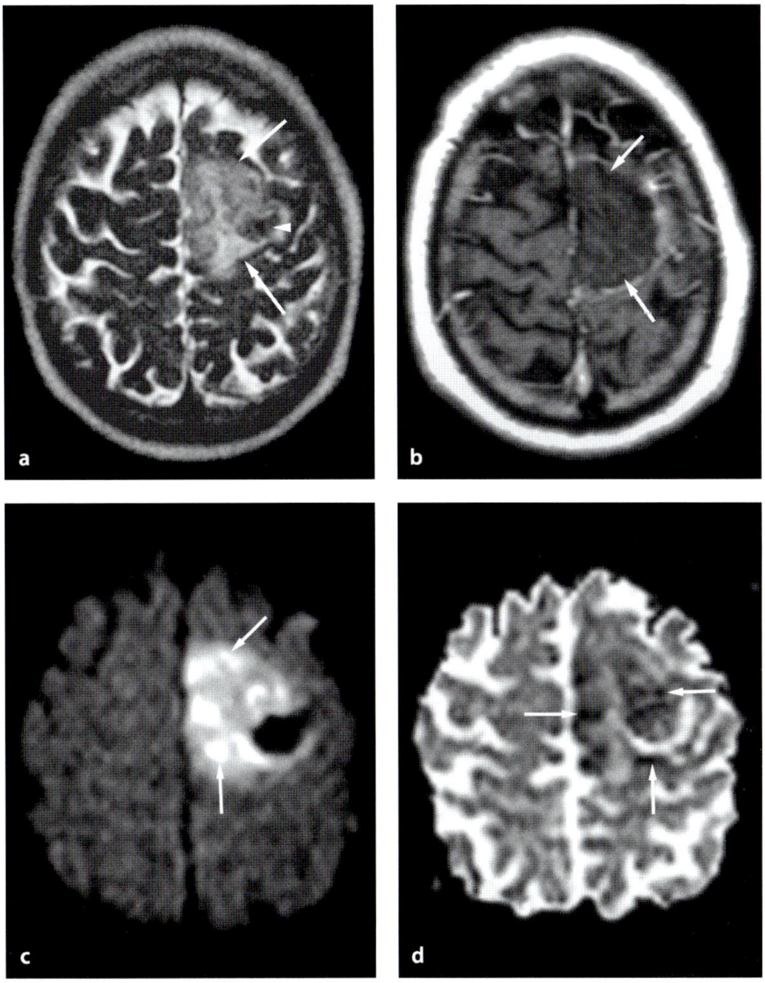

Figure 7.9 a–d

Disseminated aspergillosis in a 48-year-old female presenting with altered mental status and a history of acute lymphoblastic leukemia, post recent bone marrow transplant. **a** T2-weighted image shows a high signal intensity lesion (*arrow*) at the corticomedullary junction in the medial left frontal lobe. A hypointense spot in this lesion represents a hemorrhagic component (*arrowhead*). **b** Gadolinium-enhanced T1-weighted image reveals no enhancement within this lesion (*arrows*). **c** DW image shows this lesion containing areas of very high signal intensity (*arrows*), indicating a vasculitis mediated acute septic infarction. **d** ADC shows heterogeneous decreased signal (*arrows*) in the lesion, confirming a conglomerate of acute septic infarctions. (From [49])

7.4.10 Drug-Induced Vasculitis, Including Illicit Drugs

Some drugs, such as chemotherapeutic agents (e.g. sulfonamide, thiouracil) and illicit drugs (e.g. cocaine), can cause vasculitis [26]. Stroke can occur soon after administration of illicit drugs by an intravenous, oral or nasal route. Cocaine, heroin, amphetamine and other sympathomimetic drugs are most commonly implicated. The diagnosis of "vasculitis" depends on the pathological findings, not on the angiographic findings, which are usually non-specific and may simply indicate vasospasm induced by these drugs.

Cocaine use has emerged as an important cause of cerebrovascular events in young adults [27]. Vasculitic changes can be present on angiography, but the significance of these changes has been debated. However, elevated sedimentation rate and biopsy changes of vasculitis have been documented. MR angiography may reveal irregularity of the intracerebral vessels and DW imaging is useful for the detection of acute ischemic changes (Fig. 7.10).

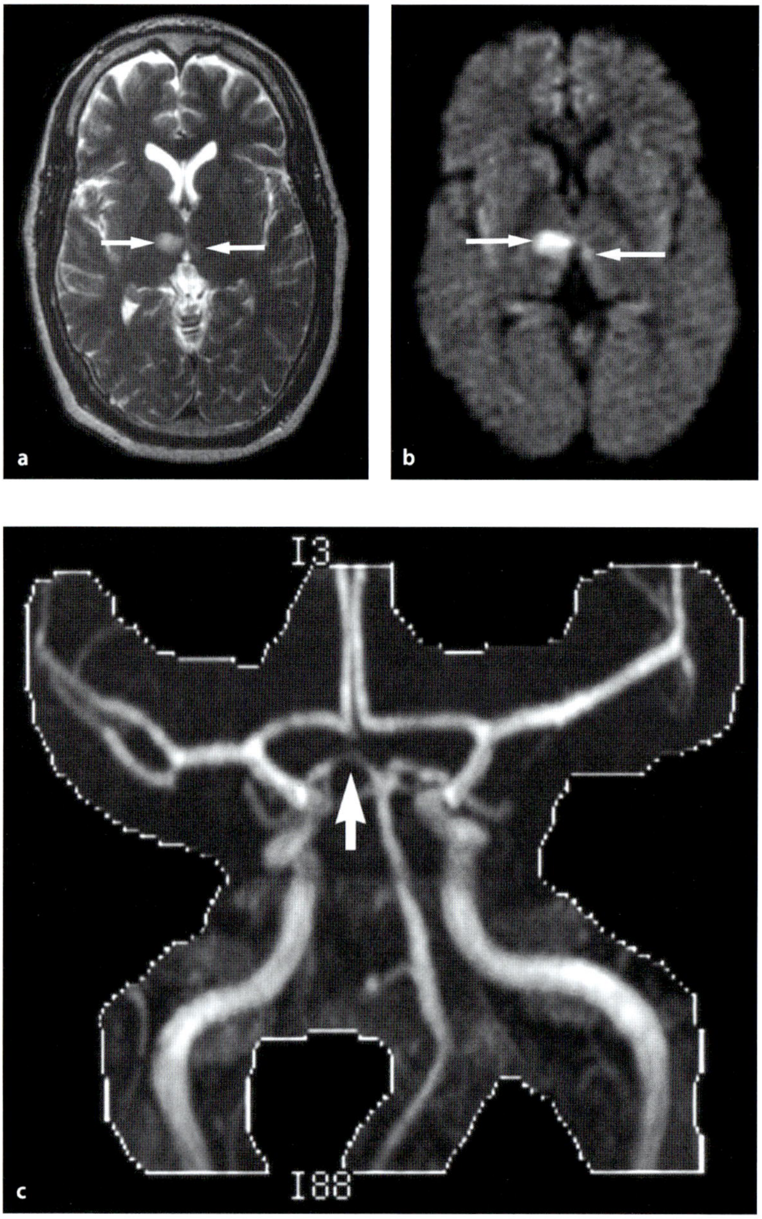

Figure 7.10 a–c

Cocaine-induced vasculopathy in a 41-year-old man with dysarthria. **a** T2-weighted image shows bilateral hyperintense lesions in medial thalami (*arrows*). **b** DW image shows the lesion in the right thalamus as hyperintense, indicating an acute infarct. **c** MR angiography shows stenosis of the right posterior cerebral artery (*arrow*). Biopsy was not performed and therefore the term cocaine-induced vasculopathy was used. (From [49])

7.5 Vasculopathy of the CNS

Vasculopathy is caused by a wide variety of underlying conditions such as degenerative, metabolic, inflammatory, embolic, coagulative and functional disorders [1]. This presentation focuses on vasculopathies that mimic vasculitis, but have no inflammation in the wall of the blood vessel (Fig. 7.11).

7.5.1 Systemic Lupus Erythematosus

Involvement of the CNS occurs in 14–75% of patients with systemic lupus erythematosus (SLE) [4]. Pathologically, microinfarcts and small vessel vasculopathy are the most common. Vasculopathy affects predominantly the arterioles and capillaries, resulting in vessel tortuosity, vascular hyalinization, endothelial proliferation and perivascular inflammation or gliosis.

Figure 7.11

Spectrum of CNS vasculopathy. (Modified from [10])

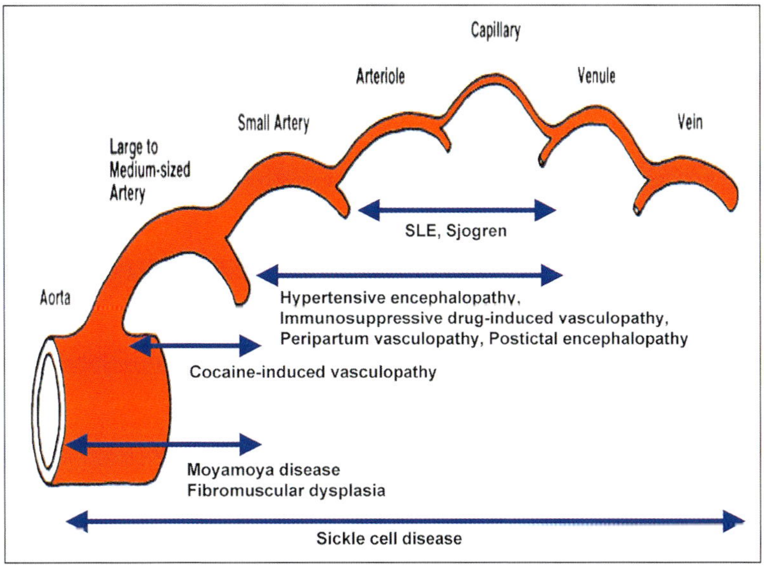

Figure 7.12 a–d

Systemic lupus erythematosus in a 39-year-old woman with recurrent episodes of stroke, who presented with fever and disturbance of consciousness. **a** T2-weighted image shows hyperintense lesions in the right thalamus, internal capsule, putamen, subcortical white matter, and the left internal capsule (*arrows*). **b** Gadolinium-enhanced T1-weighted image reveals marked enhancement of the lesion in the right side, suggesting blood–brain barrier breakdown (*arrows*). **c** DW image shows a slightly hyperintense lesion in the right thalamus but an isointense lesion in the right putamen and white matter (*arrows*). There is a linear hyperintense lesion in the right internal capsule (*long thin arrow*). A subtle hyperintense lesion in the left internal capsule is also seen (*arrowhead*). **d** The ADC map shows increased ADC of the lesion in the right side (*short thick arrows*), representing vasogenic edema. Increased ADC of the lesion in the left internal capsule (*arrowhead*) represents an old infarct. Decreased ADC is seen in the lesion in the right internal capsule (*long thin arrow*), presumably representing acute microinfarcts. (From [51])

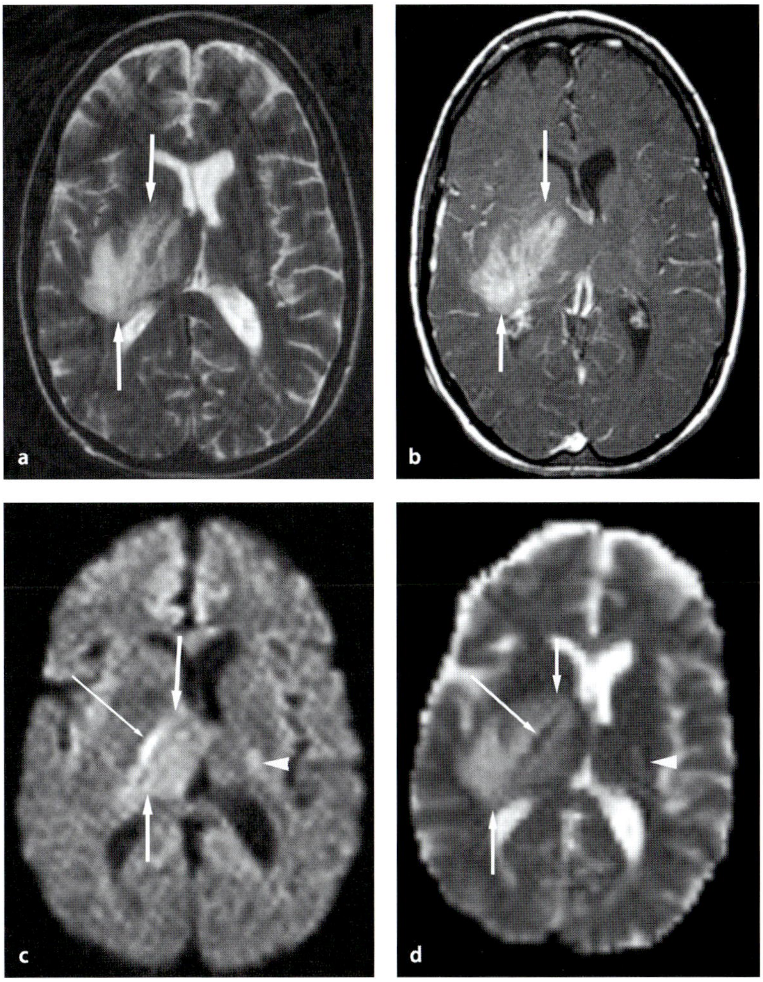

True vasculitis is very rare (0–7%). This vasculopathy may be related to both acute inflammation and ischemia [28]. In recent reports, the mechanism of vasculopathy in CNS involvement of SLE has been attributed to intravascular activation of a complement, which leads to adhesion between neutrophils and/or platelets and endothelium, resulting in leukothrombosis in the microvasculature (Shwartzman phenomenon) [29].

In this vasculopathy, despite widespread microvascular occlusions, parenchymal damage is minimal and potentially reversible. Sibbit et al. reported that up to 38% of CNS lesions in SLE were potentially reversible on MR imaging [30]. MR angiography and conventional angiography may provide additional information concerning vascular abnormalities. DW imaging shows primarily two patterns of parenchymal lesions with acute or subacute CNS symptoms: one is an acute or subacute infarction, and the other is vasogenic edema with or without microinfarcts (Fig. 7.12) [5]. CNS involvement in SLE is also due to associated uremia, hypertension, infection, Libman–Sacks endocarditis, and corticosteroid or immunosuppressive therapy.

7.5.2 Moyamoya Disease

Moyamoya disease is a rare, non-inflammatory vasculopathy of the intracranial vessels of unknown cause, which is found predominantly in East Asia. It has a bimodal age presentation, the first in childhood (first decade) and the second in adults (fourth decade). Endothelial thickening of the cellular fibrous tissue, the main pathological finding, leads to chronic progressive arterial stenosis of the circle of Willis and eventually to infarctions. In the adult form, the presenting symptom is often intracranial hemorrhage, usually intraparenchymal. The stenosis or occlusion of the supraclinoid portion of the internal carotid artery should be bilateral, but unilateral lesions can be included as "probable" cases of Moyamoya disease. DW imaging is useful in evaluating cerebral ischemia in Moyamoya disease (Fig. 7.13) [31].

7.5.3 Sickle Cell Disease

About 5–8% of patients with sickle cell disease develop symptomatic cerebrovascular disease. The risk of stroke is greatest during thrombotic crises and during the first 15 years of life [32]. Approximately 75% of strokes are the result of an occlusion of the large arteries at the base of the brain. Cortical and white matter watershed ischemia is common. This vasculopathy can be similar to Moyamoya disease. Occlusion of the small cortical branches, which leads to ischemia of deep white matter, accounts for 25% of cerebral infarctions. These lesions are thought to be related to peripheral vaso-occlusive events in which the arteriole or postcapillary venule is the major site of sickle cell adhesion [32]. If progression is associated with neurologic dysfunction, strong consideration should be given to place the patient on a long-term transfusion program. DW imaging is useful in detecting active ischemic changes and in differentiating them from chronic ischemic changes (Fig. 7.14).

7.5.4 Posterior Reversible Encephalopathy Syndrome

Posterior reversible encephalopathy syndrome is a remarkably heterogeneous group of disorders, related to hypertensive encephalopathy, severe preeclampsia/eclampsia, immunosuppressive drug or interferon neurotoxicity, uremia and thrombotic thrombocytopenic purpura [33]. The clinical symptoms are headache, altered mental status, seizures and visual loss. Posterior reversible encephalopathy syndrome is due to dysfunction of a cerebrovascular autoregulatory system, and a vasculopathy of small vessels, the arterioles. Brain perfusion is maintained by the autoregulatory system of the small arteries and arterioles that have myogenic and neurogenic components. Since the vertebrobasilar system and posterior cerebral arteries are sparsely innervated by sympathetic nerves, the occipital lobes and other posterior brain regions are susceptible to a breakthrough of the autoregulation in case of a sudden elevation of the systemic blood pressure. Endothelial damage, which can attenuate the myogenic response of the autoregulatory system, is hypothesized to be the cause of the posterior reversible encephalopathy syndrome.

Other names for the posterior reversible encephalopathy syndrome are reversible posterior leukoencephalopathy syndrome, or posterior leukoencephalopathy syndrome [34, 35]. The lesion can also involve the cerebral cortex and another name for posterior reversible encephalopathy syndrome is occipital parietal encephalopathy [36]. Similar lesions can also be seen in the frontal lobes, basal ganglia, brain stem

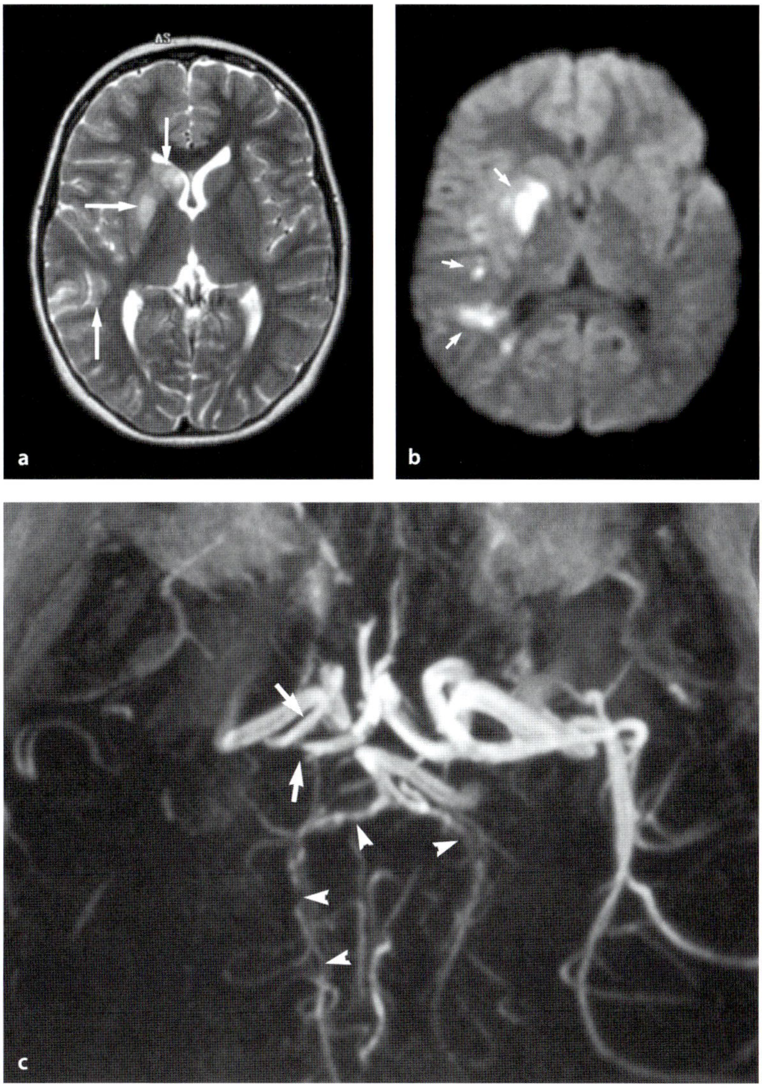

Figure 7.13 a–c

Probable Moyamoya disease in a 7-year-old girl with left hemiparesis. **a** T2-weighted image shows mild high signal lesions in the right basal ganglia and parieto-occipital cortex. **b** DW image clearly shows these lesions as high signal intensity, representing acute infarcts. **c** MR angiography shows occlusion of the right middle cerebral artery and stenosis of the right internal carotid artery and bilateral posterior cerebral arteries (*arrows*).

and cerebellum. In addition, the posterior reversible encephalopathy syndrome may not be entirely reversible, as infarction and hemorrhage may develop.

7.5.5 Hypertensive Encephalopathy

The primary cause of hypertensive encephalopathy is thought to be fluid extravasation through the interstitium, resulting from overdistension of the distal small cerebral vessels (breakthrough of autoregulation) causing vasogenic edema [37]. Ischemic processes can be triggered by vasospasm of the cerebral arteriole in response to a severe increase in blood pressure (overregulation). This will usually result in infarctions. Hypertensive encephalopathy is a clinical syndrome in which morphological and clinical phenomena are not correlated to each other. However, plasma proteins, including fibrin, are deposited in the walls of small arteries (hypertensive vasculopathy). This process leads to destruction of smooth muscle cells (fibrinoid necrosis).

The most common abnormality is seen in bilateral parieto-occipital subcortical white matter. However, these lesions can occur in the gray matter but can also involve the frontal lobes, basal ganglia, thalamus, cerebellum and brain stem. They are potentially reversible; however, if left untreated, permanent neurologic deficits or even death may occur as a result of ensuing cerebral infarction or hemorrhage. The prognosis probably depends on the extent of cytotoxic edema, which may be seen in severe cases. DW im-

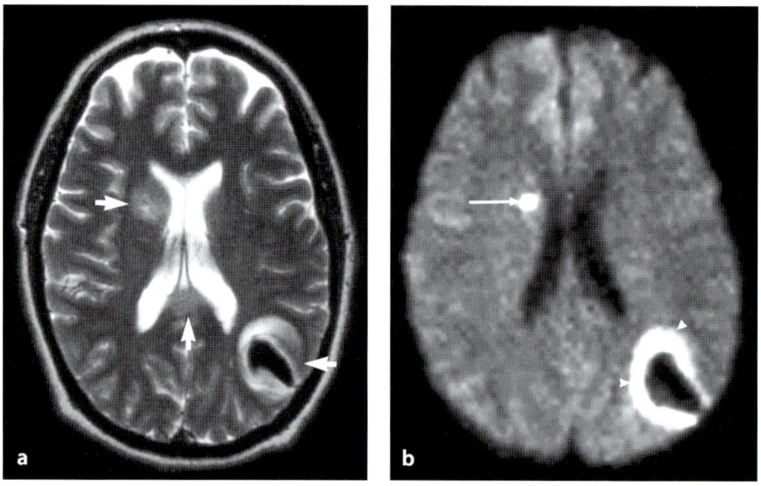

Figure 7.16 a,b

Eclampsia in a 30-year-old woman with seizures. **a** T2-weighted image shows high signal intensity lesions in the right corona radiata, posterior corpus callosum and left parieto-occipital region (*arrows*). The lesion in the left parieto-occipital region has a central very low signal intensity, representing hemorrhage. **b** On DW image, a small infarct in the right corona radiata (*arrow*) and a hemorrhagic infarct in the left parieto-occipital region (*arrowheads*) are shown as hyperintense associated with decreased ADC (not shown). The lesion in the posterior corpus callosum represents vasogenic edema

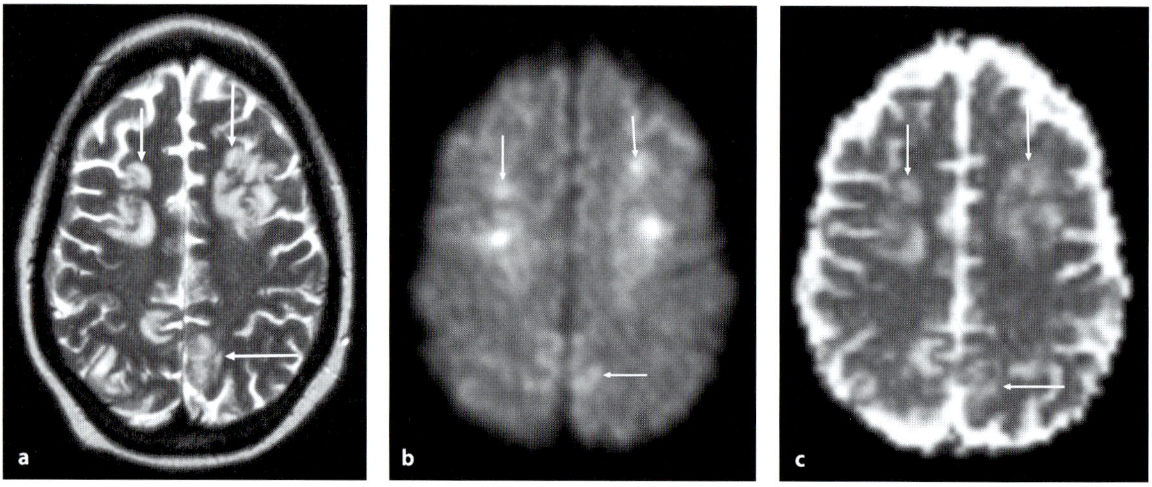

Figure 7.17 a–c

Tacrolimus neurotoxicity in a 42-year old woman with confusion after liver transplantation. **a** T2-weighted image shows high signal intensity lesions in the bilateral fronto-parieto-occipital subcortical white matter (*arrows*). **b** On DW image, these lesions show slightly hyperintense or isointense signal intensity (*arrows*). **c** ADC map shows increased ADC, which with hyperintense lesions (*arrows*) on DW image indicates T2 shine-through effect. These lesions were resolved on follow-up MRI (not shown)

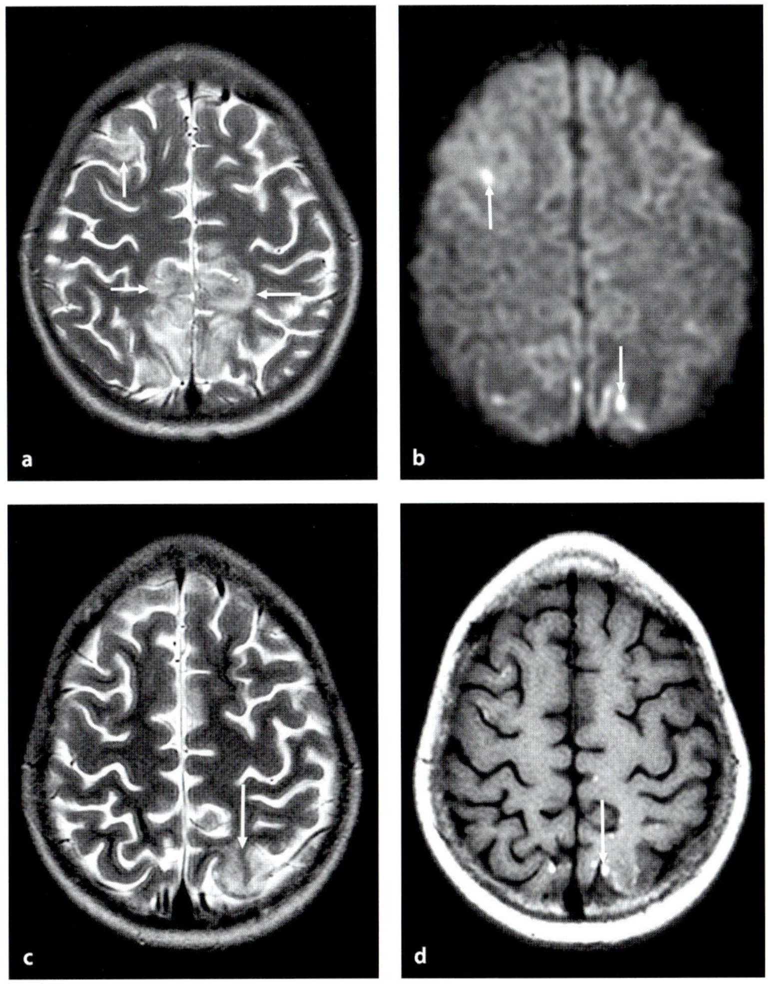

Figure 7.18 a–d

Uremic encephalopathy in a 10-year-old girl with seizure. She improved after dialysis. **a** T2-weighted image shows hyperintense lesions in bilateral parietal and right frontal subcortical white matter and cortex (*arrows*). **b** DW image reveals punctate foci of high signal within T2 high signal area (*arrow*) in the right frontal lobe and an additional bright lesion in the left occipital lobe. **c** Follow-up T2-weighted image obtained 5 days later shows incomplete resolution of left parietal high signal area (*arrow*). **d** T1-weighted image reveals punctate foci of high signal corresponding to high signal area on the initial DW image (*arrow*). (From [53])

Magnetic resonance imaging shows signal changes within the cortex and subcortical white matter in the occipital, posterior temporal, parietal and frontal lobes (Fig. 7.17). Non-transplant patients or those with total body irradiation develop white matter lesions, whereas those conditioned with chemotherapy develop mixed cortical and white matter lesions [43].

7.5.8 Uremic Encephalopathy and Hemolytic Uremic Syndrome

Uremic encephalopathy is the name given to a brain syndrome that occurs in patients with renal failure. The pathogenesis is unknown, but it has been hypothesized that it may be caused by various toxins associated with uremia (elevated parathyroid hormone level, hypercalcemia, and other metabolic abnormalities) [44]. MR imaging usually shows reversible bilateral symmetric white matter lesions. The lesions can also involve the basal ganglia or cortex. DW imaging usually shows iso- or slightly hyperintense lesions with increased ADC, mainly representing vasogenic edema (Fig. 7.18). Follow-up MRI can show cortical laminar necrosis on T1-weighted images.

Hemolytic uremic syndrome is defined as a multiorgan disease characterized by the triad of microangiopathic hemolytic anemia, thrombocytopenia and uremia. CNS complications are commonly seen (20–50%) [45]. Hemolytic uremic syndrome caused by O-157 *Escherichia coli* enterocolitis can potentially result in fatal CNS complications in infants and children. MR imaging sometimes shows irreversible lesions in the basal ganglia or cortex, representing infarction or cortical laminar necrosis. DW imaging can show these lesions as hyperintense with decreased ADC (Fig. 7.19).

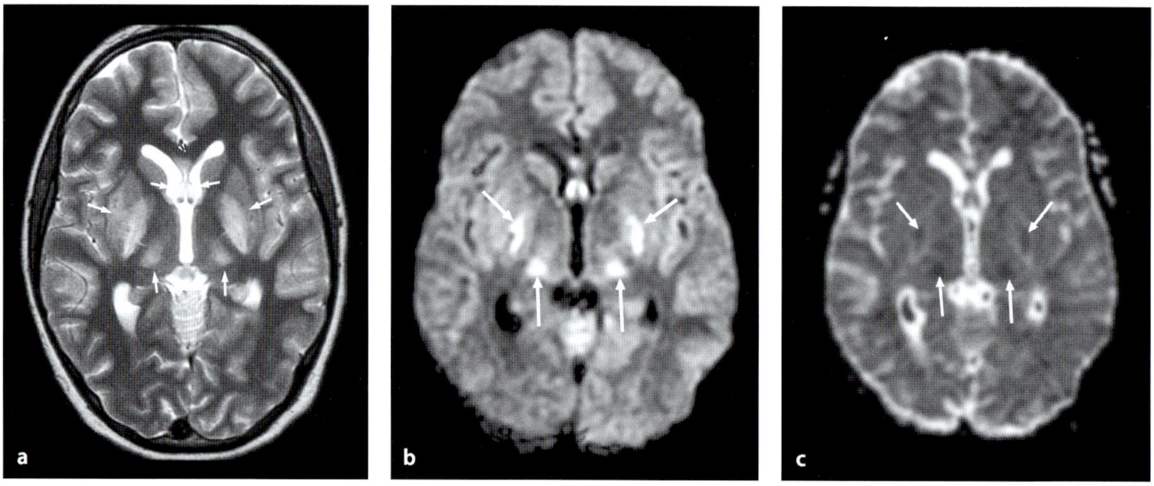

Figure 7.19 a–c

Hemolytic uremic syndrome in a 12-year-old girl with bloody diarrhea. **a** T2-weighted image shows high signal intensity lesions in the bilateral basal ganglia, thalami and fornices (*arrows*). **b** DW image shows these lesions as hyperintensity. **c** ADC map shows these lesions as decreased ADC

7.5.9 Thrombotic Thrombocytopenic Purpura

Thrombotic thrombocytopenic purpura is a multisystem vasculopathy characterized by microangiopathic hemolytic anemia, thrombocytopenia, renal involvement, fluctuating neurologic manifestations and fever [46]. Neuropathology shows hyaline thrombosis and occlusion of capillaries and arterioles without surrounding inflammatory reaction, which results in infarcts and petechial hemorrhages. MR findings are variable, ranging from punctate white matter lesions to posterior reversible encephalopathy syndrome, multifocal gray matter edema, infarction and hemorrhage. DW imaging can differentiate between these lesions (Fig. 7.20), which is important since some of the lesions seen on T2-weighted images may disappear following treatment with plasma exchange [47].

7.5.10 Cerebral Amyloid Angiopathy

Cerebral amyloid angiopathy is characterized by deposition of homogeneous eosinophilic material in the media and adventitia of arterioles and small arteries of the cortex and leptomeninges. The usual neurologic presentation is spontaneous hemorrhage. Gardien-techo T2*-weighted image is sensitive to detect microangiopathy-related microbleeds. DW image and the b0 image can show such micobleeds as low signal intensity spots [48].

7.6 Conclusion

Diffusion-weighted imaging is useful in detecting acute infarctions and in differentiating diseases that show a cytotoxic edema from those with vasogenic edema in patients with CNS vasculitis and vasculopathy. Overall, lesions demonstrating cytotoxic edema are usually irreversible and those with vasogenic edema are often reversible. Thus, DW imaging is an important tool to establish a prognosis and to determine the best treatment.

Figure 7.20 a–d

Thrombotic thrombocytopenic purpura in a 53-year-old woman with altered mental status and seizures. **a** Computed tomography shows a hemorrhage in the right parieto-occipital area (*arrow*) and a wedged-shaped area of low density more peripherally in the right occipito-parietal area (*arrowheads*). **b** T2-weighted and **c** FLAIR images show corresponding hyperintense lesions in the deep white matter and right occipito-parietal area (*arrowheads*). Hemorrhage is seen as a hypointense lesion (*arrow*). **d** On DW image, deep white matter lesions are isointense, mainly representing vasogenic edema, while a cortical lesion in the right occipito-parietal area is hyperintense, representing acute infarction (*arrowheads*). Hemorrhage is hypointense with a hyperintense rim (*arrow*)

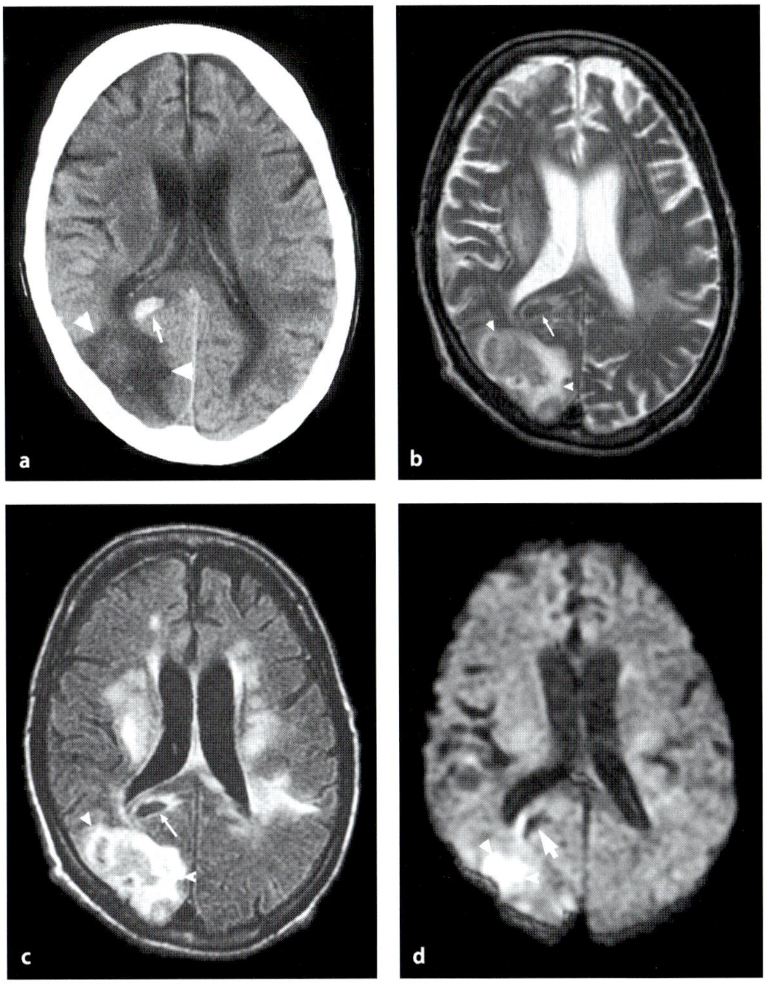

References

1. Berlit P (1994) The spectrum of vasculopathies in the differential diagnosis of vasculitis. Semin Neurol 14:370–379
2. Lie JT (1997) Classification and histopathologic spectrum of central nervous system vasculitis. Neurologic Clinics 15:805–819
3. Ebisu T, Naruse S, Horikawa Y, Ueda S, Tanaka C, Uto M, Umeda M, Higuchi T (1993) Discrimination between different types of white matter edema with diffusion-weighted MR imaging. J Magn Res Imaging 3:863–868
4. Aisen AM, Gabrielsen TO, McCune WJ (1985) MR imaging of systemic lupus erythematosus involving the brain. AJR Am J Roentgenol 144:1027–1031
5. Moritani T, Shrier DA, Numaguchi Y, Takahashi C, Yano T, Nakai K, Zhong J, Wang HZ, Shibata DK, Naselli SM (2001) Diffusion-weighted MR imaging of CNS involvement in systemic lupus erythematosus. Academic Radiology 8:741–753
6. Kocer N, Islak C, Siva A, Saip S, Akman C, Kantarci O, Hamuryudan V (1999) CNS involvement in neuro-Behçet syndrome: an MR study. AJNR Am J Neuroradiol 20:1015–1024
7. Woolfenden AR, Hukin J, Poskitt KJ, Connolly MB (1998) Encephalopathy complicating Henoch–Schonlein purpura: reversible MRI changes. Pediatr Neurol 19:74–77
8. Hinchey J, Chaves C, Appignani B, Breen J, Pao L, Wang A, Pessin MS, Lamy C, Mas JL, Caplan LR (1996) A reversible posterior leukoencephalopathy syndrome. N Engl J Med 334:494–500
9. Parisi JE, Moore PM (1994) The role of biopsy in vasculitis of the central nervous system. Semin Neurol 14:341–347
10. Jennette JC, Falk RJ, Milling DM (1994) Pathogenesis of vasculitis. Semin Neurol 14:291–299
11. Valente RM, Conn DL (1994) Current therapies for systemic vasculitis. Semin Neurol 14:380–386
12. Pomper MG, Miller TJ, Stone JH, et al. (1999) CNS vasculitis in autoimmune disease: MR imaging findings and correlation with angiography. AJNR Am J Neuroradiol 20:75–85
13. Lie JT (1994) Primary (granulomatous) angiitis of the central nervous system: a clinicopathologic analysis of 15 new cases and a review of the literature. Hum Pathol 23:164–171
14. Harris KG, Tran DD, Sickels WJ, Cornell SH, Yuh WT (1994) Diagnosing intracranial vasculitis: the role of MR and angiography. AJNR Am J Neuroradiol 15:317–330

15. Campi A, Benndorf G, Filippi M, Reganati P, Martinelli V, Terreni MR (2001) Primary angiitis of the central nervous system: serial MRI of brain and spinal cord. Neuroradiology 43:599–607
16. Ay H, Sahin G, Saatci I, Soylemezoglu F, Saribas O (2002) Primary angiitis of the central nervous system and silent cortical hemorrhages. AJNR Am J Neuroradiol 23:1561–1563
17. Moore P (1989) Diagnosis and management of isolated angitis of the central nervous system. Neurology 39:167–173
18. Brack A, Martinez-tabada V, Stanson A, et al. (1999) Disease pattern in cranial and large-vessel giant cell arteritis. Arthritis Rheum 42:311–317
19. Yamada I, Nakagawa T, Himeno Y, et al. (1998) Takayasu arteritis: evaluation of the thoracic aorta with CT angiography. Radiology 209:103–109
20. Provenzale JM, Allen NB (1996) Neurologic findings in polyarteritis nodosa. AJNR Am J Neuroradiol 17:1119–1126
21. Sehgal M, Swanson JW, DeRemee RA, et al. (1995) Neurologic manifestations of Churg-Strauss syndrome. Mayo Clin Proc 70:337–341
22. Liou HH, Liu HM, Chiang IP, et al. (1997) Churg-Strauss syndrome presented as multiple intracerebral hemorrhage. Lupus 6:279–282
23. Murphy JM, Gomez-Anson FB, Gillard JH, et al. (1999) Wegener granulomatosis: MR imaging findings in brain and meninges. Radiology 213:794–799
24. Castillo M (1994) Magnetic resonance imaging of meningitis and its complications. Top Magn Reson Imaging 6:53–58
25. DeLone DR, Goldstein RA, Petermann G, et al. (1999) Disseminated aspergillosis involving the brain: distribution and imaging characteristics. AJNR Am J Neuroradiol 20: 1597–1604
26. Lexa FJ (1995) Drug-induced disorders of the central nervous system. Semin Roentgenol 30:7–17
27. Gradon JD, Wityk R (1995) Diagnosis of probable cocaine-induced cerebral vasculitis by magnetic resonance angiography. South Med J 88:1264–1266
28. Hess DC (1997) Cerebral lupus vasculopathy. Mechanism and clinical relevance. Ann NY Acad Sci 823:154–168
29. Belmont HM, Abramson SB, Lie JT (1996) Pathology and pathogenesis of vascular injury in systemic lupus erythematosus. Interactions of inflammatory cells and activated endothelium. Arthritis Rheum 39:9–22
30. Sibbit WL, Sibbit RR, Griffey RH, et al. (1989) Magnetic resonance and computed tomographic imaging in the evaluation of acute neuropsychiatric disease in systemic lupus erythematosus. Ann Rheum Dis 48:1014–1022
31. Yamada I, Himeno Y, Nagaoka T, et al. (1999) Moyamoya disease: evaluation with diffusion-weighted and perfusion echo-planar MR imaging. Radiology 212:340–347
32. Moran CJ, Siegel MJ, DeBaun MR (1998) Sickle cell disease: imaging of cerebrovascular complications. Radiology 206:311–321
33. Casey SO, Sampaio RC, Michel E, et al. (1998) Posterior reversible encephalopathy syndrome: utility of fluid-attenuated inversion recovery MR imaging in the detection of cortical and subcortical lesions. AJNR Am J Neuroradiol 21:1199–1206
34. Hinchey J, Chaves C, Appignani B, et al. (1996) A reversible posterior leukoencephalopathy syndrome. N Engl J Med 334:494–500
35. Ay H, Buonanno FS, Schaefer PW, Le DA, et al. (1998) Posterior leukoencephalopathy without severe hypertension: utility of diffusion-weighted MRI. Neurology 51:1369–1376
36. Pavlakis SG, Frank Y, Kalina P, et al. (1997) Occipital-parietal encephalopathy: a new name for an old syndrome. Pediatr Neurol Feb;16:145–148
37. Schwartz RB, Mulkern RV, Gudbjartsson H, et al. (1998) Diffusion-weighted MR imaging in hypertensive encephalopathy: clues to pathogenesis. AJNR Am J Neuroradiol 19:859–862
38. Schwartz RB, Feske SK, Polak JF, et al. (2000) Preeclampsia-eclampsia: clinical and neuroradiographic correlates and insights into the pathogenesis of hypertensive encephalopathy. Radiology 217:371–376
39. Ursell MR, Marras CL, Farb R, et al. (1998) Recurrent intracranial hemorrhage due to postpartum cerebral angiopathy: implications for management. Stroke 29:1995–1998
40. Digre KB, Varner MW, Osborn AG, et al. (1993) Cranial magnetic resonance in severe preeclampsia vs eclampsia. Arch Neurol 50:399–406
41. Schaefer PW, Mulkern RV, Gudbjartsson H, et al. (1997) Diffusion-weighted imaging discriminates between cytotoxic and vasogenic edema in a patient with eclampsia. Stroke 28:1082–1085
42. Coley SC, Porter DA, Calamante F, et al. (1999) Quantitative MR diffusion mapping and cyclosporine-induced neurotoxicity. AJNR Am J Neuroradiol 20:1507–1510
43. Bartynski WS, Zeigler Z, Spearman MP, Lin L, Shadduck RK, Lister J (2001) Etiology of cortical and white matter lesions in cyclosporin-A and FK-506 neurotoxicity. AJNR Am J Neuroradiol 22:1901–1914
44. Okada J, Yoshikawa K, Matsuo H, et al. (1991) Reversible MRI and CT findings in uremic encephalopathy. Neuroradiology 33:524–526
45. Schmidt S, Gudinchet F, Meagher-Villemure K, et al. (2001) Brain involvement in haemolytic-uraemic syndrome: MRI features of coagulative necrosis. Neuroradiology Jul;43(7):581–585
46. Bakshi R, Shaikh ZA, Bates VE, et al. (1999) Thrombotic thrombocytopenic purpura: brain CT and MRI findings in 12 patients. Neurology 52:1285–1288
47. D'Aprile P, Farchi G, Pagliarulo R, et al. (1994) Thrombotic thrombocytopenic purpura: MR demonstration of reversible brain abnormalities. AJNR Am J Neuroradiol 15:19–20
48. Fazekas F, Kleinert R, Roob G, et al. (1999) Histopathologic analysis of foci of signal loss on gradient-echo T2*-weighted MR images in patients with spontaneous intracerebral hemorrhage: evidence of microangiopathy-related microbleeds. AJNR Am J Neuroradiol. 20: 637–42.
49. Moritani T, Hiwatashi A, Shrier DA, et al. (2003) CNS vasculitis and vasculopathy Efficacy and usefulness of diffusion-weighted echo-planar MR imaging. Clinn Imaging. In press.
50. Hiwatashi A, Garber T, Moritani T, Kinoshita T, Westesson P-L (2003) Diffusion-weighted MR Imaging of neuro_Behcet disease: a case report Neuroradiology 45: 468–471.
51. Moritani T, Shrier DA, Numaguchi Y, et al. (2001) Diffusion-weighted MR imaging of CNS involvement in systemic lupus erythematosus. Academic Radiology 8: 741–753
52. Moritani T, Numaguchi Y, Lerner NB, et al. (2003) Sickle Cell Cerebrovascular Disease: usual and unusual findings on MR Imaging and MR Angiography. Clin Imaging. In press
53. Kinoshita T, Moritani T, Shrier DA, et al. (2003) Diffusion-weighted MR imaging of posterior reversible leukoencephalopathy syndrome. A pictorial essay. Clin Imaging. 27: 307–15

Epilepsy

8.1 Definition

Epilepsy is a chronic brain disorder, which has a wide spectrum of underlying causes. It is characterized by recurrent seizures due to excessive discharge of cerebral neurons (epileptic seizures) and is associated with a variety of clinical and laboratory manifestations [1]. Epileptic seizures are defined as the clinical manifestation of abnormal excessive neuronal activity in cerebral gray matter.

8.2 Classification

The international classification of epileptic seizures is useful to describe the patients' symptoms, but this is often only the first step in the diagnostic process (Table 8.1) [2]. If the clinical characteristics are associated with a recognizable group of features, such as age of onset, genetic background and course, they may constitute an epileptic syndrome. The international classification of the epilepsy and epileptic syndromes classifies epilepsy as localized or generalized, and idiopathic or symptomatic (Table 8.2) [2].

Table 8.1. Classification of epileptic seizures

1. Partial (focal, local) seizures
   - Simple partial seizures
   - Complex partial seizures
   - Partial seizures evolving to secondary generalized seizures
2. Generalized seizures (convulsive or non-convulsive)
   - Absence seizures
   - Myoclonic seizures
   - Clonic seizures
   - Tonic seizures
   - Tonic–clonic seizures
   - Atonic seizures
3. Unclassified epileptic seizures

Table 8.2. Classification of epilepsies and epileptic syndromes

1. Localization-related epilepsies and syndromes
   - Idiopathic with age-related onset
   - Symptomatic (temporal, frontal, parietal or occipital epilepsy)
2. Generalized epilepsies and syndromes
   - Idiopathic with age-related onset
   - Idiopathic or symptomatic (West syndrome, Lennox–Gestaut syndrome)
   - Symptomatic (early myoclonic encephalopathy, specific syndromes)
3. Epilepsies and syndromes undetermined as to whether they are focal or generalized
   - Neonatal seizures
4. Special syndromes
   - Febrile convulsions
   - Drug-induced seizures
   - Eclampsia
   - Chronic progressive epilepsia partialis continua of childhood

8.3 Mechanisms and Pathophysiology of Epilepsy

The fundamental mechanisms of epilepsy can be studied at the basic level of the molecular environment of the cell. This involves cellular systems such as membrane channels, neuronal systems, cell populations and cell-to-cell interactions.

Several neurotransmitter systems are important for the mechanisms of epilepsy [3]. Glutamate is an excitatory synaptic transmitter in cerebral cortex and hippocampus. The primary receptors are divided into two groups: N-methyl-D-aspartate (NMDA) receptors and non-NMDA receptors [4]. Excessive release of glutamate activates the NMDA receptor, depolarizes the postsynaptic terminals, and induces changes in membrane function and ionic homeostasis. NMDA receptors, which have a high density in the hippocampus and other parts of the limbic system, can be related to regions that are susceptible to excitotoxic damage [5]. In some models of epileptogenesis it has been possible to block bursts of discharges by NMDA channel antagonists [6].

Another neurotransmitter related to epilepsy is gamma-aminobutyric acid (GABA), which is an inhibitor of discharges. Loss of GABAergic inhibition can generate epileptiform activity, which is believed to be important in chronic epilepsy [3].

The onset of epileptiform activity presupposes that an epileptic focus must exist where the seizure originates. The epileptic focus may exist in a discrete group of neurons, such as the CA1 pyramidal cells in the hippocampus, or over a larger region, including hippocampus and entorhinal cortex [5]. The seizure attack is associated with abnormal discharges in populations of synchronously active neurons. The spread of the seizure may be associated with an accumulation of extracellular substances, such as potassium ions or excitotoxic amines, and electrical gating mechanisms among neurons.

8.4 Magnetic Resonance Imaging of Epilepsy

Magnetic resonance imaging is widely used to evaluate patients with seizures and can detect structural brain abnormalities that may give rise to the epileptic disorder, e.g. stroke, anoxic injury, trauma, tumor, infections, demyelination and congenital anomalies. The most common cause of seizures in patients older than 45 years is stroke [7]. Intracranial tumors are often associated with epileptic seizures, and trauma is the most common cause of seizures in patients aged 15–24 years [8]. Up to 5% of epilepsy cases have a history of central nervous system infections [9].

Magnetic resonance imaging has become very useful for the diagnosis of an important cause of

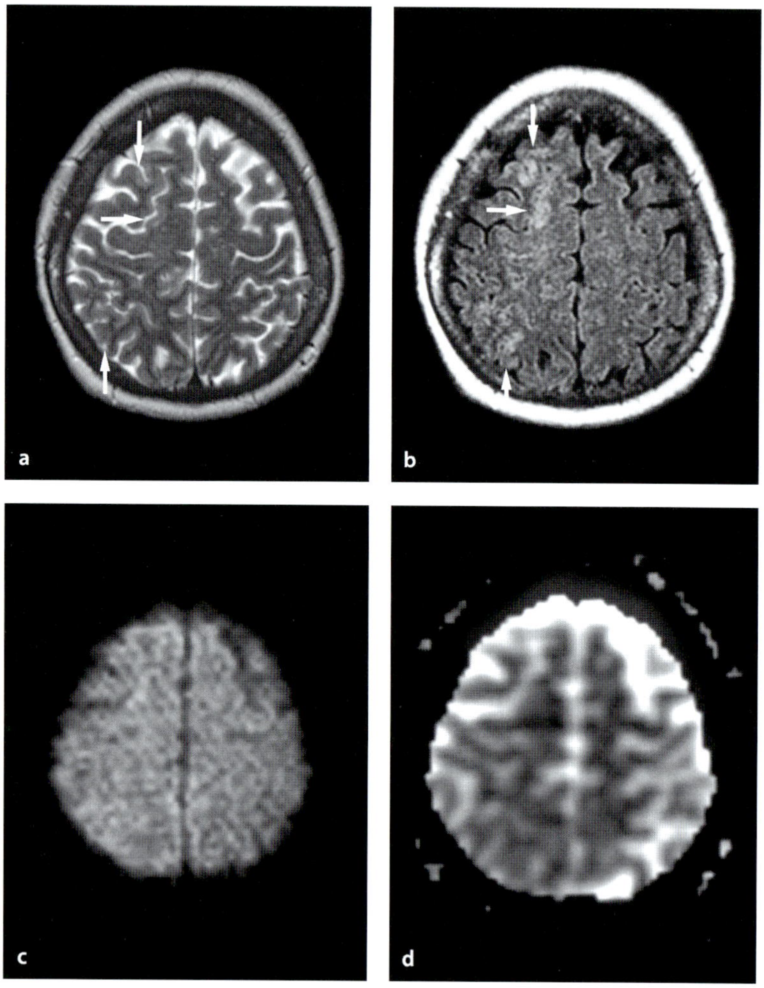

Figure 8.1 a–d

Postictal cerebral lesion following partial seizure evolving to generalized seizure in a 60-year-old woman. **a** T2-weighted and **b** FLAIR images 24 h after seizures show hyperintense lesions (*arrows*) in the right fonto-parietal cortex. **c** DW image and **d** ADC map show these lesions as isointense, probably representing mild vasogenic edema. These lesions were resolved on the 20-day follow-up MRI (not shown) seizure, namely mesial temporal (hippocampal) sclerosis [10–14]. Most seizure activity resolves within a few minutes without persistent neurologic deficits. However, transient hemiparesis (Todd's paralysis), sensory loss, persistent altered mental status, or aphasia occasionally occurs in epilepsy. MR imaging can show secondary effects of seizures on the brain [15–26]. For example, mesial temporal sclerosis, which may occur as a primary lesion, may also arise as a lesion secondary to status epilepticus [19, 21].

8.4.1 Diffusion-Weighted Imaging in Epilepsy

On routine MR imaging, signal alterations related to ictal or postictal status can be misdiagnosed as infarctions, tumorous conditions, inflammatory diseases or demyelinating diseases. This may occur because routine MR sequences will not separate vasogenic edema from cytotoxic edema. Such misdiagnoses may result in unnecessary invasive treatment. Diffusion-weighted (DW) imaging is helpful in eval-

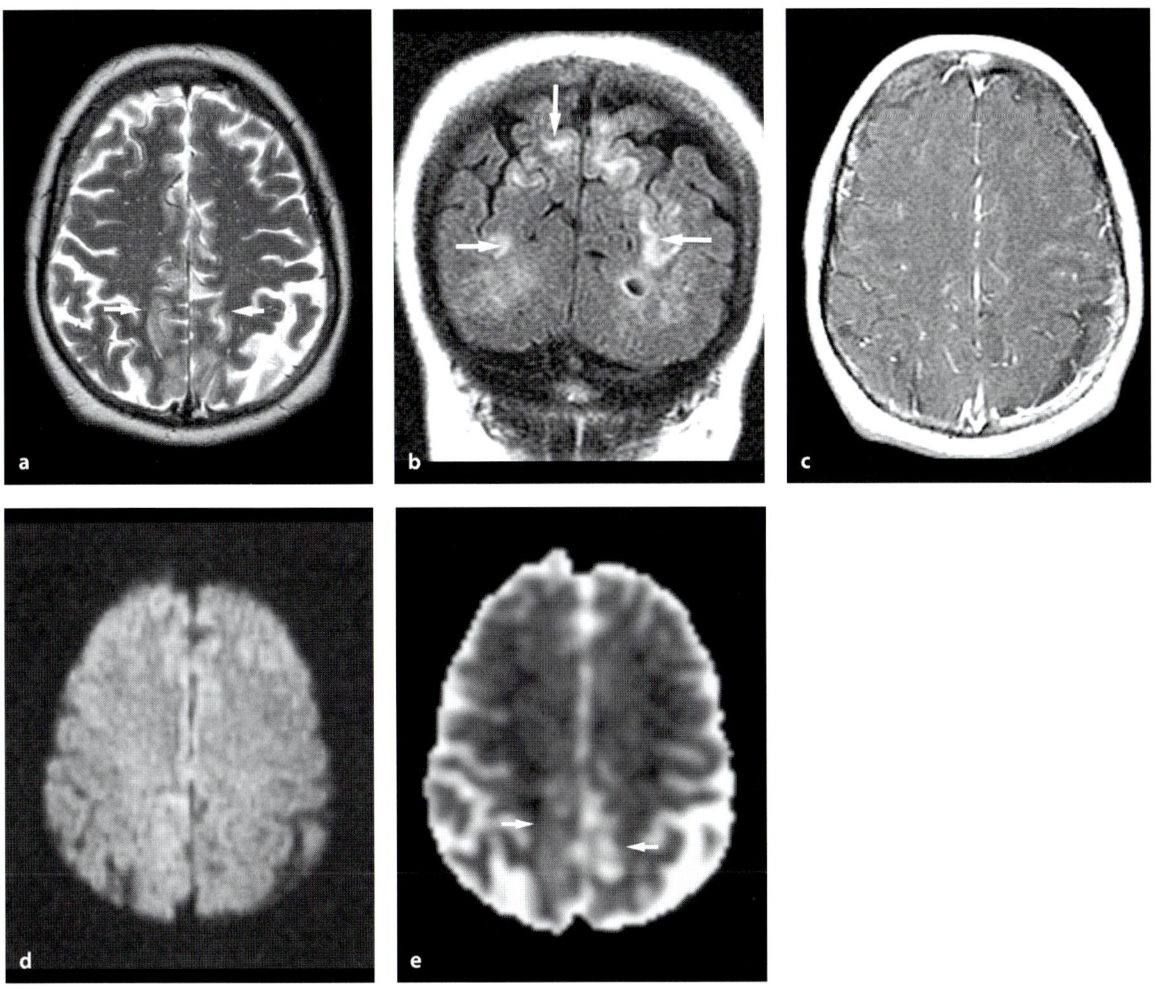

Figure 8.2 a–e

Postictal cerebral lesion following partial seizure in a 64-year-old woman with generalized seizure secondary to hyperglycemia. **a** T2-weighted image 24 h after seizures shows hyperintense lesions in bilateral medial fronto-parietal cortex (*arrows*). **b** Coronal FLAIR image also shows cortical and subcortical hyperintense lesions (*arrows*). **c** Gadolinium-enhanced T1-weighted image shows no abnormal enhancement. **d** DW image shows these lesions as isointense with increased ADC (**e**) (*arrows*), representing vasogenic edema

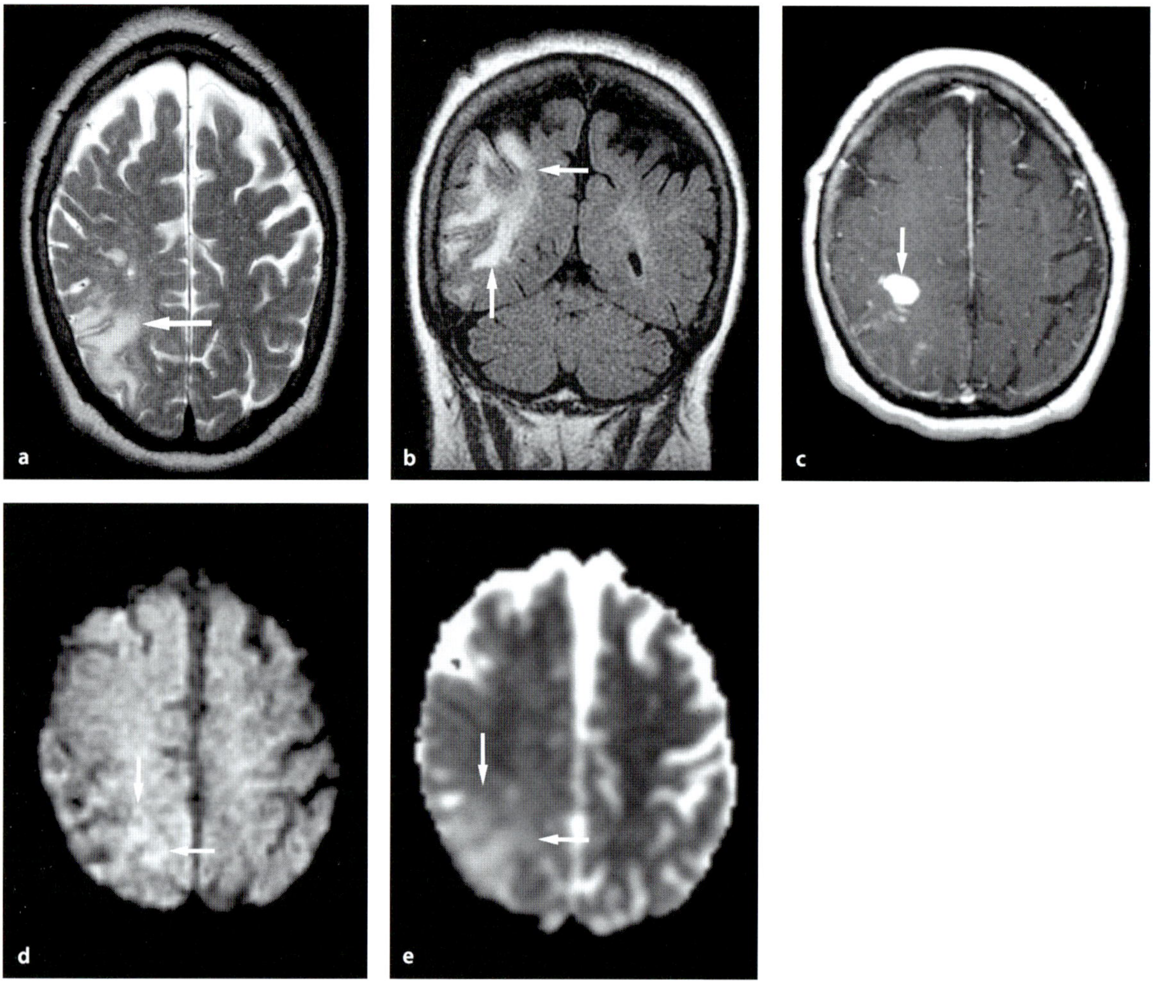

Figure 8.3 a–e

Postictal cerebral lesion following partial seizures with or without secondary generalization in a 75-year-old woman. **a** T2-weighted image at 4 days after seizures shows hyperintense lesions in the right fronto-parietal region (*arrow*). **b** Coronal FLAIR image shows cortical and subcortical hyperintense lesions (*arrows*). **c** Gadolinium-enhanced T1-weighted image shows a dense enhancing nodule (*arrow*) and leptomeningeal enhancement. **d** DW image shows these lesions as isointense to slightly hyperintense (*arrows*), with increased ADC (*arrows*) (**e**). Brain biopsy was performed and proved the lesions to be acute ischemic changes

uating epilepsy, as it will detect cytotoxic edema and can differentiate between cytotoxic and vasogenic edema in ictal and postictal lesions of the brain.

8.4.2 Magnetic Resonance Signal Alterations in Epilepsy

Variable MR findings have been reported in the brain in the ictal or postictal stage [15–26]. These findings include transient increase in T2 signal and swelling of the cerebral cortex, subcortical white matter and medial temporal lobe (including hippocampus), thalamus, claustrum and cerebellar hemispheres (Figs.

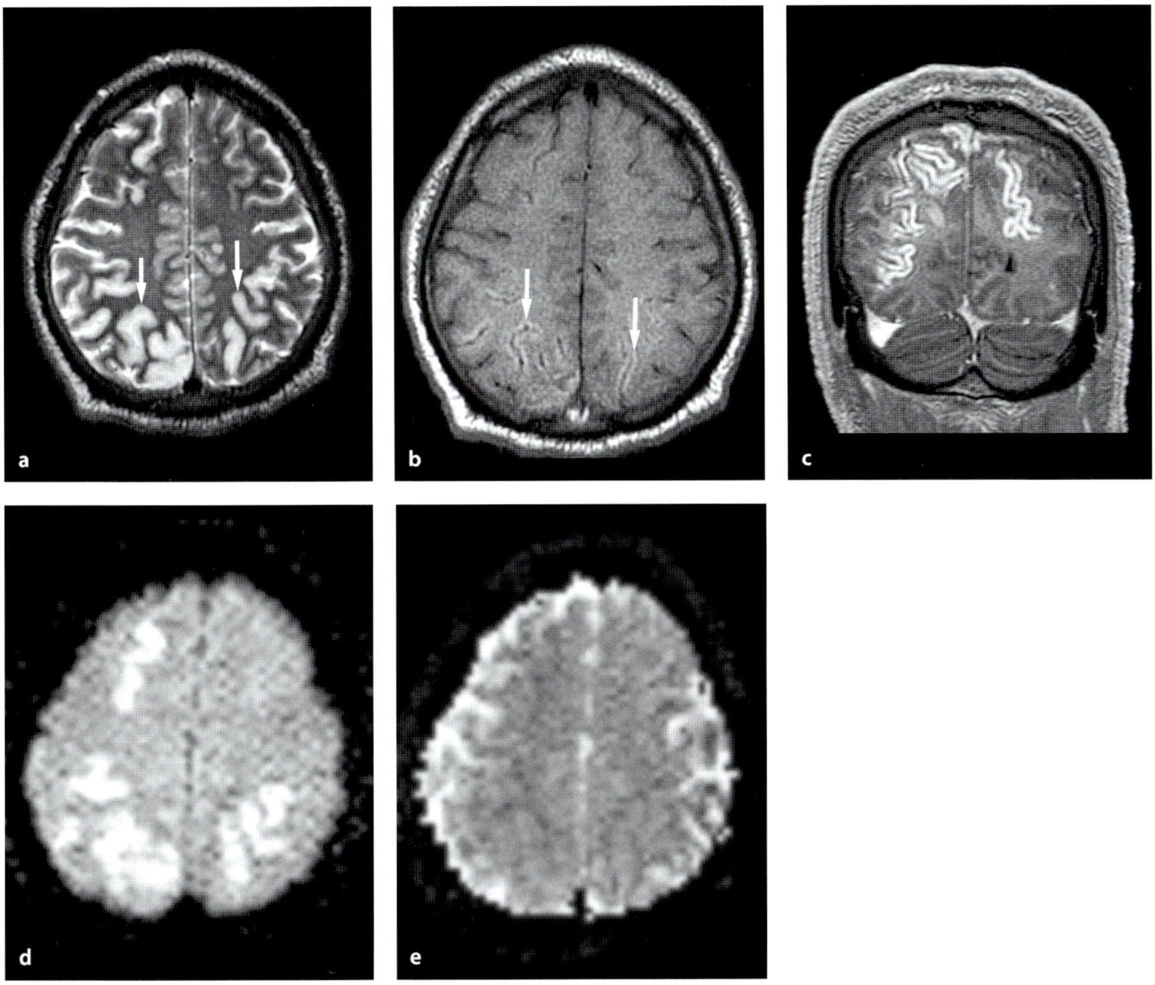

Figure 8.4 a–e

Postictal cerebral lesion following generalized tonic–clonic seizure in a 42-year-old male with alcohol withdrawal. **a** T2-weighted image shows hyperintense lesions (*arrows*) in bilateral fronto-parietal cortex. **b** T1-weighted image 8 days after the seizure shows gyriform high signal intensity in these lesions, representing cortical laminar necrosis (*arrows*). **c** Coronal gadolinium-enhanced T1-weighted image shows gyriform enhancement of these lesions. **d** DW image shows corresponding regions of hyperintensity. **e** ADC map shows almost normal ADC values of these lesions, suggesting a subacute phase of ischemia

8.1, 8.2, 8.3, 8.4, 8.5, 8.6 and 8.7). Cortical or subcortical lesions can be uni- or bilateral and they are predominantly found in the fronto-parietal regions. Whether periictal lesions are reversible or irreversible on MR imaging seems to depend on the duration and/or severity of the seizures.

8.4.3 Ictal Stage to Periictal Stage

During ictus there is an increase in metabolism (oxygen and glucose) in the seizure focus. This hypermetabolic state results in consumptive hypoxia, hypercarbia and lactic acidosis, which impair vascular autoregulation in the affected areas of cortex, leading to vasogenic edema and disruption of the blood–brain barrier [22]. If the seizures are not too prolonged, the

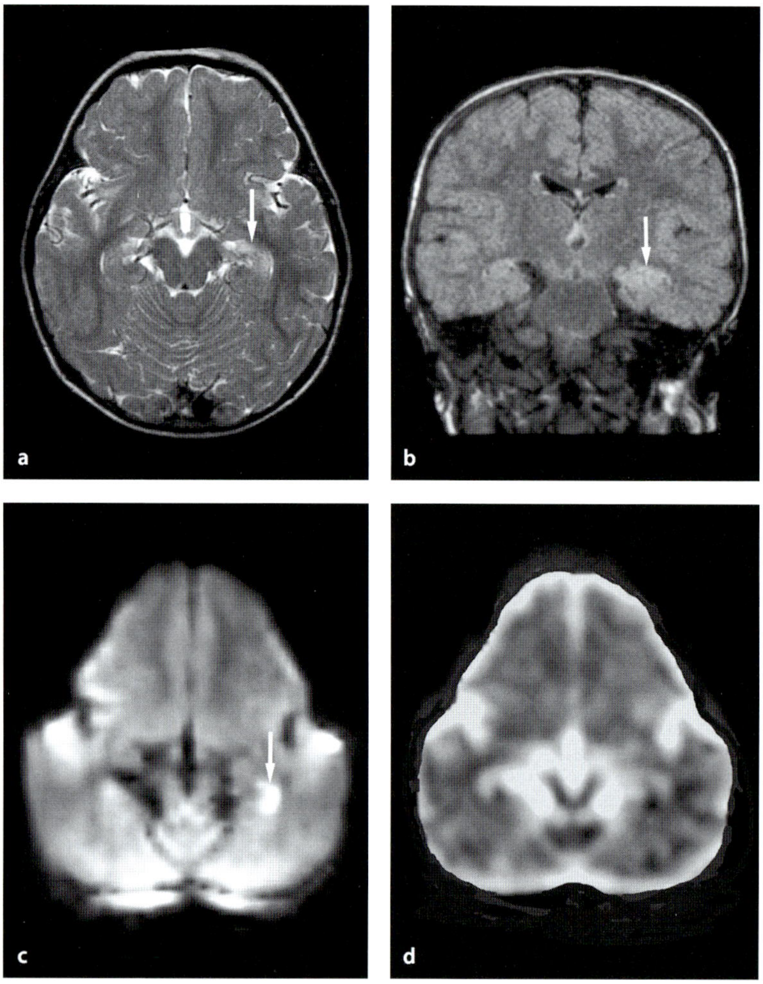

Figure 8.5 a–d

Hippocampal lesion following generalized tonic–clonic seizure in a 2-year-old female with febrile seizure leading to generalized tonic–clonic seizure. **a** T2-weighted and **b** FLAIR images 3 days after seizure show hyperintensity in the left hippocampus (*arrows*). **c** DW image shows increased signal in the left hippocampus (*arrows*) associated with normal or slightly decreased ADC (**d**), suggesting a subacute phase of the postictal hippocampal lesion

periictal brain lesions will only show transient T2 hyperintensity, mainly representing vasogenic edema. However, if the seizures are severe or prolonged, cytotoxic edema can develop. This is often seen in patients with generalized tonic–clonic seizure or status epilepticus (Figs. 8.4, 8.5, 8.6 and 8.7). Whether these lesions show enhancement or not depends on the degree of blood–brain barrier disruption (Figs. 8.2 and 8.3).

Vasogenic edema in periictal brain lesions has variable signal intensity on DW imaging, which is associated with an increase in ADC (Figs. 8.1, 8.2 and 8.3). As mentioned above, DW imaging is useful in detecting and differentiating cytotoxic from vasogenic edema.

8.4.4 Status Epilepticus

Status epilepticus is defined as continuous seizure activity that lasts longer than 30 minutes, or two or more sequential seizures that together last longer than 30 minutes and without full recovery of consciousness between the seizure attacks [28]. This is a serious event and should be treated aggressively, as the mortality associated with status epilepticus is about 8% in children and 30% in adults [28, 29].

In status epilepticus, neuronal injury is thought to result primarily from an excitotoxic mechanism mediated by intrinsic neuronal seizure activity. This is supported by the effect of kainic acid (an excitotoxic analog of glutamate), as shown in animal studies [30, 31]. Neuronal seizure activity will increase the release of glutamate from the pre-synaptic terminal of neuronal axons. The released glutamate crosses the

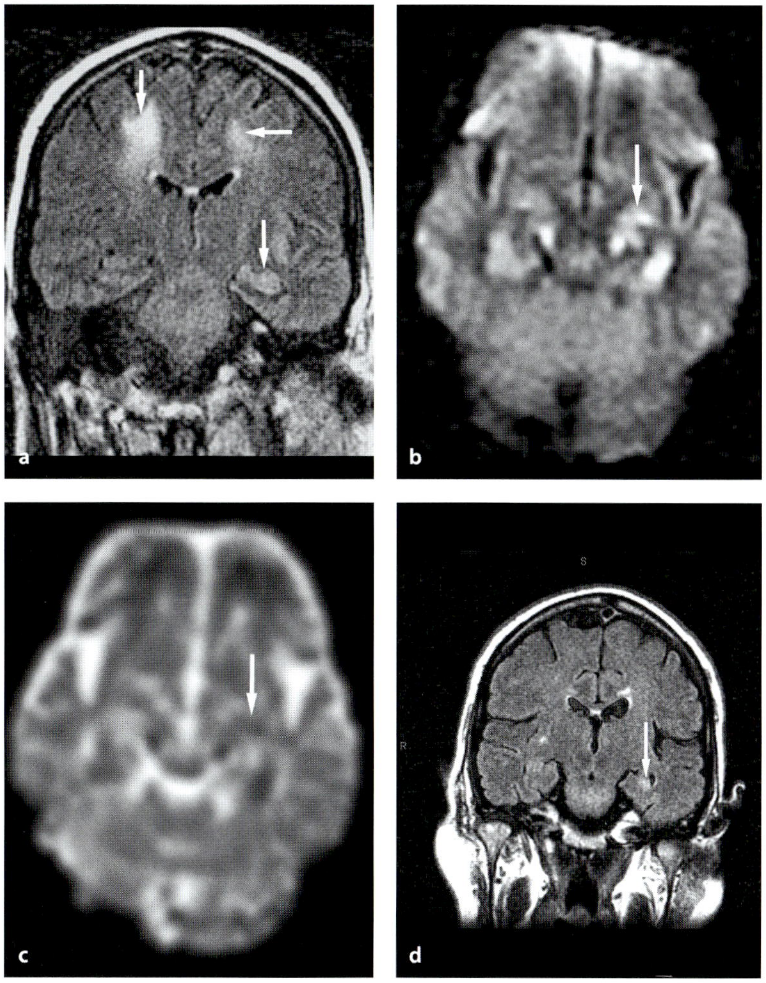

Figure 8.6 a–d

Hippocampal lesion following status epilepticus in a 55-year-old male with prolonged seizures associated with tacrolimus toxicity. **a** FLAIR image at 2 days after seizure shows high signal areas in the bilateral frontal deep white matter and the left hippocampus (*arrows*). **b** DW image (*y axis*) shows high signal in the left hippocampus (*arrow*). **c** ADC (*y axis*) shows low signal intensity (*arrow*) consistent with cytotoxic edema. **d** 7 months follow-up coronal FLAIR image shows the resolution of white matter changes and residual high signal intensity with atrophy in the left hippocampus (*arrow*)

synaptic cleft to bind to NMDA receptors of the postsynaptic neurons, resulting in cytotoxic edema. Cytotoxic edema is also seen in the astrocytes as an acute phase of reactive astrocytosis [21].

8.4.5 Cytotoxic Edema in Status Epilepticus

Cytotoxic edema following status epilepticus can be at least partially reversible [22], as compared to cerebral ischemia, where these changes are usually irreversible. In cerebral ischemia, a significant compromise of blood supply leads to irreversible failure of energy metabolism. In sustained seizures, there is an increased cerebral metabolism with an increase in cerebral blood flow. This will maintain the energy state of the neuron provided there is sufficient oxygen supply.

The parts of the human brain that are most vulnerable include parts of the hippocampus (the CA1and CA3 segments, and the hilus), amygdala, pyriform cortex, thalamus, cerebellum, and cerebral cortex. NMDA receptors are predominantly located in the CA1 of the hippocampus and layers 3 and 4 in the cerebral cortex [32, 33]. The Purkinje cell loss of the cerebellum, seen in severe epilepsy, may be explained by an increased demand for inhibition, resulting in GABA depletion and subsequent influx of calcium into neurons [34]. Unilateral hemispheric involvement is occasionally seen in status epilepticus.

Transient and reversible MR signal changes have been reported in patients following status epilepticus [16, 19–23]. On the other hand, other lesions have been proven irreversible, resulting in selective neuronal necrosis, gliosis and delayed neuronal death with subsequent atrophy [24–26] (Figs. 8.6 and 8.7). Following

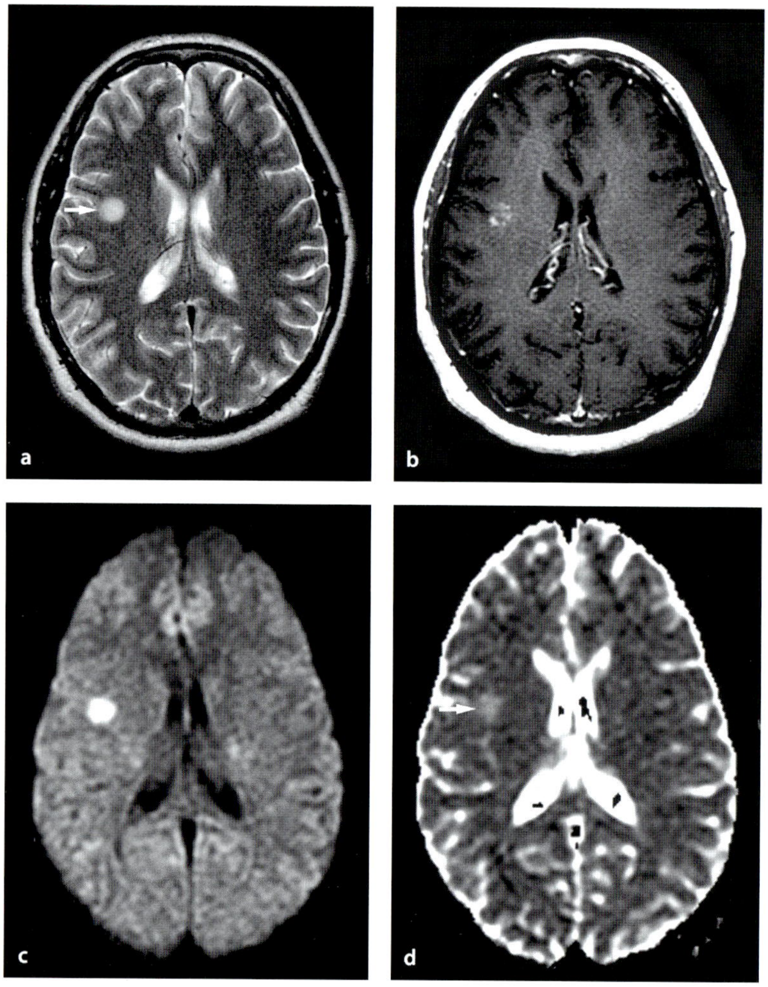

Figure 9.1 a–d

Multiple sclerosis in a 28-year-old man presenting with visual problems. **a** T2-weighted image shows a hyperintense lesion in the right frontotemporal region (*arrow*). **b** Gadolinium T1-weighted image shows mild enhancement of this lesion, representing an active plaque. **c, d** DW image (**c**) shows a hyperintense lesion associated with increased ADC (**d**), T2 shine-through (*arrow*)

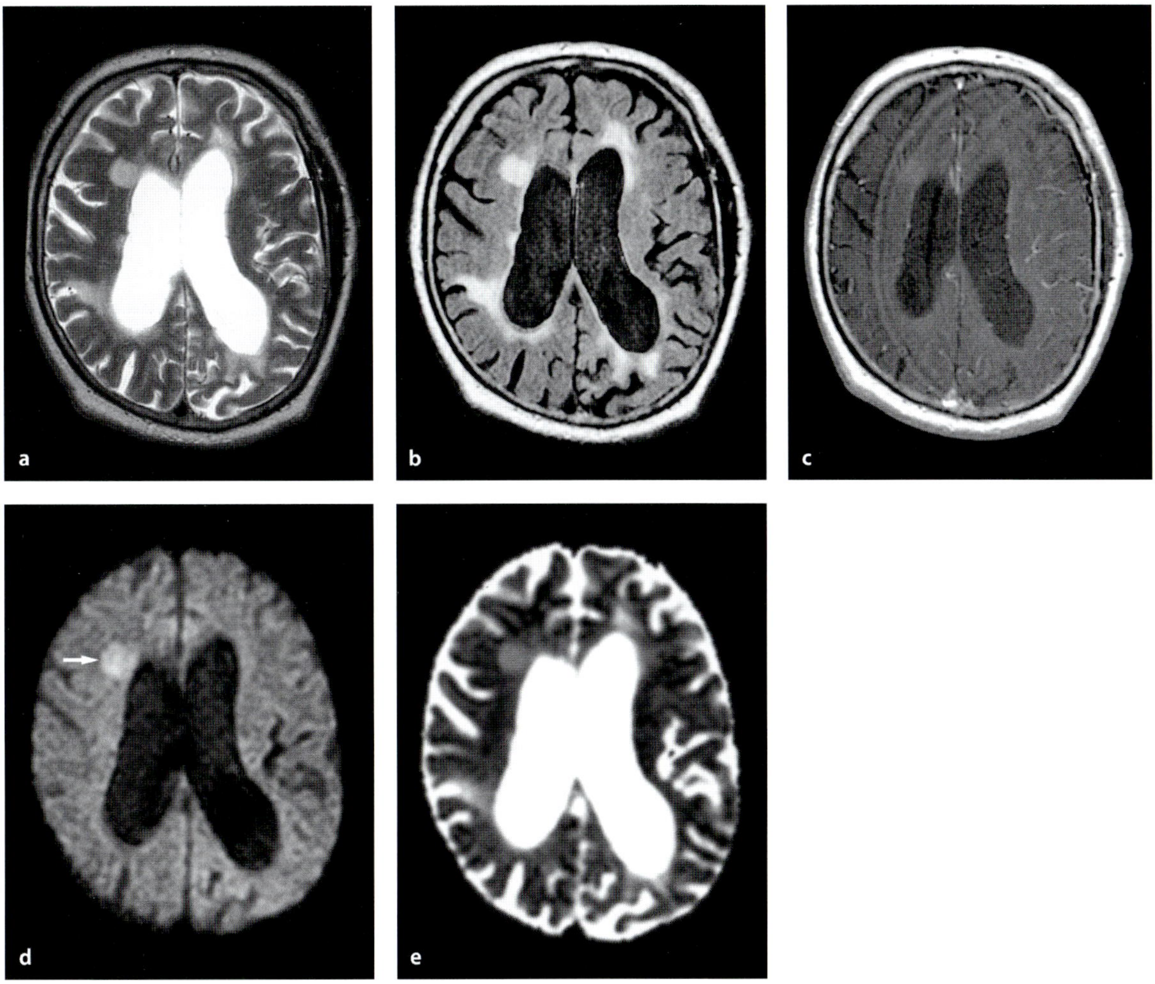

Figure 9.2 a–e

Multiple sclerosis in a 59-year-old man with a long history of recurrent seizures. T2-weighted (**a**) and FLAIR (**b**) images show multiple periventricular hyperintense lesions with ventricular dilatation. **c** On gadolinium T1-weighted image with magnetization transfer contrast, there is no enhancement of these lesions, representing relatively chronic plaques. **d** DW image shows a right frontal lesion as mildly hyperintense (*arrow*). **e** ADC is increased (T2 shine-through) and the other periventricular lesions are isointense with increased ADC (T2 washout)

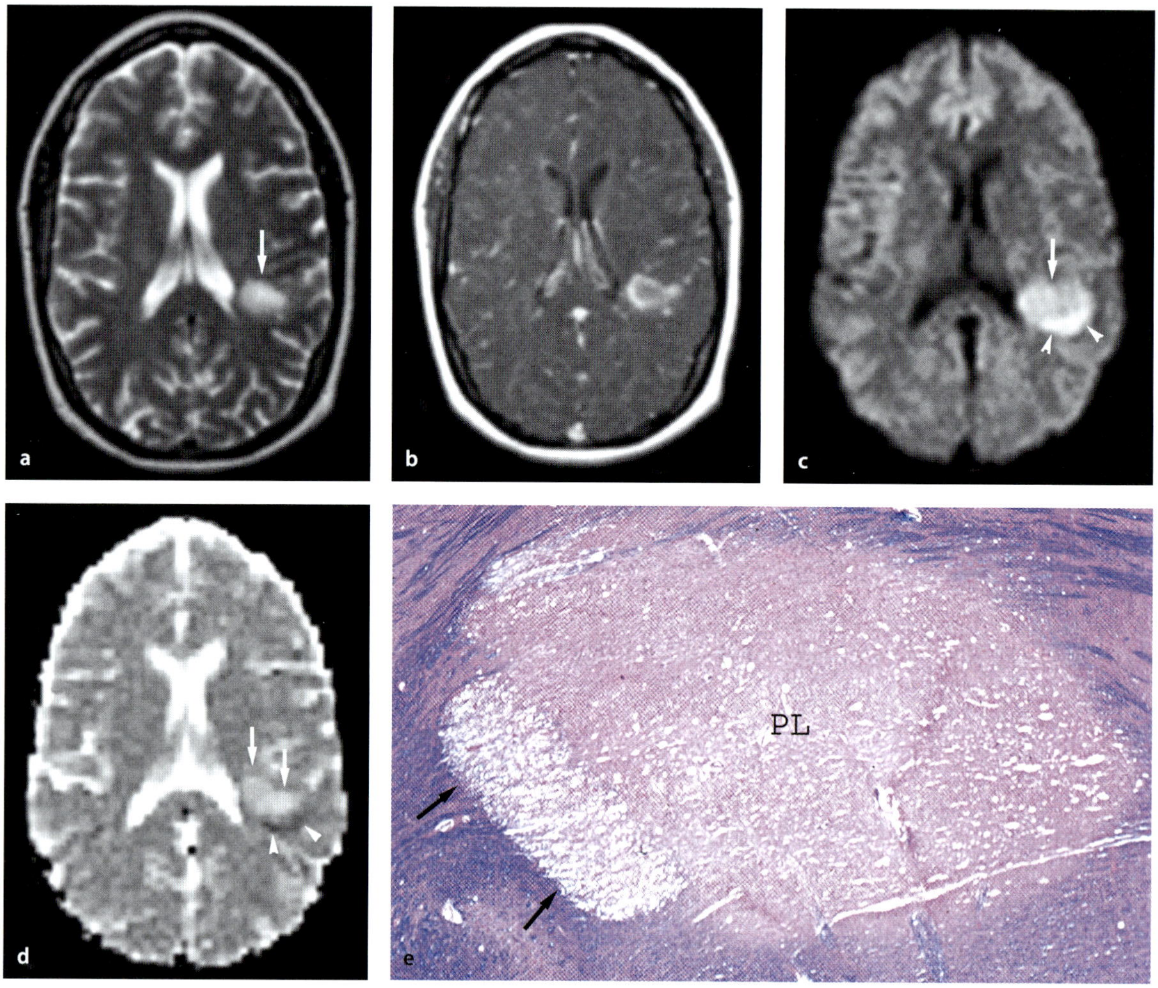

Figure 9.3 a–e

Multiple sclerosis in a 36-year-old woman presenting with subacute onset of progressive aphasia. **a** T2-weighted image shows a hyperintense lesion in the left periventricular white matter (*arrow*). **b** Gadolinium T1-weighted image with magnetization transfer contrast shows rim enhancement of this lesion. **c** DW image shows combination of a moderately hyperintense (*arrow*) and a significantly hyperintense lesion (*arrowheads*). **d** The moderately hyperintense lesion on DW image with increased ADC may represent demyelination (*arrows*), and the very hyperintense lesion on DW image with decreased ADC may represent intramyelinic edema (*arrowheads*). **e** Histopathology of another case shows that intramyelinic edema (*arrows*) is located in the periphery of a plaque (*PL*) (Luxol fast blue PAS stain, original magnification ×40). (From [33])

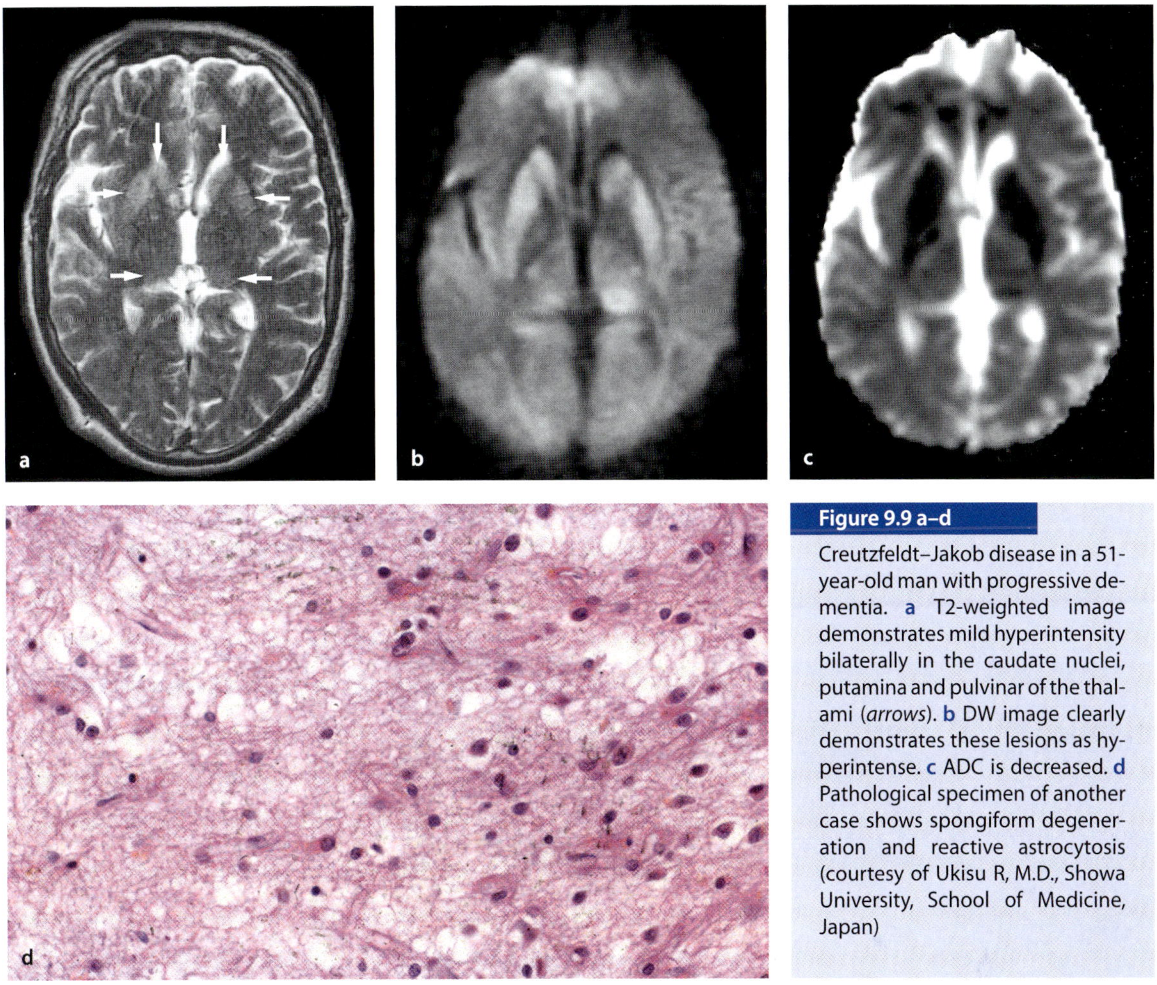

Figure 9.9 a–d

Creutzfeldt–Jakob disease in a 51-year-old man with progressive dementia. **a** T2-weighted image demonstrates mild hyperintensity bilaterally in the caudate nuclei, putamina and pulvinar of the thalami (*arrows*). **b** DW image clearly demonstrates these lesions as hyperintense. **c** ADC is decreased. **d** Pathological specimen of another case shows spongiform degeneration and reactive astrocytosis (courtesy of Ukisu R, M.D., Showa University, School of Medicine, Japan)

9.2.3 Amyotrophic Lateral Sclerosis

Amyotrophic lateral sclerosis affects middle-aged patients and is characterized by progressive muscle weakness, limb and truncal atrophy associated with bulbar signs and symptoms. The disease progression is relentless and half of the patients are dead within 3 years. MR images of amyotrophic lateral sclerosis are characterized by high T2 signal along the large myelinated pyramidal tract fibers in the posterior limb of the internal capsule and cerebral peduncles. On DW imaging there is typically increased ADC and decreased fractional anisotropy in the corticospinal tracts [32]. Diffusion tensor MR imaging may be useful in analyzing the extent and severity of axonal degeneration in amyotrophic lateral sclerosis (Fig. 9.10).

9.3 Conclusion

Magnetic resonance imaging with DW and ADC is useful in characterizing demyelinating and degenerative lesions of the brain. These imaging techniques can increase specificity and improve our understanding of the pathophysiology of these diseases.

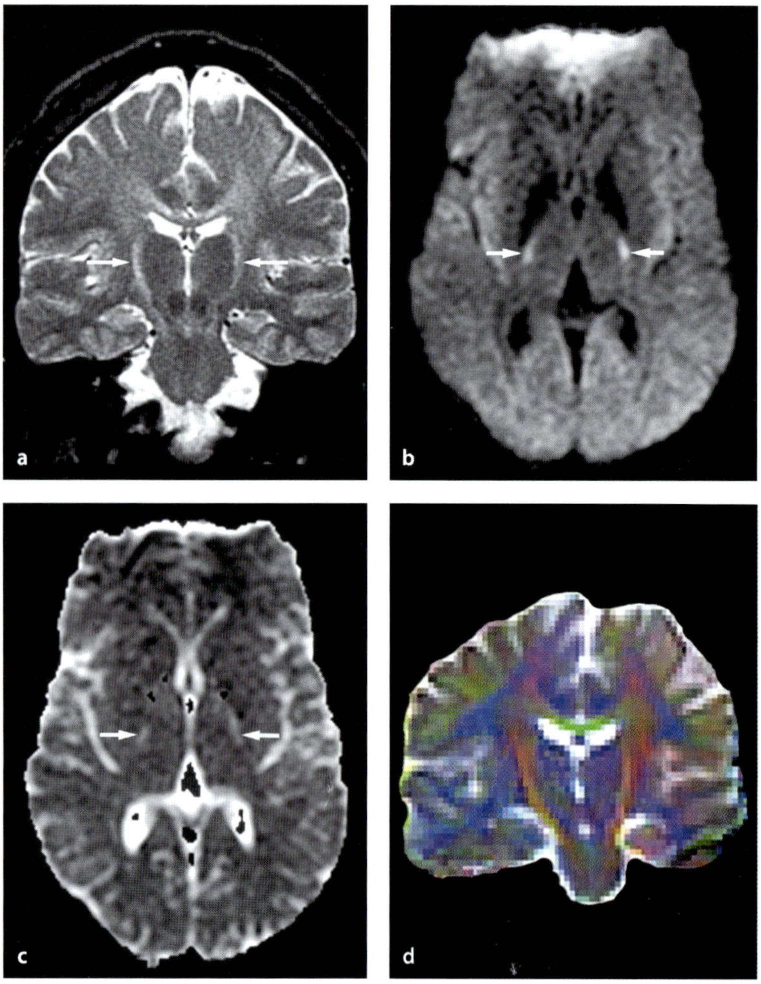

Figure 9.10 a–d

Amytrophic lateral sclerosis in a 27-year-old male with progressive weakness and dysphagia. **a** Coronal spin-echo T2-weighted image shows bilateral symmetrical hyperintensity along the corticospinal tract (*arrows*), extending into the white matter of the motor area. **b** DW image shows mild hyperintensity in bilateral corticospinal tracts. **c** ADC is increased (*arrows*). Hyperintensity and distortion in the frontal region is due to susceptibility artifact from air in the frontal sinuses. **d** Coronal diffusion tensor image with color mapping reveals decreased anisotropy along bilateral corticospinal tracts

References

1. Filippi M, Inglese M (2001) Overview of diffusion-weighted magnetic resonance studies in multiple sclerosis. J Neurol Sci 186 Suppl 1:S37–S43
2. van Walderveen MA, Lycklama A Nijeholt GJ, et al. (2001) Hypointense lesions on T1-weighted spin-echo magnetic resonance imaging: relation to clinical characteristics in subgroups of patients with multiple sclerosis. Arch Neurol 58:76–81
3. Loevner LA, Grossman RI, McGowan JC, Ramer KN, Cohen JA (1995) Characterization of multiple sclerosis plaques with T1-weighted MR and quantitative magnetization transfer. AJNR Am J Neuroradiol 16:1473–1479
4. Loevner LA, Grossman RI, Cohen JA, Lexa FJ, Kessler D, Kolson DL (1995) Microscopic disease in normal-appearing white matter on conventional MR images in patients with multiple sclerosis: assessment with magnetization-transfer measurements. Radiology 196:511–515
5. Horsfield MA, Larsson HB, Jones DK, Gass A (1998) Diffusion magnetic resonance imaging in multiple sclerosis. J Neurol Neurosurg Psychiatry 64 Suppl 1:S80–S84
6. Guo AC, MacFall JR, Provenzale JM (2002) Multiple sclerosis: diffusion tensor MR imaging for evaluation of normal-appearing white matter. Radiology 222:729–736
7. Tievsky AL, Ptak T, Farkas J (1999) Investigation of apparent diffusion coefficient and diffusion tensor anisotropy in acute and chronic multiple sclerosis lesion. Am J Neuroradiol 20:1491–1499
8. Roychowdhury S, Maldjian JA, Grossman RI (2000) Multiple sclerosis: comparison of trace apparent diffusion coefficients with MR enhancement pattern of lesions. AJNR Am J Neuroradiol 21:869–874
9. Castriota Scanderbeg A, Tomaiuolo F, Sabatini U, Nocentini U, Grasso MG, Caltagirone C (2000) Demyelinating plaques in relapsing-remitting and secondary-progressive multiple sclerosis: assessment with diffusion MR imaging. Am J Neuroradiol 21:862–868
10. Rovira A, Pericot I, Alonso J, Rio J, Grive E, Montalban X (2002) Serial diffusion-weighted MR imaging and proton MR spectroscopy of acute large demyelinating brain lesions: case report. Am J Neuroradiol 23:989–994
11. Inglese M, Salvi F, Iannucci G, Mancardi GL, Mascalchi M, Filippi M (2002) Magnetization transfer and diffusion tensor MR imaging of acute disseminated encephalomyelitis. Am J Neuroradiol 23:267–272
12. Harada M, Hisaoka S, Mori K, Yoneda K, Noda S, Nishitani H (2000) Differences in water diffusion and lactate production in two different types of postinfectious encephalopathy. J Magn Reson Imaging 11:559–563
13. Bernarding J, Braun J, Koennecke HC (2002) Diffusion- and perfusion-weighted MR imaging in a patient with acute demyelinating encephalomyelitis (ADEM). J Magn Reson Imaging 15:96–100
14. Ohta K, Obara K, Sakauchi M, Obara K, Takane H, Yogo Y (2001) Lesion extension detected by diffusion-weighted magnetic resonance imaging in progressive multifocal leukoencephalopathy. J Neurol 248:809–811
15. Castillo M, Mukheriji SK (1999) Early abnormalities related to postinfarction wallerian degeneration: evaluation with MR diffusion-weighted imaging. JCAT 23:1004–1007
16. Kang DW, Chu K, Yoon BW, Song IC, Chang KH, Roh JK (2000) Diffusion-weighted imaging in wallerian degeneration. J Neurol Sci 178:167–169
17. Ogawa T, Okudera T, Inugami A, et al. (1997) Degeneration of the ipsilateral substantia nigra after striatal infarction: evaluation with MR imaging. Radiology 204:847–851
18. Kinoshita T, Moritani T, Shrier DA, et al. (2002) Secondary degeneration of the Substantia Nigra and Corticospinal tract after hemorrhagic middle celebral artery infarction; Diffusion-weighted MR findings. Magnetic Resonance in Medical Sciences 1: 175–198
19. Nakase M, Tamura A, Miyasaka N, et al. (2001) Astrocytic swelling in the ipsilateral substantia niagra after occlusion of the middle cerebral artery in rats. Am J Neuroradiol 22:660–663
20. Johnson RT, Gibbs CJ Jr (1998) Creutzfeldt–Jakob disease and related transmissible spongiform encephalopathies. N Engl J Med 339:1994–2004
21. Brown P, Preece M, Brandel JP, et al. (2000) Iatrogenic Creutzfeldt–Jakob disease at the millennium. Neurology 55:1075–1081
22. Lucassen PJ, Williams A, Chung WCJ, et al. (1995) Detection of apoptosis in murine scrapie. Neuroscience Letters 198:185–188
23. Zeidler M, Sellar RJ, Collie DA, et al. (2000) The pulvinar sign on magnetic resonance imaging in variant Creutzfeldt–Jakob disease. Lancet 355:1412–1418
24. Molloy S, O'Laoide R, Brett F, Farrell M (2000) The "pulvinar" sign in variant Creutzfeldt–Jakob disease. Am J Roentgenol 175:555–556
25. Haik S, Brandel JP, Oppenheim C, et al. (2002) Sporadic CJD clinically mimicking variant CJD with bilateral increased signal in the pulvinar. Neurology 58:148–149
26. Demaerel P, Baert AL, Vanopdenbosch, et al. (1997) Diffusion-weighted magnetic resonance imaging in Creutzfeldt–Jakob disease. Lancet 349:847–848
27. Bahn MM, Parchi P (1999) Abnormal diffusion-weighted magnetic resonance images in Creutzfedlt–Jakob disease. Arch Neurol 56:577–583
28. Mittal S, Farmer P, Kalina P, Kingsley PB, Halperin J (2002) Correlation of diffusion-weighted magnetic resonance imaging with neuropathology in Creutzfeldt–Jakob disease. Arch Neurol 59:128–134
29. Murata T, Shiga Y, Higano S, Takahashi S, Mugikura S (2002) Conspicuity and evolution of lesions in Creutzfeldt–Jakob disease at diffusion-weighted imaging. AJNR Am J Neuroradiol 23:1164–1172
30. Dearmond MA, Kretzschmar HA, Prusiner SB (2002) Prion diseases. In: Graham DI, Lantos PL (eds) Greenfield's neuropathology, 7th edn, pp 273–323
31. Matoba M, Tonami H, Miyaji H, Yokota H, Yamamoto I (2001) Creutzfeldt–Jakob disease: serial changes on diffusion-weighted MRI. J Comput Assist Tomogr 25:274–277
32. Ellis CM, Simmons A, Jones DK, et al. (1999) Diffusion tensor MRI assesses corticospinal tract damage in ALS. Neurology 53:1051–1058
33. Moritani T (2002) Classification of brain edema. In Koredewakaru Diffusion MRI, Tokyo: Shujunsha. pp 128–137

Toxic and Metabolic Disease

In collaboration with J. Zhong

Computed tomography (CT) of patients with toxic and metabolic disease is generally non-specific, showing decreased attenuation of lesions in the white matter. Routine MR imaging is informative, but usually also non-specific, with T2 prolongation of those lesions. The white matter is often diffusely and symmetrically involved, as are the basal ganglia and/or brain stem. Diffusion-weighted (DW) imaging can add to the diagnostic information and improve the understanding of the pathophysiology of various white matter abnormalities, such as dysmyelination and demyelination in toxic and metabolic diseases.

10.1 Toxic Disease

10.1.1 Chemotherapy-Induced Leukoencephalopathy

Intrathecal or intravenous methotrexate, with or without radiation therapy, can cause diffuse white matter changes [1]. MR imaging shows diffuse or multifocal white matter lesions that are hyperintense on T2-weighted image. DW imaging show these white matter changes as diffuse hyperintensity with decreased apparent diffusion coefficient (ADC) in the white matter, even before conventional MR imaging can detect the lesions (Fig. 10.1). Pathologically these lesions represent intramyelinic and axonal swelling.

High-dose chemotherapy including carmustine (BCNU), cyclophosphamide, cisplatin, 5-fluorouracil (5-FU) and carmofur can also cause diffuse white matter disease. The lesions are hyperintense on T2-weighted images as well as on DW images, and ADC is decreased [2–4] (Fig. 10.2). Chemotherapeutic agents such as 5-FU and carmofur can have direct toxic effects on myelin, which causes intramyelinic edema [5]. Chemotherapy-associated leukoencephalopathy can be fatal and early diagnosis and discontinuation of the offending drug is therefore often necessary.

10.1.2 Heroin-Induced Spongiform Leukoencephalopathy

The inhalation of black-market heroin vapors (pyrolysate) as well as intravenous consumption of heroin can lead to toxic leukoencephalopathy [6]. The leukoencephalopathy is pathologically characterized by spongiform degeneration of the white matter as a result of fluid accumulation within the myelin sheaths (intramyelinic edema). Electron microscopy shows vacuoles between the myelin lamellae by splitting of the intraperiod lines [7]. CT and MR imaging show abnormalities in the cerebral and cerebellar white matter, cerebral peduncles, corticospinal tracts, lemniscus medialis and solitary tracts [8]. The accumulation of restricted fluid between the layers of myelin lamellae may cause hyperintensity on DW imaging with decreased ADC [9] (Fig. 10.3). Because the myelin itself and the blood–brain barrier are intact in cases of less severe heroin-induced leukoencephalopathy, one may expect the changes in the DW signal to be reversible on follow-up MR imaging [10].

10.1.3 Cocaine, Phencyclidine Hydrochloride, Amphetamines and Related Catecholaminergics

Cocaine, phencyclidine hydrochloride, amphetamines and related catecholaminergics can cause hemorrhage or infarction due to vasculitis, vasculopathy, or acute hypertensive effects [1]. DW imaging can be useful for the detection of these lesions (see also Chap. 7).

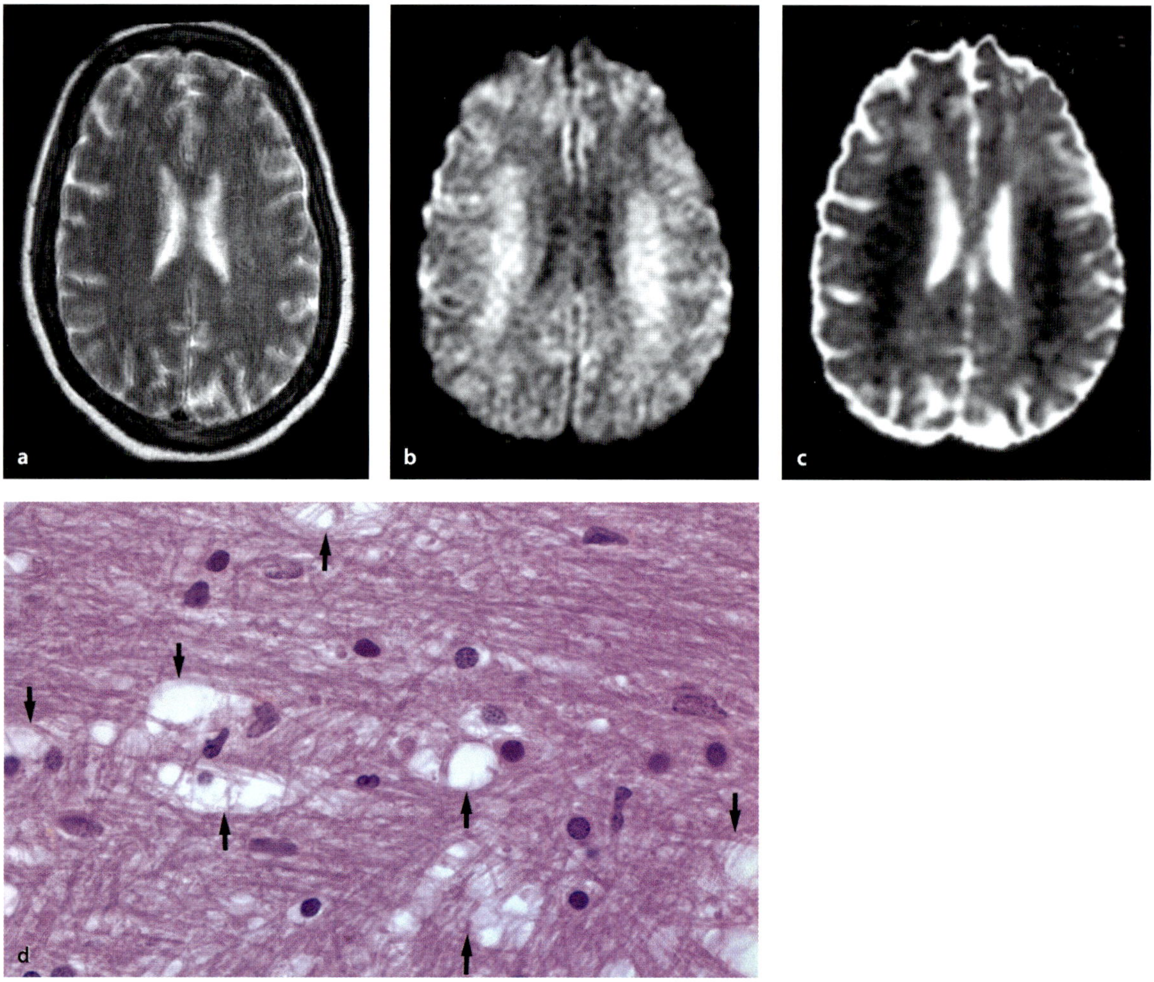

Figure 10.1 a–d

Methotrexate leukoencephalopathy in a 50-year-old female. **a** T2-weighted image does not demonstrate an appreciable abnormality in the white matter. **b** DW image shows diffuse hyperintensity in the bilateral corona radiata extending into the central semiovale. **c** ADC map shows diffuse white matter lesions as decreased ADC, which represents pure intramyelinic edema. **d** Pathological specimen shows spongiform change representing intramyelinic edema (*arrows*) diffusely in white matter. Astrocytes and axons are relatively spared (hematoxylin–eosin stain, original magnification ×200)

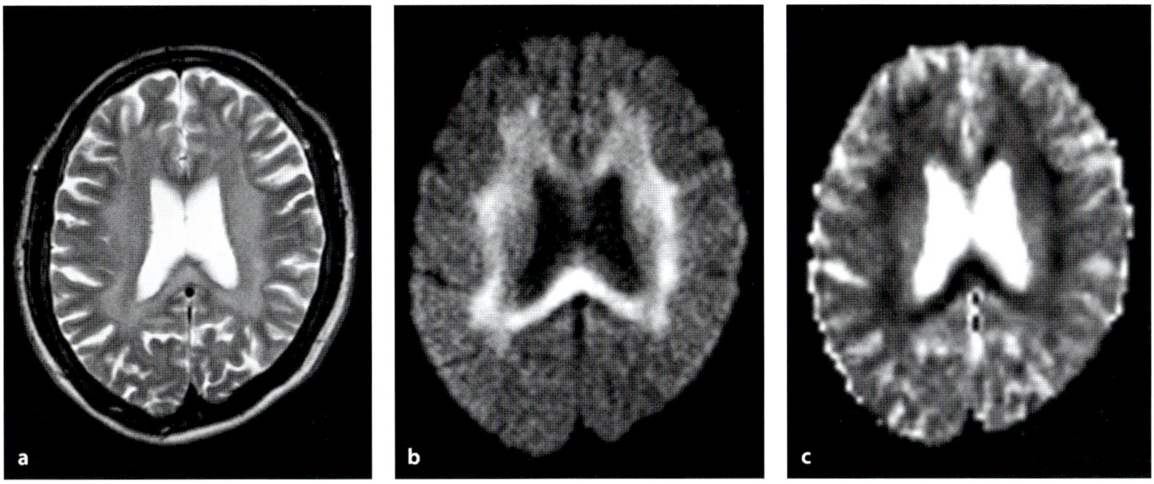

Figure 10.2 a–c

Carmofur leukoencephalopathy in a 58-year-old female. **a** T2-weighted image shows diffuse hyperintensity in the periventricular white matter including the corpus callosum. **b, c** DW image shows these lesions as hyperintense with decreased ADC, presumably related to intramyelinic edema

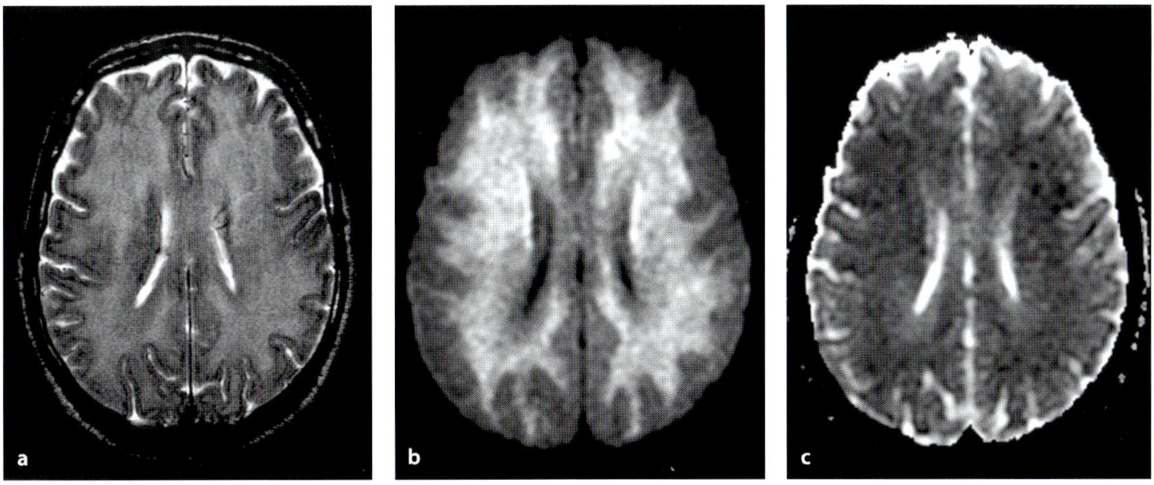

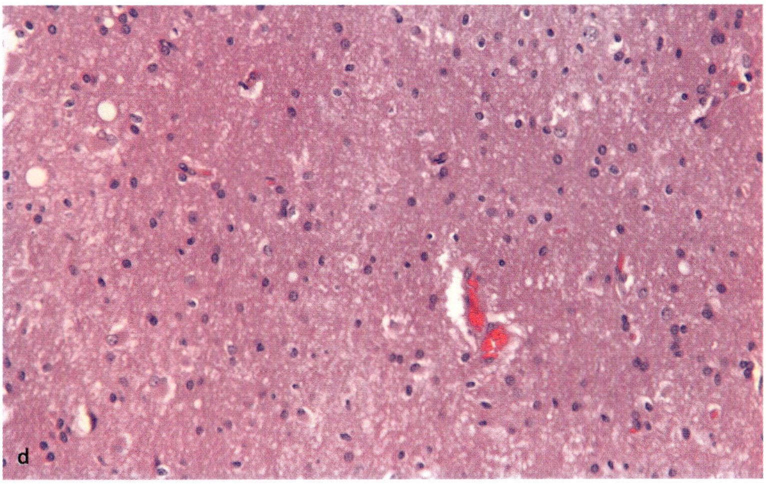

Figure 10.3 a–d

Heroin-induced leukoencephalopathy in a 55-year-old male. **a** T2-weighted image shows diffuse hyperintensity in the white matter including U fibers. **b, c** DW image shows these lesions as diffusely hyperintense with mildly decreased ADC. **d** Pathology shows intramyelinic edema and reactive astrogliosis, consistent with the subacute phase of heroin induced leukoencephalopathy (hematoxylin–eosin stain, original magnification ×200)

10.1.4 Central Pontine Myelinolysis and Extrapontine Myelinolysis

Central pontine myelinolysis (CPM) and extrapontine myelinolysis (EPM) represent destruction of myelin sheaths in characteristic places within the brain stem and cerebrum. The most common location is the central part of the basis pontis, followed by a combined type with central and extrapontine areas of myelinolysis. Isolated EPM is rare [11]. The basal ganglia, caudate nucleus, thalamus, geniculate bodies, internal and external capsules and gray–white matter junction are possible sites of EPM. The synonyms include osmotic myelinolysis and osmotic demyelination syndrome [12, 13].

Pathological findings include destruction of myelin sheaths, though the nerve cells and axons are relatively spared. The underlying etiology and pathogenesis are unknown, but the hypotheses include osmotic endothelial injury, excessive brain dehydra-

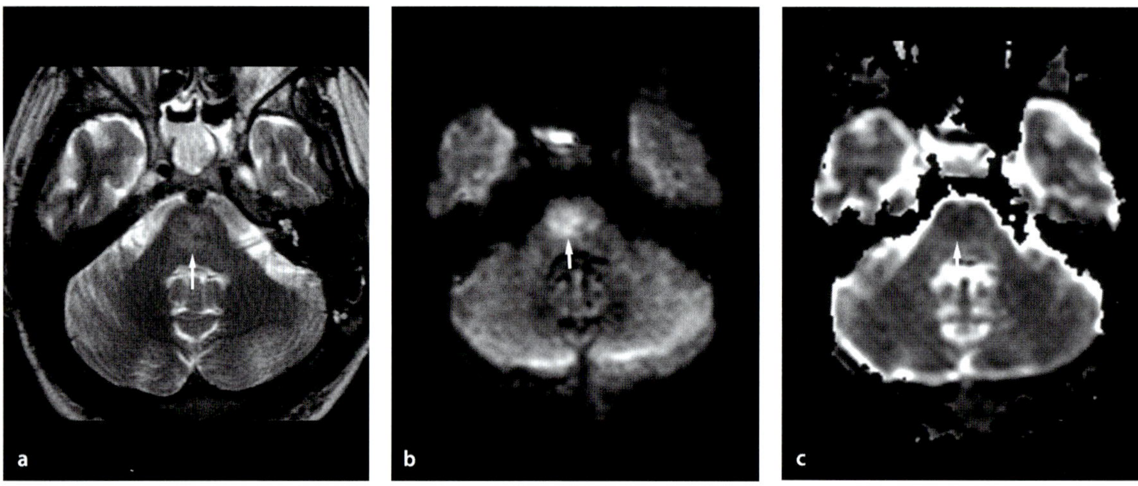

Figure 10.4 a–c

Central pontine myelinolysis in a 33-year-old male. **a** T2-weighted image shows a hyperintense lesion in the center of the pons (*arrow*). **b, c** DW image shows this lesion as hyperintense with decreased ADC

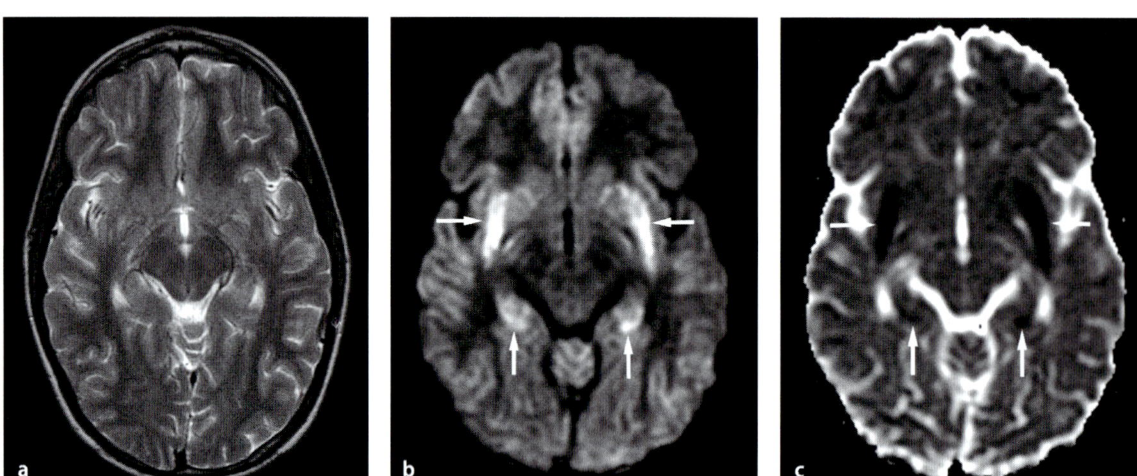

tion, and metabolic compromise [14]. The most common osmotic insult is a rapid correction of hyponatremia. However, CPM and EPM can also occur in normo- or hypernatremic states in patients with chronic alcoholism, post-liver transplantation, malnutrition and AIDS [15–17].

Magnetic resonance imaging has a fundamental role in the diagnosis and discloses hyperintense lesions on T2-weighted images, with or without enhancement on gadolinium-enhanced T1-weighted images. DW imaging can be useful in detecting the lesions in the early phase as hyperintense with decreased ADC, which represents cytotoxic edema (Figs. 10.4 and 10.5) [18]. Cytotoxic edema in CPM and EPM can occur not only in myelin sheaths, but also in neurons, axons and astrocytes [19]. The clinical outcome of CPM and EPM is highly variable, and both fatal and clinically reversible cases may be associated with this kind of cytotoxic edema.

10.1.5 Wernicke Encephalopathy

Thiamine (vitamin B1) deficiency can cause Wernicke encephalopathy, characterized by confusion, ataxia, and abnormal eye movements. Without thiamine, the Krebs and pentose phosphate cycles cannot metabolize glucose [20]. Cellular homeostasis will soon fail and the midline gray matter degenerates. Pathologic findings include decreased myelination, edema, astrocytic swelling, and necrosis in the mamillary bodies, thalamic and hypothalamic nuclei, periaqueductal gray matter, walls of the third and floor of the fourth ventricle, and less commonly, caudate, frontal, and parietal cortex.

Magnetic resonance imaging shows hyperintense lesions of these areas on T2-weighted images. They may or may not show enhancement on T1-weighted images following contrast agent injection [21]. With intravenous thiamine treatment, these lesions may

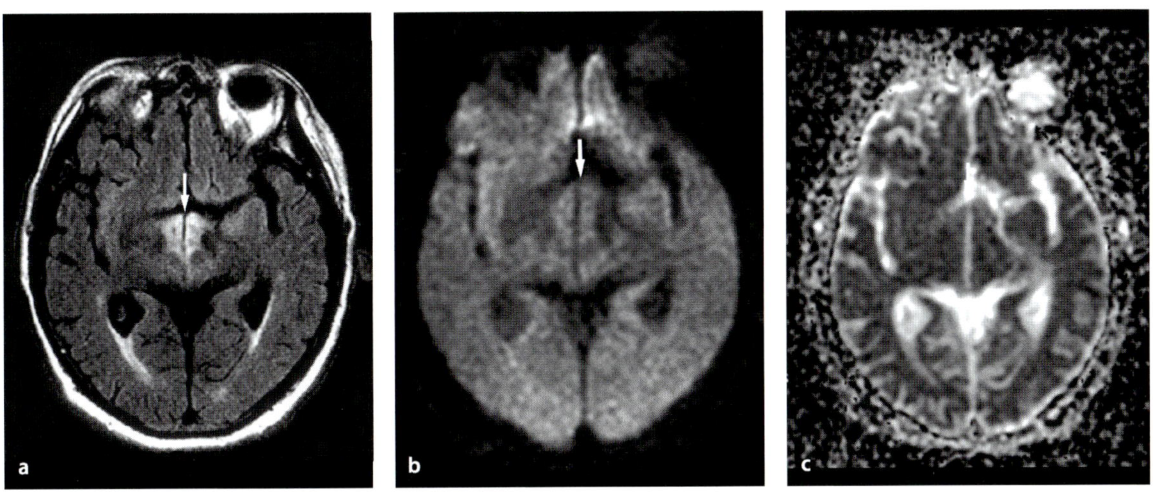

Figure 10.6 a–c

Wernicke encephalopathy in a 75-year-old male. **a** FLAIR shows a symmetrical hyperintense lesion in the hypothalamus (*arrow*). **b, c** DW image shows isointense lesions with increased ADC in the hypothalamus, which may represent vasogenic edema (*arrow*)

Figure 10.5 a–c

Extrapontine myelinolysis in an 11-year-old male. **a** On T2-weighted image shows no appreciable abnormality in the external capsules and hippocampi. **b, c** DW image demonstrates bilateral symmetrical hyperintense lesions with decreased ADC in the external capsules and hippocampi (*arrows*), representing cytotoxic edema

◀

dissipate. DW imaging shows these lesions as hyperintense with decreased or increased ADC. Lesions with decreased ADC are thought to represent cytotoxic edema of neurons or astrocytes, while lesions with increased ADC may represent vasogenic edema (Fig. 10.6) [22–24]. Both types of lesion can be reversible [25].

10.1.6 Marchiafava–Bignami Disease

Marchiafava–Bignami disease is a fatal disorder characterized by demyelination of the corpus callosum, often associated with chronic alcoholism [26]. The genu of the corpus callosum is more frequently involved, but the degeneration can extend throughout the entire corpus callosum. Occasionally, optic chiasm and the visual tracts, putamen, anterior commissure, cerebellar peduncles, cortical gray matter and U-fibers may be involved. Clinical signs include seizures, impairment of consciousness, and signs of interhemispheric disconnection, but they are nonspecific.

The corpus callosum appears hypoattenuated on CT and hyperintense on T2-weighted and fluid-attenuated inversion-recovery (FLAIR) images, which is essential to confirm the diagnosis. These lesions can be partially reversible with treatment [27]. DW imaging shows lesions in the early phase as hyperintense with decreased ADC [28] representing cytotoxic edema, mainly in the myelin sheaths (intramyelinic edema). In the subacute phase, the lesions are hyperintense on DW imaging with increased ADC representing demyelination or necrosis (Fig. 10.7).

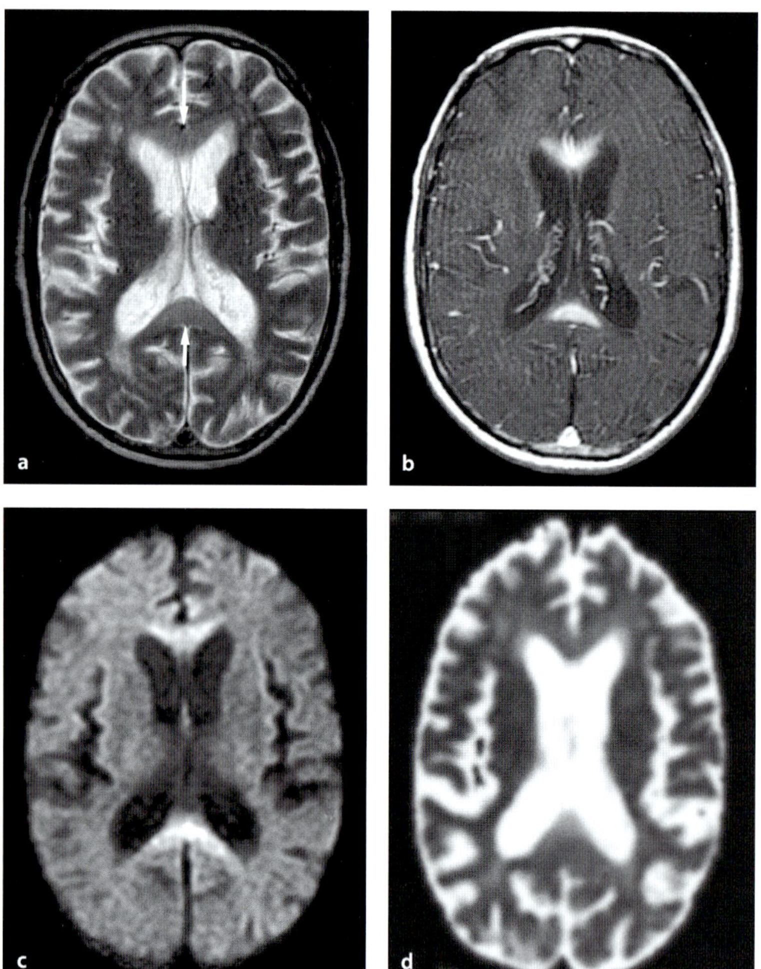

Figure 10.7 a–d

Marchiafava–Bignami disease in a 58-year-old male. **a** T2-weighted image shows hyperintensity in the anterior and posterior corpus callosum (*arrows*) and in the periventricular white matter. **b** Gadolinium-enhanced T1-weighted image with magnetization transfer contrast reveals enhancing lesions in the anterior and posterior corpus callosum. **c** DW image shows hyperintense lesions with increased ADC in the corpus cal-losum, representing demyelination and necrosis in the subacute phase

10.2 Metabolic Disease

10.2.1 Mitochondrial Encephalopathy

Mitochondrial encephalopathies are a heterogeneous group of disorders affecting primarily the central nervous system and skeletal muscles. Two main hypotheses attempt to explain the cerebral lesions: (1) metabolic damage of the endothelium, which leads to small-vessel occlusion and secondary neuronal death and (2) mitochondrial dysfunction, which results in anaerobic metabolism and neuronal death from acidosis [29].

T2-weighted images occasionally show increased signal intensity in the gray and white matter, which usually does not follow vascular territories. Proton MR spectroscopy is useful in the diagnosis by detecting elevated lactate peak. DW imaging often shows the stroke-like lesions in mitochondrial encephalopathy with lactic acidosis (MELAS) as hyperintense. They have increased or normal ADC, which presumably represents vasogenic edema [29–32] (Fig. 10.8). However, decreased ADC in these lesions representing cytotoxic edema can be observed [33–35] (Fig. 10.9).

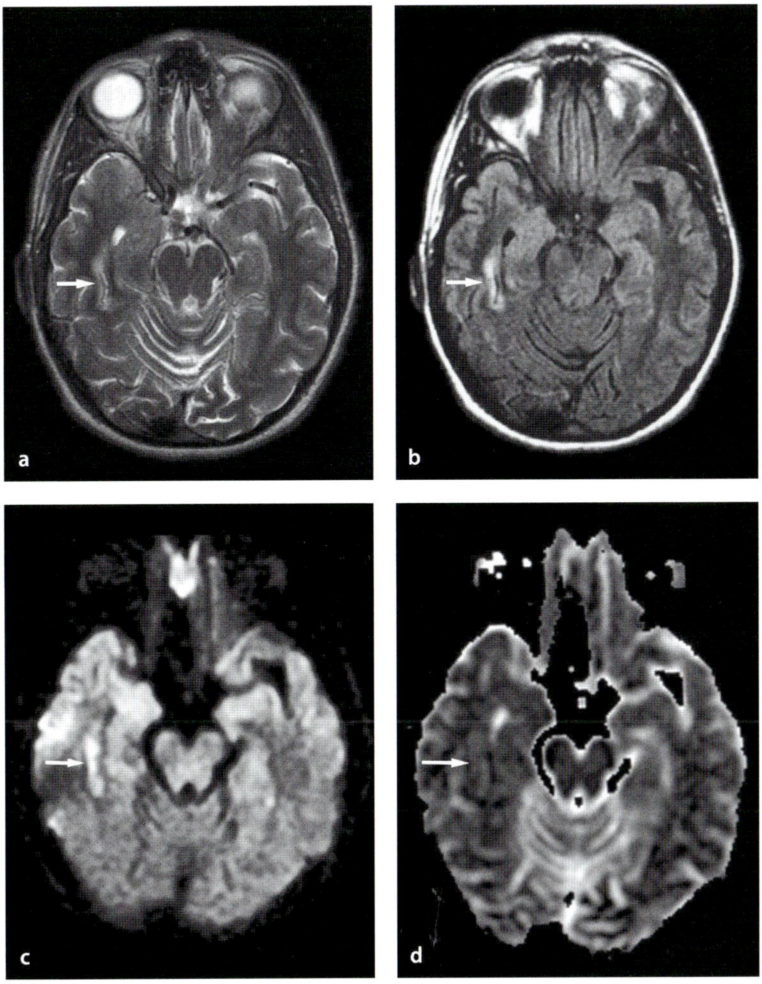

Figure 10.8 a–d

Mitochondrial encephalopathy with lactic acidosis and stroke (MELAS) in a 27-year-old female. **a** T2-weighted and FLAIR images show a hyperintense lesion in the right temporal lobe (*arrow*). **b, c** DW image shows this lesion as hyperintense with increased ADC (*arrow*), mainly representing vasogenic edema

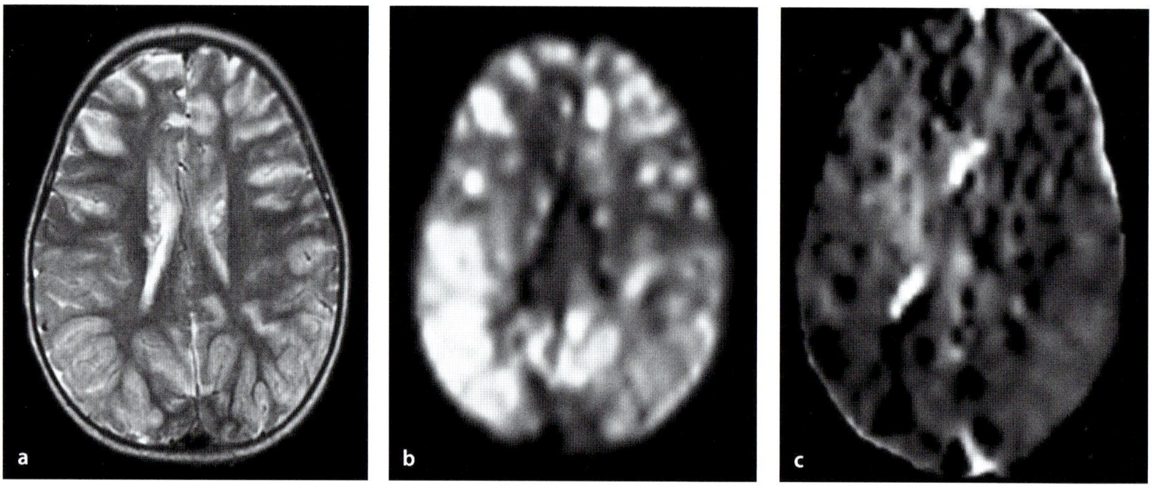

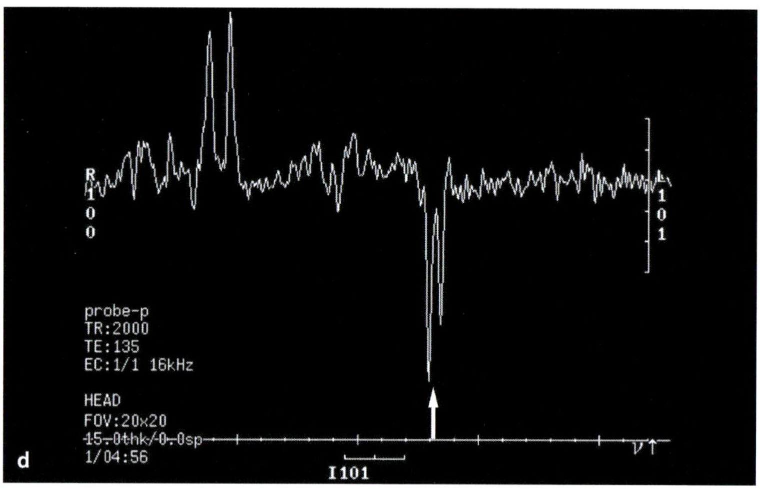

Figure 10.9 a–d

Mitochondrial encephalopathy in a 20-month-old female. **a** T2-weighted image shows multiple and diffuse hyperintense lesions in both hemispheres. **b, c** DW image shows multiple high signal intensity lesions associated with decreased ADC. **d** Proton MR spectroscopy (TE 135 ms) reveals low N-acetyl aspartate and large inverted lactate peaks (*arrow*)

10.2.2 Phenylketonuria

Phenylketonuria is an autosomal recessive disorder caused by a deficiency of phenylalanine hydroxylase. It is the most common congenital disorder of amino acid metabolism. Untreated patients typically develop mental retardation, seizures, growth retardation, hyper-reflexia, eczematous dermatitis, and hyperpigmentation. Pathologic findings include delayed or defective myelination, diffuse white matter vacuolation, demyelination, and gliosis [36].

Magnetic resonance imaging shows hyperintense lesions on T2-weighted images in the periventricular parietal and occipital regions, and in more severe cases extending to the frontal and subcortical white matter [37]. DW imaging shows these lesions as hyperintense with decreased ADC, which presumably represents intramyelinic edema and astrocytic swelling [38] (Fig. 10.10). These lesions can becompletely reversible on follow-up MRI when dietary control has been instituted.

10.2.3 Other Metabolic Diseases and Leukodystrophies

Diffusion-weighted imaging is thought to be useful to differentiate between demyelinating and dysmyelinating disorders [39, 40]. Decreased ADC in the white matter has been reported in Canavan disease [41] (Chap. 14), adrenoleukodystrophy [42], metachromatic leukodystrophy [43], L-2 hydroxyglutaric aciduria [44] (Fig. 10.11), and infantile neuronal dystrophy [45]. The cause of the decrease in ADC in these

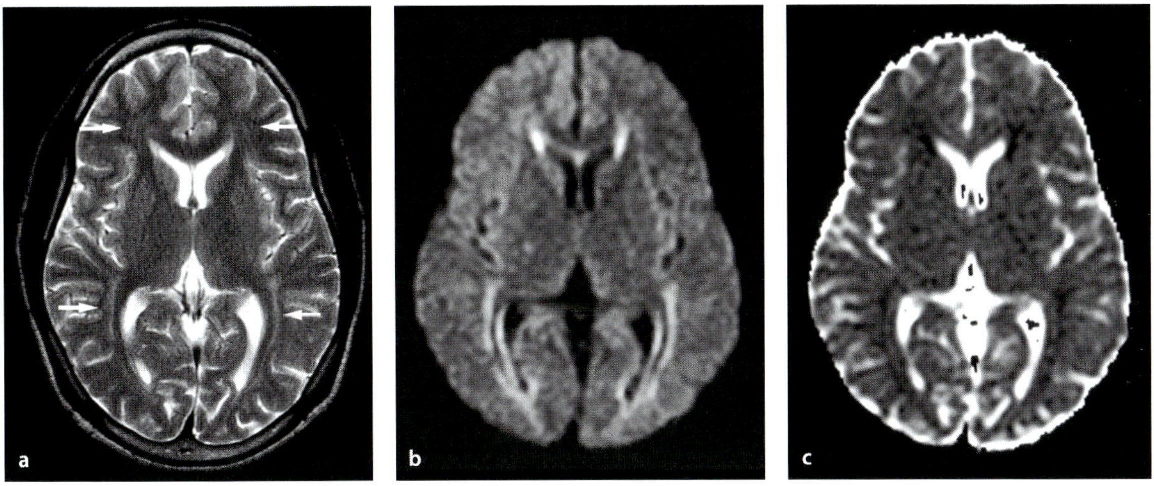

Figure 10.10 a–c

Phenylketonuria in a 36-year-old male. **a** T2-weighted image shows hyperintense lesions in the periventricular white matter (*arrows*). **b, c** DW image shows these lesions as hyperintense with decreased ADC, presumably representing intramyelinic edema

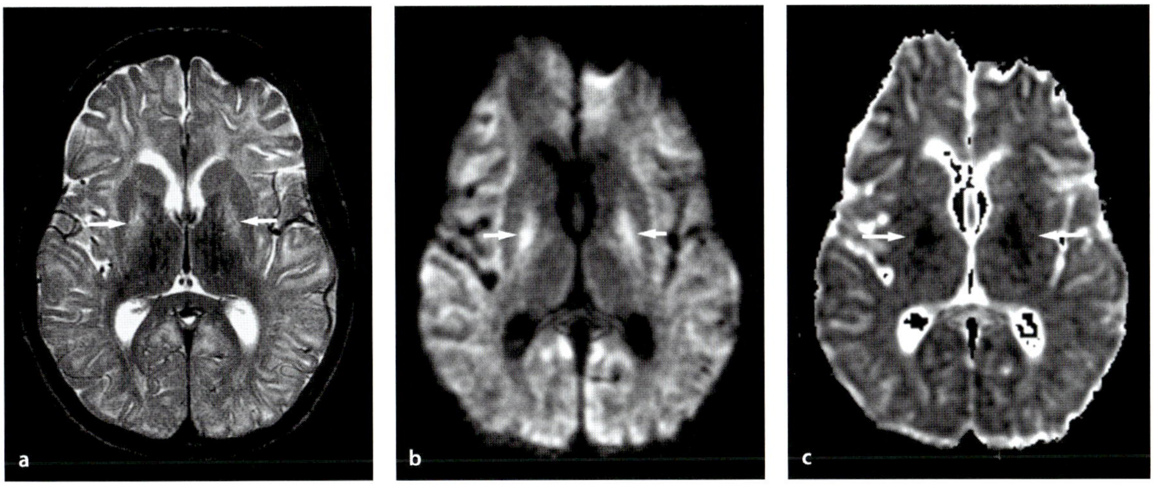

Figure 10.11 a–c

Glutaric aciduria in a 13-year-old male. **a** T2-weighted image shows symmetrical hyperintense lesions in bilateral globus pallidum (*arrows*) and diffusely in the white matter. **b, c** DW image shows these lesions as hyperintense; however, ADC is mildly decreased in the globus pallidum (*arrows*), and mildly increased in the white matter diseases seems to be intramyelinic edema [46], or axonal swelling associated with impaired myelin sheaths or axons. On the other hand, increased or normal ADC in the white matter is seen on Pelizaeus–Merzbacher disease [47] (Chap. 14), Krabbe disease, van de Knaap disease (Fig. 10.12) and leukoencephalopathy with vanishing white matter (Chap. 14). Diffusion tensor MR imaging can be useful in differential diagnosis of leukodystrophies [39, 42, 48].

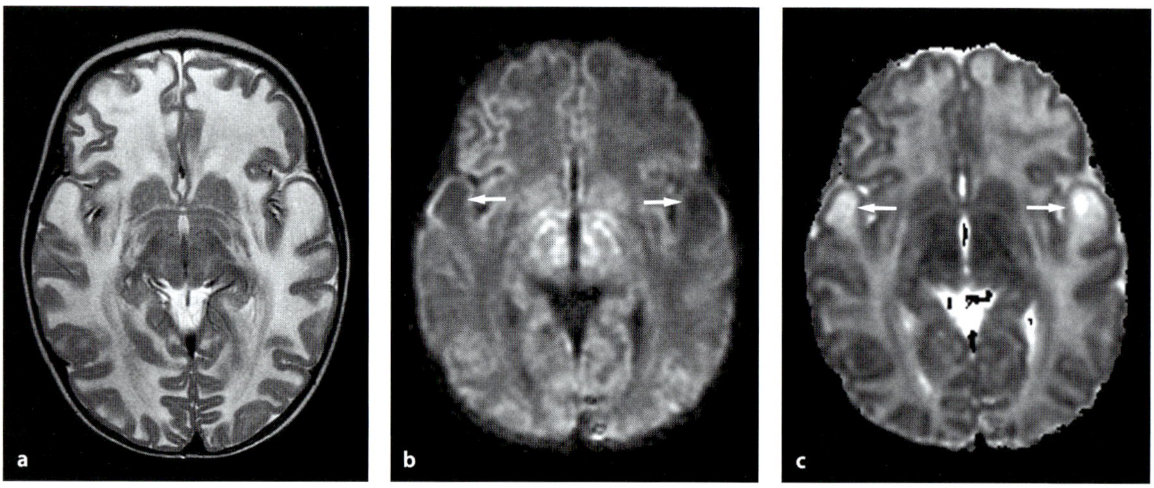

Figure 10.12 a–c

van der Knaap disease in a 10-month-old male with megalencephaly. **a** T2-weighted image shows diffuse hyperintensity in the white matter. **b, c** DW image shows diffuse hypointensity with increased ADC, especially prominent in symmetrical subcortical cysts in the temporal lobes (*arrows*)

References

1. Lexa FJ (1995) Drug-induced disorders of the central nervous system. Seminars in Roentgenol 30:7–17
2. Brown MS, Stemmer SM, Simon JH, et al. (1998) White matter disease induced by high-dose chemotherapy: longitudinal study with MR imaging and proton spectroscopy. AJNR Am J Neuroradiol 19:217–221
3. Fujikawa A, Tsuchiya K, Katase S, Kurosaki Y, Hachiya J (2001) Diffusion-weighted MR imaging of Carmofur-induced leukoencephalopathy. Eur Radiol 11:2602–2606
4. Tha KK, Terae S, Sugiura M, et al. (2002) Diffusion-weighted magnetic resonance imaging in early stage of 5-fluorouracil-induced leukoencephalopathy. Acta Neurol Scand 106:379–386
5. Matsumoto S, Nishizawa S, Murakami S, et al. (1995) Carmofur-induced leukoencephalopathy: MRI. Neuroradiology 37:649–652
6. Maschke M, Fehlings T, Kastrup O, Wilhelm HW, Leonhardt G (1999) Toxic leukoencephalopathy after intravenous consumption of heroin and cocaine with unexpected clinical recovery. J Neurol 246:850–851
7. Wolters EC, van Wijngaarden GK, Stam FC, et al (1982) Leucoencephalopathy after inhaling "heroin" pyrolysate. Lancet 2:1233–1237
8. Tan TP, Algra PR, Valk J, Wolters EC (1994) Toxic leukoencephalopathy after inhalation of poisoned heroin: MR findings. Am J Neuroradiol 15:175–178
9. Chen CY, Lee KW, Lee CC, Chin SC, Chung HW, Zimmerman RA (2000) Heroin-induced spongiform leukoencephalopathy: value of diffusion MR imaging. J Comput Assist Tomogr 24:735–737
10. Barnett MH, Miller LA, Reddel SW, Davies L (2001) Reversible delayed leukoencephalopathy following intravenous heroin overdose. J Clin Neurosci 8:165–167
11. Gocht A, Colmant HJ (1987) Central pontine and extrapontine myelinolysis: a report of 58 cases. Clin Neuropathol 6:262–270
12. Sterns RH, Riggs JE, Schochet SS Jr (1986) Osmotic demyelination syndrome following correction of hyponatremia. N Engl J Med 314:1535–1542
13. Ho VB, Fitz CR, Yoder CC, Geyer CA (1993) Resolving MR features in osmotic myelinolysis (central pontine and extrapontine myelinolysis. Am J Neuroradiol 14:163–167
14. Norenberg MD (1983) A hypothesis of osmotic endothelial injury. A pathogenetic mechanism in central pontine myelinolysis. Arch Neurol 40:66–69
15. Mascalchi M, Cincotta M, Piazzini M (1993) Case report: MRI demonstration of pontine and thalamic myelinolysis in a normonatremic alcoholic. Clin Radiol 47:137–138
16. Rodriguez J, Benito-Leon J, Molina JA, Ramos A, Bermejo F (1998) Central pontine myelinolysis associated with cyclosporin in liver transplantation. Neurologia 13:437–440
17. Miller RF, Harrison MJ, Hall-Craggs MA, Scaravilli F (1998) Central pontine myelinolysis in AIDS. Acta Neuropathol 96:537–540
18. Cramer SC, Stegbauer KC, Schneider A, Mukai J, Maravilla KR (2001) Decreased diffusion in central pontine myelinolysis. AJNR Am J Neuroradiol 22:1476–1479
19. Anderson AW, Zhong J, Petroff OA, et al (1996) Effects of osmotically driven cell volume changes on diffusion-weighted imaging of the rat optic nerve. Magn Reson Med 35:162–167
20. Victor M, Adams RD, Collins GH (1971) The Wernicke-Korsakoff syndrome. A clinical and pathological study of 245 patients, 82 with post-mortem examinations. Contemp Neurol Ser 7:1–206
21. Antunez E, Estruch R, Cardenal C, Nicolas JM, Fernandez-Sola J, Urbano-Marquez A (1998) Usefulness of CT and MR imaging in the diagnosis of acute Wernicke's encephalopathy. AJR Am J Roentgenol 171:1131–1137
22. Oka M, Terae S, Kobayashi R, et al. (2001) Diffusion-weighted MR findings in a reversible case of acute Wernicke encephalopathy. Acta Neurol Scand 104:178–181
23. Doherty MJ, Watson NF, Uchino K, Hallam DK, Cramer SC (2002) Diffusion abnormalities in patients with Wernicke encephalopathy. Neurology 58:655–657
24. Chu K, Kang DW, Kim HJ, Lee YS, Park SH (2002) Diffusion-weighted imaging abnormalities in wernicke encephalopathy: reversible cytotoxic edema? Arch Neurol 59:123–127
25. Rugilo CA, Roca MC, Zurru MC, Gatto EM (2003) Diffusion abnormalities and Wernicke encephalopathy. Neurology 60:727–728; author reply 727–728
26. Gambini A, Falini A, Moiola L, Comi G, Scotti G (2003) Marchiafava-Bignami disease: longitudinal MR imaging and MR spectroscopy study. Am J Neuroradiol 24:249–253
27. Gass A, Birtsch G, Olster M, Schwartz A, Hennerici MG (1998) Marchiafava-Bignami disease: reversibility of neuroimaging abnormality. J Comput Assist Tomogr 22:503–504
28. Inagaki T, Saito K (2000) A case of Marchiafava-Bignami disease demonstrated by MR diffusion-weighted image. No To Shinkei 52:633–637
29. Oppenheim C, Galanaud D, Samson Y, et al. (2000) Can diffusion weighted magnetic resonance imaging help differentiate stroke from stroke-like events in MELAS? J Neurol Neurosurg Psychiatry 69:248–250
30. Yoneda M, Maeda M, Kimura H, Fujii A, Katayama K, Kuriyama M (1999) Vasogenic edema on MELAS: a serial study with diffusion-weighted MR imaging. Neurology 53: 2182–2184
31. Yonemura K, Hasegawa Y, Kimura K, Minematsu K, Yamaguchi T (2001) Diffusion-weighted MR imaging in a case of mitochondrial myopathy, encephalopathy, lactic acidosis, and strokelike episodes. AJNR Am J Neuroradiol 22:269–272
32. Ohshita T, Oka M, Imon Y, et al. (2000) Serial diffusion-weighted imaging in MELAS. Neuroradiology 42:651–656
33. Kim HS, Kim DI, Lee BI, et al. (2001) Diffusion-weighted image and MR spectroscopic analysis of a case of MELAS with repeated attacks. Yonsei Med J 42:128–133
34. Majoie CB, Akkerman EM, Blank C, Barth PG, Poll-The BT, den Heeten GJ (2002) Mitochondrial encephalomyopathy: comparison of conventional MR imaging with diffusion-weighted and diffusion tensor imaging: case report. AJNR Am J Neuroradiol 23:813–816
35. Flemming K, Ulmer S, Duisberg B, Hahn A, Jansen O (2002) MR spectroscopic findings in a case of Alpers-Huttenlocher syndrome. AJNR Am J Neuroradiol 23:1421–1423

36. Huttenlocher PR (2000) The neuropathology of phenylketonuria: human and animal studies. Eur J Pediatr Suppl 2:S102–S106
37. Pearsen KD, Gean-Marton AD, Levy HL, Davis KR (1990) Phenylketonuria: MR imaging of the brain with clinical correlation. Radiology 177:437–440
38. Phillips MD, McGraw P, Lowe MJ, Mathews VP, Hainline BE (2001) Diffusion-weighted imaging of white matter abnormalities in patients with phenylketonuria. AJNR Am J Neuroradiol 22:1583–1586
39. Ono J, Harada K, Mano T, Sakurai K, Okada S (1997) Differentiation of dys- and demyelination using diffusional anisotropy. Pediatr Neurol 16:63–66
40. Engelbrecht V, Scherer A, Rassek M, Witsack HJ, Modder U (2002) Diffusion-weighted MR imaging in the brain in children: findings in the normal brain and in the brain with white matter diseases. Radiology 222:410–418
41. Sener RN (2003) Canavan disease: diffusion magnetic resonance imaging findings. J Comput Assist Tomogr 27:30–33
42. Eichler FS, Itoh R, Barker PB, et al. (2002) Proton MR spectroscopic and diffusion tensor brain MR imaging in X-linked adrenoleukodystrophy: initial experience. Radiology 225:245–252
43. Sener RN (2002) Metachromatic leukodystrophy: diffusion MR imaging findings. AJNR Am J Neuroradiol 23:1424–1426
44. Sener RN (2003) L-2 hydroxyglutaric aciduria: proton magnetic resonance spectroscopy and diffusion magnetic resonance imaging findings. J Comput Assis Tomo 27:38–43
45. Sener RN (2003) Diffusion magnetic resonance imaging in infantile neuroaxonal dystrophy. J Comput Assis Tomo 27:34–37
46. Ellison D, Love S (1998) Toxic injury of the CNS. In: Neuropathology, 1st edn, section 8, chapter 25. Barcelona, Mosby, pp 25.1–25.22
47. Ono J, Harada K, Sakurai K, et al. (1994) MR diffusion imaging in Pelizaeus-Merzbacher disease. Brain Dev 16:219–223
48. Guo AC, Petrella JR, Kurtzberg J, Provenzale JM (2001) Evaluation of white matter anisotropy in Krabbe disease with diffusion tensor MR imaging: initial experience. Radiology 218:809–815

Infectious Diseases

11.1 Overview of Brain Infections

Infections of the brain are caused by bacteria, virus, fungi or parasites. Bacterial infections are often related to septic emboli and extracranial infections spreading intracranially and intra-axially. This can result in cerebritis and brain abscesses. Viral infections are more diffuse and cause encephalitis and vasculitis. Toxoplasmosis, which is the most common parasitic infection of the brain, causes encephalitis and abscesses, while disseminated aspergillosis causes vasculitis-mediated infarctions resulting in extensive cerebritis and/or abscess formation.

The pathophysiology and the imaging findings vary greatly depending on the organism causing the infection. Diffusion-weighted (DW) imaging is useful for diagnosis of infectious conditions of the brain by means of differentiating vasogenic edema from cytotoxic edema [1]. DW imaging can also separate abscesses from cystic and necrotic tumors [2–6].

11.2 Bacterial Brain Abscess

Bacterial brain abscesses are potentially fatal, but can often be medically and surgically treated if detected early. Symptoms are often non-specific and vague, and imaging is therefore necessary for detection and characterization.

A brain abscess begins as a focal area of microvascular injury, usually at the gray–white matter junction or deeper in the white matter. Pathologically the initial stage of a brain abscess is a focal area of cerebritis or presuppurative encephalitis. This is characterized by early necrosis of the cerebral parenchyma, vascular congestion, petechial hemorrhage, neutrophil infiltration and vasogenic edema [7–9].

Late cerebritis is characterized by a necrotic and purulent center. This evolves to frank abscess formation, which is characterized by central pus, inflammatory granulation tissue and a fibrous capsule. The pus usually consists of both dead and viable neutrophils. Even in the chronic phase of an abscess, there is both coagulative necrosis and bacteria (Fig. 11.1).

The early phase of the brain abscess has a homogeneous, bright signal on DW imaging associated with decreased apparent diffusion coefficient (ADC) (Fig. 11.1). On follow-up DW imaging, the chronic phase of an abscess can still show hyperintensity, but ADC values are partially increased [10]. A possible explanation for the high signal on DW imaging is restriction of water mobility due to the high viscosity of coagulative necrosis and the hypercellularity of polynucleated neutrophils in the pus.

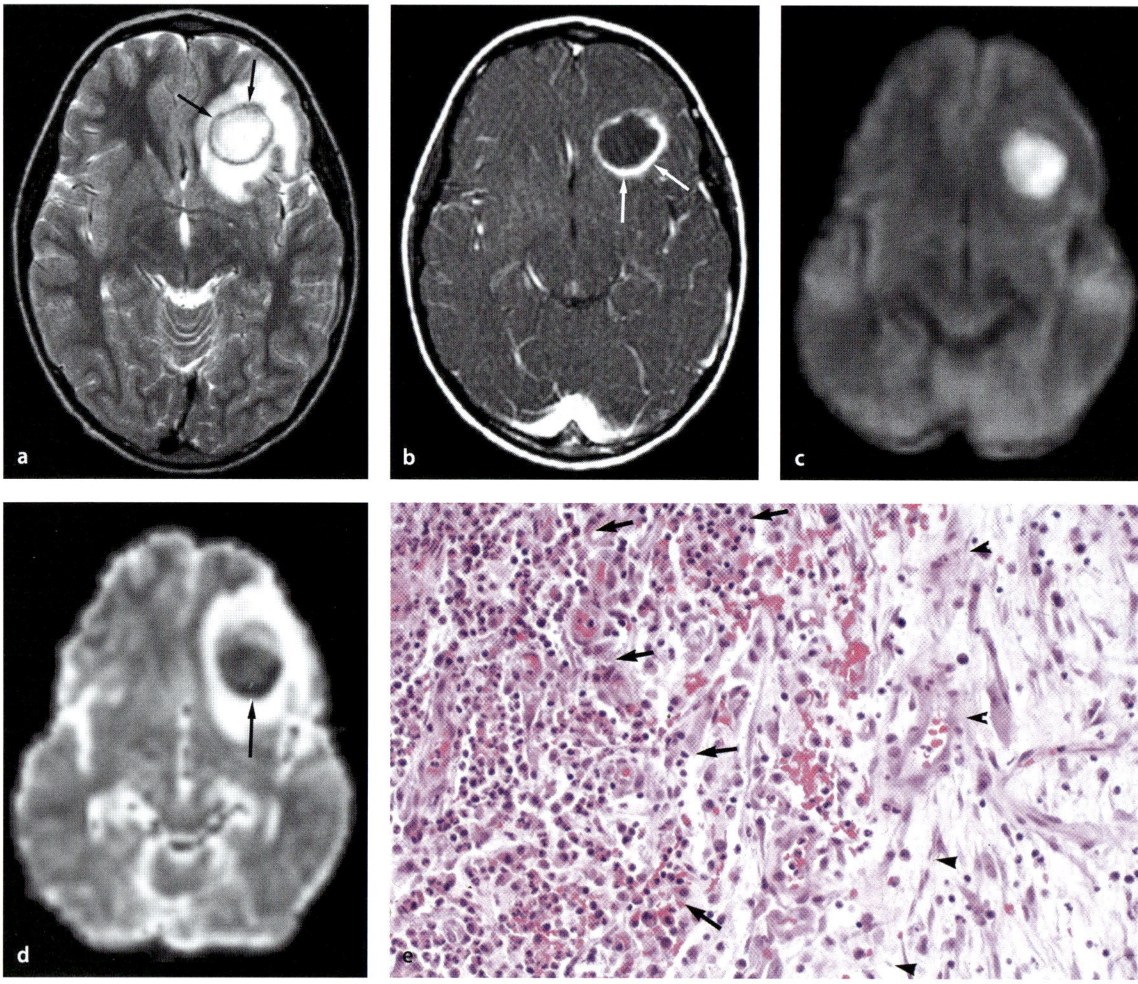

Figure 11.1 a–e

Chronic streptococcal brain abscess in a 7-year-old boy presenting with severe headache. **a** T2-weighted image shows a central hyperintense mass lesion with low signal rim (*black arrows*) and peripheral edema in the left frontal lobe. **b** Gadolinium-enhanced T1-weighted image with magnetization transfer contrast shows this mass with ring enhancement. **c** DW image shows a central cystic component as hyperintense. **d** ADC map shows decreased ADC of this component. **e** Specimen shows numerous neutrophils (*arrows*) in the pus and granulomatous fibrous capsules (*arrowheads*) in a chronic abscess

11.3 Septic Emboli

The main risk factors for brain abscesses are bacterial endocarditis and chronic suppurative intrathoracic infections [11]. If septic emboli of sufficient size are lodged in an intracerebral arterial vessel, infarction can occur. Infarctions are bright on DW imaging, with decreased ADC (Fig. 11.2). Septic infarctions usually occur in the distal cortical branch territories, while small septic emboli are characteristically found in the corticomedullary junction. It can take a few weeks and up to several months for septic emboli to develop into an abscess. Serial DW imaging is therefore often useful in patients at risk for septic encephalopathy. By means of repeated DW imaging, it is possible to identify the initial infarction and the subsequent cerebritis/abscess evolution [12]. This allows for early treatment.

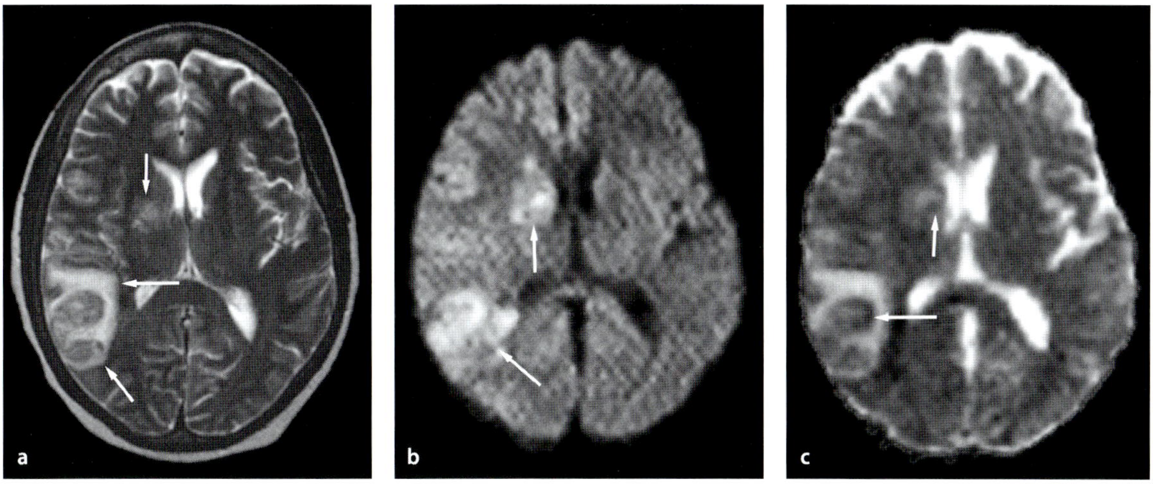

Figure 11.2 a–c

Septic emboli from staphylococcus endocarditis in a 44-year-old woman presenting with left hemiparesis of 3 days' duration. **a** T2-weighted image shows hyperintense lesions in the right basal ganglia and posteriorly in the middle cerebral artery territory (*arrows*). **b** DW image shows these lesions as hyperintense (*arrows*). **c** ADC map shows these lesions as decreased ADC, mainly representing infarcts (*arrows*)

11.4 Brain Abscess Caused by Unusual Bacteria

The classical finding in a brain abscess is a cystic lesion with marked enhancement following contrast medium injection. On DW imaging, the cyst shows a high signal and ADC is reduced. However, the characteristics of brain abscess appear to be related to the type of organism and the immunity of the host. Thus, in an immunodeficient patient with sepsis, multiple micro-abscesses are often observed. Multiple micro-abscesses of gram-negative rods may involve basal ganglia bilaterally and mimic small infarcts (Fig. 11.3) [13].

Listeria infection is often associated with parenchymal involvement, especially in the brain stem [14]. A central small abscess may be seen as high signal on DW imaging, with decreased ADC (Fig. 11.4).

11.4.1 Differential Diagnosis

Diffusion-weighted imaging can discriminate brain abscesses from cystic or necrotic tumors, which is often difficult with conventional MR imaging (Fig. 11.5) [2–6]. Abscesses and tumor necrosis are both generally bright on DW images, but the abscesses characteristically have low ADC, whereas the ADC in tumors varies. There are exceptions and central necrosis of a tumor or metastasis can occasionally show the same characteristics with hyperintensity on DW imaging with low ADC (Fig. 11.6) [15–17]. Sterile and liquefactive coagulative necrosis, hemorrhage and viscous mucinous components are possible causes for this finding.

Pure coagulative necrosis typically develops after radiofrequency thalamotomy [18]. The lesion often shows hyperintensity on DW imaging, with decreased ADC (Fig. 11.7). Although imaging characteristics are very similar to those of an abscess, the history of the patient and the symptomatology can usually help to differentiate post-surgical lesions from abscesses.

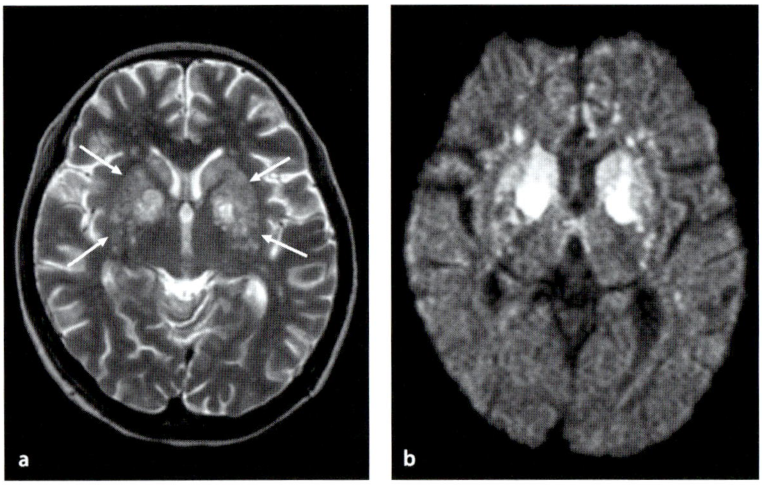

Figure 11.3 a,b

Brain abscesses due to gram-negative rods bacteria in a 28-year-old woman with Crohn's disease and long-term steroid use presenting with headache. Blood culture showed gram-negative rods. **a** T2-weighted image shows multiple small hyperintense lesions in bilateral basal ganglia (*arrows*) and white matter. **b** DW image shows these lesions as very hyperintense with decreased ADC (not shown). This finding mimics multiple small infarcts but it represents multiple small abscesses

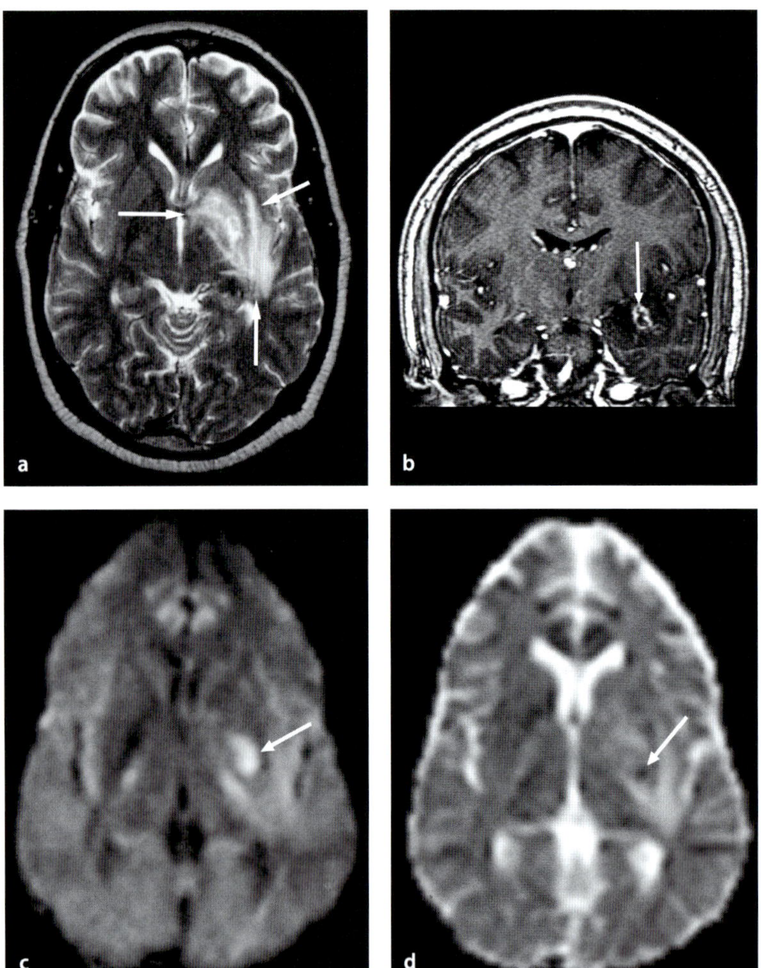

Figure 11.4 a–d

Listeria meningoencephalitis after bone marrow transplantation for chronic myeloblastic leukemia in a 31-year-old man presenting with a 2-day history of severe headache. **a** T2-weighted image shows hyperintense lesions in the left basal ganglia, internal capsule and white matter in the temporal lobe, representing encephalitis (*arrows*). **b** Coronal gadolinium-enhanced T1-weighted image shows an irregular ring-enhancing lesion (*arrow*). **c** DW image shows a central cystic component as hyperintense with decreased ADC (**d**) (*arrow*), representing an abscess

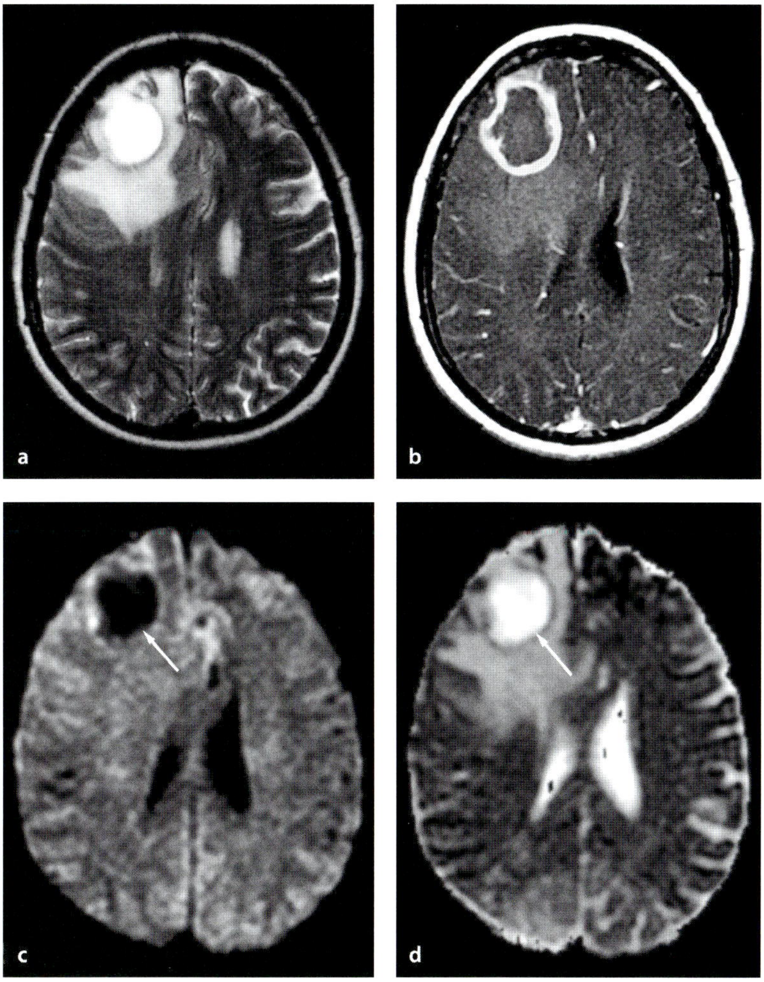

Figure 11.5 a–d

Glioblastoma multiforme in a 69-year-old woman. **a** T2-weighted image shows a central hyperintense mass lesion with low signal rim and peripheral edema in the right frontal lobe. **b** Gadolinium-enhanced T1-weighted image with magnetization transfer contrast shows this mass with irregular ring enhancement. **c** DW image shows a central cystic component as hypointense (*arrow*). **d** ADC map shows increased ADC of this cystic component (*arrow*)

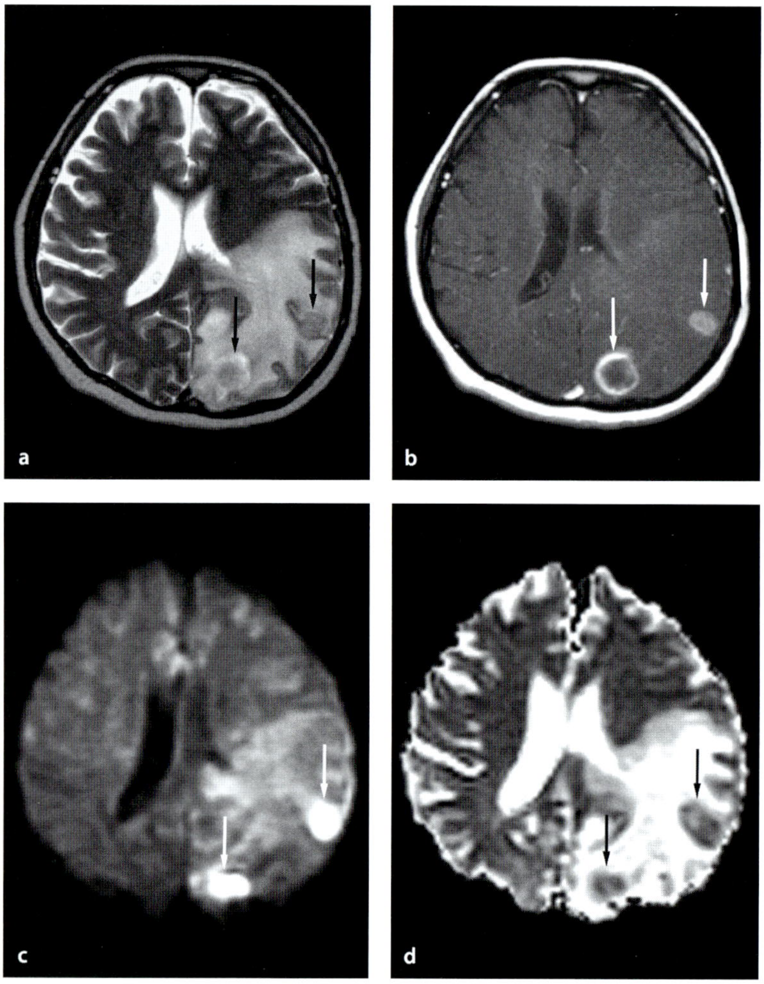

Figure 11.6 a–d

Brain metastasis from adenocarcinoma of the lung in a 44-year-old man. **a** T2-weighted image shows multiple mass lesions and peripheral edema in the left parieto-occipital lobe. **b** Gadolinium-enhanced T1-weighted image with magnetization transfer contrast shows these lesions with irregular ring enhancement (*arrows*). **c** DW image shows central cystic components as hyperintense (*arrows*). **d** ADC map shows relative low ADC values of cystic components, presumably representing coagulative necrosis or mucinous substance (*arrows*)

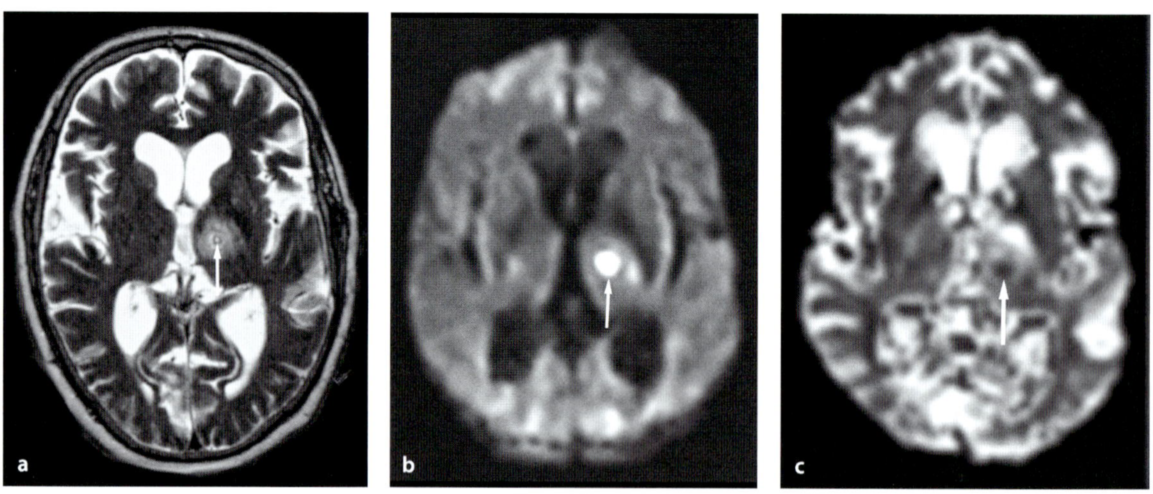

11.5 Bacterial Abscess in the Extra-Axial Space

Infections can enter the extra-axial spaces by a variety of mechanisms, including direct spread from an adjacent focus, retrograde septic thrombophlebitis, hematogenous seeding and sequela of purulent leptomeningitis [19]. Abscesses can occur in epidural, subdural (Fig. 11.8), leptomeningeal (Fig. 11.9) or intraventricular spaces (Fig. 11.10) [20, 21]. Wherever they occur, DW imaging shows pus in the abscess as hyperintense, with relatively low ADC values. An exception to the rule can be found in some cases of extraaxial pus collections where the ADC may from low to high. Regions of increased ADC may represent dilution of pus with CSF.

11.5.1 Differential Diagnosis

Hematomas and epidermoids occasionally have imaging characteristics similar to those of an extra-axial abscess on DW imaging. The hematomas often show hyperintensity with decreased ADC, probably because of the hypercellularity or hyperviscosity (Fig. 11.11) [22, 23]. Epidermoids also show an extra-axial hyperintense lesion on DW imaging (Fig. 11.12) caused by high viscosity keratohyalin- containing materials, which are arranged in layers, as in an onion bulb. Moreover, they are rich in cholesterol crystals. ADC maps usually show slightly increased ADC as compared with normal brain parenchyma [24].

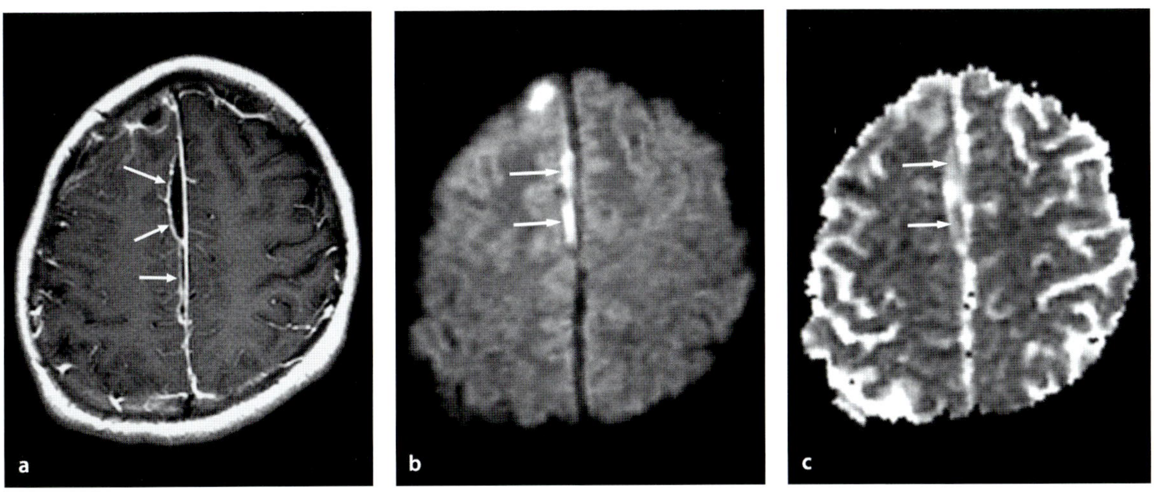

Figure 11.8 a–c

Subdural empyema in a 16-year-old male. **a** Gadolinium-enhanced T1-weighted image with magnetization transfer contrast shows rim-enhancing lesions along the falx (*arrows*). Mild meningeal enhancement is also seen. **b** DW image shows these lesions as hyperintense, representing subdural abscesses (*arrows*). **c** ADC map shows these lesions as relatively decreased ADC (*arrows*) compared with CSF (Courtesy of Morikawa M, M.D., Nagasaki University, School of Medicine, Japan)

Figure 11.7 a–c

Coagulative necrosis 4 weeks after radiofrequency thalamotomy for essential tremor in a 70-year-old man. **a** T2-weighted image shows a hyperintense lesion with mild peripheral edema in the left thalamus (*arrow*). **b** DW image shows a central cystic component as hyperintense. **c** ADC map shows this component as decreased ADC, probably representing coagulative necrosis. This lesion was decreased in size on follow-up MRI without antibiotic treatment

◀

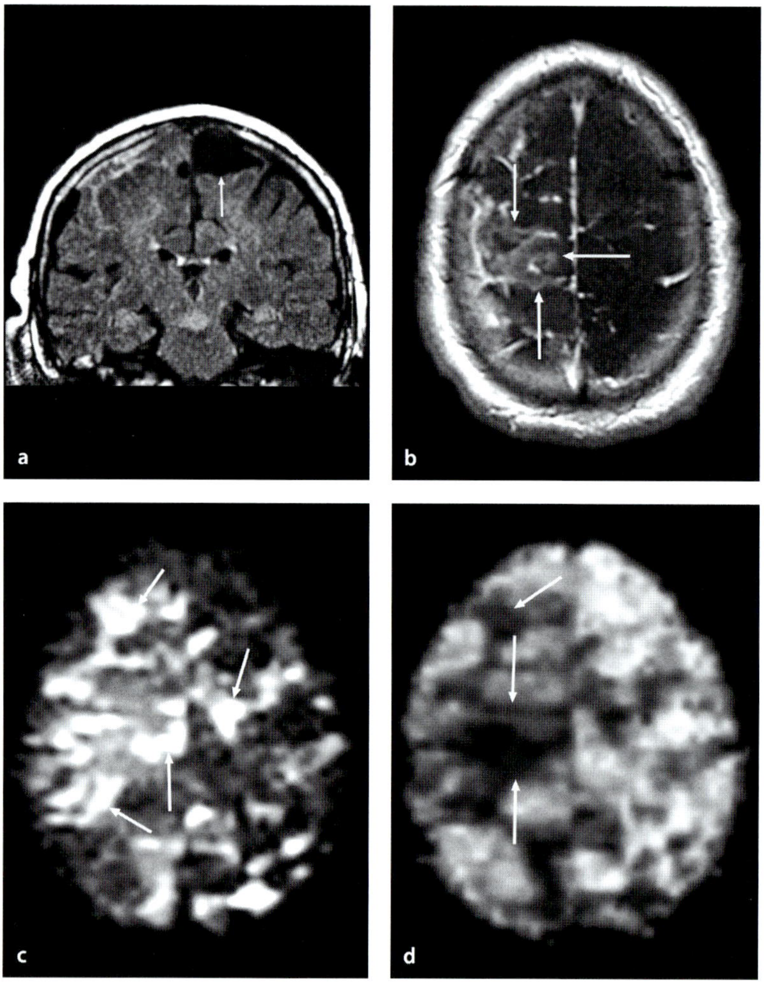

Figure 11.9 a–d

Purulent leptomeningitis in a 77-year-old man. Pneumococcus was proven by CSF examination. **a** Coronal fluid-attenuated inversion-recovery image shows hyperintense lesions in subarachnoid space in the right frontoparietal region (*arrow*). **b** Gadolinium-enhanced T1-weighted image with magnetization transfer contrast shows irregular enhancement of these lesions (*arrows*). **c** DW image shows these lesions as hyperintense, representing purulent leptomeningitis (*arrows*). **d** ADC map shows these lesions as decreased ADC (*arrows*)

Figure 11.10 a–d

Purulent ventriculitis after surgery in a 60-year-old man. **a** Gadolinium-enhanced T1-weighted image with magnetization transfer contrast shows linear enhancement along the ventricle (*arrows*). There are post-surgical changes in the left frontal lobe and extracranially on the left. **b** T2-weighted image shows a fluid–fluid level in bilateral lateral ventricles (*arrows*) and postoperative changes in the left frontal lobe (*arrowhead*). **c** DW image shows the fluid (*arrows*) as hyperintense, representing purulent ventriculitis, and postoperative changes in the left frontal lobe (*arrowhead*). **d** ADC map shows these lesions as decreased ADC

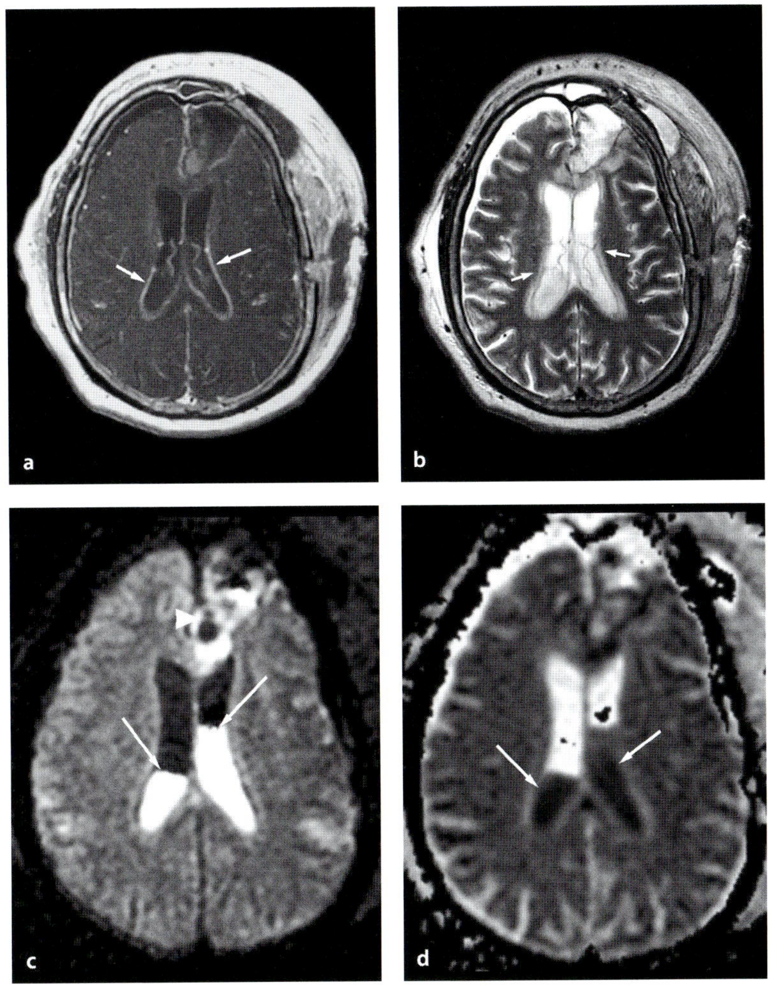

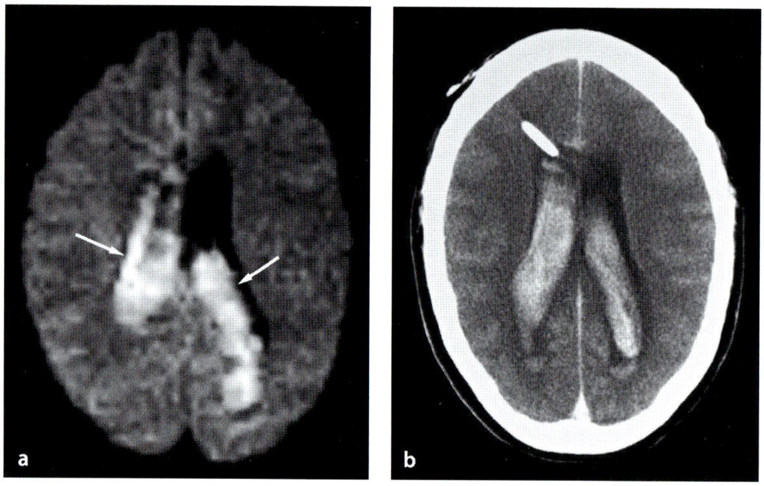

Figure 11.11 a,b

Intraventricular hemorrhage due to arteriovenous malformation in a 29-year-old man. **a** DW image shows hyperintense lesions (*arrows*) with decreased ADC (not shown), similar to the finding seen in purulent intraventriculitis. **b** Computed tomography shows the high density of intraventricular hemorrhage

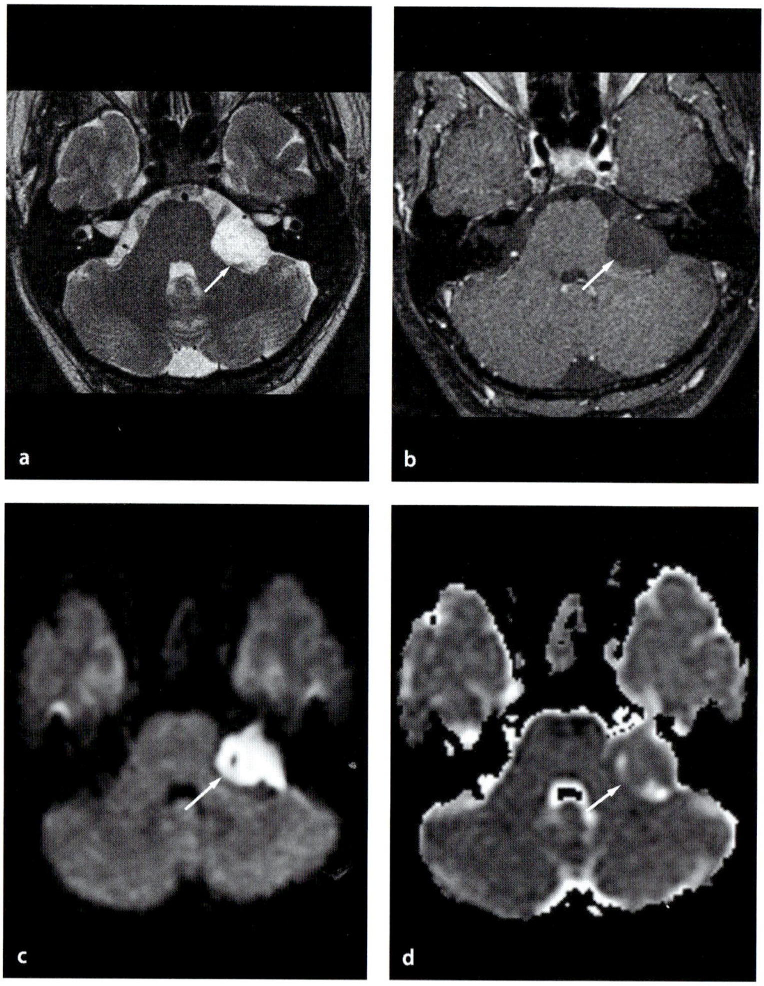

Figure 11.12 a–d

Epidermoid in a 29-year-old male. **a** T2-weighted image shows a hyperintense mass (*arrow*) in the left cerebello-pontine angle. **b** Gadolinium-enhanced T1 weighted image with magnetization transfer contrast shows the lesion (*arrow*) as hypointense with no enhancement. **c** DW image shows this lesion (*arrow*) as hyperintense. **d** ADC map shows almost similar ADC value to the cerebellar parenchyma

11.6 Bacterial Vasculitis

Cerebral infarctions secondary to meningitis are well documented in the pediatric age group. Thus, *Streptococcus* B meningitis in neonates can result in infarction, a condition that is rare in adults. The blood–brain/blood–cerebrospinal fluid (CSF) barrier and the mechanical integrity of meninges are immature in young infants, which may explain the preponderance for these changes to occur in babies and young infants [25]. Both arteries and veins can be involved by infection via the perivascular space. Pial arteriolar occlusion causes infarction of the subpial cortex (Fig. 11.13). Although vasculitic changes occur early, they only become prominent resulting in an infarction by the second to third week after onset of the meningeal infection. DW imaging is useful in early detection of the infarction, which is bright on DW imaging, with low ADC.

11.7 Toxoplasmosis

Toxoplasma abscesses consist of ischemic necrosis of the brain tissue associated with sclerosing endoarteritis and a variety of inflammatory reactions. Toxoplasma abscesses are more commonly seen in immunocompromised patients, such as patients with acquired immunodeficiency syndrome (AIDS). In fact, toxoplasma abscesses in AIDS patients were once so common that they were morphologically grouped into three subtypes: (1) poorly circumscribed areas of necrosis (necrotizing abscess), (2) a central area of coagulative necrosis surrounded by macrophages and organism (organizing abscess), and (3) well-demarcated cystic spaces (chronic abscess) [26].

Diffusion-weighted imaging of a toxoplasma abscess may show a variety of signal characteristics, which probably is a reflection of the different pathologic/morphologic subtypes (Fig. 11.14). Abscesses with different characteristics are often seen in the same patient. Central necroses in toxoplasma abscesses do not contain as many inflammatory cells as regular bacterial abscesses.

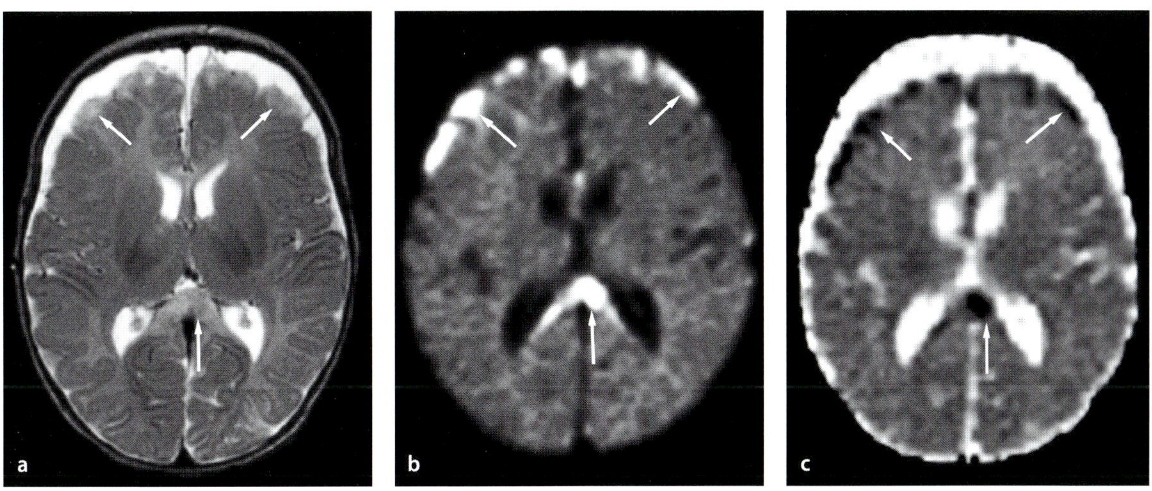

Figure 11.13 a–c

Group B *Streptococcus* meningitis in a 5-week-old boy. **a** T2-weighted image shows bifrontal subdural effusion and hyperintense lesions in bilateral frontal superficial cortices and posterior corpus callosum (*arrows*). **b** DW image shows these lesions (*arrows*) as hyperintense. **c** ADC map shows these lesions (*arrows*) as decreased ADC, representing acute infarcts of the subpial cortex

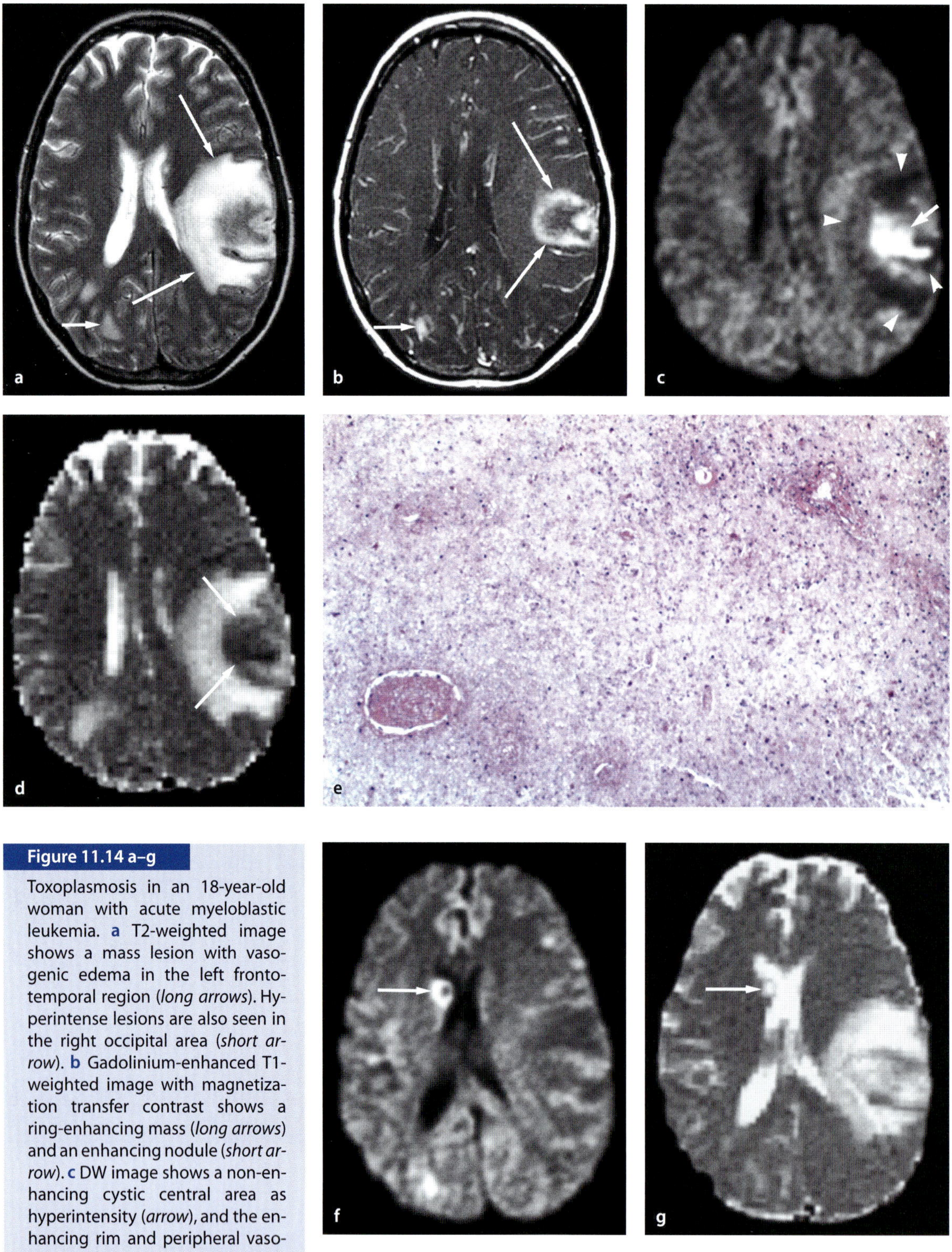

Figure 11.14 a–g

Toxoplasmosis in an 18-year-old woman with acute myeloblastic leukemia. **a** T2-weighted image shows a mass lesion with vasogenic edema in the left fronto-temporal region (*long arrows*). Hyperintense lesions are also seen in the right occipital area (*short arrow*). **b** Gadolinium-enhanced T1-weighted image with magnetization transfer contrast shows a ring-enhancing mass (*long arrows*) and an enhancing nodule (*short arrow*). **c** DW image shows a non-enhancing cystic central area as hyperintensity (*arrow*), and the enhancing rim and peripheral vasogenic edema as hypointensity (*arrowheads*). **d** ADC map shows decreased ADC of this cystic component (*arrows*). **e** Biopsy specimen shows coagulative necrosis of this cystic component (hematoxylin–eosin stain). **f**, **g** DW image shows a small cystic lesion (*arrow*) as hypointense with increased ADC (**g**) in the right caudate nucleus (*arrow*), which may indicate the different phase of toxoplasma abscess. Left fronto-temporal mass and multiple hyperintense nodules are also seen

11.7.1 Differential Diagnosis

The main differential diagnosis is lymphoma. In AIDS patients, central nervous system lymphoma is often associated with central necrosis. This is occasionally liquefactive, presumably related to hypoxia or apoptosis of tumor cells. The enhancing portion of the lesion usually shows hyperintensity on DW imaging, with relatively low ADC due to hypercellularity of lymphomas (Fig. 11.15). A central necrosis usually shows hypointensity on DW imaging, with increased ADC.

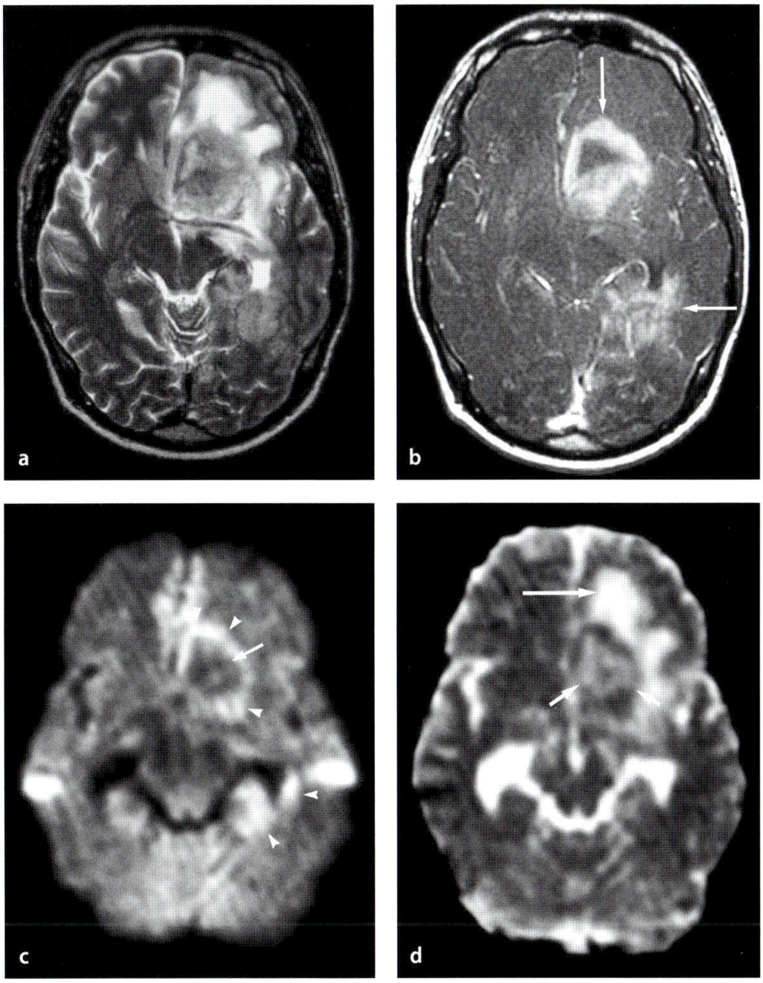

Figure 11.15

AIDS-related lymphoma in a 23-year-old man. **a** T2-weighted image shows a necrotic mass in the left frontal lobe extending into the temporo-occipital lobe. **b** Gadolinium-enhanced T1-weighted image magnetization transfer contrast shows these lesions (*arrows*) with irregular ring enhancement. **c** DW image shows enhancing solid components as hyperintensity (*arrowheads*), and a necrotic component as hypointensity (*arrow*). **d** ADC map shows central necrosis as increased ADC (*long arrow*) and the solid components (*short arrows*) as relatively low ADC compared to the central necrosis

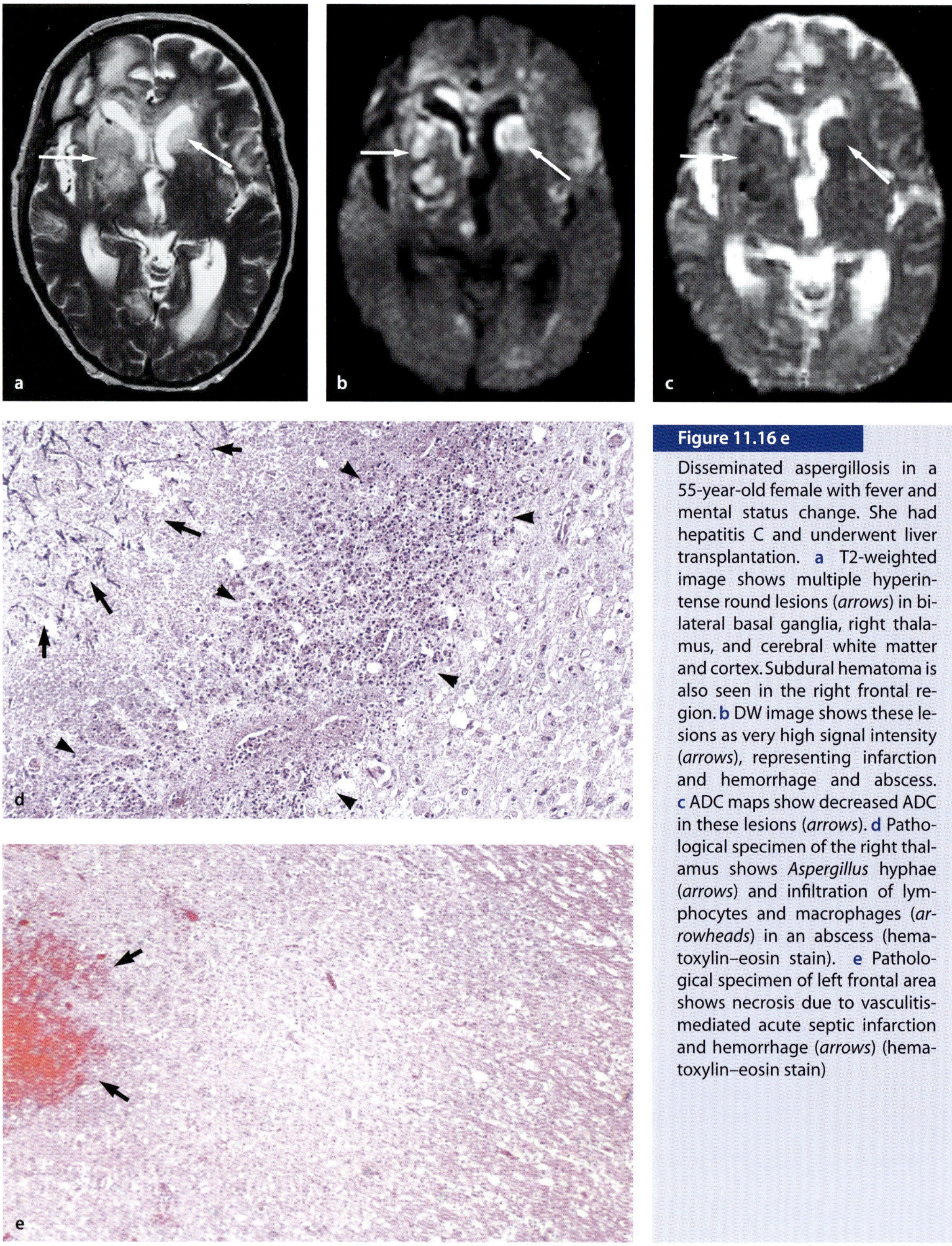

Figure 11.16 e

Disseminated aspergillosis in a 55-year-old female with fever and mental status change. She had hepatitis C and underwent liver transplantation. **a** T2-weighted image shows multiple hyperintense round lesions (*arrows*) in bilateral basal ganglia, right thalamus, and cerebral white matter and cortex. Subdural hematoma is also seen in the right frontal region. **b** DW image shows these lesions as very high signal intensity (*arrows*), representing infarction and hemorrhage and abscess. **c** ADC maps show decreased ADC in these lesions (*arrows*). **d** Pathological specimen of the right thalamus shows *Aspergillus* hyphae (*arrows*) and infiltration of lymphocytes and macrophages (*arrowheads*) in an abscess (hematoxylin–eosin stain). **e** Pathological specimen of left frontal area shows necrosis due to vasculitis-mediated acute septic infarction and hemorrhage (*arrows*) (hematoxylin–eosin stain)

11.8 Disseminated Aspergillosis

In disseminated aspergillosis, *Aspergillus* infiltrates and destroys the internal elastic lamina of cerebral arteries and causes vasculitis [27]. This leads initially to acute infarction or hemorrhage. Most of the process extends into the surrounding tissue as cerebritis and eventually evolves into an abscess. Infection of already infarcted brain tissue is often aggressive with rapid progression. DW imaging is useful for early detection of this vasculopathy-mediated septic infarction and abscess (Fig. 11.16). New antifungal therapies (triazole) have made effective treatment possible.

11.9 Herpes Encephalitis

Pathologically, herpes encephalitis has both cytotoxic and vasogenic edema associated with massive tissue necrosis and petechial or even confluent hemorrhage. DW imaging is more sensitive than conventional MR imaging in detecting early changes of herpes encephalitis (Fig. 11.17) [28, 29]. This is important for early diagnosis and treatment. Herpes encephalitis typically affects the temporal lobes, which occasionally can make the detection of lesions in the middle cranial fossa difficult on DW imaging because of susceptibility artifacts.

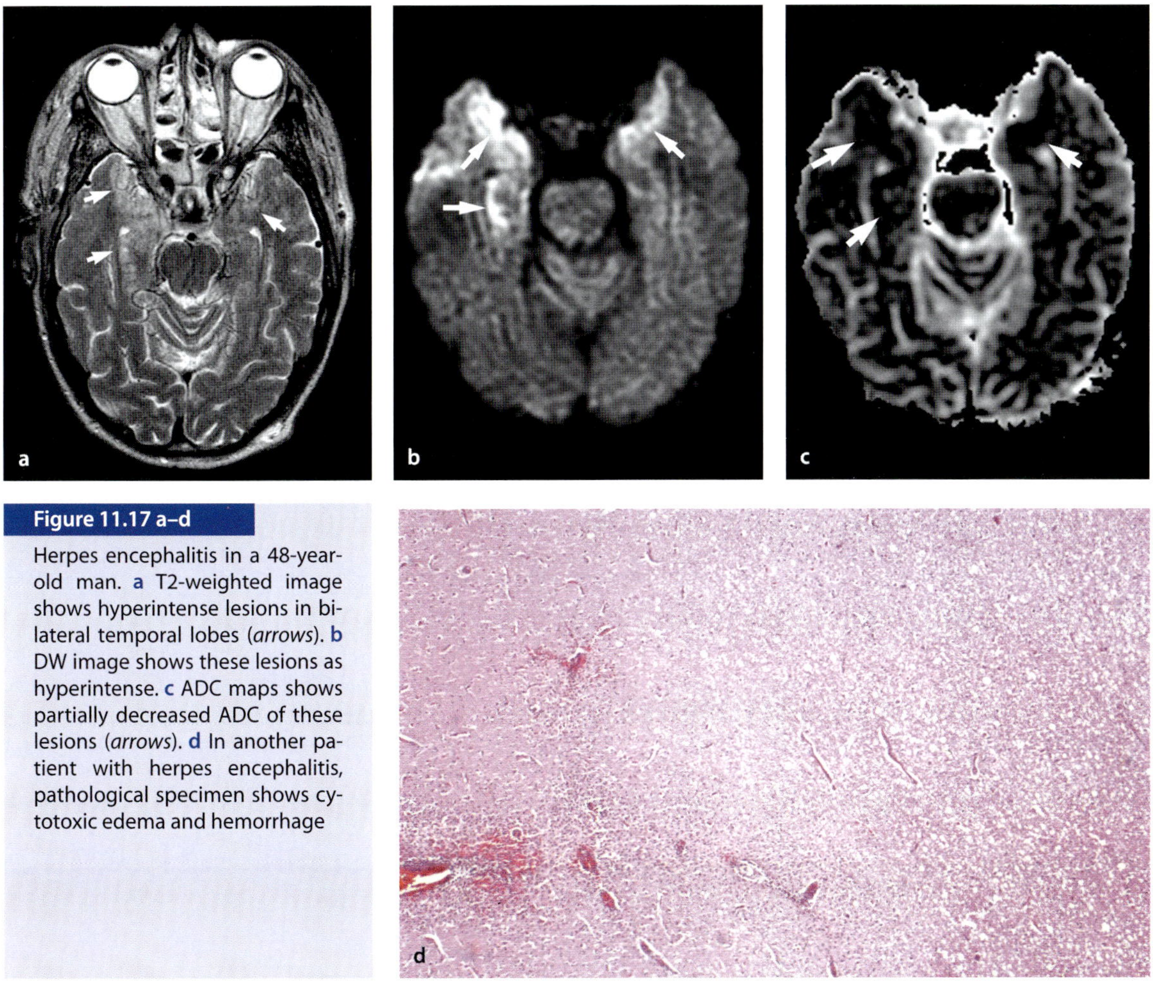

Figure 11.17 a–d

Herpes encephalitis in a 48-year-old man. **a** T2-weighted image shows hyperintense lesions in bilateral temporal lobes (*arrows*). **b** DW image shows these lesions as hyperintense. **c** ADC maps shows partially decreased ADC of these lesions (*arrows*). **d** In another patient with herpes encephalitis, pathological specimen shows cytotoxic edema and hemorrhage

11.10 Human Immunodeficiency Virus Infection

The pathological hallmark of human immunodeficiency virus (HIV) encephalopathy is multinucleated giant cells in the white matter [30]. This results in myelin pallor and rarefaction. MR imaging typically shows diffuse periventricular white matter lesions, but brain stem and basal ganglia can also be involved. DW imaging usually shows mild hyperintensity with decreased ADC which is secondary to a T2 shine through effect (Fig. 11.18).

Cerebral infarction in HIV patients is common and has been seen on MR imaging in up to 18% of the patients [31]. The infarctions are caused by opportunistic infections, drug use and primary HIV vasculitis. AIDS-related bilateral basal ganglia lesions are reported to be numerous microinfarcts on post-mortem neuropathological examination [32]. DW imaging can show numerous hyperintense lesions in bilateral basal ganglia, presumably representing microinfarcts (Fig. 11.19).

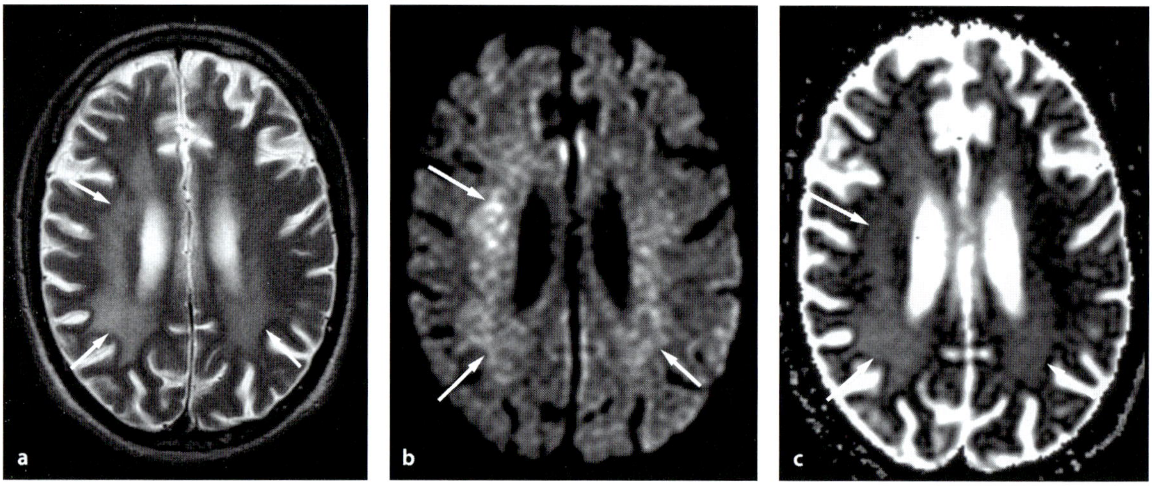

Figure 11.18 a–c

HIV encephalopathy in a 60-year-old man. **a** T2-weighted image shows periventricular hyperintense lesions (*arrows*). **b** DW image shows these lesions as mild hyperintensity (*arrows*). **c** ADC map shows these lesions as increased ADC. Mild hyperintensity on DW imaging is due to T2 shine-through effect (*arrows*)

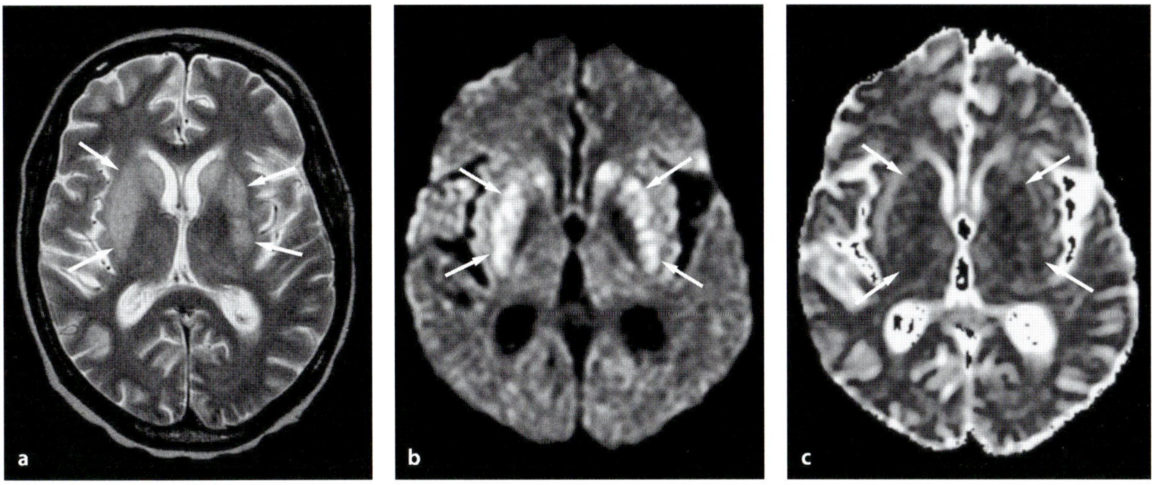

Figure 11.19 a–c

AIDS-related bilateral basal ganglia lesions in a 38-year-old female with multiple small infarcts, probably associated with HIV vasculitis. **a** T2-weighted image shows hyperintense lesions in bilateral caudate nuclei and putamina (*arrows*). **b** DW image demonstrates these lesions as numerous hyperintense spots, probably representing microinfarcts associated with HIV infection and/or drug abuse (*arrows*). **c** ADC map shows these lesions as partially decreased ADC (*arrows*)

References

1. Ebisu T, Naruse S, Horikawa Y, et al. (1993) Discrimination between different types of white matter edema with diffusion-weighted MR imaging. JMRI 3:863–868
2. Ebisu T, Tanaka C, Umeda M, et al. (1996) Discrimination of brain abscess from necrotic or cystic tumors by diffusion-weighted echo planar imaging. Magn Reson Imaging 14:1113–1116
3. Kim YJ, Chang KH, Song IC, et al. (1998) Brain abscess and necrotic or cystic brain tumor: discrimination with signal intensity on diffusion-weighted MR imaging. AJR Am J Roentgenol 171:1487–1490
4. Desprechins B, Stadnik T, Koerts G, et al. (1999) Use of diffusion-weighted MR imaging in differential diagnosis between intracerebral necrotic tumors and cerebral abscesses. AJNR Am J Neuroradiol 20:1252–1257
5. Noguchi K, Watanabe N, Nagayashi T, et al. (1999) Role of diffusion-weighted echo-planar MRI in distinguishing between brain abscess and tumor: a preliminary report. Neuroradiology 41:171–174
6. Castillo M (1999) Imaging brain abscess with diffusion-weighted and other sequences. AJNR Am J Neuroradiol 20:1193–1194
7. Gray F, Nordmann P (1997) Bacterial infections. In: Graham DI, Lantos PL (eds) Greenfield's neuropathology, vol 2, 6th edn, pp 114–129
8. Falcone S, Post MJD (2000) Encephalitis, cerebritis, and brain abscess: pathophysiology and imaging findings. Neuroimaging Clin North Am 10:333–353
9. Hatta S, Mochizuki H, Kuru Y, et al. (1994) Serial neurological studies in focal cerebritis. Neuroradiology 36:285–288
10. Ketelslegers E, Duprez T, Ghariani S, et al. (2000) Time dependence of serial diffusion-weighted imaging features in a case of pyogenic brain abscess. JCAT 24:478–481
11. Bakshi R, Wright PD, Kinkel PR, et al. (1999) Cranial magnetic resonance imaging findings in bacterial endocarditis: the neuroimaging spectrum of septic brain embolization demonstrated in twelve patients. J Neuroimaging 9:78–84
12. Höllinger P, Zürcher R, Schroth G, et al. (2000) Diffusion magnetic resonance imaging findings in cerebritis and brain abscess in a patient with septic encephalopathy. J Neurol 247:232–234
13. Nagase T, Wada S, Nakamura R, et al. (1995) Magnetic resonance imaging of multiple brain abscesses of bilateral basal ganglia. Internal Medicine 34:554–558
14. Southwick FS, Purich DL (1996) Intracellular pathogenesis of listeriosis. New Eng J Med 334:770–776
15. Holtås S, Geijer B, Strömblad LG, et al. (2000) A ring-enhancing metastasis with central high signal on diffusion-weighted imaging and low apparent diffusion coefficients. Neuroradiology 42:824–827
16. Hartman M, Jansen O, Heiland S, et al. (2001) Restricted diffusion within ring enhancement is not pathognomonic for brain abscess. AJNR Am J Neuroradiol 22:1738–1742
17. Tung GA, Evangelista P, Rogg JM, et al. (2001) Diffusion-weighted MR imaging of rim-enhancing brain masses: is markedly decreased water diffusion specific for brain abscess? AJR Am J Roentgenol 177:709–712

18. Fredman DP, Goldman HW, Flanders AE (1997) MR imaging of stereotaxic palidotomy and thalamotomy. AJR Am J Roentgenol 169:894–896
19. Ackerman LL, Traynelis VC (1999) Dural space infection: cranial subdural empyema and cranial epidural abscess. In: Osenbach RK, Zeidman SM, (eds) Infections in neurological surgery. Philadelphia: Lippincott-Raven, pp 85–99
20. Ramsay DW (2000) Diffusion-weighted imaging of cerebral abscess and subdural empyema. AJNR Am J Neuroradiol 21:1172
21. Rana S, Albayram S, Lin DDM, et al. (2002) Diffusion-weighted imaging and apparent diffusion coefficient maps in a case of intracerebral abscess with ventricular extension. AJNR Am J Neuroradiol 23:109–112
22. Ebisu T, Tanaka C, Umeda M, et al. (1997) Hemorrhagic and nonhemorrhagic stroke: diagnosis with diffusion-weighted and T2-weighted echo-planar MR imaging. Radiology 203:823–828
23. Atlas SW, Dubois P, Singer MB, et al. (2000) Diffusion measurements in intracranial hematomas: implications for MR imaging of acute stroke. AJNR Am J Neuroradiol 21:1190–1194
24. Tsuruda JS, et al. (1990) Diffusion-weighted MR imaging of the brain: value of differentiating between extraaxial cysts and epidermoid tumors. AJNR Am J Neuroradiol 11:925–931
25. Moodley M, Bullock MR (1985) Severe neurological sequelae of childhood bacterial meningitis. S Afr Med J 68: 566–570
26. Navia BA, Petito CK, Gold JWM, et al. (1986) Cerebral toxoplasmosis complicating the acquired immune deficiency syndrome: clinical and neurological findings in 27 patients. Ann Neurol 19:224–238
27. DeLone DR, Goldstein RA, Petermann G, et al. (1999) Disseminated aspergillosis involving the brain: distribution and imaging characteristics. AJNR Am J Neuroradiol 20: 1597–1604
28. Sener RN (2001) Herpes simplex encephalitis: diffusion MR imaging findings. Comput Med Imaging Graph 25:391–397
29. Tsuchiya K, Katase S, Yoshino A, et al. (1999) Diffusion-weighted MR imaging of encephalitis. AJR Am J Roentgenol 173:1097–1099
30. Flowers CH, Mafee MF, Crowell R, et al. (1990) Encephalopathy in AIDS patients: evaluation with MR imaging. AJNR Am J Neuroradiol 11:1235–1245
31. Connor MD, Lammie GA, Bell JE, et al. (2000) Cerebral infarction in adult AIDS patients: observations from the Edinburgh HIV autopsy cohort. Stroke 31:2117–2126
32. Meltzer CC, Wells SW, Becher MW, et al. (1998) AIDS-related MR hyperintensity of the basal ganglia. AJNR Am J Neuroradiol 19:83–89

Trauma

12.1 Introduction

Head injuries are the most common cause of death and permanent disability in the early decades of life. They vary widely in their etiology, pathophysiology, clinical presentation, and optimal treatment strategies. Traumatic brain injuries are classified in two main categories: focal and diffuse brain injuries [1]. Focal brain injuries usually result from direct impact force to the head, like cerebral contusions and epidural hematomas. Diffuse brain injuries are caused by sudden changes in movement of the head, usually rotational accelerations, which result in a variety of injuries, ranging from a brief cerebral concussion to extensive diffuse axonal injuries (DAI).

Computed tomography (CT) is imperative in patients with focal and diffuse injuries, especially when hemodynamically or neurologically unstable [2]. However, CT is often false negative or underestimates contusions shortly after trauma, and DAIs are often not detected. Conventional MR imaging has higher detection sensitivity with regard to these lesions because of its greater sensitivity for edema [3].

Once it was thought that edema following traumatic brain injury was vasogenic, but recent experimental studies using diffusion-weighted (DW) imaging have shown that edema after head trauma consists of both vasogenic and cytotoxic edema [4–9]. Since DW imaging is also very sensitive in detecting small lesions of cytotoxic edema and can differentiate cytotoxic from vasogenic edema, it has become especially useful in the evaluation and staging of patients with DAI.

12.2 Diffuse Axonal Injury

Diffuse axonal injury results from a diffuse shearing-strain deformation causing change in shape of brain tissue from unequal movement of adjacent tissues that differ in density and rigidity [1]. Patients with DAI more often than with other types of primary brain injuries show severe impairment of consciousness at impact.

Pathologically, injury related to DAI is always more extensive microscopically than at gross examination [10]. Microscopically, shearing injuries initially produce multiple, characteristic axonal bulbs, or retraction balls, as well as numerous foci of perivascular hemorrhages.

The origin behind cytotoxic edema in DAI seems to be related to an excitotoxic mechanism, in particular glutamate [1, 11, 12]. Damage at the node of Ranvier will result in a traumatic defect in the axonal membrane. This defect causes excessive neurotransmitter release with increase in intracellular calcium ions, as in brain ischemia, which leads to axonal and glial cell swelling (cytotoxic or neurotoxic edema). These changes can eventually lead to axonal degeneration or necrosis with microglial and astrocytic reactive changes. Accumulation of hemosiderin-laden macrophages is also seen in the chronic phase.

12.2.1 Location

Common locations of DAI are at the gray–white matter junctions (Fig. 12.1), in the corpus callosum (Fig. 12.2) and at the dorsolateral aspect of the upper brain stem (Fig. 12.3). DAI may be confined to the white matter of the frontal and temporal lobes in mild head trauma [13]. With more severe rotational acceleration, lesions are also seen in the lobar white matter as well as in the posterior half of the corpus callosum. In cases with even greater trauma, lesions will also be found in the anterior corpus callosum, and the dor-

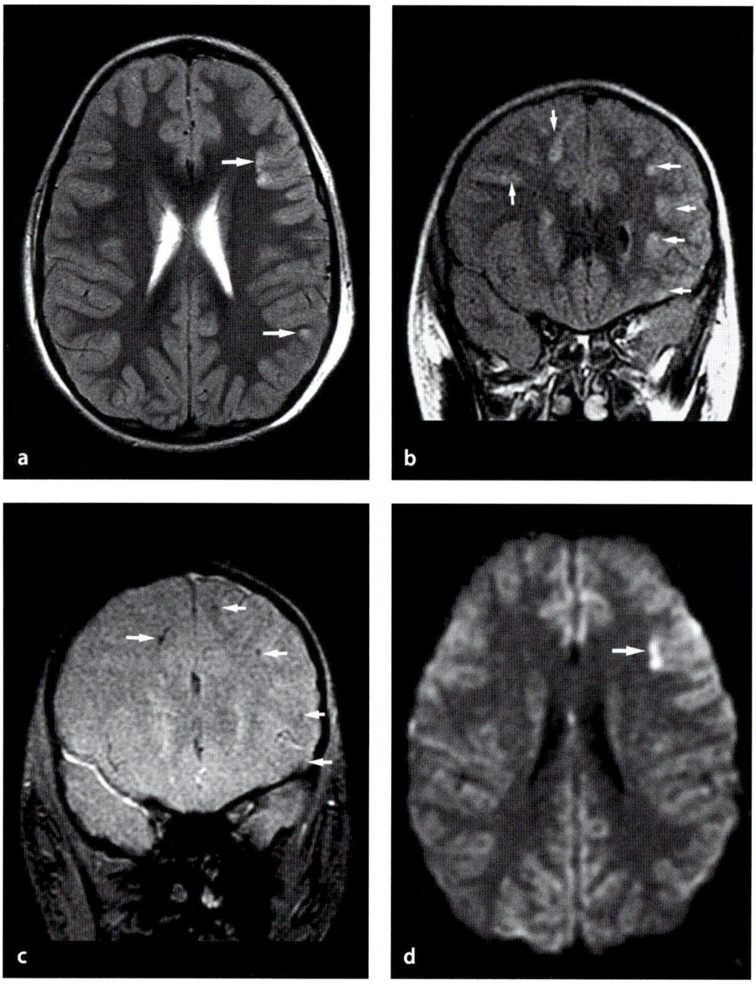

Figure 12.1 a–d

Fig. 12.1a–d. Diffuse axonal injury in gray–white matter junction in a 7-year-old male after a motor vehicle accident. T2-weighted (**a**) and coronal FLAIR (**b**) images show multiple hyperintense lesions in the gray–white matter junction of bilateral frontoparietal lobes (*arrows*). **c** Coronal GRE image shows multiple small hemorrhages as low signal in these lesions (*arrows*). **d** DW image demonstrates diffuse axonal injury as high signal intensity (*arrow*) with decreased ADC (not shown), representing cytotoxic edema

solateral aspects of the midbrain and upper pons. Occasionally, DAI lesions occur in the parietal and occipital lobes, internal and external capsules (Fig. 12.2), basal ganglia (Fig. 12.4), thalamus (Fig. 12.2), fornix (Fig. 12.5), septum pellucidum, and cerebellum (Fig. 12.6). Intraventricular hemorrhage can accompany these findings. They have the same mechanical origin and are due to disruption of the subependymal plexus of capillaries and veins that lie along the ventricular surface of the corpus callosum, fornix, and septum pellucidum [14].

Figure 12.3 a–c

Diffuse axonal injury in the brain stem in a 28-year-old male after a motor vehicle accident. **a** DW image shows a hypointense lesion with a hyperintense rim in the dorsolateral aspect of the midbrain, representing a hemorrhagic lesion of diffuse axonal injury (*arrow*). **b** ADC map shows decreased ADC of this lesion (*arrow*). This might be due to a paramagnetic susceptibility artifact. **c** Coronal GRE image clearly shows hemorrhagic lesions as hypointense in the brain stem (*arrow*) and in the right frontoparietal region

▶

Figure 12.2 a–d

Diffuse axonal injury in the corpus callosum, internal capsule and thalamus in a 29-year-old female after a motor vehicle accident. T2-weighted (**a**) and FLAIR (**b**) images show multiple hyperintense lesions in the anterior and posterior corpus callosum, internal capsules and left thalamus (*arrows*). **c** DW image demonstrates these lesions as high signal intensity with decreased ADC (**d**)

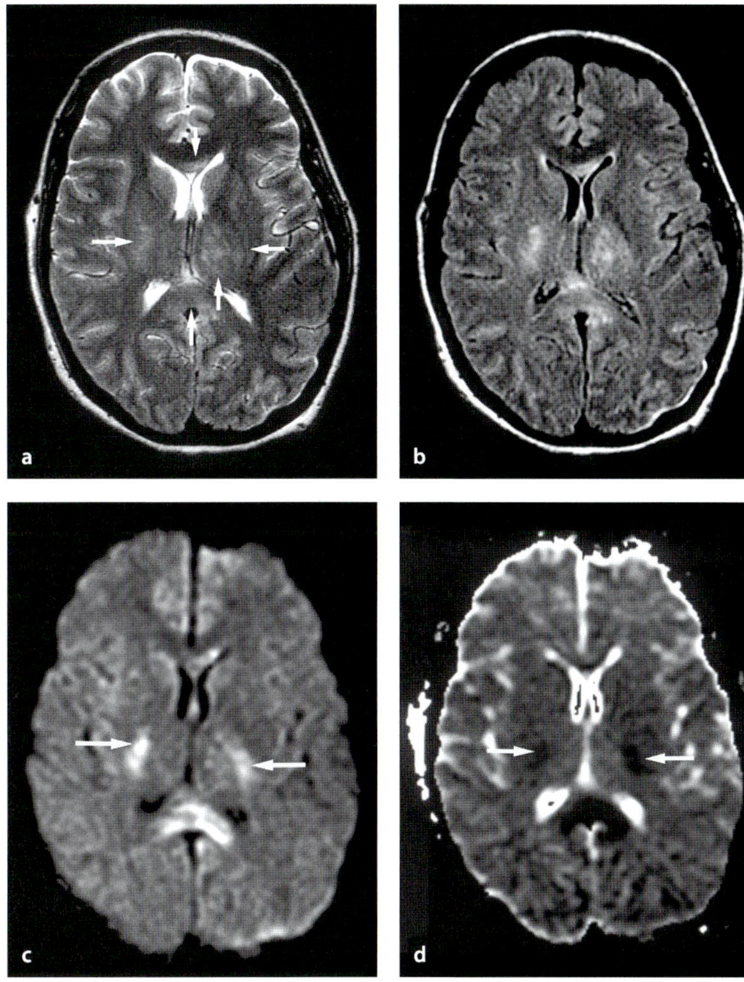

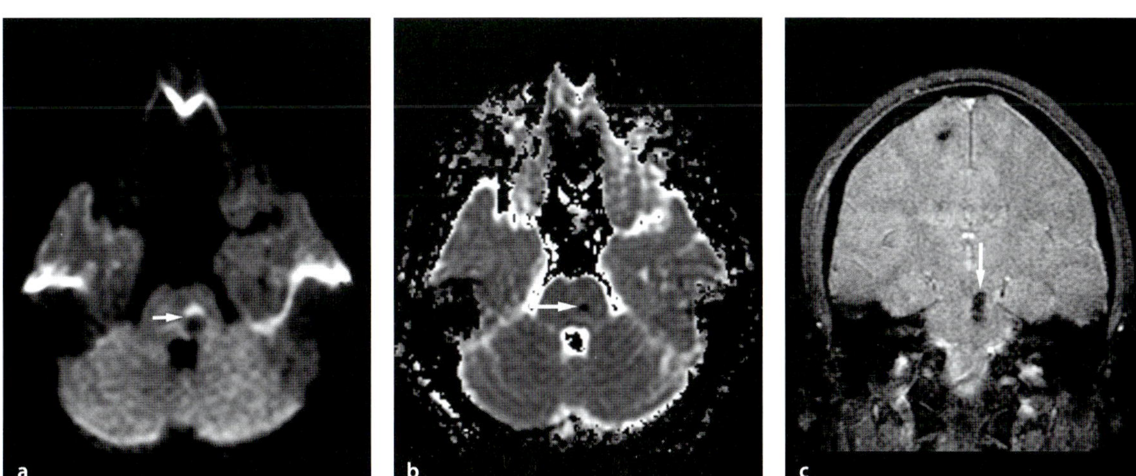

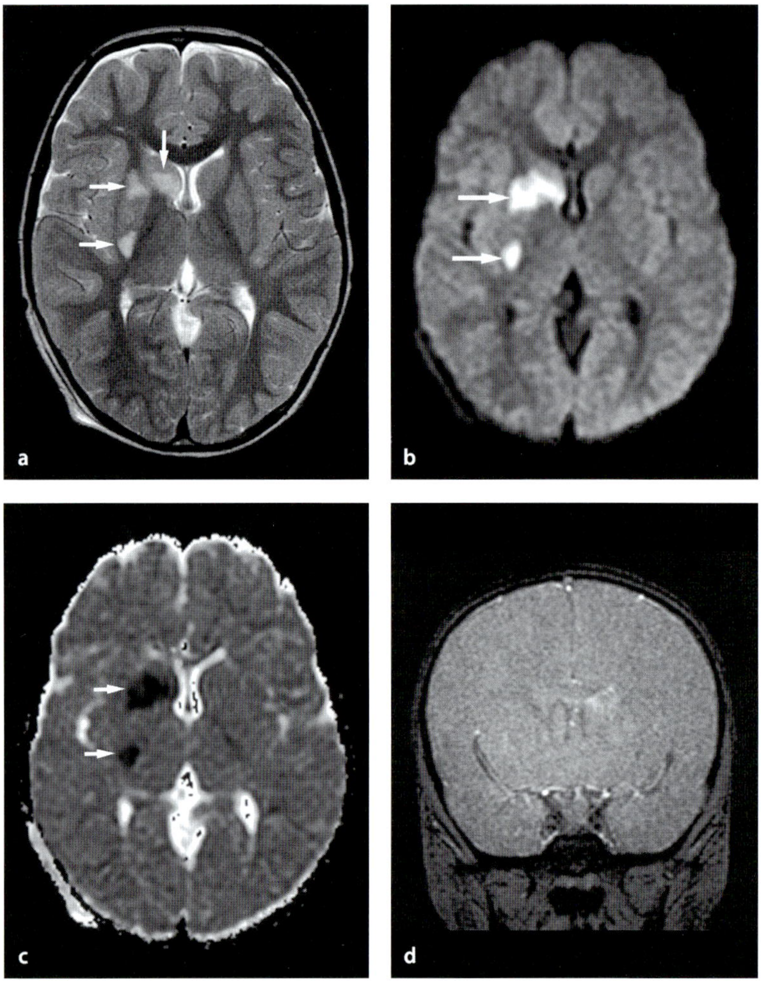

Figure 12.4 a–d

Diffuse axonal injury in the basal ganglia in a 3-year-old male after a motor vehicle accident. **a** T2-weighted image shows hyperintense lesions in the right lentiform and caudate nucleus (*arrows*). **b, c** DW imaging shows these lesions as hyperintense with decreased ADC (*arrows*). **d** Coronal GRE image clearly shows no hemorrhagic foci in these lesions

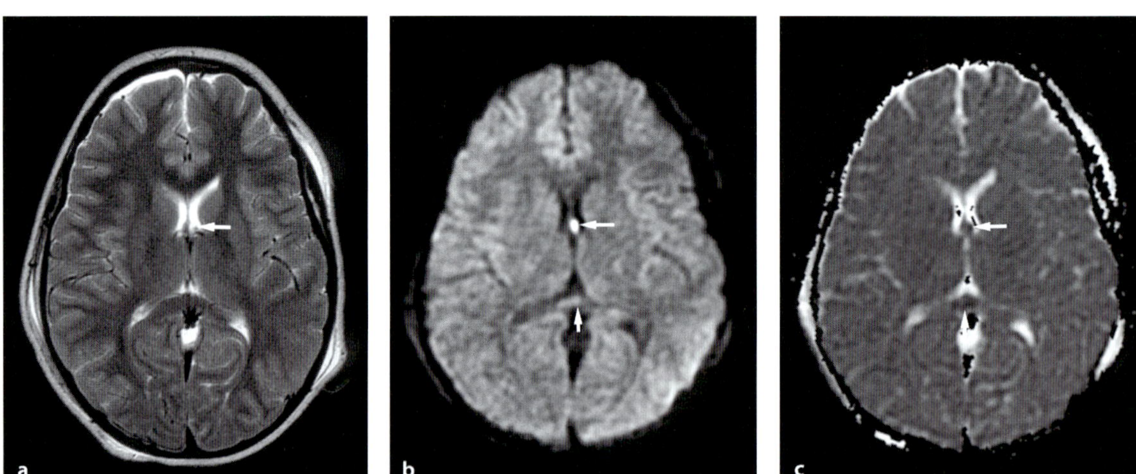

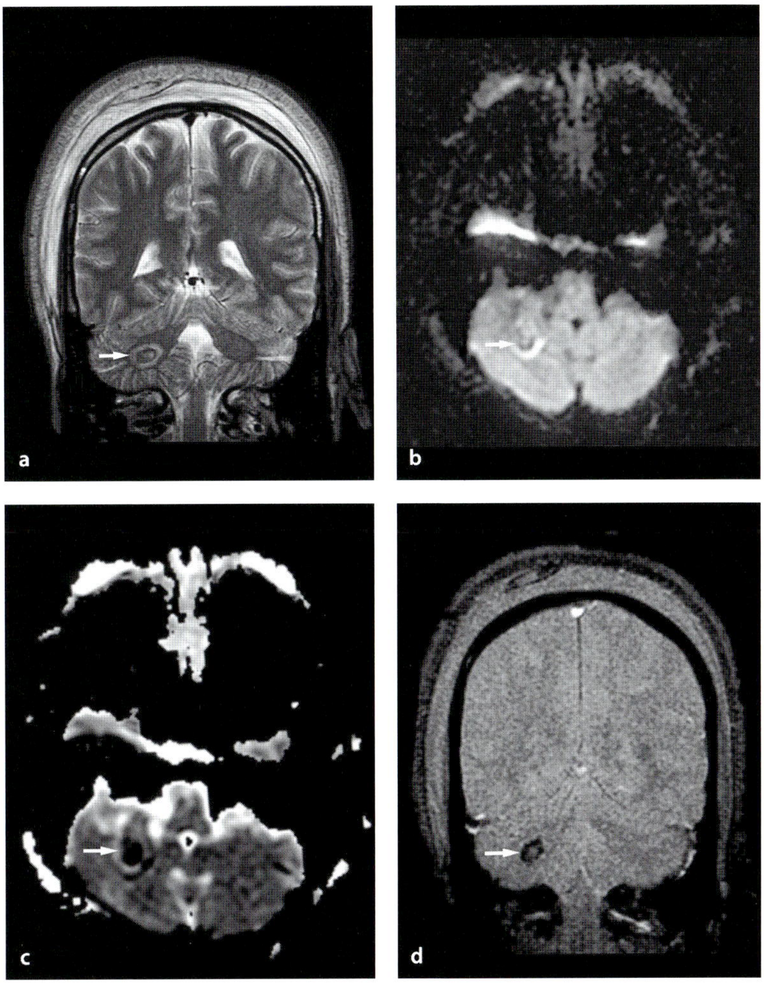

Figure 12.6 a–d

Diffuse axonal injury in the cerebellum of an 18-year-old male after a motor vehicle accident. **a** T2-weighted image shows a hypointense lesion in the right middle cerebellar peduncle (*arrow*). **b** DW image shows a hypointense lesion with a hyperintense rim, representing a hemorrhagic lesion (*arrow*). **c** ADC map reveals decreased ADC in this lesion (*arrow*). This may be due to a paramagnetic susceptibility artifact. **d** Coronal GRE image clearly demon- strates hemorrhagic lesions as hypointense (*arrow*)

Figure 12.5 a–c

Diffuse axonal injury in the fornix of an 11-year-old female after a motor vehicle accident. **a** On T2-weighted image, it is difficult to detect a small hyperintense lesion in the fornix (*arrow*). **b, c** DW image shows the lesion in the fornix and posterior corpus callosum as hyperintense with decreased ADC (*arrows*)

◀

12.2.2 Computed Tomography and MR Imaging

Few DAI lesions are visible with CT. Only large lesions or those that are grossly hemorrhagic are seen. MR imaging has been proven to be more sensitive for detection as well as for characterization of DAI lesions [2]. Conventional MR imaging shows multiple, small, deeply situated elliptical lesions that spare the overlying cortex. Fluid-attenuated inversion-recovery (FLAIR) images are more sensitive than T2-weighted images to detect small hyperintense lesions adjacent to the cerebrospinal fluid, such as in the fornix and septum pellucidum [15]. These lesions are, moreover, often accompanied by small, petechial hemorrhages. They occur in 10–30% of all DAI lesions [16] and are best appreciated on T2*-weighted gradient-echo (GRE) images because of their susceptibility effects [17]. However, even these MRI sequences are thought to underestimate the true extent of DAI.

12.2.3 Diffusion-Weighted Imaging

Diffusion-weighted imaging measures a unique physiologic parameter, movement of water in the tissue, which allows for identification of DAI lesions that may not be visible on T2/FLAIR or T2*-weighted GRE images. DAI lesions on DW imaging are hyperintense and associated with decreased apparent diffusion coefficient (ADC) [17–22]. The precise mechanisms underlying the diffusion changes associated with DAI are unknown. Cytotoxic edema, which seems to be the cause of reduced ADC in ischemic brain injury, can also occur in the early phase of DAI. However, reduced ADC could also be due to the development of retraction balls and concomitant cytoskeletal collapse along the severed axons [21].

The time course of the ADC abnormality seems to be different from that of ischemic brain injury. Prolonged decrease in ADC, over 2 weeks, was occasionally observed in DAI [18], and cytotoxic edema in the corpus callosum can be partially reversible on follow-up imaging using T2-weighted sequences [22]. However, axonal and glial cell swelling in DAI is thought to be mainly due to excitotoxic mechanisms. It can be a slower or reversible form of cellular swelling than that seen in ischemic brain injuries [6]. Hemorrhagic components, which often accompany these brain injuries, will affect the signal intensity on DW images.

12.3 Brain Contusion

Brain contusions are defined as traumatic injuries to the cortical surface of the brain [1]. They are caused by direct contact between the skull and the brain parenchyma. Compared with DAI, contusions tend to be larger, more superficial, more ill defined and more likely to contain areas of hemorrhage. Cytotoxic edema in brain contusions is also related to excitotoxic mechanisms [23].

12.3.1 Location

Common locations of brain contusions are in the temporal and frontal lobes, especially along their anterior, lateral, and inferior surfaces (Fig. 12.7). The parietal occipital lobes, hippocampus (Fig. 12.8), cerebellar hemisphere, vermis and cerebellar tonsils (Fig. 12.9) are less frequently involved [2].

Figure 12.8 a–c

Brain contusion in the hippocampus in an 11-year-old female after a motor vehicle accident. **a** FLAIR image shows a hyperintense lesion in the left hippocampus (*arrow*). **b** DW image shows this lesion as hyperintense. **c** ADC is decreased in the left hippocampus and left side of the brain stem (*arrows*), representing mainly cytotoxic edema

▶

Figure 12.7 a–d

Brain contusion in the frontal lobe in a 37-year-old male after a motor vehicle accident. **a** On CT obtained after evacuation of epidural hematoma, it is difficult to delineate the extent of a mass lesion in the right frontal lobe (*arrows*). **b** T2-weighted image delineates the extent of the edematous brain contusion (*arrows*). **c** DW image shows heterogeneous signal intensity in these lesions, representing mixed vasogenic and cytotoxic edema with hemorrhagic necrotic tissues (*arrows*). **d** ADC map reveals mixed increase and relative decrease of ADC (*arrows*) in these lesions

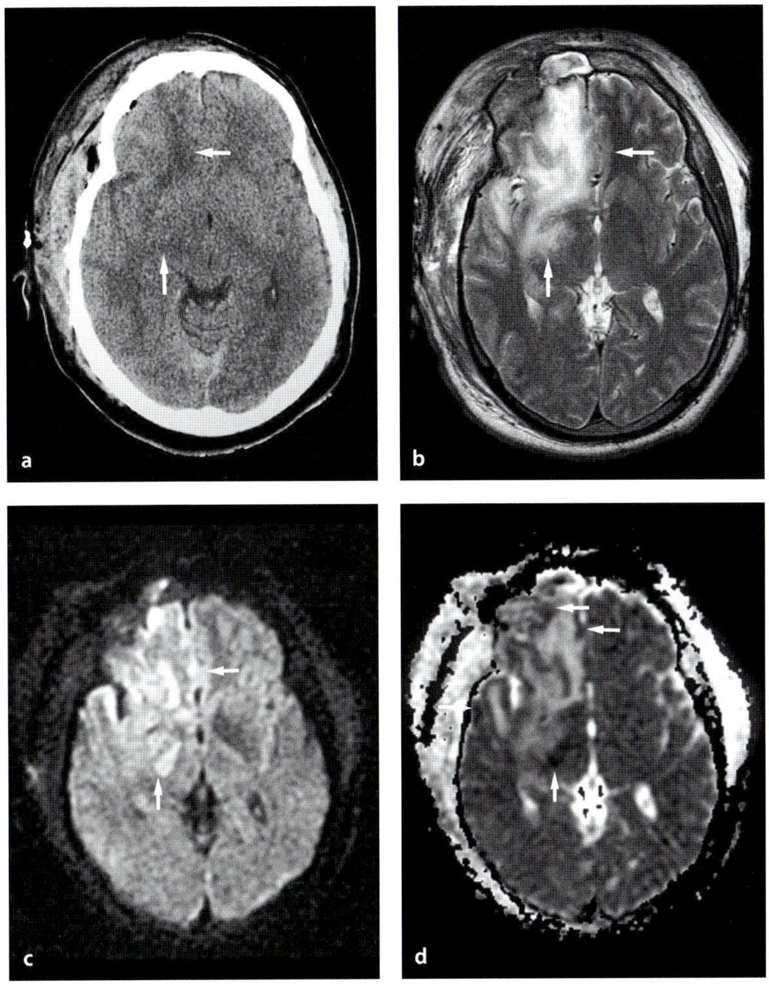

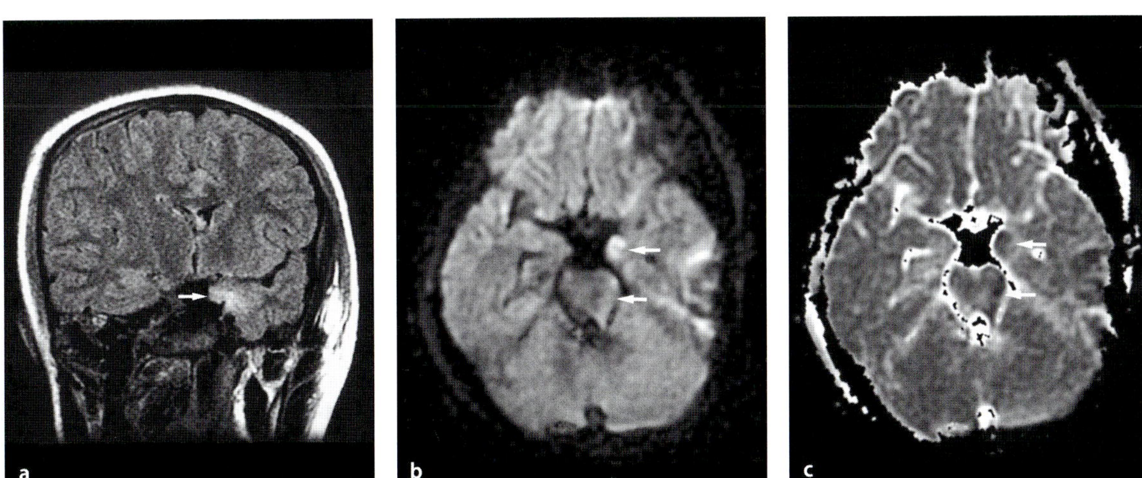

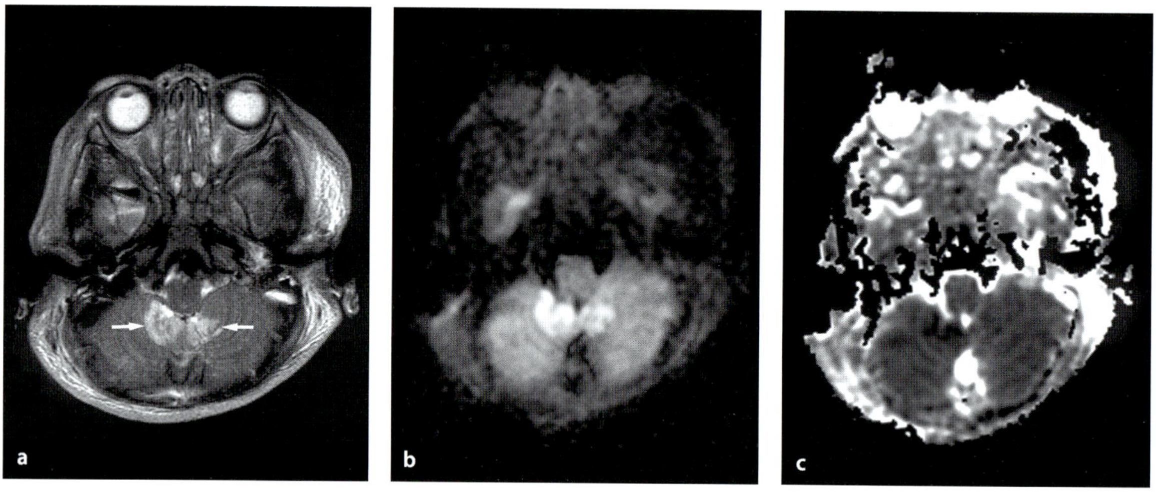

Figure 12.9 a–c

Brain contusion in the cerebellar tonsil in an 11-year-old female after a motor vehicle accident. **a** T2-weighted image shows hyperintense lesions in the cerebellar tonsils (*arrows*). **b** DW image shows this lesion as hyperintense. **c** ADC is partially decreased

12.3.2 Computed Tomography and MR Imaging

Contusions are often difficult to identify on CT obtained shortly after trauma unless they are large or contain areas of hemorrhage [2]. Initial CT will often show only faint areas of low attenuation, sometimes mixed with a few tiny areas of petechial hemorrhage. MRI is considerably more sensitive than CT for early detection and evaluation of their extent.

12.3.3 Diffusion-Weighted Imaging Findings

Brain contusions are sometimes associated with a non-hemorrhagic mass effect, which progresses rapidly after the trauma. Edema in brain contusions is heterogeneous, composed of cytotoxic and vasogenic edema [24], which can be demonstrated by DW imaging. Kawamata et al. reported a specific DW imaging finding of brain contusions [25, 26]. On DW imaging, the contusion is shown as a low intensity core, with increased ADC, surrounded by a rim of a high intensity, with decreased ADC. This suggests that intra- and extracellular components undergo disintegration and homogenization within the central area, whereas cellular swelling is predominant in the peripheral area.

12.4 Hemorrhage Related to Trauma

Traumatic hemorrhages result from injury to a cerebral vessel (artery, vein or capillary) [2]. Subdural hematomas originate from disruption of the bridging cortical veins, which are vulnerable to rapid stretching. Epidural hematoma can have either an arterial or a venous sinus origin, typically associated with a skull fracture. Traumatic intracerebral hematomas result from a shear-strain injury involving arteries, veins or capillaries. Traumatic subarachnoid hemorrhage is usually seen after severe head trauma and may as such accompany brain contusion or DAI. Hemorrhages can also represent a disruption of intracranial arteries, especially arteries of the vertebrobasilar system.

12.4.1 Computed Tomography and MR Imaging

Computed tomography is the modality of choice for the initial evaluation of traumatic brain hemorrhages, as it is a fast examination technique, is widely available, has no contraindications and relatively accurately depicts most hematomas [2].

Magnetic resonance imaging is usually not the primary imaging technique and findings will then depend on the stage of degradation of the hemoglobin at the time of examination. However, in most instances MR imaging is extremely helpful to detect hematomas, especially along the vertex and skull base, and can in certain questionable cases differentiate between subdural and epidural hematomas [2]. T2*-weighted GRE and FLAIR images seem to be more sensitive to detect hemorrhage than conventional spin-echo imaging [27–30].

12.4.2 Diffusion-Weighted Imaging

Diffusion-weighted imaging findings of subdural and epidural hematomas have not been well described in the literature. Depending on the age of the hematoma, DW imaging will vary in signal intensity (Fig. 12.10). Gradient-echo sequences are better in detecting hematomas, including subdural and epidural hematomas, than DW imaging [30]. Although often difficult to detect [31, 32], DW imaging can occasionally depict a subarachnoid hemorrhage as a hyperintense signal (Fig. 12.11). The benefit of DW imaging is probably for the detection of underlying or associated parenchymal lesions. For example, subarachnoid hemorrhage will often cause vasospasm of the intracranial arteries, which can result in brain ischemia. Mass effect secondary to subdural or epidural hematomas, which is closely related to morbidity and mortality, is due to a combination of the hematoma, underlying parenchymal edema and diffuse cerebral swelling.

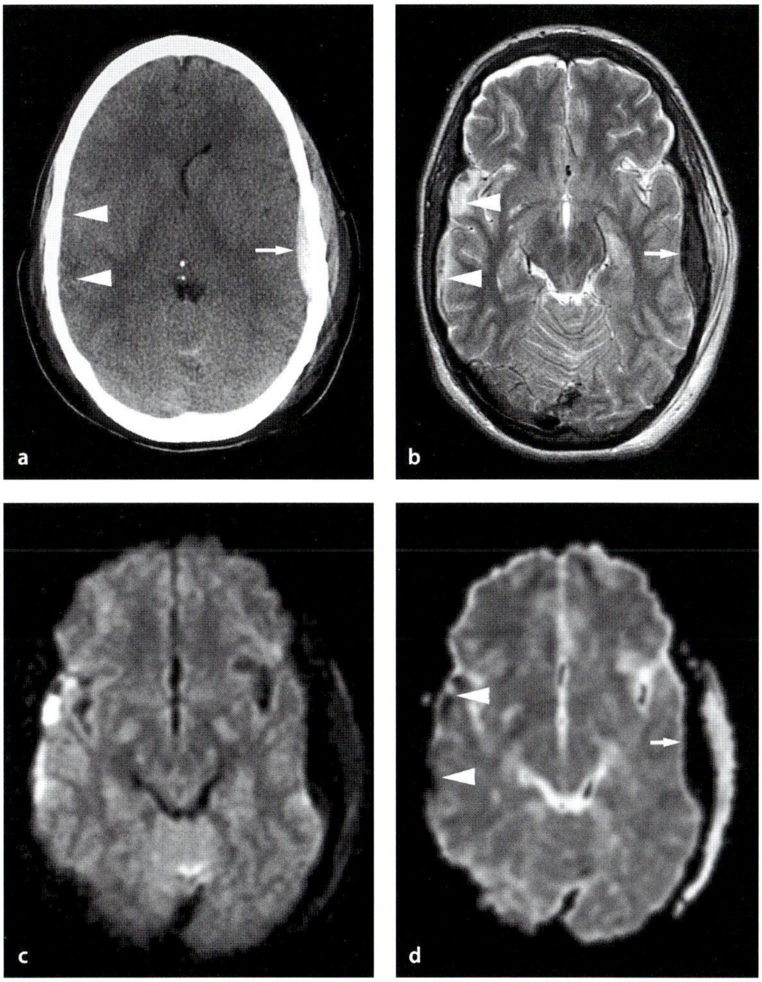

Figure 12.10 a–d

Epidural and subdural hematoma in a 26-year-old male after a motor vehicle accident. **a** CT shows a left epidural hematoma (*arrow*) but it is difficult to depict the isodense small subdural hematoma in the right side (*arrowheads*). **b** T2-weighted image shows the left epidural hematoma (*arrow*) as a hypointense lesion and the right subdural hematoma as partially hypointense lesions (*arrowheads*). **c** DW image shows the epidural hematoma as very hypointense due to deoxy-hemoglobin, and the subdural hematoma as very hyperintense presumably due to high viscosity or hypercellularity of hematoma. **d** ADC map shows hypointensity due to loss of pixels with background masking in the left epidural hematoma (*arrow*). ADC map also shows decreased ADC in the right subdural hematoma (*arrowheads*)

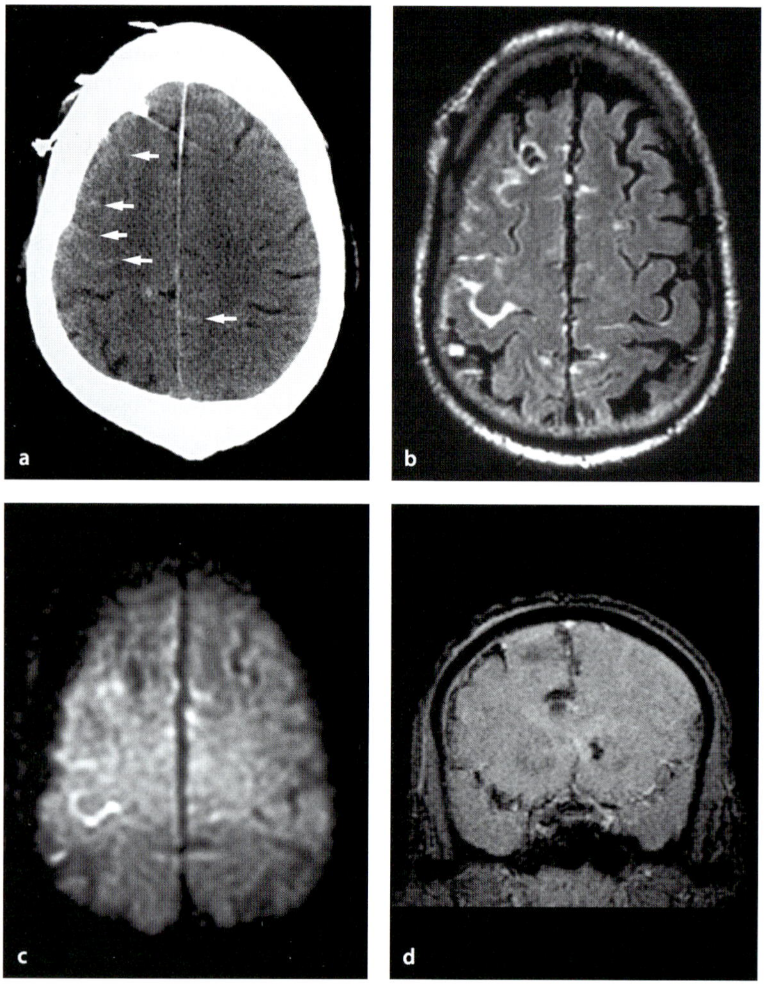

Figure 10.11 a–d

Subarachnoid hemorrhage in a 68-year-old male with ruptured aneurysm of the right middle cerebral artery bifurcation. **a** Postoperative CT shows subtle high density of subarachnoid space in the right frontoparietal area (*arrows*). **b** FLAIR image shows subarachnoid hemorrhage as hyperintensity. **c** DW image also shows subarachnoid hemorrhage as hyperintensity with mildly increased ADC (not shown). **d** Coronal GRE shows the hemorrhage as low signal intensity

12.5 Vascular Injuries

Traumatic arterial and venous injuries (dissections, lacerations, occlusions, pseudoaneurysm, arteriovenous fistulas) are more prevalent than generally believed [2]. Many asymptomatic lesions probably escape detection, and others are recognized several days to months after the injury (Fig. 12.12). CT is useful to detect skull base fractures and CT angiography may help to evaluate the vascular injuries. However, a combination of MR imaging and MR angiography is probably the most efficacious way to screen high-risk patients for traumatic vascular injuries, especially if combined with DW imaging, which is very sensitive to detect small and early ischemic lesions secondary to traumatic vascular injuries. Still, one has to acknowledge that conventional angiography continues to be the gold standard in the evaluation of known or suspected traumatic arterial lesions.

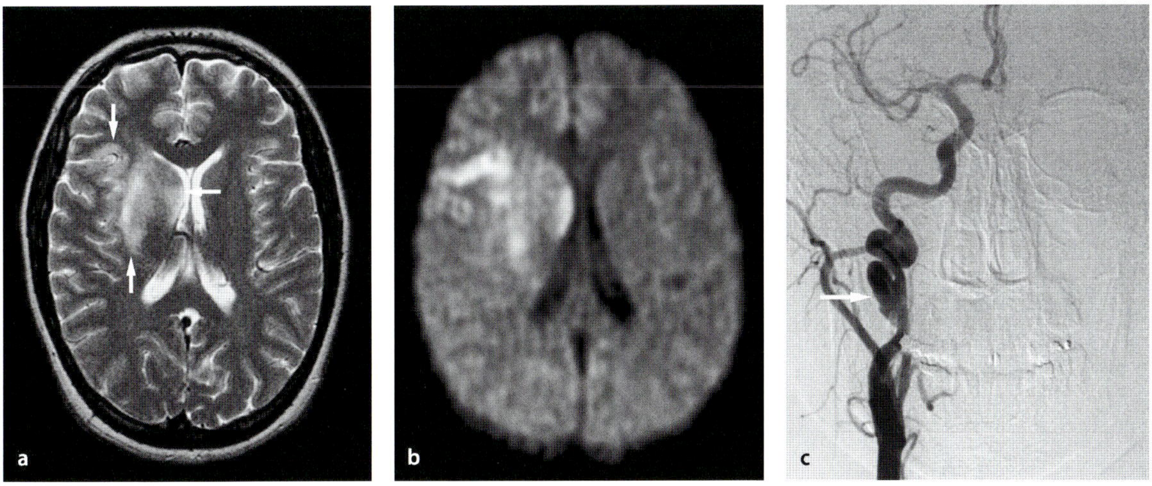

Figure 12.12 a–c

Cerebral infarction after carotid artery dissection with pseudoaneurysm in a 20-year-old female after a motor vehicle accident. **a** T2-weighted image shows hyperintense lesions in the right middle cerebral artery territory including the right basal ganglia (*arrows*). **b** DW image also shows these lesions as hyperintense with decreased ADC (not shown), representing acute infarction. **c** Conventional angiogram shows pseudoaneurysm of the right carotid artery (*arrow*)

References

1. Gennarelli TA (1993) Mechanisms of brain injury. J Emerg Med 11 Suppl 1:5–11
2. Gentry LR (1994) Imaging of closed head injury. Radiology 191:1–17
3. Kelly AB, Zimmerman RD, Snow RB, Gandy SE, Heier LA, Deck MD (1988) Head trauma: comparison of MR and CT-experience in 100 patients. AJNR Am J Neuroradiol 9:699–708
4. Alsop DC, Murai H, Detre JA, McIntosh TK, Smith DH (1996) Detection of acute pathologic changes following experimental traumatic brain injury using diffusion-weighted magnetic resonance imaging. J Neurotrauma 13:515–521
5. Hanstock CC, Faden AI, Bendall MR, Vink R (1994) Diffusion-weighted characterization, edema, diffusion aging differentiates ischemic tissue from traumatized tissue. Stroke 25:843–848
6. Barzo P, Marmarou A, Fatouros P, Hayasaki K, Corwin F (1997) Contribution of vasogenic and cellular edema to traumatic brain swelling measured by diffusion-weighted imaging. J Neurosurg 87:900–907
7. Ito J, Marmarou A, Barzo P, Fatouros P, Corwin F (1996) Characterization of edema by diffusion-weighted imaging in experimental traumatic brain injury. J Neurosurg 84:97–103
8. Albensi BC, Knoblach SM, Chew BG, O'Reilly MP, Faden AI, Pekar JJ (2000) Diffusion and high resolution MRI of traumatic brain injury in rats: time course and correlation with histology. Exp Neurol 162:61–72
9. Assaf Y, Beit-Yannai E, Shohami E, Berman E, Cohen Y (1997) Diffusion- and T2-weighted MRI of closed-head injury in rats: a time course study and correlation with histology. Magn Reson Imaging 15:77–85
10. Gennarelli TA, Graham DI (1998) Neuropathology of the head injuries. Semin Clin Neuropsychiatry 3:160–175
11. Faden AI, Demediuk P, Panter SS, Vink R (1989) The role of excitatory amino acids and NMDA receptors in traumatic brain injury. Science 244:798–800
12. Katayama Y, Becker DP, Tamura T, Ikezaki K (1990) Cellular swelling during cerebral ischaemia demonstrated by microdialysis in vivo: preliminary data indicating the role of excitatory amino acids. Acta Neurochir Suppl (Wien) 51:183–185
13. Mittl RL, Grossman RI, Hiehle JF, Hurst RW, Kauder DR, Gennarelli TA, Alburger GW (1994) Prevalence of MR evidence of diffuse axonal injury in patients with mild head injury and normal head CT findings. AJNR Am J Neuroradiol 15:1583–1589
14. Gentry LR, Thompson B, Godersky JC (1988) Trauma to the corpus callosum: MR features. Am J Neuroradiol 9:1129– 1138
15. Ashikaga R, Araki Y, Ishida O (1997) MRI of head injury using FLAIR. Neuroradiology 39:239–242
16. Yanagawa Y, Tsushima Y, Tokumaru A, Un-no Y, Sakamoto T, Okada Y, Nawashiro H, Shima K (2000) A quantitative analysis of head injury using T2*-weighted gradient-echo imaging. J Trauma 49:272–277
17. Huisman TA, Sorensen AG, Hergan K, Gonzalez RG, Schaefer PW (2003) Diffusion-weighted imaging for the evaluation of diffuse axonal injury in closed head injury. J Comput Assist Tomo 27:5–11
18. Liu AY, Maldjian JA, Bagley LJ, Sinson GP, et al. (1999) Traumatic brain injury: diffusion-weighted MR imaging findings. AJNR Am J Neuroradiol 20:1636–1641
19. Rugg-Gunn FJ, Symms MR, Barker GJ, Greenwood R, Duncan JS (2001) Diffusion imaging shows abnormalities after blunt head trauma when conventional magnetic resonance imaging is normal. J Neurol Neurosurg Psychiatry 70:530–533
20. Hergan K, Schaefer PW, Sorensen AG, Gonzalez RG, Huisman TA (2002) Diffusion-weighted MRI in diffuse axonal injury of the brain. Eur Radiol 12:2536–2541
21. Arfanakis K, Haughton VM, Carew JD, Rogers BP, Dempsey RJ, Meyerand ME (2002) Diffusion tensor MR imaging in diffuse axonal injury. AJNR Am J Neuroradiol 23:794–802
22. Takayama H, Kobayashi M, Sugishita M, Mihara B (2000) Diffusion-weighted imaging demonstrates transient cytotoxic edema involving the corpus callosum in a patient with diffuse brain injury. Clin Neurol Neurosurg 102:135–139
23. Maeda T, Katayama Y, Kawamata T, Yamamoto T (1998) Mechanisms of excitatory amino acid release in contused brain tissue: effects of hypothermia and in situ administration of CO2+ on extracellular levels of glutamate. J Neurotrauma 15:655–664
24. Umeda M, Yamaki T, Tanaka C, et al. (1996) Analysis of brain edema in experimental cerebral contusion using diffusion weighted MRI and ADC value. Neurotraumatology 19:79–83
25. Kawamata T, Katayama Y, Mori T, Aoyama N, Tsubokawa T (2002) Mechanisms of the mass effect of cerebral contusion: ICP monitoring and diffusion MRI study. Acta Neurochir Suppl 81:281–283
26. Kawamata T, Katayama Y, Aoyama N, Mori T (2000) Heterogeneous mechanisms of early edema formation in cerebral contusion: diffusion MRI and ADC mapping study. Acta Neurochir Suppl 76:9–12
27. Melhem ER, Jara H, Eustace S (1997) Fluid-attenuated inversion recovery MR imaging: identification of protein concentration thresholds for CSF hyperintensity. Am J Roentgenol 169:859–862
28. Singer MB, Atlas SW, Drayer BP (1998) Subarachnoid space disease: diagnosis with fluid-attenuated inversion-recovery MR imaging and comparison with gadolinium-enhanced spin-echo MR imaging–blinded reader study. Radiology 208:417–422
29. Dechambre SD, Duprez T, Grandin CB, Lecouvet FE, Peeters A, Cosnard G (2000) High signal in cerebrospinal fluid mimicking subarachnoid haemorrhage on FLAIR following acute stroke and intravenous contrast medium. Neuroradiology 42:608–611
30. Taoka T, Yuh WT, White ML, Quets JP, Maley JE, Ueda T (2001) Sulcal hyperintensity on fluid-attenuated inversion recovery MR images in patients without apparent cerebrospinal fluid abnormality. AJR Am J Roentgenol 176: 519–524
31. Lin DD, Filippi CG, Steever AB, Zimmerman RD (2001) Detection of intracranial hemorrhage: comparison between gradient-echo images and b0 images obtained from diffusion-weighted echo-planar sequences. AJNR Am J Neuroradiol 22:1275–1281
32. Wiesmann M, Mayer TE, Yousry I, Medele R, Hamann GF, Bruckmann H (2002) Detection of hyperacute subarachnoid hemorrhage of the brain by using magnetic resonance imaging. J Neurosurg 96:684–689

Brain Neoplasms

In collaboration with A. Hiwatashi

13.1 Introduction

Routine MR imaging is the most sensitive method of detecting tumors of the brain. It is, however, not specific enough to determine the histologic nature of most tumors. Diffusion-weighted (DW) imaging can differentiate between tumor and infection [1–51] and can provide information about tumor cellularity, thereby helping in the characterization and grading of brain tumors. This chapter will demonstrate DW imaging characteristics of intracranial tumors.

13.2 Gliomas

The signal intensity of gliomas on DW images is variable and depends mainly on their T2 and apparent diffusion coefficient (ADC) values [1–22]. Thus, some gliomas are hyperintense on DW images with decreased ADC, which reflects a reduced volume of the extracellular space. Other gliomas have a normal or increased ADC, that is the DW signal is a T2 shine-through effect (Figs. 13.1, 13.2, 13.3, 13.4, 13.5, 13.6 and 13.7).

13.2.1 High-Grade Tumors

It has been reported that high-grade gliomas typically are hyperintense on DW images with decreased ADC [3, 8–10, 12, 13, 19]. High tumor cellularity is probably the major determinant of the decreased ADC values in high-grade brain tumors [3, 8, 12, 18]. Other studies have suggested that ADC correlates not only with tumor cellularity, but also with total nuclear area and tumor grade [8, 12, 13, 17, 19], with high-grade tumors having high cellular density and decreased ADC. Other studies have correlated areas of decreased ADC and found it to be associated with increased choline on MR spectroscopy. Choline is a marker for cell membrane turnover [7, 9].

Although there is a general principle of high-grade gliomas having high DW signal with decreased ADC, there are still controversies regarding how well DW imaging can differentiate between high-grade primary brain tumors, lymphoma and metastasis. Krabbe et al. reported that both the contrast-enhancing portions and the peritumoral edema of metastasis have higher ADC than high-grade gliomas [6]. In the individual case the distinction between metastasis and high-grade glioma is often difficult to make, as some high-grade gliomas also have high ADC [18, 46]. For lymphomas, Guo et al. reported that ADC was generally lower than in high-grade gliomas. This could be useful to differentiate the two [16], but in the clinical situation there is often overlap between lymphoma and high-grade lymphoma. In our experience, the ADC of lymphoma ranges between 0.51 and 0.71×10^{-3} mm^2/s, whereas that of high-grade gliomas ranges between 0.58 and 0.89×10^{-3} mm^2/s. Lymphomas tend to have lower ADC values because of a higher nuclear to cytoplasmic ratio [16]. These are general principles, but in practical clinical work it is often difficult to distinguish between lymphomas, metastasis and high-grade gliomas, even with the most sophisticated ADC maps [4, 10–13, 18].

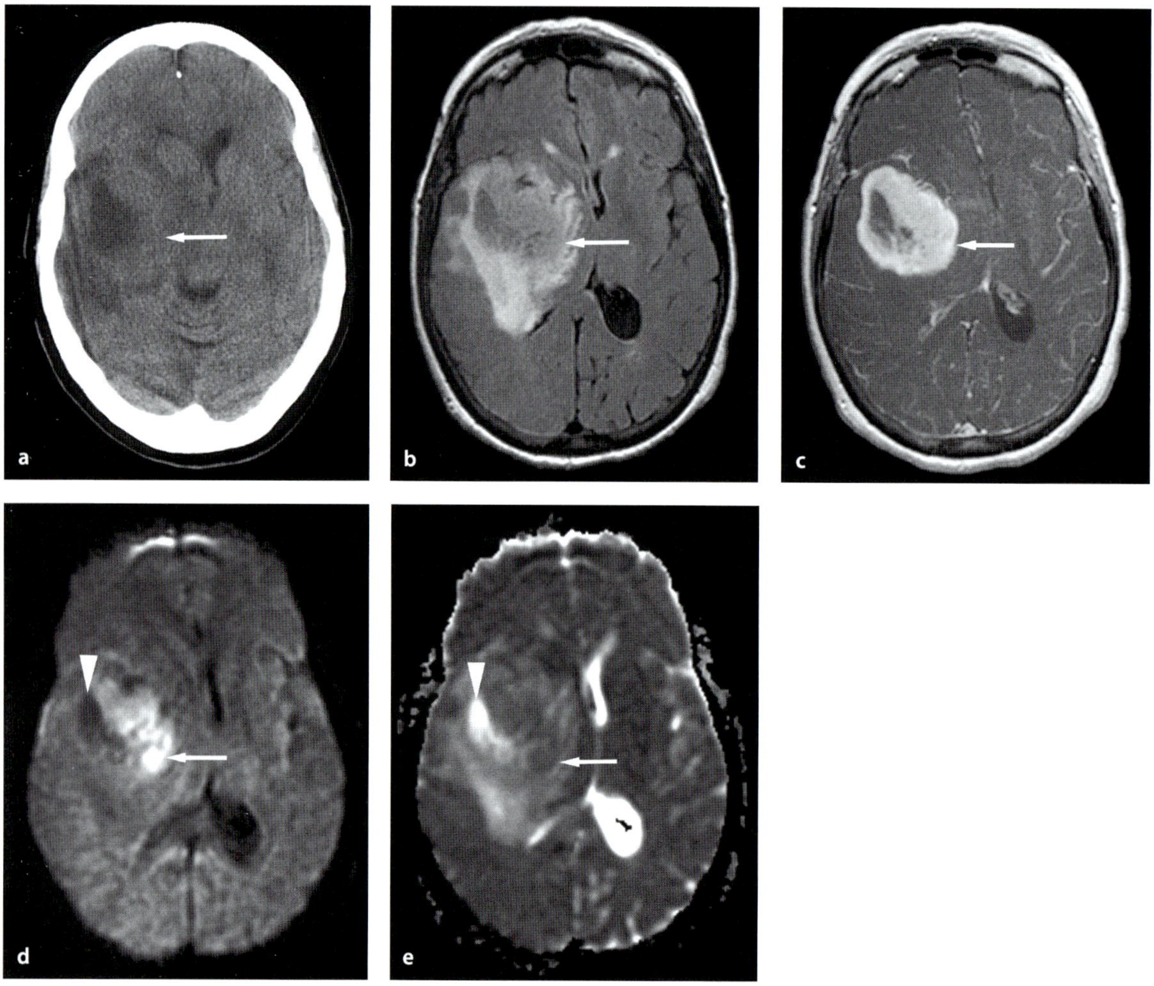

Figure 13.1 a–e

Glioblastoma multiforme in a 69-year-old female with left-side weakness. **a** Unenhanced computed tomography shows a heterogeneous iso- to hypodense lesion in the right temporal lobe (*arrow*). **b** Fluid-attenuated inversion-recovery image shows the heterogeneous hyperintense lesion in the right temporal lobe (*arrow*). **c** Gadolinium-enhanced T1-weighted image shows heterogeneous enhancement in the mass (*arrow*). **d** DW image shows hyperintensity in the enhancing portion of the mass (*arrow*) and hypointensity in the cystic/necrotic portion of the mass (*arrowhead*). **e** ADC map shows heterogeneous hypointensity ($0.74–0.85\times10^{-3}$ mm^2/s; *arrow*) in the enhancing portion of the mass compared to the surrounding vasogenic edema. Cystic/necrotic portion of the mass (*arrowhead*) is hyperintense. These findings may correspond to the high cellularity of the enhancing tumor and increased diffusibility of the cystic/necrotic portion

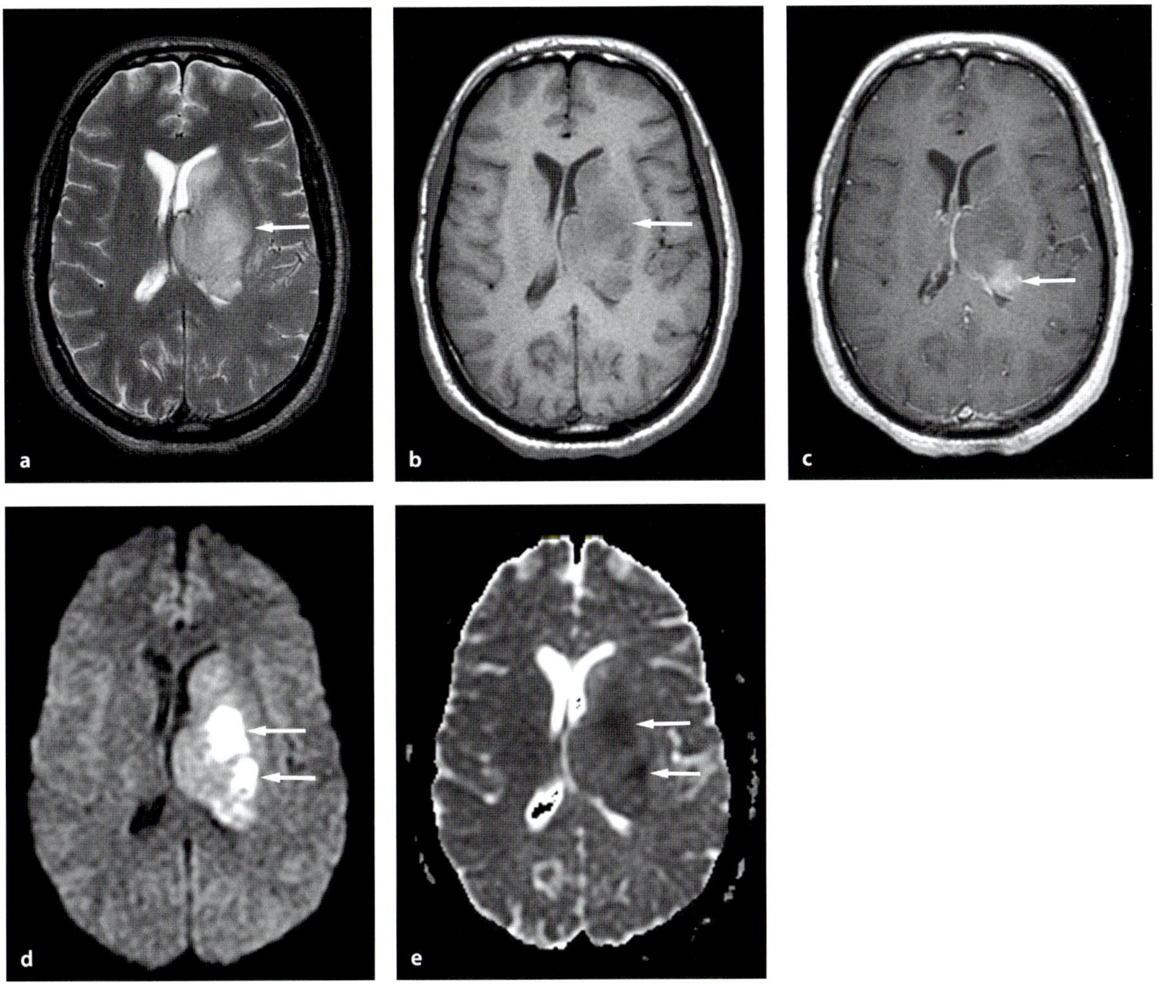

Figure 13.2 a–e

Glioblastoma multiforme in a 51-year-old male with right-side weakness. **a** T2-weighted image shows a hyperintense mass in the left basal ganglia and thalamus (*arrow*). **b** T1-weighted image shows hypointensity in the lesion (*arrow*). **c** Gadolinium-enhanced T1-weighted image shows heterogeneous enhancement in the posterior portion of the tumor (*arrow*). **d** DW image shows hyperintensities (*arrows*). The areas of marked hyperintensity on DW image does not show contrast enhancement in this patient. **e** ADC map shows heterogeneous hypointensity in the lesion ($0.58–0.89\times10^{-3}$ mm^2/s; *arrows*)

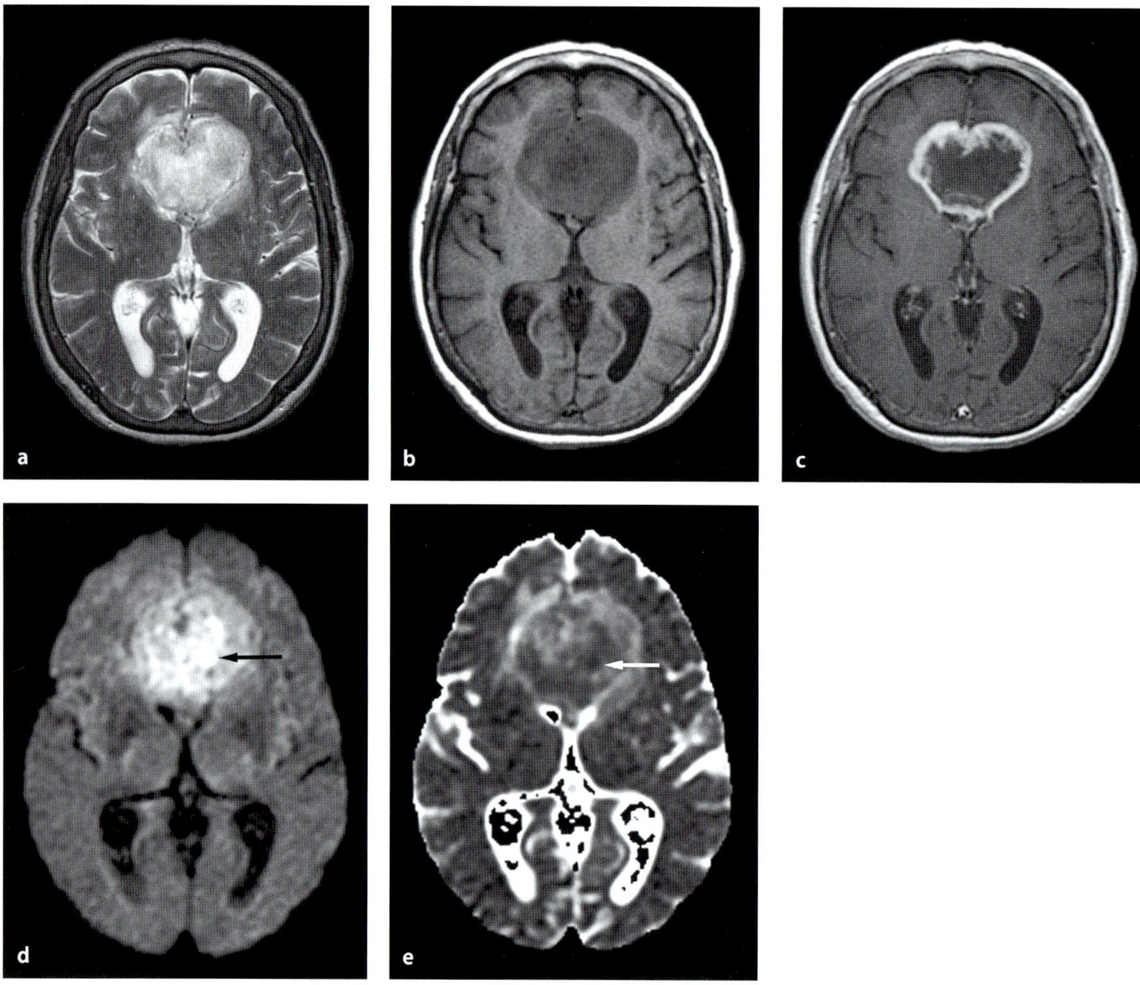

Figure 13.3 a–e

Glioblastoma multiforme in an 80-year-old female with personality change. **a** T2-weighted image shows a hyperintense mass, which involves the genu of corpus callosum (butterfly tumor). **b** T1-weighted image shows hypointensity in the lesion. **c** Gadolinium-enhanced T1-weighted image shows irregular ring-like enhancement of the tumor. **d** DW image shows hyperintensity (*arrow*). **e** ADC map shows heterogeneous intensity in the mass. Note the hypointensity in the center of the lesion (0.65×10^{-3} mm^2/s; *arrow*). These findings may correspond to the cellularity of the tumor

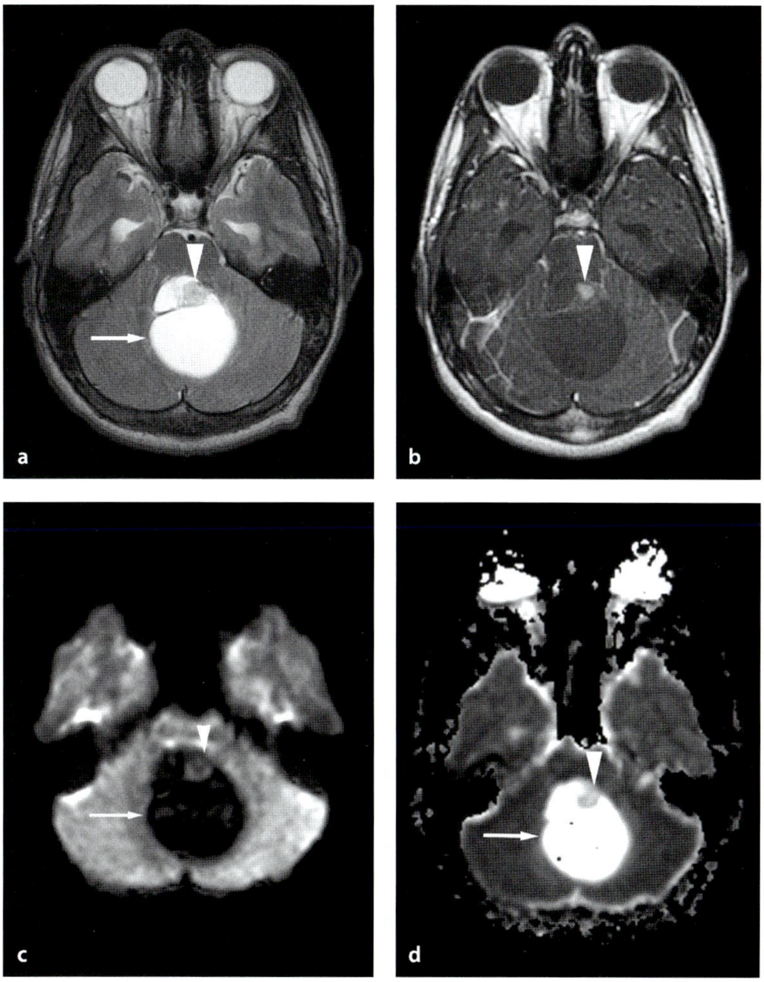

Figure 13.4 a–d

Juvenile pilocytic astrocytoma in a 14-year-old male with headache. **a** T2-weighted image shows a hyperintense mass in the cerebellum (*arrow*). There is a mural nodule in the tumor (*arrowhead*). **b** Gadolinium-enhanced T1-weighted image shows enhancement in the nodule (*arrowhead*). **c** DW image shows hypointensity in the cystic component (*arrow*) and mild hypointensity in the nodule (*arrowhead*). **d** ADC map shows hyperintensity in the cystic component (*arrow*) and mild hyperintensity in the nodule (1.18×10^{-3} mm^2/s; *arrowhead*) of the mass

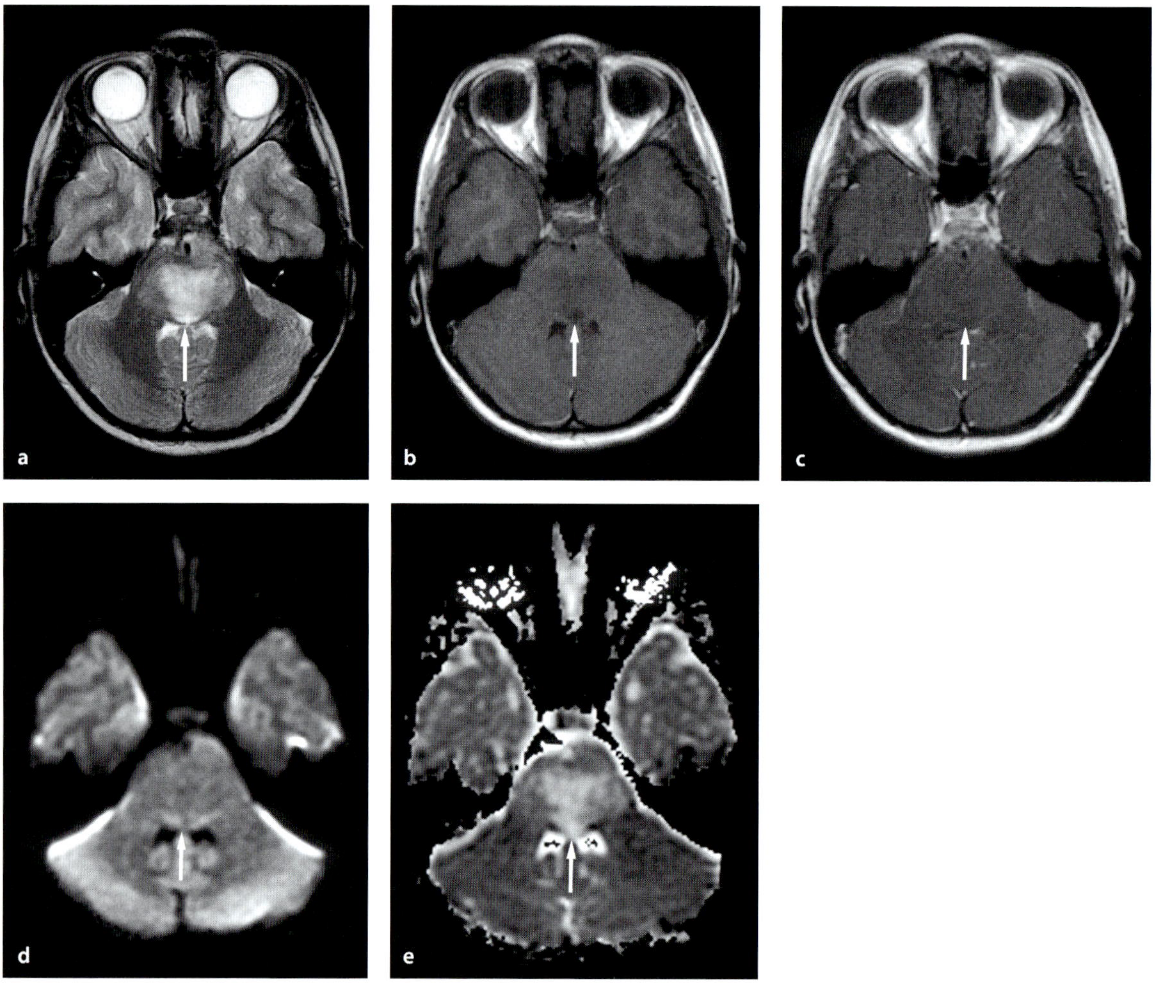

Figure 13.5 a–e

Brain stem glioma in an 8-year-old female with headache. **a** T2-weighted image shows a hyperintense lesion (*arrow*) with surrounding edema in pons. **b** T1-weighted image shows the hypointense lesion (*arrow*). **c** Gadolinium-enhanced T1-weighted image shows no significant enhancement. **d** DW image shows isointensity in the lesion (*arrow*). **e** ADC map shows hyperintensity in the lesion ($0.85–1.17 \times 10^{-3}$ mm^2/s; *arrow*). The isointensity on DW image is caused by a balance between increased T2 and ADC

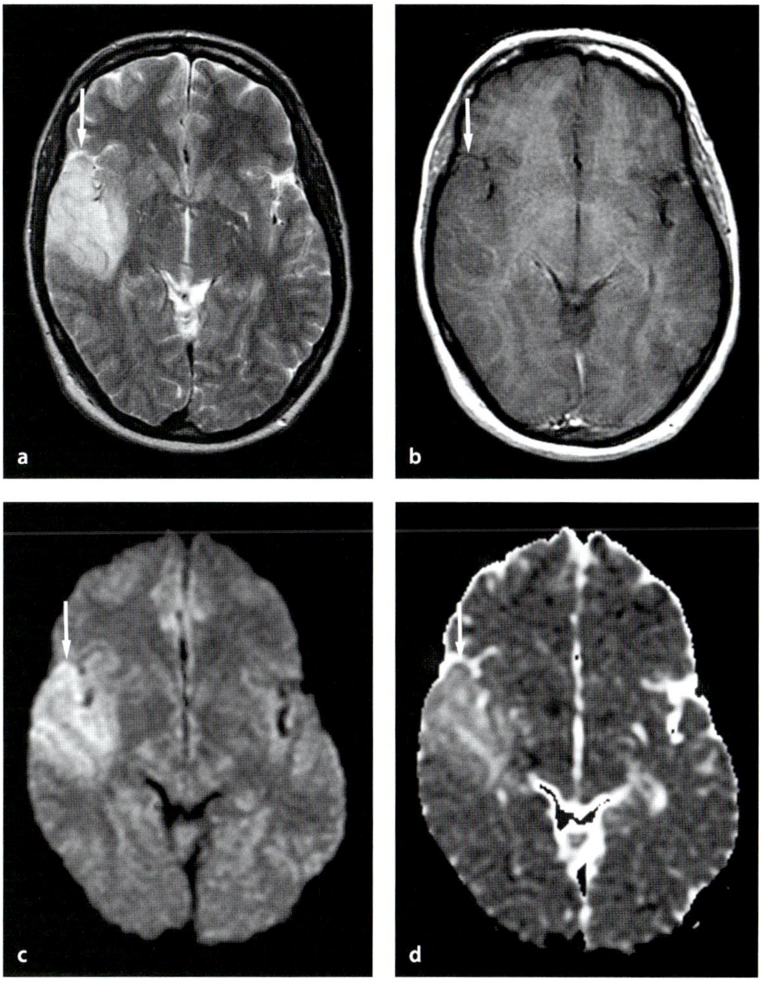

Figure 13.6 a–d

Low-grade oligoastrocytoma in a 48-year-old female with seizures. **a** T2-weighted image shows a hyperintense lesion in the right temporal lobe (*arrow*). **b** Gadolinium-enhanced T1-weighted image shows a slightly hypointense lesion and no enhancement (*arrow*). **c** DW image shows hyperintensity (*arrow*). **d** ADC map shows hyperintensity in the lesion (0.98–1.19×10^{-3} mm^2/s; *arrow*)

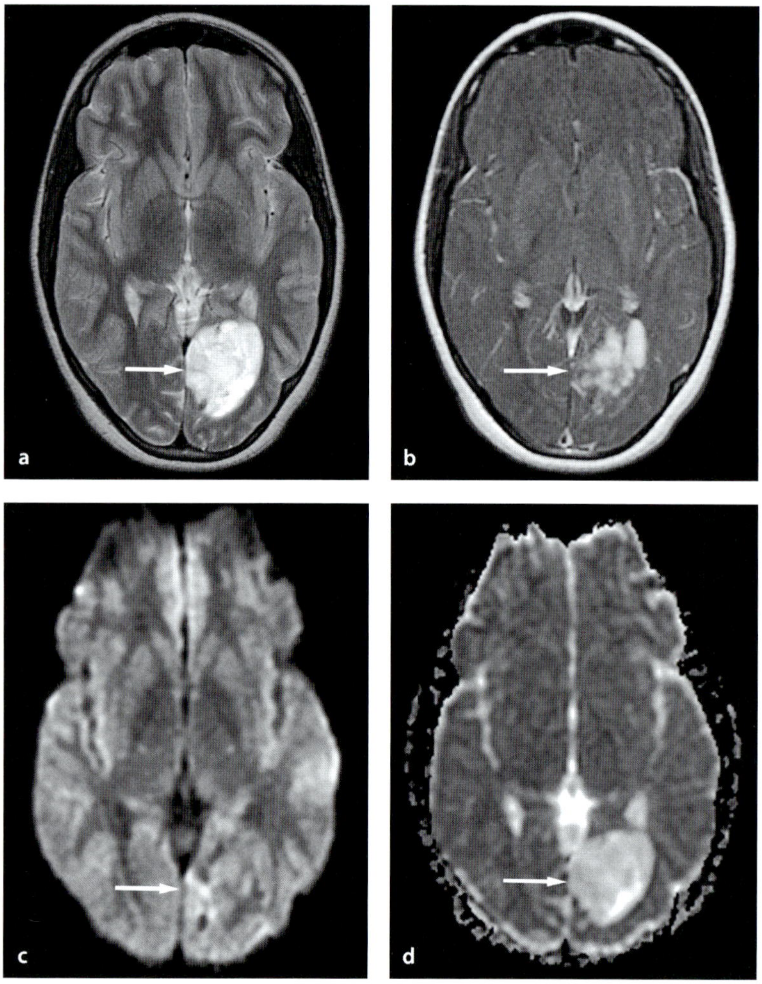

Figure 13.7 a–d

Ganglioglioma in a 12-year-old female with seizures. **a** T2-weighted image shows a hyperintense lesion in the left occipital lobe (*arrow*). **b** Gadolinium-enhanced T1-weighted image shows heterogeneous enhancement (*arrow*). **c** DW image shows heterogeneous mild hyperintensity in the lesion (*arrow*). **d** ADC map shows hyperintensity in the lesion ($0.98–1.35\times10^{-3}$ mm^2/s; *arrow*)

13.2.2 Peritumoral Infiltration

The value of DW imaging for the delineation of peritumoral invasion in primary brain tumors is controversial. Some authors have suggested that ADC is useful to determine the extent of tumor invasion [3, 19], but most of the recent studies have shown that it is not possible to determine accurately the degree of peritumoral infiltration by DW imaging and ADC mapping [4, 10–12, 18, 19]. The poor delineation is probably due to the conjoined effects of T2 and ADC on DW images. For tumors that are biologically different, such as glioblastomas and meningiomas, it has been reported that ADC, T1 and fractional anisotropy of the enhancing tumor and its peritumoral edema are markedly different [15]. Future studies will show if diffusion tensor imaging can add more information about tumor infiltration.

13.2.3 Treatment Response

Diffusion-weighted imaging has been attempted to follow response to treatment and disease progress. For example, animal studies have shown a tendency to an increase in ADC during treatment, followed by a return to the pretreatment level during recurrent tumor growth [20–22]. There are no published human studies, but the preliminary reports have confirmed the observations in animals. Thus, recent studies have shown that radiation necrosis has a higher ADC than tumor recurrence [23].

13.3 Epidermoid Tumors and Arachnoid Cysts

Epidermoid tumors are benign neoplasms of ectodermal origin with stratified squamous epithelium and keratinaceous debris [13, 24–31]. They are hyperintense on DW images, with decreased ADC. The ADC of epidermoid tumors has been reported to be lower than that of cerebrospinal fluid and cqual to or higher than that of brain parenchyma [24–26, 28, 29, 31]. Therefore, the hyperintensity of epidermoid tumors on DW images is primarily caused by a T2 shine-through effect (Fig. 13.8).

Arachnoid cysts have a similar appearance on routine MR imaging as epidermoid tumors, but it is well known that DW imaging can distinguish the two [24–31]. Arachnoid cysts are hypointense on DW images as a result of free diffusion and in general their DW signal characteristics are similar to those of cerebrospinal fluid (Fig. 13.9) [24–26].

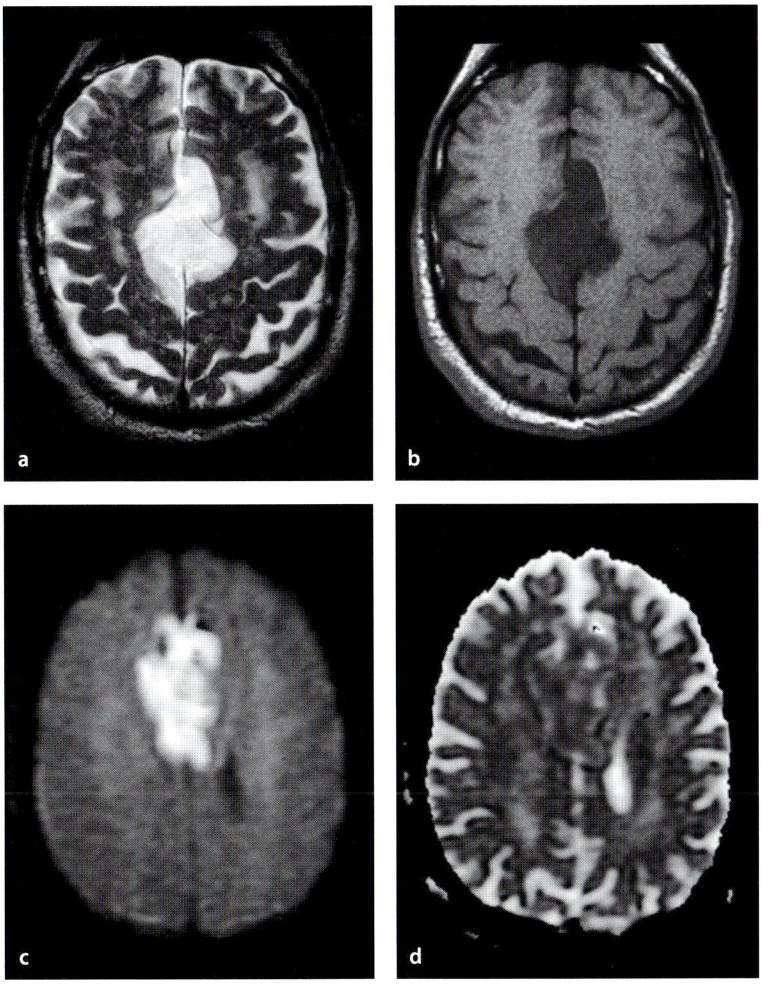

Figure 13.8 a–d

Epidermoid tumor in a 9-year-old female without symptoms. **a** T2-weighted image shows a hyperintense mass near the falx. **b** T1-weighted image shows the hypointense lesion. **c** DW image shows hyperintensity in the lesion, which is caused by both increased T2 and restricted diffusibility. **d** ADC map shows heterogeneous hypointensity in the lesion, consistent with restricted diffusion. ADC was not calculated because raw data were not available

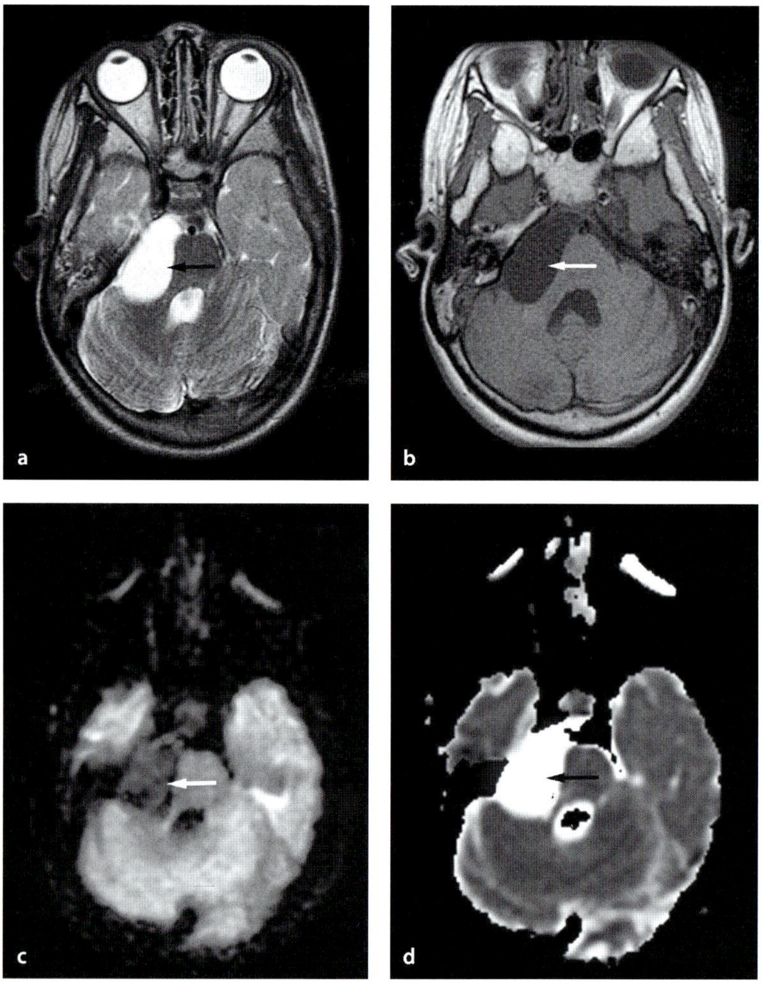

Figure 13.9 a–d

Arachnoid cyst in a 9-year-old female with developmental delay. **a** T2-weighted image shows a large hyperintense lesion in the right cerebellopontine angle (*arrow*). **b** T1-weighted image shows hypointensity in the lesion (*arrow*). **c** DW image shows hypointensity (*arrow*). **d** ADC map shows hyperintensity in the lesion due to increased diffusibility ($3.07–3.12 \times 10^{-3}$ mm^2/s; *arrow*)

13.4 Primitive Neuroectodermal Tumors

Primitive neuroectodermal tumors are a group of histologically similar tumors that occur mostly in children. They include embryonal, largely undifferentiated tumors, such as medulloblastoma, neuroblastoma, pineoblastoma, ependymoblastoma and medulloepithelioma. These tumors have a high cellular density and are typically hyperintense on DW images, with decreased ADC [32–35]. The hyperintensity on DW images and decreased ADC is caused by their dense cellularity and high nuclear to cytoplasmic ratio (Fig. 13.10).

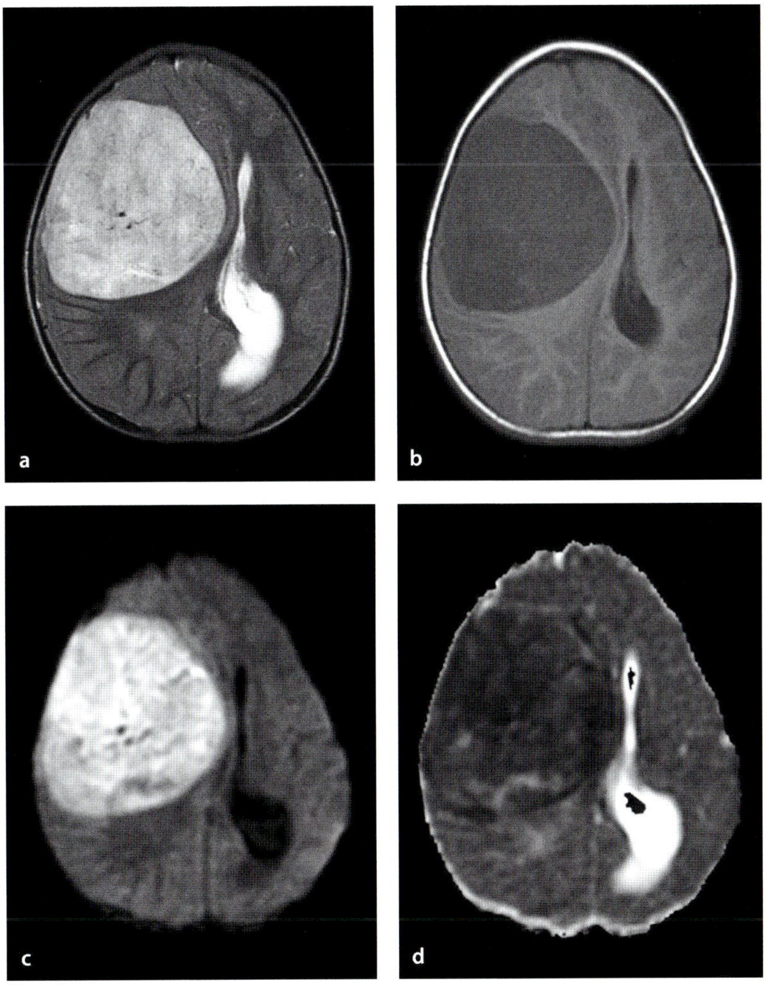

Figure 13.10 a–d

Primitive neuroectodermal tumor in a 20-month-old female with lethargy and nausea. **a** T2-weighted image shows a well-demarcated and heterogeneous intense mass in the right frontal lobe. **b** T1-weighted image shows heterogeneous hypointensity in the lesion. **c** DW image shows hyperintensity. **d** ADC map shows heterogeneous hypointensity (0.54–0.74×10^{-3} mm^2/s) in the lesion, which may represent hypercellularity

13.5 Meningiomas

The signal characteristics of meningiomas on DW images are variable [11, 12, 18, 36, 37]. Most benign meningiomas are isointense on DW images and ADC maps, but some are slightly hyperintense on both DW images and ADC maps (Fig. 13.11). Malignant or atypical meningiomas have markedly increased signal intensity on DW images and decreased ADC due to a high tumor cellularity [12, 18, 36]. However, other factors, such as multifocal areas of necrosis, numerous abnormal mitoses and cytologic pleomorphism may also cause the high DW signal in atypical and malignant meningiomas (Fig. 13.12).

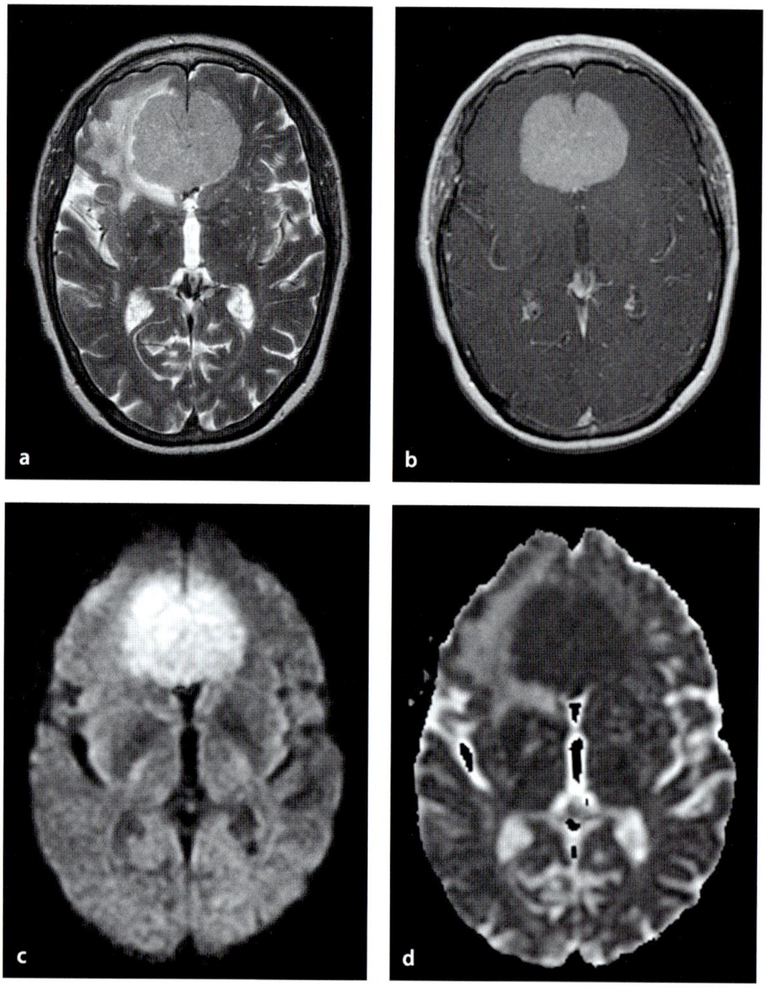

Figure 13.11 a–d

Benign meningioma (syncytial meningioma) in a 72-year-old female with a visual disturbance. **a** T2-weighted image shows a slightly hyperintense mass near the frontal aspect of the falx. **b** Gadolinium-enhanced T1-weighted image shows homogeneous enhancement. **c** DW image shows hyperintensity in the lesion. **d** ADC map shows mild hypointensity (0.73–0.78×10^{-3} mm^2/s)

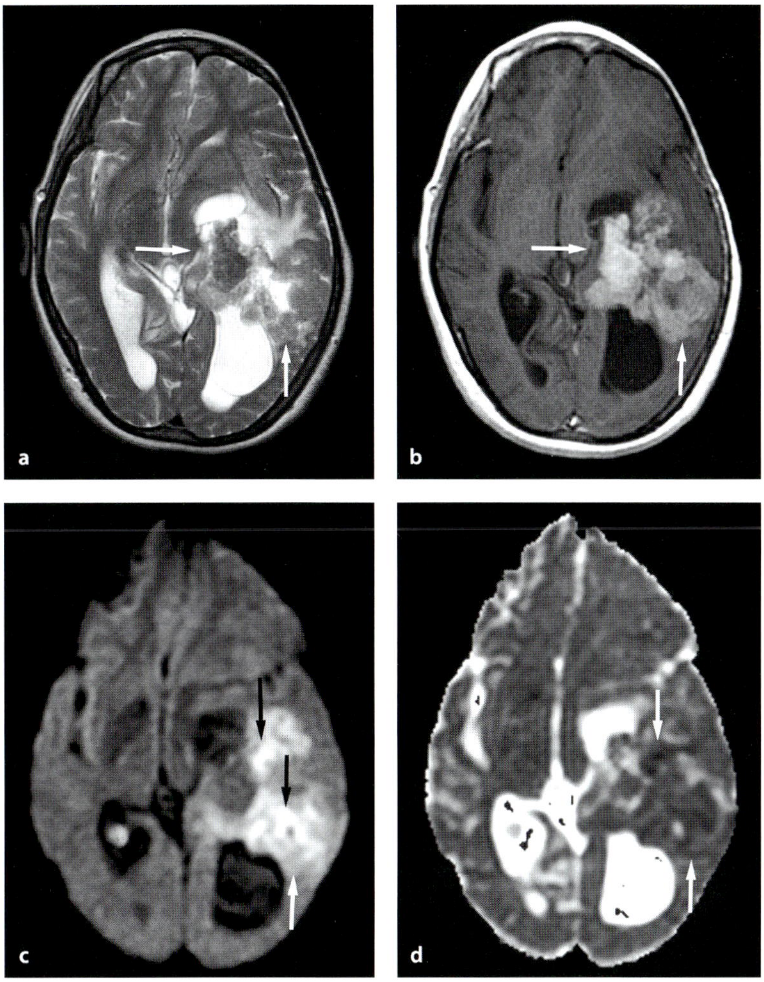

Figure 13.12 a–d

Atypical meningioma in a 45-year-old female with headache. **a** T2-weighted image shows a heterogeneous intense mass in the temporal lobe (*arrows*). **b** Gadolinium-enhanced T1-weighted image shows heterogeneous enhancement (*arrows*). **c** DW image shows heterogeneous hyperintensity (*arrows*). **d** ADC map shows hypointensity, especially in the right side of the mass (0.51×10^{-3} mm^2/s; *arrows*)

13.6 Malignant Lymphomas

Primary lymphomas of the central nervous system are rare malignant tumors, accounting for less than 1% of all brain tumors (Fig. 13.13). However, the incidence has been increasing subsequent to the spread of the acquired immunodeficiency syndrome [18, 38–41].

A majority of lymphomas are iso- or hypointense on T2-weighted images [18]. Most lymphomas show homogeneous enhancement in immunocompetent patients, but in immunosuppressed patients a rim enhancement is a more common finding [18, 38–40].

The enhancing components of lymphomas are generally hyperintense on DW images [16, 18, 41]. The ADC of lymphomas is often lower than that in high-grade gliomas [16]. As mentioned above, this corresponds to the hypercellularity and might help in differentiating between lymphomas and high-grade gliomas [16].

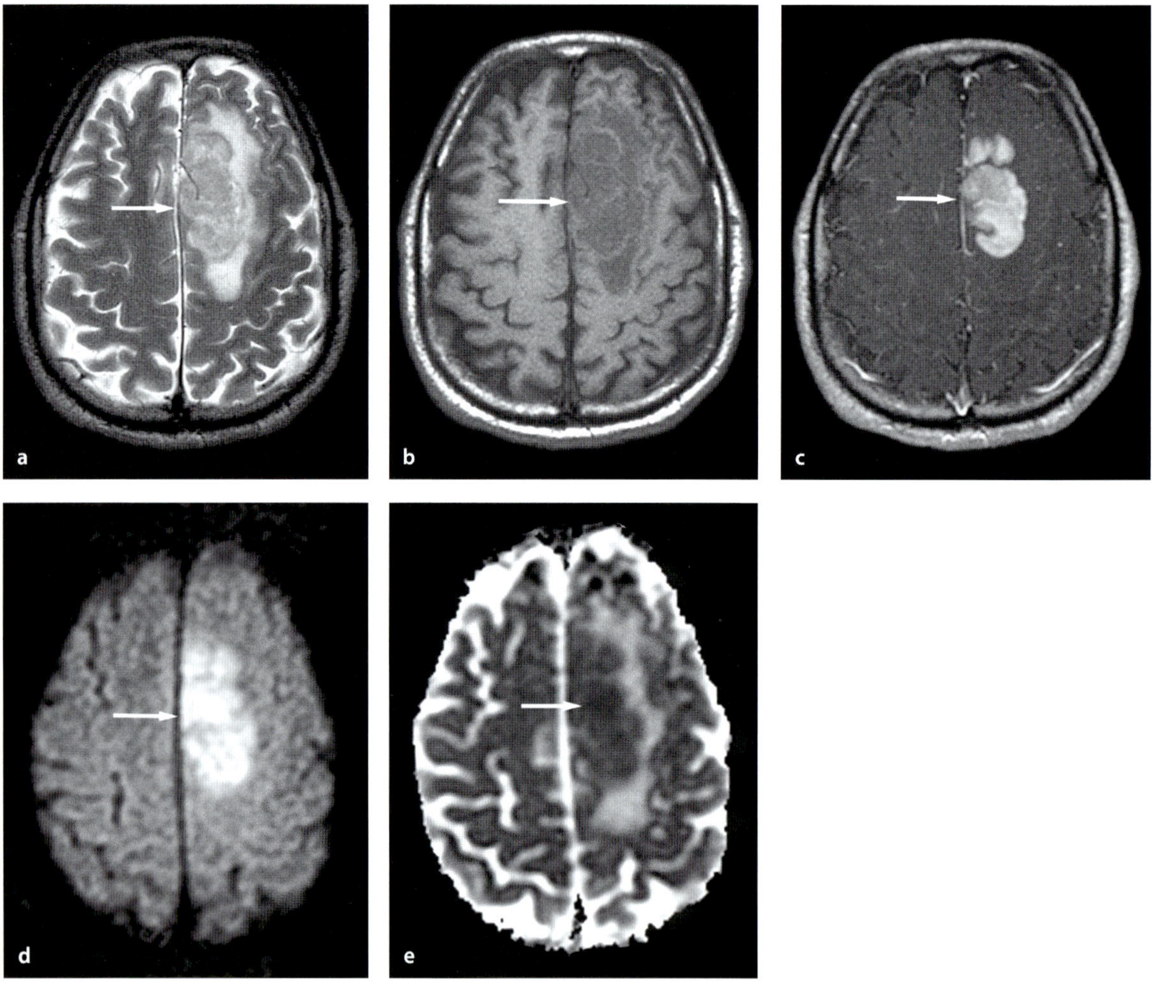

Figure 13.13 a–e

Lymphoma (diffuse large B cell type) in a 64-year-old male with seizure. **a** T2-weighted image shows a slightly hyperintense mass (*arrow*) with surrounding edema in the left frontal lobe. **b** T1-weighted image shows the hypointense mass (*arrow*) in the left frontal lobe. **c** Gadolinium-enhanced T1-weighted image shows the heterogeneously enhancing mass (*arrow*) in the left frontal lobe. **d** DW image shows hyperintensity in the lesion (*arrow*). **e** ADC map shows hypointensity in the lesion ($0.51–0.71 \times 10^{-3}$ mm^2/s; *arrow*)

13.7 Craniopharyngiomas

Craniopharyngiomas typically show a combination of contrast enhancement, cyst formation and calcifications (Fig. 13.14). Sener et al. reported hyperintensity on DW images, with increased ADC, corresponding to an increased diffusibility in the tumor [42] and a T2 shine-through effect.

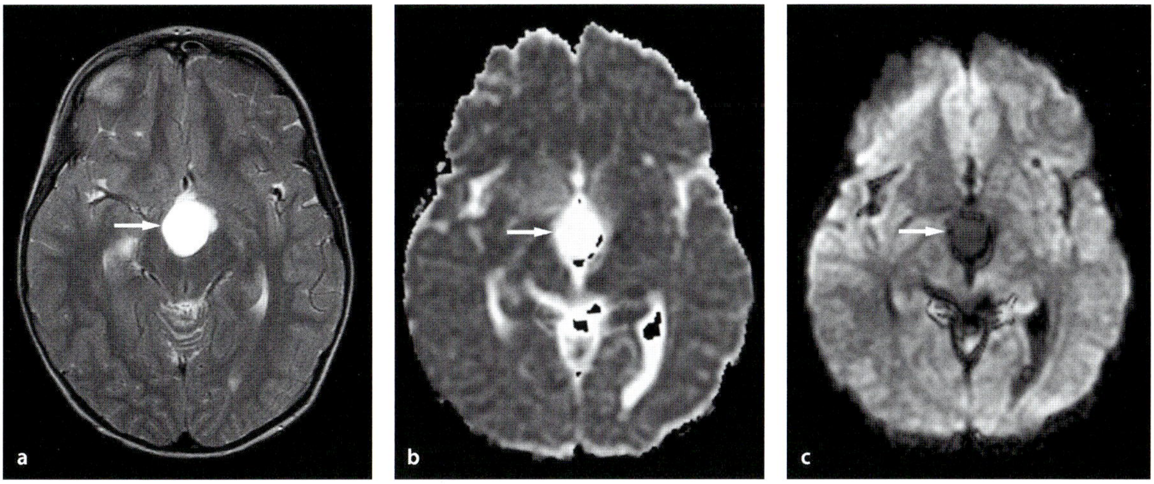

Figure 13.14 a–c

Craniopharyngioma in an 8-year-old male with panhypopituitarism. **a** T2-weighted image shows a hyperintense mass (*arrow*) in the suprasellar region. **b** DW image shows hypointensity in the mass (*arrow*). **c** ADC map shows hyperintensity ($2.25–2.38 \times 10^{-3}$ mm^2/s; *arrow*)

13.8 Metastases

The signal intensity of non-necrotic components of metastases on DW images is variable and depends on their T2 and ADC [2, 6, 11, 12, 18, 19, 30, 43–51]. DW imaging findings of solid components of metastasis are similar to those of gliomas, probably reflecting the cellularity of the primary tumor. The cellularity is a major determinant of their DW signal intensity [6, 11, 18, 30].

The common signal intensity of necrotic/cystic components of cerebral metastases may relate to an increase in free water, showing hypointensity on DW images and increased ADC. However, in the presence of extracellular met-hemoglobin and/or increased viscosity, DW images can show hyperintensity with decreased ADC (Figs. 13.15 and 13.16) [18, 43–51]. This condition is rare, but it should be considered as one differential diagnosis of pyogenic abscesses.

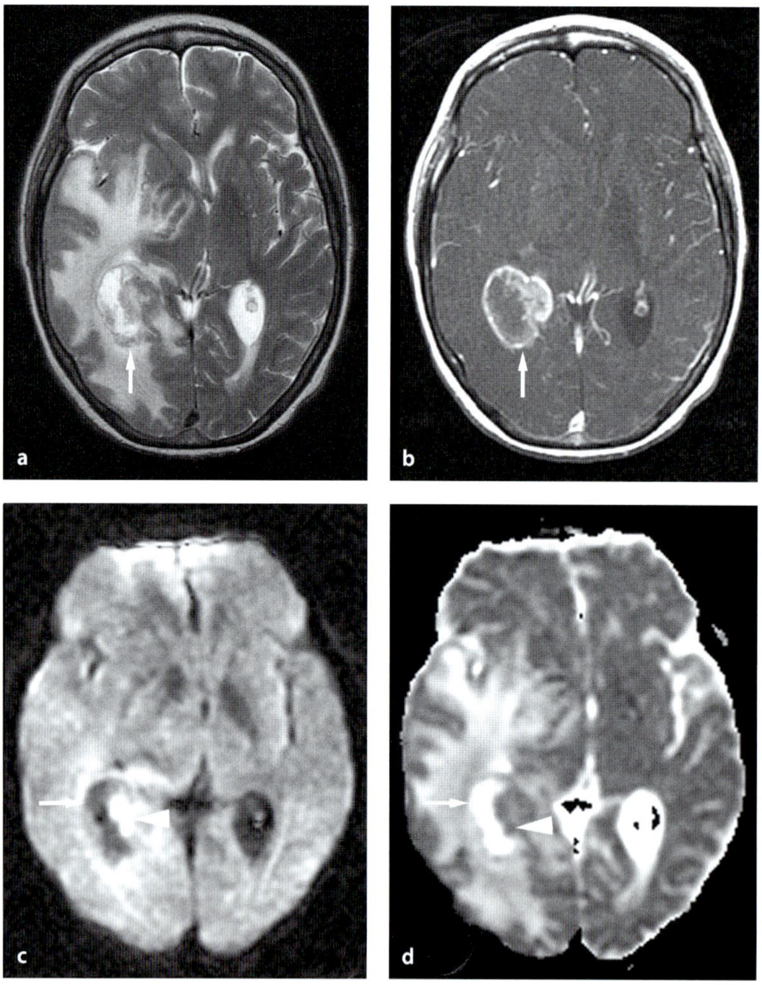

Figure 13.15 a–d

Metastasis (lung cancer) in a 59-year-old female with adenocarcinoma of the lung. **a** T2-weighted image shows heterogeneous intensity of a mass (*arrow*) with surrounding edema in the right temporal lobe. **b** Gadolinium-enhanced T1-weighted image shows heterogeneous ring-like enhancement (*arrow*). **c** DW image shows hyperintensity in the solid portion (*arrowhead*) and hypointensity in the cystic/necrotic portion (*arrow*). **d** ADC map shows hypointensity in the solid portion ($0.97–1.00\times10^{-3}$ mm^2/s; *arrowhead*) and hyperintensity ($2.21–2.35\times10^{-3}$ mm^2/s; *arrow*). Hyperintensity in the surrounding edema, consistent with vasogenic edema, is also seen

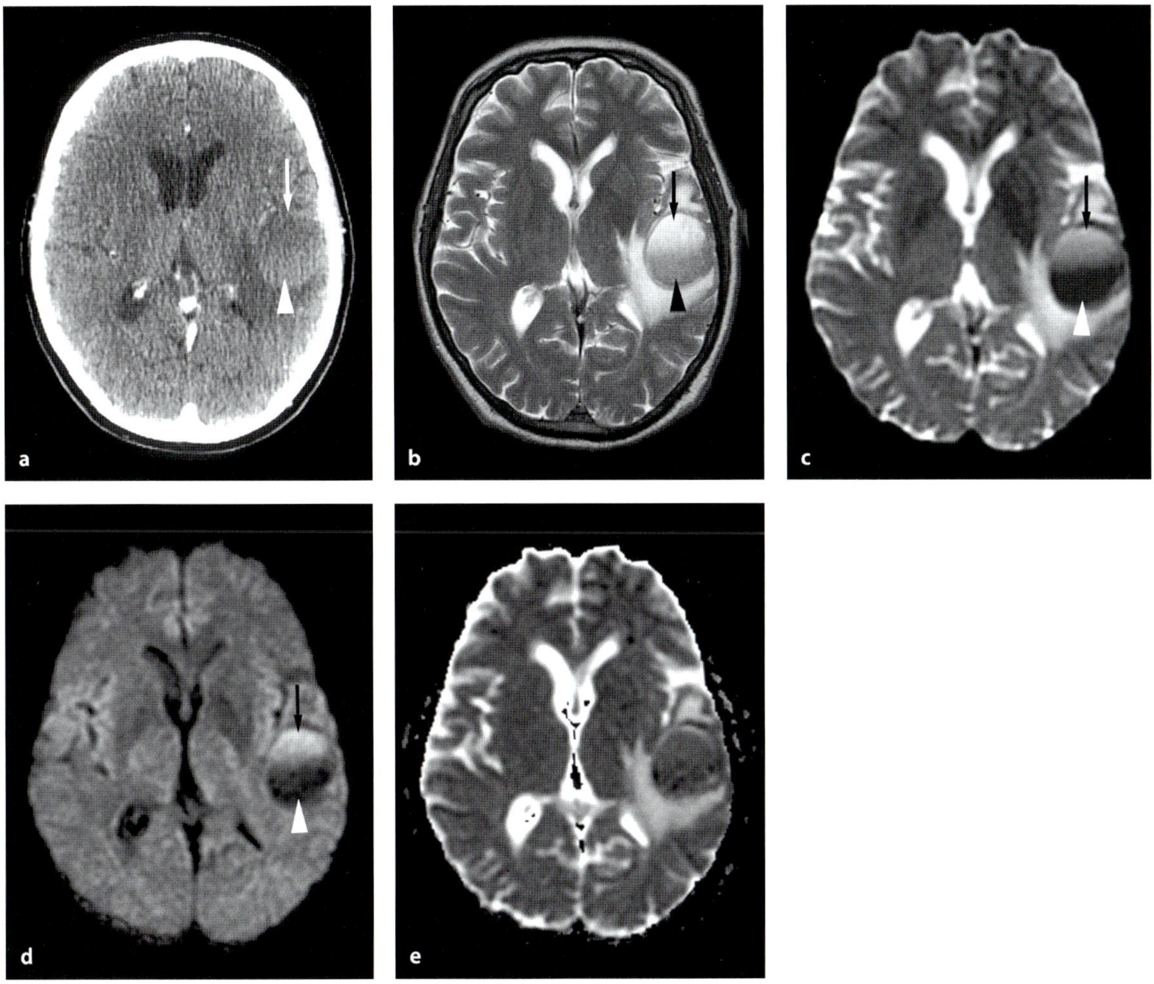

Figure 13.16 a–e

Metastasis (melanoma) in a 56-year-old male with metastatic melanoma. **a** Postcontrast computed tomography shows a heterogeneous density mass, which shows hypodensity (*arrow*) in the anterior portions and hyperdensity (*arrowhead*) in the posterior portions. **b** T2-weighted image shows the heterogeneous intense mass, which contains anterior hyperintense portion (*arrow*) and posterior hypointense portion (*arrowhead*) with surrounding edema in the left temporal lobe. **c** b_0 image also shows anterior hyperintense portion (*arrow*) and posterior hypointense portion (*arrowhead*). **d** DW image again shows anterior hyperintense portion (*arrow*) and posterior hypointense portion (*arrowhead*). **e** ADC map shows iso- to hypointensity in the lesion. The accurate calculation of ADC in the hemorrhage is difficult

13.9 Conclusion

Diffusion-weighted imaging can provide valuable information about tumor cellularity and help in the characterization and grading of tumors of the brain. In most situations, it is difficult to differentiate between specific tumors and to determine tumor infiltration. Future studies will show whether diffusion tensor imaging can improve our ability to characterize and grade brain tumors on imaging studies.

References

1. LeBihan D, Breton E, Lallemand D, Grenier P, Cabanis E, Laval-Jeantet M (1986) MR imaging of intravoxel incoherent motions: application to diffusion and perfusion in neurologic disorders. Radiology 161:401–407
2. Hajnal JV, Doran M, Hall AS, et al. (1991) MR imaging of anisotropically restricted diffusion of water in the nervous system: technical, anatomic, and pathologic considerations. J Comput Assist Tomogr 15:1–18
3. Tien RD, Felsberg GJ, Friedman H, Brown M, MacFall J (1994) MR imaging of high-grade cerebral gliomas: value of diffusion-weighted echoplanar pulse sequences. AJR Am J Roentgenol 162:671–677
4. Eis M, Els T, Hoehn-Berlage M, Hossmann KA (1994) Quantitative diffusion MR imaging of cerebral tumor and edema. Acta Neurochir Suppl (Wien) 60:344–346
5. Brunberg JA, Chenevert TL, McKeever PE, et al. (1995) In vivo MR determination of water diffusion coefficients and diffusion anisotropy: correlation with structural alteration in gliomas of the cerebral hemispheres. AJNR Am J Neuroradiol 16:361–371
6. Krabbe K, Gideon P, Wagn P, Hansen U, Thomsen C, Madsen F (1997) MR diffusion imaging of human intracranial tumours. Neuroradiology 39:483–489
7. Gupta RK, Sinha U, Cloughesy TF, Alger JR (1999) Inverse correlation between choline magnetic resonance spectroscopy signal intensity and the apparent diffusion coefficient in human glioma. Magn Reson Med 41:2–7
8. Sugahara T, Korogi Y, Kochi M, et al. (1999) Usefulness of diffusion-weighted MRI with echo-planar technique in the evaluation of cellularity in gliomas. J Magn Reson Imaging 9:53–60
9. Gupta RK, Cloughesy TF, Sinha U, et al. (2000) Relationships between choline magnetic resonance spectroscopy, apparent diffusion coefficient and quantitative histopathology in human glioma. J Neurooncol 50:215–226
10. Castillo M, Smith JK, Kwock L, Wilber K. (2001) Apparent diffusion coefficients in the evaluation of high-grade cerebral gliomas. AJNR Am J Neuroradiol 22:60–64
11. Stadnik TW, Chaskis C, Michotte A, et al. (2001) Diffusion-weighted MR imaging of intracerebral masses: comparison with conventional MR imaging and histologic findings. AJNR Am J Neuroradiol 22:969–976
12. Kono K, Inoue Y, Nakayama K, et al. (2001) The role of diffusion-weighted imaging in patients with brain tumors. AJNR Am J Neuroradiol 22:1081–1088
13. Gauvain KM, McKinstry RC, Mukherjee P, et al. (2001) Evaluating pediatric brain tumor cellularity with diffusion-tensor imaging. AJR Am J Roentgenol 177:449–454
14. Sinha S, Bastin ME, Whittle IR, Wardlaw JM (2002) Diffusion tensor MR imaging of high-grade cerebral gliomas. AJNR Am J Neuroradiol 23:520–527
15. Bastin ME, Sinha S, Whittle IR, Wardlaw JM (2002) Measurements of water diffusion and T1 values in peritumoural oedematous brain. Neuroreport 13:1335–1340
16. Guo AC, Cummings TJ, Dash RC, Provenzale JM (2002) Lymphomas and high-grade astrocytomas: comparison of water diffusibility and histologic characteristics. Radiology 224:177–183
17. Yang D, Korogi Y, Sugahara T, et al. (2002) Cerebral gliomas: prospective comparison of multivoxel 2D chemical-shift imaging proton MR spectroscopy, echoplanar perfusion and diffusion-weighted MRI. Neuroradiology 44:656–666
18. Stadnik TW, Demaerel P, Luypaert RR, et al. (2003) Imaging tutorial: differential diagnosis of bright lesions on diffusion-weighted MR images. Radiographics 23:E7–7
19. Bulakbasi N, Kocaoglu M, Ors F, Tayfun C, Ucoz T (2003) Combination of single-voxel proton MR spectroscopy and apparent diffusion coefficient calculation in the evaluation of common brain tumors. AJNR Am J Neuroradiol 24:225–233
20. Chenevert TL, McKeever PE, Ross BD (1997) Monitoring early response of experimental brain tumors to therapy using diffusion magnetic resonance imaging. Clin Cancer Res 3:1457–1466
21. Chenevert TL, Stegman LD, Taylor JM, et al. (2000) Diffusion magnetic resonance imaging: an early surrogate marker of therapeutic efficacy in brain tumors. J Natl Cancer Inst 92:2029–2036
22. Mardor Y, Roth Y, Lidar Z, et al. (2001) Monitoring response to convection-enhanced taxol delivery in brain tumor patients using diffusion-weighted magnetic resonance imaging. Cancer Res 61:4971–4973
23. Asao C, Korogi Y, Kitajima M Nishimura R, Baba Y, Kochi M, Yamashita Y (2003) Diffusion-weighted MR imaging of radiation-induced brain injury: the usefulness for differentiation from tumor recurrence. American Society of Neuroradiology 41st annual meeting 2003 proceedings 19:16
24. Tsuruda JS, Chew WM, Moseley ME, Norman D (1990) Diffusion-weighted MR imaging of the brain: value of differentiating between extraaxial cysts and epidermoid tumors. AJNR Am J Neuroradiol 11:925–931
25. Tsuruda JS, Chew WM, Moseley ME, Norman D (1991) Diffusion-weighted MR imaging of extraaxial tumors. Magn Reson Med 19:316–320
26. Maeda M, Kawamura Y, Tamagawa Y (1992) Intravoxel incoherent motion (IVIM) MRI in intracranial, extraaxial tumors and cysts. J Comput Assist Tomogr 16:514–518
27. Laing AD, Mitchell PJ, Wallace D (1999) Diffusion-weighted magnetic resonance imaging of intracranial epidermoid tumours. Australas Radiol 43:16–19
28. Dechambre S, Duprez T, Lecouvet F, Raftopoulos C, Gosnard G (1999) Diffusion-weighted MRI postoperative assessment of an epidermoid tumour in the cerebellopontine angle. Neuroradiology 41:829–831

29. Chen S, Ikawa F, Kurisu K, Arita K, Takaba J, Kanou Y (2001) Quantitative MR evaluation of intracranial epidermoid tumors by fast fluid-attenuated inversion recovery imaging and echo-planar diffusion-weighted imaging. AJNR Am J Neuroradiol 22:1089–1096
30. Bergui M, Zhong J, Bradac GB, Sales S (2001) Diffusion-weighted images of intracranial cyst-like lesions. Neuroradiology 43:824–829
31. Annet L, Duprez T, Grandin C, Dooms G, Collard A, Cosnard G (2002) Apparent diffusion coefficient measurements within intracranial epidermoid cysts in six patients. Neuroradiology 44:326–328
32. Kotsenas AL, Roth TC, Manness WK, Faerber EN (1999) Abnormal diffusion-weighted MRI in medulloblastoma: does it reflect small cell histology? Pediatr Radiol 29:524–526
33. Klisch J, Husstedt H, Hennings S, von Velthoven V, Pagenstecher A, Schumacher M (2000) Supratentorial primitive neuroectodermal tumours: diffusion-weighted MRI. Neuroradiology 42:393–398
34. Wilke M, Eidenschink A, Muller-Weihrich S, Auer DP (2001) MR diffusion imaging and 1H spectroscopy in a child with medulloblastoma. A case report. Acta Radiol 42:39–42
35. Erdem E, Zimmerman RA, Haselgrove JC, Bilaniuk LT, Hunter JV (2001) Diffusion-weighted imaging and fluid attenuated inversion recovery imaging in the evaluation of primitive neuroectodermal tumors. Neuroradiology 43:927–933
36. Filippi CG, Edgar MA, Ulug AM, Prowda JC, Heier LA, Zimmerman RD (2001) Appearance of meningiomas on diffusion-weighted images: correlating diffusion constants with histopathologic findings. AJNR Am J Neuroradiol 22:65–72
37. Bitzer M, Klose U, Geist-Barth B (2002) Alterations in diffusion and perfusion in the pathogenesis of peritumoral brain edema in meningiomas. Eur Radiol 12:2062–2076
38. Jack CR Jr, Reese DF, Scheithauer BW (1986) Radiographic findings in 32 cases of primary CNS lymphoma. AJR Am J Roentgenol 146:271–276
39. Hochberg FH, Miller DC (1988) Primary central nervous system lymphoma. J Neurosurg 68:835–853
40. Poon T, Matoso I, Tchertkoff V, Weitzner I Jr, Gade M (1989) CT features of primary cerebral lymphoma in AIDS and non-AIDS patients. J Comput Assist Tomogr 13:6–9
41. Chang L, Ernst T (1997) MR spectroscopy and diffusion-weighted MR imaging in focal brain lesions in AIDS. Neuroimaging Clin N Am 7:409–426
42. Sener RN, Dzelzite S, Migals A (2002) Huge craniopharyngioma: diffusion MRI and contrast-enhanced FLAIR imaging. Comput Med Imaging Graph 26:199–203
43. Noguchi K, Watanabe N, Nagayoshi T, et al. (1999) Role of diffusion-weighted echo-planar MRI in distinguishing between brain abscess and tumour: a preliminary report. Neuroradiology 41:171–174
44. Park SH, Chang KH, Song IC, Kim YJ, Kim SH, Han MH (2000) Diffusion-weighted MRI in cystic or necrotic intracranial lesions. Neuroradiology 42:716–721
45. Holtas S, Geijer B, Stromblad LG, Maly-Sundgren P, Burtscher IM (2000) A ring-enhancing metastasis with central high signal on diffusion-weighted imaging and low apparent diffusion coefficients. Neuroradiology 42:824–827
46. Tung GA, Evangelista P, Rogg JM, Duncan JA 3rd (2001) Diffusion-weighted MR imaging of rim-enhancing brain masses: is markedly decreased water diffusion specific for brain abscess? AJR Am J Roentgenol 177:709–712
47. Hartmann M, Jansen O, Heiland S, Sommer C, Munkel K, Sartor K (2001) Restricted diffusion within ring enhancement is not pathognomonic for brain abscess. AJNR Am J Neuroradiol 22:1738–1742
48. Geijer B, Holtas S (2002) Diffusion-weighted imaging of brain metastases: their potential to be misinterpreted as focal ischaemic lesions. Neuroradiology 44:568–573
49. Chang SC, Lai PH, Chen WL, et al. (2002) Diffusion-weighted MRI features of brain abscess and cystic or necrotic brain tumors: comparison with conventional MRI. Clin Imaging 26:227–236
50. Lai PH, Ho JT, Chen WL, Hsu SS, Wang JS, Pan HB, Yang CF (2002) Brain abscess and necrotic brain tumor: discrimination with proton MR spectroscopy and diffusion-weighted imaging. AJNR Am J Neuroradiol 23:1369–1377
51. Guzman R, Barth A, Lovblad KO, et al. (2002) Use of diffusion-weighted magnetic resonance imaging in differentiating purulent brain processes from cystic brain tumors. J Neurosurg 97:1101–1107

Pediatrics

14.1 Water Content of the Pediatric Brain

The water content of the pediatric brain is considerably higher than that of the adult brain. This makes it more difficult to diagnose ischemic and other lesions in pediatric patients using computed tomography (CT) and MR imaging. Diffusion-weighted (DW) imaging is sensitive to alteration in diffusion of water molecules, and this technique can help overcome some of these difficulties [1]. DW imaging is primarily useful for detecting and characterizing ischemic lesions, but also for evaluation of myelinization by demonstrating anisotropy of the white matter earlier than conventional MR imaging [2, 3].

14.2 Normal Structures

Diffusion-weighted imaging characteristics of a normal brain in young infants are different from those in adults [4]. Apparent diffusion coefficient (ADC) values in both gray and white matter of newborns are considerably higher than in adults. This reflects the high water content of the pediatric brain [5]. For the same reason, the deep white matter in the newborn normally shows hypointensity on DW imaging associated with increased ADC (Fig. 14.1). With increasing age, there is a relative decrease in water content of a pediatric brain. This is more evident in the white matter than in the gray matter. The decrease in water content is caused by myelinization replacing water during normal white matter development [2].

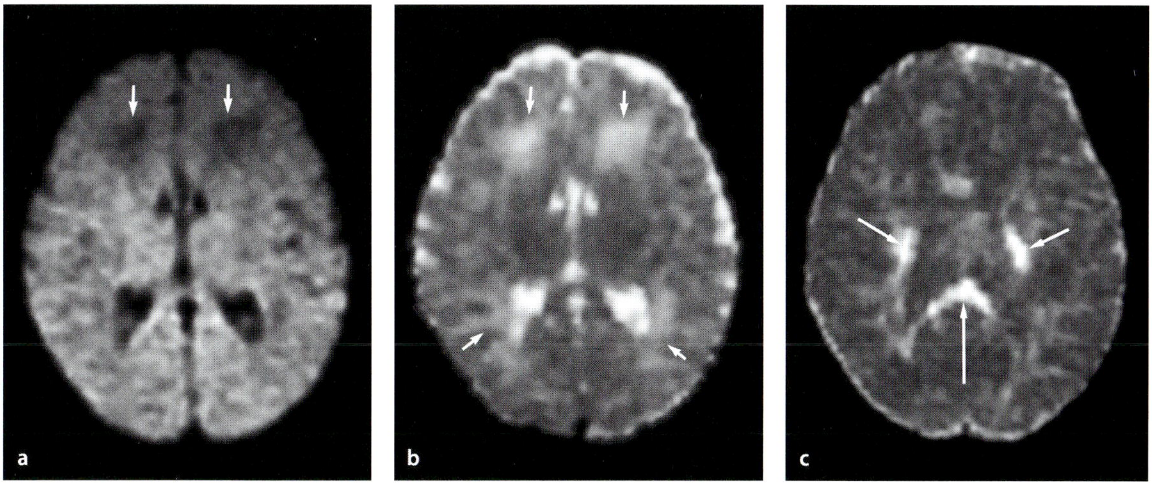

Figure 14.1 a–c

Normal pediatric DW imaging in a 2-day-old boy. **a** Low signal intensity and **b** increased ADC in the deep white matter are normal in this age group (*arrows*). **c** Fractional anisotropy map demonstrates anisotropy along the anterior and posterior limbs of the internal capsules (*arrows*), the corpus callosum (*long arrow*), and the temporo-parieto-occipital white matter earlier than regular T1- and T2-weighted images

14.3 Anisotropy

Normally anisotropy is much less evident in the immature brain than in the adult. One exception is corpus callosum, where anisotropy is already visible by DW imaging as early as the 28th gestational week. This occurs although the corpus callosum is composed of non-myelinated fibers. The phenomenon has been called premyelination anisotropy [3]. The anisotropic effect in the immature brain is thought to be related to structural changes of the axonal membrane. In general, anisotropy in the white matter of newborns is lower than in adults (Fig. 14.1). The anisotropic pattern can vary depending on the irregularity of axonal orientation as well as the degree of myelination.

14.4 Infarction and Ischemia

Ischemic infarctions in children are uncommon when compared with adults and they have different etiologies. They can be caused by thrombosis, embolism, arterial dissection, vasculitis, Moyamoya disease, sickle cell disease, child abuse, etc. (Figs. 14.2, 14.3, 14.4 and 14.5) [6, 7]. Hyperacute and acute infarctions are characterized by cytotoxic edema. Vasogenic edema occurs later and is typically seen in the subacute phase. DW imaging is useful for early detection of infarction in children, but also to differentiate between acute/subacute infarctions and chronic infarctions or ischemic gliosis.

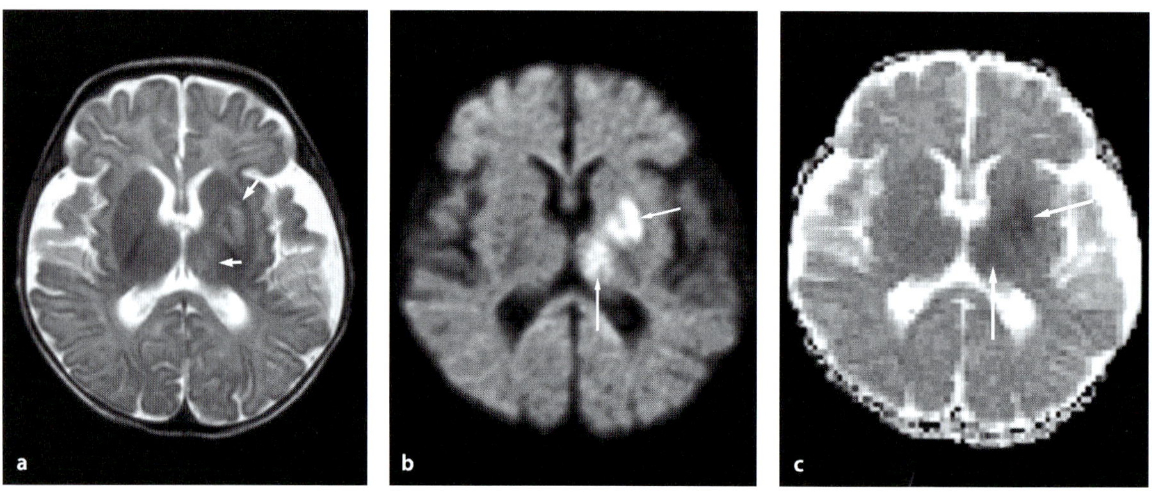

Figure 14.2 a–c

Cerebral infarction due to embolism in a 3-month-old boy. He had Down syndrome and ventricular septal defect. **a** T2-weighted image shows mild high signal lesions in the left putamen and thalamus (*arrows*). **b** DW image shows these lesions as high signal intensity (*arrows*). **c** ADC is decreased (*arrows*), consistent with acute infarcts

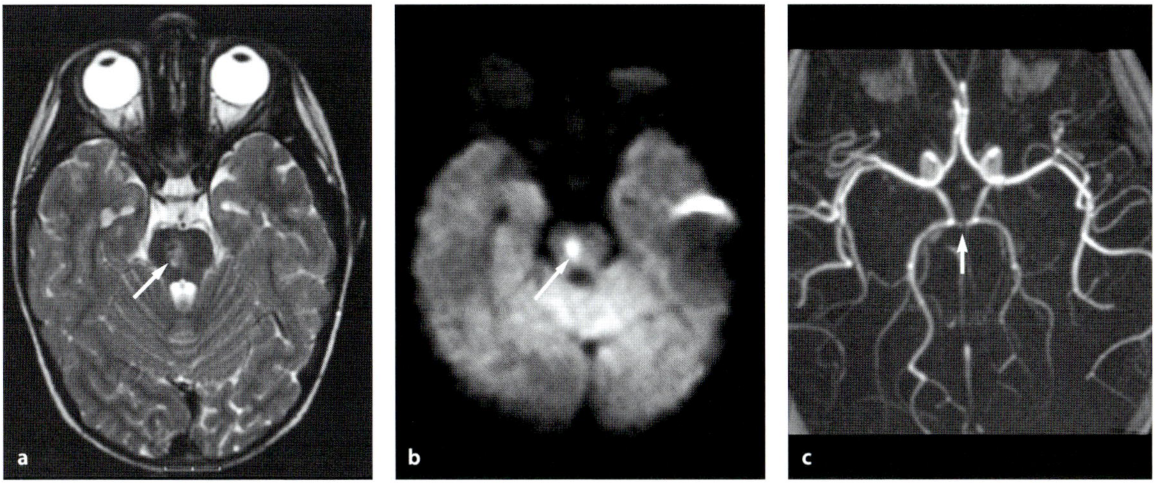

Figure 14.3 a–c ▲

Dissection of the vertebrobasilar arteries and infarction in a 4 year-old girl. **a** T2-weighted image shows high signal lesion in the right medial side of the pons (*arrow*). **b** DW image shows this lesion as hyperintense with decreased ADC (not shown), representing an acute infarct (*arrow*). **c** On MR angiography the vertebrobasilar arteries are absent due to dissection of the vertebrobasilar arteries (*arrow*). The posterior cerebral arteries are supplied from the anterior circulation

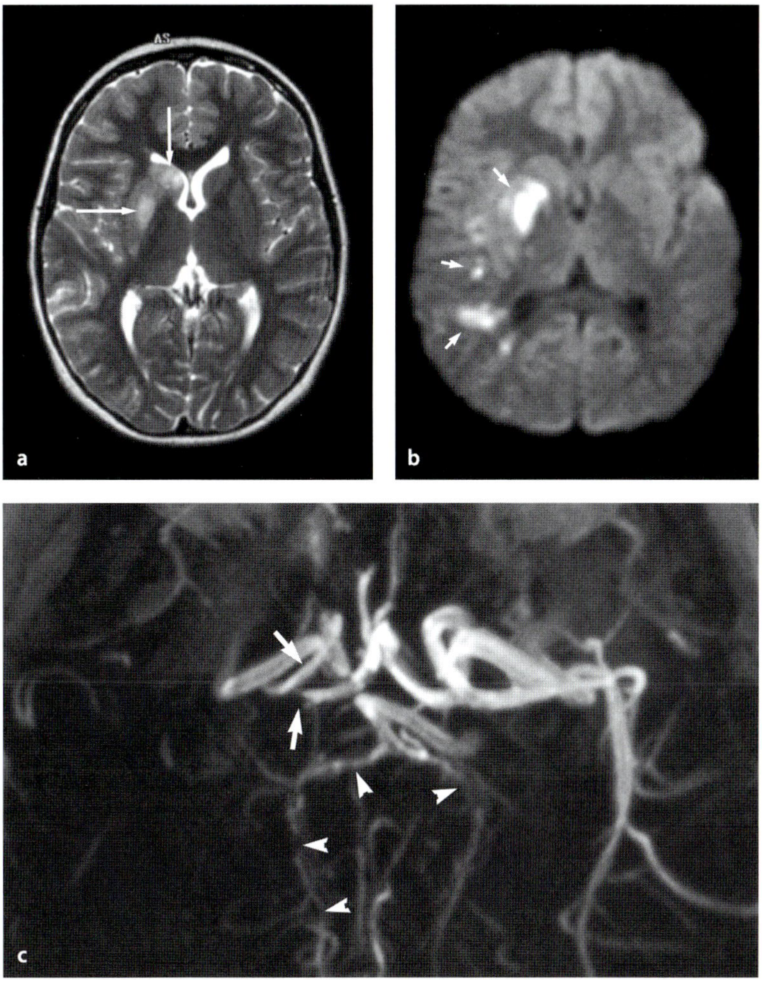

Figure 14.4 a–c

"Probable" Moyamoya disease in a 7-year-old girl. **a** T2-weighted image shows mild high signal lesions in the right basal ganglia. **b** DW image shows ischemic lesions not only in the basal ganglia but also in the parieto-occipital region as high signal, representing acute infarcts (*arrows*). **c** MR angiography shows occlusion of the right middle cerebral artery and stenosis of the right internal carotid artery (*arrows*) and bilateral posterior cerebral arteries (*arrowheads*).

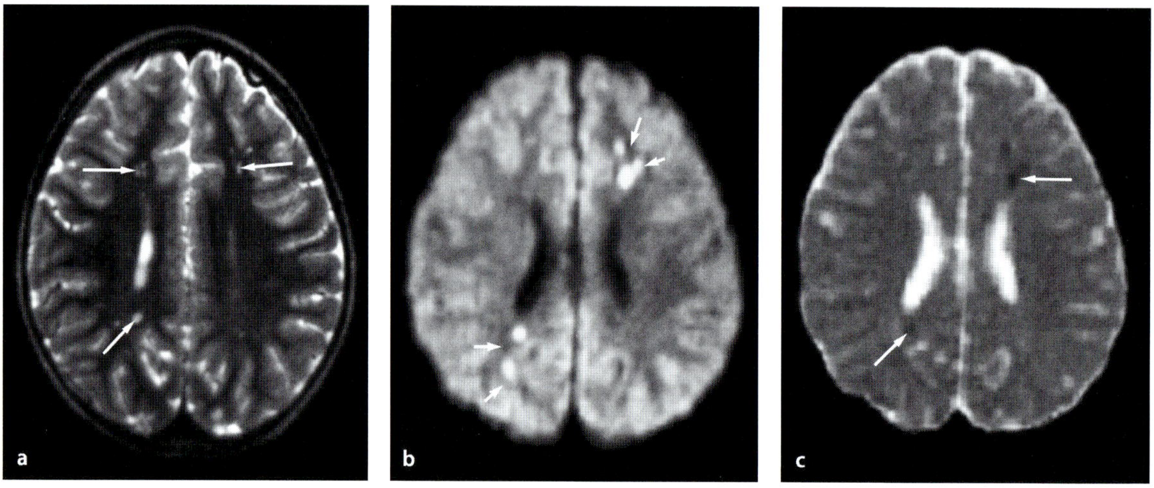

Figure 14.5 a–c

Sickle cell disease in a 4-year-old boy presenting with severe headache. **a** T2-weighted image shows multiple high signal spots in the white matter (*arrows*). **b** DW image shows some of these spots as very high signal intensity, representing small acute infarcts (*arrows*). **c** ADC map shows decreased ADC of these lesions consistent with small ischemic lesions (*arrows*). (From [49]).

14.4.1 Moyamoya Disease

Moyamoya disease is a chronic cerebrovascular occlusive disease of unknown origin that occurs predominantly in East Asia. In children it is characterized by progressive arterial stenosis with cerebral infarctions. The stenosis involves primarily the circle of Willis and the supraclinoid portion of the internal carotid arteries. Typically the internal carotid arteries are occluded bilaterally. In so-called "probable" Moyamoya disease, there is unilateral occlusion of one of the carotid arteries in its supraclinoid portion. DW imaging is useful for early detection of cerebral ischemia in this disease (Fig. 14.4) [6].

14.4.2 Sickle Cell Disease

About 5–8% of patients with sickle cell disease develop symptomatic cerebrovascular disease [7]. The risk of stroke is greatest during thrombotic crises and during the first 15 years of life. Stenosis or occlusion of both large and small vessels can cause cerebral infarction. Sickle cell disease results in vasculopathy, which in many respects is similar to Moyamoya disease. Cortical and white matter watershed ischemia is common; however, patients with sickle cell disease often also demonstrate multiple ischemic white matter lesions. These lesions can occur in spite of normal MR angiography and conventional angiography (Fig. 14.5). They are thought to be due to small vessel ischemic gliosis, similar to what is seen in small vessel disease of older patients.

14.4.3 Hypoxic Ischemic Encephalopathy

Hypoxic ischemic encephalopathy is the result of decreased global perfusion or oxygenation. It is generally due to neonatal anoxia, hypoglycemia, suffocation, cardiac arrest or child abuse. Whether produced by ischemia, anoxia or hypoglycemia, it is remarkably similar to infarcted brain in its appearance.

Diffusion-weighted imaging often depicts acute or subacute ischemic lesions when MR imaging and CT scans are normal or show only subtle abnormalities (Fig. 14.6) [8–13]. Diffuse hyperintensity on DW imaging with decreased ADC in the corpus callosum, and along the pyramidal tract in the internal capsules and the brain stem is occasionally seen. This is pre-

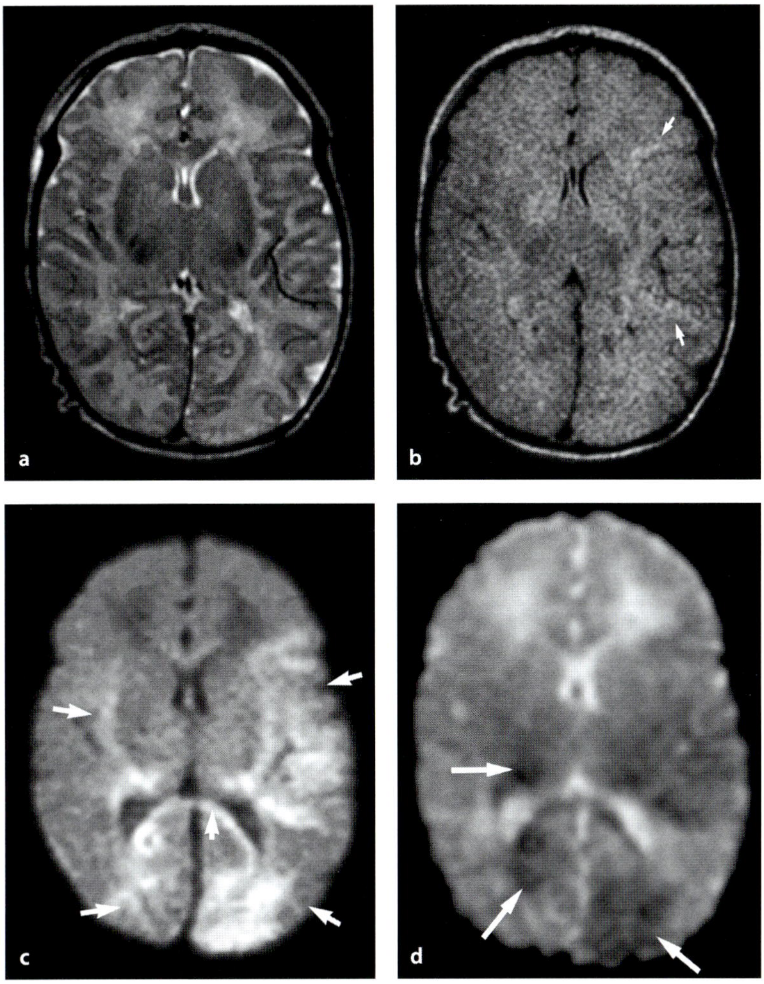

Figure 14.6 a–d

Hypoxic ischemic encephalopathy secondary to intrauterine cerebrovascular accident in a 2-day-old term girl. **a** T2-weighted image appears normal. **b** Fluid-attenuated inversion-recovery (FLAIR) image shows partially slightly high signal in the white matter (*arrows*). **c** DW image shows bilateral hyperintense lesions in the temporo-occipital cortices, white matter and corpus callosum (*arrows*), representing ischemic lesion and cytotoxic edema. **d** ADC map shows corresponding decreased ADC values

sumably due to cytotoxic edema of the glial cells, axons and myelin sheaths [14]. The prognosis of hypoxic ischemic encephalopathy depends on the extension of the cytotoxic edema, which is seen as hyperintensity on DW imaging. Hypoxic ischemic injury with cytotoxic edema is usually irreversible. DW imaging is helpful in establishing both the diagnosis and the prognosis, but also in the management of hypoxic ischemic encephalopathy.

14.5 Trauma

14.5.1 Battered Child Syndrome

In the USA, there are an estimated 3,000 deaths per year from non-accidental injuries. The pathogenesis of these brain parenchyma injuries is unknown, but unmyelinated white matter may be more vulnerable to shearing stress [15]. Young infants have a relatively large head, weak neck muscles and a thin skull, making them extremely vulnerable to traumatic injuries. In experimental studies of acute subdural hematomas in the infant rat, the glutamate concentration in the extracellular fluid of cortex was increased more than seven times over the basal level [16]. This suggests that the primary increased release of glutamate from the pre-synaptic terminal following traumatic stimuli and the primary decreased re-uptake of glutamate from the synapse following hypoxic or ischemic events are related to brain parenchymal injuries.

Histologic similarities have been observed in child abuse victims and infants with hypoxic ischemic encephalopathy. However, a history of apnea suggesting hypoxic–ischemic injury was only found in 57% of the child abuse cases. In a neuropathology study it was noted that diffuse axonal injuries were rare among child abuse victims, only seen in three out of 53 cases [17].

The distribution of widespread parenchymal injuries is often not related to the vascular territories or the location and size of acute subdural hematomas on CT and MRI (Fig. 14.7). However, it can be difficult to detect brain parenchyma injuries on CT as well as on routine MRI.

Diffusion-weighted imaging has a significant role in recognizing the extent of brain parenchymal injury. The parenchymal lesions can be unexpectedly extensive and caution is needed to window DW imaging optimally (Fig. 14.8). Quantifying the ADC value is especially useful to detect extensive parenchyma abnormalities.

The severity of DW imaging abnormality correlates with the patient's outcome [18]. However, MR spectroscopy may be better in evaluating the severity of trauma, as this will show decreased *N*-acetyl aspartate (NAA) (decreased neuronal activity), increased lactate (metabolic acidosis) and increased glutamate/glutamine (increased extracellular glutamate); the degree of these changes seems to be related to the severity of brain damage (Fig. 14.7) [19]. This grading may become important in the future since neuroprotective effects have been reported with several kinds of selective glutamate receptor antagonists in animal studies [20–22].

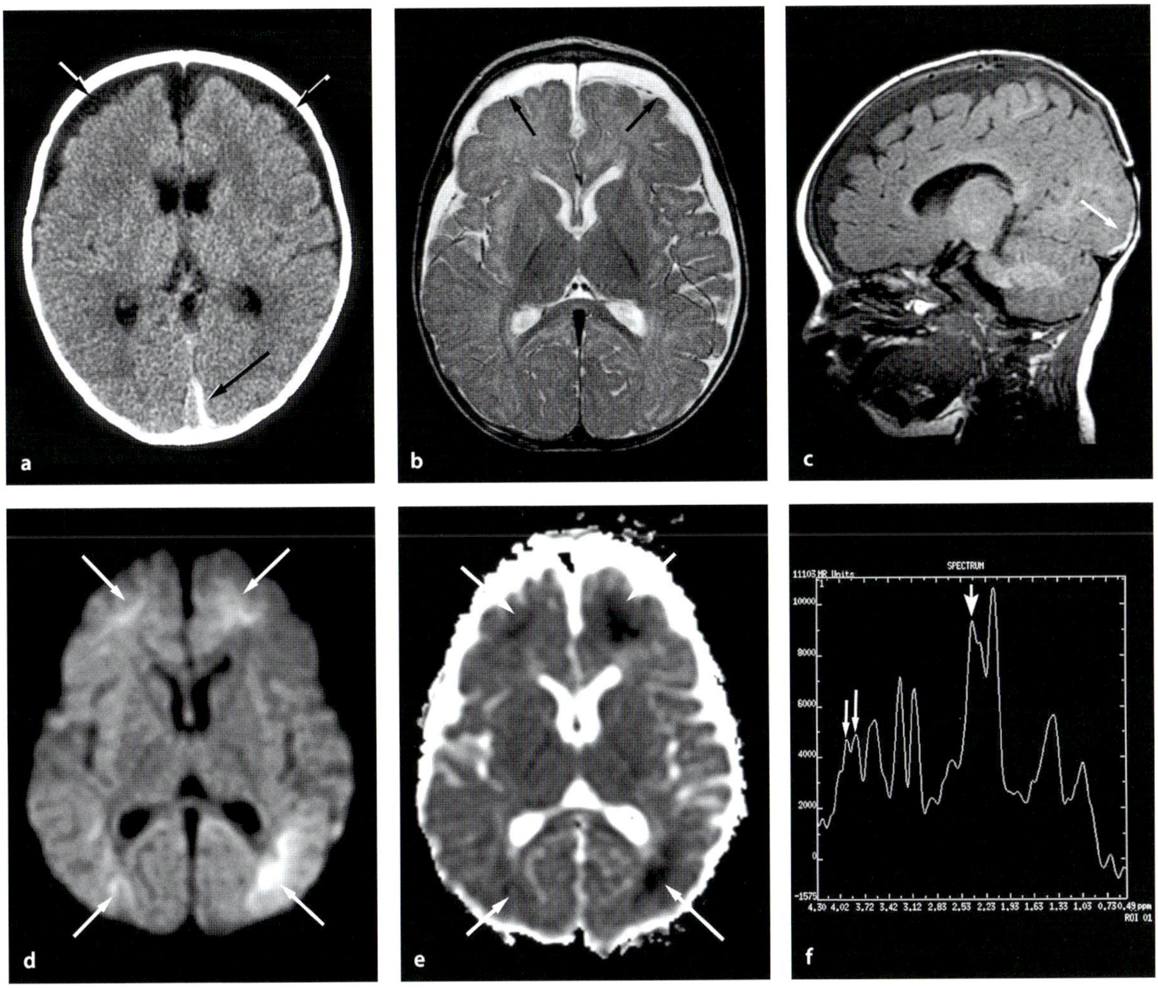

Figure 14.7 a–f

Battered child syndrome in a 6-month-old boy. **a** CT shows high density area representing acute subdural hematoma in the left occipital region (*long arrow*), and bilateral subdural fluid collections in the frontal region (*short arrows*). **b** T2-weighted image also shows bilateral subdural fluid collection and no apparent abnormalities in the brain parenchyma. **c** Sagittal T1-weighted image shows acute subdural hematoma as small linear hyperintensity in the occipital area (*arrow*). **d**, **e** DW image shows the extent of parenchymal abnormality as hyperintense lesions with decreased ADC in bilateral fronto-parieto-occipital white matter (*arrows*). This distribution is related to the hypoxic–ischemic encephalopathy rather than the subdural hematoma. **f** MR spectroscopy (TE 30 ms) shows an increased glutamate/glutamine peak (*arrow*) that may represent increased glutamate release or decreased glutamate re-uptake

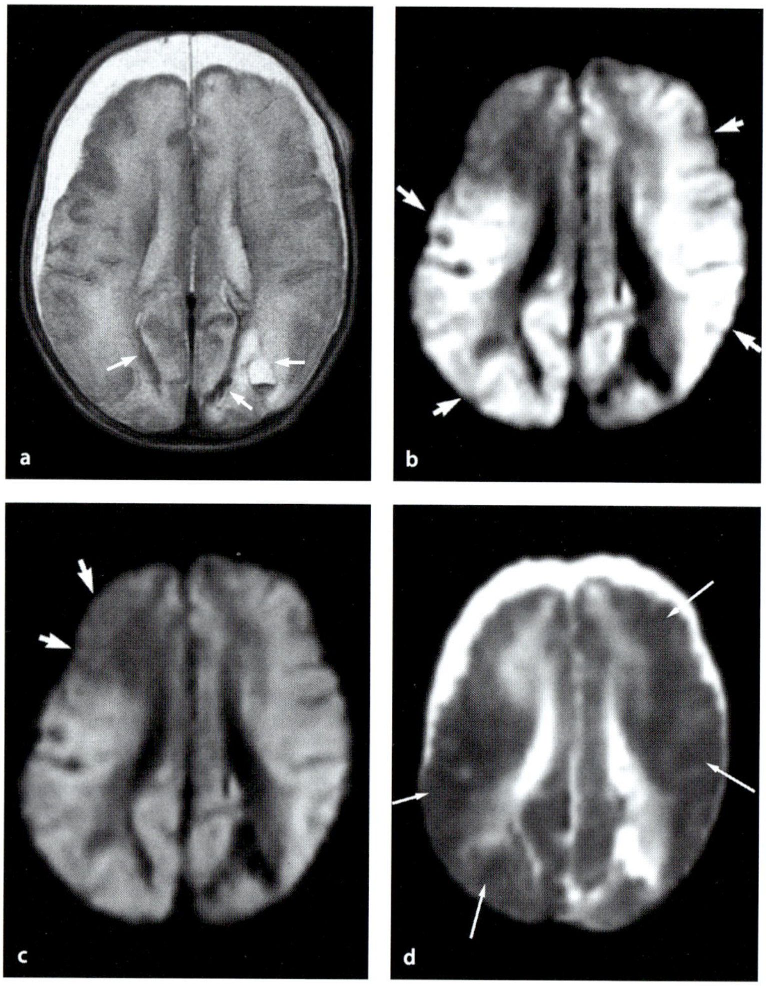

Figure 14.8 a–d

Battered child syndrome in a 2-month-old boy. **a** T2-weighted image shows intracranial hemorrhages with shearing injury (*arrows*) and bilateral chronic subdural hematomas in the bilateral frontal region. **b** DW image shows diffusely increased signal in both hemispheres (*arrows*) with sparing of only the right frontal area. **c** DW image filmed with incorrect window and level setting, suggesting wrongly that the low signal in the right frontal area is abnormal (*arrows*), when indeed this is the only normal portion of the brain. Correct window and levels are critical, as is comparison with findings on other sequences. **d** ADC values are decreased ($0.31 \times 10^{-3}/mm^2$ per s) in the diffuse parenchymal abnormalities (*arrows*)

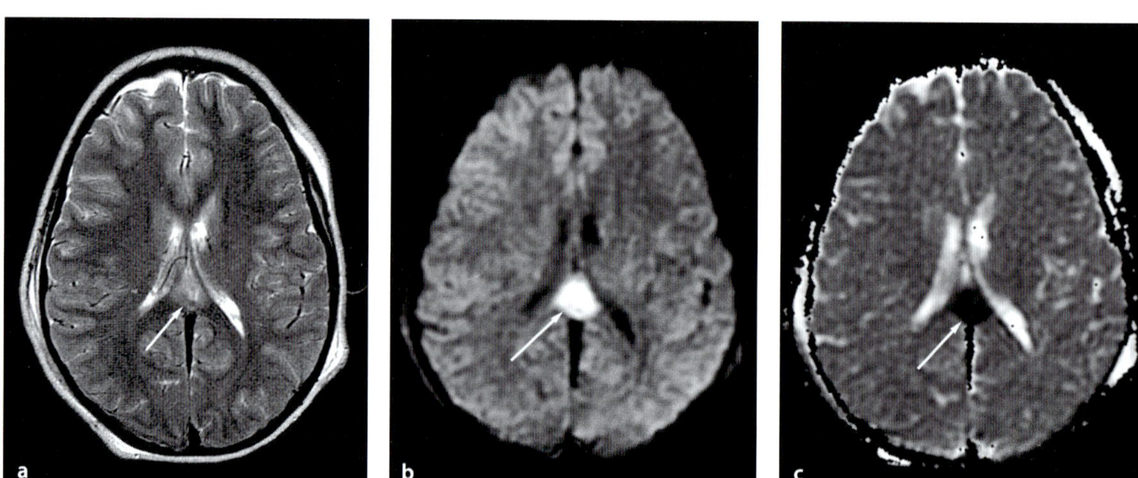

14.5.2 Diffuse Axonal Injury and Brain Contusion

It was once considered that edema following brain contusion or diffuse axonal injury was vasogenic. Recent experimental studies using DW imaging have shown that edematous regions following injury consist of both vasogenic and cytotoxic edema [23, 24]. DW imaging shows diffuse axonal injury as hyperintense, presumably due to cytotoxic edema (Fig. 14.9). It should be noted, however, that hemorrhagic components often accompany these brain injuries, which will affect the signal intensity on DW imaging. Brain contusions near the skull base are also often overlooked on DW imaging due to susceptibility artifacts.

14.6 Encephalopathies

14.6.1 Mitochondrial Encephalopathy

Mitochondrial encephalopathies show various patterns of central nervous system involvement. These include multiple infarcts in the cortex, white matter or basal ganglia. Infarcts are related to the type of mitochondrial encephalopathy and do not usually follow vascular territories. Other features of mitochondrial encephalopathy are spongy encephalopathy or demyelination with intramyelinic edema [25] and atrophy. DW imaging is useful for detecting the active ischemic changes (Fig. 14.10) [26].

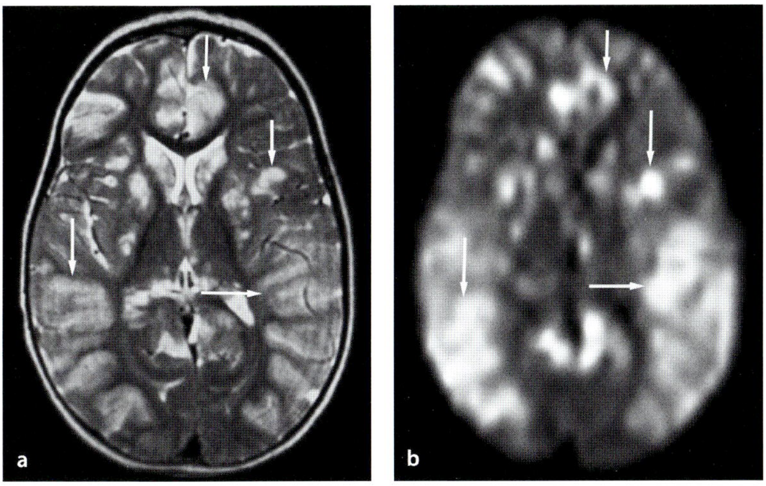

Figure 14.10 a,b

Mitochondrial encephalomyopathy in a 20-month-old girl with seizures. **a** T2-weighted image shows multiple high signal lesions in bilateral basal ganglia and fronto-parieto-occipital cortex not corresponding to a vascular territory (*arrows*). **b**, **c** DW image shows multiple high signal intensity lesions associated with decreased ADC (*arrows*)

Figure 14.9 a–c

Diffuse axonal injury in an 11-year-old boy injured in a motor vehicle accident. **a** T2-weighted image shows a mildly hyperintense lesion in the corpus callosum (*arrow*). **b** DW image demonstrates this lesion as hyperintense (*arrow*). **c** ADC map shows decreased ADC of this lesion (*arrow*), probably representing cytotoxic edema associated with diffuse axonal injury

◀

14.6.2 Acute Necrotizing Encephalopathy

Acute necrotizing encephalopathy is a clinicopathological entity recently separated from acute encephalopathy of unknown etiology. It frequently occurs in East Asia. The hallmark of acute necrotizing encephalopathy is multiple, bilateral symmetric brain lesions showing necrosis, petechiae and cytotoxic edema without inflammatory cell infiltration [27]. These lesions are mainly seen in the thalamus and tegmentum (Fig. 14.11). The prognosis is generally poor.

14.6.3 Hypertensive Encephalopathy

Hypertensive encephalopathy occurs most often secondary to renal diseases in children. It also occurs in children treated for myeloproliferative disorders [28]. In children, convulsions are often accompanied by severe headache and restlessness. The most common abnormality on MR imaging is bilateral high signal intensity in parieto-occipital subcortical white matter. These lesions can also occur in the frontal lobes and gray matter, including basal ganglia, thalamus, cerebellum and brain stem. The mechanism of the disease is thought to be vasogenic edema from failure of autoregulation and/or a cytotoxic edema triggered by severe vasospasm. DW imaging can distinguish irreversible ischemic changes from reversible conditions with vasogenic edema (Fig. 14.12) [29].

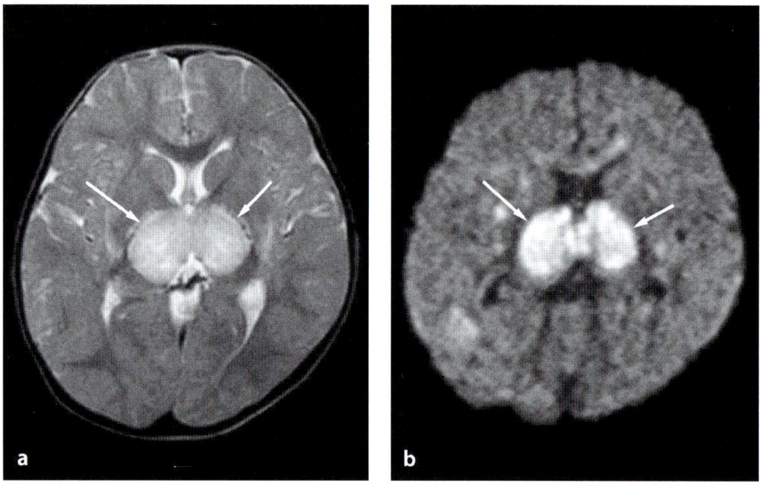

Figure 14.11 a,b

Acute necrotizing encephalopathy in a 1-year-old boy with seizure. **a** T2-weighted image shows multiple hyperintense lesions in bilateral thalami and right temporo-occipital region (*arrows*). **b** DW image also shows these lesions (*arrows*) as hyperintense associated with decreased ADC (not shown). (Courtesy of Ida M, M.D. Ekara Municipal Hospital, Japan)

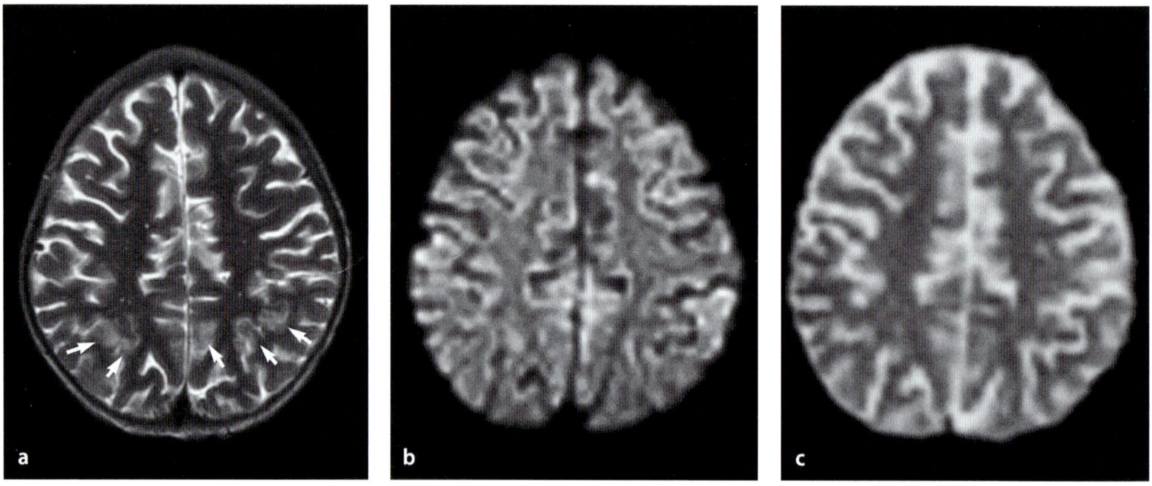

14.7 Infections

14.7.1 Encephalitis

Diffusion-weighted imaging can detect early encephalitic changes [30] and is generally more sensitive than conventional MR imaging. Herpes encephalitis demonstrates pathologically severe edema including both cytotoxic and vasogenic edema and massive tissue necrosis with petechial or confluent hemorrhage. Herpes simplex type 1 encephalitis in older children and adults usually involves the medial temporal lobe, inferior frontal lobes and insula (Fig. 14.13). Neonatal herpes simplex type 2 encephalitis involves the cortex and white matter extensively (Fig. 14.14). The early detection by DW imaging is valuable for early institution of treatment

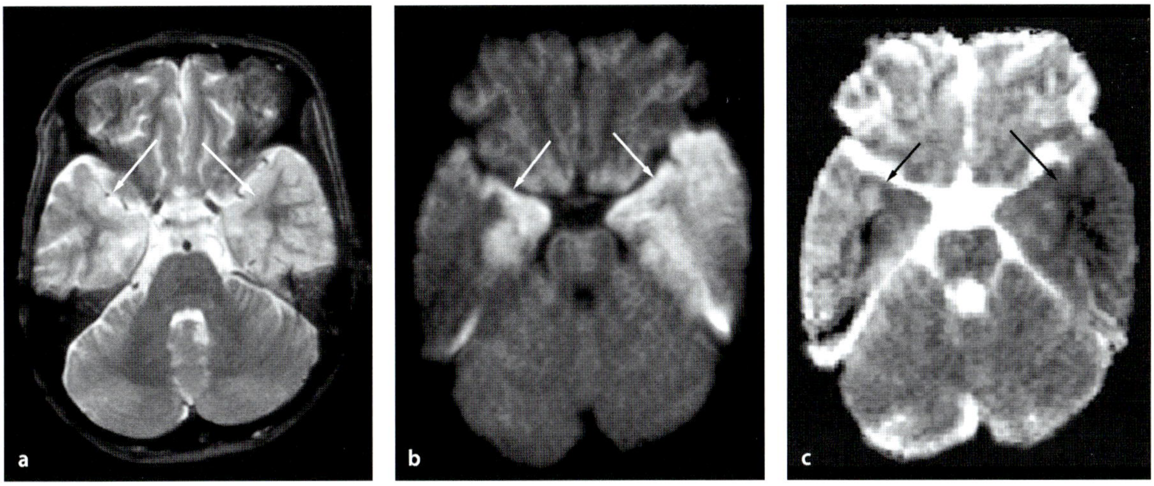

Figure 14.13 a–c

Herpes simplex type 1 encephalitis in an 11-year-old boy. **a** T2-weighted image shows hyperintense lesions in bilateral temporal lobes (*arrows*). **b** DW image clearly shows these lesions as hyperintense (*arrows*). **c** ADC map shows decreased ADC of these lesions (*arrows*)

Figure 14.12 a–c

Hypertensive encephalopathy in a 5-year-old girl with acute lymphoblastic leukemia during chemotherapy presenting with seizure. **a** T2-weighted image shows hyperintense lesions in bilateral parieto-occipital cortex (*arrows*). This high signal seems to be due to subtle vasogenic edema rather than frank ischemia. **b, c** DW image and ADC map show no apparent abnormal signal intensities

◄

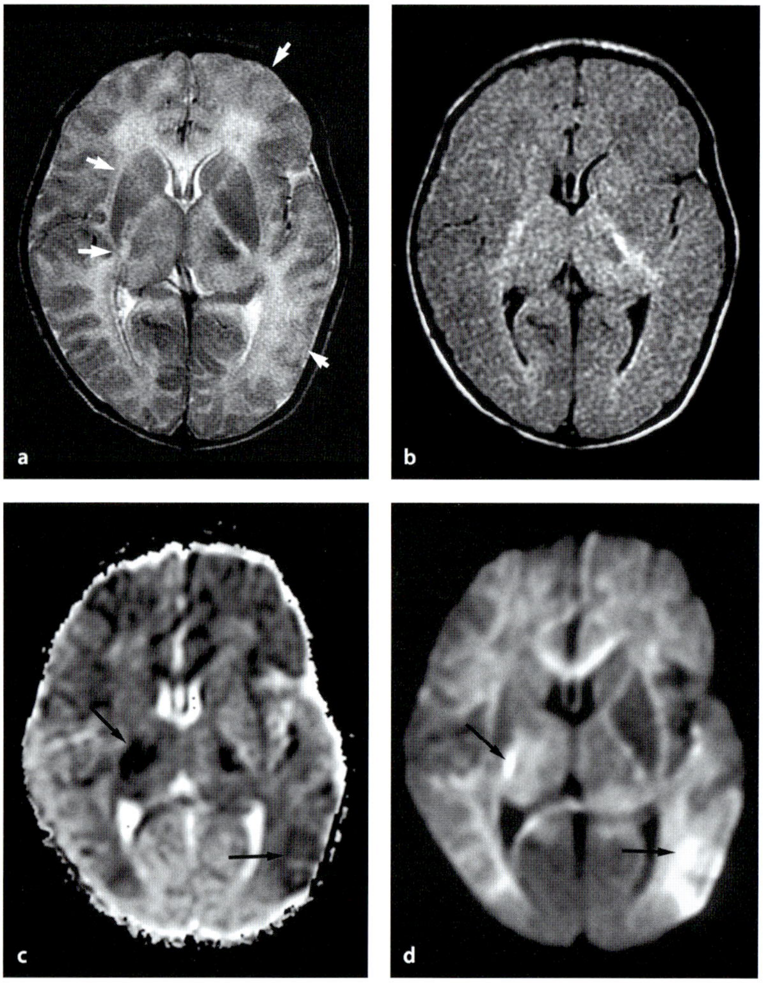

Figure 14.14 a–d

Herpes simplex type 2 encephalitis in a 2-week-old girl. **a** T2-weighted images show asymmetric hyperintense lesions in the thalami and right basal ganglia and cerebral cortices (*arrows*). The precise extent of the lesions is difficult to determine. **b** FLAIR image appears normal. **c, d** DW image shows asymmetric but extensive hyperintense lesions (*arrows*) with decreased ADC in the thalamus and gray and white matter of both hemispheres

14.7.2 Brain Abscess

Abscesses in the brain are potentially fatal, but may be successfully treated by early medical or surgical intervention. DW imaging can discriminate a brain abscess from a cystic or necrotic tumor, which is often difficult with conventional MR imaging [31]. The brain abscess shows very high signal on DW imaging associated with decreased ADC (Fig. 14.15). Pus usually consists of both dead and still viable neutrophils, along with necrotic tissue and bacteria, as well as exuded plasma. A possible explanation for the high signal on DW imaging is limited water mobility, presumably due to the high viscosity of continuous coagulative necrosis and hypercellularity of polynucleated neutrophils in the pus.

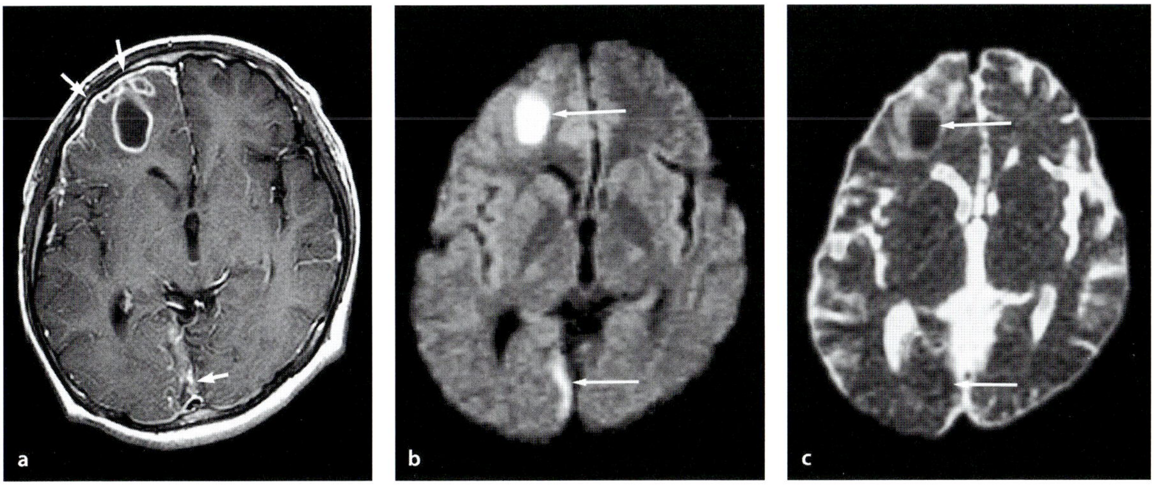

Figure 14.15 a–c

Brain abscess and subdural abscess in a 16-year-old male with high fever and headache. **a** Gadolinium-enhanced T1-weighted image shows thin-walled ring or rim-enhancing lesions with pachymeningeal enhancement in the right frontal lobe and right medial occipital region (*arrows*). **b** DW image shows cystic component of the right frontal and occipital lesion as very hyperintense (*arrows*). **c** ADC map shows decreased ADC in these cystic components (*arrows*). (Courtesy of Morikawa M, M.D., Nagasaki University, School of Medicine, Japan)

14.8 Brain Tumor

The signal intensity on DW imaging and the ADC values of brain tumors are variable and related to the architecture of the tumor. Malignant brain tumors, such as medulloblastoma (Fig. 14.16), primitive neuroectodermal tumor, glioma and lymphoma [32–35], often show high signal intensity on DW imaging associated with decreased ADC. The decreased ADC value is caused by increased intracellular water, hypercellularity and/or decreased extracellular water in tumor interstitium. Other brain tumors, either benign or malignant, show hyperintensity associated with increased ADC, indicating a T2 shine-through effect (Fig. 14.17).

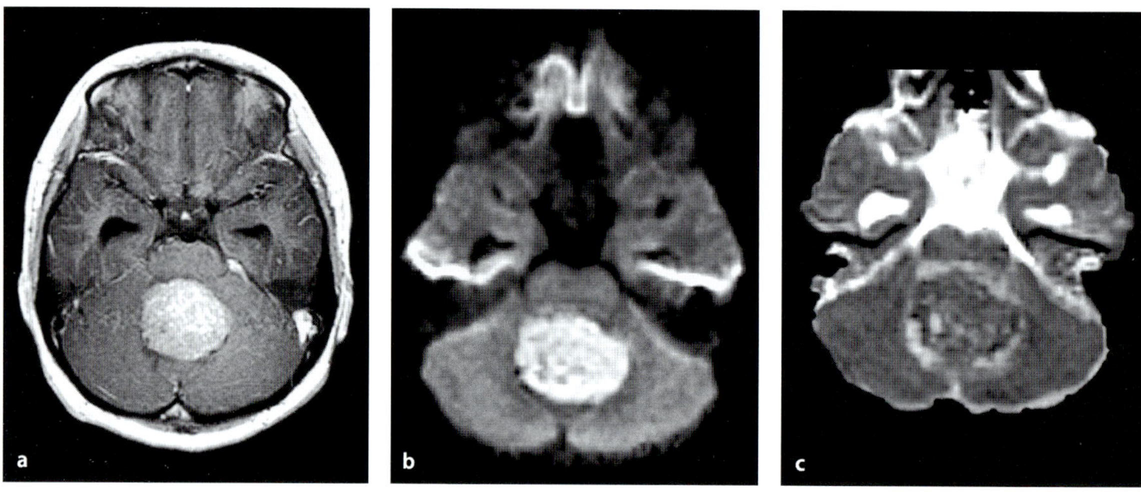

Figure 14.16 a–c

Medulloblastoma in a 10-year-old boy. **a** Gadolinium-enhanced T1-weighted image shows a solid mass with enhancement in the cerebellar vermis. **b** DW image shows this solid mass as hyperintense. **c** ADC map shows decreased ADC of this mass. This is due to high cellular density causing restricted diffusion. (Courtesy of Morikawa M, M.D., Nagasaki University, School of Medicine, Japan)

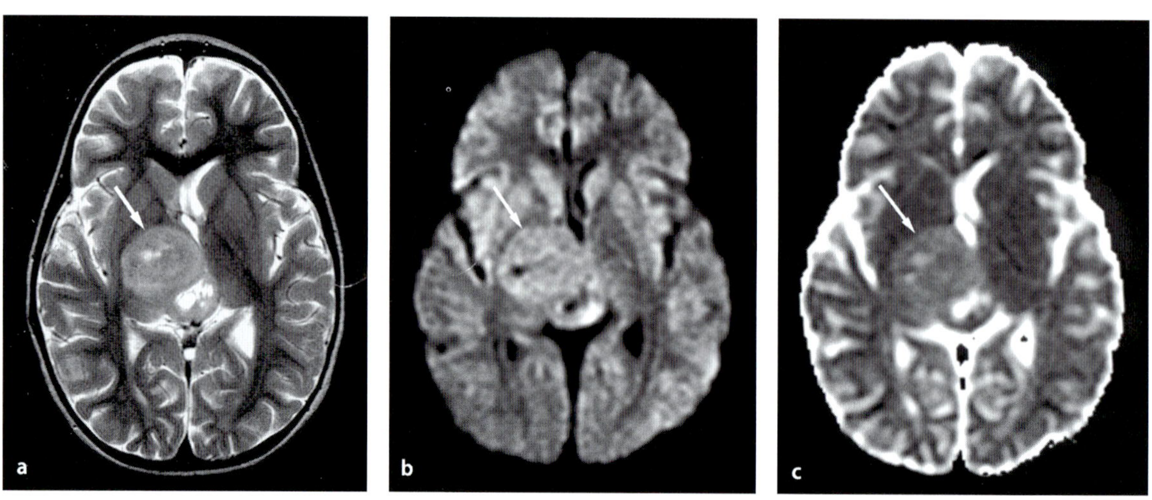

14.9 Dysmyelination and Demyelination

Diffusion-weighted imaging is useful in differentiating between white matter diseases in pediatric patients.

14.9.1 Pelizaeus–Merzbacher Disease

Pelizaeus–Merzbacher disease, one of the typical dysmyelination disorders, pathologically shows hypo- or amyelination and spares the axon. Despite hypo- or amyelination in Pelizaeus–Merzbacher disease, DW imaging shows normal diffusional anisotropy in the white matter (Fig. 14.18) [36]. This finding suggests that anisotropy is primarily related to structural changes of the axonal membrane in the immature brain.

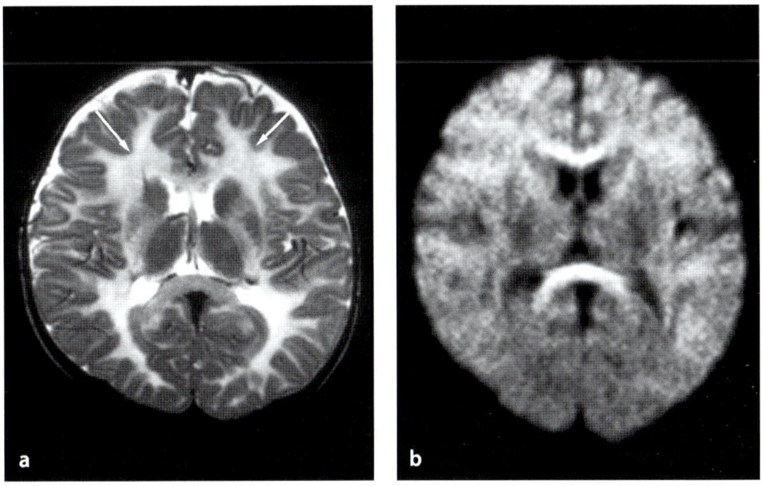

Figure 14.18 a,b

Pelizaeus–Merzbacher disease in a 7-year-old boy with developmental delay. **a** T2-weighted image shows diffuse white matter hyperintensity extending into the U-fibers due to hypo- or amyelination (*arrows*). **b** DW image (*z axis*) demonstrates isointensity in the white matter and essentially normal anisotropy in the corpus callosum. ADC values are diffusely slightly increased in the white matter (not shown) (From [46]).

Figure 14.17 a–c

Pilocytic astrocytoma in a 4-year-old girl. **a** T2-weighted images show a hyperintense mass lesion in the right thalamus (*arrow*). **b** DW image shows this solid mass as hyperintense (*arrow*). **c** ADC map shows slightly increased ADC of this mass. Hyperintensity on DW imaging is due to T2 signal effect, which is called T2 shine-through effect

◀

14.9.2 Vanishing White Matter Disease

Leukoencephalopathy with vanishing white matter is a newly recognized dysmyelinating disease of unknown etiology. In the late stage of this disease there is demyelination or hypomyelination and extensive cystic degeneration of the white matter associated with reactive changes. DW imaging shows low signal intensity, presumably representing cystic degeneration (Fig. 14.19).

14.9.3 Metabolic or Toxic Leukoencephalopathies

Some metabolic or toxic leukoencephalopathies such as phenylketonuria (Fig. 14.20), adrenoleukodystrophy, metachromatic leukodystrophy, L-2 hydroxyglutaric acidurig (Chap. 10), infantile neuronaldystrophy and Canavan's disease (Fig. 14.21) can show hyperintense lesions associated with decreased ADC on DW imaging [37–42]. One possible explanation for the hyperintense lesion is intramyelinic edema, which is one form of cytotoxic edema selectively occurring in myelin sheaths.

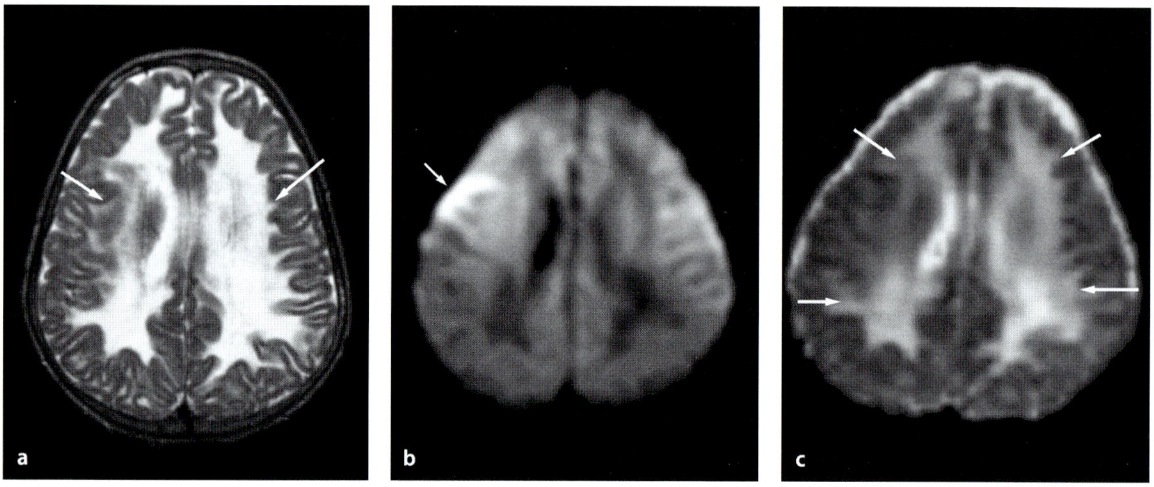

Figure 14.19 a–c

Vanishing white matter disease in an 8-year-old boy with progressive leukoencephalopathy. **a** T2-weighted image shows diffuse white matter hyperintensity extending into the U-fibers (*arrows*). **b** DW image demonstrates diffuse low signal intensity in the white matter. High signal intensity along right frontal lobes is magnetic susceptibility artifact due to the field heterogeneity (*arrow*). **c** ADC map shows diffusely increased ADC in the white matter, presumably due to cystic changes (*arrows*)

14.9.4 Multiple Sclerosis

Multiple sclerosis occasionally occurs in pediatric patients. Hyperintense plaques on DW imaging with decreased ADC are rare in multiple sclerosis (Fig. 14.21) [39]. When present, a possible explanation for the hyperintense, cytotoxic multiple sclerosis plaques may be intramyelinic edema.

14.9.5 Osmotic Myelinolysis

Osmotic myelinolysis also occurs in pediatric patients and can result in hyperintense lesions on DW imaging associated with decreased ADC (Fig. 14.22) [40].

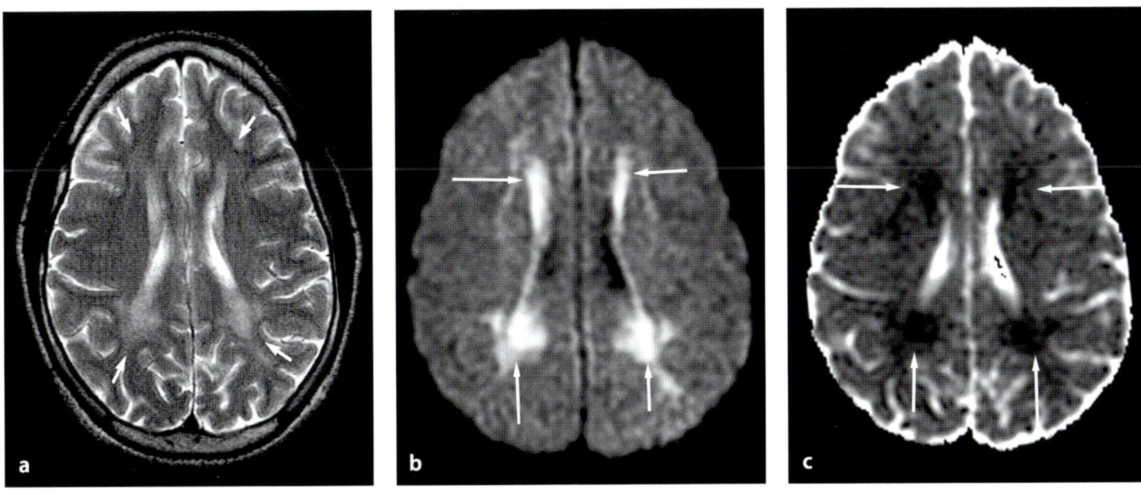

Figure 14.20 a–c

Phenylketonuria in a 36-year-old male **a** T2-weighted image shows hyperintense lesions in the periventricular white matter (*arrows*). **b** DW image shows these lesions as hyperintense (*arrows*). **c** These hyperintense lesions have decreased ADC representing cytotoxic edema, especially intramyelinic edema (*arrows*)

Figure 14.21 a,b

Canavan disease. 15-month-old boy presented with delayed psychomotor development seizures, and spasticity. **a** T2-weighted image reveals high signal in peripheral white matter, globi pallidi and thalami. Note diffuse atrophy and thinning of the cortex. **b** DW image shows high signal changes in the peripheral white matter and globi pallidi with mildly decreased ADC partially (not shown). (From Prof. Dr R. Nuri Sener, Department of Radiology, Ege University Hospital, Turkey, Journal of Computer Assisted Tomography 27, 2003, [42] with permission)

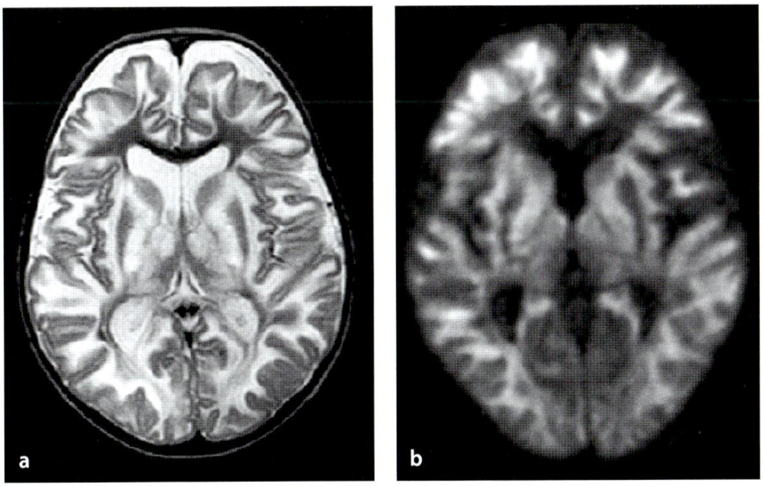

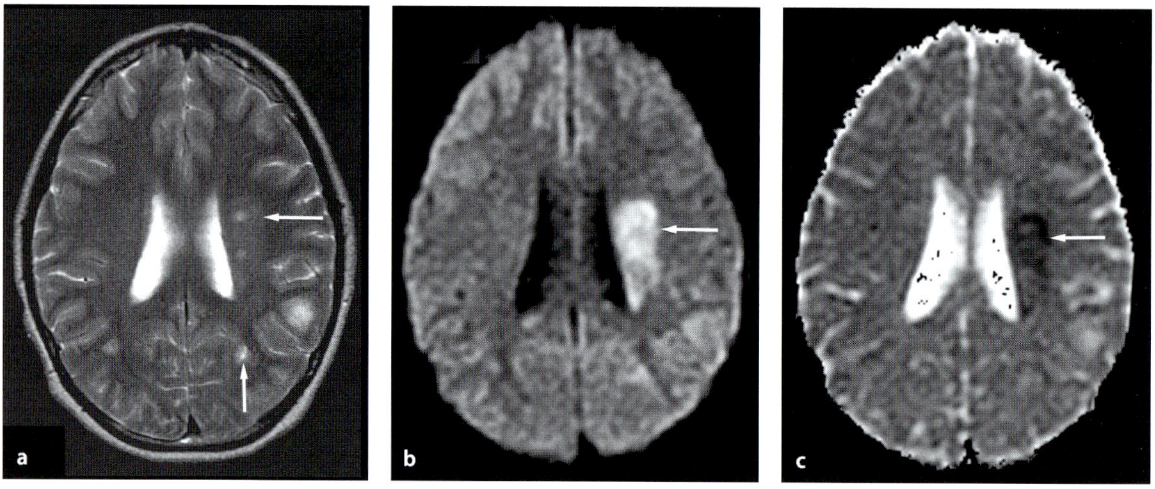

Figure 14.22 a–c

Acute multiple sclerosis in a 13-year-old female. **a** T2-weighted image shows multiple hyperintense lesions in the periventricular and subcortical white matter (*arrows*). **b, c** DW image shows a hyperintense plaque with decreased ADC (*arrows*) in the left corona radiata. This is a cytotoxic plaque, presumably mainly composed of intramyelinic edema

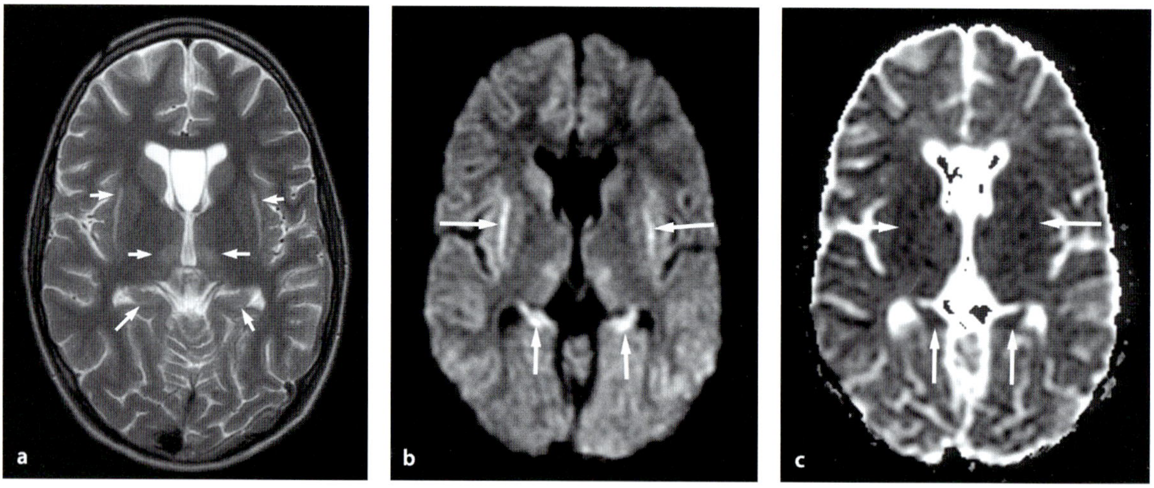

Figure 14.23 a–c

Osmotic myelinolysis (extrapontine myelinolysis) in an 11-year-old boy. **a** T2-weighted image shows symmetrical multiple slightly hyperintense lesions in the external capsules, hippocampi and medial thalami (*arrows*). **b, c** DW image shows these lesions (*arrows*) as hyperintense with decreased ADC, representing cytotoxic edema

14.10 Conclusion

Diffusion-weighted imaging plays an important role in the diagnosis of various pathological conditions in the pediatric brain, which has a high water content . DW imaging can also depict acute or subacute ischemic changes in children when conventional MR imaging or CT is normal or shows only subtle abnormalities. DW imaging is useful in differentiating white matter diseases and in differentiating between tumor and abscess. The calculation of ADC maps or fractional anisotropic images is quantitative and demonstrates the water content or anisotropy more precisely than DW imaging. However, recognition of imaging pitfalls is important for optimal interpretation of DW imaging.

References

1. Roley HA, Grant PE, Roberts TPL (1999) Diffusion MR imaging. Theory and applications. Neuroimaging Clin N Am 9:343–361
2. Neil JJ, Shiran SI, McKinstry RC, et al. (1998) Normal brain in human newborns: apparent diffusion coefficient and diffusion anisotropy measured by using diffusion tensor MR imaging. Radiology 209:57–66
3. Wimberger DM, Roberts TP, Barkobich AJ, et al. (1995) Identification of "premyelination" by diffusion-weighted MRI. JCAT 19:28–33
4. Tanner SF, Ramenghi LA, Ridgway JP, et al. (2000) Quantitative comparison of intrabrain diffusion in adults and preterm and term neonates and infants. AJR Am J Roentgenol 174:1643–1649
5. Morris MC, Zimmerman RA, Bilaniuk LT, et al. (1999) Changes in brain water diffusion during childhood. Neuroradiology 41:929–934
6. Yamada I, Himeno Y, Nagaoka T, et al. (1999) Moyamoya disease: evaluation with diffusion-weighted and perfusion echo-planar MR imaging. Radiology 212:340–347
7. Moran CJ, Siegel MJ, DeBaun MR (1998) Sickle cell disease: imaging of cerebrovascular complications. Radiology 206:311–321
8. Phillips MD, Zimmerman RA (1999) Diffusion imaging in pediatric hypoxic ischemia injury. Neuroimaging Clin N Am 9:41–52
9. Robertson RL, Ben-Sira L, Barnes PD, et al. (1999) MR line-scan diffusion-weighted imaging of term neonates with perinatal brain ischemia. AJNR Am J Neuroradiol 20:1658–1670
10. Johnson AJ, Lee BC, Lin W (1999) Echoplanar diffusion-weighted imaging in neonates and infants with suspected hypoxic-ischemic injury: correlation with patient outcome. AJR Am J Roentgenol 172:219–226
11. Inder T, Huppi PS, Zientara GP, et al. (1999) Early detection of periventricular leukomalacia by diffusion-weighted magnetic resonance imaging techniques. J Pediatr 134:631–634
12. Forbes KP, Pipe JG, Bird R (2000) Neonatal hypoxic-ischemic encephalopathy: detection with diffusion-weighted MR imaging. AJNR Am J Neuroradiol 21:1490–6
13. Wolf RL, Zimmerman RA, Clancy R, Haselgrove JH (2001) Quantitative apparent diffusion coefficient measurements in term neonates for early detection of hypoxic-ischemic brain injury: initial experience. Radiology 218:825–33
14. Rumpel H, Nedelcu J, Aguzzi A, et al. (1997) Late glial swelling after acute cerebral hypoxic-ischemia in the neonatal rat: a combined magnetic resonance and histochemical study. Pediatr Res 42:54–59
15. Chen CY, Zimmerman RA, Rorke LB (1999) Neoroimaging in child abuse: a mechanism-based approach. Neuroradiology 41:711–722
16. Bullock R, Butcher SP, Chen MH, et al. (1991) Correlation of the extracellular glutamate concentration with extent of blood flow reduction after subdural hematoma in the rat. J Neurosurg 74:794–802
17. Geddes JF, Hackshaw AK, Vowles GH, et al. (2001) Neuropathology of inflicted head injury in children. Patterns of brain damage. Brain 124:1290–1298
18. Suh DY, Davis PC, Hopkins KL, et al. (2001) Nonaccidental pediatric head injury: diffusion-weighted imaging findings. Neurosurgery 49:309–320
19. Holshouser BA, Ashwal S, Luh G, et al. (1997) Proton MR spectroscopy after central nervous system injury: outcome prediction in neonates, infants, and children. Radiology 202:487–496
20. Duhaime AC, Gennarelli LM, Boardman C (1996) Neuroprotection by dextromethorphan in acute experimental subdural hematoma in the rat. J Neurotrauma 13:79–84
21. Ikonomidou C, Qin Y, Labruyere J, Kirby C, et al. (1996) Prevention of trauma-induced neurodegeneration in infant rat brain. Pediatr Reseach 39:1020–1027
22. Smith SL, Hall ED (1998) Tirilazad widens the therapeutic window for riluzole-induced attenuation of progressive cortical degeneration in an infant rat model of the shaken baby syndrome. J Neurotrauma 15:707–719
23. Barzo P, Marmarou A, Fatouros P, et al. (1997) Contribution of vasogenic and cellular edema to traumatic brain swelling measured by diffusion-weighted imaging. J Neurosurg 87:900–907
24. Liu AY, Maldjian JA, Bagley LJ, Sinson GP, et al. (1999) Traumatic brain injury: diffusion-weighted MR imaging findings. AJNR Am J Neuroradiol 20:1636–1641
25. Ellison D, Love S (1998) Toxic injury of the CNS. In: Neuropathology, 1st edn. London: Mosby 25:5
26. Yonemura K, Hasegawa Y, Kimura K, Minematsu K, Yamaguchi T (2001) Diffusion-weighted MR imaging in a case of mitochondrial myopathy, encephalopathy, lactic acidosis, and strokelike episodes. AJNR Am J Neuroradiol 22:269–72
27. Yagishita A, Nakano I, Ushioda T, Otsuki N, Hasegawa A (1995) Acute encephalopathy with bilateral thalamotegmental involvement in infants and children: imaging and pathology findings. AJNR Am J Neuroradiol 16:439–447

28. Cooney MJ, Bradly WG, Symko SC, et al. (2000) Hypertensive encephalopathy: complication in children treated for myeloproliferative disorders: report of three cases. Radiology 214:711–716
29. Schwartz RB, Mulkern RV, Gudbjartsson H, et al. (1998) Diffusion-weighted MR imaging in hypertensive encephalopathy: clues to pathogenesis. AJNR Am J Neuroradiol 19:859–862
30. Tsuchiya K, Katase S, Yoshino A, et al. (1999) Diffusion-weighted MR imaging of encephalitis. AJR Am J Roentgenol 173:1097–1099
31 Ebisu T, Tanaka C, Umeda M, et al. (1996) Discrimination of brain abscess from necrotic or cystic tumors by diffusion-weighted echo planar imaging. Magn Reson Imaging 14:1113–1116
32. Kotsenas AL, Roth TC, Manness WK, et al. (1999) Abnormal diffusion-weighted MRI in medulloblastoma: does it reflect small cell histology? Pediatr Radiol 29:524–526
33. Erdem E, Zimmerman RA, Haselgrove JC, Bilaniuk LT, Hunter JV (2001) Diffusion-weighted imaging and fluid attenuated inversion recovery imaging in the evaluation of primitive neuroectodermal tumors. Neuroradiology Nov; 43(11):927–933
34. Tien RD, Felsberg GJ, Friedman H, et al. (1993) MR imaging of high-grade cerebral gliomas: value of diffusion-weighted echoplanar pulse sequence. AJR Am J Roentgenol 162:671–677
35. Guo AC, Cummings TJ, Dash RC, Provenzale JM (2002) Lymphomas and high-grade astrocytomas: comparison of water diffusibility and histologic characteristics. Radiology 224:177–183
36. Ono J, Harada K, Sakurai K, et al. (1994) MR diffusion imaging in Pelizaeus-Merzbacher disease. Brain Dev 16:219–223
37. Ono J, Harada K, Mano T, et al. (1997) Differentiation of dys- and demyelination using diffusion anisotropy. Pediatr Neurol 16:63–66
38. Eichler FS, Itoh R, Barker PB, et al. (2002) Proton MR spectroscopic and diffusion tensor brain MR imaging in X-linked adrenoleukodystrophy; initial experience. Radiology 225: 245–252
39. Sener RN (2002) Metachromatic leukodystrophy: diffusion MR imaging findings. AJNR Am J Neuroradiol 23: 1424–1426
40. Sener RN (2003) L-2 hydroxyglutaric aciduria: proton magnetic resonance spectroscopy and diffusion magnetic resonance imagimg findings. J Comput Assis Tomogr 27: 38–43
41. Sener RN (2003) Diffusion magnetic resonance imaging in infantile neuroaxonal dystrophy. J Comput Assis Tomogr 27: 34–37
42. Sener RN (2003) Canavan disease: diffusion magnetic resonance imaging findings. J Comput Assist Tomogr 27:30–33
43. Rovira A, Pericot I, Alonso J, Rio J, et al. (2002) Serial diffusion-weighted MR imaging and proton MR spectroscopy of acute large demyelinating brain lesions: case report. AJNR Am J Neuroradiol Jun-Jul; 23(6):989–994
44. Cramer SC, Stegbauer KC, Schneider A, et al. (2001) Decreased diffusion in central pontine myelinolysis. AJNR Am J Neuroradiol Sep; 22(8):1476–1479
45. Moritani T, Numaguchi Y, Lerner NB, et al. (2003) Sickle Cell Cerebrovascular Disease: usual and unusual findings on MR Imaging and MR Angiography, Clin Imaging. In press.
46. Moritani T, Shrier DA, Numaguchi Y, et al. (2000) Diffusion-weighted echo-planar MR imaging: clinical applications and pitfalls – a pictorial essay. Clin Imaging. 24: 181–192

How to Use This Book

In collaboration with A. Hiwatashi

Most textbooks related to imaging are organized into chapters based on the disease or the conditions that are described. This traditional way of organizing an imaging book, namely from the disease to the images, provides a firm structure and allows the author or editor to present all the imaging characteristics of a specific disease condition with its typical, atypical and specific features in one place. This book is no different and is also organized in this traditional way.

For the clinician using the book as an aid to solve a clinical case, this traditional approach is not practical. To match the imaging characteristics of your specific clinical case you essentially have to go through the entire book to find all the images that match the imaging features of your specific patient. To overcome this problem of a disease-oriented imaging book, we developed this chapter: here we turned the organization around, proceeding from the images to the disease/diagnosis. In this chapter we have prepared the material based on the imaging characteristics and grouped all conditions with similar imaging features together in seven tables. We used DW imaging, ADC and T2 characteristics to create seven tables with conditions that appeared similar on MR imaging.

The clinician can go directly to this chapter, determine if the lesion in question has a high, intermediate or low DW imaging signal intensity and then determine the same with regard to the ADC and T2 characteristics. He or she can then go to the table in Chap 15 that lists conditions with these imaging features. Each table is essentially a list of differential diagnosis for conditions with similar imaging characteristics. When combined with the knowledge of patient symptomatology and demographic criteria, the radiologist will be able to narrow the differential diagnosis to a few conditions.

These tables take into account that the same condition may have variable imaging characteristics; for this reason, several conditions are listed in more than one table. Moreover, within the tables there are variations as to the degree of a specific imaging feature, which allows the clinician to match his/her clinical case to the best table and condition. Chapter 15 makes direct reference to other chapters of the book, where a full description is then provided.

Table 1

Differential diagnoses for lesions with a **high diffusion** signal associated with **low ADC** and **iso intense T2** signal

Diagnose	Reference images DWI high	 ADC low	 T2WI iso
Infarction/hypoxia			
Battered child	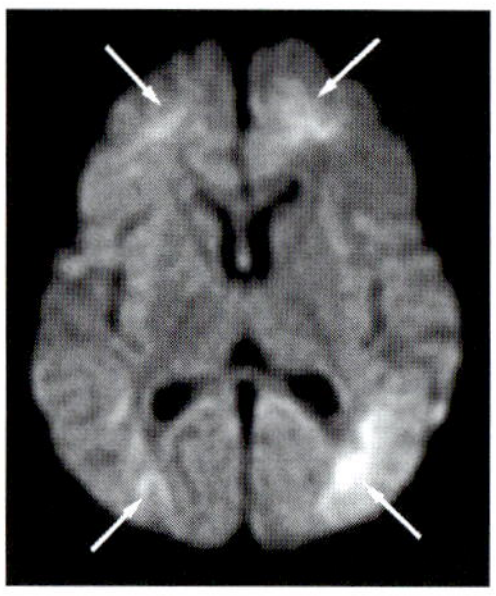 Fig. 14.7 d	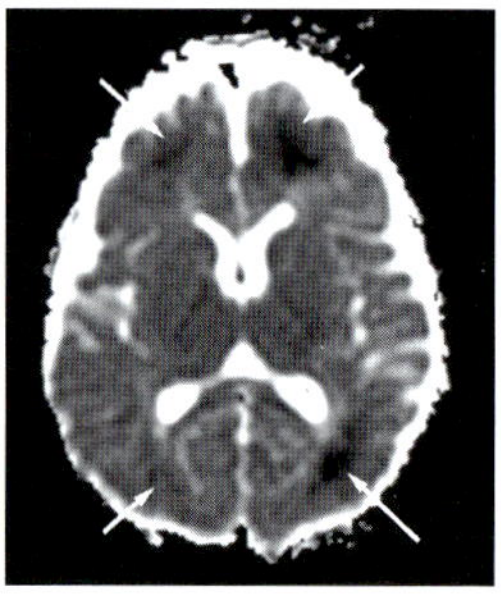 Fig. 14.7 e	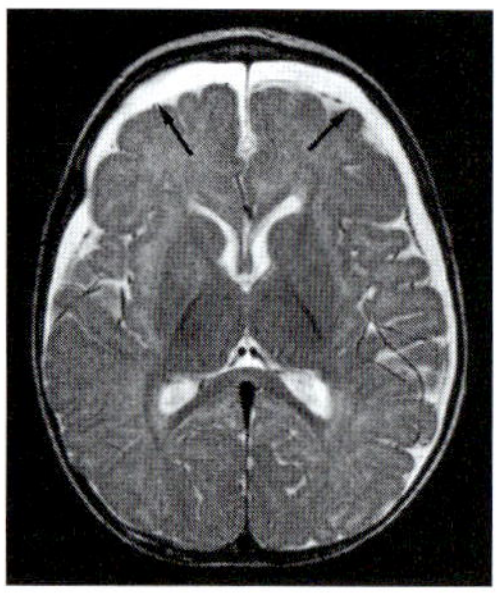 Fig. 14.7 b
Hyperacute infarction	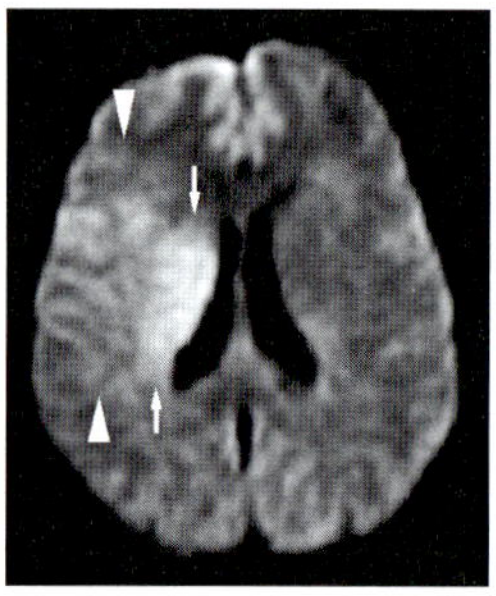 Fig. 5.1 b	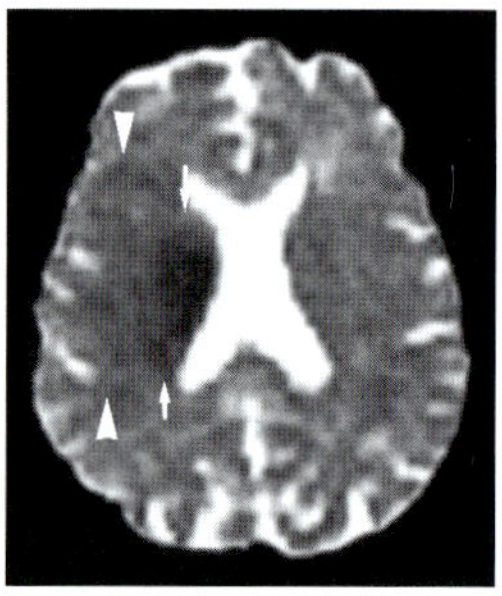 Fig. 5.1 c	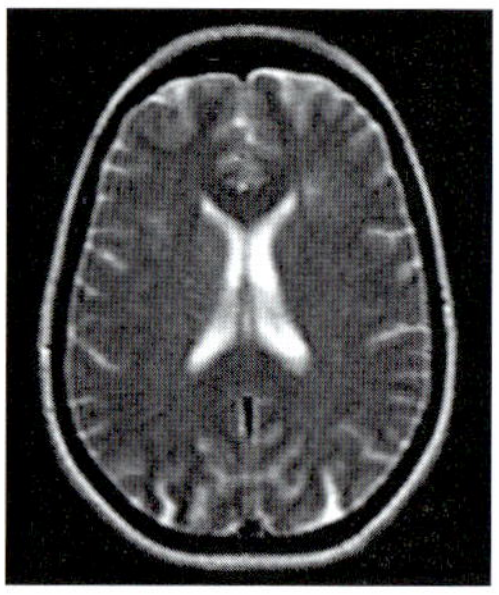 Fig. 5.1 a
Hypoxic ischemic encephalopathy	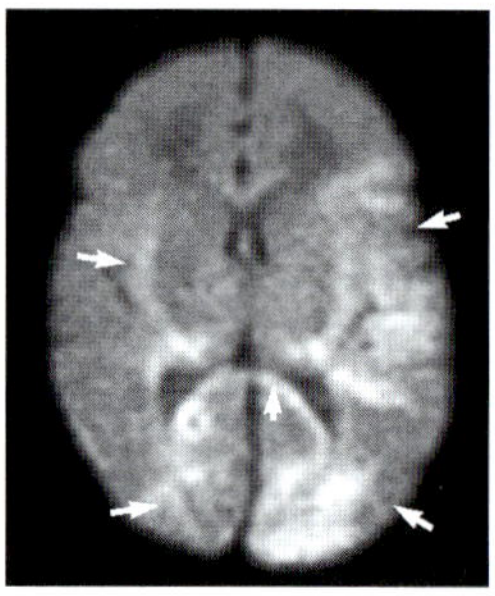 Fig. 14.6 c	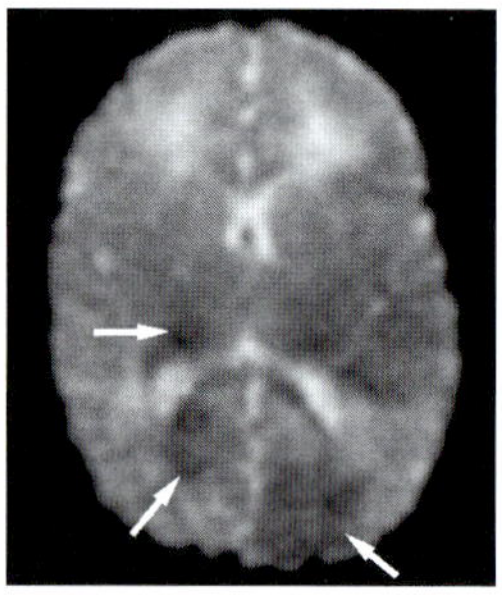 Fig. 14.6 d	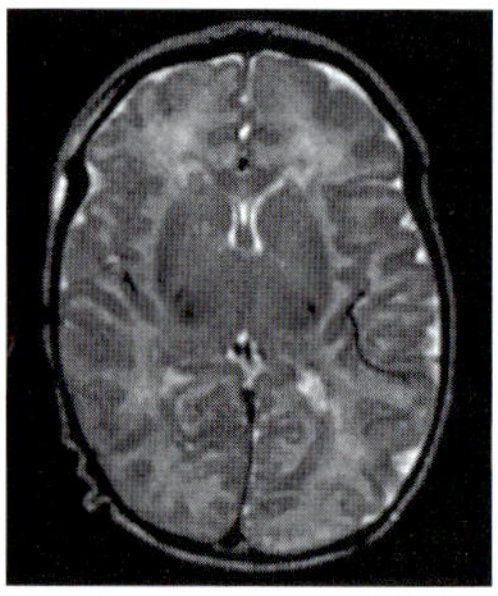 Fig. 14.6 a
Toxic/metabolic			
Methotrexate toxicity	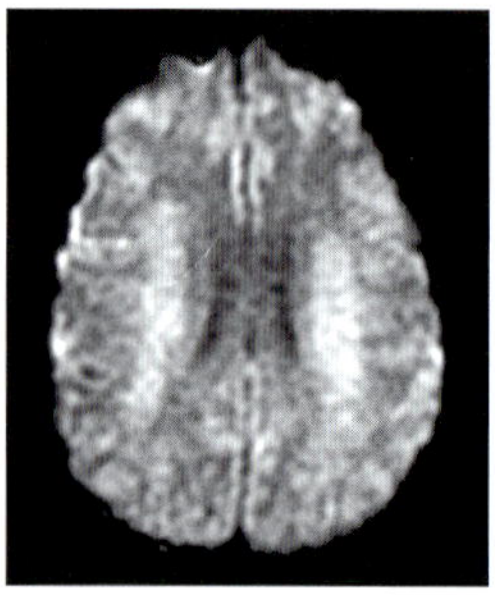 Fig. 10.1 b	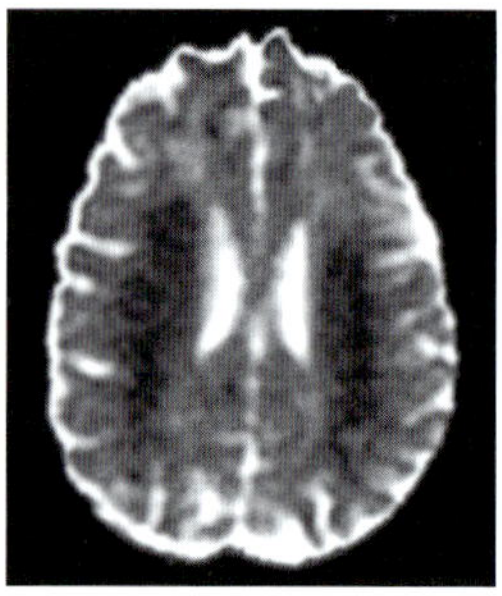 Fig. 10.1 c	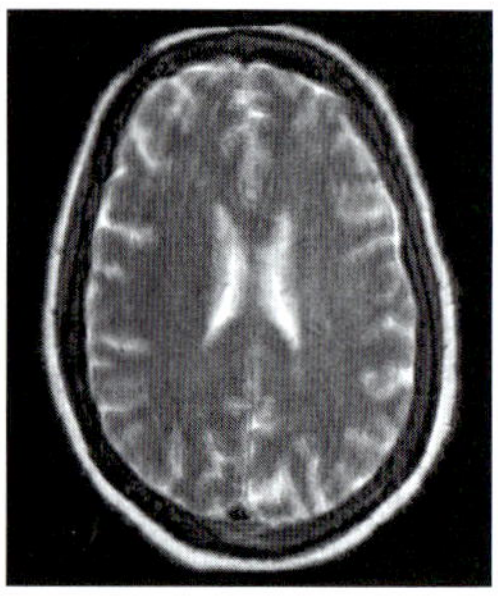 Fig. 10.1 a

Table 2

Differential diagnoses for lesions with a **high diffusion** signal associated with **iso-high ADC** and a **high intense T2** signal

Diagnose	Reference images DWI high	ADC iso-high	T2 high
Degeneration			
Amyotrophic lateral sclerosis (ALS)	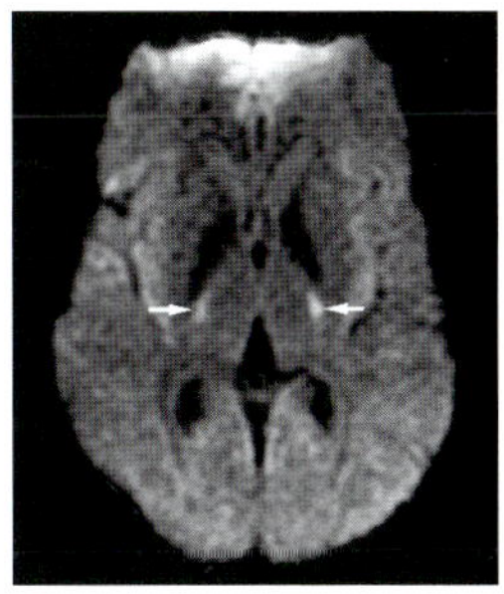 Fig. 9.10 b	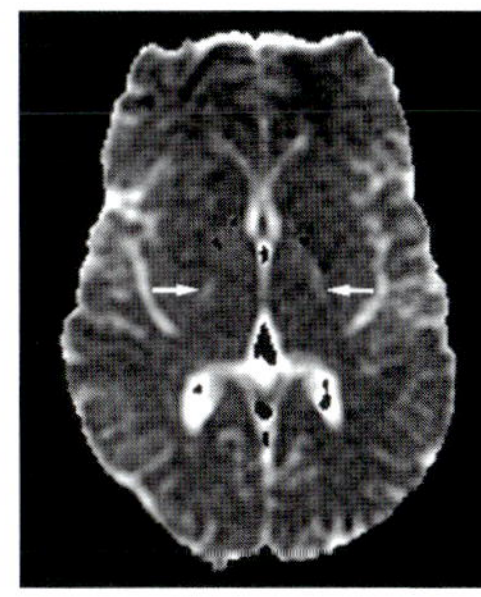 Fig. 9.10 c	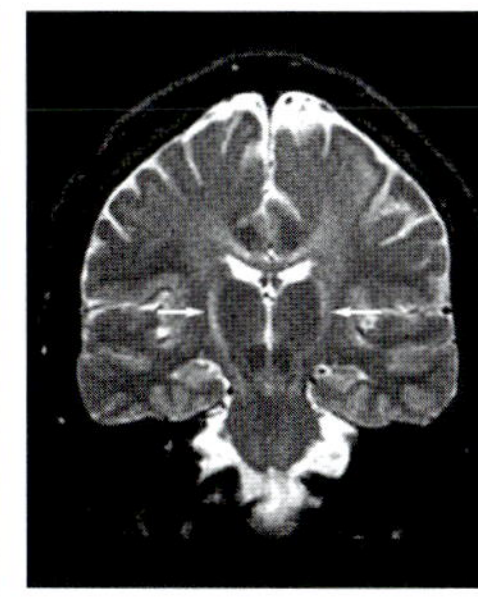 Fig. 9.10 a
Demyelination			
Acute disseminated encephalomyelopathy (ADEM)	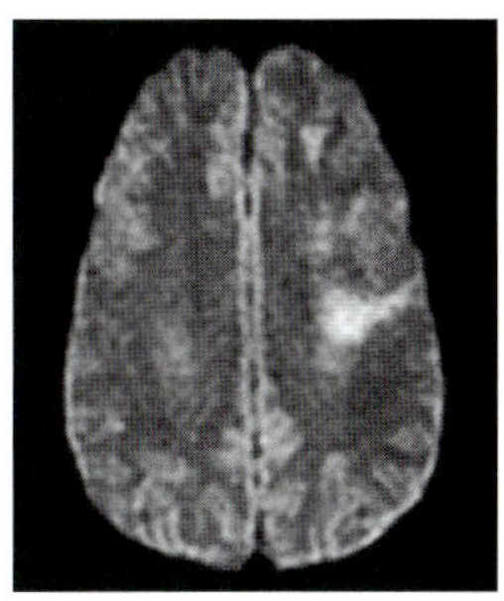 Fig. 9.5 c	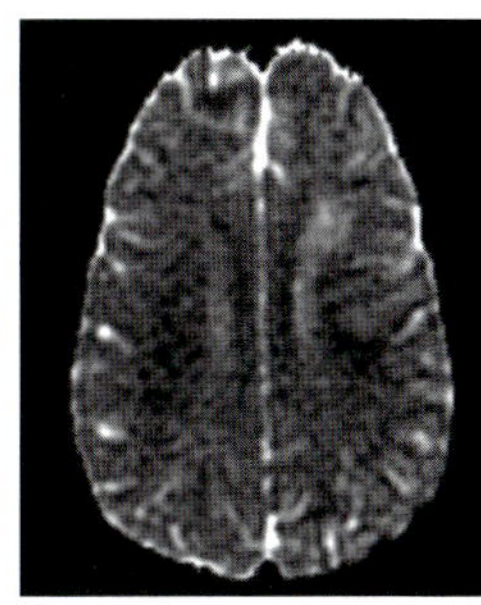 Fig. 9.5 d	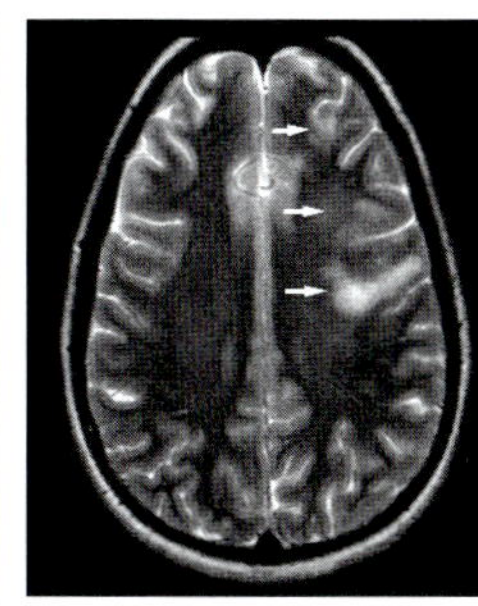 Fig. 9.5 a
Multiple sclerosis (MS)	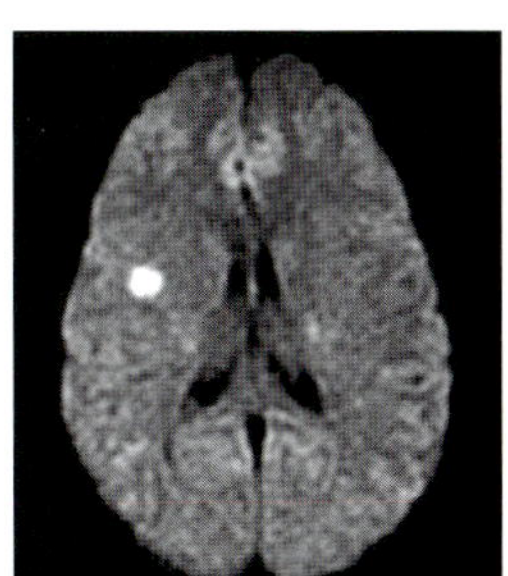 Fig. 9.1 c	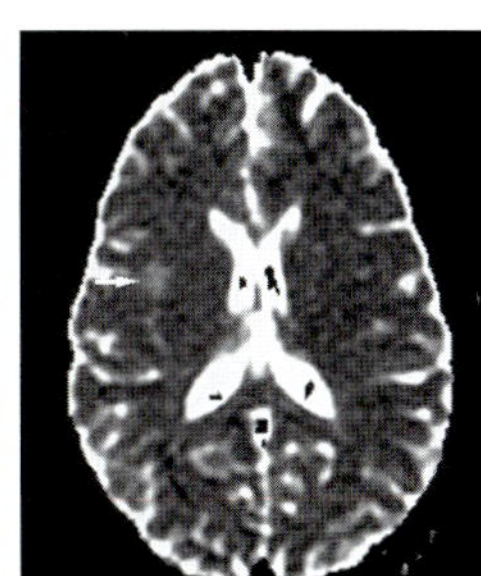 Fig. 9.1 d	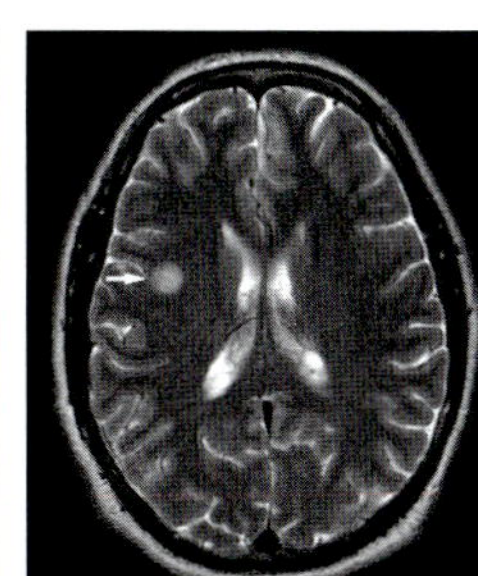 Fig. 9.1 a
Progressive multiple leukoencephalophaty (PML)	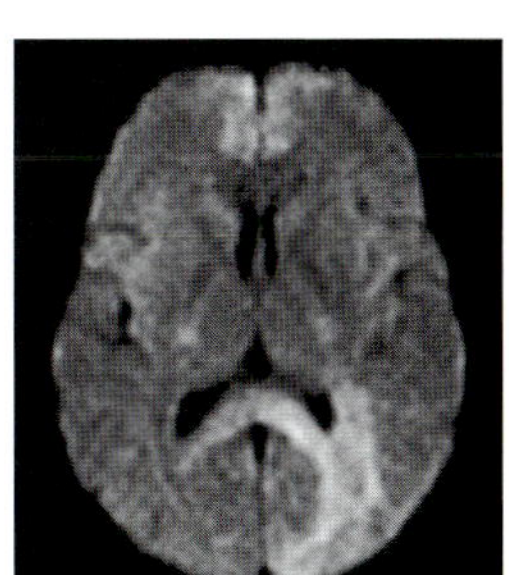 Fig. 9.6 b	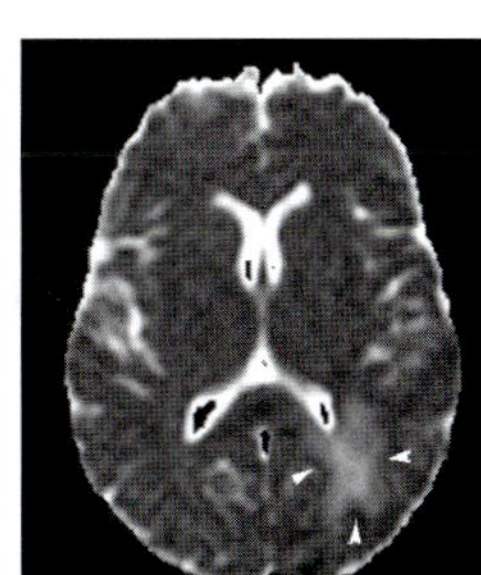 Fig. 9.6 c	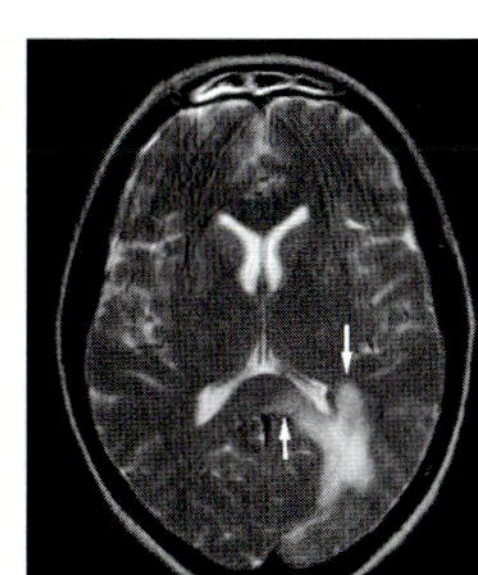 Fig. 9.6 a

Table 2

Differential diagnoses for lesions with a **high diffusion** signal associated with **iso-high ADC** and a **high intense T2** signal

Diagnose	Reference images		
	DWI high	ADC iso-high	T2 high
Epilepsy			
Postictal	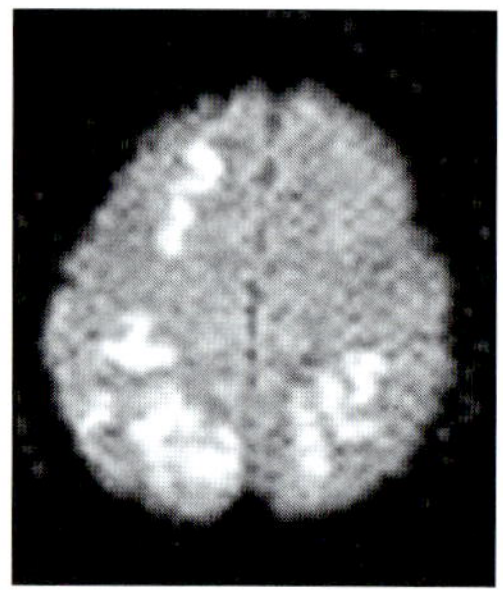 Fig. 8.4 d	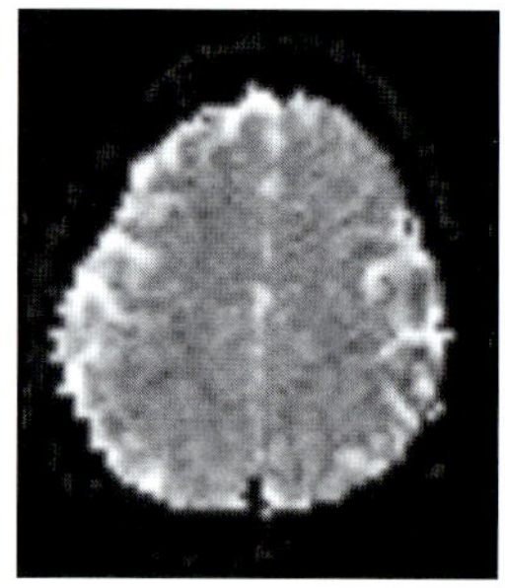 Fig. 8.4 e	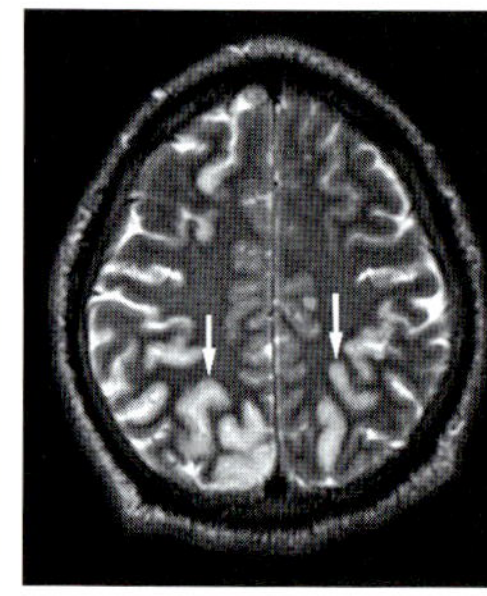 Fig. 8.4 a
Infarction			
Venous infarction	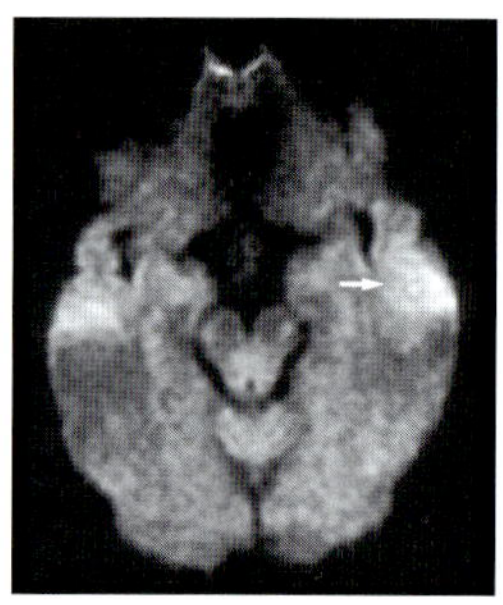 Fig. 5.7 c	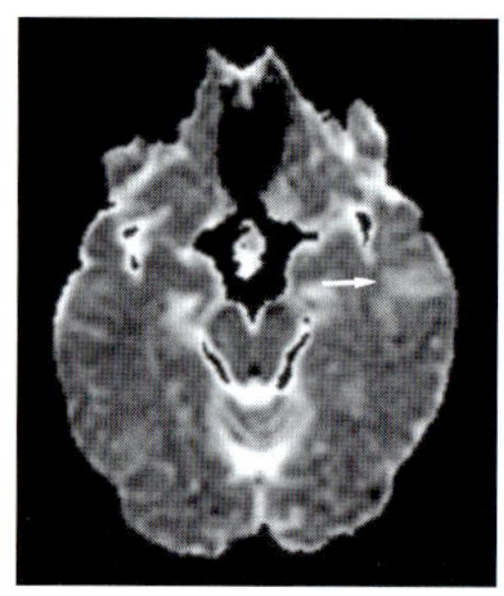 Fig. 5.7 d	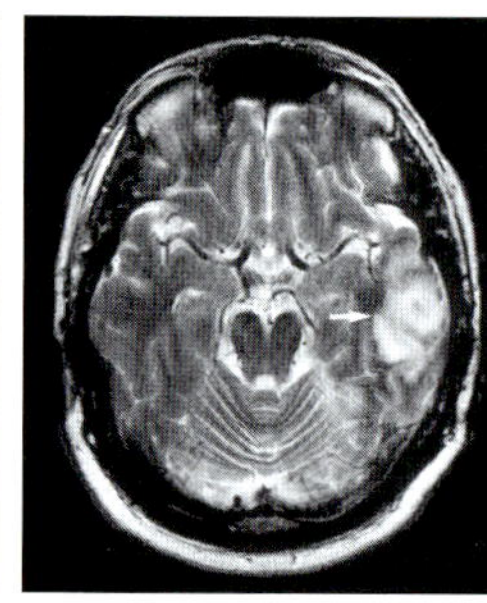 Fig. 5.7 b
Infection			
Human immunodeficiency virus (HIV)	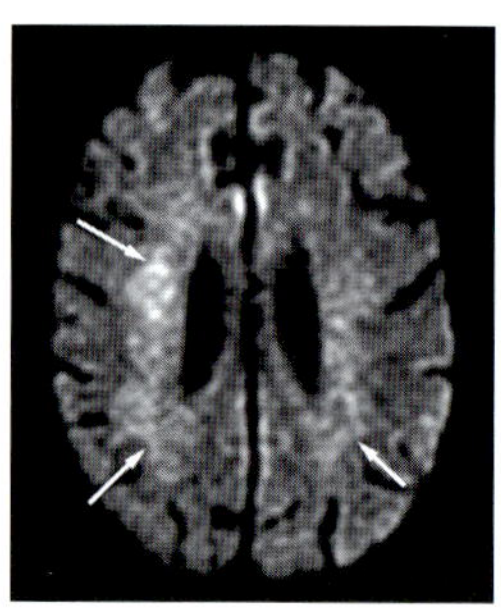 Fig. 11.18 b	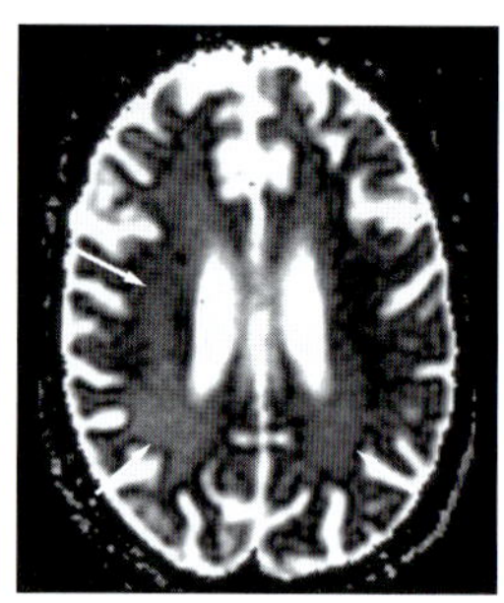 Fig. 11.18 c	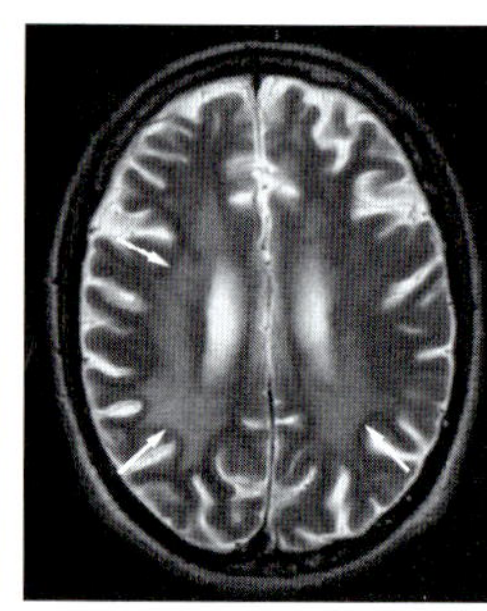 Fig. 11.18 a
Toxic/metabolic			
Marchiafava-Bignami disease	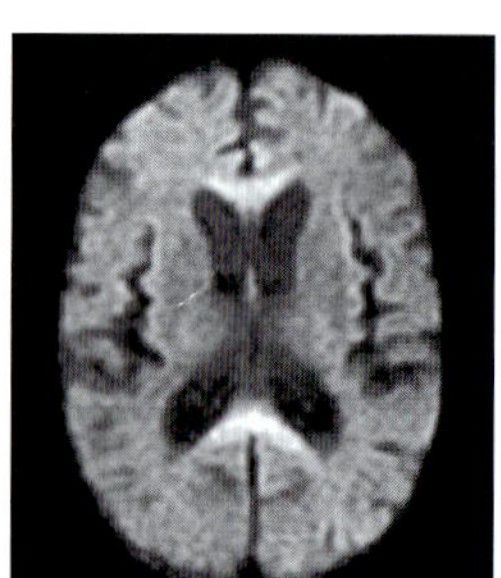 Fig. 10.7 c	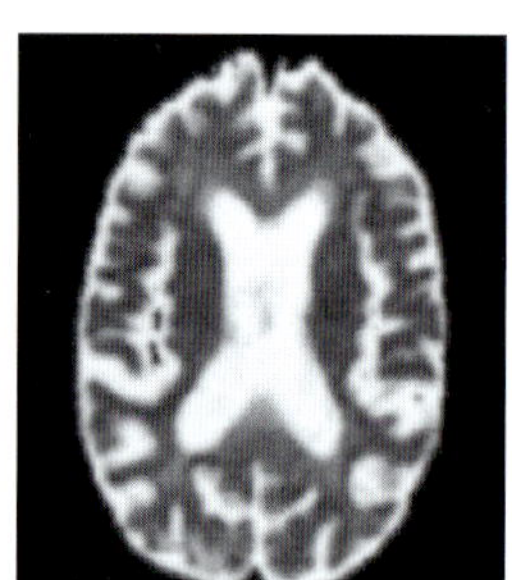 Fig. 10.7 d	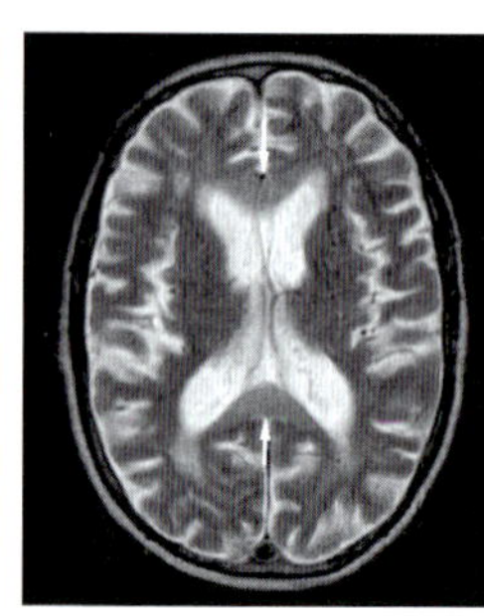 Fig. 10.7 a

Table 2

Differential diagnoses for lesions with a **high diffusion** signal associated with **iso-high ADC** and a **high intense T2** signal

Diagnose	Reference images		
	DWI high	ADC iso-high	T2 high
Mitochondrial myopathy, encephalophathy, latic acidosis and stroke-like episodes (MELAS	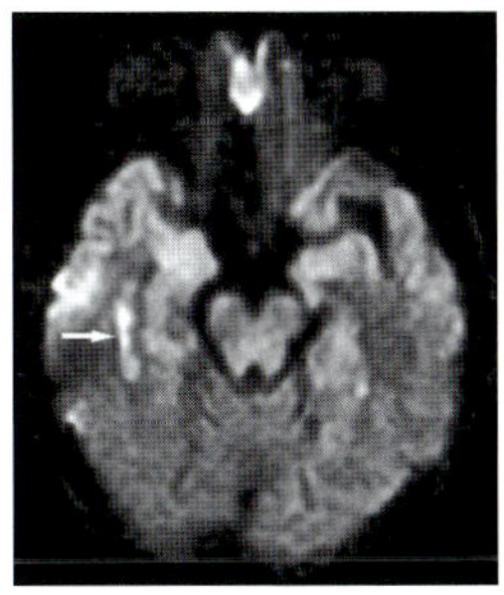 Fig. 10.8 c	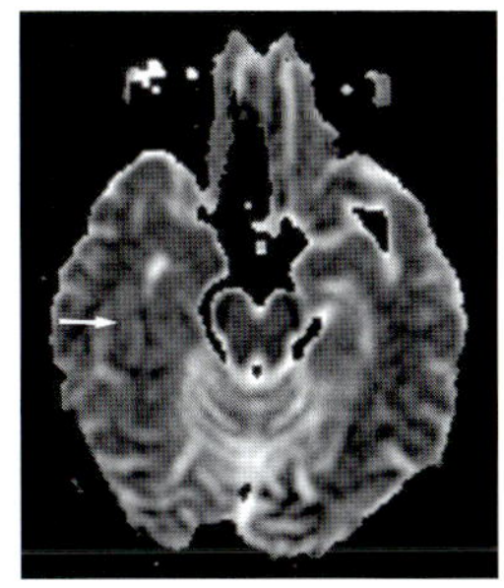 Fig. 10.8 d	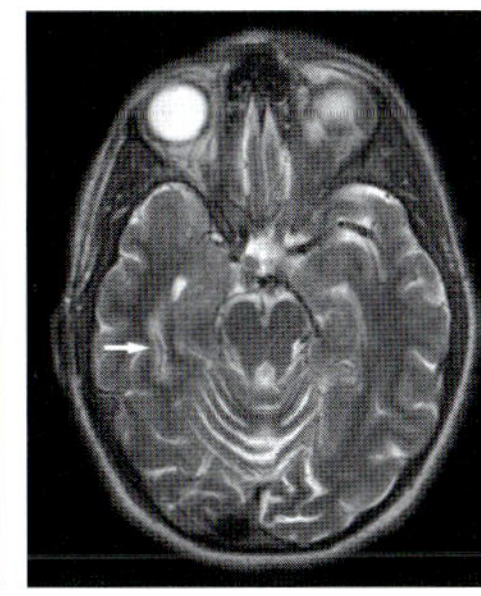 Fig. 10.8 a
Tumor			
Epidermoid	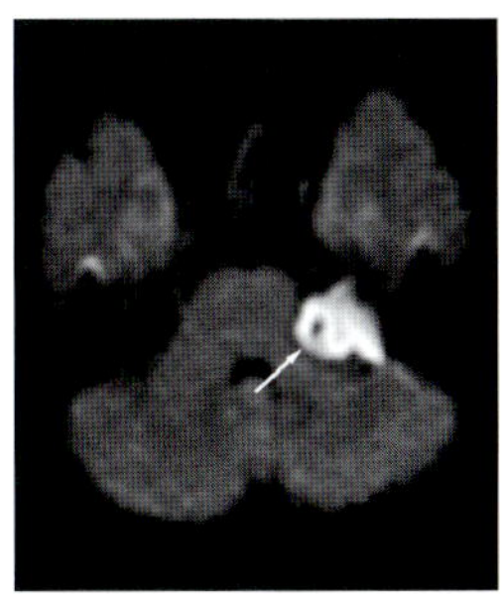 Fig. 11.12 c	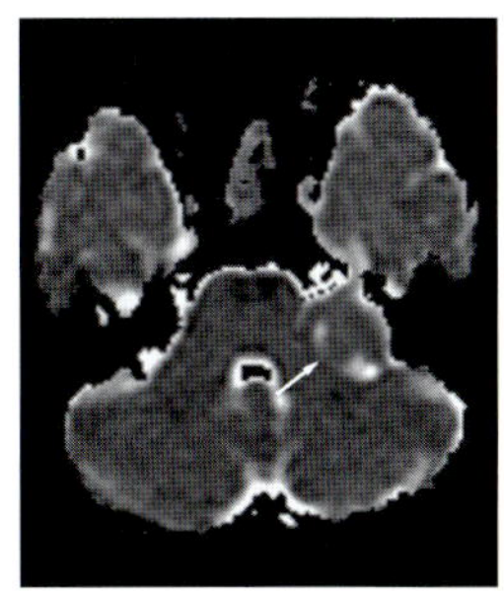 Fig. 11.12 d	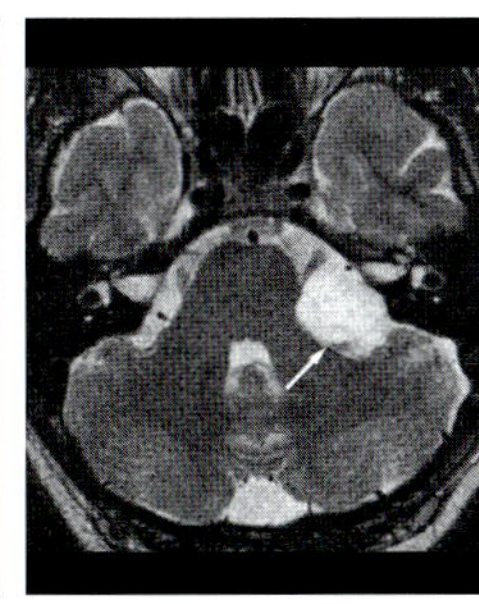 Fig. 11.12 a
Brain stem glioma	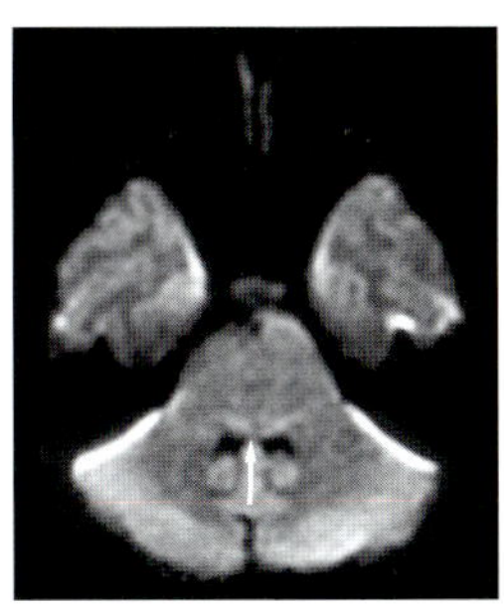 Fig. 13.5 d	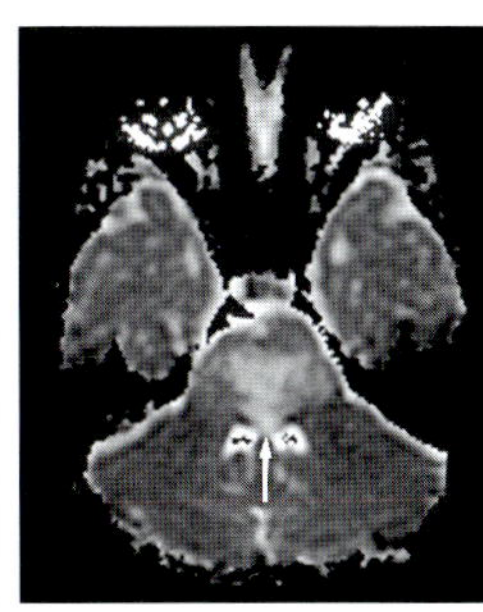 Fig. 13.5 e	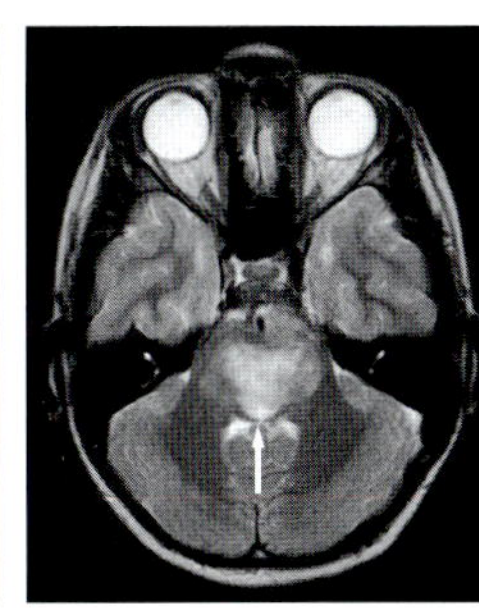 Fig. 13.5 a
Anaplastic astrocytoma	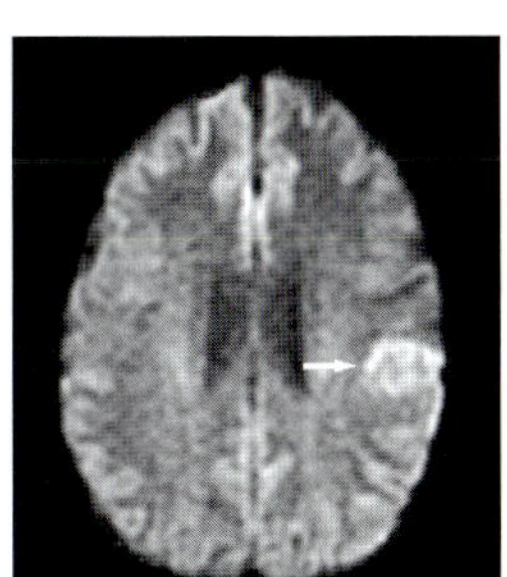 Fig. 3.2 c	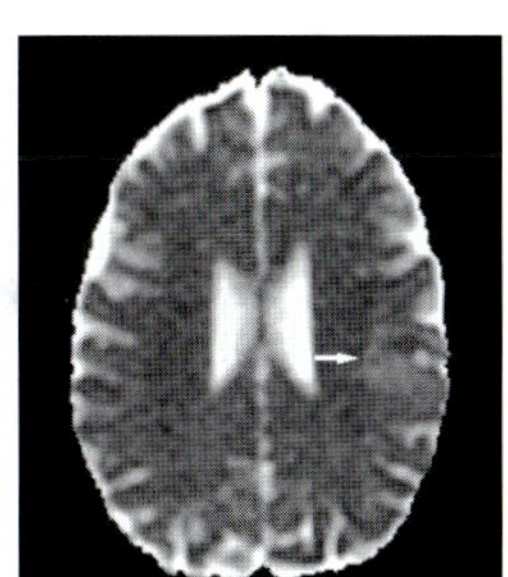 Fig. 3.2 d	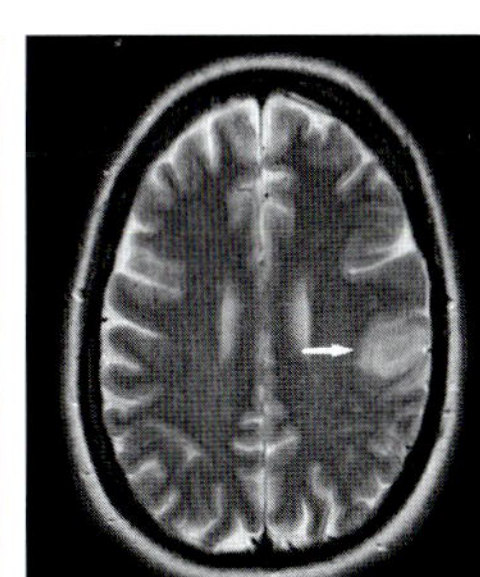 Fig. 3.2 a

Table 2

Differential diagnoses for lesions with a **high diffusion** signal associated with **iso-high ADC** and a **high intense T2** signal

Diagnose	Reference images: DWI high	ADC iso-high	T2 high
Low-grade oligoastrocytoma	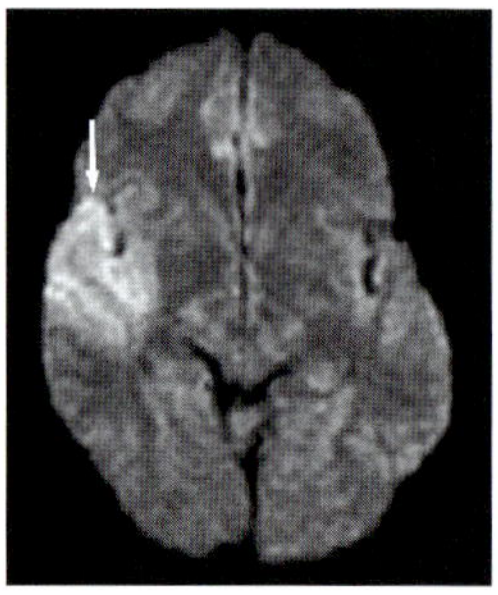 Fig. 13.6 c	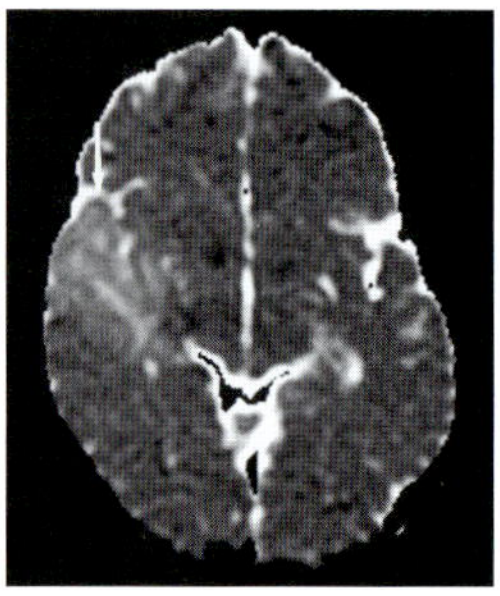 Fig. 13.6 d	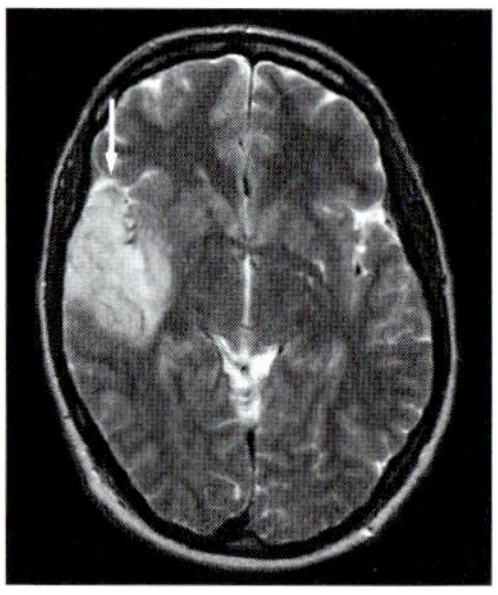 Fig. 13.6 a
Ganglioglioma	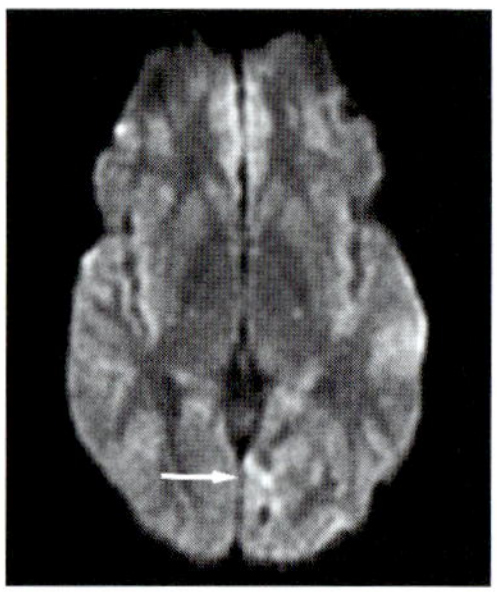 Fig. 13.7 c	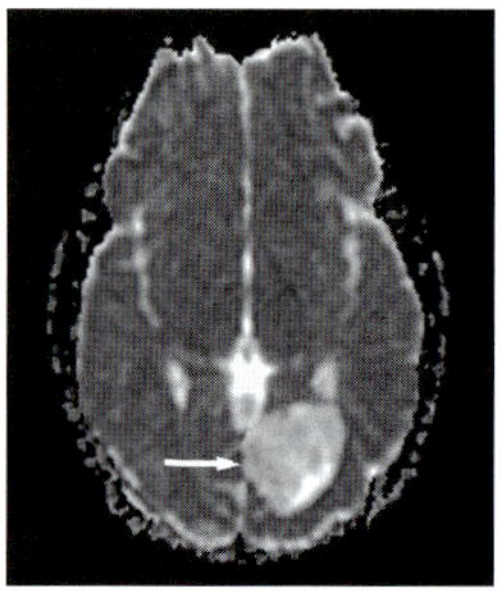 Fig. 13.7 d	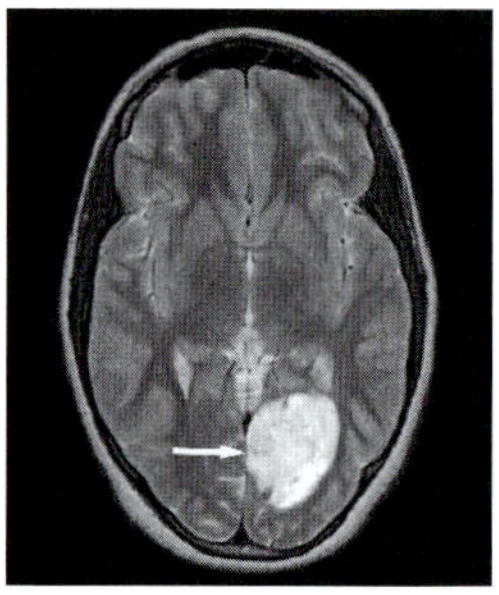 Fig. 13.7 a
Pilocytic astrocytoma	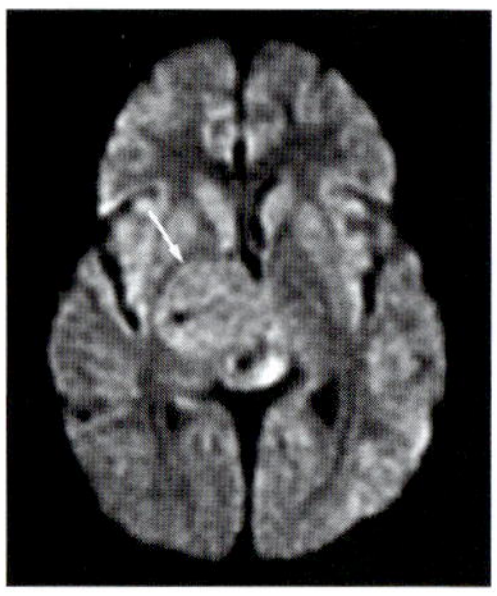 Fig. 14.17 b	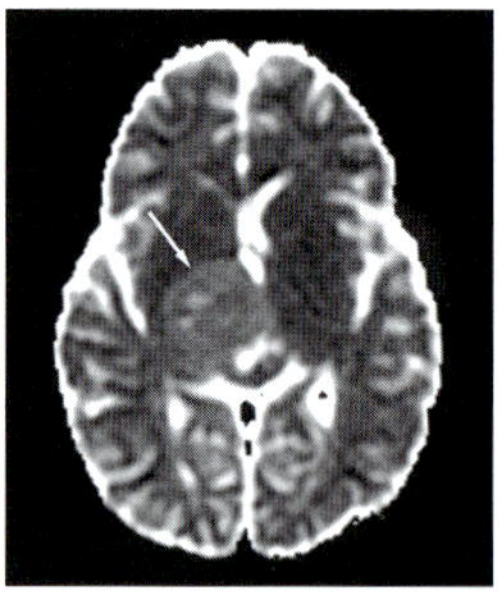 Fig. 14.17 c	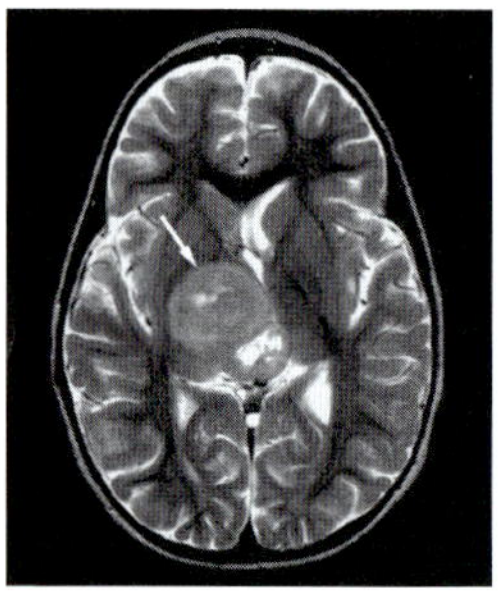 Fig. 14.17 a
Vasculitis/vasculopathy			
Primary angitis of central nervous system (PACNS)	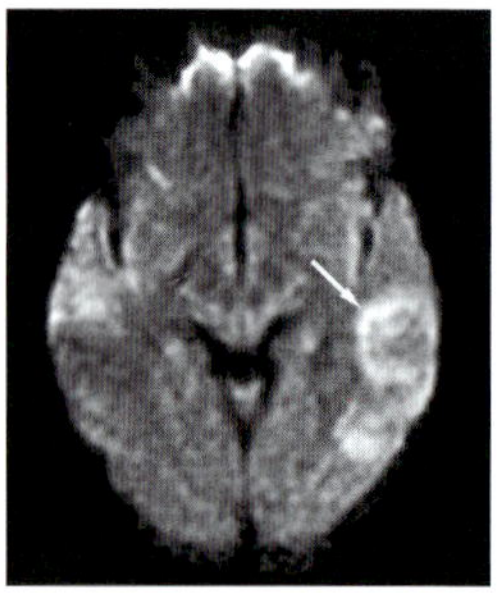 Fig. 7.2 c	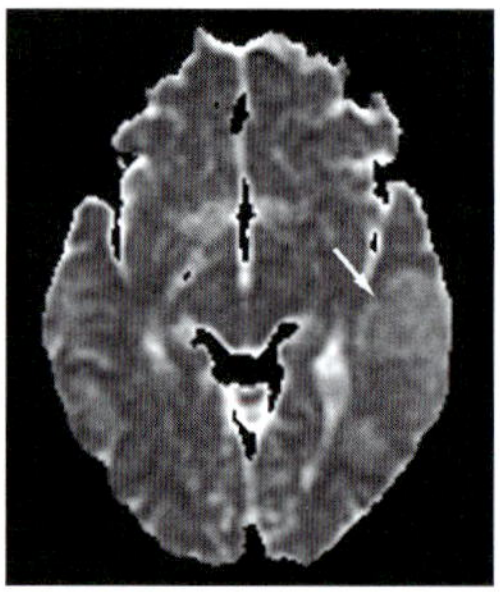 Fig. 7.2 d	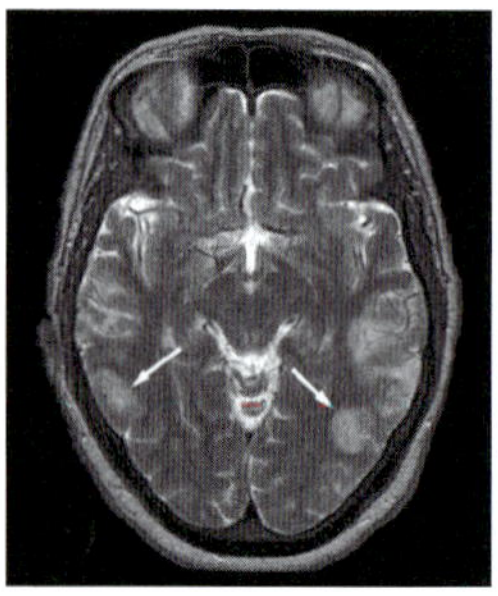 Fig. 7.2 a

Table 2

Differential diagnoses for lesions with a **high diffusion** signal associated with **iso-high ADC** and a **high intense T2** signal

Diagnose	Reference images DWI high	 ADC iso-high	 T2 high
Neuro-Behchet's disease	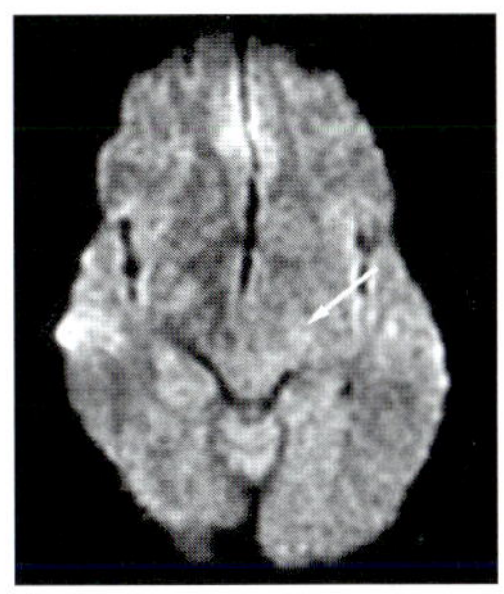 Fig. 7.7 c	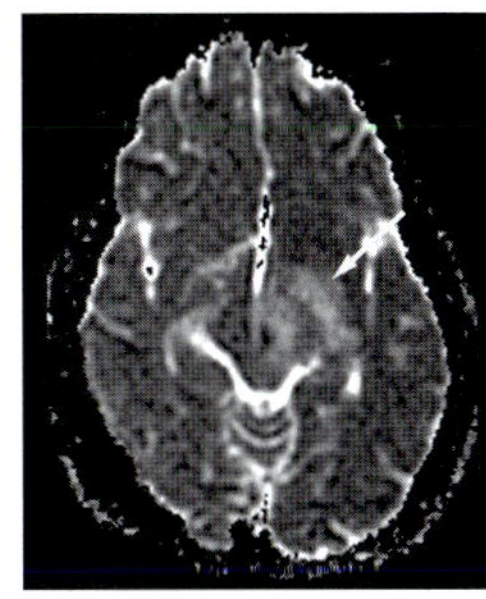 Fig. 7.7 d	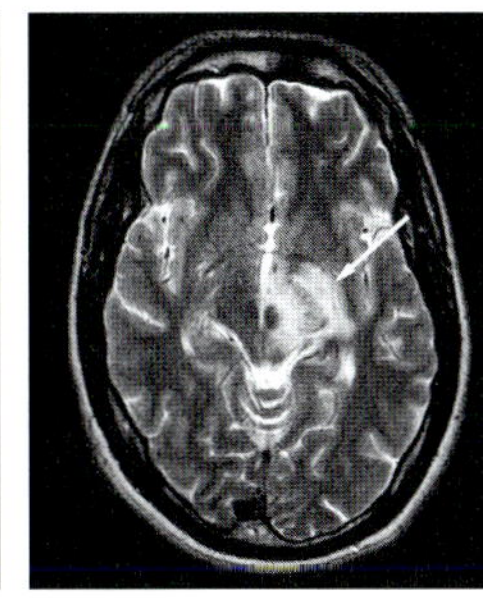 Fig. 7.7 a
Tacrolimus neurotoxicity	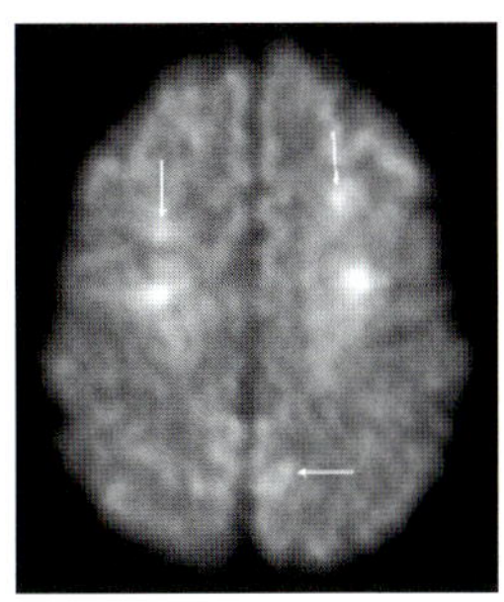 Fig. 7.17 b	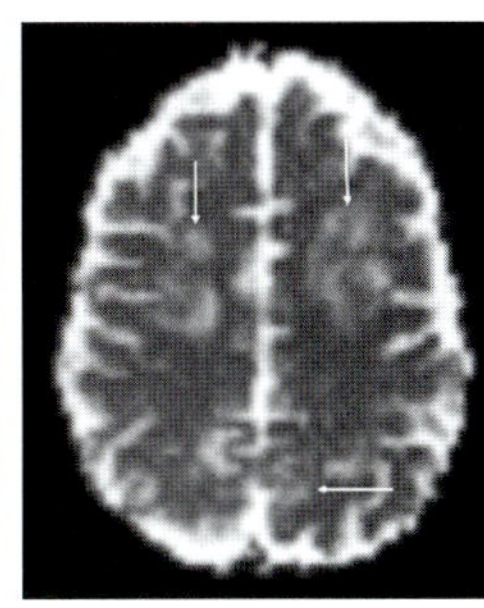 Fig. 7.17 c	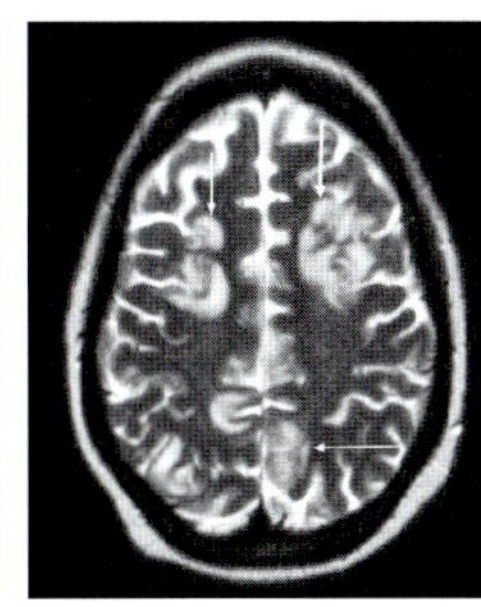 Fig. 7.17 a

Table 3

Differential diagnoses for lesions with a **high diffusion** signal associated with a **low ADC** and **high intense T2** signal

Diagnose	Reference images DWI high	ADC low	T2 high
Degeneration			
Wallerian degeneration	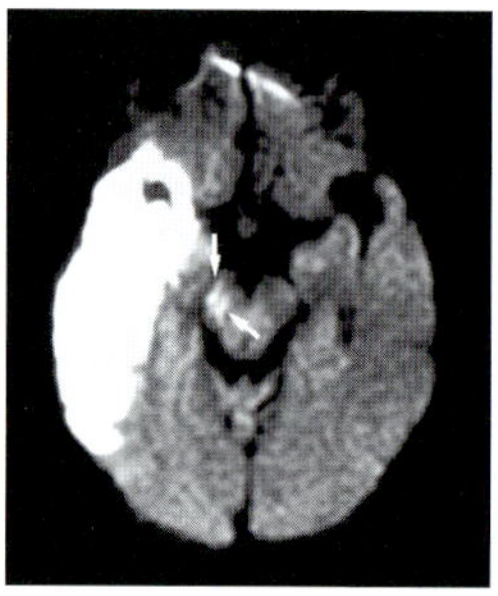 Fig. 9.7 c	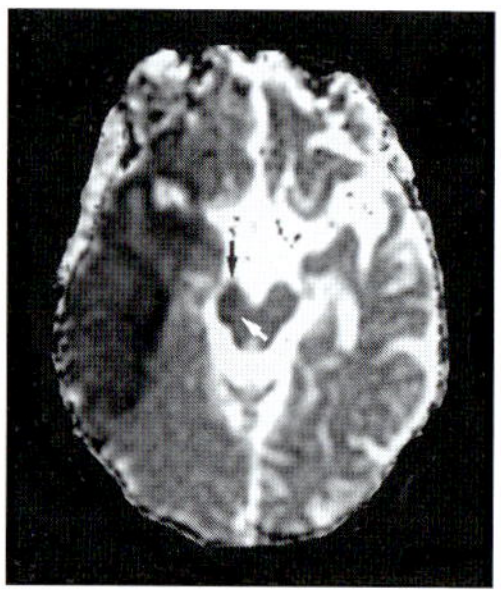 Fig. 9.7 d	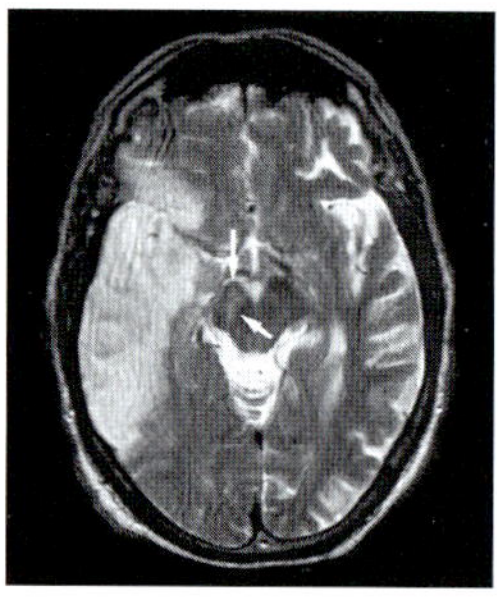 Fig. 9.7 b
Demyelination			
Multiple sclerosis (MS)	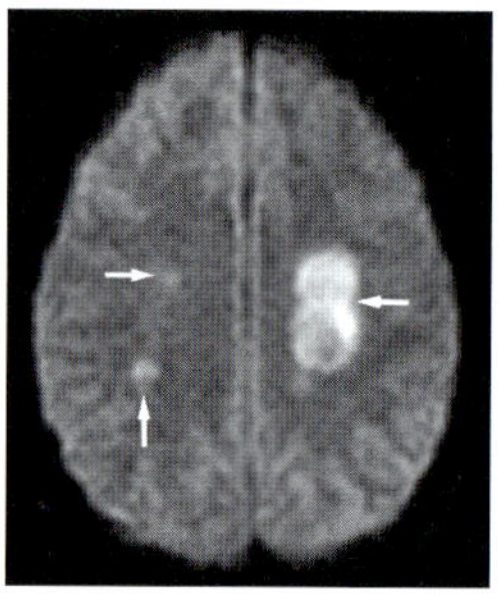 Fig. 9.4 b	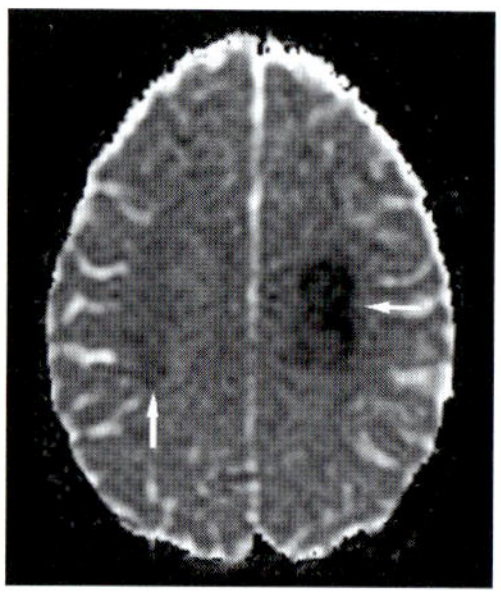 Fig. 9.4 c	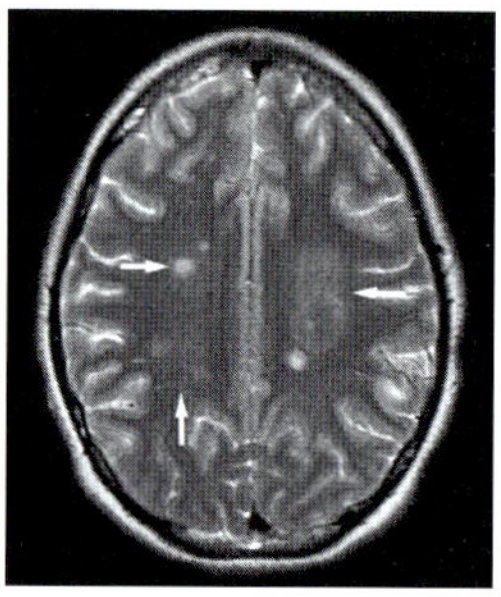 Fig. 9.4 a
Epilepsy			
Status epileptics	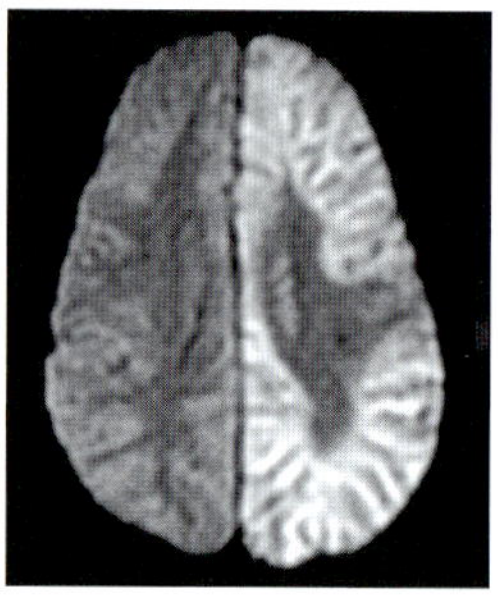 Fig. 4.8 b	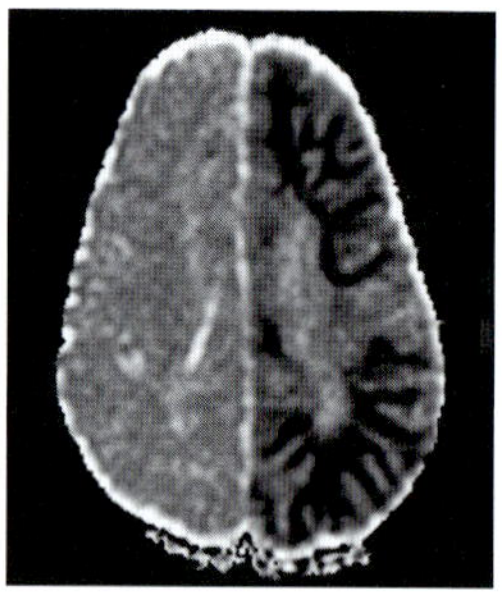 Fig. 4.8 c	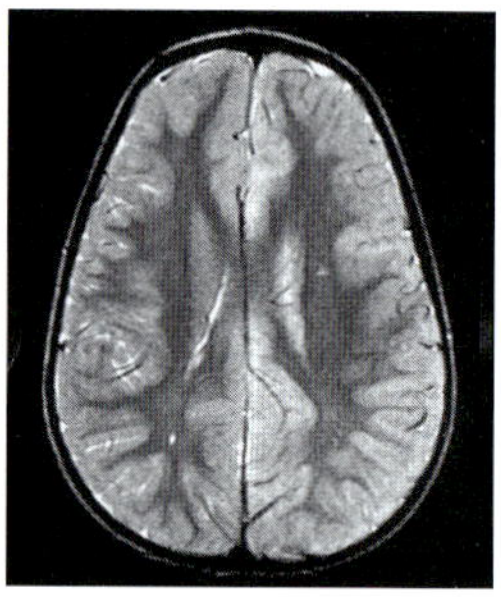 Fig. 4.8 a
Postictal	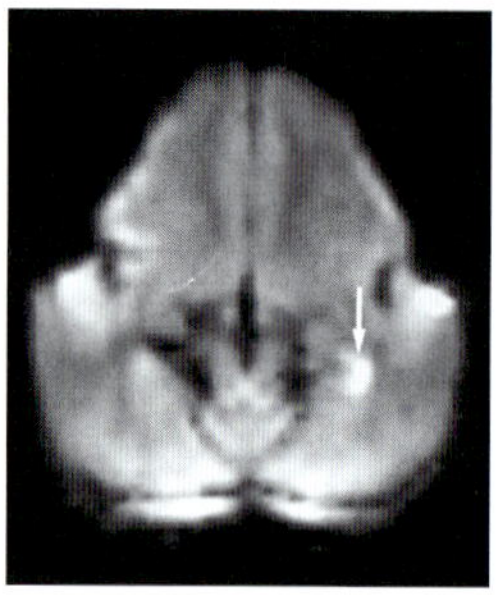 Fig. 8.5 c	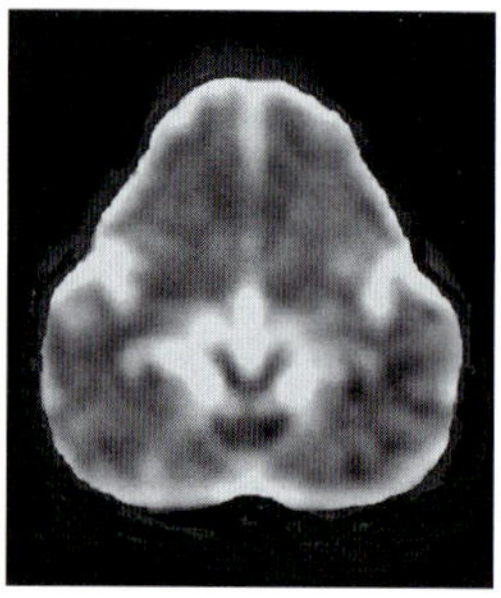 Fig. 8.5 d	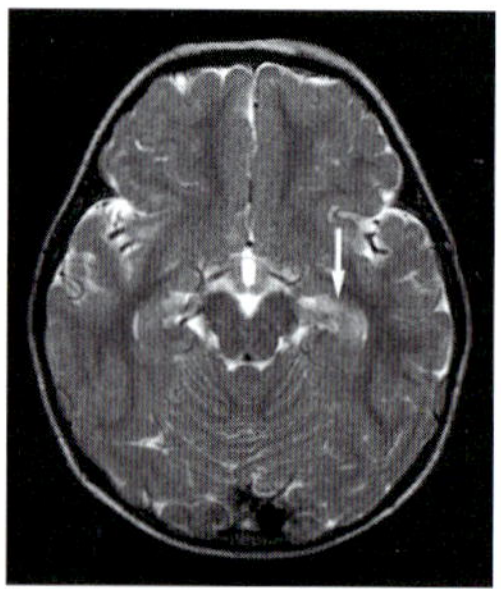 Fig. 8.5 a

Table 3

Differential diagnoses for lesions with a **high diffusion** signal associated with a **low ADC** and **high intense T2** signal

Diagnose	Reference images DWI high	 ADC low	 T2 high
Hematoma			
Late subacute hematoma (Extracellular methemoglobin)	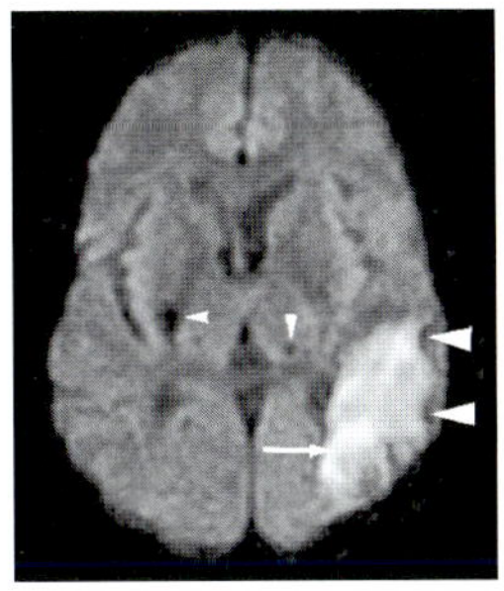 Fig. 6.4 c	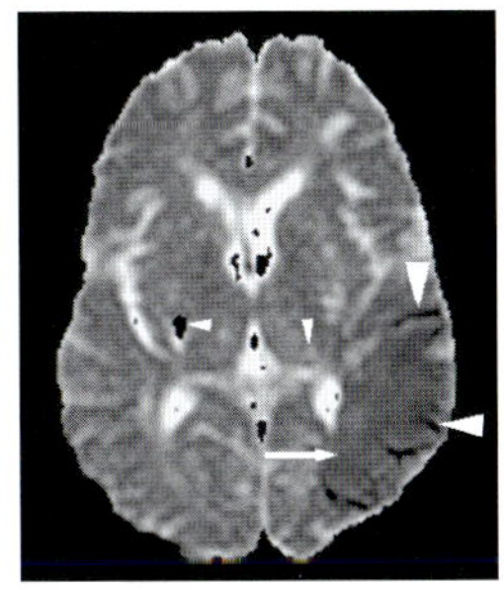 Fig. 6.4 e	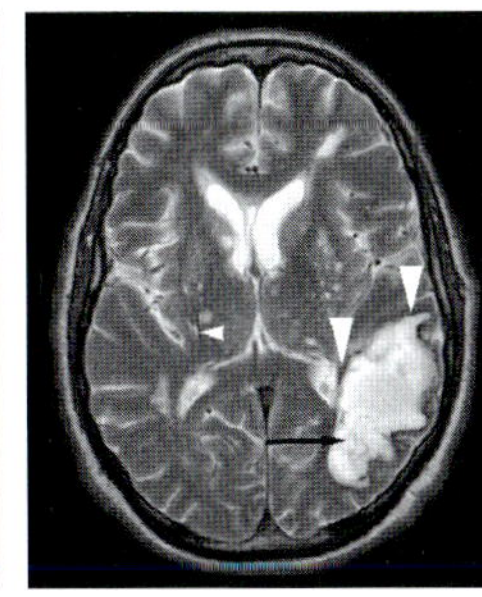 Fig. 6.4 a
Subdural hematoma (SDH)	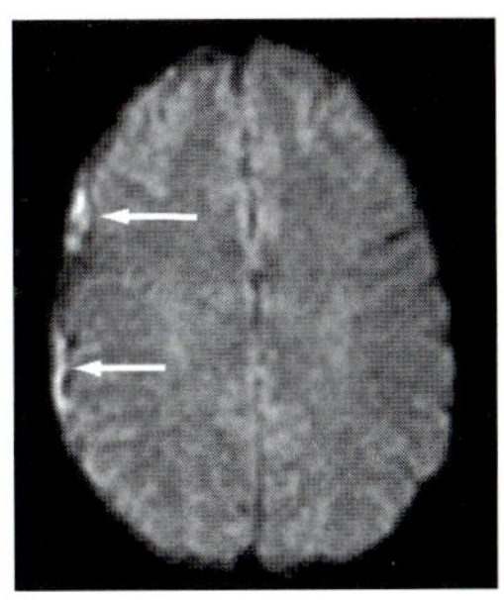 Fig. 6.6 c	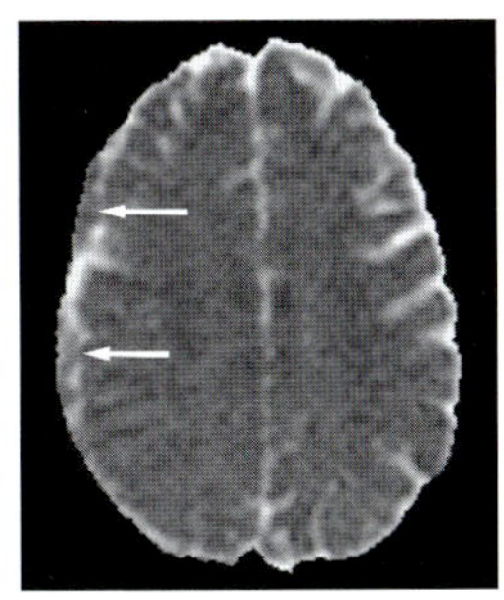 Fig. 6.6 e	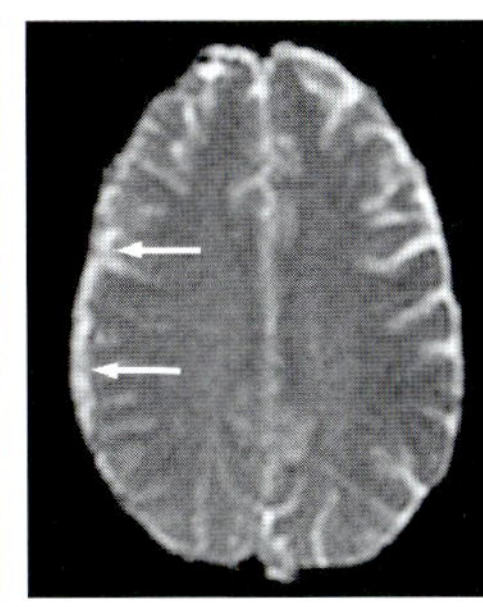 Fig. 6.6 d
Infarction/Ischemia			
Hyperacute reversible ischemia (2 h)	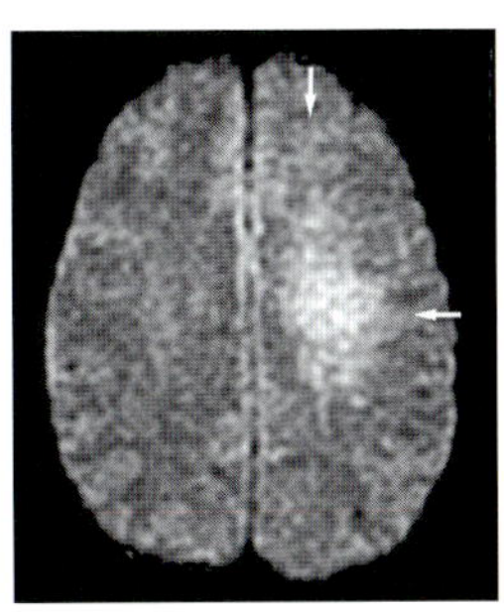 Fig. 5.5 b	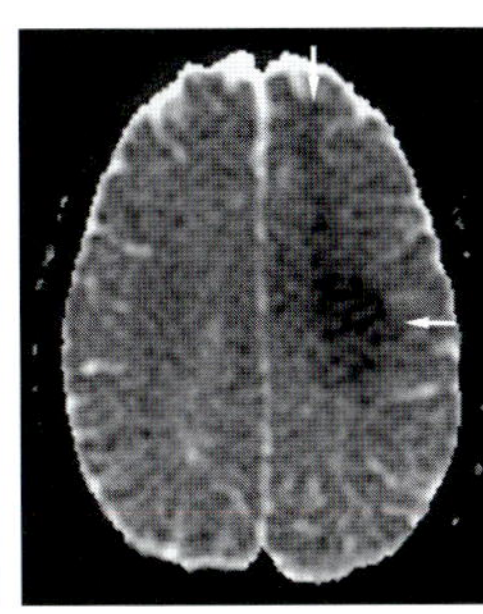 Fig. 5.5 c	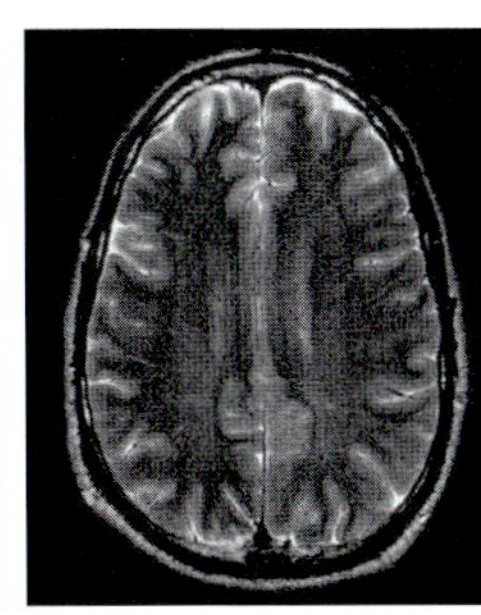 Fig. 5.5 a
Acute infarction (24 h)	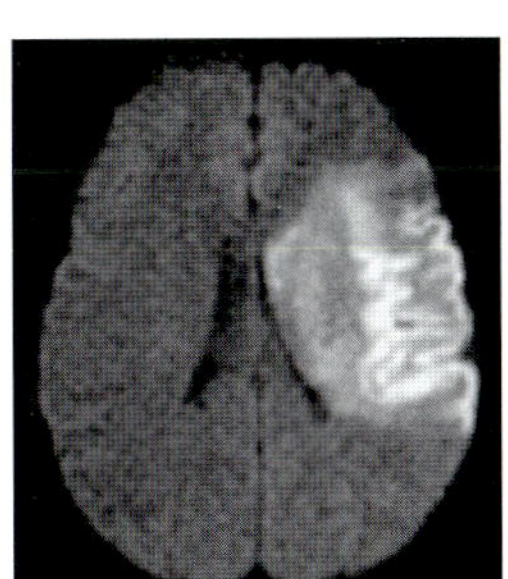 Fig. 3.3 d	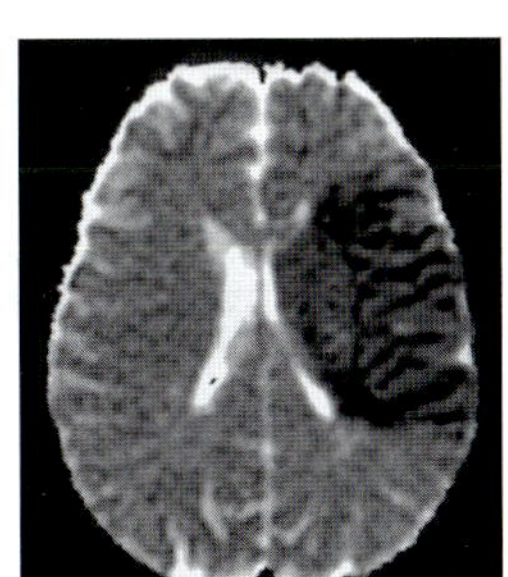 Fig. 3.3 e	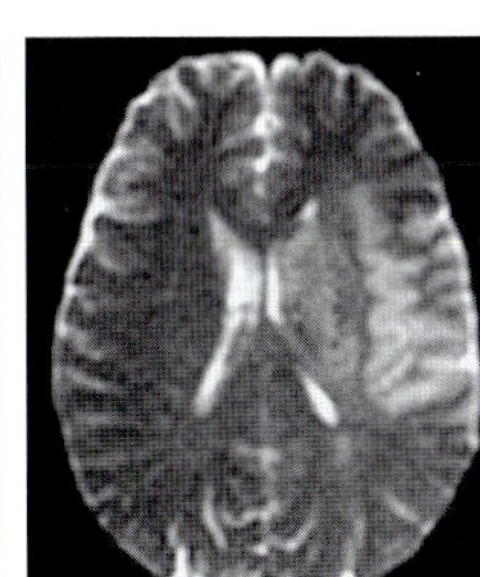 Fig. 3.3 c

Table 3

Differential diagnoses for lesions with a **high diffusion** signal associated with a **low ADC** and **high intense T2** signal

Diagnose	Reference images DWI high	 ADC low	 T2 high
Subacute infarction (10 day)	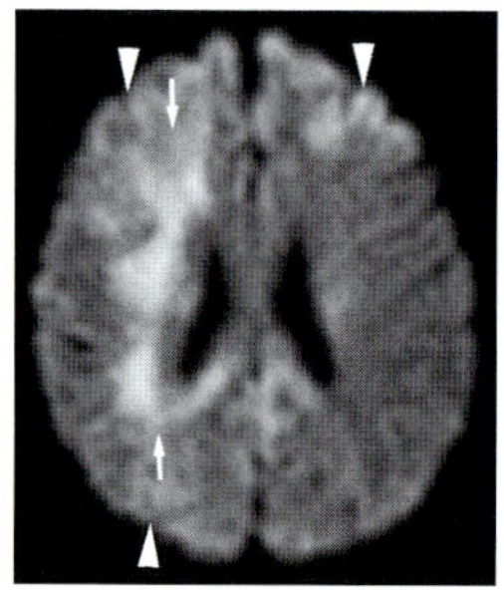 Fig. 5.3 c	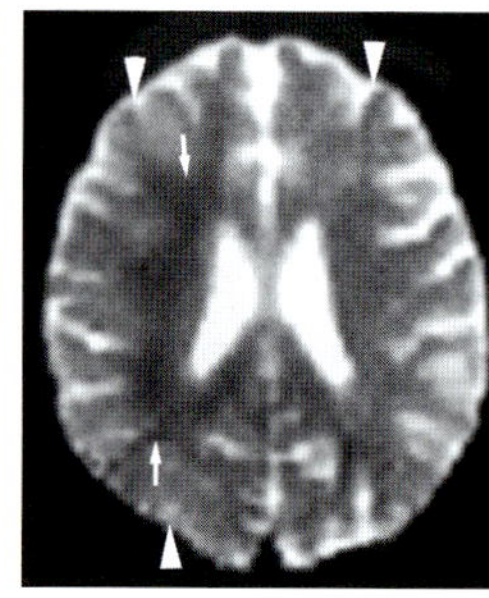 Fig. 5.3 d	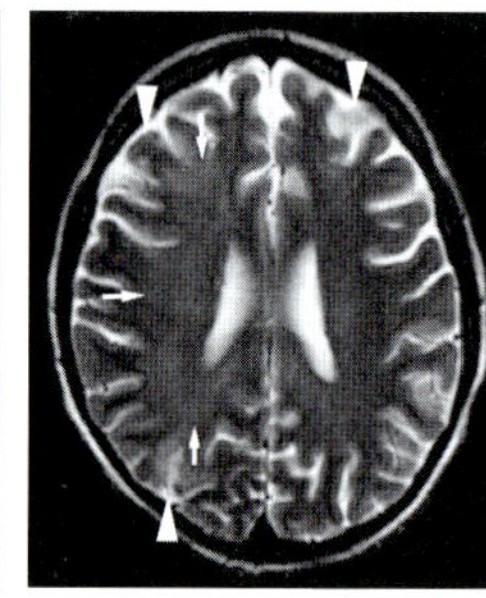 Fig. 5.3 a
Infection			
Abscess	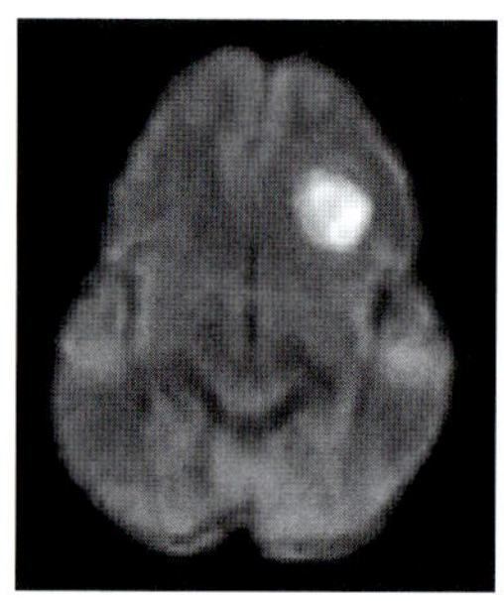 Fig. 11.1 c	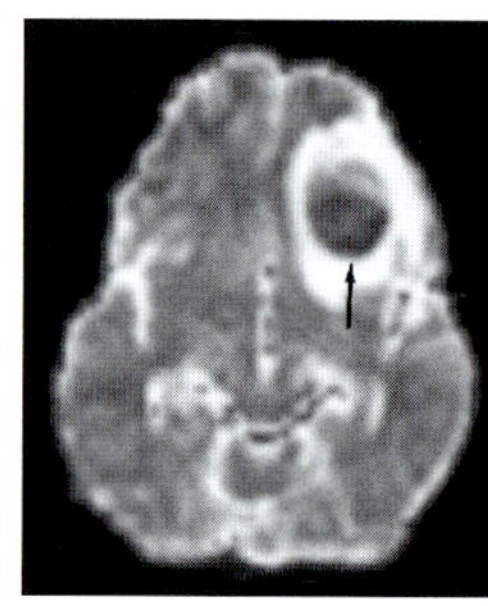 Fig. 11.1 d	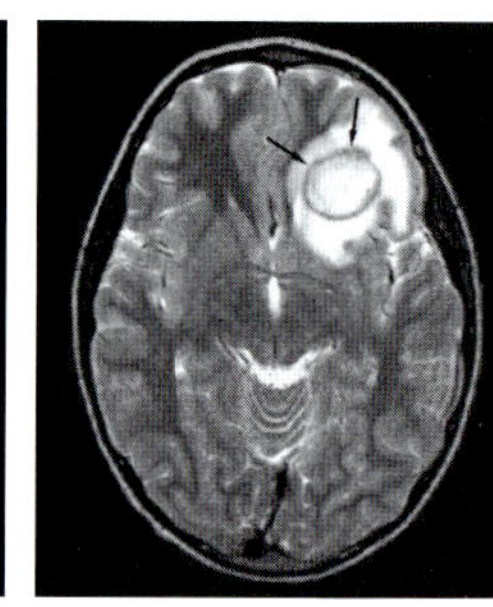 Fig. 11.1 a
Septic emboli	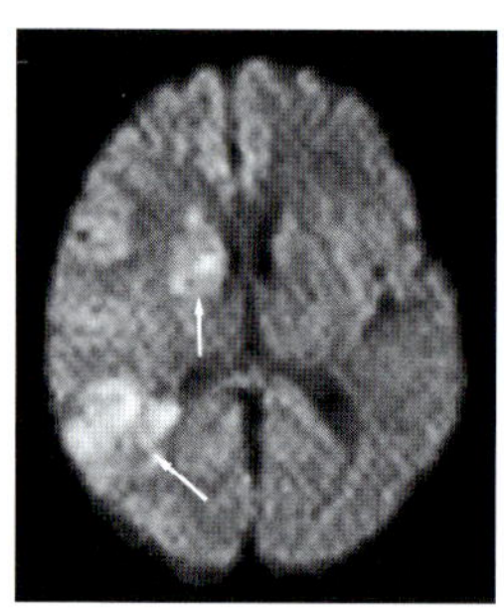 Fig.11.2 b	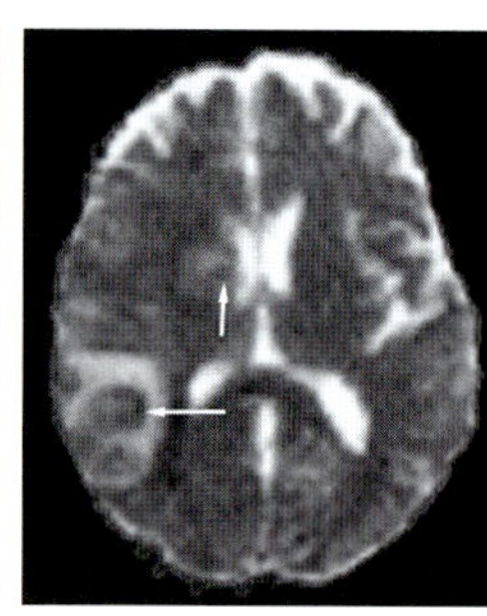 Fig.11.2 c	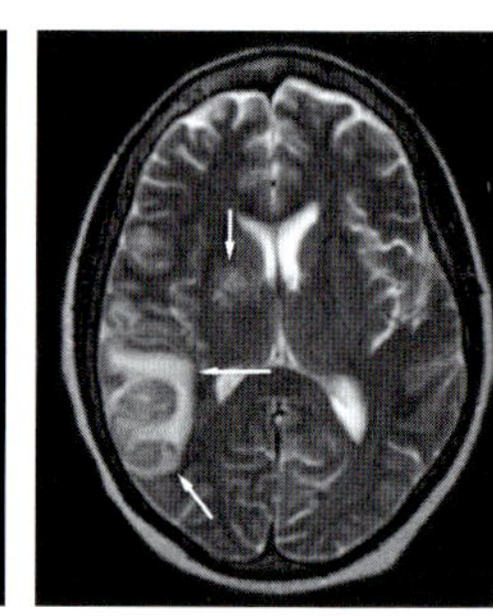 Fig.11.2 a
Ventriculitis	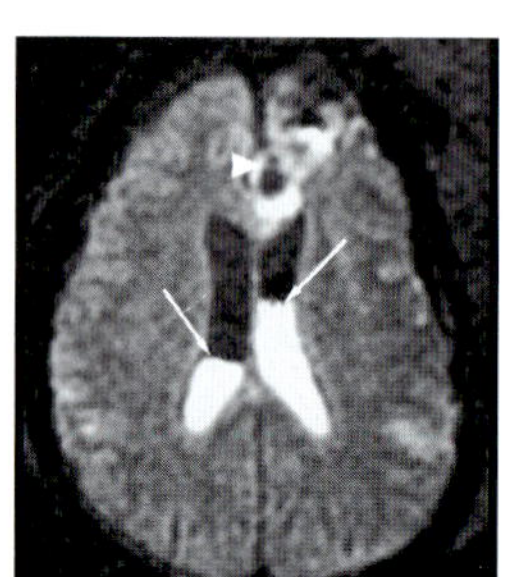 Fig. 11.10 c	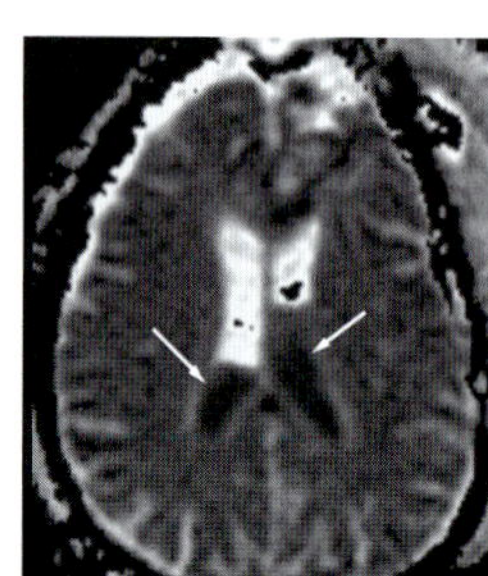 Fig. 11.10 d	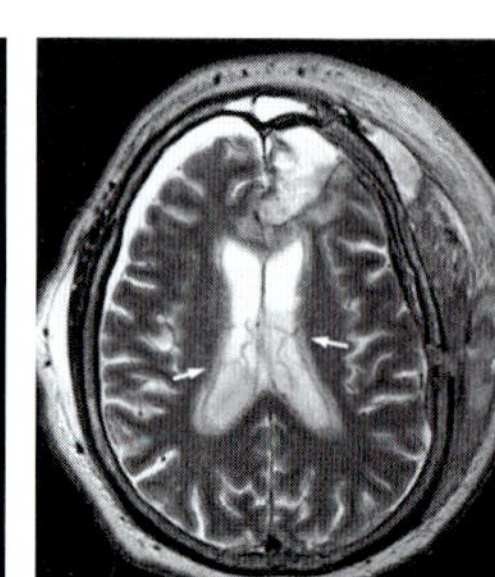 Fig. 11.10 b

Table 3

Differential diagnoses for lesions with a **high diffusion** signal associated with a **low ADC** and **high intense T2** signal

Diagnose	Reference images		
	DWI high	ADC low	T2 high
Acquired immunodeficiency syndrome (AIDS)	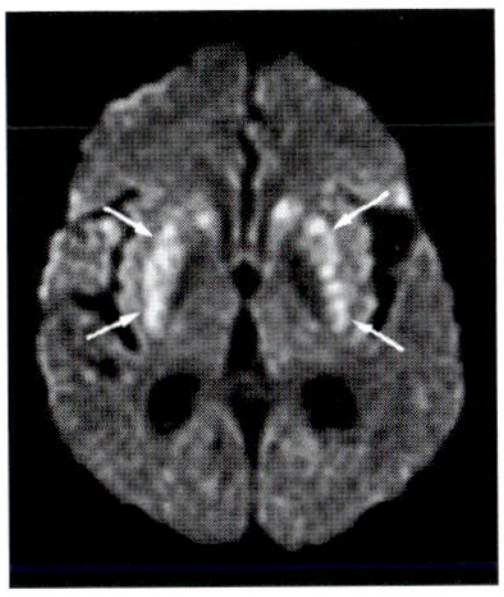 Fig. 11.19. b	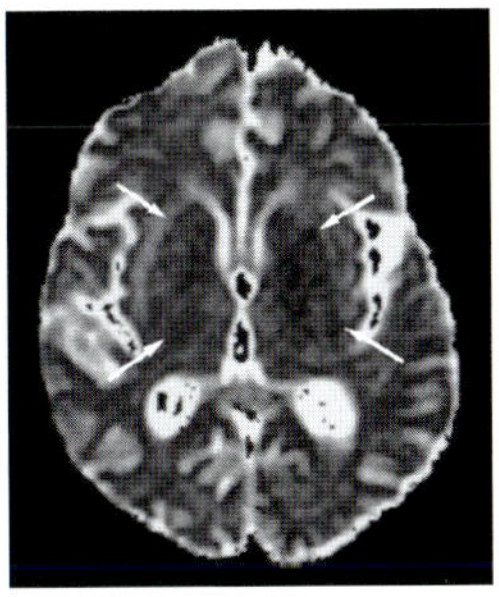 Fig. 11.19. c	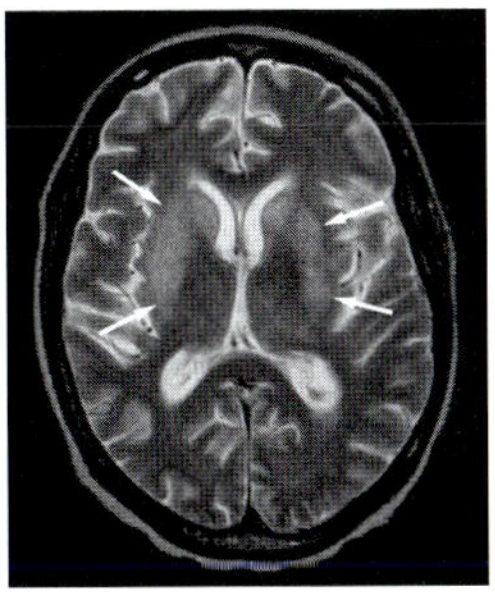 Fig. 11.19. a
Aspergillosis (disseminated)	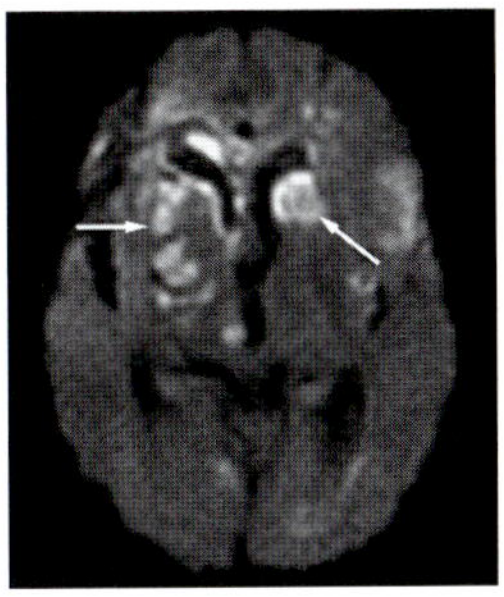 Fig.11.16 b	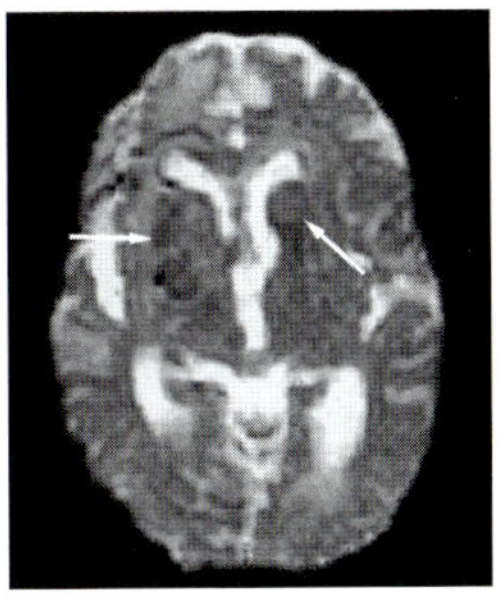 Fig.11.16 c	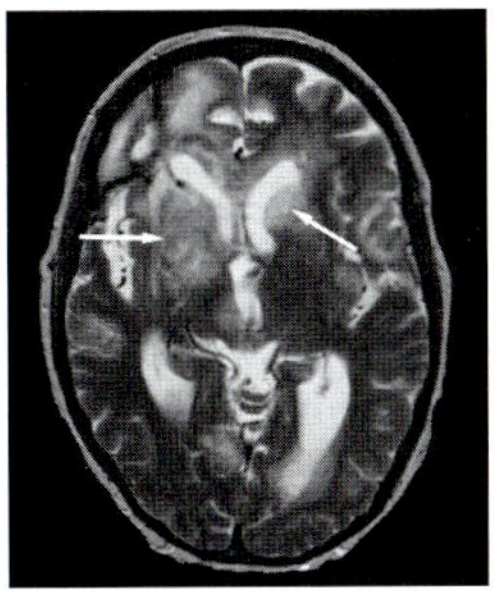 Fig.11.16 a
Creutzfeldt-Jakob disease (CJD)	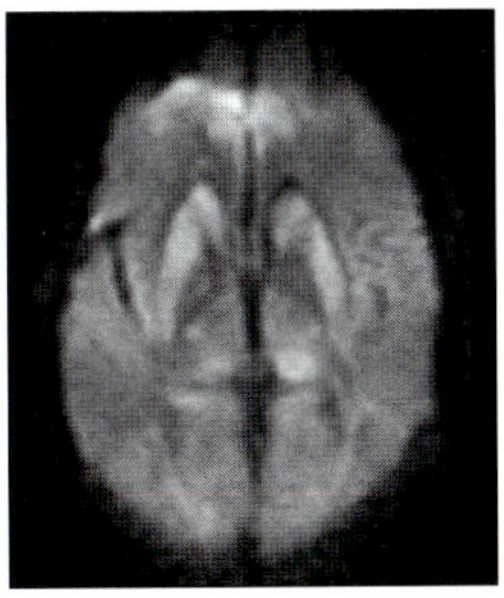 Fig. 9.9 b	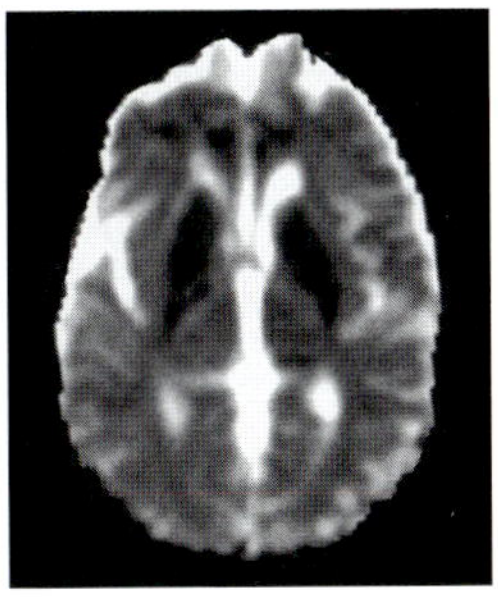 Fig. 9.9 c	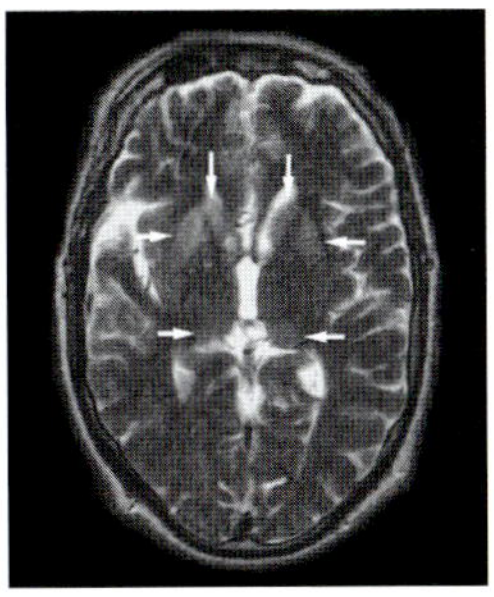 Fig. 9.9 a
Group B Streptococcus meningitis	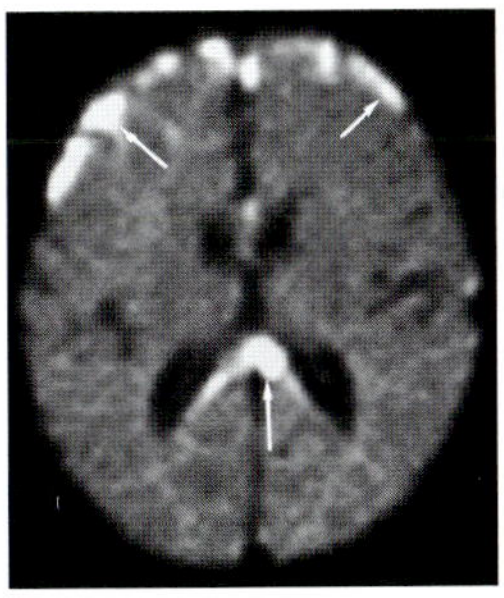 Fig. 11.13 b	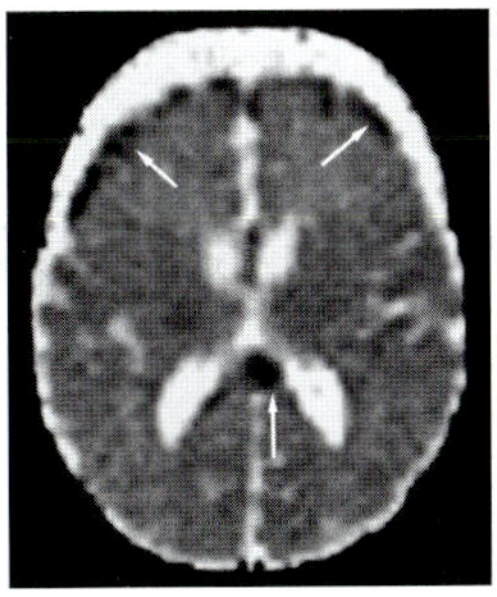 Fig. 11.13 c	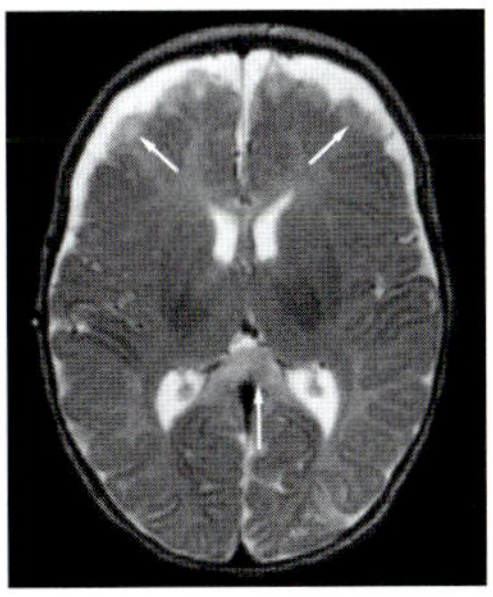 Fig. 11.13 a

Table 3

Differential diagnoses for lesions with a **high diffusion** signal associated with a **low ADC** and **high intense T2** signal

Diagnose	Reference images: DWI high	Reference images: ADC low	Reference images: T2 high
Herpes simplex virus	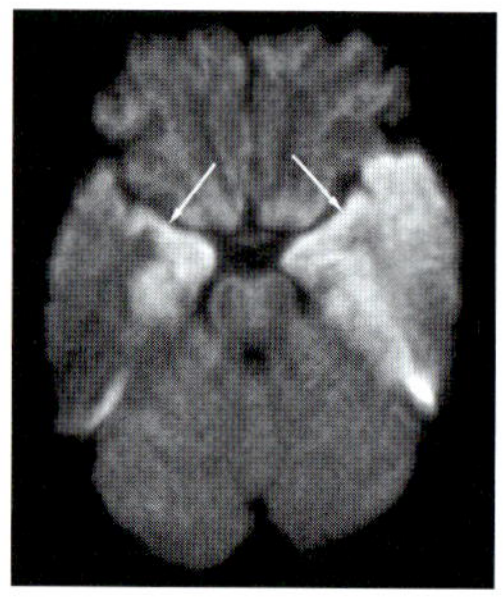 Fig. 14.13 b	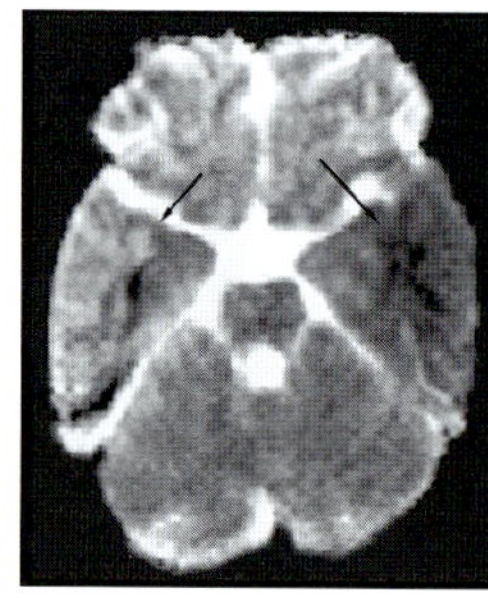 Fig. 14.13 c	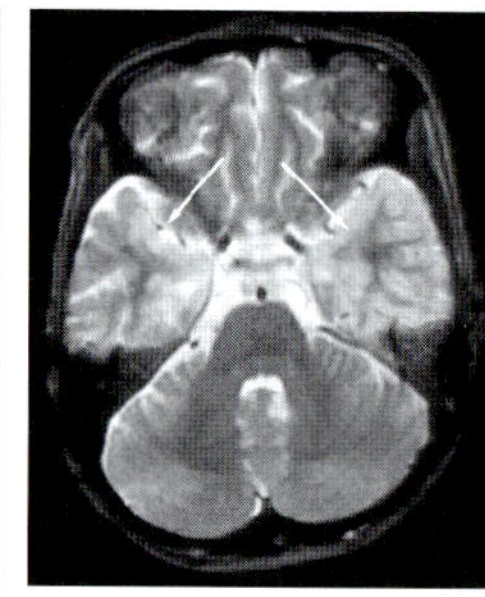 Fig. 14.13 a
Toxoplasmosis	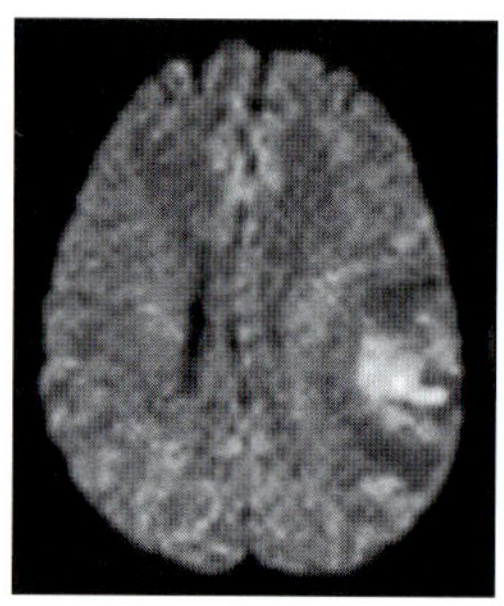 Fig. 4.16 b	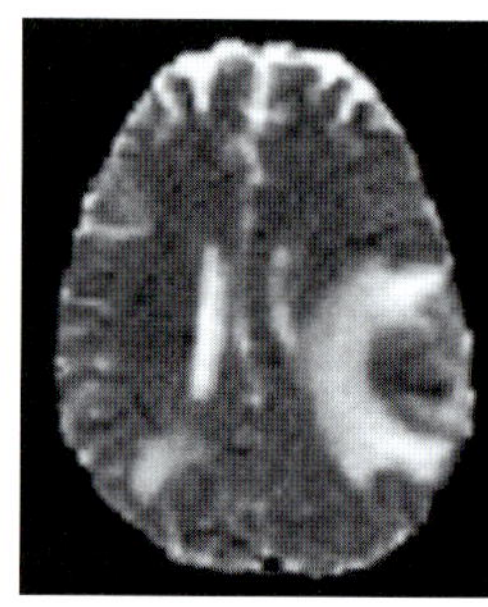 Fig. 4.16 c	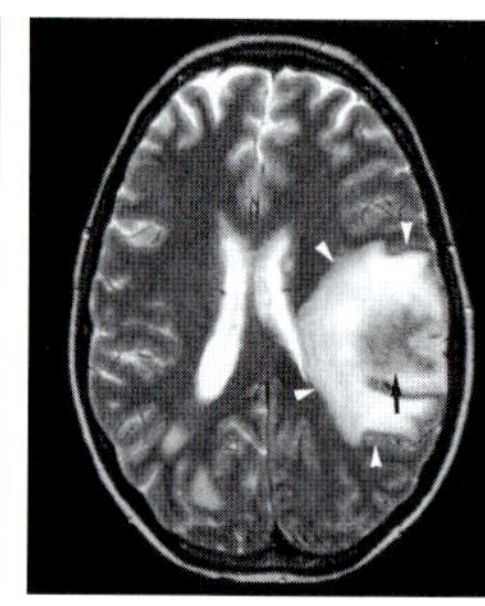 Fig. 4.16 a
Toxic/metabolic			
Carmofur	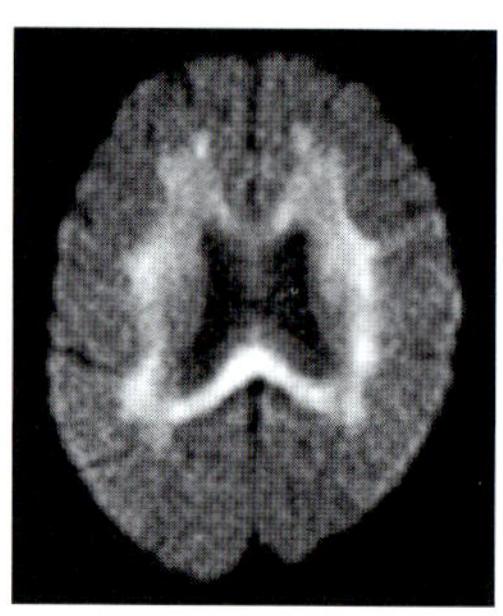 Fig.10.2 b	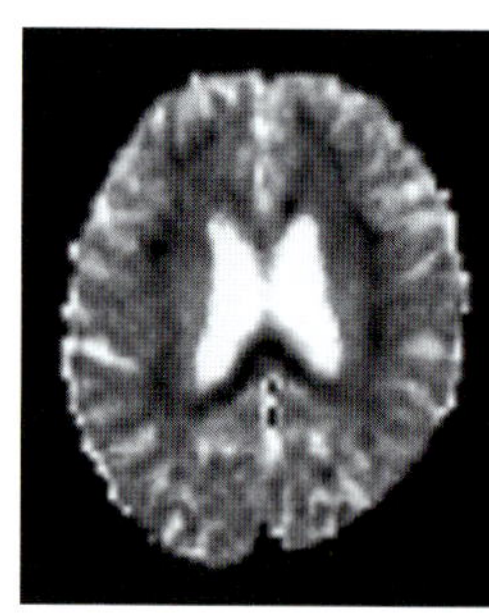 Fig.10.2 c	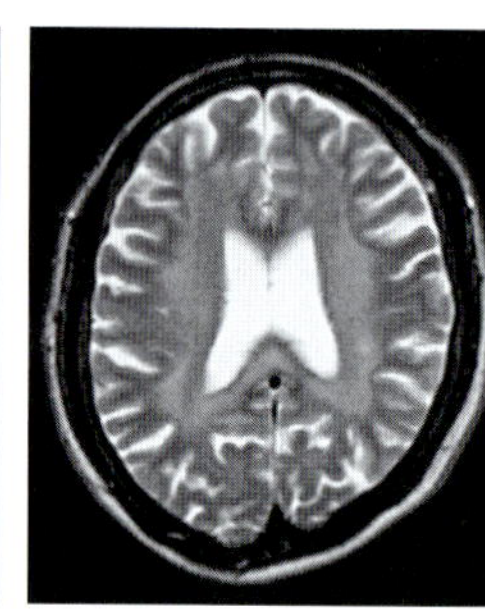 Fig.10.2 a
Heroin	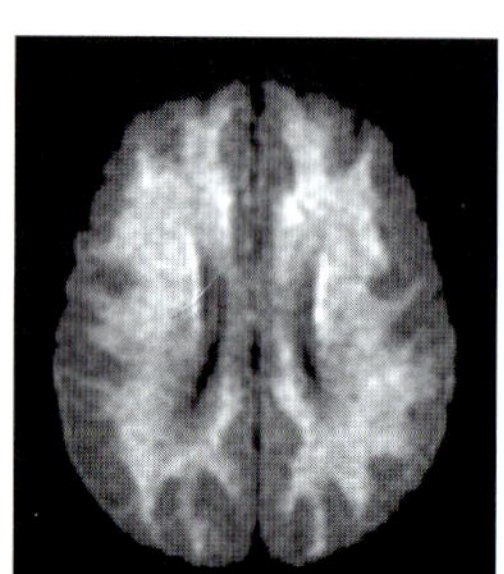 Fig. 10.3 b	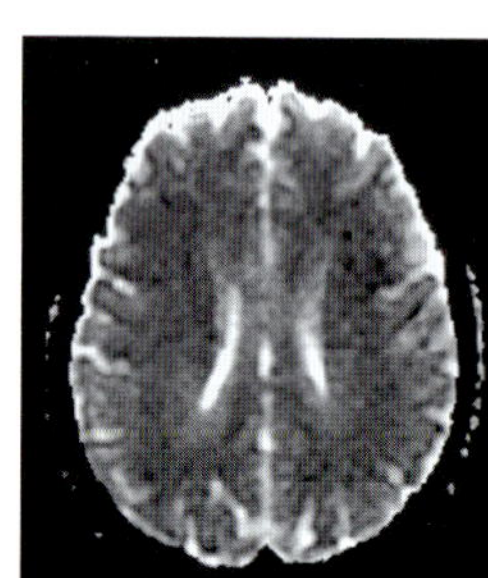 Fig. 10.3 c	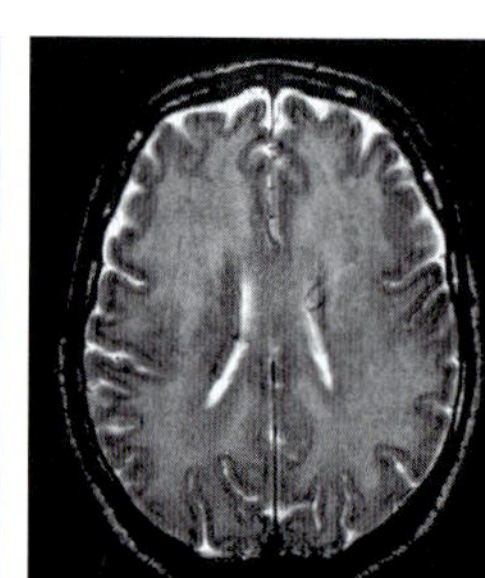 Fig. 10.3 a

Table 3

Differential diagnoses for lesions with a **high diffusion** signal associated with a **low ADC** and **high intense T2** signal

Diagnose	**Reference images**		
	DWI high	**ADC low**	**T2 high**
Central pontine myelinolysis (CPM)	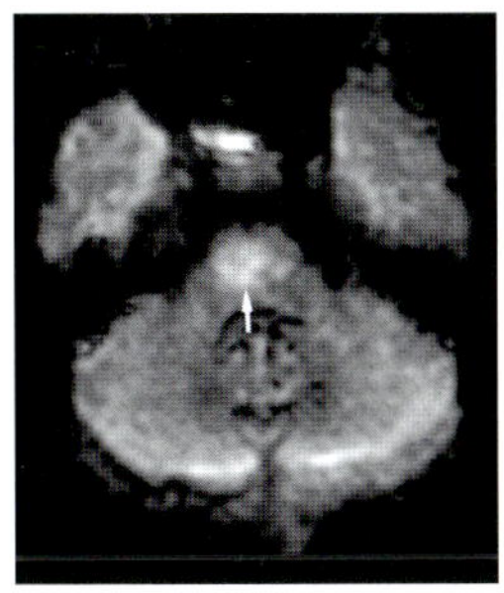 Fig. 10.4. b	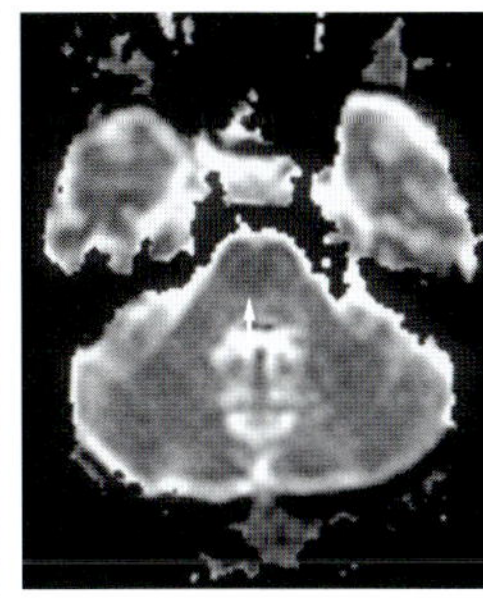 Fig. 10.4. c	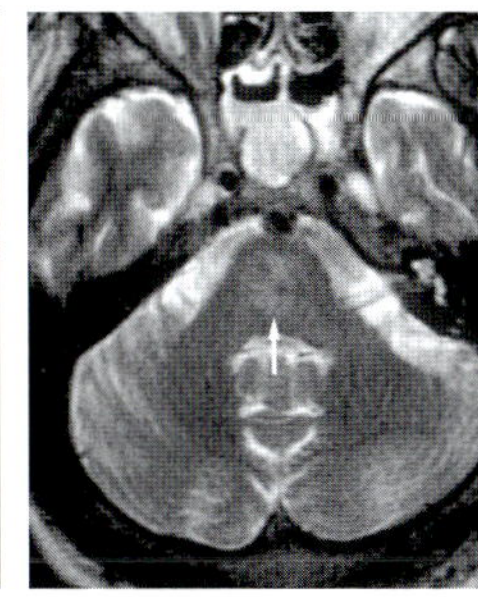 Fig. 10.4. a
Extrapotine meyelinolysis (EPM)	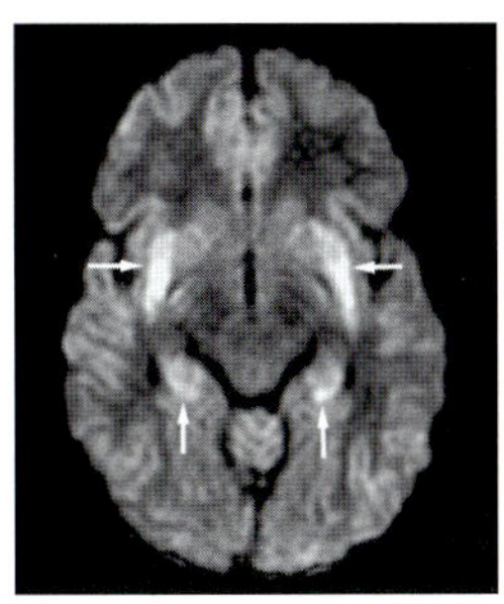 Fig. 10.5 b	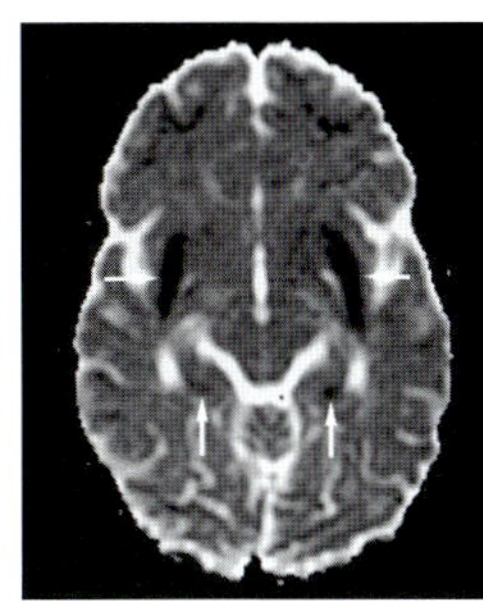 Fig. 10.5 c	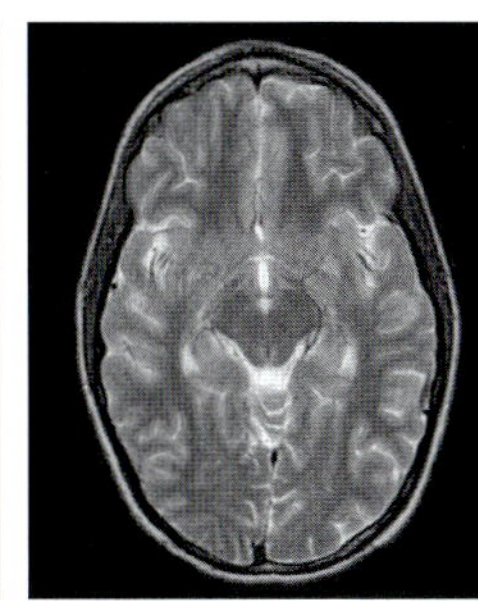 Fig. 10.5 a
Glutaric aciduria	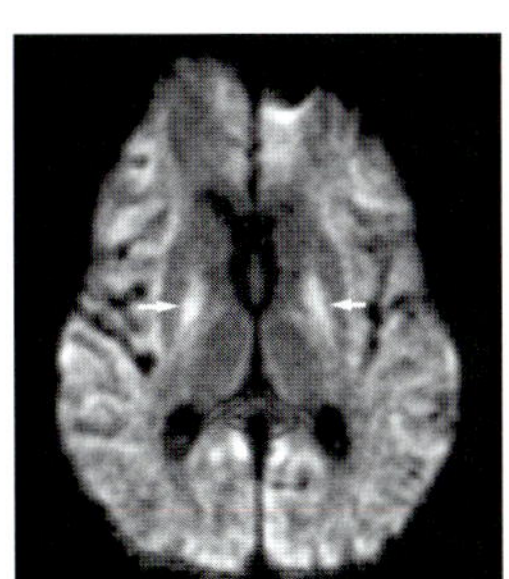 Fig.10.11 b	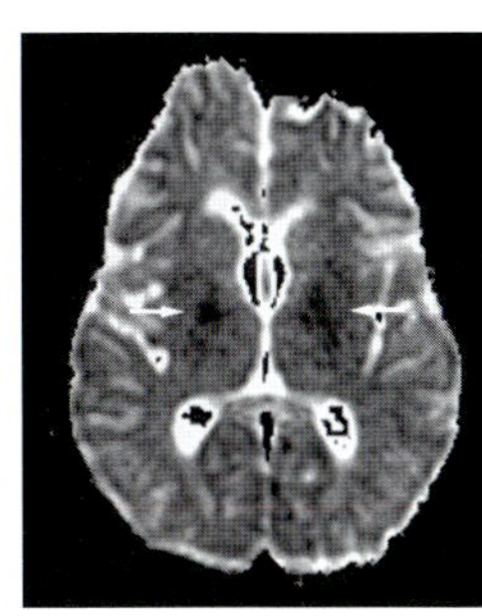 Fig.10.11 c	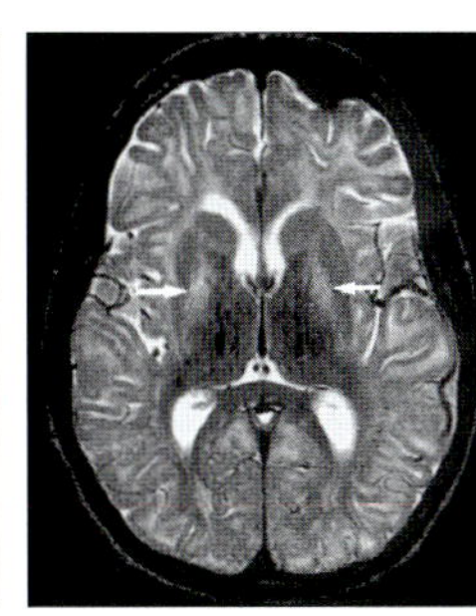 Fig.10.11 a
Mitochondrial encephalomyopathy	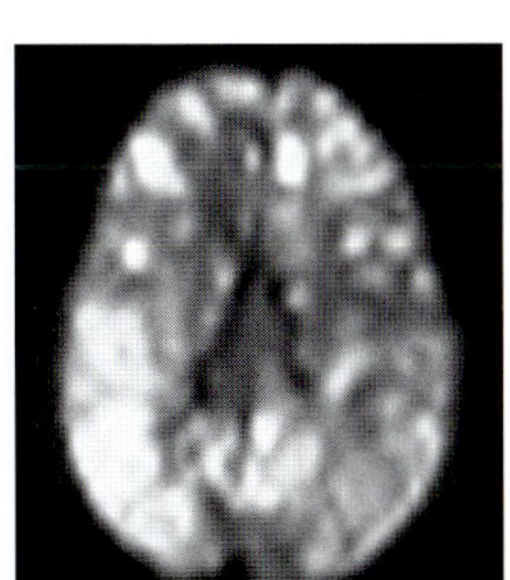 Fig. 10.9. b	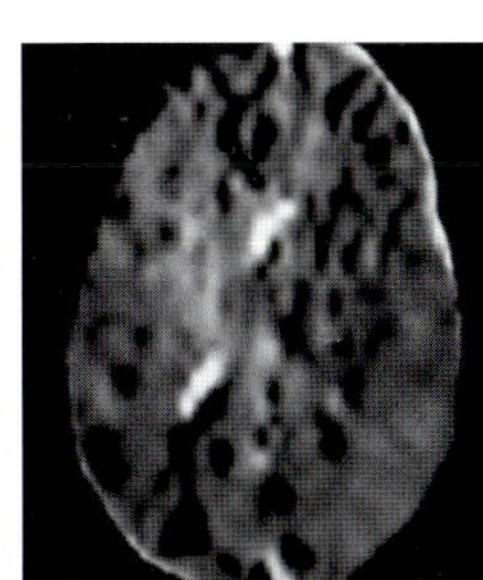 Fig. 10.9. c	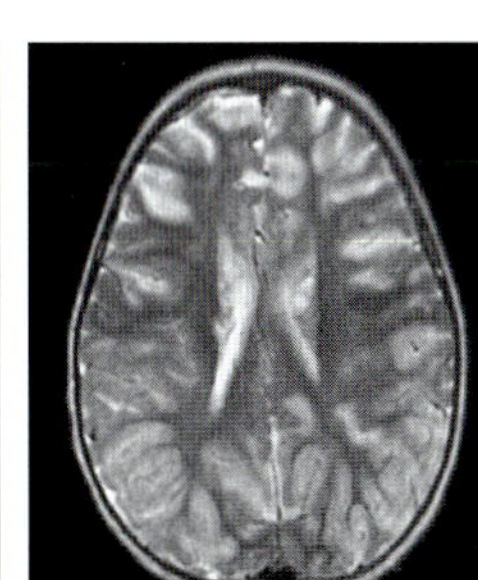 Fig. 10.9. a

Table 3

Differential diagnoses for lesions with a **high diffusion** signal associated with a **low ADC** and **high intense T2** signal

Diagnose	**Reference images** DWI high	ADC low	T2 high
Phenylketonuria (PKU)	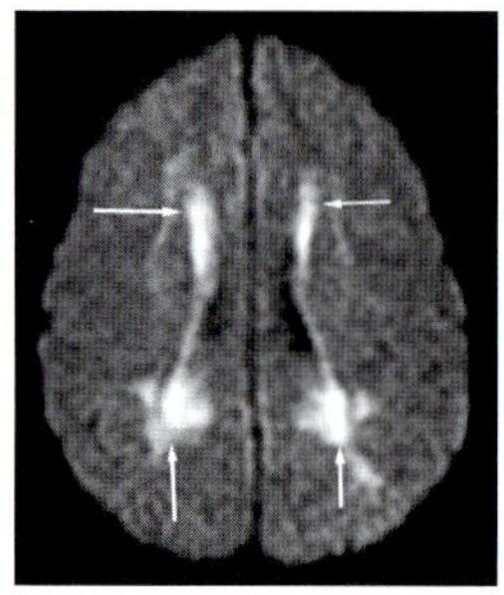 Fig. 14.20 b	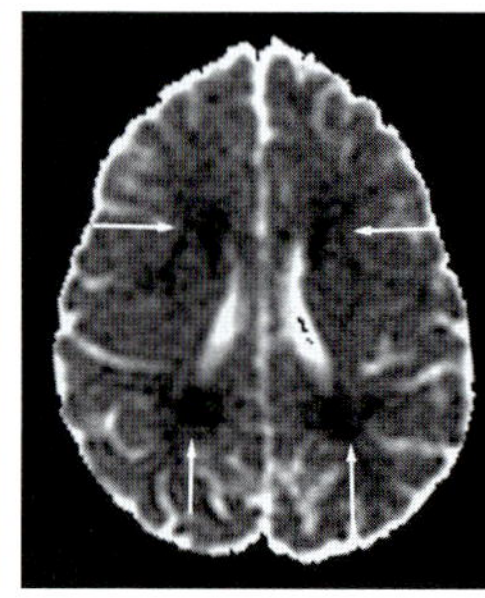 Fig. 14.20 c	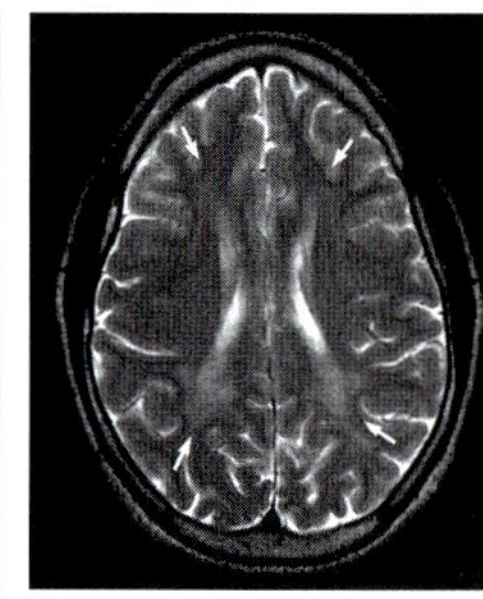 Fig. 14.20 a
Trauma			
Battered child	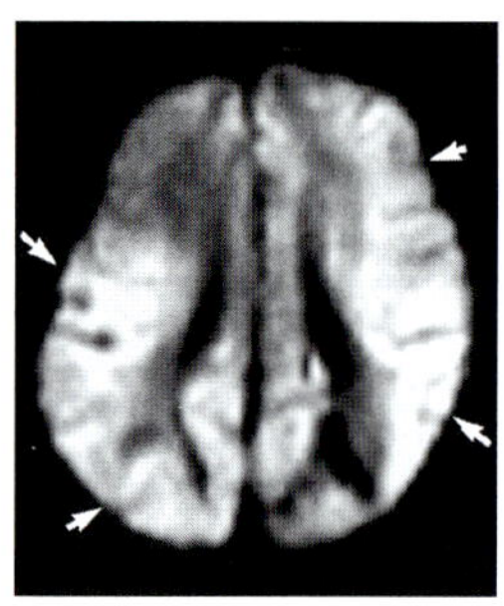 Fig. 14.8 b	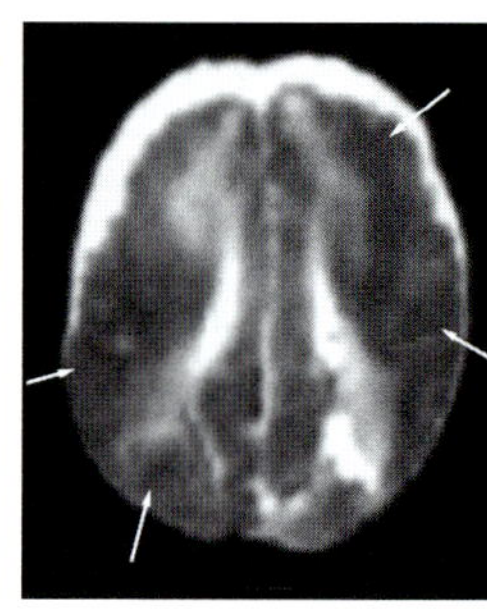 Fig. 14.8 d	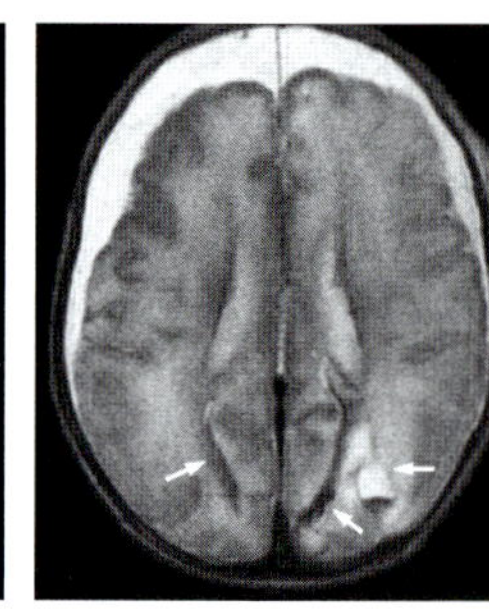 Fig. 14.8 a
Contusion	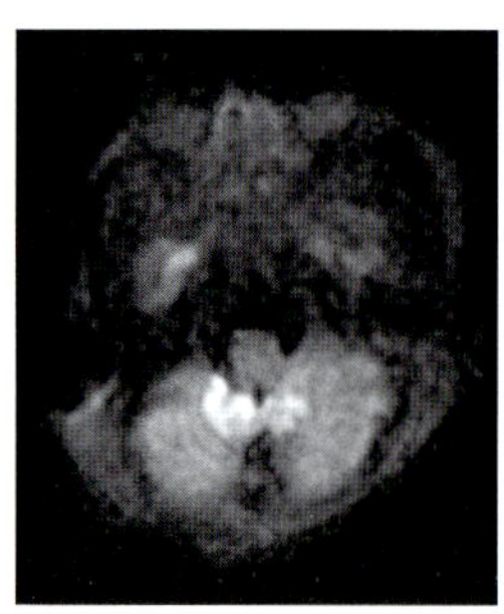 Fig.12.9 b	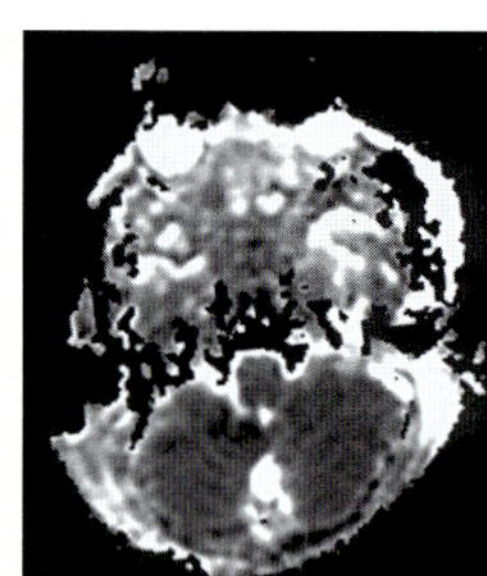 Fig.12.9 c	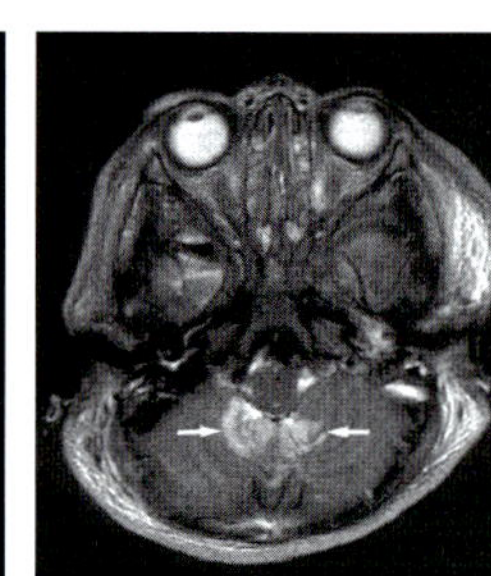 Fig.12.9 a
Diffuse axonal injury (DAI)	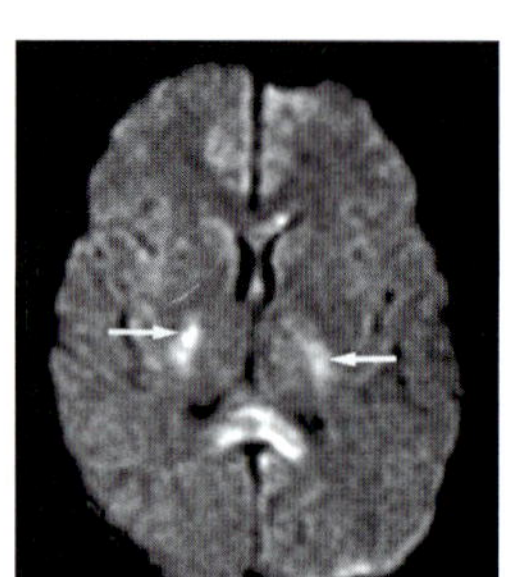 Fig. 12.2 c	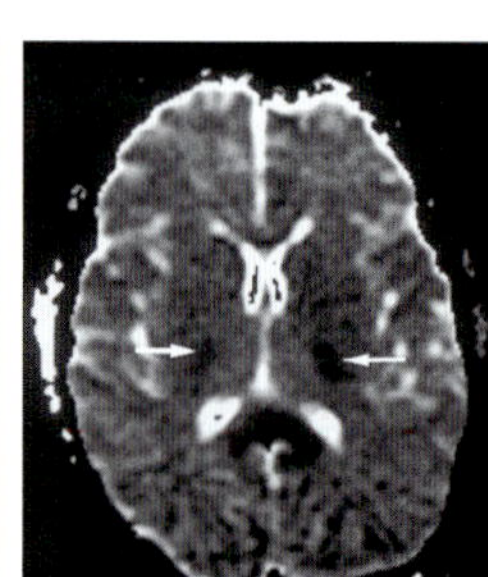 Fig. 12.2 d	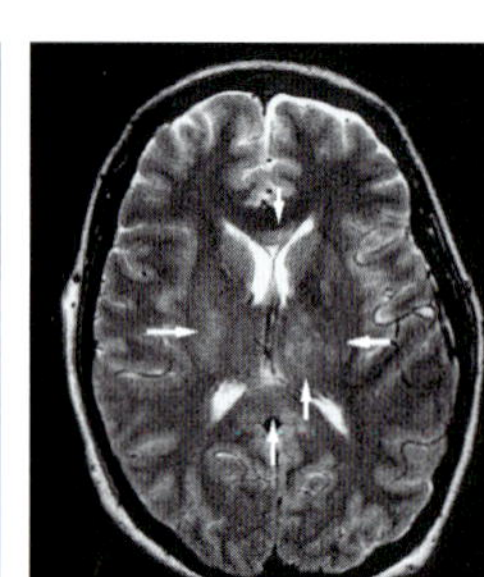 Fig. 12.2 a

Table 3

Differential diagnoses for lesions with a **high diffusion** signal associated with a **low ADC** and **high intense T2** signal

Diagnose	**Reference images**		
	DWI high	**ADC low**	**T2 high**
Tumor			
Glioblastoma multiforme (GBM) (solid)	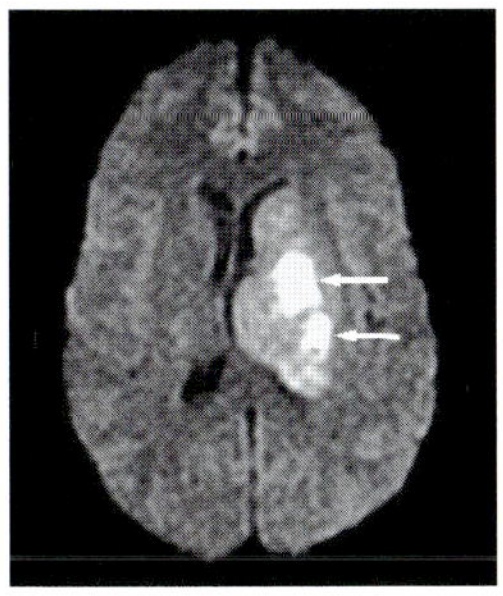 Fig. 13.2 d	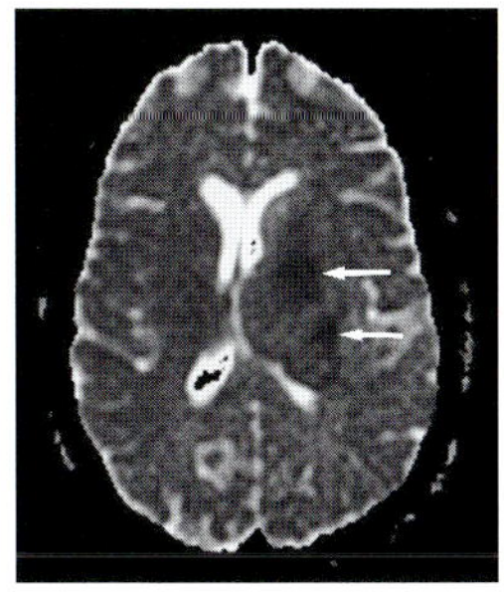 Fig. 13.2 e	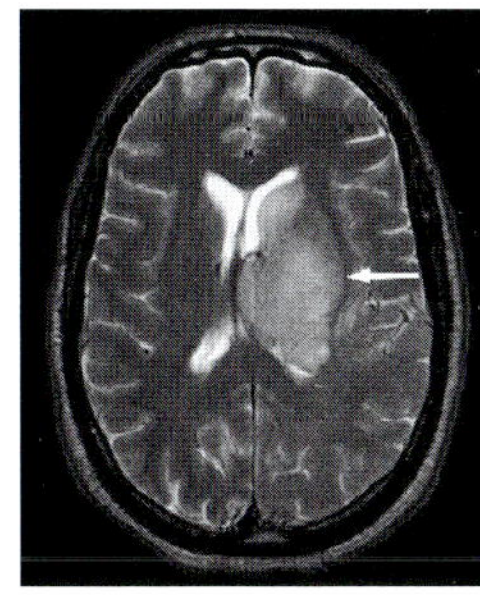 Fig. 13.2 a
Lymphoma	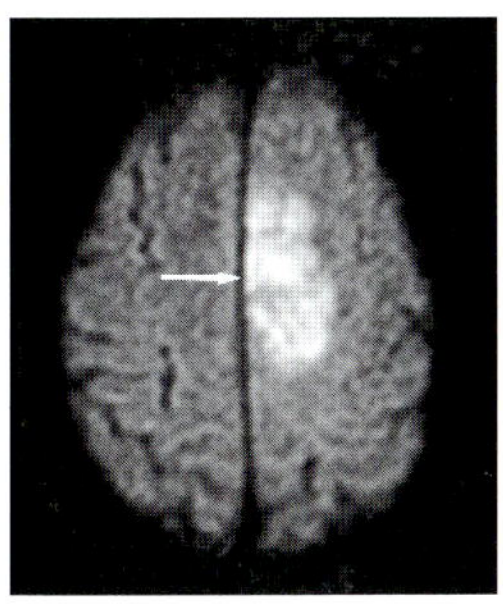 Fig. 13.13 d	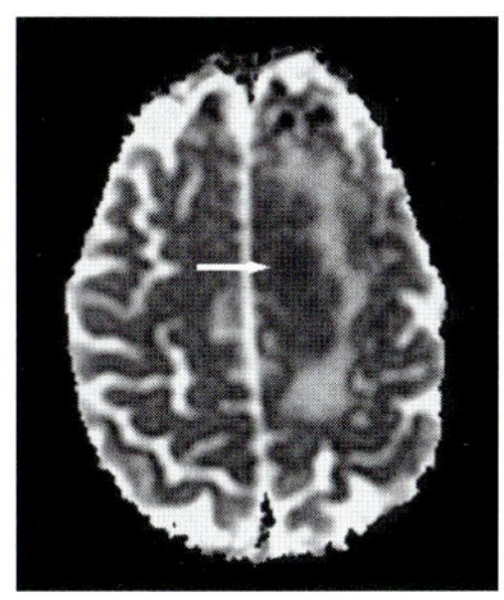 Fig. 13.13 e	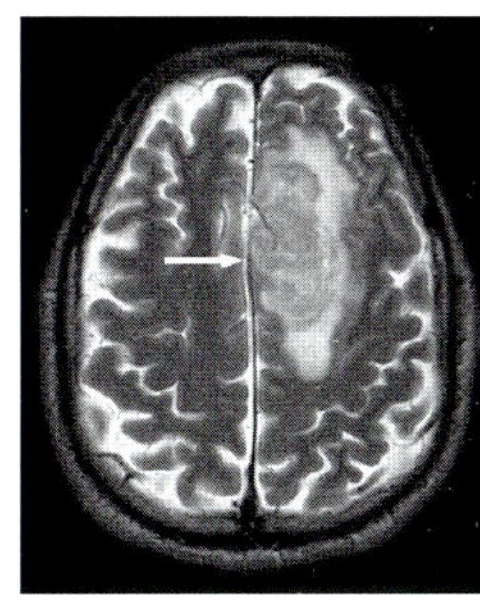 Fig. 13.13 a
Meningioma (syncytial)	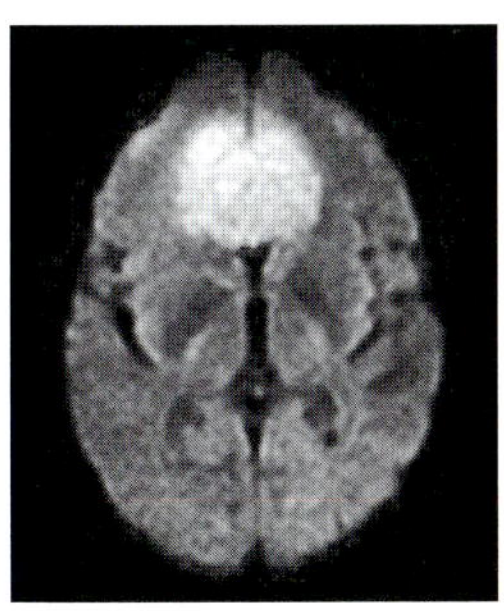 Fig.13.11 c	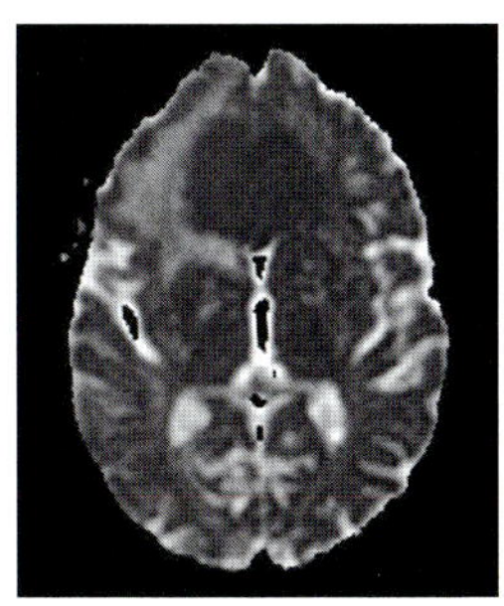 Fig.13.11 d	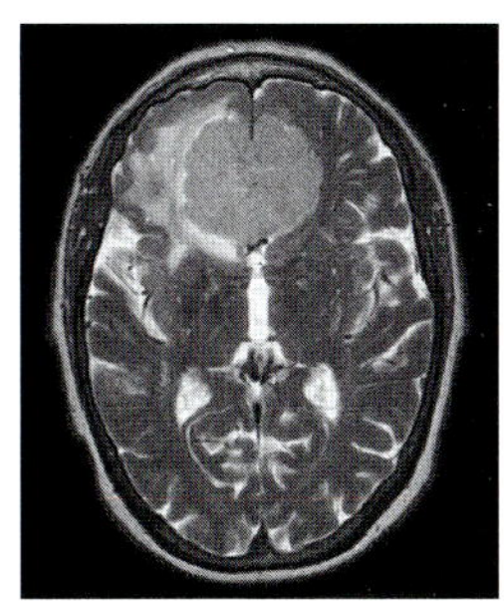 Fig.13.11 a
Meningioma (atypical)	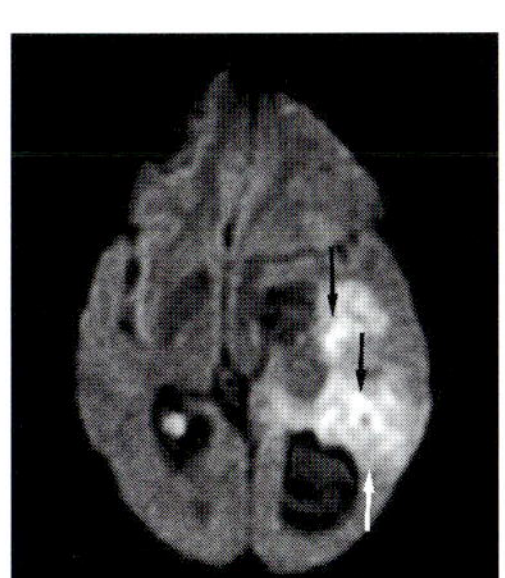 Fig. 13.12 c	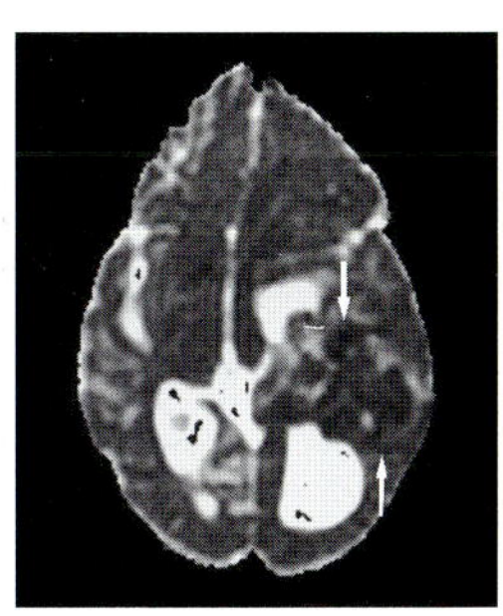 Fig. 13.12 d	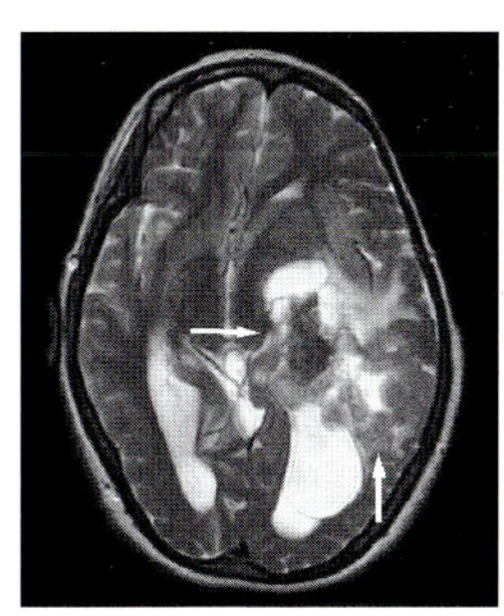 Fig. 13.12 a

Table 3

Differential diagnoses for lesions with a **high diffusion** signal associated with a **low ADC** and **high intense T2** signal

Diagnose	Reference images DWI high	ADC low	T2 high
Metastasis	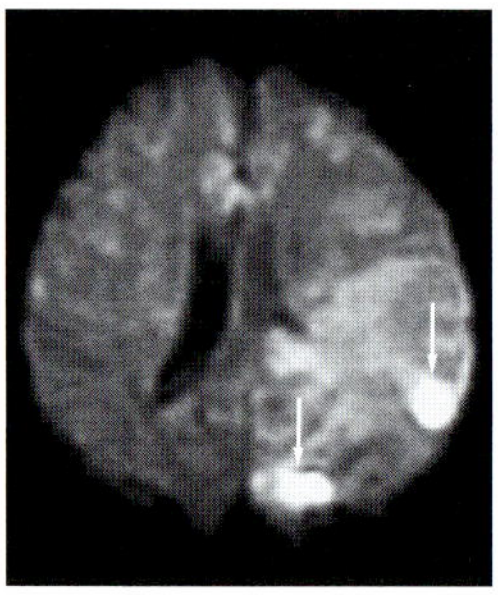 **Fig. 11.6 c**	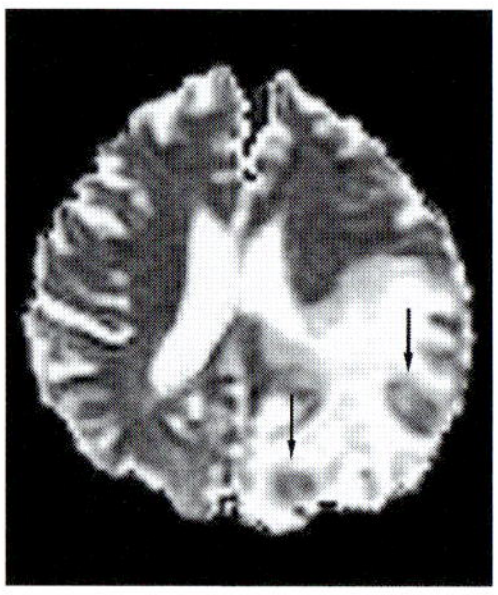 **Fig. 11.6 d**	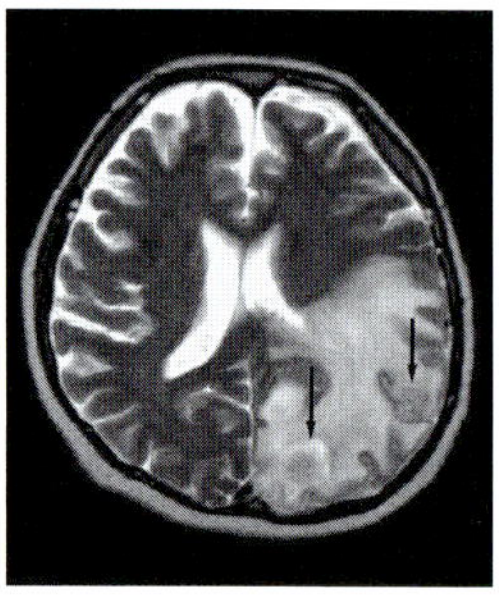 **Fig. 11.6 a**
Primitive neuroectodermal tumor (PNET)	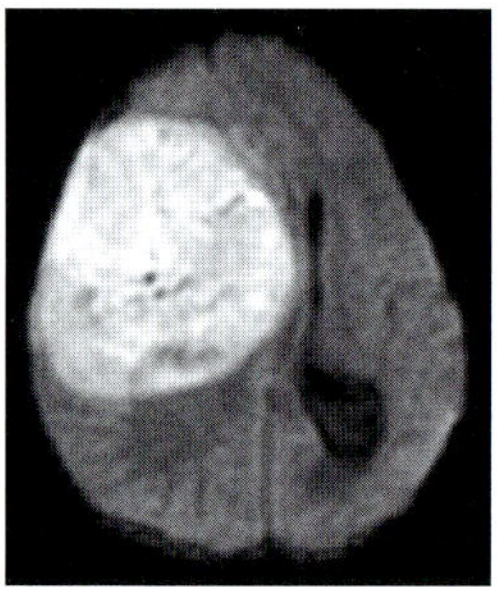 **Fig. 13.10 c**	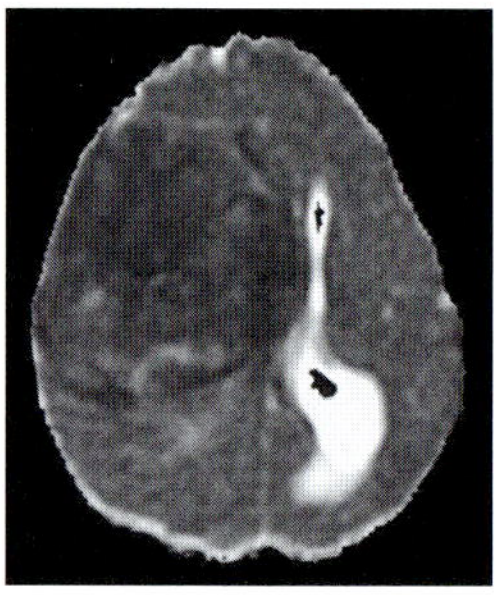 **Fig. 13.10 d**	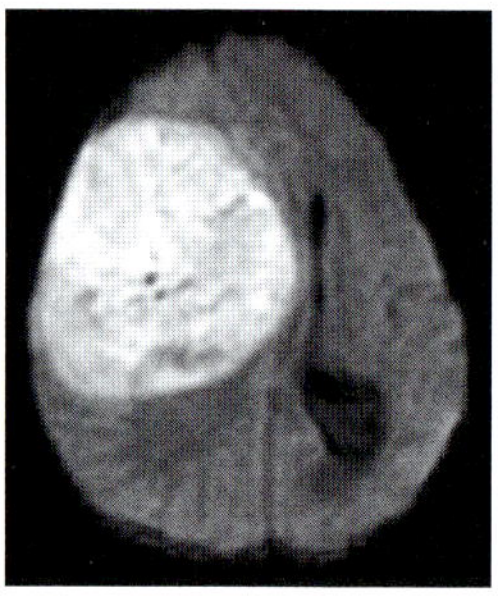 **Fig. 13.10 a**
Radiation necrosis	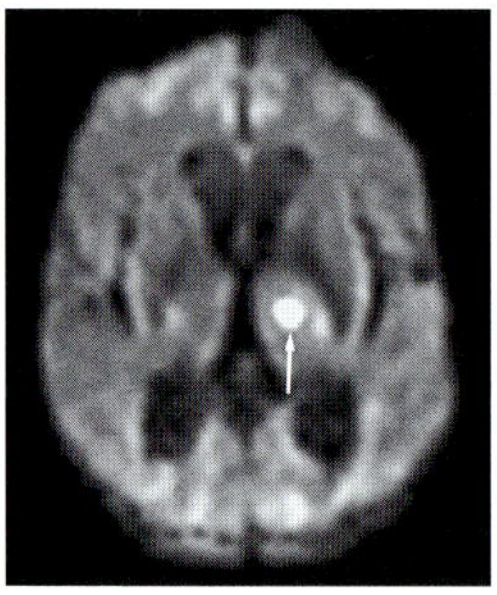 **Fig.11.7 b**	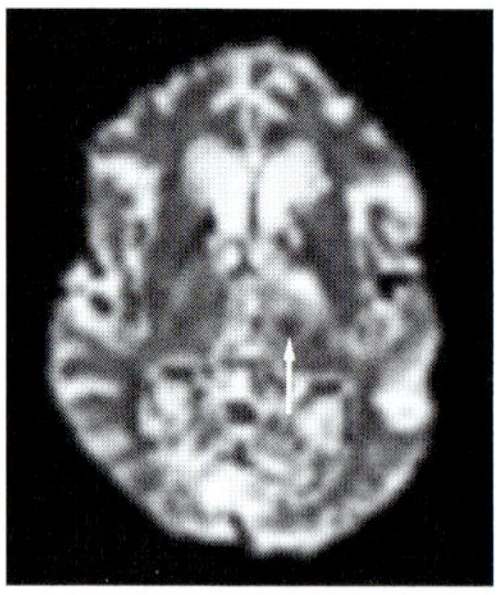 **Fig.11.7 c**	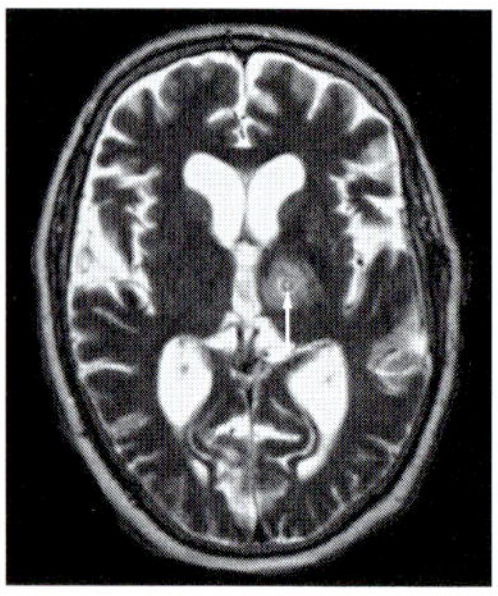 **Fig.11.7 a**
Vasculitis			
Primary angitis of central nervous system (PACNS)	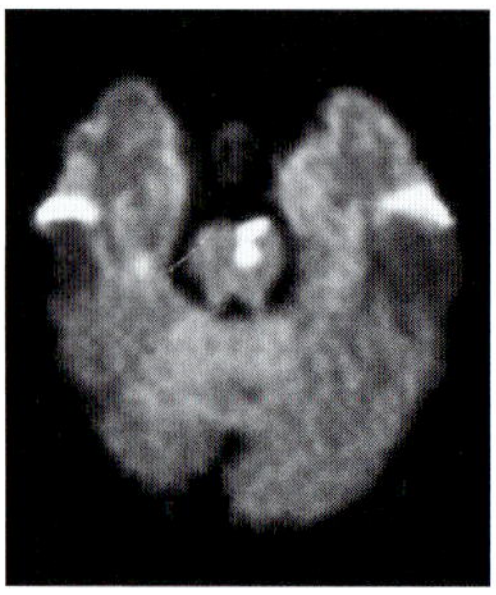 **Fig. 7.3 e**	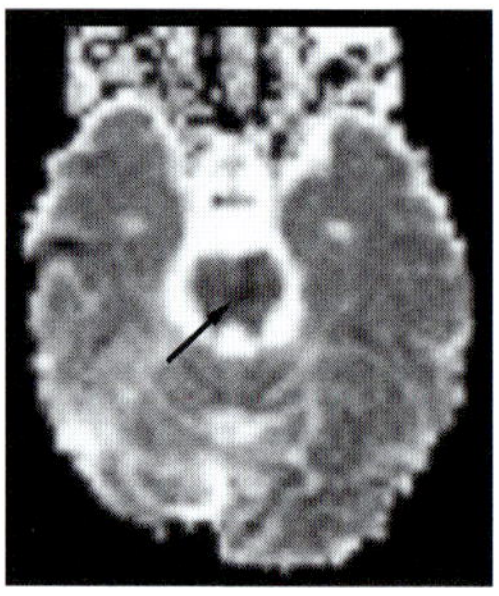 **Fig. 7.3 f**	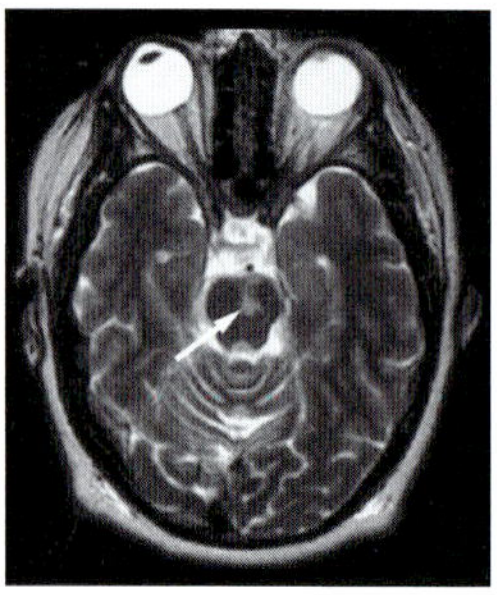 **Fig. 7.3 b**

Table 3

Differential diagnoses for lesions with a **high diffusion** signal associated with a **low ADC** and **high intense T2** signal

Diagnose	Reference images		
	DWI high	**ADC low**	**T2 high**
Pneumococcal meningitis and vasculitis	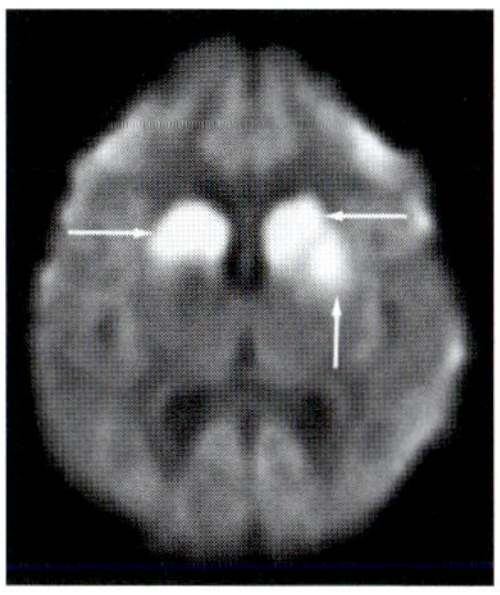 Fig. 7.8 b	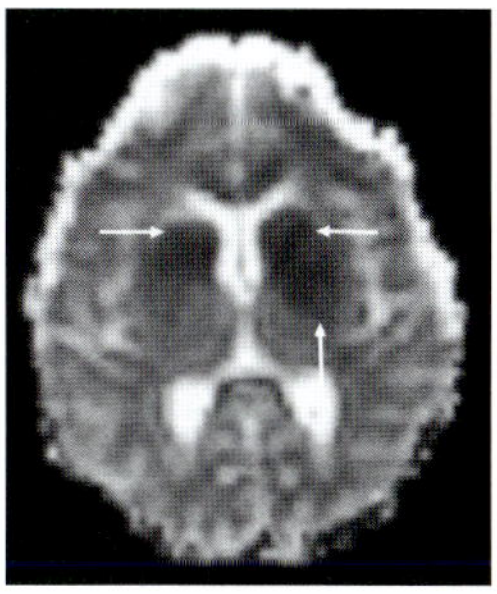 Fig. 7.8 c	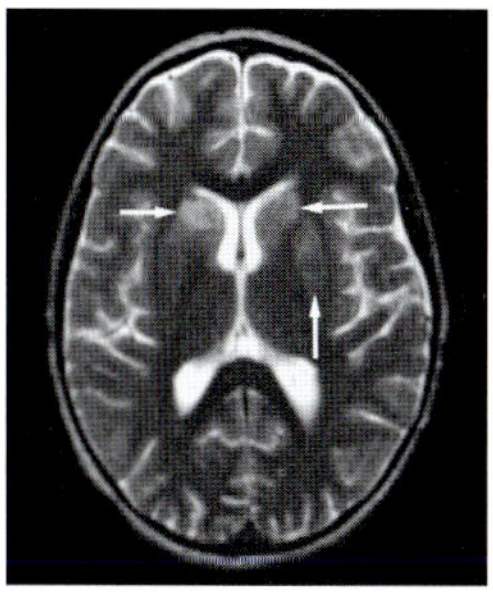 Fig. 7.8 a
Sickle cell disease	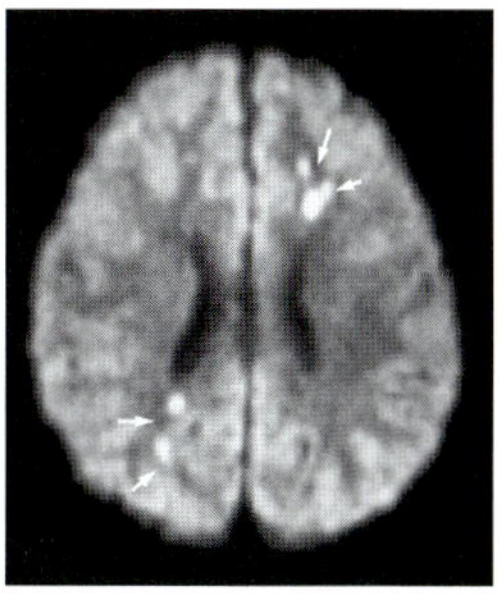 Fig. 14.5 b	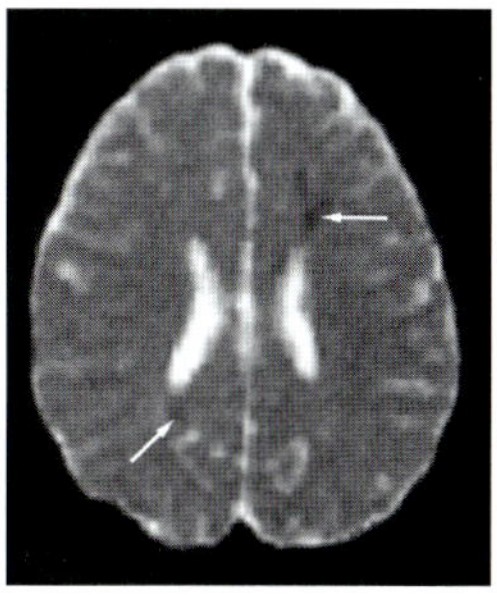 Fig. 14.5 c	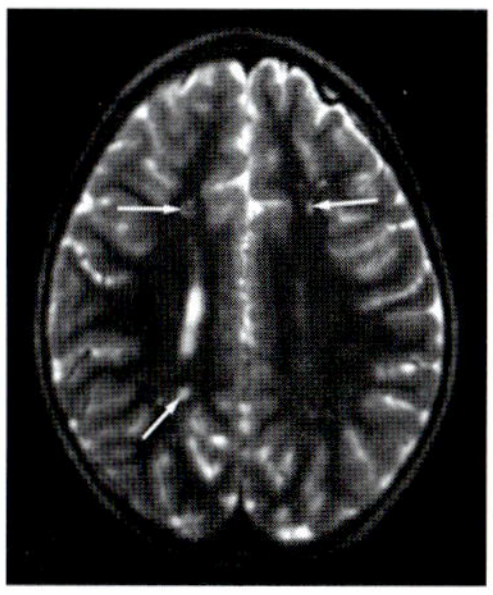 Fig. 14.5 a
Hemolytic uremic syndrome (HUS)	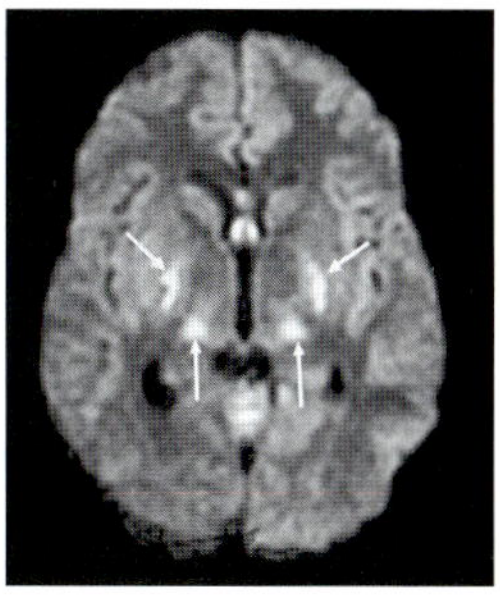 Fig.7.19 b	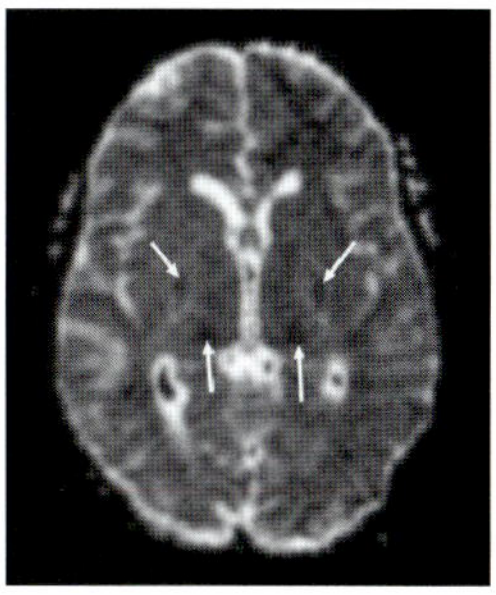 Fig.7.19 c	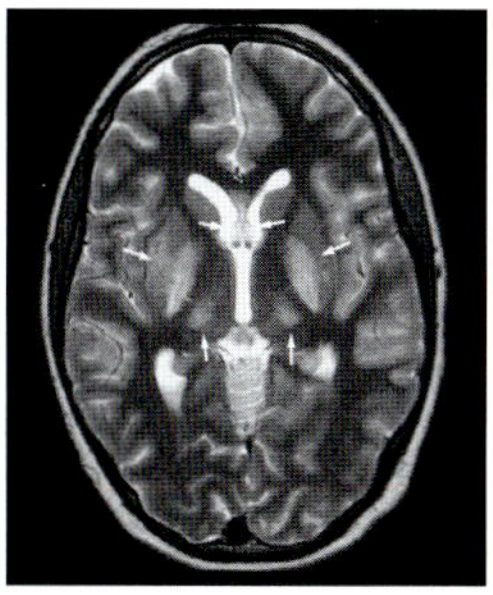 Fig.7.19 a

Table 4

Differential diagnoses for lesions with an **iso diffusion** signal associated with a **high ADC** and **high intense T2** signal

Diagnose	Reference images DWI iso	 ADC high	 T2 high
Vasculitis/Vasculopathy			
Posterior reversible encephalopathy synsdrome (PRES)	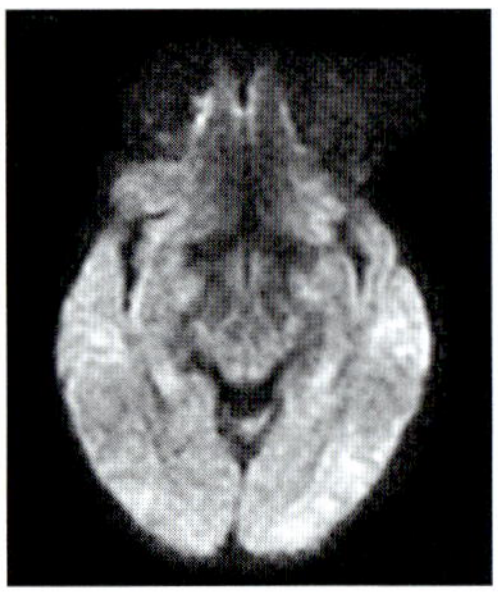 Fig. 3.4 c	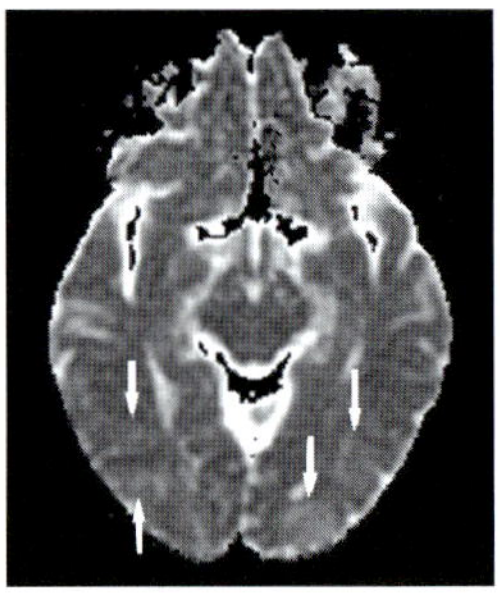 Fig. 3.4 d	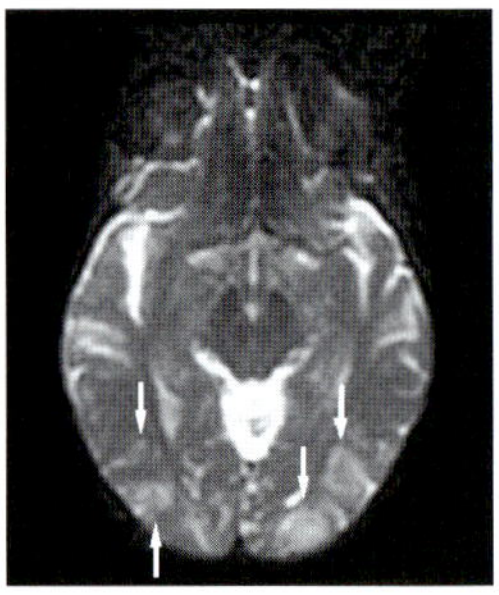 Fig. 3.4 b
Vasogenic edema (metastasis)	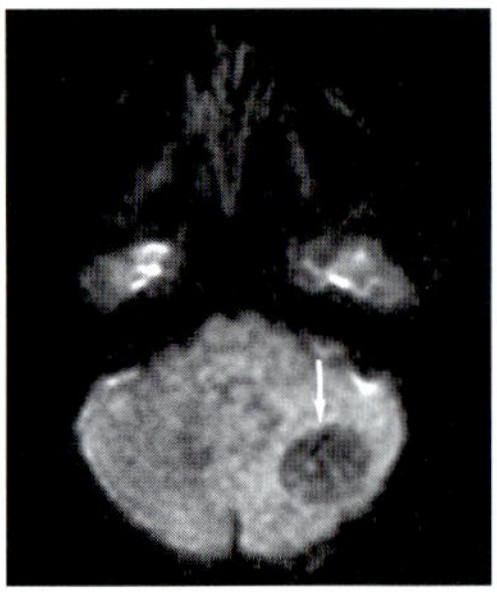 Fig. 3.5 e	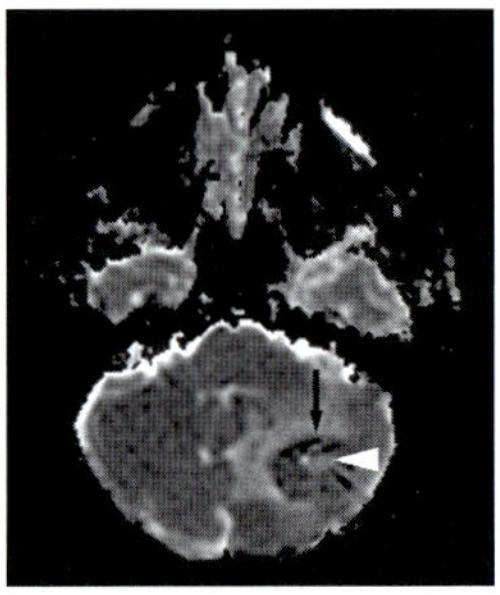 Fig. 3.5 d	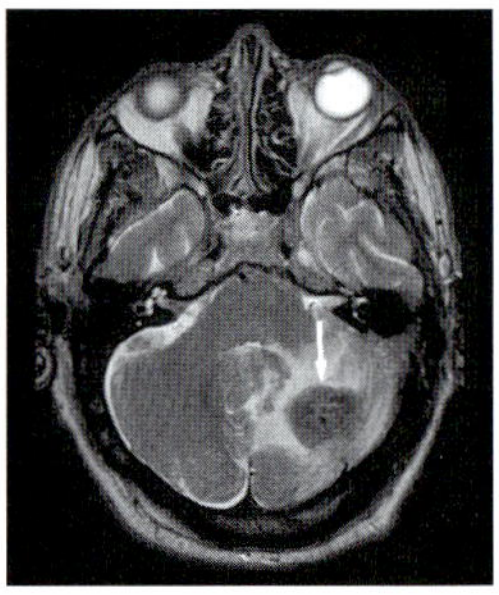 Fig. 3.5 a
Tacrolimus neurotoxicity	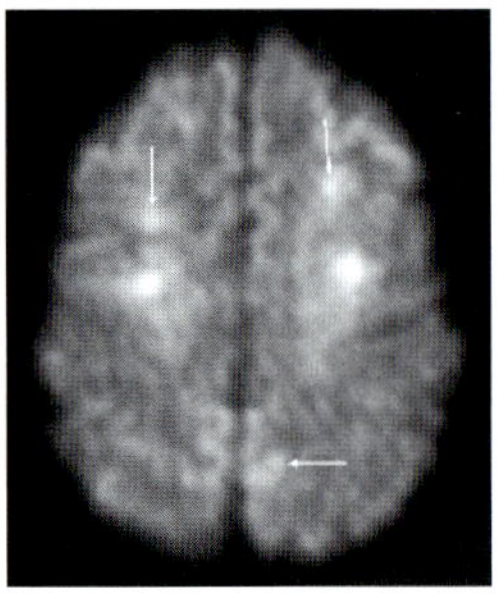 Fig. 7.17 b	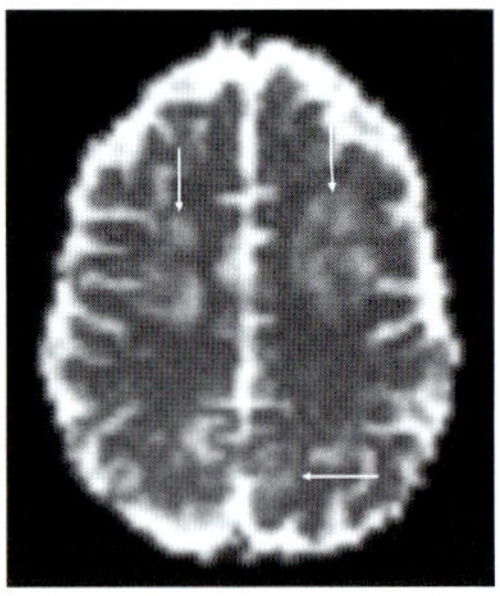 Fig. 7.17 c	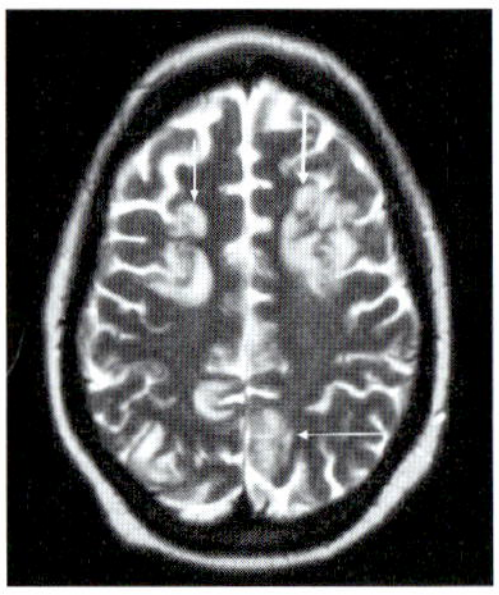 Fig. 7.17 a
Postictal	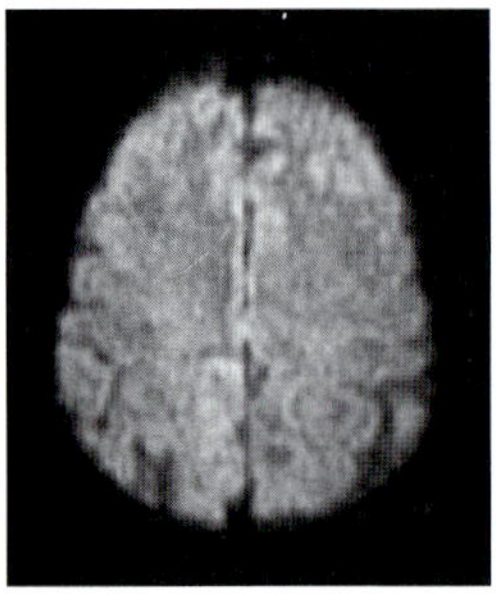 Fig. 8.2 d	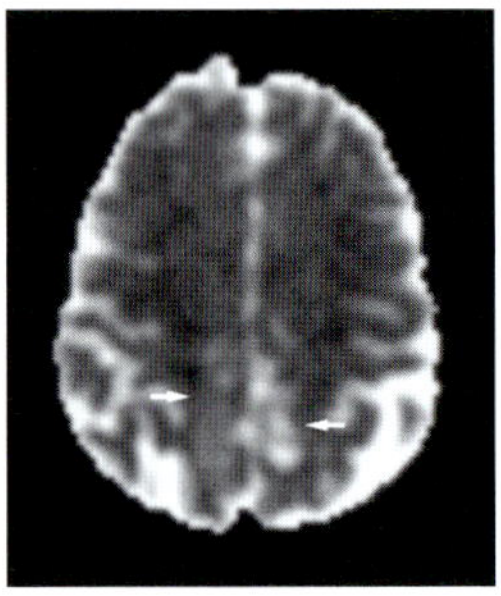 Fig. 8.2 e	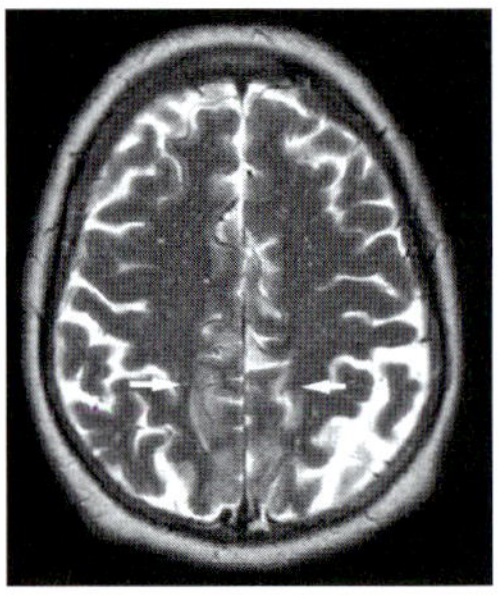 Fig. 8.2 a

Table 4

Differential diagnoses for lesions with an **iso diffusion** signal associated with a **high ADC** and **high intense T2** signal

Diagnose	Reference images		
	DWI iso	ADC high	T2 high
Metabolic disease Wernicke	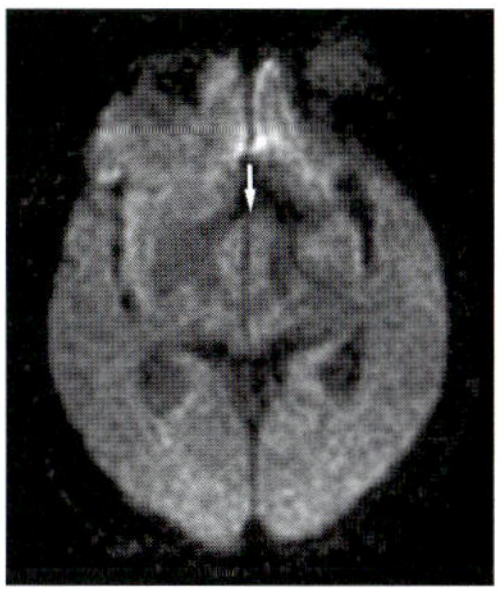 Fig. 10.6 b	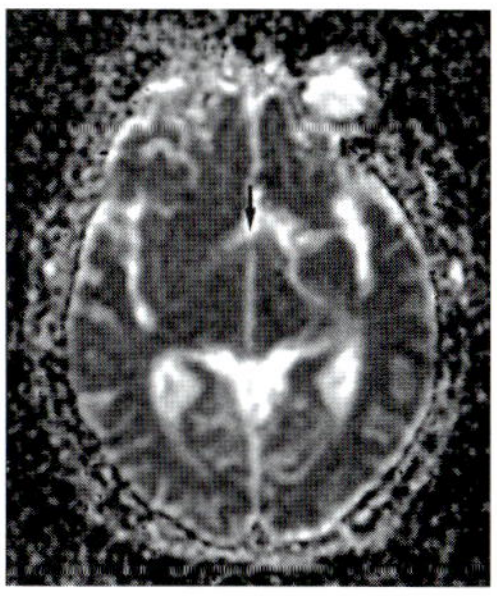 Fig. 10.6 c	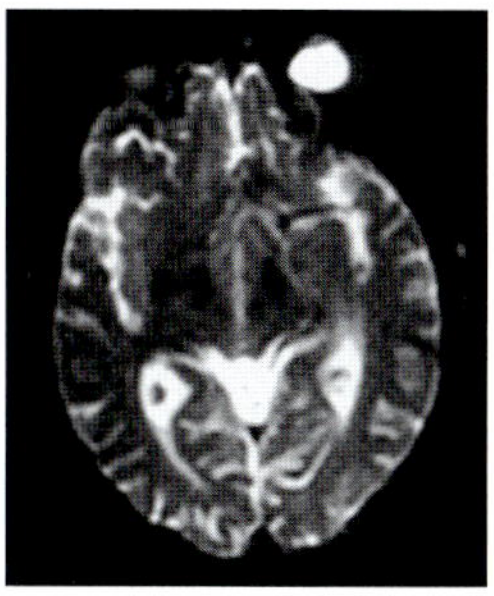 Fig. 10.6 a

Table 5

Differential diagnoses for lesions with a **low diffusion** signal associated with a **high ADC** and **high intense T2** signal

Diagnose	Reference images		
	DWI low	ADC high	T2 high
Infarction Chronic infarction (10 months)	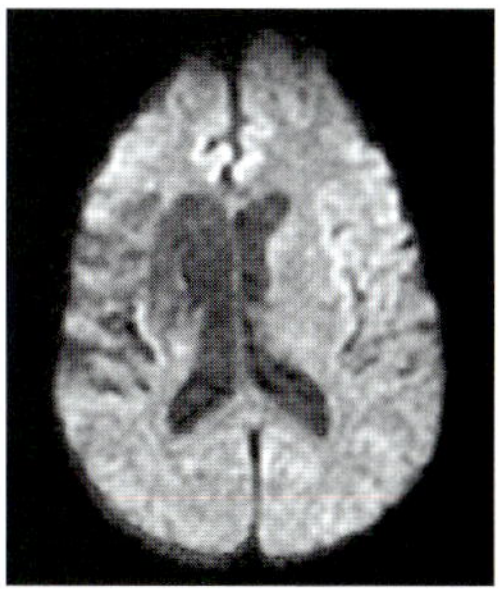 Fig. 5.4 c	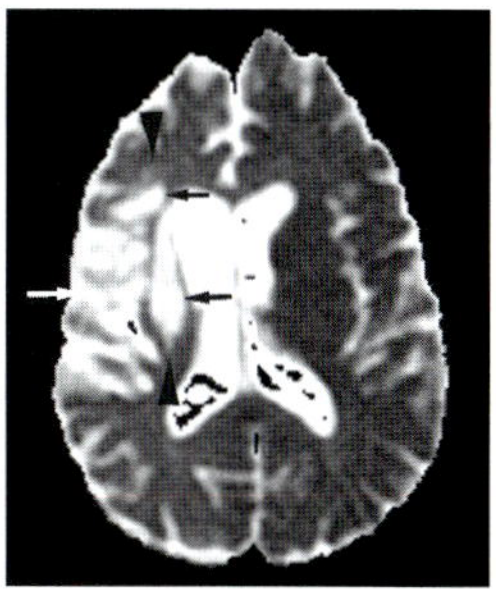 Fig. 5.4 d	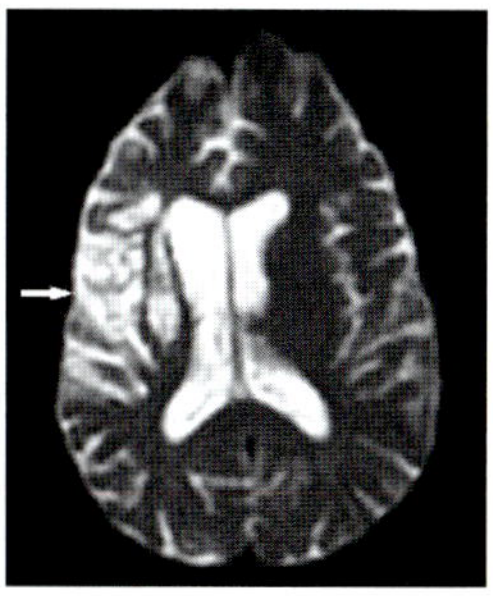 Fig. 5.4 b
Tumor GBM (necrosis)	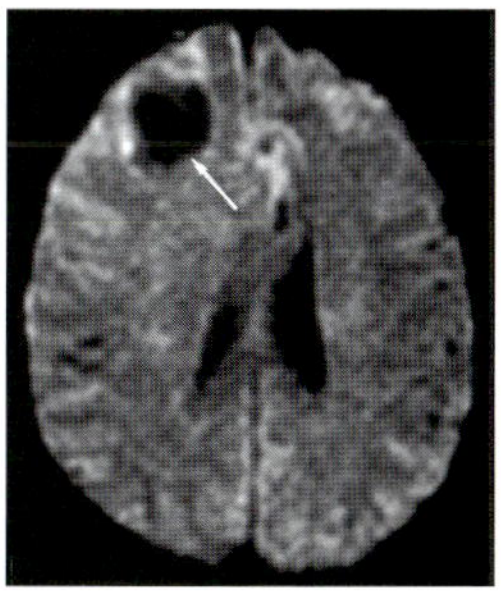 Fig. 11.5 c	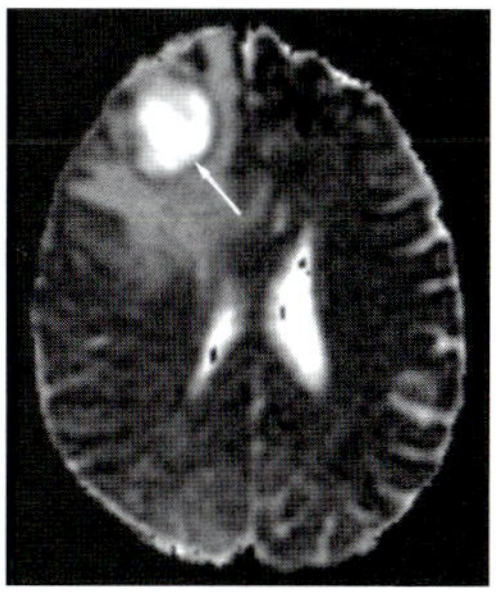 Fig. 11.5 d	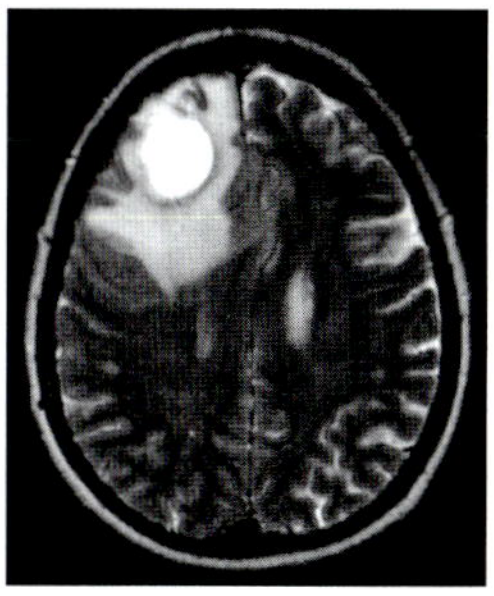 Fig. 11.5 a

Table 5

Differential diagnoses for lesions with a **low diffusion** signal associated with a **high ADC** and **high intense T2** signal

Diagnose	Reference images DWI low	 ADC high	 T2 high
Craniopharyngioma	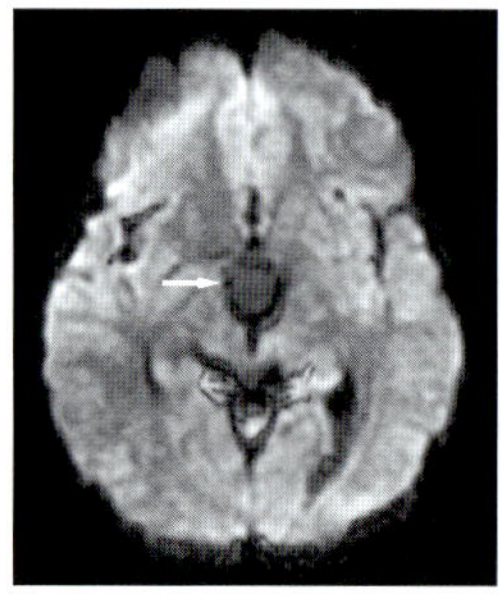 Fig. 13.14 c	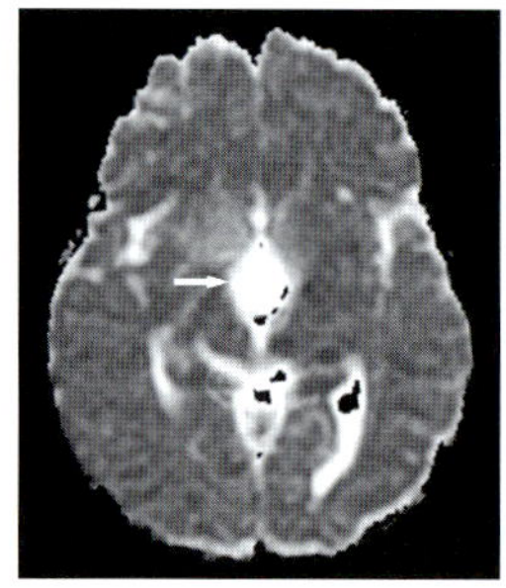 Fig. 13.14 b	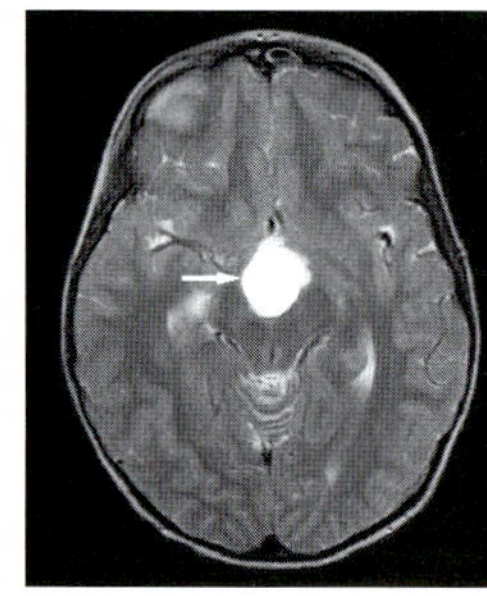 Fig. 13.14 a
Matastasis (lung)	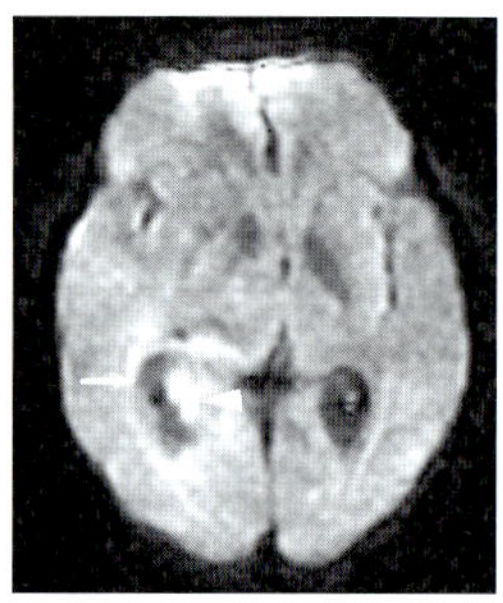 Fig. 13.15 c	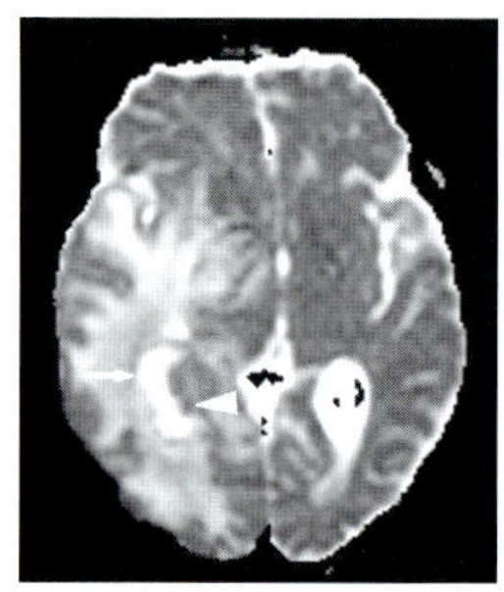 Fig. 13.15 d	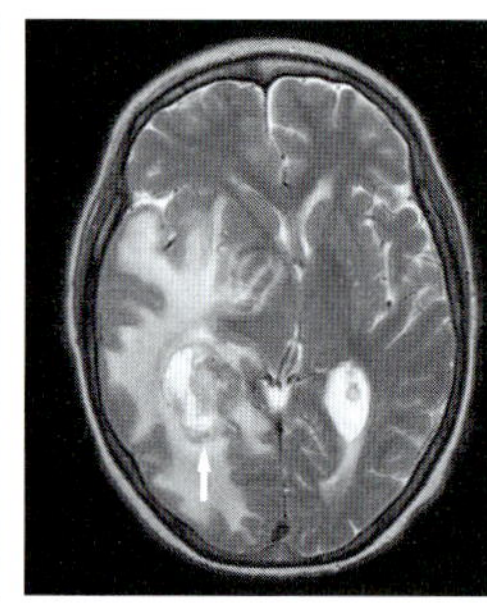 Fig. 13.15 a
Arachnoid cyst	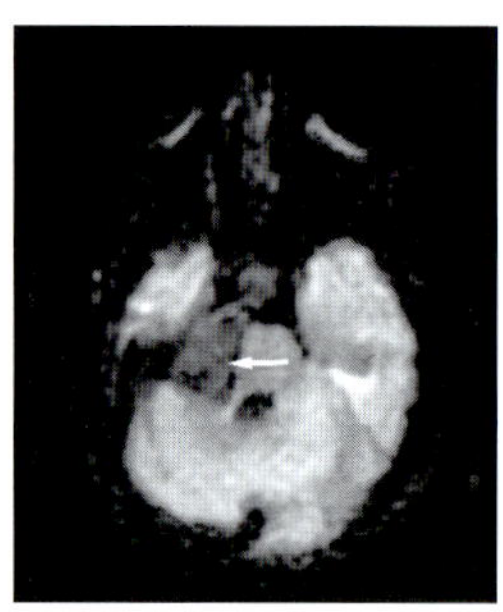 Fig. 13.9 c	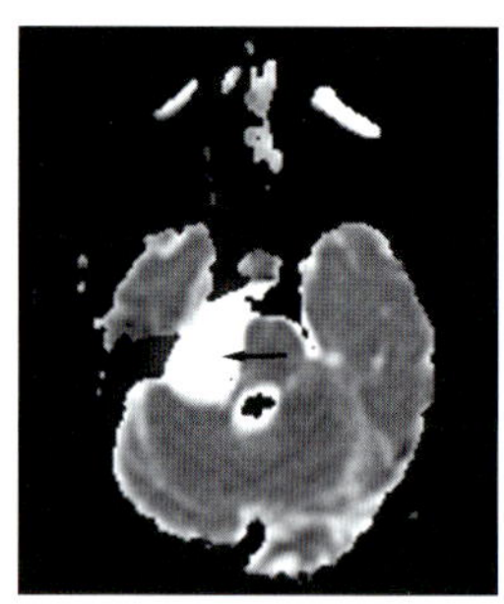 Fig. 13.9 d	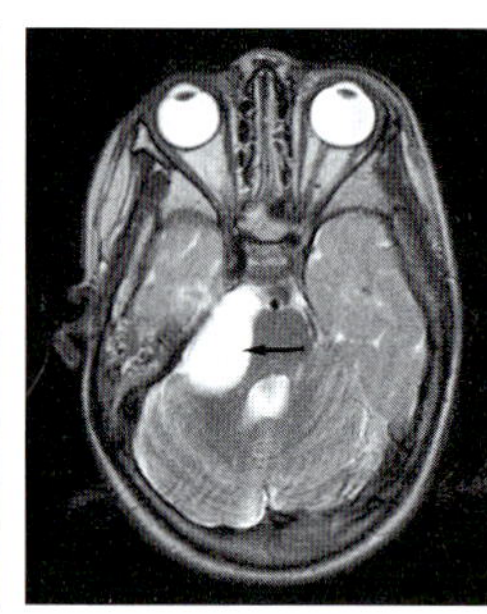 Fig. 13.9 a
Vanishing white matter disease	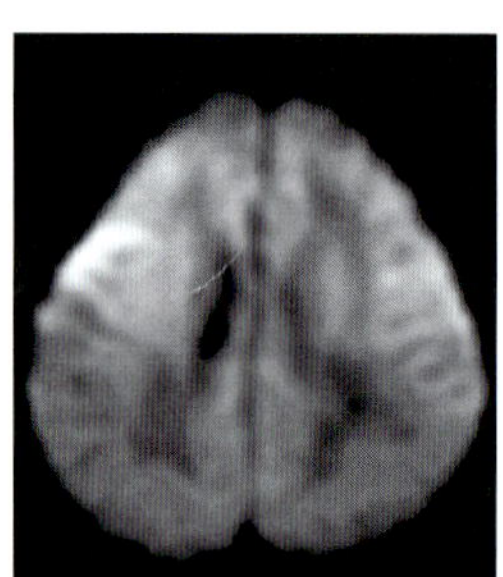 Fig. 14.19 b	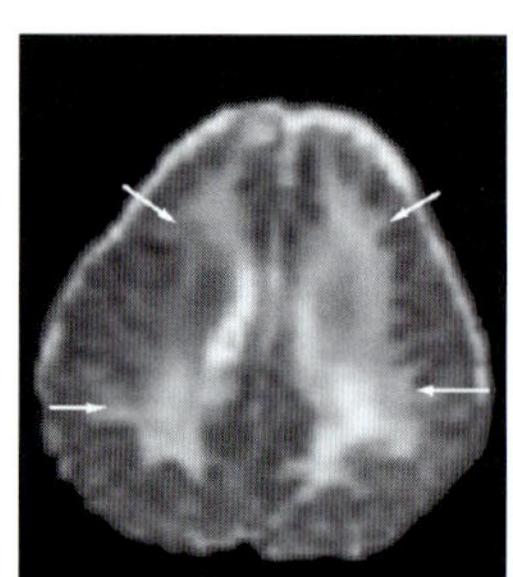 Fig. 14.19 c	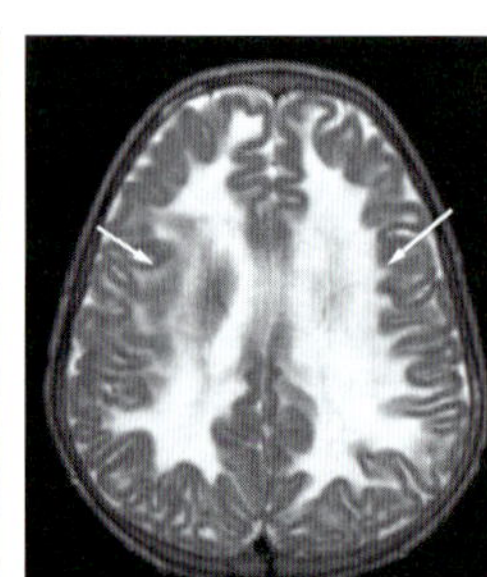 Fig. 14.19 a

Table 5

Differential diagnoses for lesions with a **low diffusion** signal associated with a **high ADC** and **high intense T2** signal

Diagnose	Reference images		
	DWI low	ADC high	T2 high
van der Knaap disease	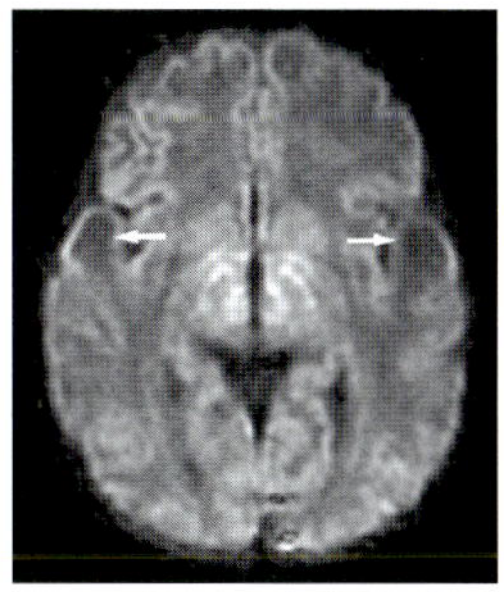 Fig. 10.12 b	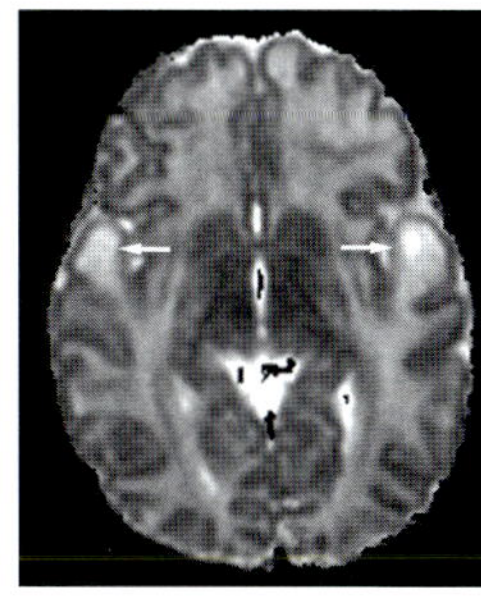 Fig. 10.12 c	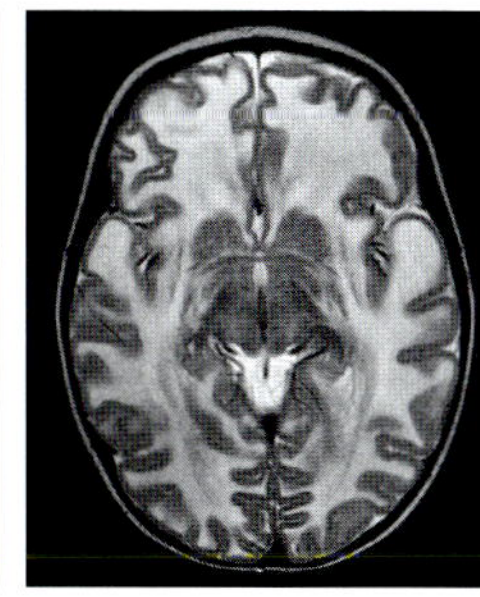 Fig. 10.12 a
Vasogenic edema (toxo)	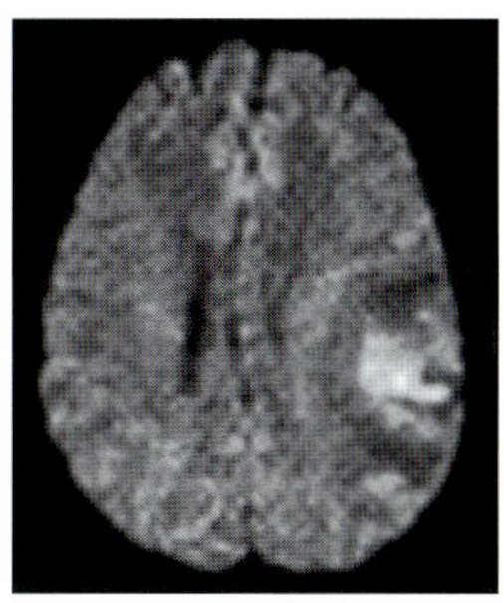 Fig. 4.16 b	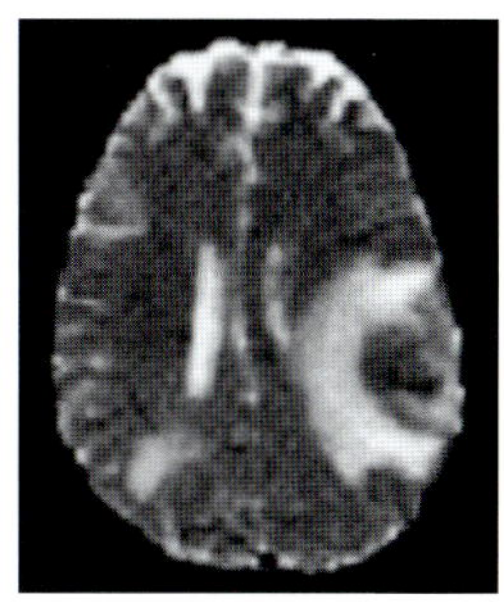 Fig. 4.16 c	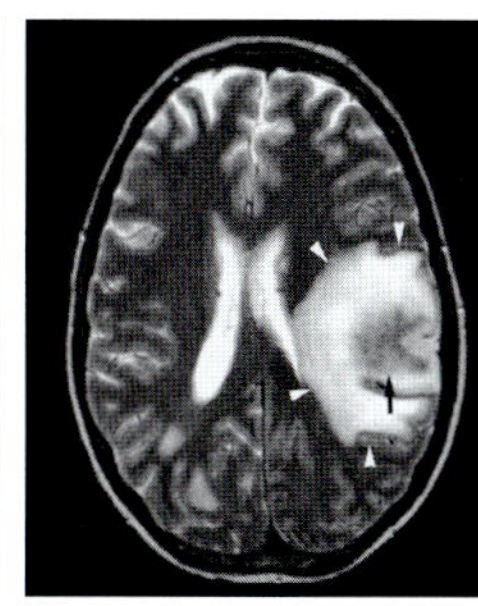 Fig. 4.16 a

Table 6

Differential diagnoses for lesions with a **low diffusion** signal associated with a **high ADC** and **iso intense T2** signal

Diagnose	Reference images		
	DWI low	ADC high	T2 iso
Normal Neonate	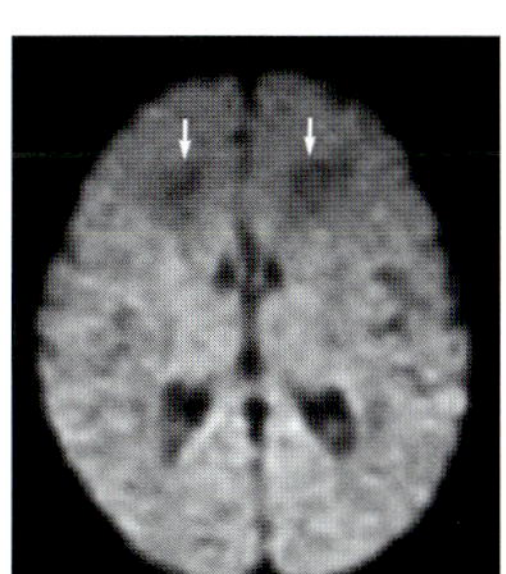 Fig. 2.3 a	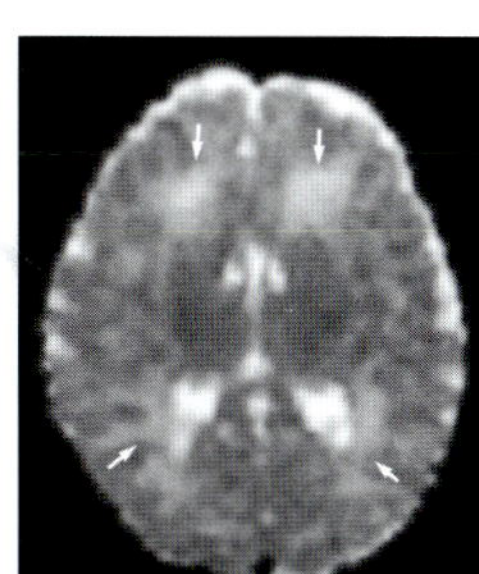 Fig. 2.3 b	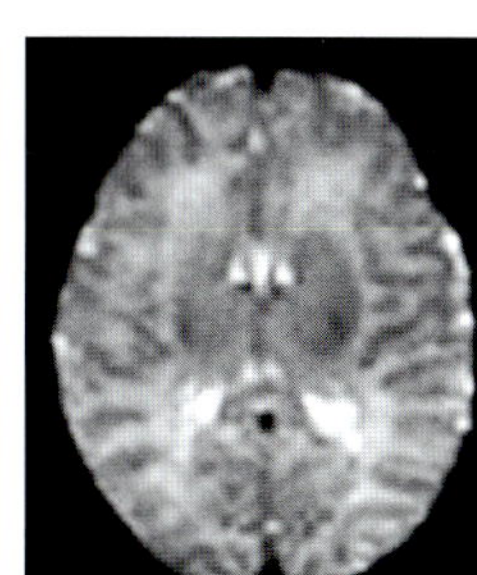 Fig. 2.3 c

Table 7

Differential diagnoses for lesions with **artifacts**

Diagnose	Reference images DWI	ADC	T2
Susceptibility/Artifact **Normal iron deposition** **Hematoma**	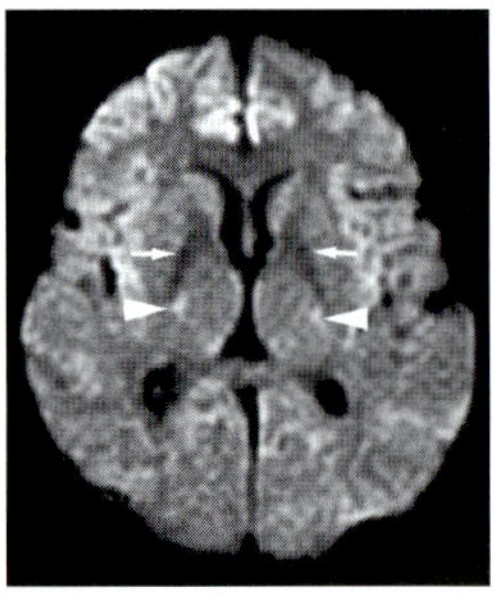 Fig. 2.1 a	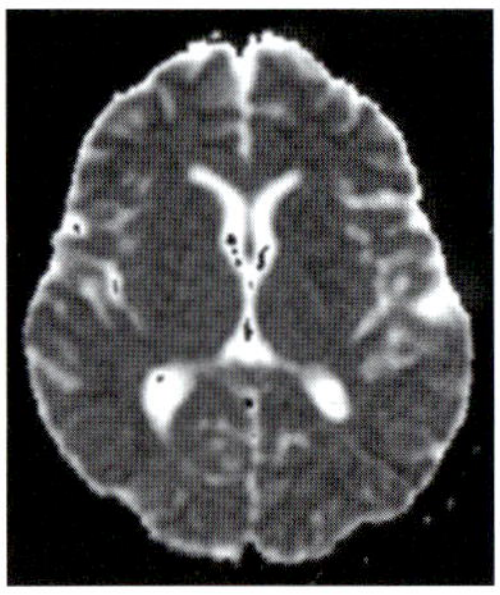 Fig. 2.1 b	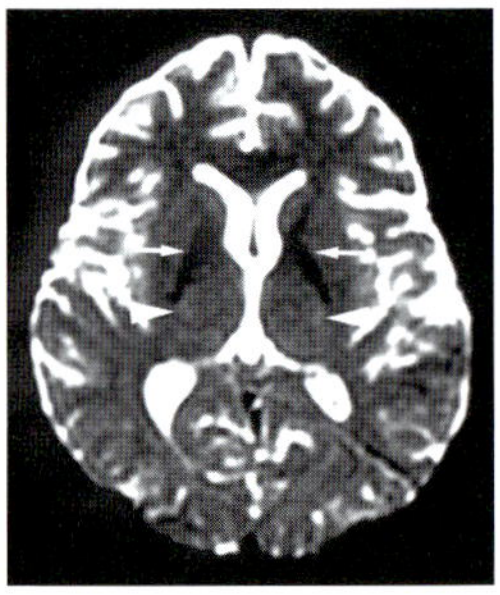 Fig. 2.1 c
Oxy/deoxy hemoglobin	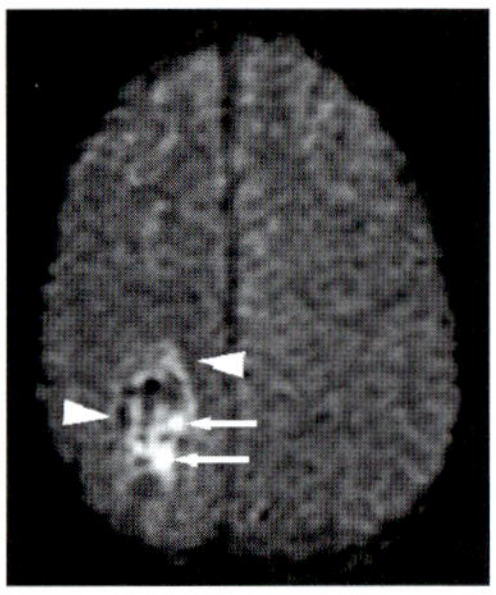 Fig. 6.2 d	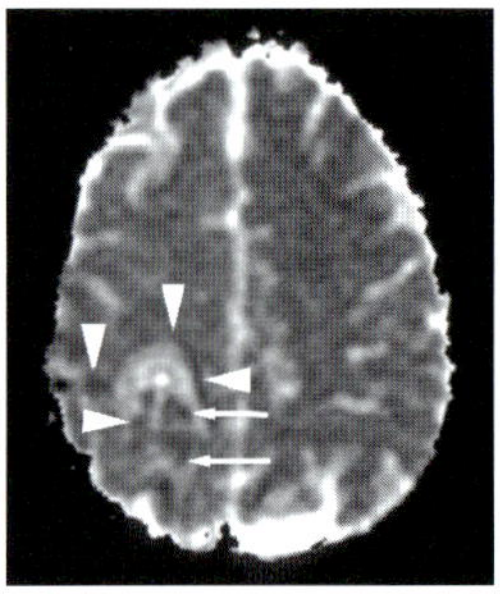 Fig. 6.2 f	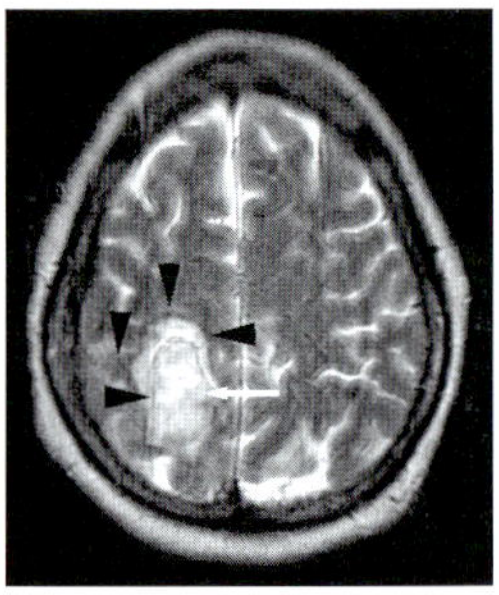 Fig. 6.2 b
Deoxy, IC met	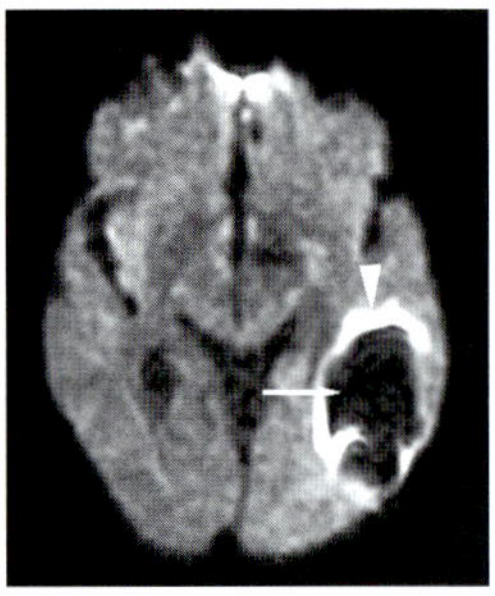 Fig. 6.3 c	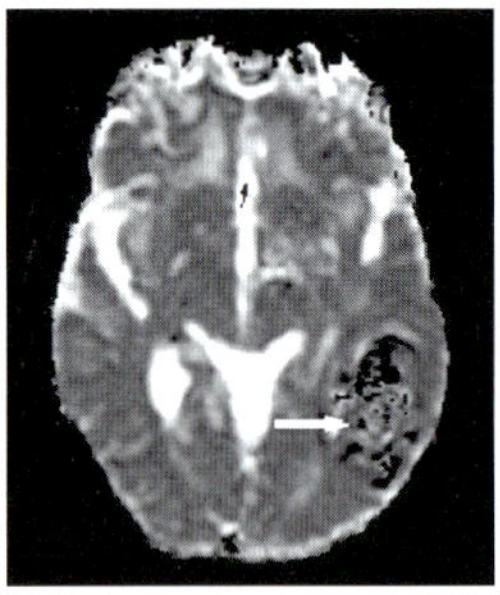 Fig. 6.3 e	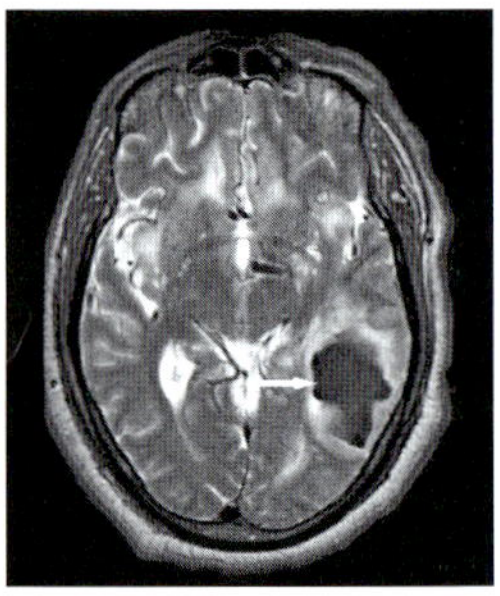 Fig. 6.3 a
Epidural hematoma (EDH)	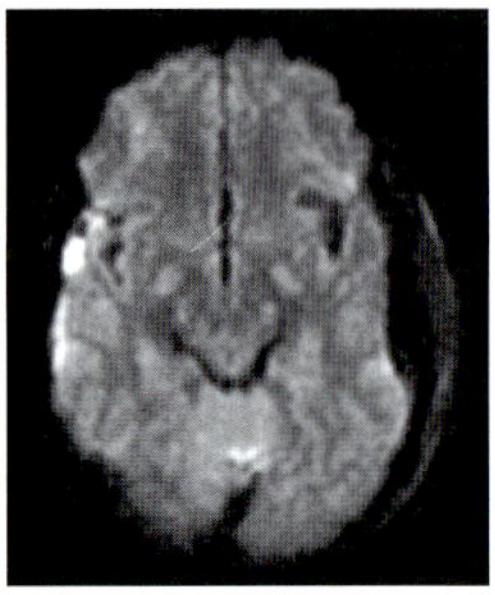 Fig. 12.10 c	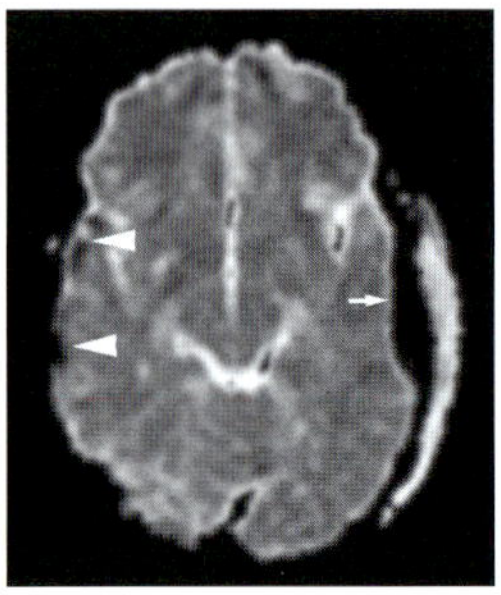 Fig. 12.10 d	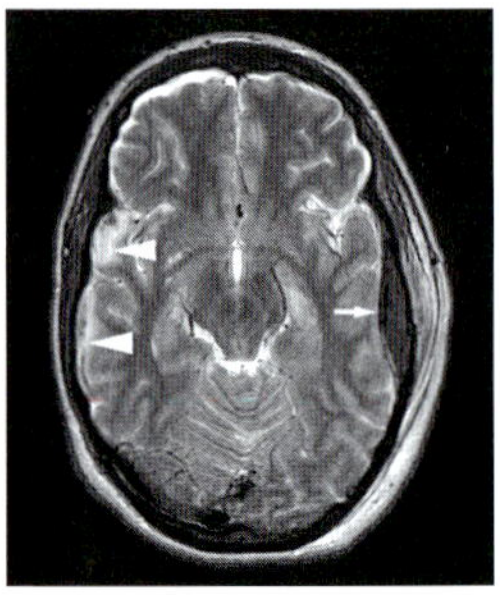 Fig. 12.10 b

Table 7

Differential diagnoses for lesions with **artifacts**

Diagnose	Reference images		
	DWI	ADC	T2
Hemosiderin/ferritin	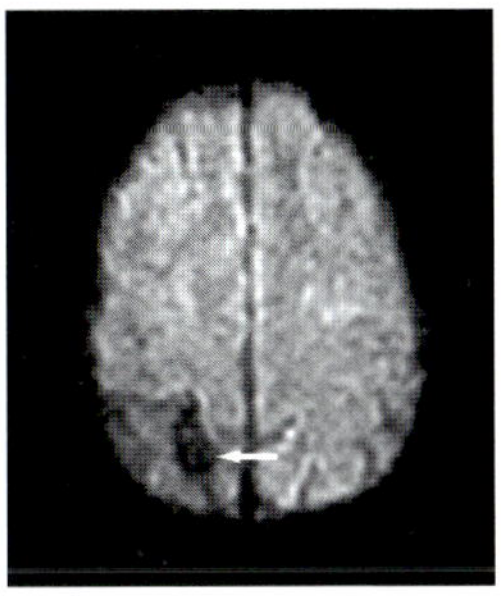 Fig. 6.2 k	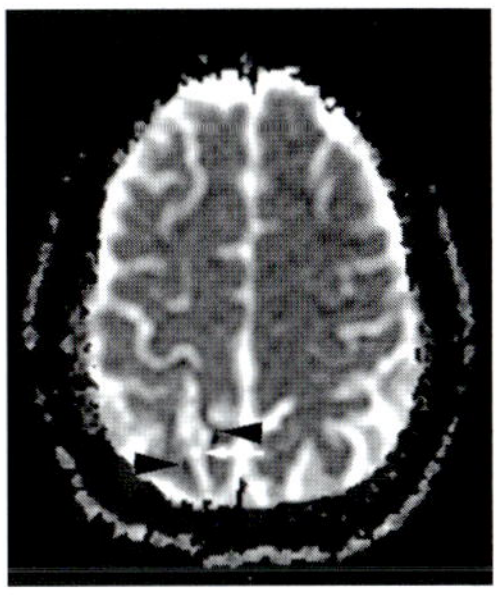 Fig. 6.2 m	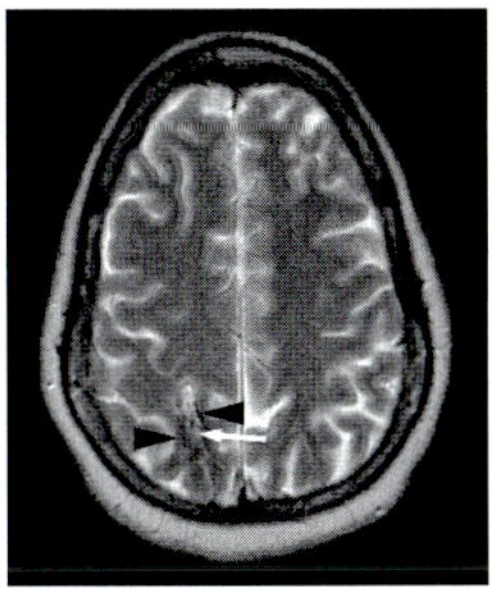 Fig. 6.2 i
Hemorrhagic infarction	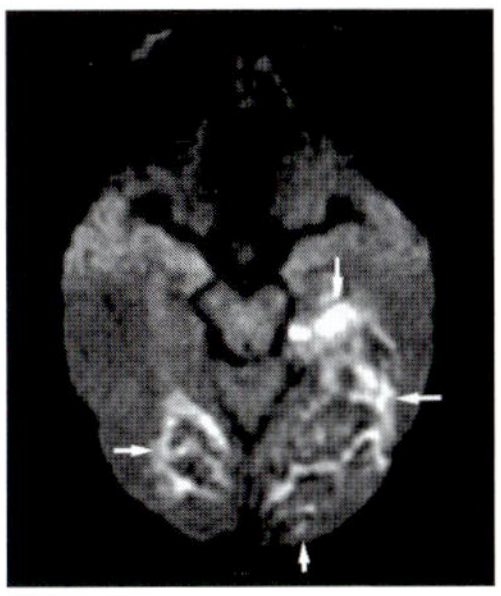 Fig. 5.11 b	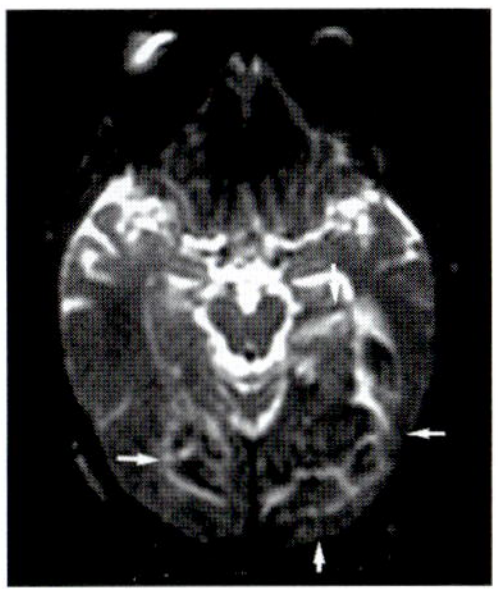 Fig. 5.11 c	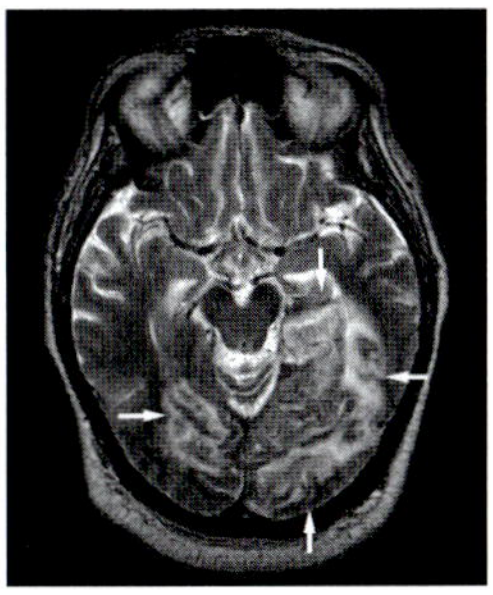 Fig. 5.11 a
Disseminated aspergillosis	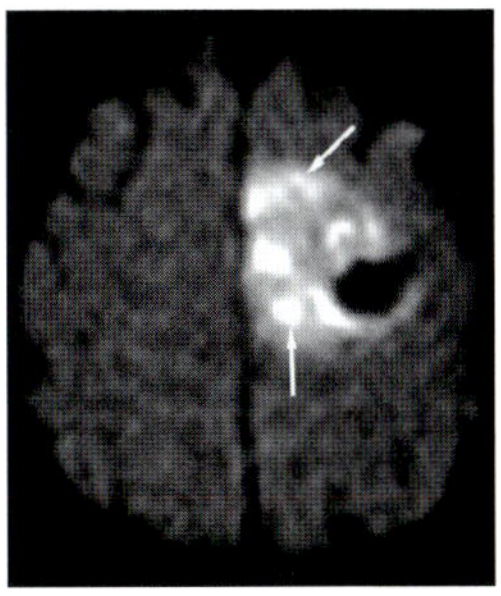 Fig. 7.9 c	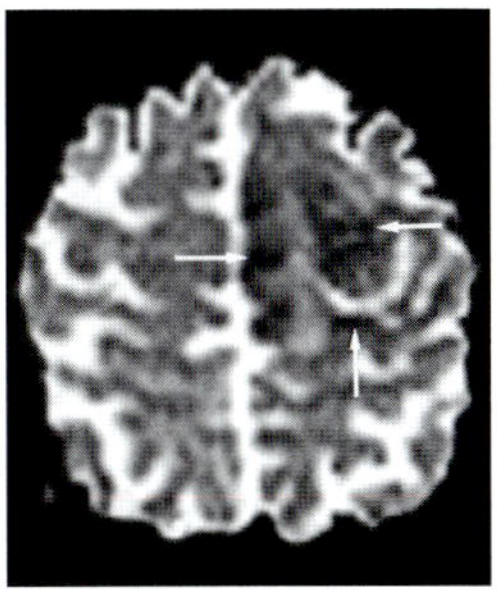 Fig. 7.9 d	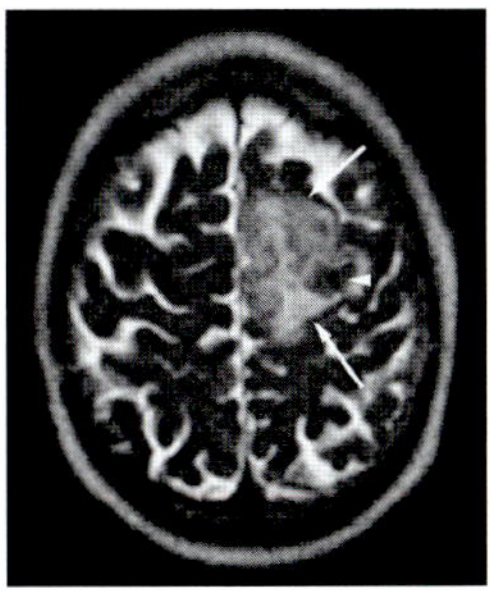 Fig. 7.9 a
Metastasis (melanoma)	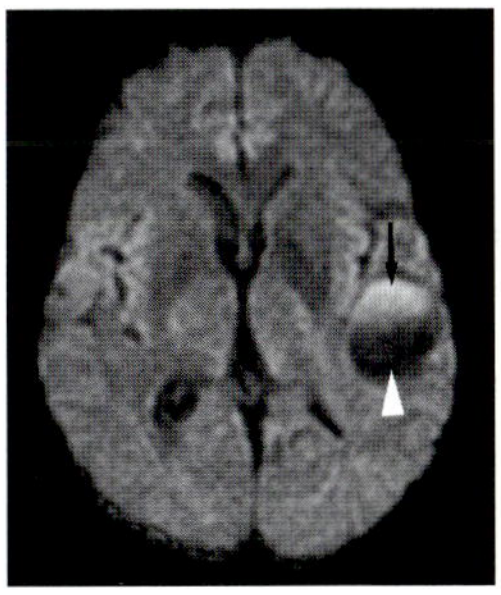 Fig. 13.16 d	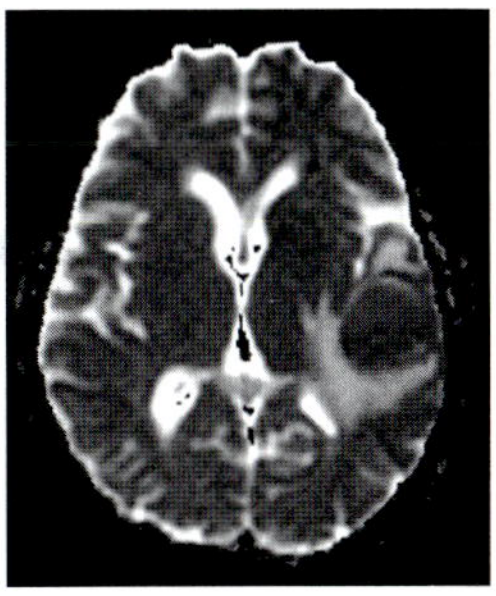 Fig. 13.16 e	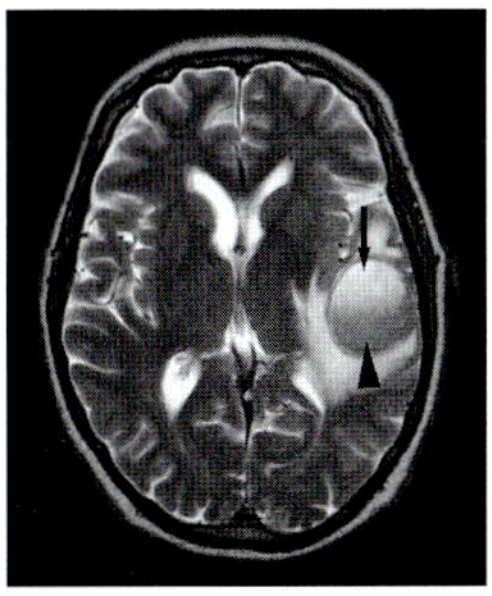 Fig. 13.16 b

Table 7

Differential diagnoses for lesions with **artifacts**

Diagnose	Reference images: DWI	ADC	T2
Contusion	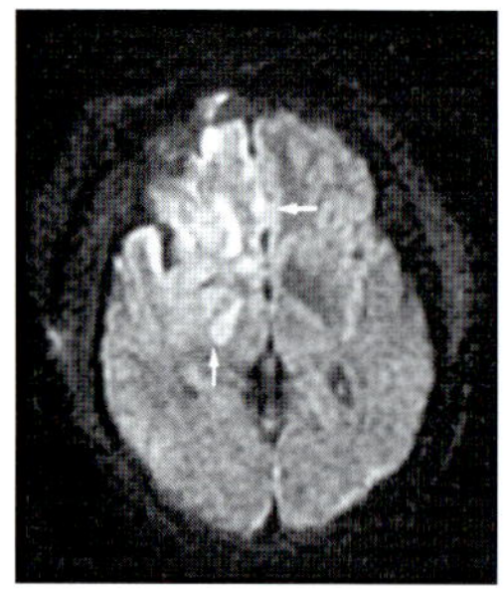 Fig. 12.7 c	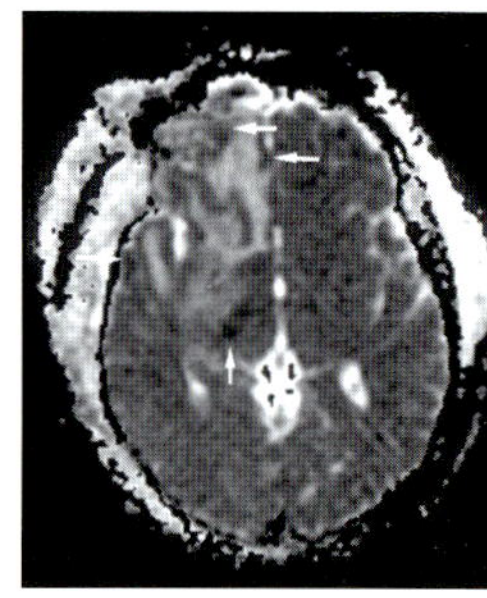 Fig. 12.7 d	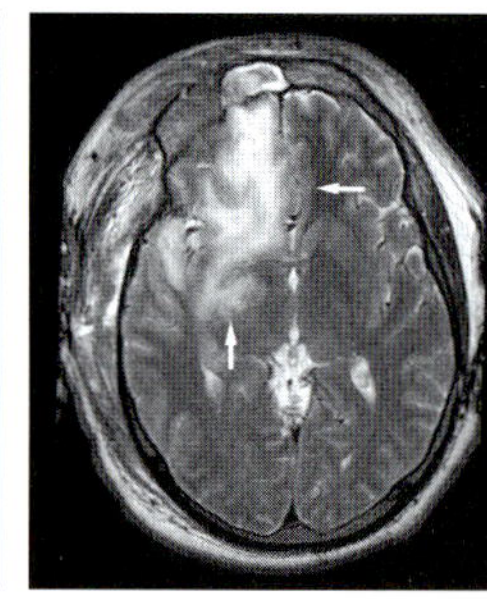 Fig. 12.7 b
Eddy current artifacts	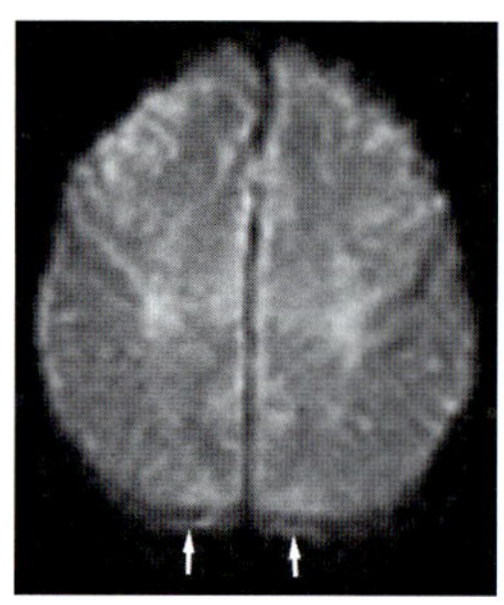 Fig. 3.7 a	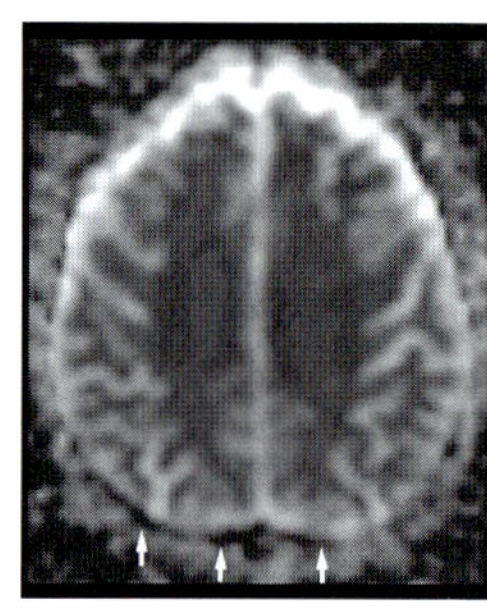 Fig. 3.7 b	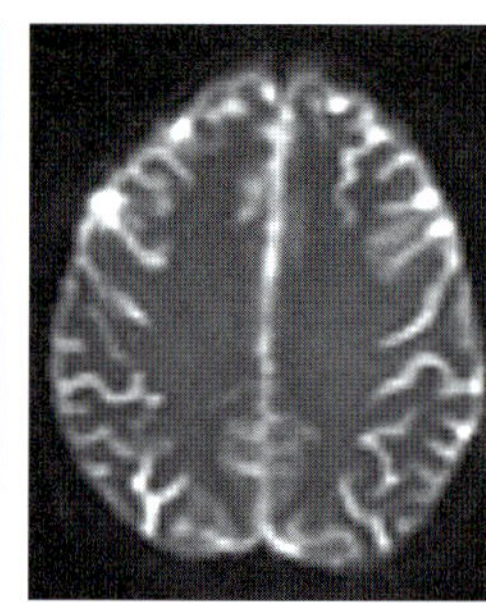 Fig. 3.7 c
N/2 ghosting artifacts	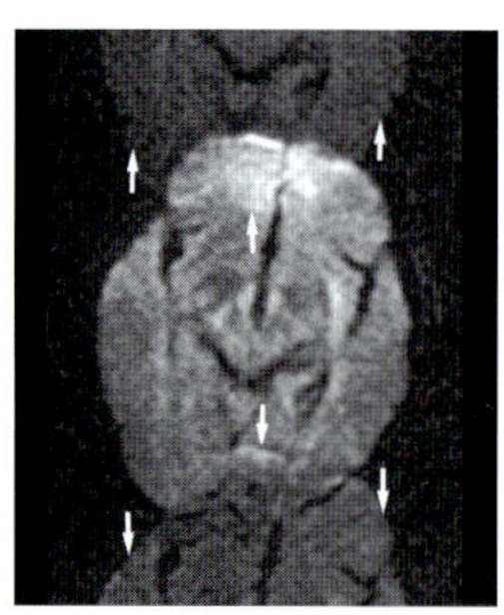 Fig. 3.10 a	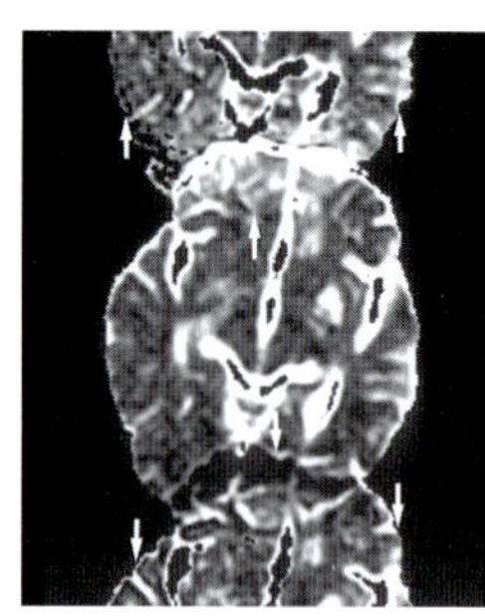 Fig. 3.10 b	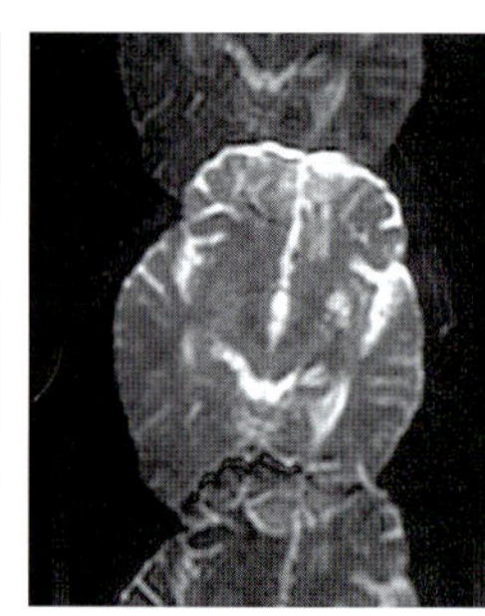 Fig. 3.10 c
Motion artifacts	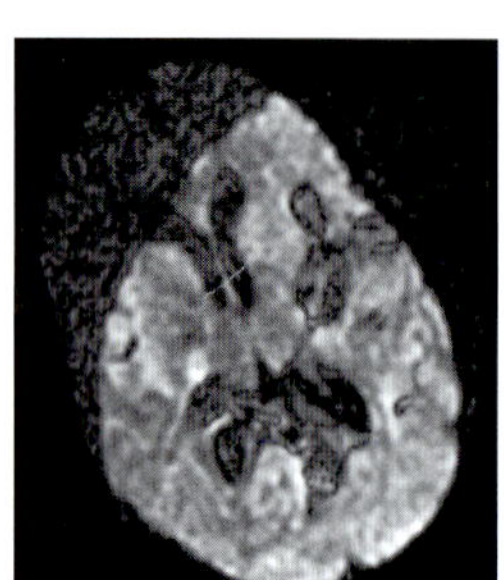 Fig. 3.12 a	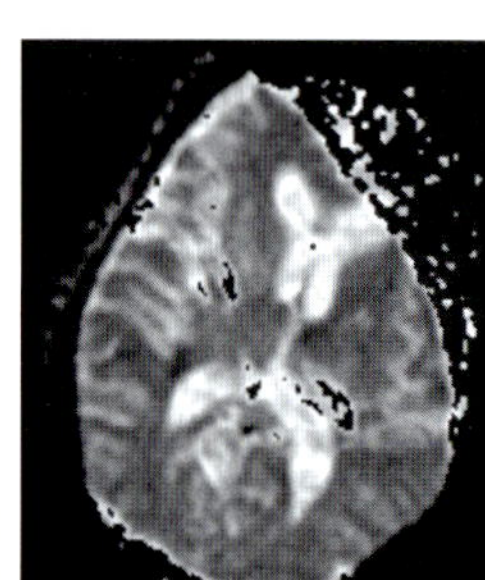 Fig. 3.12 b	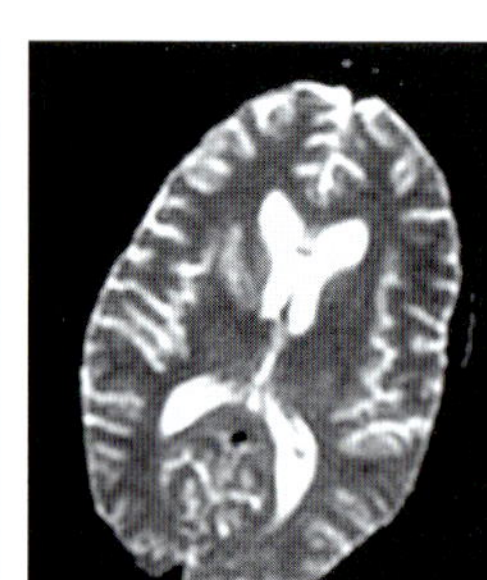 Fig. 3.12 c

Subject Index

A

abscess/brain abscess 131–132, 134, 137, 193
- bacterial 131–132, 134, 137
- subdural 193

aciduria
- glutaric 127
- L-2 hydroxyglutraric 126

ADC (apparent diffusion coefficient) 1–3, 11, 40, 43, 45, 74
- ADC maps 11
- in children 181
- pseudo-normalization of the ADC 43
- relative ADC 45

ADEM (acute disseminated encephalomyelitis) 111, 112
adrenoleukodystrophy 126
AIDS-related (*see also* HIV) 146–147
- bilateral basal ganglia lesions 147
- lymphoma 143

amnesia, transient global 45
amphetamine 83, 119–121
amyloid angiopathy, cerebral 92
amyotrophic lateral sclerosis 116
angiography
- DSA (digital subtraction angiography) 73, 77
- MRA (MR angiography) 73

angitis
- microscopic polyangitis 80
- primary, of the CNS 74–756

anisotropy/anisotropic 1, 3, 182
- diffusion, anisotropic 1, 3
- premyelination anisotropy 182

aortitis syndrome (*Takayasu's* arteritis) 78–79
apoptosis 28
apparent diffusion coefficient (*see* ADC) 1–3, 11, 40, 43, 45, 74, 181
arachnoid cysts 169–170
arteritis
- giant cell 75, 77–78
- polyarteritis nodosa 80
- *Takayasu's* (aortitis syndrome) 78–79

arthritis, scleroderma of rheumatoid arthritis 80
artifacts 16–21
- *Eddy* current artifacts 18
- motion artifacts 20–22
- N/2 ghosting artifact (nyquist ghost) 20
- susceptibility (*see there*) 16–19, 49, 66

aspartate 27
aspergillosis, disseminated 80, 144–145
Aspergillus 145
ASSET (array spatial and sensitivity encoding technique) 18
astrocytoma
- anaplastic 13
- juvenile pilocytic 165
- oligoastrocytoma 167

axonal
- diffuse axonal injury 32, 68, 149–154, 189
- swelling 25

B

bacterial
- brain abscess 131–132, 134, 137, 193
- vasculitis 141

battered child syndrome 186–188
BCNU (carmustine) 119
Behçet's disease 80–81
b-factor 1–2, 4
bleeding
- hemorrhage/hematoma (*see there*) 55–71, 140
- microangiopathy-related microbleeds 92

brain
- edema (*see there*) 25–38, 46, 64, 87, 101–103, 107
- infarcts (*see there*) 51
- infectious diseases (*see there*) 131–148
- injury (*see there*; *see also* trauma) 27, 149–160
- metastases 136, 161
- neoplasms 161–179
- pediatric brain 181
- tumors 66, 133, 161–168, 169–170

Brownian motion 1
b-values, higher 5

C

Canavan disease 126, 196
capsule, internal 9
carmofur leukencephalopathy 120
carmustine (BCNU) 119
carvernous angioma 67
catecholaminergics 119–121
caudate nucleus 9
central nervous system (*see* CNS) 74–76, 84–92
cerebellar infarct 51
cerebral
- amyloid angiopathy 92
- infarction 14, 29, 39–54, 159, 182–185
- – acute 14, 42
- – brain stem 51
- – cerebellar 51
- – chronic 44
- – hemorrhagic 53
- – hyperacute 29, 41
- – septic 132–133
- – small vessel 50
- – subacute 43
- – thromboembolic 40
- – venous 45
- – watershed 47–48
- postictal cerebral lesion 96–99
- toxoplasmosis 37, 141–142

chemical shift 20
- selective (CHESS) 20

children (*see also* pediatrics) 181–199
chorioid plexus 7, 9
Churg-Strauss disease 80
cisplatin 119
CNS (central nervous system) 74–76, 84–92
- primary angitis 74–76
- vasculopathy of the CNS 84–92

coagulative necrosis 137
cocain 83, 119
- vasculopathy, cocain-induced 84

collagen vascular disease 80
contusion of the brain 154–156, 189
corpus callosum 52, 124, 151
- focal lesion in the splenium of 103–104

craniopharnygioma 175
Creutzfeldt-Jakob disease 32, 114–115
cryoglobulinemia, essential 80

cyclophosphamide 119
cyclosporine 88
cytokines 27
cytotoxic edema 25–27, 29–35, 46, 101–103, 123, 149
- in status epilepticus 101–103

D

degenerative diseases 28, 107–117
- encephalomyelitis, acute disseminated 111
- transneuronal degenertion 113
- *Wallerian* degeneration 33, 113
delayed neuronal death 28
demyelinating diseases 107–117
- encephalomyelitis, acute disseminated (ADEM) 112
deoxy-hemoglobin 17, 56–57, 59–60
dephasing 2
diamagnetic substance 55
diffusion
- anisotropic 1, 3
- apparent diffusion coefficient (*see* ADC) 1–3, 11, 40, 43, 45, 74, 181
- free 3
- isotropic 3, 7, 11
- tensor imaging 5
diffusion-encoding 1
diffusion-weighted imaging of normal brain 7–10
dipole-dipole relaxation 62
diseases (*see* syndromes/diseases)
drug-induced vasculitis including illicit drugs 83–84
DSA (digital subtraction angiography) 73, 77
dural hemorrhage 64
- epidural 64, 156–157
- subdural 64, 156–157

E

echo-planar imaging 4
- multi-shot 18
eclampsia/preeclampsia 88, 90
Eddy current artifacts 18
edema of the brain 25–38, 46, 64, 87, 101–103, 107
- axonal swelling 25
- cytotoxic (*see there*) 25–27, 29–35, 46, 101–103, 123, 149
- interstitial 35–36
- intramyelinic 25, 30, 107, 120–121, 124, 126, 197
- subdural edema/hemorrhage 64, 156–157
- vasogenic 35, 87, 149
emboli/embolic
- septic 132–133
- thromboembolic infarction 40
embryonal tumors 171
empyema, subdural 137
encephalitis 191–192
- herpes simplex encephalitis (*see there*) 146, 191–192
- listeria meningoencephalitis 134
encephalopathy
- HIV (human immunodeficiency virus) encephalopathy (*see also* AIDS) 146–147
- hypertensive 87–89, 190
- hypoxic ischemic 30, 184–185
- mitochondrial 125–126, 189–190
- necrotizing encephalopathy, acute 190
- syndrome (*see also* leukoencephalopathy syndrome) 15, 86–87
- uremic 91
- *Wernicke* encephalopathy 123–124
ependymoblastoma 171
epidermoid 140
- tumors 169–170
epidural hemorrhage 64, 156–157
epilepsy/epileptic 31, 95–106
- cytotoxic edema 101–103
- excitotoxic mechanism 100, 154
- magnetic resonance signal alterations 98–99
- MRI 96–97
- *Rasmussen* syndrome (epilepsia partialis continua) 103
- seizures, epileptic 95
- status epilepticus 31, 100
- syndromes, epileptic 95
- - hemiplegic hemiconvulsion epilepsy syndrome 102–103
excitotoxic
- brain injury 27–28, 154
- mechanisms 28, 100, 154
exponential DW images 11

F

b-factor 1–2, 4
ferritin 55, 57, 59, 61
fibrinolytic therapy 45
- intraarterial 45
- intravenous 45
field gradients 1–2
5-fluouroracil (5-FU) 119
free radicals 27

G

GABA (gamma-aminobutyric acid) 96
ganglia, AIDS-related bilateral basal ganglia lesions (*see also* HIV) 146–147
ganglioglioma 168
giant cell arteritis 75, 77–78
glioblastoma multiforme 66, 135, 162–164
glioma 161–168, 194
- brain stem glioma 166
- ganglioglioma 168
- high-grade tumors 161–168
- peritumoral infiltration 168
- treatment response 168
gliosis, mesial temporal 102
glutamate 27–28, 95, 100
glutaric aciduria 127
ghosting artifacts 20
- N/2 ghosting artifact (nyquist ghost) 20
gradient
- factor 1
- field gradient 1–2
gray matter 7, 9
- ischemia, gray and white matter 45

H

hematoma (*see*hemorrhage/hematoma) 55–71, 140
hemiplegia/hemiplegic
- hemiplegic hemiconvulsion epilepsy syndrome 102–103
- migraine 45
hemoglobin
- deoxy-hemoglobin 17, 56–57, 59–60
- met-hemoglobin (*see there*) 17, 55, 57, 59–60
- oxy-hemoglobin 55–57
hemolytic uremic syndrome 91–92
hemorrhage, intracranial/hematoma 55–71, 140
- chronic
- - hematoma 62
- - hemorrhage 57
- epidural 64, 156–157
- hyperacute (*see there*) 55–59
- infarct 53
- intratumoral 66
- intraventricular 63, 65, 140
- microangiopathy-related microbleeds 92
- subacute 59–63
- subarachnoid 62–63, 156, 158
- subdural 64, 156–157
- transformation 53
- trauma-related 68, 156–159
hemosiderin 55, 57, 59, 61
Henoch-Schöenlein purpura 80
heroin 83
- spongiform leukoencephalopathy, heroin-induced 119, 121
herpes simplex encephalitis 145, 191–192
- type 1 191
- type 2 191–192

hippocampus/hippocampal lesion 21, 100–101
HIV (human immunodeficiency virus)
– encephalopathy (*see also* AIDS) 146–147
– vasculitis 146
L-2 hydroxyglutraric aciduria 126
hyperacute
– hematoma 55–59
– – deoxy-hemoglobin 55–57, 59
– – hemosiderin/ferritin 57, 59, 61
– – met-hemoglobin (*see there*) 17, 55, 57, 59–60
– – oxy-hemoglobin 55–57
– infarction 29, 41
hypertensive
– encephalopathy 87–89, 190
– vasculitits 80
hypoxic ischemic encephalopathy 30, 184–185

I

illicit drugs, vasculitis 83–84
image
– exponential DW images 11
– perfusion MR image 48
– perfusion-weighted image 46, 47
infantile neuronal dystrophy 126
infarction, cerebral (*see* cerebral infarction) 14, 29, 39–54, 132–133, 159, 182–185
infectious diseases 131–148
– bacterial brain abscess 131–132, 134, 137, 193
– septic emboli 132–133
– vasculitis, infectious 82–83
injury of the brain (*see* trauma/injury) 27, 149–160, 186–189
interferon 88
internal capsule 9
interstitial edema 35–36
– vasogenic 36
intracranial hemorrhage 55–71
intramyelinic
– cleft 26, 30
– edema 25, 107, 120–121, 124, 126, 197
intratumoral hemorrhage 66
intraventricular hemorrhage 63, 65, 140
ischemia/ischemic 29–30, 40, 45–49, 182–185
– encephalopathy, hypoxic ischemic 30, 184–185
– gray and white matter ischemia 45
– penumbra 29, 40, 48
– reversible 46
– transient ischemic attack 45
– venous 36, 48–49
isotropic diffusion 3, 7, 11

K

van der Knaap disease 128
Krabbe disease 128

L

lactic acidosis (*see also* MELAS) 125
leptomeningitis, purulent 138
Leukoencephalopathy
– carmofur 120
– metabolic 196
– methotraxate 120
– posterior leukoencephalopathy syndrome 86
– – occipital parietal 86
– – reversible posterior 15, 86–87
– progressive multifocal (PML) 112
– spongiform leukencephalopathy, heroin-induced 119, 121
– toxic 196
– with vanishing white matter 128, 196
leukodystrophy 126
– adrenoleukodystrophy 126
– metachromatic 126
line scan 20
listeria meningoencephalitis 134
lupus erythematosus, systemic (SLE) 84–85
lymphoma 161, 194
– AIDS-related (*see also* HIV) 143
– large B-cell type 174
– malignant 174
– primary 174

M

magnetic resonance signal alterations in epilepsy 98–99
magnetization transfer ratio (MRT) 107
Marchiafava-Bignami disease 124
medulloepithelioma 171
medulloblastoma 171, 194
melanoma 177
MELAS (mitochondrial encephalopathy with lactic acidosis and stroke) 125
meningiomas 172–173
– atypical 173
meningitis
– pneumococcal 80
– purulent leptomeningitis 138
– *Streptococcus* menigitis, group B 141
meningoencephalitis, listeria 134
mesial temporal
– gliosis 102
– sclerosis 97, 102
metabolic diseases 119–130
– leukoencephalopathy, metabolic 196
metachromatic leukodystrophy 126
metastasis of the brain 133, 176–177
met-hemoglobin 17, 55, 57, 59–60, 64–65
– extracellular 61, 64
– intracellular 17, 60, 65
methothrexate 119
– leukoencephalopathy 120
N-methyl-D-aspartate (NMDA) receptors 28, 95
microangiopathy-related microbleeds 92
microscopic polyangitis 80
migraine, hemiplegic 45
mitochondrial encephalopthy 125–126, 189–190
– with lactic acidosis and stroke (MELAS) 125
motion
– artifacts 20–22
– *Brownian* motion 1
Moyamoya disease 86–87, 183–184
– probable 97
MRI, epilepsy 96–97
MRT (magnetization transfer ratio) 107
multi-directional 4
multiple sclerosis 12, 34, 91–112, 197
multi-shot echo-planar imaging 18
myelinolysis 122–123
– central pontine 122–123
– extra pontine 122–123
– osmotic 122, 197–198

N

N/2 ghosting artifact (nyquist ghost) 20
navigator method 20
necrosis 28
– acute necrotizing encephalopathy 190
– coagulative necrosis 137
– tumor necrosis 133
neoplasms of the brain 161–179
neuroblastoma 171
neuroectodermal tumors, primitive 171, 194
neuronal
– delayed neuronal death 28
– infantile neuronal dystrophy 126
neurotoxicity, tacrolimus 90
NMDA (N-methyl-D-aspartate) receptors 28, 95
normal brain, diffusion-weighted imaging 7–10
nucleus, caudate 9
nyquist ghost (N/2 ghosting artifact) 20

O

oligoastrocytoma, low-grade 167, 195
osmotic myelinolysis 122, 197, 198
oxy-hemoglobin 55–57

P

paramagnetic 59
– susceptibility artifacts 16
pediatrics 181–199
– battered child syndrome 186–188
– pediatric brain 181
Pelizaeus-Merzbacher disease 128, 195
penumbra, ischaemic 29, 40, 48
perfusion MR image 48
perfusion-weighted image 46–47
peritumoral invasion 168
phase encoding 18
phencyclidine hydrochloride 119–121
phenylketonuria 35, 126–127, 196–197
pilocytic astrocytoma, juvenile 165, 195
pineoblastoma 171
PML (progressive multifocal leukoencephalopathy) 112
pneumococcal meningitis 80
polyarteritis nodosa 80
pontine myelinolysis
– central pontine 122–123
– extra pontine 122–123
postictal cerebral lesion 96–99
preeclampsia/eclampsia 88, 90
PROPELLER (periodically rotated overlapping parallel lines with enhanced reconstruction) 18, 20
proton-electron 62
pulse sequence 2
pulvinar sign 114
purpura
– *Henoch-Schöenlein* 80
– thrombotic thrombocytopenic 92–93
purulent
– leptomeningitis 138
– ventriculitis 139

R

radiofrequency thalamotomy 133, 137
Rasmussen syndrome (epilepsia partialis continua) 103
relaxation, dipole-dipole 62
reversible ischemia 46

S

S_0 (signal intensity) 1
scleroderma of rheumatoid arthritis 80
sclerosis
– amyotrophic lateral sclerosis 116
– mesial temporal sclerosis 97, 102
– multiple sclerosis 12, 34, 91–112, 197
SENSE (sensitive encoding) 18, 20
septic
– emboli 132–133
– infarctions 132
sickle cell disease 86, 88, 184
single-shot 4
small vessel infarcts 50
splenium of the corpus callosum, focal lesion in 103–104
SSFSE 20
status epilepticus 31, 100
Streptococcus meningitis, group B 141
stroke 39
– MELAS (*see there*) 125
subarachnoid hemorrhage 62–63, 156, 158
subdural
– abscess 193
– edema/hemorrhage 64, 156–157
– empyema 137
susceptibility
– artifacts 16–19, 49
– – magnetic 17, 66
– – paramagnetic 16
– effect 59
syndromes/diseases
– *Behçet's* 80–81
– *Canavan* 126, 196
– *Churg-Strauss* 80
– *Creutzfeldt-Jakob* 32, 114–115
– *Henoch-Schöenlein* 80
– *van der Knaap* 128
– *Krabbe* 128
– *Marchiafava-Bignami* 124
– *Moyamoya* 86–87, 183–184
– *Pelizaeus-Merzbacher* 128, 196
– *Rasmussen* 103
– *Takayasu's* 78–79
– *Wallerian* 33, 113
– *Wegener's* 80
– *Wernicke* 123
systemic lupus erythematosus (SLE) 84–85

T

T_2
– T_2 blackout 11–13
– T_2-shine-through/shine-through effect 2, 5, 11–12, 40, 43, 108–109
– T_2 washout 11, 15, 109
tacrolimus 88
– neurotoxicity 90
Takayasu's arteritis (aortis syndrome) 78–79
TE (time evolution) 1
tensor
– analysis 3
– diffusion tensor imaging 5
thalamotomy, radiofrequency 133, 137
thromboembolic infarction 40
thrombosis 39
– purpura, thrombotic thrombocytopenic 92–93
– thrombolysis/thrombolytic treatment 39
time evolution (TE) 1
tonic-clonic seizure, generalized 100
toxic diseases 119–130
– cytotoxic edema (*see there*) 25–27, 29–35, 46, 101–103, 123
– excitotoxic injury of the brain 27–28, 154
– leukoencephalopathy, toxic 196
– neurotoxicity, tacrolimus 90
toxoplasmosis 37, 141–142
transneuronal degenertion 113
– trans-synaptic 113
trauma/injury of the brain 149–160, 186–189
– brain contusion 154–156, 189
– children 185–189
– diffuse axonal injury 32, 149–154
– excitotoxic (*see there*) 27–28, 154
– hemorrhage related to trauma (*see there*) 68, 156–159
tumor
– arachnoid cysts 169–170
– craniopharnygioma 175
– embryonal 171
– epidermoid 169–170
– glioma (*see there*) 161–168, 194
– intratumoral hemorrhage 66
– lymphoma (*see there*) 143, 161, 174, 194
– meningiomas 172–173
– metastasis of the brain 133, 176–177
– necrosis 133
– neoplasms of the brain 161–179
– neuroectodermal, primitive 171, 194
– peritumoral invasion 168
two-point measurement 5

U

uremic
– encephalopathy 91
– hemolytic uremic syndrome 91–92

V

vanishing white matter disease 128, 196
vasculitis 73–94
– bacterial 141
– drug-induced 83–84
– HIV (human immunodeficiency virus) vasculitis (*see also* AIDS) 146
– hypertensive 80
– infectious 82–83

vasculopathy/vascular 73–94
- of the CNS 84–92
- cocain-induced 84
- collagen vascular disease 80
vasogenic edema 35–36, 87, 149
venous
- infarction 45
- ischemia 36, 48–49
ventricle
- intraventricular hemorrhage 63, 65, 140
- purulent ventriculitis 139

W

Wallerian degeneration 33, 113
watershed infarction 47–48
Wegener's granulomatosis 80
Wernicke encephalopathy 123–124
white matter 7, 9
- ischemia, gray and white matter 45
- leukoencephalopathy with vanishing white matter 128, 196
- myelination 9
- subcortical 9